Pharmaceutical Dosage Forms: Disperse Systems

Pharmaceutical Dosage Forms: Disperse Systems

In Three Volumes
VOLUME 3
Second Edition, Revised and Expanded

edited by

Herbert A. Lieberman
H.H. Lieberman Associates, Inc.
Livingston, New Jersey

Martin M. Rieger
M.& A. Rieger Associates
Morris Plains, New Jersey

Gilbert S. Banker
University of Iowa
Iowa City, Iowa

CRC Press is an imprint of the
Taylor & Francis Group, an **informa** business

CRC Press
Taylor & Francis Group
6000 Broken Sound Parkway NW, Suite 300
Boca Raton, FL 33487-2742

First issued in paperback 2019

ISBN-13: 978-0-8247-9842-0 (hbk)
ISBN-13: 978-0-367-40060-6 (pbk)

Visit the Taylor & Francis Web site at
http://www.taylorandfrancis.com

and the CRC Press Web site at
http://www.crcpress.com

Preface

This text completes this three-volume treatise on disperse systems. As with the other revised, second edition volumes in the series, this one has been updated and expanded to contain subject matter not included in the first edition. Volume one covers the theoretical parameters involved in formulating the many different types of disperse system products. The second volume describes the characteristics and formulations of specific types of disperse system dosage forms, including many practical examples of their formulation and stability test procedures. In the first section of this volume there is detailed information on several specialized products such as certain types of emulsions, liposomes, different polymers used in dispersions as either drugs or film carriers, biodegradable and new polymeric dispersions, and the characteristics of polymeric pharmaceutical excipients currently used in drug formulations.

In Part B of this text, there is a description of equipment selection and processing, scale-up to manufacturing, quality assurance, validation, and drug regulatory affairs related to disperse system pharmaceutical products and dosage forms.

The chapter "Specialized Pharamaceutical Emulsions" (Chapter 1) has been updated and the reference literature citations expanded. It describes emulsions within the context of systems of arrested phase separation, which may or may not be at equilibrium. A broader classification results that includes to special types of macroemulsions, multiple emulsions, microemulsions, vesicles, and liquid crystals. A section has also been added on gel emulsions. The uses and implications of these systems for drug delivery are also explored.

Much progress has been made in the development of liposomes (Chapter 2) in pharmaceutical products. In the short time between the publication of the first and second editions of the text on disperse systems, the first two liposomally formulated dosage form—an anticancer agent and an antifungal product—have been approved by the Food and Drug Administration. Additionally, liposomal-based products are gaining widespread use as preferred dosage forms for the emerging cosmeceutical field. Most exciting, however, is the prospect that liposomes may well emerge as the preferred carriers for the delivery of DNA in gene therapy. The revised chapter on liposomes, while retaining its emphasis on fundamental concepts and their general application to pharmaceutical dosage forms, describes the newest advances in liposome technology that have led to breakthroughs in introducing liposomal drug products.

"Polymeric Dispersions as Drug Carriers" (Chapter 3) presents an overview of the different methods used to prepare polymeric nanoparticles and the physicochemical methods to characterize them. The nanoparticles are formed either by various emulsification or phase separation methods or from preformed lipophilic or hydrophilic polymers of synthetic, semisynthetic, or natural origin. Important process and formulation parameters of the different preparation techniques are described. Precise characterizations of several different physico-chemical parameters are included in order to help formulate a drug product of high quality. "Aqueous Polymer Dispersions as Film Formers" (Chapter 4) offers a comprehensive review of the various aqueous polymeric dispersion systems that are currently used in coating pharmaceutical products. The mechanisms for the formation of films from aqueous disperse systems are described in depth. Several sample formulations are presented. The effects of varying the formula and process conditions on the properties of the resulting film are explained. Both the theories and the instrumental methods used to characterize the thermal, mechanical, and permeation properties of the films formed are also depicted.

"Biodegradable Nanoparticles of Poly(lactic acid) and Poly(lacto-*co*-glycolic acid) for Parenteral Administration" (Chapter 5) details the importance of the nanoparticles PLA and PEG-PLA as they are used in sustained release parenteral formulations and for drug targeting due to their chemical and physical stability and their biodegradability. Numerous preparative methods are described, with special emphasis on recently developed techniques using polyethylene glycol (PEG) and PEG-PLA copolymers because of their reduced uptake by the mononuclear phagocytic system. Topics such as physicochemical characterization, purification, freeze-drying, sterilization, and in vitro release studies are depicted. Chapter 6 "Aqueous Polymeric Dispersions" provides an overview of the properties and applications of these systems, particularly their methods of preparation and use in drug dosage forms. Applications range from new or enhanced topical products, to bioadhesive, biodegradable, and drug targeting delivery systems. Numerous polymers are described, including those that have been converted into aqueous dispersion, colloidal, near colloidal latex, or pseudolatex and readily water-dispersible powder forms. The stability of various aqueous disperse products is also presented.

"Polymeric Pharmaceutical Excipients" (Chapter 7) concentrates on the scientific, regulatory, and toxicological information for select and commonly used polymers in disperse system formulations. It provides facts concerning the present as well as the historical regulations of pharmaceutical excipients under the Food, Drug, and Cosmetic Act. There are suggestions for expanding the list of the currently acceptable excipients in order to meet the increasing demands for several new dosage delivery systems. Sample formulations of pharmaceutical disperse system dosage forms are included. This is followed by several tables, summarizing the regulatory status of the polymers and listing their manufacturers. Finally, there are product excipient monographs with detailed information on many of the polymers. "A Practical Guide to Equipment Selection and Operating Techniques" (Chapter 8) has been revised and expanded. A new technique that uses the physical characteristics of supercritical fluids to create very fine and uniform dispersions is described. Also included is a new design of cyclone-like processing equipment that can be used on dispersion products having a wide range of viscosities.

Chapter 9, "Scale-Up of Disperse Systems: Theoretical and Practical Aspects," is intended to provide the formulator with an appreciation of the complexity of the scale-up problems associated with disperse system pharmaceutical products. It prompts an awareness of the extent to which these problems may be resolved by the integration of pharmaceutical technology and chemical engineering. This chapter appears to be the first monograph in the pharmaceutical literature to address the issue of the scale-up of disperse systems from an engineering perspective.

"The Scale-Up of Dispersed Parenteral Dosage Forms" (Chapter 10) describes the specific procedures applicable for the development of dosage forms starting with basic information concerning mixing and homogenization, followed by descriptions of freeze drying and solvent removal. Several areas of practical importance, such as material compatibilities, equipment cleanliness, and sterilization-in-place procedures are explained in simple terms. Problems and pitfalls in scale-up of parenteral dosage forms are emphasized. A hypothetical case study is presented so that a novice in this field may profit from the information imparted regarding the scale-up of dispersed parenteral dosage forms.

Chapter 11, "Quality Assurance," has been revised to reflect new trends in enhancing the quality of disperse system dosage forms by adding new analytical, stability, environmental, pyrogen, sterility, and governmental regulations sections. New figures and tables have been added, and the reference list has been updated to reflect many new publications that have appeared since the first edition. "Validation of Disperse Systems" (Chapter 12) serves to fill an important gap in the pharmaceutical literature, which is replete with information concerning the validation of sterile products and solid dosage forms but is lacking in disperse system dosage forms. The essential principles of pharmaceutical process validation are introduced and a strategy is offered that can be used to validate and control the manufacture of suspensions, gels, ointments, creams, lotions, and aerosols.

"Drug Regulatory Affairs" (Chapter 13) captures important Food and Drug Administration regulatory information necessary to legally market disperse system drug products and any drug dosage form in the United States. Included in this chapter are a history of the FDA and the evolution of the drug regulations, and an overview of the various requirements for conducting clinical research and obtaining marketing approvals for new drug products (IND, NDA, ANDA). Also included are descriptions of the requirements for marketing over-the-counter drugs, the recent Prescription Drug User Fee Act (PDUFA) and its impact on FDA review practices, and the regulations concerning drug product production. These sections review current inspection practices, adulteration and misbranding, preapproval inspections, advisory committee functions, labeling and advertising, good manufacturing practices (GMP), and performance goals. This chapter is a comprehensive review of the current status of drug regulations in the United States.

As with the other books on pharmaceutical dosage forms, the approach taken in preparing this book was to choose authors recognized for their experience and expertise in their assigned subject matter. The hope of the editors is to provide theoretical and practical information on modern industrial pharmacy. Each of the books in the series will be useful to those involved in the pharmaceutical and allied health sciences, particularly those in the formulation, manufacture, and control of drug dosage forms, as well as hospital pharmacists, patent attorneys, governmental scientists in health research, and drug regulatory personnel.

The choice of topics and authors belongs to the editors. The problem for all editors of a multiauthored book is to gently coerce very busy people to finish their chapters on schedule, and to revise and polish their manuscripts with our suggestions. Despite these pressures we are grateful to all contributors for their patience and forbearance with us.

Each author and editor will be pleased if all of our efforts have resulted in a useful teaching and reference source on the practice of industrial pharmacy.

Herbert A. Lieberman
Martin M. Rieger
Gilbert S. Banker

Preface to the First Edition

Pharmaceutical Dosage Forms is the title of a series of books describing the theory and practice of product development associated with specific types of pharmaceutical products. The first three volumes describe tablets, and the two that follow are devoted to parenteral medications. This book is the first of two volumes that covers disperse system products, namely various types of liquid and semisolid emulsions, liquid and paste-like suspensions, and colloidal dispersions. A more specific product description of this type of pharmaceutical preparation would include: injectables, suppositories, aerosols, ingested and topically applied emulsions, abrasive pastes, such as toothpastes, and ingestable types of suspensions, such as liquid antacids.

Despite the diversity of disperse systems, there are unifying theoretical concepts that are applicable to all of them. Thus, the principles of rheology, of surface activity, and of the potential energy barriers to coalescence offer a systematic approach to an understanding of disperse systems. As a result, chapters with an emphasis on theory are included in the first volume of this two-volume treatise. From an understanding of these scientific fundamentals, product development formulators can develop their own approaches to the compounding of specific products. In addition, an understanding of theory and basic concepts may lead the development scientist to new ideas and technology necessary for novel drug delivery advances.

The pharmaceutical chemists, in developing a particular dosage form, must pursue many objectives. The integrity of the drug substance must be maintained; the pharmacological activity should be optimized; a uniform and consistent amount of drug must be dispensed throughout the lifetime of the product; the product must be physically and chemically stable as well as provide good user acceptance throughout its intended shelf and usage life. Successful completion of these objectives requires information on the practical aspects of formulation. Experienced practitioners have acquired this type of knowledge during many years of painstaking trials and errors. Therefore, for the second volume of this treatise, on formulation aspects, we chose as chapter contributors specialists in developing particular types of products.

A comprehensive book on disperse systems must teach two things: underlying principles and practical approaches. The novice requires detailed guidance, and the experienced formulator seeks information on new technologies and theories. Our task was to help select subject matter for this two-volume text to achieve these objectives. Thus, the book will teach the novice product development scientist the theoretical principles and practical aspects necessary for developing a disperse system product. In addition, the information is sufficiently comprehensive and extensive in order to offer knowledgeable formulators specific information that will enable them to achieve their goals.

Any multiauthor book includes redundancies. We deliberately did not strive to completely eliminate this type of duplication because authors may cover subjects from different points of view, which can help to clarify a particular subject matter. In fact, the background, experience, and bias of the various authors, in our opinion, help to contribute materially to learning and to broaden the comprehension of the subject matter for the reader.

We acknowledge the diligence with which the authors worked on their contributions. We also recognize gratefully the authors' forbearance in conforming with our numerous requests for modifications and additions. The real credit for this book belongs to each contributing author. On the other hand, the editors must assume the major responsibility for any criticism of the choice of subject matter in the text.

We sincerely hope that this book will be of help to students in the pharmaceutical sciences. In addition, we look for this book in the dosage form series to be a useful source of information for the product development specialists in the pharmaceutical industry, as well as others in related industries, academia, and government who have a need for this knowledge.

Herbert A. Lieberman
Martin M. Rieger
Gilbert S. Banker

Contents

Contributors

Eric Allémann, Ph.D. Research Associate, Pharmaceutical Technology and Biopharmacy, School of Pharmacy, University of Geneva, Geneva, Switzerland

Gilbert S. Banker, Ph.D. Professor and Dean, College of Pharmacy, University of Iowa, Iowa City, Iowa

Lawrence H. Block, Ph.D. Professor and Chair, Department of Medicinal Chemistry and Pharmaceutics, Duquesne University, Pittsburgh, Pennsylvania

Roland Bodmeier, Ph.D. Professor, Institut fur Pharmazie, Freie Universität Berlin, Berlin, Germany

Matthew Cherian, Ph.D. Director, Pharmaceutical Research, Pharmacia and Upjohn, Nerviano, Italy

Ganesh S. Deshpande, Ph.D.* College of Pharmacy, University of Iowa, Iowa City, Iowa

Isaac Ghebre-Sellassie, Ph.D. Associate Research Fellow, Parke-Davis Pharmaceutical Research, Warner-Lambert Company, Morris Plains, New Jersey

Robert Gurny, Ph.D. Professor and Head, Department of Biopharmaceutics and Pharmaceutical Technology, School of Pharmacy, University of Geneva, Geneva, Switzerland

Samir A. Hanna, Ph.D. Vice President, Worldwide Quality Control and Bulk Quality Assurance, Bristol-Meyers Squibb Company, New Brunswick, New Jersey

Vijay Kumar, Ph.D. Assistant Professor, Pharmaceutics Division, College of Pharmacy, University of Iowa, Iowa City, Iowa

Danilo D. Lasic, Ph.D. Consultant, Liposome Consultations, Newark, California

Jean-Christophe Leroux, Ph.D. Assistant Professor, School of Pharmacy, University of Montreal, Montreal, Quebec, Canada

---

**Current affiliation*: SmithKline Beecham Consumer Healthcare, Parsippany, New Jersey.

Philippe Maincent, Ph.D. Professor, Department of Pharmaceutical Technology, Faculty of Pharmacy, University of Nancy, Nancy, France

Frank Martin, Ph.D. Sequus Pharmaceuticals, Inc., Menlo Park, California

Robert A. Nash, Ph.D. Associate Professor, Department of Pharmacy and Administrative Sciences, St. John's University, Jamaica, New York

Joel B. Portnoff, M.S. Director, Pharmaceutical Research, Fujisawa USA, Melrose Park, Illinois

Joseph A. Ranucci, B.S.Pharm. Associate Scientific Leader, Technical Department, Roche Vitamins, Inc., Nutley, New Jersey

Mohammad Riaz, Ph.D. Associate Professor, Faculty of Pharmacy, University of the Punjab, Lahore, Pakistan

Morton Rosoff, Ph.D. Arnold and Marie Schwartz College of Pharmacy and Health Sciences, Long Island University, Brooklyn, New York

Dina R. Russello, B.S., R.A.C. Manager, Consumer Care Division, Department of Regulatory Affairs, Bayer Corporation, Morristown, New Jersey

Roy R. Scott, B.S., Ch.E. Manager, Department of Sales Engineering, ARDE Barinco, Inc., Norwood, New Jersey

Irwin B. Silverstein, Ph.D. Manager, Department of Quality Assurance, International Specialty Products, Wayne, New Jersey

S. Esmail Tabibi, Ph.D. Project Officer and Chemist, Pharmaceutical Resources Branch, National Cancer Institute, National Institutes of Health, Bethesda, Maryland

John P. Tomaszewski, M.S., R.A.C. Associate Director, Consumer Care Division, Department of Regulatory Affairs, Bayer Corporation, Morristown, New Jersey

Jean Wang, Ph.D. Scientist, Department of Technology, Parke-Davis Pharmaceutical Research, Warner-Lambert Company, Morris Plains, New Jersey

Norman Weiner, Ph.D. Professor, College of Pharmacy, University of Michigan, Ann Arbor, Michigan

Contents of Pharmaceutical Dosage Forms: Disperse Systems, Second Edition, Revised and Expanded, Volumes 1 and 2

edited by Herbert A. Lieberman, Martin M. Rieger, and Gilbert S. Banker

VOLUME 1

VOLUME 2

Contents of Pharmaceutical Dosage Forms: Tablets, Second Edition, Revised and Expanded, Volumes 1–3

edited by Herbert A. Lieberman, Leon Lachman, and Joseph B. Schwartz

VOLUME 1

VOLUME 2

VOLUME 3

Contents of Pharmaceutical Dosage Forms: Parenteral Medications, Second Edition, Revised and Expanded, Volumes 1–3

edited by Kenneth E. Avis, Herbert A. Lieberman, and Leon Lachman

VOLUME 1

VOLUME 2

VOLUME 3

Pharmaceutical Dosage Forms: Disperse Systems

1

Specialized Pharmaceutical Emulsions

Morton Rosoff

Arnold and Marie Schwartz College of Pharmacy and Health Sciences, Long Island University, Brooklyn, New York

I. INTRODUCTION

The definition of an emulsion invariably leads to semantic difficulties. If one considers colloidal dispersions as systems of arrested phase separation, which may or may not be at equilibrium, and places emulsions within this connotation, then a broader classification results that includes microemulsions, vesicles, and liquid crystalline systems. This extension is particularly apt for a chapter on specialized pharmaceutical emulsions, where stability of the formulations is a primary requirement for the application.

Traditionally, emulsions have served to render homogeneous substances that are normally immiscible. Compare, for example, the differences between a salad treated with oil and vinegar separately and one containing mayonnaise. In pharmacy, emulsions have been used orally, parenterally, and topically (mainly in cosmetics), as well as for diagnostic purposes. Inherent instability, nonuniformity, and lack of understanding of drug release mechanisms have been major drawbacks to more widespread use in drug delivery, but with recent interest in controlled release, that is, targeting of drugs to specific sites, and less conventional routes of administration, emulsions, both the conventional and the novel, may find increased usefulness. In addition to specific macroemulsions, multiple emulsions, microemulsions, gel emulsions, and vesicles are discussed herein.

II. MACROEMULSIONS

Ordinary emulsions comprise four possible types: "oil-in-water" (O/W), "water-in-oil" (W/O), "water-in-water" (W/W), and "oil-in-oil" (O/O). Emulsions of the W/W type are formed from aqueous solutions of mixtures of incompatible polymers. They include systems containing polysaccharides, synthetic polymers, and proteins. The O/O emulsions consist of incompatible organic solvents stabilized by block copolymers with residues of differing solubilities in the two components. Multiple emulsions are emulsions of emulsions and consist of more than one dispersed phase, usually oil or water droplets not identical with the continuous phase. Depending on their end use, emulsions vary considerably in their composition, structure, and properties. Rheologically, for example, emulsions range from Newtonian and non-Newtonian liquids, including thixotropic systems, to gel-like solids. Surfactants are necessary in formulating stable emulsions of low viscosity and may be of low or high molecular weight. The surface activity of low molecular weight surfactants is sensitive to the composition of both phases of the system, and it is necessary to have a range of surfactants to match the hydrophilic-lipophilic balance (HLB) of a particular surfactant to the emulsion composition. Unfortunately, except for phospholipids, mono- and diglycerides, and bile salts, the catalog of suitable surfactants is limited. Although their potential scope is considerable, extensive toxicological studies are required to prove the harmlessness of novel surfactants. The O/W emulsions are administered orally as well as parenterally, whereas W/O emulsions are used mainly as diffusion barrier depots when given intramuscularly.

A. Parenteral Emulsions

Injectable emulsions are used mainly for lipids or lipid-soluble materials. They are used for nutritional purposes and delivery of vaccines, and as drug carriers and diagnostic agents. For intramuscular use, paraffin oils as well as oils of vegetable origin are emulsified with nonionic surfactants such as sorbitan fatty acid esters and polyoxyethylene sorbitan fatty acid esters. Via the intravenous route, properly purified vegetable oils with egg or soya phospholipids as emulsifiers are commonly used. The subject has been well reviewed [1–3] and this section highlights salient information.

For depot preparations, where biodegradability or erosion is not desired as part of the release mechanism, a branched long-chain saturated hydrocarbon [4,5] is preferred from the viewpoint of stability and toxicity. However, other considerations depend on the type of drug incorporated and the volume of oil injected.

Of the many vegetable oils tried, soybean, safflower, and cottonseed oils are the most favored because they show the least incidence of toxic reactions and the greatest resistance to oxidation. By modifying vegetable oils enzymatically via biotechnology, structural lipids can be obtained, i.e., 1,3 specific triglycerides of differing chain lengths that influence the overall hydrophobicity and affect drug release and the rate of in vivo hydrolysis [6]. In addition to natural lecithins as emulsifiers, the polyoxyethylene-polyoxypropylene compounds (poloxamers), available commercially as Pluronics (BASF Wyandotte Corp., Parsippany, NJ) [7] are one of the few classes of nonionic surfactants found to be relatively nontoxic, and they are used alone or in combination with lecithin [8].

Table 1 gives typical compositions of soybean and egg lecithin. When pure phospholipids, phosphatidylcholine (PC) and phosphatidylethanolamine (PE) were used on their own, the resulting emulsions showed poor stability [9]. The enhanced stability produced by natural or commercial lecithin presumably stems from the presence of negatively charged phosphatides, which aid in the formation of gel-like liquid crystalline interfacial films [10,11].

Parenteral emulsions are subject to two important constraints: particle size and sterilization. For intravenous emulsions, the particle size cannot be greater than 5 μm without risking emboli in the capillaries. Average particle size for fat emulsions is less than 1 μm and is achieved by homogenization at high temperatures under high pres-

Table 1 Typical Phospholipid Composition of Lecithins (%)

Phospholipid	Egg yolk lecithin	Soybean lecithin
Phosphatidylcholine (PC)	7.3	40
Lysophosphatidylcholine (LPC)	5.8	—
Phosphatidylethanolamine (PE)	15.0	35
Lysophosphatidylethanolamine (LPE)	2.1	—
Phosphatidylinositol (PI)	0.6	20
Phosphatidic acid (PA)	—	5
Sphingomyelin (SP)	2.5	—

Source: Ref. 10.

sure [12]. The resulting homogenized product is then autoclaved at 110°C for 40 min. Surprisingly, such treatment does not affect stability but actually serves to decrease the particle size [3]. Alternatively, parenteral emulsions may be sterilized by filtration, provided the particle size is small enough to pass the filter. Sterilization of individual components and aseptic assembly is also feasible. A major application of lipid emulsions is the delivery of fat in parenteral nutrition. Fat is a concentrated source of energy and can supply essential fatty acids. Since a 30% fat emulsion is not hypertonic, the addition of other nutrients such as carbohydrates and amino acids is permitted. Table 2 shows the composition of various intravenous fat emulsions.

Instability of fat emulsions can arise from changes in particle size of the oil droplets leading to creaming and coalescence, or from changes in pH, hydrolysis of emulsifier, or oxidation of the oil. If the pH of the emulsion is maintained on the alkaline side of neutrality, the stability is satisfactory and the product may be kept at temperatures below 30°C.

The close resemblance of fat emulsion particles to chylomicrons, the natural emulsion particles that transport ingested lipophiles such as triglycerides into the lymphatic and circulatory systems, suggests that fat emulsions can be used as carriers for drugs and as targeted delivery systems [13–18].

Other advantages of such drug-containing emulsions would be their ability to be diluted in vivo in the blood or gastrointestinal tract without ensuing precipitation of solid drug particles, as well as the enhanced stability furnished by a nonaqueous environment. Many drugs, including barbituric acid, diazepam, and anesthetics, have been dissolved in the oil phase and administered by all parenteral routes [19] (Table 3). Jeppson and Ljunberg [20,21] concentrated on intravenous fat emulsions, and some of their original indications of sustained release or targeting have not been substantiated. Such emulsions can affect the metabolism of the drug, which can in turn affect clearance of the particles from the blood [22]. Positive prolonged release effects have, however, been noted for progesterone [23] and corticosteroids [24]. Intralipid has been used to incorporate valinomycin, an antitumor agent [25], and it was found that a 20-fold lower dose was sufficient to produce the same effects as an aqueous suspension. Amphotericin B administered as a lipid emulsion, in comparison with a 5% aqueous dextrose solution, gave a lower incidence of fever and nephrotoxicity and was as effective [26]. The addition of drugs or electrolytes to fat emulsions can cause instability and breaking of the emulsion, and carrier formulations may have to be specially formulated.

Table 2 Composition of Some Intravenous Fat Emulsions (% w/v)

Trade name	Oil phase	Emulsifier	Other components
Intralipid[a]	Soybean 10 or 20	Egg lecithin 12, soybean lecithin 20	Glycerol 25
Lipofundin S[b]	Soybean 10 or 20	Soybean lecithin 0.75 or 1.5	Xylitol 5
Liposyn[c]	Safflower 10	Egg lecithin 1.2	Glycerol 2.5
Nutrifundin[b]	Soybean 3.8	Soybean lecithin 0.38	Xylitol 10 amino acids 6
Trivemil[d]	Soybean 3.8	Soybean lecithin 0.7	Sorbitol 10 amino acids 6
Lipofundin[b]	Cottonseed 10	Soybean lecithin 0.75	Sorbitol 5 tocopherol 0.058
SR 695[e]	Cottonseed 15	Pluronic F-68 0.3, polyethylene glycol monopalmitate 1.2	Dextrose 5

[a]Travenol.
[b]Braun.
[c]Abbott.
[d]Egic.
[e]Southern Regional Research Laboratory.

Table 3 Composition of Diazepam Lipid Emulsion

Diazepam	0.5 g
Soybean oil	15.0 g
Acetyl monoglycerides	5.0 g
Egg yolk phosphatides	1.2 g
Glycerol	2.5 g
Distilled water, q.s. ad	100.0 mL

Source: Ref. 19.

Poloxamers (nonionic block copolymers) have been used as emulsifiers for fat emulsions and appear to slow the clearance of emulsion particles from the plasma as compared to phospholipid-stabilized formulations [27,28]. The presence of liposomal structures (lipid vesicles) or other non-monolayer structures has been observed [29–31] in phosphatide-stabilized emulsions and may affect the physical and biological behavior of such emulsion systems [32].

Emulsion uptake by the reticuloendothelial system (RES) has slowed progress in parenteral drug delivery. Inclusion of polyethylene glycol (PEG) derivatives in emulsion formulations prolongs circulation time presumably by sterically preventing binding of serum opsonins to the emulsion particle surface, thus reducing affinity for the RES [33,34]. Similar results have been reported for other steric stabilizers [35].

The introduction of a nonabsorbable oil into the intestinal lumen in the form of an emulsion can alter the normal absorption of lipophiles. Intervention in this manner may hinder the absorption of dietary lipophilic toxins and possibly reduce tissue depots of toxic lipophiles stored in adipose and other tissues. Kriegelstein et al. deduced, by infusing Intralipid in animals, that blood levels of lipophilic drugs were reduced [36].

B. Fluorocarbon Emulsions

The discovery that animals can survive breathing oxygen-saturated silicones and certain perfluorinated liquids such as the isomers of perfluorotetrahydrofuran (FC-75) suggested that such inert substances might be useful as intravascular O_2 and CO_2 transporting agents [37]. Figure 1 gives structures of some of the newer candidates for artificial blood substitutes. Perfluorocarbons (PFCs) cannot be injected intravenously in the pure state as they lead to formation of emboli. Moreover, they do not dissolve water-soluble materials such as electrolytes, glucose, and other nutrients, so these cannot be administered at the same time. The alternative is to formulate emulsions of the O/W type. Fluosol-DA, (Green Cross Corp., Osaka, Japan), the existing standard used in human studies, is an aqueous emulsion of perfluorodecalin (14% w/v) and perfluorotripropylamine (6% w/v); Pluronic F-68 (5.4% w/v) and egg yolk phospholipids (0.8% w/v) serve as emulsifiers. The interfacial surfactant layer was found to have negligible effect on O_2 transfer through the O/W phase boundary in PFC emulsions [38]. Additives such as small amounts of inorganic salts and other ingredients confer characteristics similar to those of blood plasma. The average particle size is 100–200 nm, and since the long-term stability is low, the product must be kept refrigerated. Fluorinated surfactants have been proposed as potentially superior emulsifying agents [39], although Davis et al. [40] questioned the rationale for their effectiveness, particularly where

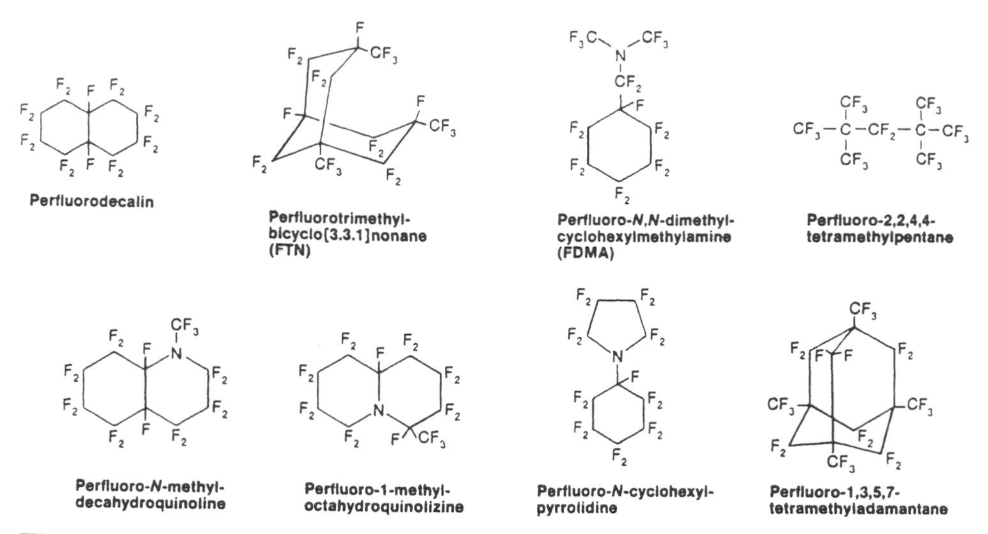

Fig. 1 Structures of oxygen-carrying perfluorochemicals.

emulsion instability is due to the Kelvin effect (i.e., large particles growing at the expense of the smaller ones) [41]. The main degradation pathway is thought to be molecular diffusion rather than coalescence [42]. The stabilizing effect of surfactant molecules on coalescence of droplets has been attributed to their structural rheological properties at the interface [43]. Thus, lower adhesion of the hydrophobic portion of the surfactant with the nonpolar phase would result in greater interaction with the aqueous medium at the interface and enhanced viscosity and strength of the adsorption layer. The greater adhesion energy of perfluorinated surfactants for PFC oils would, on this view, tend to make such emulsions less stable than when a poloxamer is used [44]. Natural egg yolk phospholipids (EYP) and Pluronic F-68 are limited in their adaptability to each therapeutic application and a series of perfluorinated analogs of phosphatidtyl choline [45], and perfluoralkylated derivatives of disaccharides [46] as surfactants or cosurfactants with EYP have been developed to overcome these limitations, but potential toxicity is a problem.

Toxicity and clearance rates from the body are important considerations, and, as has been noted, emulsions that are stable in vitro tend to be slowly eliminated in vivo, and vice versa [39]. This may be related to the lipid solubility of the fluorocarbons, the more lipid-soluble material rapidly passing through cell membranes and out of the body. Table 4 compares the solubility parameters of fluorocarbons with those of related compounds.

Table 4 Solubility Parameters

Compound	δ (cal/min)$^{1/2}$
Fluorocarbons	6–7
Hydrocarbons	7–9
Water	23.4
Oxygen	5.7

From the viewpoint of toxicity and stability, it is desirable that the emulsion particles be small (<1 μm). Microemulsions, which are fluid, transparent, and thermodynamically stable microheterogeneous systems, eminently satisfy these criteria and offer an alternative to commercial macroemulsions currently being used. Besides their long-term stability and spontaneous formation, they provide better solubilization of fluorocarbons, and consequently of molecular oxygen, than macroemulsions [47]. Perfluorinated surfactants are required for formulating microemulsions from perfluorinated oils because of the mutual phobicity of hydrocarbons and fluorocarbons [48]. Fluorinated microemulsions have been prepared using fluorinated ionic surfactants [49], fluorinated polyoxyethylene surfactants [47], and a mixture of hydrogenated and fluorinated nonionic surfactants [50–54]. "Waterless" fluorinated microemulsions as media for oxidation reactions have also been obtained by using formamide in place of water [55]. Unfortunately, these perfluorinated surfactants are inclined to be eliminated slowly and to liberate toxic fluoride ions, making them bioincompatible. By synthesizing high oxygen capacity fluorinated oils containing varying proportions of hydrogenated and fluorinated portions in the molecule, it was possible to produce microemulsions using nonionic surfactants of the poloxamer type (which are biocompatible) [56], although the toxicity of the oil derivatives was not established.

Second-generation emulsions, such as Oxygent, use perfluorooctyl bromide (PFOB) as the oil [57] at higher concentrations of fluorochemical, thus enhancing the potential for gas transport not only of oxygen but also of nitrogen, which can protect against gas embolization. These PFOB emulsions stabilized with egg yolk lecithin as an emulsifier were stable over four years at 50°C [58]. The superiority of lecithins to artificial surfactants is believed to be due to fluorocarbon-filled vesicles [59].

An interesting question is what effect concomitantly administered drugs have during treatment of patients with perfluorocarbon blood substitutes. Little is known about the solubility, pharmacokinetics, and possible interactions of other chemicals with perfluorocarbons. Tremper et al. [60] showed that there is a significantly increased blood-gas partition coefficient of inhalation anesthetics in perfluorocarbon and perfluorocarbon-blood mixtures. Following penicillin exchange transfusion in rats with Fluosol-DA, the half-life was prolonged [61], possibly due to blockage of the rate-determining step in renal elimination. The possibility of altered pharmacokinetics prompts caution in the administration of drugs to patients treated with fluorocarbon emulsions [62].

The use of perfluorocarbon emulsions as vehicles to transport drugs to specific target sites is being explored. Where oxygenation mediates the action of radiotherapy or of cytotoxic drugs, fluorocarbon emulsions may enhance the anticancer effects [63]. Oxygen can be imaged by NMR through the effect on the spin-relaxation time of ^{19}F. Oxygen-carrying contrast agents, which can be taken up by the liver, spleen, or tumors, may prove useful in angiography.

C. Radiopaque Agents

Radiopaque substances are used in conjunction with x-rays to visualize body organs and structures. Oil-in-water emulsions of iodized oil, iophendylate, and ethiodized oil (Ethodiol) injected into the peritoneal cavity of rats were found to give excellent outlines of the peritoneal cavity and in some cases of special structures such as the spleen [64]. Intrathecal injections of iophendylate emulsions into dogs and into the carpal joint of a horse showed good contrast and demonstrated good mixing with the respective body fluids. Intravenous administration in humans of ethiodized fat emulsions with a particle diameter of 1–4 μm gave superior results in computerized tomography of hepatomas [65].

As mentioned previously, perfluorocarbon emulsions, particularly those containing a bromide atom, have been found to be useful in angiography and bronchography. Perfluorooctyl bromide emulsions stabilized with Pluronic F-68 were assessed for radiographic quality after administration in animals and man [66]. The flow properties of these non-Newtonian systems may play a role in the fineness of details observed in radiographs. When combined with F-alkyl iodides, stability was maintained and the radiopaque effect was enhanced [67].

D. Water-in-Oil Emulsions

Water-in-oil emulsions administered by subcutaneous or intramuscular routes are usually designed to exert a depot effect or delayed release of drug where the barrier to diffusion is the oil continuous phase. This is particularly exemplified in their use as adjuvants that increase the immune response induced by an antigen. Mineral oil [68] as well as low molecular weight hydrocarbons and sesame and peanut oils [69,70], have been used as the oil phases in emulsions consisting of aqueous antigen and nonionic emulsifiers such as Arlacel A (ICI Americas, Wilmington, DE) [70] or zwitterionic lecithin [71]. Erodability of the oil is an advantage in preventing abscess formation at the site of injection and in enhancing release by gradual breakdown. The mechanism of adjuvant action is not well elucidated. In some cases the effect is attributable to the nature of the surfactant's ability to adsorb to cell membranes, to cause aggregation of various cell types involved in the immune response, or delay absorption of the antigen. The effectiveness of some formulations as a function of the amount of surfactant and oil used suggests that particle size may also be important [72].

E. Oral Emulsions

Davis [73] reviewed the renewed interest in emulsions as drug delivery systems that may modify the bioavailability of orally ingested drugs. Early work describing the more complete absorption of drugs in O/W emulsions administered orally as compared to those administered with an aqueous suspension, dealt with sulfonamides [74] and indole [75]. Griseofulvin suspended in the oil phase of an O/W emulsion was found to give enhanced bioavailability, presumably due to the effect of metabolic products of the oil on gastric emptying time [76]. This view of the effect of the oil has been further elaborated by Bates and Sequeira [77] and by Palin et al. [78] (Fig. 2). The influence of the HLB of surfactants on ephedrine absorption and release from mineral oil emulsions administered orally to dogs was observed by Lin et al. [79].

In an attempt to explain the mechanism of intestinal absorption of drugs from O/W emulsions, Kakemi and coworkers [80,81], using both in vitro and in vivo models, concluded that drugs are absorbed mainly via the aqueous phase; the absolute volume of the aqueous phase is critical for poorly oil-soluble drugs ($K < 1$); the amount of drug in the aqueous phase is important for $K > 1$, where K is the O/W partition coefficient of the drug. When the initial amount of drug in the aqueous phase is large in comparison with the total amount or when little drug exists in the aqueous phase initially, the rate limiting step for absorption is drug transfer from oil to water rather than from oil to membrane. In intermediate states the kinetics of absorption is due to a combination of these processes and is no longer first order. Ogata et al. [82] took into ac-

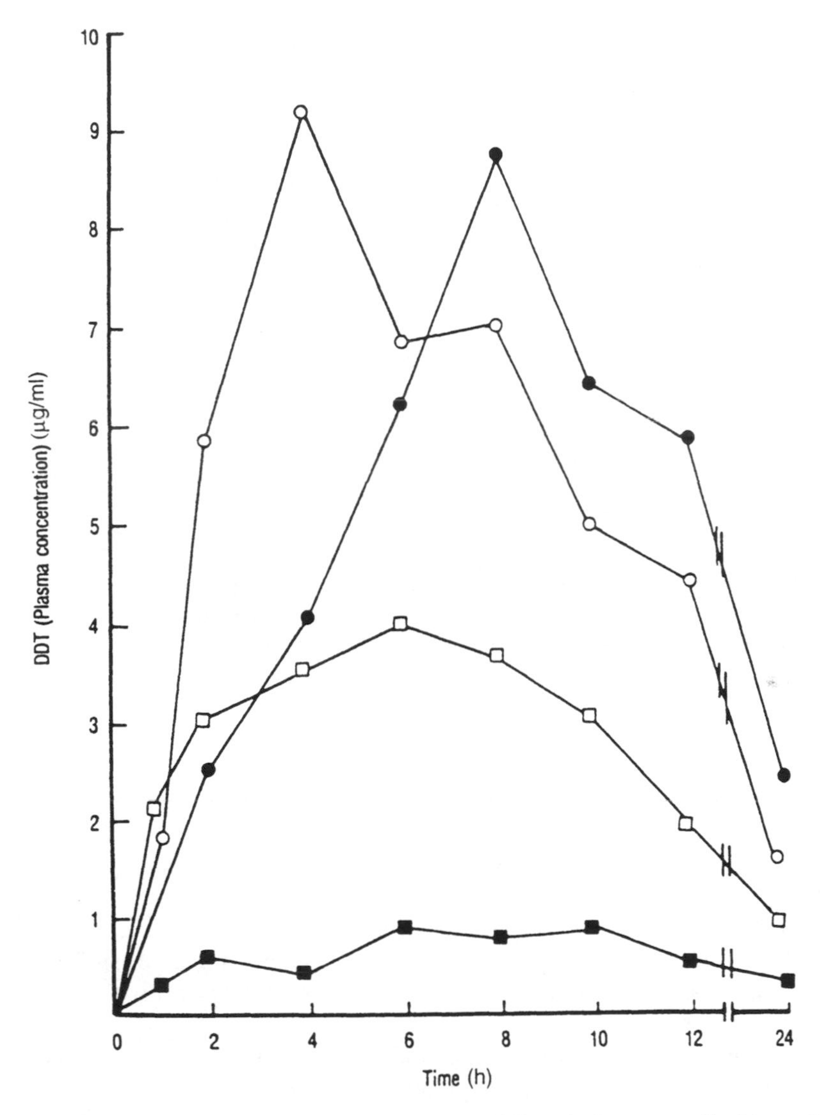

Fig. 2 The effect of emulsification and the difference between a liquid paraffin emulsion and a vegetable oil emulsion: mineral oil, squares (solid square = oil solution, open squares = emulsion); arachis oil, circles (solid circle = oil solution, open circle = emulsion). DDT orally administered to rats. Emulsifier, Tween 80, 6P6 v/v. (From Ref. 73.)

count the competing influence of micelles. As shown in Fig. 3, if adsorption takes place from a micellar phase, (b), oil droplets will compete with micelles for adsorption to the mucosal surface. On the other hand, if absorption is from free drug (a) in the aqueous phase, (a) adsorption of oil droplets will not inhibit absorption. In this scheme, possible binding of drug molecules to oil droplets is not considered, and the role of micellar absorption has not been resolved. In contrast to these results, Noguchi et al. [83] view the absorption process for an oil-soluble drug as comprising adsorption of oil droplets to the mucosal surface, partitioning into membrane lipids, release into inner compartments, and transport via the portal or lymphatic systems.

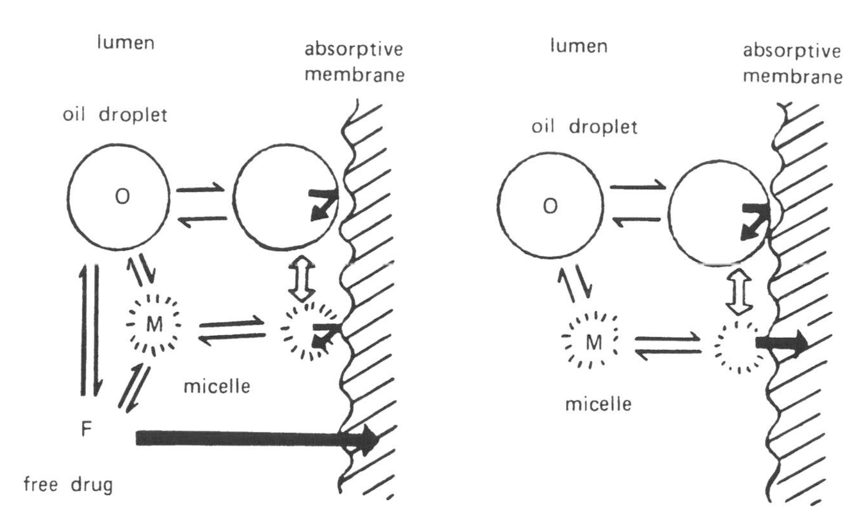

Fig. 3 Schematic of the relationship between absorption and drug distribution in an O/W emulsion, where O, M, and F represent drug in oil droplets, micelles, and aqueous phase, respectively. Arrows represent equilibria and two-headed arrows competition between states. (From Ref. 82.)

Bloedow [84] described the first stage of absorption of oils as hydrolysis at the oil-water interface followed by solubilization into bile salt micelles, which are assumed to penetrate into the villi of the small intestine; a drug could thus be carried along with its lipid solvent. The lipid component of lipid-bilesalt mixed micelles or perhaps of the finely divided emulsion state, is thought to enhance the fluidity and permeability of the mucosal membrane to poorly absorbed drugs [85].

Given the various ways for absorption to take place, Amstrong and James [86] outlined the parameters influencing the choice of route and method of absorption from lipid vehicles; they are solubility of drug in vehicle, aqueous solubility of drug, nature of lipid, and nature of drug.

It appears that highly lipophilic drugs are absorbed by a lipid transport process, and the less lipophilic drugs are first dissolved in the aqueous luminal fluid. The lipid solvents such as oils would promote absorption of the former drugs, and the less hydrophobic solvents such as PEG would further uptake of the latter. The importance of the lipid vehicle has been pointed out by Bloedow and Hayton [87], who suggest that digestible lipids function as carriers for lipophilic drugs, whereas nonpolar, nondigestible lipids retard the absorption of lipid-soluble drugs.

Lipid microemulsions for improving drug dissolution and oral absorption have been reviewed [88]. Water-in-oil microemulsions appear to enhance the absorption of water-soluble drugs and protect proteins, for example, from degradation [89–91]. Drugs derivatized with cholesterol to achieve hydrophobicity could be formulated for delivery in phospholipid microemulsions [92]. Two advantages these have over vesicles as drug delivery systems are the greater physical stability in plasma and the resistance of the internal phase to leaching.

Self-emulsifying systems are isotropic mixtures of oil and surfactant that form fine O/W emulsion droplets when introduced in aqueous phases under conditions of gentle agitation [93]. The rapid emulsification that takes place in the aqueous contents of the stomach can provide a large interfacial area for partitioning of drugs subject to dissolution-rate-limited absorption. Such formulations are an alternative to more traditional oral ones for lipophilic compounds.

Use of the potential of oral emulsions for promoting the bioavailability of drugs depends on a greater appreciation of the mechanisms underlying drug release and absorption as well as the ability to control stability. Offsetting these limitations are the greater availability of emulsifiers (hydrocolloids, nonionics) than exists for parenteral emulsions and the wide range of vegetable oils. The GRAS (generally recognized as safe) listings of the U.S. Food and Drug Administration provide a useful index [94].

1. Lymphatic Absorption

Drug delivery via lymphatic circulation, although slow, has the advantage of avoiding first-pass effects and attaining high concentrations of antimetastatic agents along lymphatic pathways [95]. Delivery of bleomycin into lymph by intraperitoneal and intramuscular injection of W/O and O/W emulsions was found to be more effective than similar injections of aqueous solutions in the treatment of malignant lymphomas [96]. The W/O emulsions gave better absorption than the O/W, since bleomycin is water soluble. Nevertheless, enough drug was bound to the oil droplets to improve the absorption into lymph over the aqueous formulation. To overcome stability problems, emulsions have been spray-dried and then reconstituted [97]. Use was made of this technique for bleomycin emulsions [96].

The emulsifier can greatly influence the amount of drug bound to oil droplets as in the case of O/W emulsions of mitomycin C, where gelatin gave better results than polysorbate 80 [98]. When high concentrations of gelatin (20%) were used in forming W/O emulsions of antineoplastic drugs, the internal aqueous phase was replaced by gelatin microspheres approximately 1.5 μm in diameter [99]. Maximum concentrations of drug in the lymph nodes were higher for the microsphere-oil emulsion than for the W/O formulation [100,101]. Probably both emulsion types form multiple emulsions on injection, and the greater stability and multiplicity of the internal droplets of the gelatin emulsions enhance lymphatic transport and prolongation of release [102]. In Fig. 4 the concentration-time course of bleomycin in the regional lymph nodes shows the enhanced and sustained delivery of the microsphere emulsion in comparison to intravenous and topical injection of aqueous solutions [103]. The beneficial effect of a bleomycin microsphere-in-oil emulsion for treatment of lymphangioma has been reported [104].

III. MULTIPLE EMULSIONS

The term "multiple emulsion" has been used to describe emulsion systems in which drops of the dispersed phases contain smaller droplets that have the same composition as the external phase. The potential of these systems for application in pharmacy and separation technology has generated increased attention, and comprehensive reviews have dealt with their preparation, stability, and pharmaceutical uses [105–107].

A schematic illustration of a W/O/W emulsion drop is shown in Fig. 5; photomicrographs of three different types appear in Fig. 6.

Multiple emulsions, which appear to form most readily when an emulsion is inverting from O/W to W/O or vice versa, are inherently unstable. By proper choice of

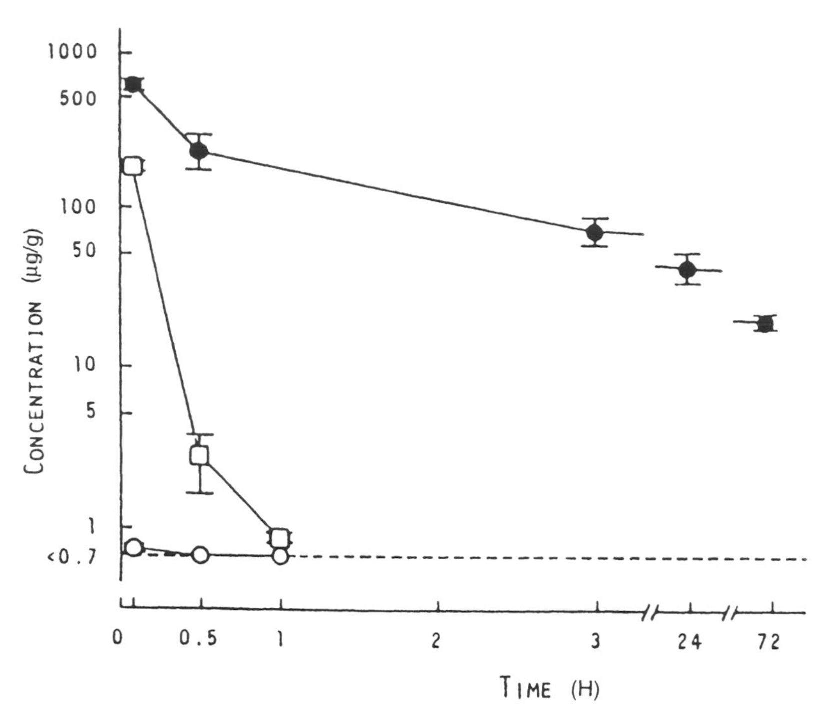

Fig. 4 Concentration of bleomycin in the regional lymph node of rabbits after intravenous injection of aqueous solution (open circles), topical injection of aqueous solution (open squares), and topical injection of gelatin microsphere-in-oil emulsion (solid circles). (From Ref. 103.)

surfactants the primary emulsion can be formed using, for example, a lipophilic, low-HLB surfactant. The primary emulsion can then be emulsified with a hydrophilic, high-HLB surfactant, using low shear, to produce a W/O/W emulsion of reasonable stability (Fig. 7). The reverse procedure could be followed to prepare an O/W/O emulsion. Microporous glass membranes with a narrow range of pore sizes can facilitate preparation of W/O/W emulsions by application of pressure [108].

Although W/O/W emulsions have many of the attributes of W/O emulsions, their lower viscosity, derived from water as the external phase, makes them easier to inject. Adjuvant effects have been reported to be improved [109,110] compared to W/O emulsions or aqueous solutions of antigen. Similar increases in effectiveness of anticancer drug delivery using multiple emulsions have been observed [111,112]. The most prom-

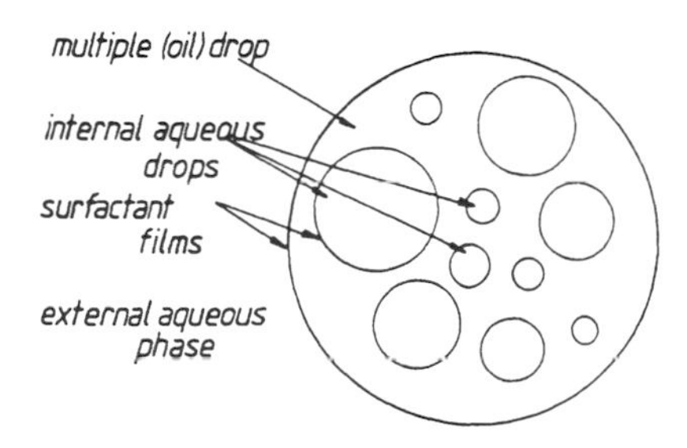

Fig. 5 Structure of a W/O/W emulsion drop.

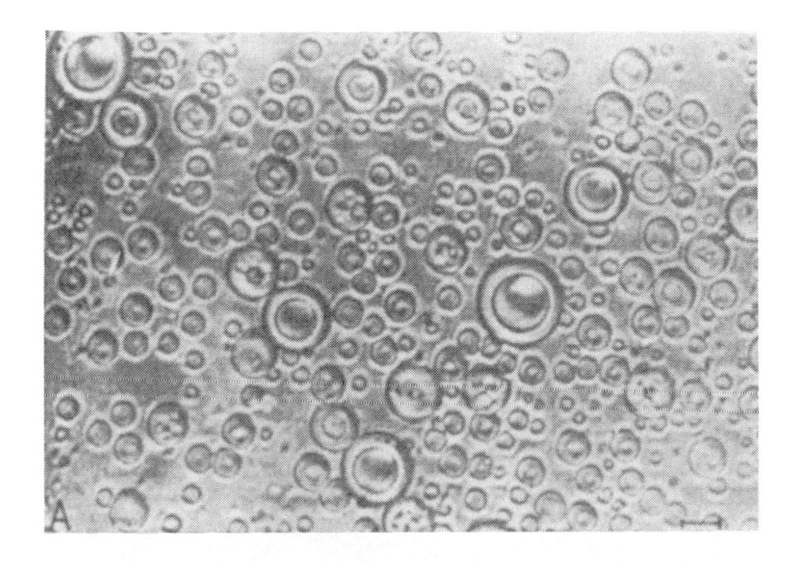

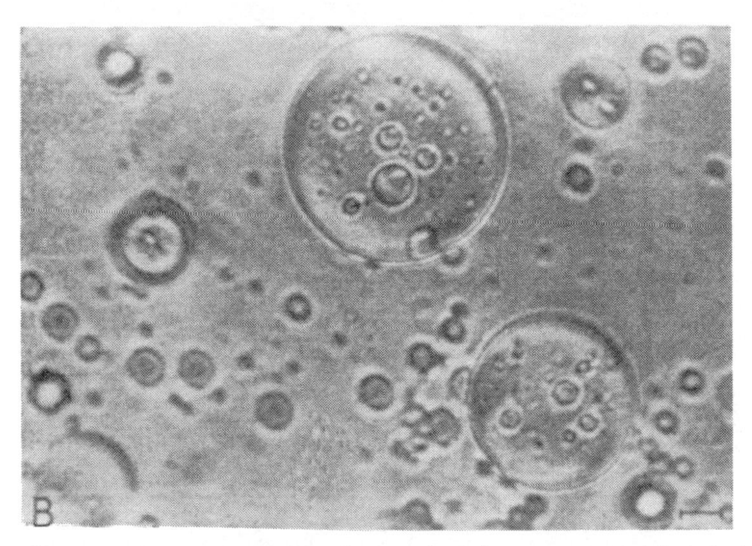

Fig. 6 Micrographs of water-isopropyl myristate-water emulsions designated types A, B, and C according to Florence and Whitehill [105] Type A drops contain one large internal droplet. Type B contain several small internal droplets. Type C consist of large drops in which many smaller internal droplets are entrapped.

ising use of multiple emulsions is in the area of sustained-release drug formulations, since the oil layer between two aqueous phases can behave like a membrane controlling solute release [113].

"Liquid membrane" emulsions of the O/W/O type have been used to separate hydrocarbons where the aqueous phase serves as the membrane and a solvent as the external phase [114]. The system W/O/W, on the other hand, can extract contaminants from wastewater, which acts as the external phase [115]. A multiple-emulsion formulation has also been proposed for the treatment of drug overdose [116]. An aqueous base-in-oil system was designed [117] to remove acidic drugs such as barbiturates and

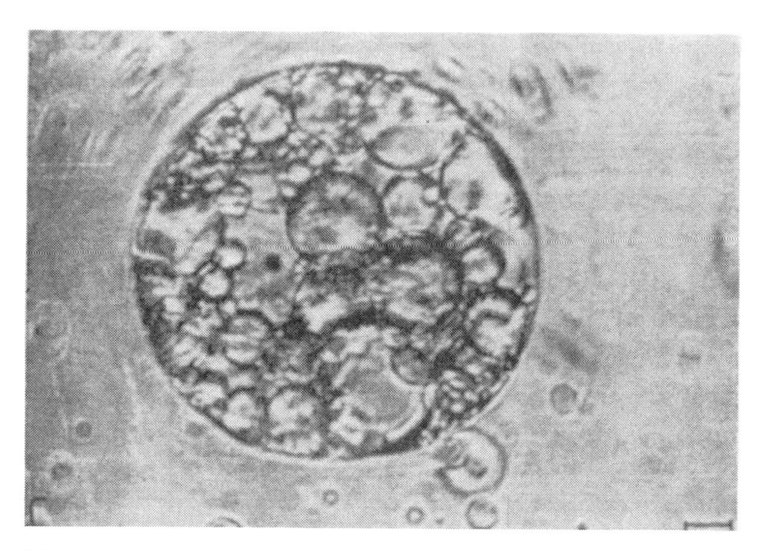

Fig. 6 Continued

salicylates from the gastrointestinal tract by entrapping unionized drug permeating through the oil membrane into the inner basic phase, where it is converted to an oil-insoluble anion. Other trapping mechanisms that could be envisioned include antibodies, enzymes [118], and activated adsorbents. The rate of in vitro drug removal was observed to be quite rapid; under favorable conditions up to 95% of phenobarbital was removed in 5 min, and acetylsalicylic acid was extracted even faster. The extraction kinetics were first order, with the viscosity of the oil phase as the rate-determining parameter.

A novel technique for taste-masking of drugs using multiple emulsions has been proposed [119]. By dissolving drug in the inner aqueous phase of a W/O/W emulsion under conditions of good shelf stability, the formulation is designed to release drug through the oil phase in the presence of gastric fluid. The use of taste-masking and slow release of an antimalarial drug from a W/O/W multiple emulsion has been described [120].

Related to multiple emulsions are the "polyaphrons," a term coined by Sebba [121]. Polyaphrons are three-liquid-phase dispersions, the internal phase being stabilized by encapsulation in a thin, aqueous soapy film. The film, as in a soap bubble, behaves as though it were a phase distinct from bulk water. Polyaphrons exhibit a foamlike character in which the oil-encapsulated cells, the "aphrons," aggregate to form stable polyhedral structures. Dispersions containing up to 97% of dispersed oil phase within a continuous structure that contains only 3% water is claimed to be achievable. Proposed applications are safety fuels, enzyme immobilization, cleansers, soil removers, and cosmetics.

A. Stability of Multiple Emulsions

Figure 8 summarizes some of the breakdown pathways that can occur in W/O/W emulsions. Rupture of the oil layer and consequent loss of the internal aqueous drops can result if the concentration of the primary emulsifier is depleted by transfer through the oil layer to the external aqueous component. Partitioning effects as well as mixed mi-

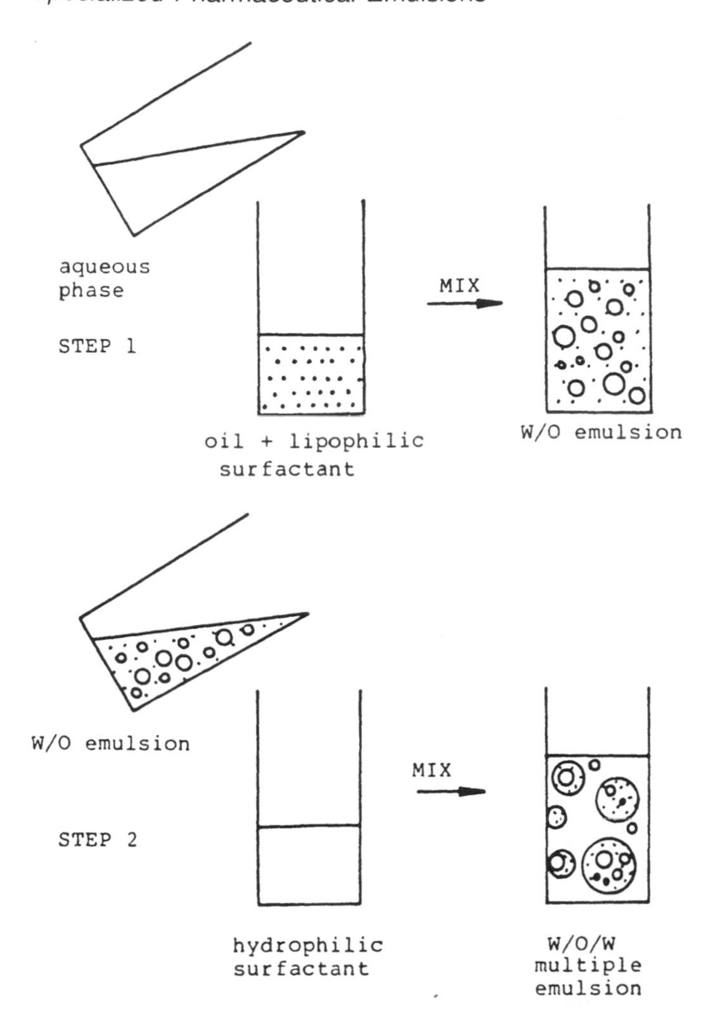

Fig. 7 Two-stage method of formulation of a W/O/W multiple emulsion.

celle formation of the two surfactants may be involved, suggesting that there is an optimum ratio of primary to secondary surfactant for greater yields of preparation and high stability [122]. Emulsifier migration is responsible for a decrease in the effective HLB of the second emulsifier, that is, an increase in the optimal HLB proportional to the concentration of primary emulsifier. At a fixed concentration of primary emulsifier, the HLB shift is inversely proportional to the concentration of secondary emulsifier. From these linear relationships it is possible to predict the optimal HLB if the two emulsifier concentrations and the required HLB of the oil are known [123]. Inversion of W/O/W emulsions to O/W emulsions occurs if the HLB of the total emulsifiers approaches the required HLB of the oil or if the droplet size, as a result of increasing secondary emulsifier concentration, becomes too small to contain the internal water droplets [122,124].

An important criterion for the pathway being followed is the lowering of the interfacial free energy [125]. Coalescence of multiple drops (pathway a in Fig. 8) pro-

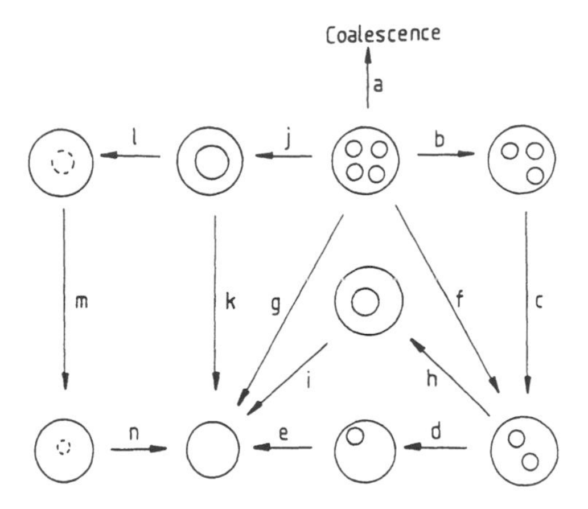

Fig. 8 Breakdown processes that may occur in multiple emulsions [105]. Pathway a, coalescence of multiple oil drops with other oil drops, single or multiple. Pathways b, c, d, and e, expulsion of single internal droplets. Pathways f and g, expulsion of more than one internal droplet. Pathways h and j, coalescence of internal droplets before being expelled (i and k). Pathways l, m, and n, shrinkage of internal droplets due to diffusion of water through the oil phase.

duces a large decrease in the water-oil interfacial area, whereas coalescence of the small internal droplets (pathway jk) would not contribute significantly to breakdown.

Notwithstanding the complexity of a theoretical analysis, some attempt has been made to understand the stability of multiple emulsions in terms of interactions. Some of these are depicted in Fig. 9. It was calculated, for example, that the reduction in diameter of a multiple emulsion drop as a result of expulsion of internal droplets by rupture of the oil membrane leads to a significant diminution of the van der Waals forces of attraction between multiple emulsion droplets (A, in Fig. 9) [125]. From a consideration of the free energy change on expulsion of an internal aqueous droplet, emulsions that contain only a few small droplets would be expected to be more stable than the ones containing a single large droplet or many smaller droplets. The greater attraction

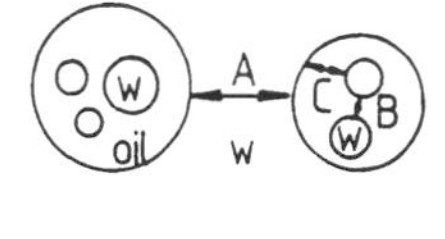

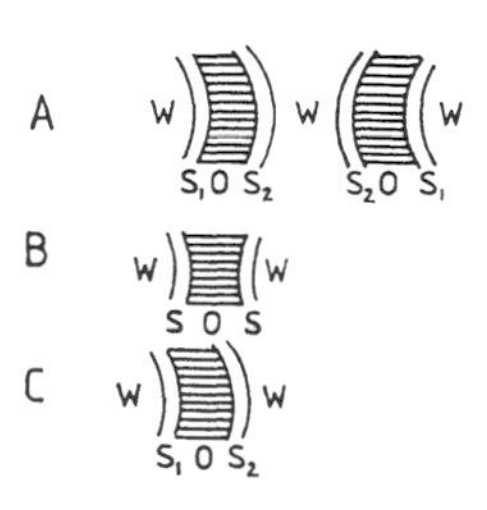

Fig. 9 Possible interactions A, B, and C between phases in multiple emulsions (top). Schematic of structures of oil (O) and water (W) phases and surfactant films (S1, S2) involved in attractive and repulsive interactions. (From Ref. 106.)

of internal water droplets for the continuous phase than for each other (C vs. B) predicts that a decrease in the size of the external droplet for a given W/O primary emulsion would lead to decreased stability of the internal droplets. Nevertheless, theoretical interpretations are incomplete and are complicated by the often observed stabilizing presence of liquid crystalline phases at the water-oil and oil-water interfaces of both W/O/W and O/W/O multiple emulsions [126]. Based on the permeability coefficients of water across the oil membrane, Matsumoto [127], inferred the similarity of the mechanism of water permeation to lipid membrane systems and concluded that the structure of the oil layers consists of multilamellar leaflets of hydrophobic emulsifier. Such liquid crystalline phases are known to contribute to the stability of ordinary emulsions [128,129] and even to form secondary droplets, that is, aggregates of emulsion droplets surrounded by closed lamellae in the continuous phase, which improve the stability and water-retaining ability of the emulsion [130]. A microemulsion in equilibrium with excess oil or water when blended with the conjugate phase may form multiple emulsions of the type: M.E./oil/M.E. and oil/M.E./oil [131]. The emulsion stability will be related to the microstructure that surrounds the emulsion droplets.

A significant feature of multiple emulsions as compared to ordinary emulsions is the osmotic gradient for the passage of water across the oily membrane. Swelling or shrinking of the internal droplets can influence the stability of the multiple emulsion. Electrolytes, proteins, sugars, and drugs engender osmotic gradients and can affect viscosity, which is sensitive to the volume of the dispersed aqueous phase. Rupture of the oil layer causes the volume fraction of the disperse aqueous phase to disappear and the viscosity to fall, thus permitting the evaluation of the stability of W/O/W emulsions by following changes in viscosity with time [132]. Using this technique, it was found that addition of sodium chloride to the aqueous inner phase promoted aggregate formation on aging, but various organic acids such as acetic acid, citric acid, and ascorbic acid actually rupture the oil layer [90]. An isotonic condition can be attained by the addition of glucose or sucrose to both aqueous phases.

B. Drug Release

Stability and drug release are closely interrelated. Coalescence of the internal aqueous phase and rupture of the oil droplets will release drug molecules directly into the dispersion medium. On the other hand, diffusion across the thin oil membranes is a possibility if the oil layer is not considered to be semipermeable (i.e., permeable to solvent molecules only). Water permeability coefficients have been calculated [125,133], assuming that true osmotic flow occurs in W/O/W systems. There is evidence, however, that factors other than osmotic gradients affect the flow of water molecules in multiple emulsions. Matsumoto et al. [133] showed that the permeability coefficient decreases with an increase in concentration of glucose in the aqueous phase. Davis and Burbage [134] found that in larger multiple drops a critical concentration of sodium chloride is needed to effect shrinkage, and Tomita et al. [135] suggested that viscosity changes in multiple emulsions could be explained by assuming that solutes, as well as water, can permeate the oil layer. The release of naltrexone hydrochloride from W/O/W emulsions was investigated in developing a sustained release dosage form for narcotic antagonism [113]. Release rates from the internal aqueous phase were decreased by the addition of increasing amounts of sodium chloride and sorbitol to the internal aqueous phase. Results of a study confirming that migration of electrolytes from the internal to the external aqueous phase is controlled by diffusion rather than breakdown of droplets [136] imply that for a dosage form it is not sufficient to obtain stable emulsions without also controlling diffusion.

Ionized or unionized materials may be transported across the oil phase by swollen or reverse micelles [124,137], or water may diffuse across regions of thin bilayers of surfactants at the oil-water interface [137,138]. Another possibility is facilitated diffusion [139], in which a carrier molecule interacts, as in ion pairing, with the drug in the internal phase or the membrane to render it lipophilic. Carrier molecules are essentially conserved because of their low aqueous solubility; they diffuse from one interface to the other alternately picking up and releasing drug in a zero-order fashion. Finally, and perhaps most important, there is diffusion of the unionized portion of the drug through the oil layer. For lipid spheres of radii of less than 1 mm encapsulated by films, however, it is calculated that interfacial transfer rather than free diffusion may be the rate-controlling step [140], and release over short periods of time will be effectively zero-order.

C. Multiple Emulsion Stabilization

Multiple emulsions accentuate the properties of ordinary emulsions that are promising for pharmacy, such as the ability to compartmentalize drugs, but they aggravate such pitfalls as instability. Attempts to block the principal routes of breakdown of multiple emulsions by gelation of internal or external phases were undertaken by Florence and Whitehill [141]. Gelation was accomplished by cross-linking acrylamide monomers or poloxamer emulsifiers by γ-irradiation. In the internal phase, W/O/W systems with polyacrylamide gel resembled the gelatin microsphere oil-water emulsions discussed earlier [101,142]. To overcome the undesirable exposure of drug to irradiation, poloxamer surfactants have been developed that gel by interfacial polymerization [143]. Interaction of polymer and surfactant at the interfacial boundary to create a barrier to coalescence is yet another approach [144,145]. Nonionic surfactants and macromolecules such as bovine serum albumin have been used to form a complex interfacial membrane that prolonged the release of drug from the internal aqueous phase of W/O/W droplets. With poloxamer nonionics, the droplets were found to be encapsulated in aqueous surfactant membranes similar to "polyaphrons" [121,146]. Reinforcement of the inner aqueous phase with co-emulsifiers has been proposed [147]. Incorporation of co-emulsifiers into the external water phase may induce structure formation surrounding the multiple droplets and improves the temperature stability of W/O/W emulsions [148]. Comb-graft hydrophobic copolymeric surfactants enable the formation of small droplets of increased stability [149].

Surfactant migration and osmotic gradients are largely responsible for the notably poor physical stability of multiple emulsions. To counteract the physical instability of emulsions, especially multiple emulsions, formulations containing emulsifying agents have been gelled, dried [150], and, using sucrose as a carrier, converted to glasses [151]. Such solid-state dispersions obviate the need for a second emulsifying agent and thus promote increased stability of the multiple emulsion formed when adding an aqueous phase to the solid.

Microencapsulation by forming multiple emulsions of immiscible polymer solutions has been described by Morris and Warburton [152]. Adsorbed polymers at the interfacial boundaries provide a barrier to coalescence, and microcapsules composed of three layers of polymer material have been formed. Proteins have been microencapsulated by double emulsion evaporation in a matrix of polylactide [153]. The internal structure of the microparticles formed can be varied from a multivesicular to a matrixlike structure. Such microporous materials may have applications in controlled drug delivery.

IV. GEL EMULSIONS

Highly concentrated W/O emulsions that form in the water-rich regions of water/nonionic surfactant/oil systems have been termed gel emulsions [154,155]. The internal structure of gel emulsions can be of the multiple emulsion type, or of aphrons when the content of dispersed phase is so high that close packing of spherical liquid droplets is not possible and the droplets change to a polyhedral shape to fill up the available space. These emulsions can be formulated with a large amount of water ($> 99\%$ w/v) and very low surfactant concentrations ($< 0.5\%$ w/v). They are structurally foamlike, with a gel appearance, and are transparent or white depending on composition and temperature. At equilibrium these gels can be considered as W/O emulsions in a W/O microemulsion or reverse (W/O) swollen micelles as the continuous phase. For a not too high water content these systems are equivalent to a dispersion of water globules in a pure oil phase surrounded by a surfactant monolayer. Depending on the oil/surfactant ratio, liquid crystalline phases may also be present [156]. Highly concentrationed W/O gel emulsions can be spontaneously formed from microemulsions or oil-swollen aqueous micellar solutions with a quick change in temperature. During the temperature change, the spontaneously self-organizing structure goes from a water continuous microemulsion to a reverse micelle via vesicle, lamellar (L_α) liquid crystalline and bicontinuous (L_3) structures [157]. In the bicontinuous structure, oil and water form two intertwined continuous networks separated by a film of surfactant continuous in three dimensions. The conversion of an O/W polyaphron to a W/O microemulsion on exposure of the polyaphrons to alcohol has also been observed [158]. Apparently, the alcohol affects interface fluidity, resulting in the aggregate structure that is formed.

Both normal and reverse surfactant-based gel emulsions are of intrinsic theoretical interest and have potential for use as formulations in pharmaceutical, cosmetic, food, chemical engineering, and other technologies. They are useful in combining dissimilar polymers into multiphase blends due to the existence of phase domains [159]. Diffusion and release of solute molecules through the interfaces of gel emulsions have been studied [160].

Microemulsion-based gels have been formed by dissolving gelatin in W/O microemulsions [161]. The gelatin molecules undergo a coil to helix transition forming helical linkages via H-bonds, creating a nanodroplet network or gel. Lecithin organogels can be made by adding a small amount of water to lecithin reverse micelles or W/O microemulsions [162]. The droplets form rodlike aggregates, and at a critical water concentration large cylindrical reverse micelles form a network that can dissolve proteins as well as drugs and bioactives. Domains of aggregated units equivalent to microemulsions in the cubic phase are typical binary phase gels—transparent viscous solutions of rodlike micelles [163]. They possess unique viscoelastic properties, so that when struck with a soft object they produce metallic sounds, and therefore have been called "ringing gels." Gels can also be formed where the assemblies are lamellar or hexagonal.

V. MICROEMULSIONS

Microemulsions have been treated separately in this volume, but since they are often considered, perhaps inaccurately, as specialized emulsions they are briefly discussed. If a definition is needed, the one of Danielsson and Lindman [164] may seem appropriate: "A microemulsion is defined as a system of water, oil and amphiphile which is a single optically isotropic and thermodynamically stable liquid solution."

This interpretation lays stress on phase equilibria of systems that may often contain four or more components. It excludes any description of aggregate structure, which may be variable, and eliminates thermodyamically unstable systems such as ordinary emulsions and transparent emulsions containing very small droplets of long-lived kinetic stability. Included are solubilized, normal micellar solutions, reverse micelles, cores or droplets of water or oil, and bicontinuous structures, in which neither oil nor water can be said to surround the other [165] (see Fig. 10). This definition should be widened, however, to include metastable states, spontaneous emulsions of long-lived kinetic stability [166]. Spontaneous emulsions that cannot be broken appear frequently in the vicinity of a microemulsion domain in phase diagrams [167], and the distinction between macroemulsions and microemulsions, solely on the basis of thermodynamic stability, is not discerning enough. Does the kinetic stability of spontaneous long-lived coarse emulsions involve the same or a different set of interactions than that of microemulsions? There is no difference in nature between swollen micelles and microemulsion droplets and they can be considered synonymous. The properties of the internal phase of swollen micelles evolve smoothly toward those of a bulk phase without any well-defined transition.

The phenomenological approach using phase diagrams has been extensively studied [168] and has yielded a comprehensive knowledge of how to formulate microemulsions. Figure 11 depicts a way to map the phase behavior of a four-component system, and Fig. 12 shows a planar section representing a reverse micelle or W/O microemulsion. Microemulsions can be in equilibrium with excess oil, excess water, or both (Figs. 13 and 14).

Liquid crystalline phases are associated with microemulsions and often obtrude on their regions of formation (Fig. 15). Kahlweit et al. [169], in particular, has systematized features of the phase diagrams by associating microemulsions with critical phenomena.

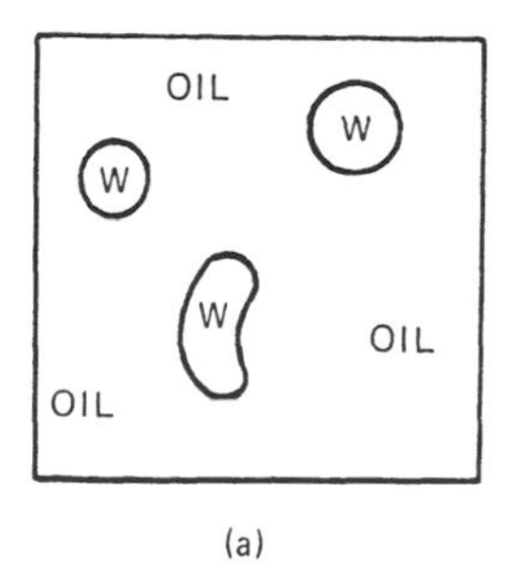

(a)

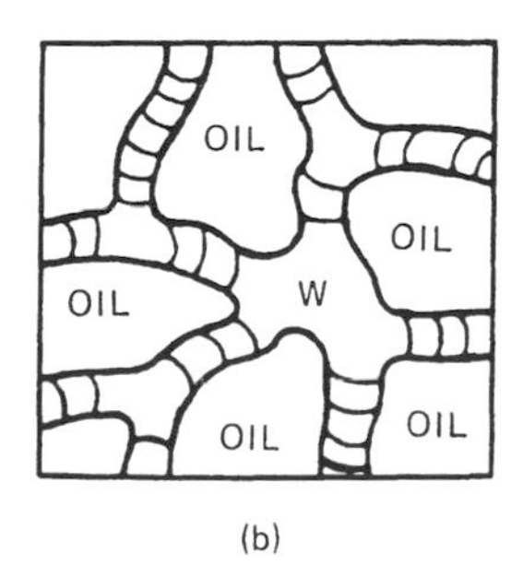

(b)

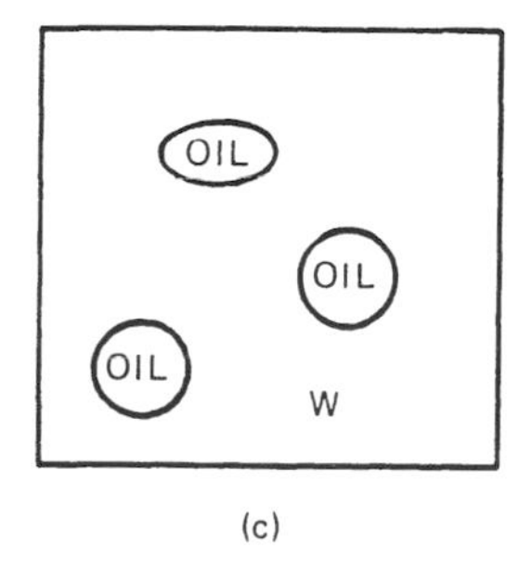

(c)

Fig. 10 (a) and (c) At low concentrations of oil and water, swollen micelles may be formed. (b) A network of water tubes in an oil matrix at intermediate ratios of water to oil—the bicontinuous structure. (Surfactant film is represented as a continuous sheet.)

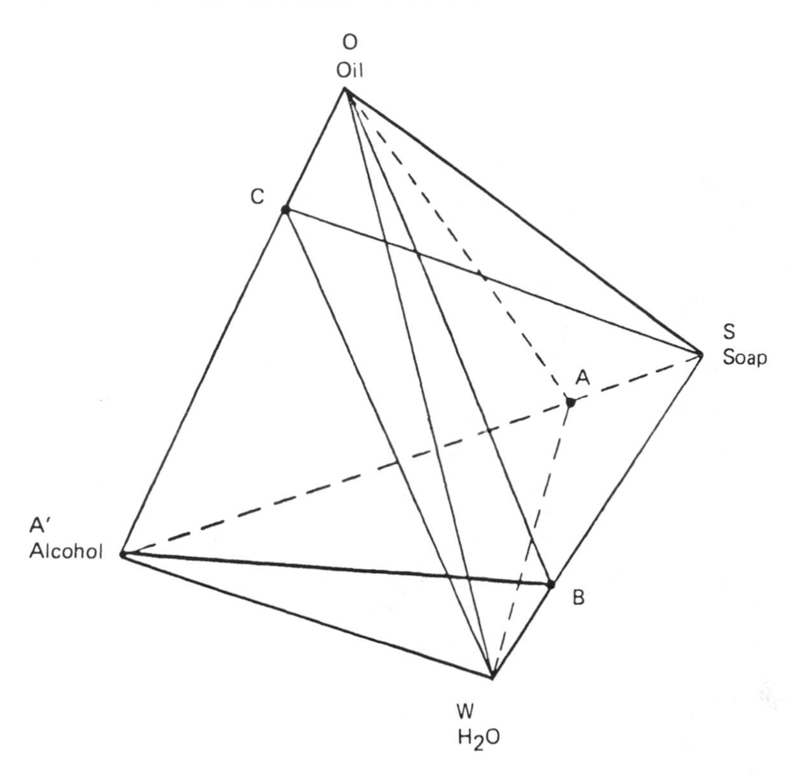

Fig. 11 Phase behavior of a four-component system. Plane sections correspond to two components kept in a fixed ratio.

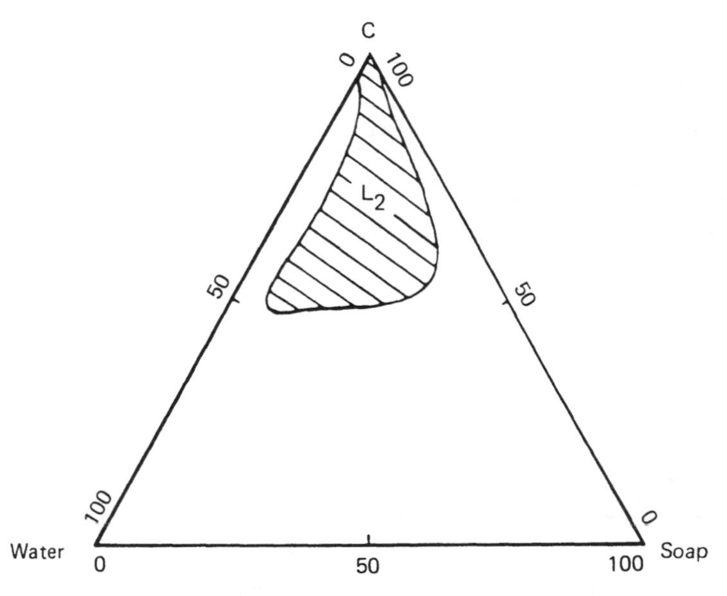

Fig. 12 Water-in-oil microemulsion for a system of potassium oleate, water, and a fixed ratio of 1-hexanol to hydrocarbon (C). This planar section corresponds to triangle OSW of Fig. 11.

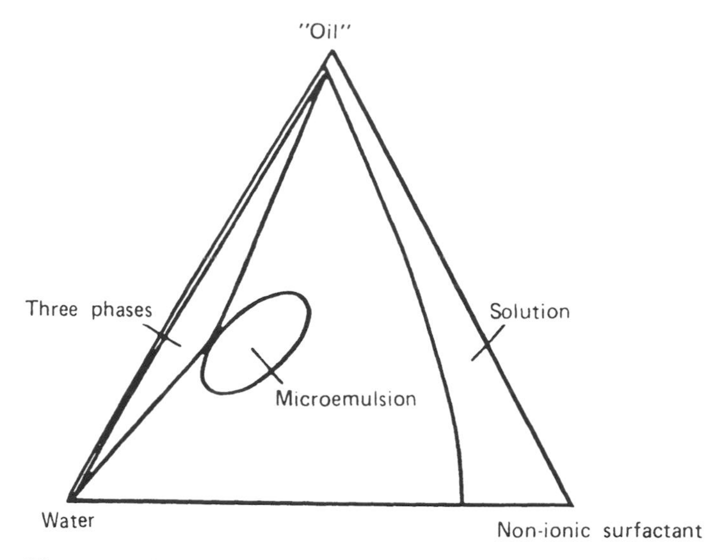

Fig. 13 Microemulsion region in equilibrium with excess oil and water phases.

Devices of this sort can reveal relationships, allow qualitative generalizations, and considerably shorten the tedious trial-and-error method of formulation, but they do not tell how the phase equilibria are controlled, why one type of microstructure is formed over another, what the origins of their stability are, and where in the phase diagram different morphologies are to be found [131]. Monophasic areas in a phase diagram do not correspond in surfactant systems to one single type of structure, and transformations from one structure to another can occur without leaving a monophase area.

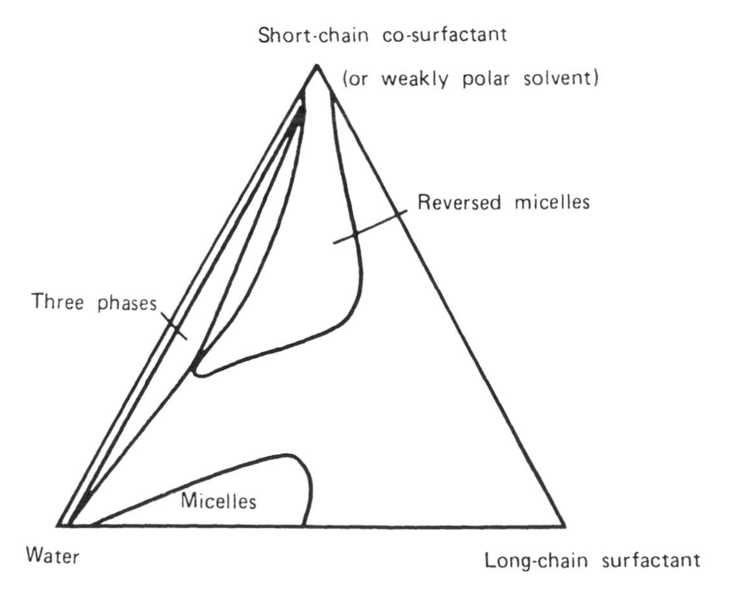

Fig. 14 Equilibrium between W/O micelles and excess aqueous and cosurfactant phases. Hydrocarbon may be dissolved into the reverse micelle solution to form microemulsions containing large amounts of water.

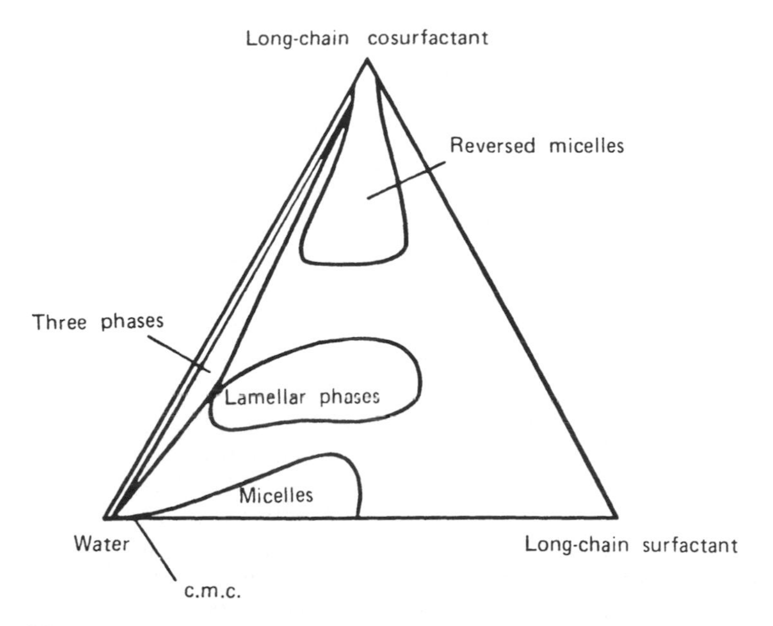

Fig. 15 Equilibrium between liquid crystalline phase and aqueous and organic solutions.

The pioneering description by Schulman of an O/W microemulsion invoked a droplet model and was based on surface tension [170,171]. The surfactant is assumed to saturate the interfaces between oil and water [172] rather than building up micelles in one of the two phases, and all the oil is to be found inside the microemulsion droplets. The interfacial tension at the droplet-solvent interface is given by:

$$\gamma = \gamma_0 - \pi \tag{1}$$

where
γ_0 = surface tension in the absence of surfactant
π = spreading pressure of the mixed film.

The condition for stability is $\gamma = 0$. Surfactant and cosurfactant contribute to the spreading pressure; the cosurfactant and the oil penetrate the interface, affecting its curvature, rendering it more disordered and flexible. The molecular structure of the surfactant and cosurfactant as well as their concentrations thus determine the structures that are formed through their influence on the surface pressure and the O/W surface tension. This simple approach avoids a detailed consideration of oil-and-water surfactant interactions by including them in the surface tension concept.

When $\gamma \approx 0$, contributions to the stability of the entropy of mixing of the droplets and the energy due to curvature become important and refinements must be introduced. Ruckenstein and Chi [173], using the surface tension concept, also included the dispersion entropy of the droplets and the reduction of the chemical potential of surfactant and cosurfactant in the bulk phase by adsorption at the interface. The latter two negative contributions may exceed the positive free energy arising from the lower interfacial tension produced by using a cosurfactant and lead to thermodynamically stable emulsions. Overbeek et al. [174], in an analogous thermodynamic treatment, arrived at similar conclusions but noted that it may be necessary to augment the interfacial free energy with the curvature contributions from bending and flexibility of the interface. These phenomonological theories involve cancellation of large and opposite contributions that make the system very sensitive to slight changes in external conditions.

The nature of the microemulsion that is formed (i.e., the phase limits of O/W or W/O microemulsions) depends on the structure and geometric packing of the polar heads and hydrocarbon tails of the surfactant molecules. Israelachvili [175] and Mitchell and Ninham [176] showed that the nature of the aggregate unit that is formed depends on the packing ratio υ/a_0l_c, where υ is the partial molecular volume of the surfactant, a_0 is the area of the head group of the surfactant, and l_c is the maximum chain length that the chains can assume at a given temperature in the sense that chain lengths above l_c cause a sudden increase in the free energy of the system. The magnitude of this ratio, which is related to the HLB, and the spontaneous curvature of the monolayer predict inverted micelles and bilayer structures (0.5–1), inverse micelles (> 1), and spherical micelles when small (< 0.33). The packing ratio is sensitive to pH, temperature, and ionic strength. Addition of cosurfactant has the effect of increasing υ without increasing l_c, acting as "padding" for separating the charged groups [177]. Alcohols, for example, with chain length equal to or greater than the chain length of the oil, do not induce microemulsions. Electrolytes or an increase in temperature reduce the area per head group, increasing the packing ratio. The latter effect predicts that the solubilization of oil in nonionic surfactants will increase with temperature as observed, until the packing ratio reaches a value of 1 and phase inversion takes place. With further temperature increases the ratio exceeds 1, and W/O microemulsions would be expected. The packing ratio only accounts for geometry and must be used with caution since different structures can coexist.

Since many microemulsions are continuously stable in going from oil-rich to water-rich compositions, the role of the cosurfactant cannot be entirely reconciled to its effect on packing. A concentration effect has been proposed whereby cosurfactant migrates to regions of strong curvature, increasing the flexibility of the interface and facilitating the formation of bicontinuous regions. By focusing on the interface between oil and water and its random fluctuations, Jouffroy et al. [178] calculated an entropy of mixing based on the number of ways in which the interface can be partitioned in space rather than by the ways in which the droplets can be distributed.

In this random structure model, at low water content (volume fraction of water $\phi_\omega < 0.15$) the water phase is discontinuous (the same for the oil phase at low oil content). But, over a large concentration range, both oil and water phases are continuous. A progressive change from the droplet to the random model can be envisaged as the dispersed phase concentration, and attractive interactions between droplets increase. Without cosurfactant, lamellar phases may form due to interface inflexibility.

The various theoretical treatments have attempted with some success to explain the following phenomena: the ultralow interfacial tensions; the requirement of greater amounts of cosurfactant for O/W microemulsions than for W/O ones; the relatively greater stability of W/O microemulsions; the effect of electrolyte in promoting W/O microemulsions; the greater solubilization of water in W/O microemulsions at the largest possible droplet radius; the tendency toward monodispersity of droplet sizes; phase equilibria between microemulsions and excess oil or water or both, as well as the structure of the middle-phase microemulsion in equilibrium with the latter; and the dynamic properties of the interface.

A. Pharmaceutical Applications

The impetus for increased understanding of the structures and properties of microemulsions comes from the practical applications, both actual and potential. Solubilization of water-insoluble drugs in micelles or transparent systems of thermodynamically stable droplets, formed spontaneously, is clearly an advantage over conventional emulsion formulations. The feasibility of contacting oil- and water-soluble reactants at a large interface with varying microenvironments allows the possibility of enhancing heterogeneous reactions such as with lipids and water-soluble enzymes [179,180]. The structure of the interface (i.e., "hard sphere" or bicontinuous) may influence diffusion-controlled reactions and their rates. Dynamic properties of the interface such as rapid solubilization-desolubilization can give large increases in the rate of transfer of solute, and microemulsions have been proposed as liquid membrane carriers to transport lipophilic substances through an aqueous medium, or inversely to carry hydrophilic substances across a lipoidal medium [181]. As liquid membrane extractors, microemulsions have advantages of stability over multiple emulsions [182], and have been adapted to enzymatic synthesis, where protein can easily be adsorbed to or released from the W/O droplets [183].

One of the difficulties in realizing the potential of microemulsions as drug delivery systems through solubilization of drugs is the narrow range of acceptable surfactants and cosurfactants and the high concentrations usually required. It is apparent that more research needs to be done on the use of lipids or surfactants based on natural products. Some attempts have been made to overcome these limitations by using hydrophobic nonionic cosurfactants rather than pharmaceutically nonacceptable long-chain alcohols [184]. Jayakrishnan et al. [185] attempted to formulate microemulsions with low concentrations of pharmaceutical nonionic surfactants and isopropanol and solubilized hydrocortisone, a drug with low aqueous and lipid solubilities. Friberg and Li [186] investigated the preparation of microemulsions using esters instead of the usual hydrocarbon oils. Detergentless microemulsions, advantageous for carrying out enzyme reactions, consisting of *n*-hexane-water-2-propanol can be formed [187], as well as waterless microemulsions in which a nonaqueous polar compound replaces water [188]. Birefringement microemulsions have been reported in both aqueous [189] and nonaqueous systems [190]. They bear resemblance to lyotropic liquid crystalline systems and may be examples of long-lived kinetically stable emulsions containing liquid crystalline material.

Of novel interest is the existence of microemulsions that can be cooled continuously into the glassy state [191]. The microemulsion state apparently facilitates the formation of glasses from many liquids that would not ordinarily do so. This should provide opportunities for study of the microemulsion state as well as for applications to drug delivery systems.

Microemulsions can be used to produce ultrafine nanoparticles with sharp particle size distributions by precipitation reaction when two microemulsions of the same composition but containing different reactants are mixed. In these W/O emulsions the water droplets behave as microreactors, the nanoparticle size being a function of the droplet size.

Kemken et al. describe an interesting application of an O/W microemulsion for transdermal delivery. A supersaturated system of an apolar drug is formed in a microemulsion and, as the water concentration increases on occlusion, the drug's thermodynamic activity increases. The increased drug flux was found to correlate with the decreasing solubility [192].

Extensive interfaces and the ability to compartmentalize large amounts of material on a microscopic level within the same medium are provided by microemulsions. This can have profound effects on ionization potentials, oxidation-reduction properties, dissociation constants, reaction rates, and photochemistry. The applications to such pharmaceutical questions as drug stability and absorption have yet to be thoroughly explored. A disadvantage peculiar to micellar systems (somewhat less so for microemulsions) is the lability of their interfaces. Temperature, electrolyte concentration, composition, and the presence of proteins and other biological substances can alter structures and solubilization capacity. Polymerization of the micelle or microemulsion membrane is a way to eliminate this instability by forming a capsule wall. The drug can be encapsulated, serving as a reservoir, and since the usual size of the particles lies between 30 and 350 nm, they have been named nanocapsules [193]. Such carrier systems can be administered both parenterally and orally for sustained release [194]. Incorporation of antigen protects the antigen from degradation, antibody formation is increased [195], and lymphocyte proliferation is stimulated [196]. Intracellular or transcellular transport of drugs to specific sites by endocytosis of nanocapsules has been shown with Actinomycin D, an anticancer drug [197]. Migration of nanocapsules containing drugs through the intestinal epithelium to the blood or lymphatics is also considered possible [194]. Polymerization of microemulsions of monomers yields monodisperse microlatices that can be used as carriers for pharmaceuticals [198,199].

VI. VESICLES

Among the organized surfactant structures—monolayers, micelles, microemulsions, bilayers, and vesicles—the latter have attracted great interest as potential drug carriers of greater specificity and duration of action. The major forces governing the self-assembly of these aggregates derive from the hydrophobic interactions between the hydrocarbon tails, which cause the molecules to associate, and the hydrophilic nature of the head groups, which dictates that they remain in contact with water.

Without invoking a detailed analysis of the complex interactions between hydrocarbon chains and polar head groups, geometrical considerations predict the structures into which amphiphiles will assemble [200]. Molecules that cannot pack into smaller micellar structures because of bulky hydrocarbon chains that increase the surface area above the optimal value, or because of a small optimal head group area a_0, at the hydrocarbon-water interface will form vesicles or flexible bilayers when the value of υ/a_0l_c lies between 0.5 and 1. This means that for the same head group area and maximum chain length l_c, the hydrocarbon volume υ, must be twice that of micelle-forming molecules, for which the packing ratio is normally in the range 0.33–0.5. Surfactants with two hydrocarbon chains will therefore generally form bilayers. This will also be true for single-chain lipids with very small head groups. Closed spherical bilayers will be favored over infinite planar bilayers (packing ratio $\approx$ 1) as long as the head group areas do not exceed their optimal value. When the ratio exceeds 1, such molecules cannot pack into bilayers because their head group area is too small and they will form inverted structures or precipitate out of the solution (see Fig. 16).

Phospholipids, a main constituent of biological membranes, have a high tendency to aggregate to bilayers both in the crystalline state and in aqueous dispersions. Un-

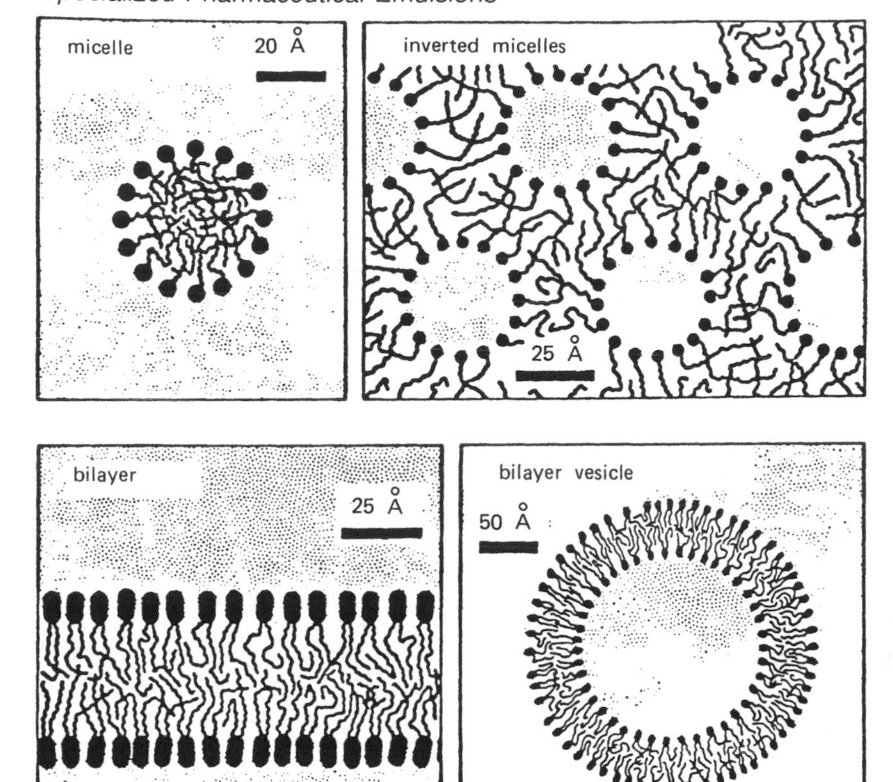

Fig. 16 Association structures of amphiphiles such as surfactants and lipids in water, which can transform one to another by changing solution conditions, such as electrolyte or lipid concentration, pH, or temperature.

charged or zwitterionic lipids, which form lamellar liquid crystalline phases in water, give rise to a two-phase system consisting of multilamellar liposomes and excess water. These structures consist of onionlike stacks of swollen bilayers ranging from 1 to 10 μm in size and can be sonicated to form small unilamellar vesicles surrounded by a single bilayer. Charged lipids and neutral lipids doped with a charged amphiphile spontaneously form dispersions of unilamellar vesicles with diameters ranging from 0.1 to 2 μm. In excess ion concentrations, charged lipids exhibit the swelling properties of neutral lipids. Small vesicle dispersions are generally not thermodynamically stable and have a tendency to aggregate and fuse, separating into a lamellar phase and dilute aqueous solution. Some combinations of nonionic surfactants spontaneously form reverse vesicles [201].

A. Methods of Liposome Preparation

The original method of Bangham et al. [202] for the preparation of liposomes involves the deposition of a thin lipid film from an organic solvent medium on the walls of a container, followed by agitation with an aqueous solution of the material to be encapsulated. If the agitation is carried out above the gel-liquid phase transition temperature of the phospholipid and the film is kept thin, multilamellar vesicle (MLVs) form spontaneously. Drug-trapping efficiency of these MLVs is low, and the method cannot be easily scaled up.

Ether injection techniques [203] are based on the infusion of a solution of lipids in ether into an aqueous solution of the material to be encapsulated at a temperature high enough for the solvent to be rapidly evaporated. Large unilamellar vesicles (LUVs) are produced, but the encapsulating efficiency is not high and the material to be entrapped requires possible detrimental contact with organic solvent.

Small unilamellar vesicles (SUVs) are formed by first dispersing a phospholipid and a surfactant in an aqueous solution of the material to be encapsulated. The dispersion of mixed micelles is then exhaustively dialyzed to remove surfactant, and a homogeneous dispersion of SUVs is obtained [204]. The degree of encapsulation is low due to loss of material during dialysis, except in the case of macromolecules, and significant residual surfactant is present. Gel filtration can be used to form homogeneous populations of unilamellar liposomes, and this step more rapidly and completely removes detergents and unbound drug at the same time.

The reverse micelle method [205] starts with the dispersion of an aqueous solution of the material to be encapsulated in a volatile organic solvent containing the lipid mixture. This step is followed by redispersing in an aqueous buffer medium to form a W/O/W multiple emulsion. On evaporation of the solvent, SUVs are obtained.

The reverse phase evaporation method [206] is based on the transformation of an emulsion to a liposomal dispersion. The initial formation of a W/O emulsion of the aqueous drug solution in a lipid-containing organic solvent is followed by the removal of solvent by rotary evaporation. Drug encapsulating efficiency is high (approximately 50%), and in addition to SUVs, MLVs and LUVs may be obtained under suitable conditions [207]. A recent technique called microfluidization is based on impinging two fluidized streams at high velocity. The method claims greater uniformity, smaller sizes, and high capture rates, up to 75% [208].

B. Liposomes as Drug Carriers

The number of works devoted to the biomedical applications of liposome vesicles is voluminous and growing [209–212]. There are several reasons for this popularity: liposomes obtained from natural phospholipids are biocompatible and biodegradable; a wide range of pharmaceutical agents can be incorporated into the inner aqueous or lipid phases; drugs so compartmentalized are sequestered from the external environment and cause less undesirable reactions; and drugs may apparently be introduced into the cells by fusion and endocytosis.

Important considerations in the preparation of lipid vesicles are the lipid surface charge, vesicle size, and aqueous volume of the liposomes. Relatively high ratios of aqueous space to lipid are required for entrapment of water-soluble drugs, and high ratios of lipid to aqueous space for lipophilic drugs. One of the problems has been the low efficiency of loading of hydrophilic drug molecules and the rapid leakage of weakly hydrophobic molecules after encapsulation, although the use of pH and concentration gradients to force entrapment with high efficiency and retention appears promising [213]. Liposome size can also have an important effect on liposome disposition in vivo. Large uni- or multilamellar liposomes are cleared rapidly from the circulation on intravenous injection and are targeted mainly to the reticuloendothelial system, thus affording the possibility of drug delivery to organs such as the liver and spleen as well as circulating monocytes. Small unilamellar liposomes remain in the blood circulation for some time and have the opportunity to sustain delivery of therapeutic agents to cells within the circulation. Targeting of liposomes to specific cell sites by incorporating homing ligands into the liposome membrane that bind selectively to determinants on target cells has not been very successful. The overriding preference of liposomes for the reticuloendothelial system has not been easy to overcome, and circulating liposomes have difficulty crossing blood vessel walls. A new generation of sterically stabilized liposomes, however, where the lipid bilayer contains lipids conjugated with ethylene glycol, have two orders of magnitude greater blood-circulating lifetimes than conventional liposomes [214]. These so-called stealth liposomes, since they are less subject to nonspecific uptake, also lend themselves to targeting by attaching ligands such as antibodies [215].

Where treatment involves the reticuloendothelial system, as in leishmaniasis, liposome-encapsulated drugs are safer and far more effective than unencapsulated drugs [216]. Liposomes injected subcutaneously or intramuscularly show slow clearance from the injection site followed by absorption into lymphatics. The slow clearance permits high levels of drugs to be sustained as in depot preparations. Direct drug delivery of liposome-containing corticosteroids to joints of arthritic patients has been clinically successful [217]. Potential applications of the feature of lymphatic uptake include delivery of anticancer agents for treatment of lymph node metastases [218] and delivery of radioimaging agents in lymph nodes [219]. The localization of intravenously administered liposomes within phagocytes has been used to activate macrophages to render them cytotoxic for tumor cells by systemic administration of liposomes containing immunomodulators [220]. Liposomes also find wide use in the field of immunology as carriers of antigens and adjuvants, and in the analysis of antibodies [221].

The use of liposomes containing large molecules to improve the oral absorption of drugs has been disappointing and offers no advantage over conventional emulsions. Direct administration to the nose or lungs as an aerosol formulation, however, holds promise [222]. Topical application of liposomes on the skin produced higher levels of drug in the dermis and epidermis and reduced drug levels systemically [153,223]. Moisturizing agents and water encapsulated in liposomes [225] were reported to exhibit enhanced delivery due to the heightened substantivity of phospholipid vesicles for keratin in the stratum corneum. Handjani-Vila et al. [226] prepared vesicles termed "niosomes" from synthetic surfactants, dicetyl phosphate, and cholesterol, that facilitated the uptake of various topically applied substances, such as humectants, sunscreens, and tanning agents, by the skin. When applied to the eye, liposomal preparations encapsulating idoxuridine [227], penicillin G [228], and inulin and epinephrine [229] showed greater corneal uptake than the unencapsulated material. Liposome preparations containing antibiotics applied topically to the eye resulted in sustained release; 55% of the original liposomal drug concentration was present in the ocular cavity after 24 h, compared to zero concentration for the control after 3 h [230].

1. Stability of Liposomes

An important aspect in the use of liposomes as drug delivery systems is their stability. This includes the chemical stability of lipids and entrapped drugs, the physical stability of the liposomal membrane and the liposome size in terms of leakage of drug and half-life, and, finally, the interaction of liposomes in vivo with blood components, which may lead to their transformation or breakdown.

Increase in size and leakage are a function of lipid constitution and entrapped material [231], SUVs being generally less stable than liposomes with smaller bilayer curvature. For storage purposes, it is possible to overcome these problems by freeze-drying aqueous liposome suspensions or lyophilizing the liposome components from a suitable organic solvent [232]. For hydrophilic drugs or for unilamellar vesicles such approaches may not be feasible. Addition of cholesterol to the bilayer and cryoprotective agents such as lactose can be beneficial in maintaining physical structure and decreasing leakage rates during freeze-drying and freeze/thaw cycles [233]. Membrane-permeable drugs can also be immobilized by internal gelation or precipitation [234].

Sterilization of liposomes is limited to filtration by Millipore filters (Millipore Corp., Bedford, MA) through pore sizes no greater than 0.22 μm, since other procedures would destroy the integrity of the bilayer. Liposomes larger than 0.5 μm cannot be sterilized by this method as they leak when forced through the requisite filters. For human testing, all materials must be germ free as well as pyrogen free, and the entire formulation process must be done under sterile conditions.

Proteins, particularly serum lipoproteins, have a marked effect on liposomes, causing alteration of membrane permeability and even disruption of the bilayer structure by removal of phospholipid. Cholesterol has a stabilizing influence [235], presumably because of its ability to tighten the packing of phospholipid molecules in bilayers that are fluid [236], thus making it difficult for proteins to penetrate the liposomal membrane. As has been mentioned previously, stealth liposomes inhibit interactions with plasma proteins and provide a steric barrier to greatly improve biological stability, thus furthering the concept of liposomes as sustained release systems.

Liposomes have been encapsulated to increase the physical stability, in W/O emulsions [237] and with calcium alginate hydrogel as the encapsulating matrix [238]. The latter system showed a lag-and-burst type of release suitable for a vaccine carrier.

C. Synthetic Vesicles

The inherent instabilities of natural phospholipid vesicles and the difficulty of entrapment of solutes has led to a search for simpler systems of synthetic surfactants that pack readily into bilayers. It was found that lipid bilayer vesicles can be formed from various kinds of synthetic single-chain, double-chain, and triple-chain amphiphiles having cationic, anionic, and nonionic polar head groups [239]. Formation of cationic dialkyldiethylammonium halide [240]; dialkyl phosphate, sulfonate, and carboxylate [241]; and zwitterionic surfactant vesicles [242] has been reported. Single-chain ammonium amphiphiles bearing two charged end groups and a rigid segment consisting of biphenyl groups are also able to form monolayer assemblies [243] (Fig. 17a). Vesicles prepared from lipid analogous dialkyl compounds [244], dialkyl polyoxyethylene ethers [245], and niosomes made from a single-alkylchain nonionic surfactant with or without cholesterol [246] are representative of the neutral types. All these models simulate the behavior of liposomes in their fluidity, osmotic activity, and entrapment of solutes but give no improvement in stability or reduction of fragility over the liposome. In vivo experiments carried out with niosomes containing methotrexate showed that they were effective in maintaining drug levels in the blood; as with liposomes, large amounts of vesicles were taken up in the liver [247]. Probably because of their relative toxicity ionic vesicles have not been tested in this way. Niosomes have also been administered via the topical route [248].

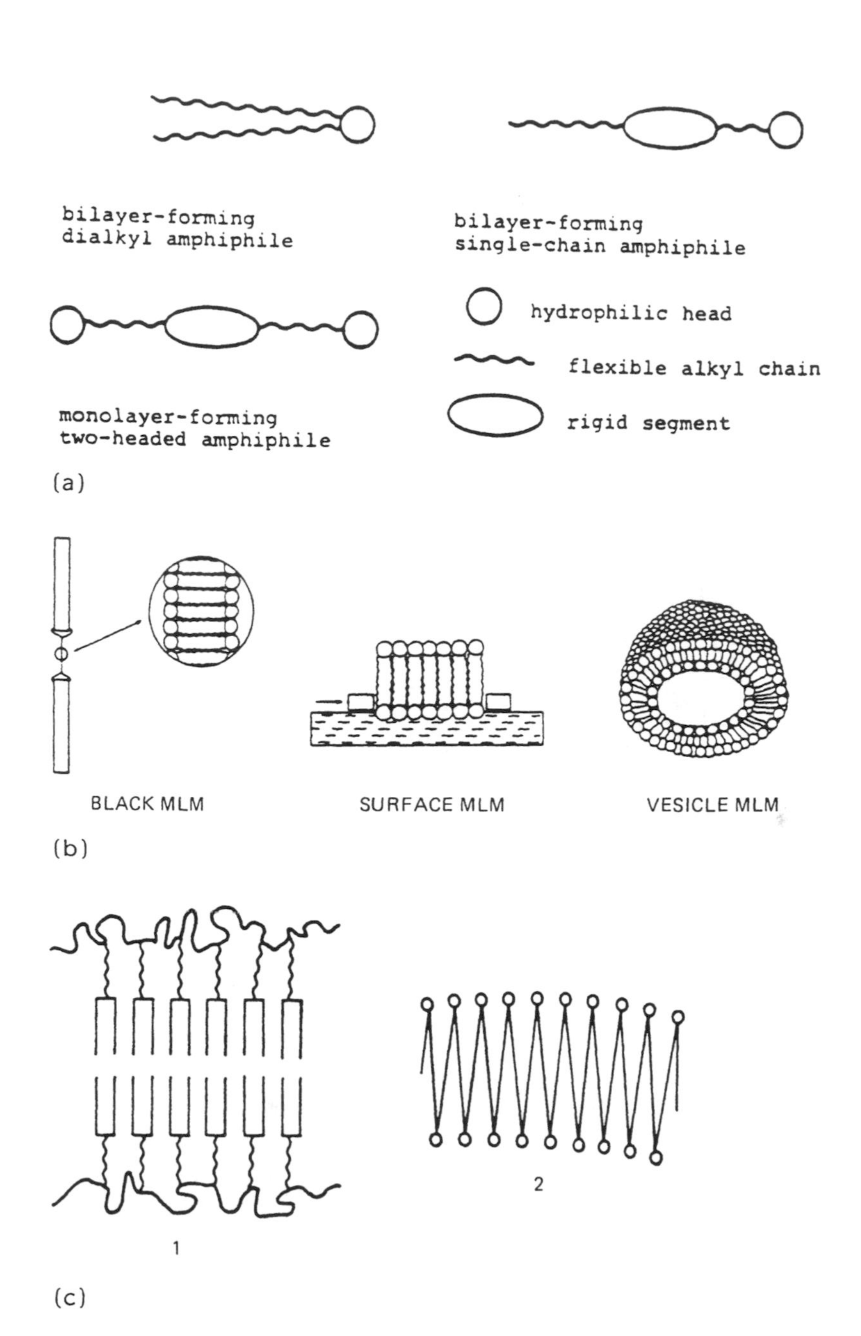

Fig. 17 (a) Double-chain and single-chain amphiphiles, which assemble spontaneously to form stable monolayers and bilayers. Rigid segments are elongated aromatic groups that have liquid crystalline properties (from Ref. 239); (b) bolaamphiphiles and their high molecular weight aggregates, which form monolayer liquid membranes (MLMs) (from Ref. 255); (c) schematic polymeric bilayer and monolayer membranes: (1) copolymer; (2) ionene polymer,

$$\left[\begin{array}{c} CH_3 \\ | \\ N^+ \text{—}(CH_2)_n \\ | \\ CH_3 \end{array}\right] Br^-$$

(from Ref. 251); (d) methods of synthesizing polymeric membranes (x = polymerizable group): (1), (2), and (3), polymerization with no effect on head group properties; (4), polymerization with preservation of chain fluidity (from Ref. 255).

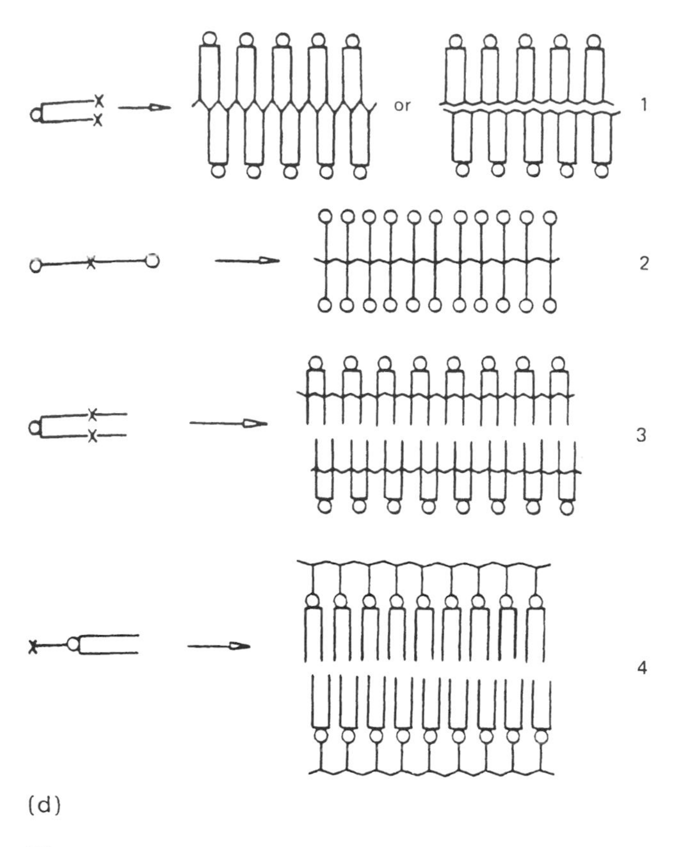

Fig. 17 Continued

Bolaamphiphiles [249], which consist of a chain of hydrophobic groups connecting two hydrophilic head groups, form monolayered liquid membranes that may curve to form vesicles (Fig. 17b). Although no liquid crystal formation has been observed, these vesicles are fluid above 10°C and can be stable for periods longer than a year [250]. Their utility lies in their thinness (pore formation) and ability to dissolve in biological membranes and interact with biological polyelectrolytes.

1. Polymeric Vesicles

Water-soluble copolymers containing hydrophilic units and dialkyl groups connected to the polymer backbone as well as ionene polymers give characteristic bilayer or monolayer vesicles through aggregation of the side chains in the former case and by chain folding in the latter [251]. (See Fig. 17c.)

Attempts to stabilize bilayer vesicles by polymerization have been reported by several groups [240,252–254]. Surfactants that readily form lamellar phases in water are synthesized to contain polymerizable groups, such as diacetylene, methacryl, styryl, or vinyl groups. Polymerization may be induced by exposure to ultraviolet light, γ-radiation, or by a free-radical mechanism. Depending on the location of the polymerizable group, vesicles can be "zipped up" at the inner and/or outer surfaces or within the hydrophobic layer as shown in Fig. 17d. The covalent linking of the individual lipid molecules to any desired level confers enhanced stability, the ability to control bilayer fluidity, and thus the permeability of the bilayer to drug release. Polymeric vesicles are also of interest in that the surface polymeric network resembles the cytoskeleton of living cells.

Two recent developments illustrate the potential applications of polymeric vesicles. Gros et al. [255] envisage the incorporation into polymeric vesicles carrying anticancer agents of a "cork" of destabilizing areas, which can be "uncorked" at a therapeutic site by a trigger-mediated process such as an enzyme reaction or alteration of pH. In an analogous innovation, Okahata [256] developed a lipid bilayer-corked capsule membrane that combines the characteristics of a microcapsule with those of lipid bilayer vesicles. Multiple bilayers are deposited parallel to the membrane plane of porous, ultrathin, nylon membranes. The lipid bilayer wall is strengthened by the nylon membrane, and the permeability of the nylon capsule is controlled by changes in the physical state of the lipid bilayer, which can be mediated by temperature, pH, photoirradiation, electric field, and interaction with divalent cations. Capsule membranes of this type, having a trigger-responsive lipid bilayer permeability valve, may be effective in controlled drug delivery devices. A related, novel method of drug delivery, termed the "molecular drill," involves stretching of a unilamellar vesicle adsorbed to a porous substrate, thereby increasing the membrane's permeability [257]. Molecules contained inside the vesicle wall that could not otherwise cross the membrane are released into the substrate. The reverse process of delivery of an adsorbed drug to a cell membrane may also be applicable.

D. DNA Transfer

The burgeoning development of recombinant DNA technology and the possibility of gene therapy on the cellular and genetic level has enhanced the need for efficient delivery of DNA into cells. Liposomes have shown utility as carriers or vectors of genetic information. DNA has been successfully encapsulated in liposomes [258]. Complexation of negatively charged DNA with cationic lipid-containing liposomes increased transfection many fold [259] when administered systemically, and promising results were also obtained from the inhalation of an aerosol of DNA-liposome complexes (genosomes) [260]. The nature of the supramolecular aggregates that are formed has not yet been precisely elucidated, but appears to result from a combination of liposome aggregation, fusion, and condensed DNA via a process of bridging flocculation [261].

VII. OTHER PHASE STRUCTURES

Lamellar liquid crystals (L_α) have been used in ointment formulations [262]. They are thermodynamically stable, being equilibrium structures, and can solubilize large amounts of hydrophobic or polar solutes. Diffusion rates of lipophilic drugs in lecithin-water vehicles were higher than from aqueous suspensions, as the solutes were located between the lipophilic molecules rather than the hydrophilic layers (translamellar route). The diffusion coefficient of a drug within a lamellar phase is about one or two orders of magnitude less than that in solution, which may be advantageous for topical delivery. The lamellar liquid crystalline phases are stable only for rather well-defined cosurfactant/surfactant ratios of nonionic or zwitterionic surfactants. Addition of an ionic surfactant to an aqueous solution containing enough surfactant to form an L_α or L_3 phase, transforms the bilayers into gels of densely packed multilamellar vesicles [263].

The L_3 phase (bicontinuous) is a locally flat bilayer structure swollen by a thin oil film separating large water domains. Interconnecting passages between lamellar sheets make for a spongy structure. The L_3' phase is the reverse phase, where the extensive domains are oil (Fig. 10b).

Cubic phases are thermodynamically stable states exiating in a variety of surfactant and lipid systems. They are optically clear, isotropic, and gel-like, and may be bicontinuous. Polymerization in bicontinuous cubic phases results in a cross-linked permanent polymer matrix of porous material having a bicontinuous microstructure [264]. A reversible micellar to cubic gel phase transition at body temperature of a greater than 20 wt% aqueous solution of Pluronic F-127 can be used for sustained drug delivery [265]. Monoolein forms a cubic phase of high solubilization capacity that is biodegradable and bioadhesive [266]. When dispersed with lecithin and Pluronic F-127 in water, the resultant cubic phase emulsion, or "cubosomes," is of the parenteral nutrition type.

Hexagonal phases are viscous and birefringent and can be used similarly to L_α phases, although the release pattern may be different. A transition from the lamellar phase to a reverse hexagonal phase triggered by weakly acidic pH is involved in membrane fusion and release of vesicle contents. The trick of formulating to effect phase transitions in situ to a desirable microstructure is a recurring one in drug delivery.

Lipospheres consist of water-dispersible microparticles containing a solid hydrophobic fat core, mainly triglycerides, stabilized by a surface layer of phospholipid molecules (lecithin). In the polymeric lipospheres, biodegradable polymers such as polylactide substitute for the fat. Originally developed for parenteral drug delivery [267], they have also been used in formulating vaccines [268].

Fibrous molecular assemblies consisting of helical bilayers as well as rolled up bilayer scrolls have been reported [269]. These micellar and vesicular elongated rods can be lyophilized and stored, unlike micelles and vesicles, which tend to degrade on dehydration.

VIII. CONCLUSION

Microemulsions, liposomes, vesicles, other phase structures, and polymerized versions of these—nanoparticles and synthetic vesicles—may be regarded as variants of emulsions. The structural chemistry of such noncovalent aggregates can be just as kaleidoscopic as covalent polymers, i.e., proteins and nucleic acids. Such microheterogeneous systems have great potential for new pharmaceutical applications, particularly drug delivery. Self-assembled systems can ultimately be controlled and manipulated as in molecular synthesis. Using tailor-made synthetic molecules coupled with a thorough knowledge of their phase relationships, suitable microstructures can be designed based on the end goal.

Nevertheless, conventional emulsions have been known for a long time and are clinically accepted. They are versatile, serving as drug carriers, nutritional media, blood substitutes, and diagnostic agents that can be produced technologically on a large scale. Their stability is reasonable, and they offer large drug-carrying capacity. In certain cases, emulsions bear the characteristics of the more novel systems. Lipid emulsions, for example, resemble liposomes in many respects, both in structure and in in vivo effects. At high emulsifier concentrations, emulsions may also contain micellar aggregates or liquid crystalline structures.

Multiple emulsions are a degree more complex than ordinary emulsions but provide multicompartments and possibilities for prolonged release. Further understanding of their morphology and stability in relation to the surfactants used in their formulation as well as the mechanisms of release will contribute to designing more effective dosage forms.

REFERENCES

1. S. S. Davis, J. Hadgraft, and K. J. Palin, Medical and pharmaceutical applications of emulsions. In: *Encyclopedia of Emulsion Technology*, Vol. 2 (P. Becher, ed.), Marcel Dekker, New York, 1985, p.159.
2. S. S. Davis, *J. Clin. Pharm.*, 1:11 (1976).
3. P. K. Hansrani, S. S. Davis, and J. Groves, *J. Parenter. Sci. Technol.*, 37:145 (1983).
4. B. I. Wilner, M. A. Ekers, H. D. Troutman, F. W. Trader, and I. W. McLean, *J. Immunol.*, 91:210 (1963).
5. D. E. S. Stewart-Tull, T. Shimono, S. Kotani, and B. A. Knights, *Int. Arch. Allergy Appl. Immunol.*, 52:118 (1976).
6. L. Illum, R. West, and C. Washington, *Int. J. Pharm.*, 54:41 (1989).
7. I. R. Schmolka, *Fed. Proc.*, 29:1717 (1970).
8. A. R. Vela, O. L. Hartwig, M. Atik, R. R. Marrero, and I. Cohn, *Am. J. Clin. Nutr.*, 26:80 (1965).
9. S. S. Davis, *J. Hosp. Pharm.* (Suppl.), 149:165 (1974).
10. L. Rydhag, *Fette Seife Anstrichm.*, 81:168 (1979).
11. S. S. Davis, In: *Advances in Clinical Nutrition* (I. D. A. Johnson, ed.). MTP, Lancaster, England, 1983, p. 213.
12. R. P. Geyer, D. M. Watkin, W. P. Waddell, F. R. Olsen, and F. J. Stare, *J. Am. Oil Chem. Soc.*, 32:365 (1955).
13. T. Yamaguchi and Y. Mizushima, *Crit. Rev. Ther. Drug Carr. Syst.*, 11:215 (1994).
14. R. J. Prankerd, and V. J. Stella, *J. Parent. Sci. Technol.*, 44:139 (1990).
15. T. Yamaguchi, K. Nishizak, S. Itai, H. Hayashi, and H. Oshima, *Pharm. Res.*, 12:1273 (1995).
16. T. Yamaguchi, N. Tanabe, Y. Fukishima, T. Dasu, and H. Hayashi, *Chem. Pharm. Bull.*, 42:646 (1964).
17. Y. Mizushima and K. Hoshi, *J. Drug Targeting*, 1:93 (1993).
18. S. Benita and M. Y. Levy, *J. Pharm. Sci.*, 82:1069 (1993).
19. R. Jeppson, *Acta Pharm. Suec.*, 9:81 (1972).
20. R. Jeppson and S. Ljunberg, *Acta Pharm. Suec.* 4, 9:199 (1972)
21. R. Jeppson and S. Ljunberg, *Acta Pharm. Suec.*, 10:129 (1973).
22. E. J. Jesmok, G. M. Walsh, W. Ditzler, and E. F. Woods, *Proc. Soc. Exp. Biol.*, 162:458 (1979).
23. B. Meyerson, *Horm. Behav.*, 3:1 (1972).
24. Y. Mizushima, T. Hamano, and K. Kokoyama, *J. Pharm. Pharmacol.*, 34:49 (1982).
25. A. J. Repta, In: *Topics in Pharmaceutical Sciences* (D. D. Breimer and P. Speiser, eds.). Elsevier, New York, 1981, p. 181.
26. P. Sorkin, H. Nagar. A. Weinbroum, A. Setton, E. Israetil, A. Scarlatt, A. Silberberg, V. Rudick, Y. Kluger, and P. Halpern, *Crit. Care Med.*, 24:13116 (1996).
27. R. Jeppson and S. Ljunberg, *Acta Pharm. Suec.*, 36:312 (1975).
28. S. Jeppson and S. Rossner, *Acta Pharmacol. Toxicol.*, 37:134 (1975).
29. M. J. Groves, M. W. Wineberg, and A. P. R. Brain, *J. Dispersion Sci. Technol.*, 6:237 (1985).
30. J. Li and K. D. Caldwell, *J. Pharm. Sci*, 83:1586 (1994).
31. K. Wesesen and T. Wehler, *J. Pharm. Sci.*, 81:777 (1992).

32. H. Hedeman, H. Brønstad, A. Müllertz, and S. Frokjaer, *Pharm. Res.*, 13:725 (1996).
33. F. Liu and D. Liu, *Pharm. Res.*, 12:1060 (1995).
34. J. T. Wheeler, V. F. Wong, S. M. Ansell, D. Masin, and M. B. Bally, *J. Pharm. Sci.*, 83:1558 (1994).
35. B. B. Lundberg, B. C. Mortimer, and T. G. Redgrave, *Int. J. Pharm.*, 134:118 (1996).
36. J. Kriegelstein, A. Meffert, and D. H. Niemeyer, *Experientia*, 30:926 (1974).
37. L. C. Clark, Jr., and F. Gollan, *Sci.*, 152:1755 (1966).
38. K. Ju, J. F. Lee, and W. B. Arminger, *Biotech. Bioeng.*, 37:505 (1991).
39. R. P. Geyer, In: *Drug Design*, Vol. 7 (A. J. Ariens, ed.). Academic Press, New York, 1976, p. 1.
40. S. S. Davis, H. P. Round, and T. S. Purewall, *J. Colloid Interface Sci.*, 82:307 (1981).
41. A. S. Kabalnov and E. D. Shchukin, *Adv. Colloid Interface Sci.*, 38:69 (1992).
42. C. Varescon, C. Arlen, M. LeBlanc, and J. G. Riess, J. Chim. Phys., 58:2111 (1989).
43. E. D. Schukin, E. A. Amelina, K. N. Makarova, A. M. Parfenova, N. A, Safronova, and L. L. Gervits, *Kolloidn. Zh.*, 46:1211 (1984).
44. E. A. Amelina, E. D. Schukin, K. N. Makarova, A. M. Parfenova, N. A. Safronova, E. A. Golubeva, and L. L. Gervits, *Kolloidn. Zh.*, 42:1209 (1984).
45. C. Santaella, P. Vierling, and J. G. Riess, *Biomat. Art. Cells Immob. Technol.*, 19:1055 (1991).
46. S. Abouhilale, J. Greiner, and J. G. Riess, *J. Am. Oil Chem. Soc.*, 69:1 (1992).
47. G. Mathis, P. Leempoel, J. C. Ravey, C. Selve, and J.-J. Delpeuch, *J. Am. Chem. Soc.*, 106:6162 (1984).
48. P. Mukurjee, *J. Am. Oil Chem. Soc.*, 59:573 (1982).
49. E. Oliveros, M. T. Maurette, and A. M. Braun, *Helv. Chim. Acta*, 66:1183 (1983).
50. H. L. Rosano and W. F. Gerbacia, German Patent 2 319 971 (1973).
51. G. Serratrice, L. Matos, J. J. Delpuech, and A. Cambon, *J. Chim. Phys.*, 87:1969 (1990).
52. C. Cecutti, A. Novelli, I. Rico, and A. Lattes, *J. Dispersion Sci. Technol.*, 11:115 (1990).
53. A. Lattes and I. Rico-Lattes, *Art. Cells Blood Subs. Immob. Biotechnol.*, 22:1007 (1994).
54. K. V. Schubert and E. W. Kaler, *Colloids Surfaces*, 84:90 (1994).
55. I. Rico and A. Lattes, *J. Colloid Interface Sci.*, 102:285 (1984).
56. C. Cecutti, I. Rico, and A. Lattes, *J. Dispersion Sci. Technol.*, 7:307 (1986).
57. J. G. Riess, *Biomat. Art. Cells Immob. Biotechnol.*, 20:183 (1992).
58. J. G. Riess, J. L. Dalfors, G. K. Hanuk, D. H. Klein, M. P. Krafft, T. J. Pelura, and E. G. Schutt, *Biomat. Art. Cells Immob. Biotechnol.*, 20:839 (1992).
59. U. Gross, H. Reichelt, and J. Draffehn, *Biomat. Art. Cells Immob. Technol.*, 19:1055 (1991).
60. K. Z. Tremper, T. Zaccari, B. F. Cullen, and S. M. Hufstedler, *Anesth. Analg.*, 63:690 (1984).
61. G. R. Hodges, T. S. Reed, C. E. Hignite, and W. R. Snodgrass, *Prog. Clin. Biol. Res.*, 122:430 (1983).
62. J. Matsumoto, J. Bianchine, and R. Thompson, *Proc. West. Pharmacol. Soc.*, 26:403 (1983).
63. J. Horada, H. Sieki, H. Ohmura, and S. Myagati, *Prog. Chem. Biol. Res.*, 122:439 (1983).
64. A. L. Kunz, R. E. Lewis, and G. L. Sperandio, *Bull. Parenter. Drug Assoc.*, 19:13 (1965).
65. J. P. Dutcher, P. J. Haney, N. O. Whitley, R. Finley, P. Pearl, M. S. Didolkar, and P. H. Wiernik, *J. Clin. Oncol.*, 2:118 (1984).
66. A. S. Arambulo, M. S. Liu, A. L. Rogen, G. Dobben, and D. M. Long, *Drug Dev. Commun.*, 1:73 (1974–75).
67. V. Sanchez and J. Greiner, *J. Fluorine Chem.*, 75:217 (1995).
68. Q. N. Myrvik, *Ann. N. Y. Acad. Sci.*, 221:324 (1974).
69. C. H. Stuart-Harris, *Bull. WHO*, 41:617 (1969).
70. M. R. Hilleman, *Prog. Med. Virol.*, 8:131 (1966).
71. J. A. Reynolds, *Infect. Immunol.*, 28(B):937 (1980).

72. E. Yarkoni, M. S. Meltzer, and H. J. Rapp, *Int. J. Cancer*, 19:818 (1977).
73. S. S. Davis, *Pharm. Technol.*, 5:71 (1981).
74. S. E. Svenson, W. F. Delorenzo, R. Engelberg, M. Spodner, and L. O. Randall, *Antiobiot. Med.*, 11:148 (1956).
75. J. Wagner, E. Gerard, and D. Kaiser, *Clin. Pharmacol. Ther.*, 7:610 (1966).
76. P. Carrigan and T. Bates, *J. Pharm. Sci.*, 62:1476 (1973).
77. T. R. Bates and J. Sequeira, *J. Pharm. Sci.*, 64:793 (1975).
78. K. Palin, S. S. Davis, A. J. Phillips, D. Whalley, and C. G. Wilson, *J. Pharm. Pharmacol. Suppl.*, 32:62P (1980).
79. G. M. Lin, F. Sadik, W. F. Gilmore, and J. H. Fincher, *J. Pharm. Sci.*, 63:666 (1974).
80. K. Kakemi, H. Sezaki, S. Muranishi, H. Ogata, and S. Isemura, *Chem. Pharm. Bull.*, 20:708 (1972).
81. K. Kakemi, H. Sezaki, S. Uranishi, H. Ogata, and K. Giga, *Chem. Pharm. Bull.*, 20:715 (1972).
82. H. Ogata, K. Kakemi, S. Muranishi, and H. Sezaki, *Chem. Pharm. Bull.*, 23:707 (1975).
83. T. Noguchi, K. Taniguchi, S. Muranishi, and H. Sezaki, *Chem. Pharm. Bull.*, 25:434 (1977).
84. D. C. Bloedow, Bioavailability of lipophilic drugs in lipid vehicles, Ph.D. thesis, Washington State University, 1974, p. 40.
85. N. Muranishi, Y. Nakajima, M. Kinugawa, S. Muranishi, and H. Sezaki, *Int. J. Pharm.*, 4:281 (1980).
86. N. A. Armstrong and K. C. James, *Int. J. Pharm.*, 6:195 (1980).
87. D. C. Bloedow and W. L. Hayton, *J. Pharm. Sci.*, 65:328 (1976).
88. P. P. Constanides, *Pharm. Res.*, 12:1561 (1995).
89. P. P. Constanides, J. P. Scalart, C. Lancaster, J. Marcello, G. Marks, H. Ellens, and P. L. Smith, *Pharm. Res.*, 11:1385 (1994).
90. X. H. Zhou, and A. Li Wanpu, *Int. J. Pharm.*, 75:117 (1991).
91. J. E. Talmadge, *Adv. Drug Del. Rev.*, 10:247 (1993).
92. J. L. Murtha and H. Y. Ando, *J. Pharm. Sci.*, 83:1222 (1994).
93. C. W. Pouton, *Int. J. Pharm.*, 27:335 (1985).
94. T. Furia, ed., *Handbook of Food Additives*, Chemical Rubber Co., Cleveland, 1968, p. 626.
95. T. Takahashi, *Crit. Rev. Ther. Drug Carr. Syst.*, 2:245 (1986).
96. Y. Nakamoto, M. Hashida, S. Muranishi, and H. Sezaki, *Chem. Pharm. Bull.*, 23:3125 (1975).
97. H. Richter and K. Steiger-Trippi, *Pharm. Acta Helv.*, 36:322 (1961).
98. Y. Nakamoto, M. Fujiwara, T. Naguchi, T. Kimura, S. Muranishi, and H. Sezaki, *Chem. Pharm. Bull.*, 23:2232 (1975).
99. M. Hashida, M. Egawa, S. Murawishi, and H. Sezaki, *J. Pharmacokin. Biopharm.*, 5:225 (1977).
100. M. Hashida, S. Muranishi, and H. Sezaki, *Chem. Pharm. Bull.*, 25:2410 (1977).
101. M. Hashida, Y. Takahashi. S. Muranishi, and H. Sezaki, *J. Pharmacokin. Biopharm.*, 5:241 (1977).
102. M. Hashida, T. Yoshioka, S. Muranishi, and H. Sezaki, *Chem. Pharm. Bull.*, 28:1009 (1980).
103. M. Hashida, S. Muranishi, H. Sezaki, N. Tanigwa, K. Satomura, and Y. Hikasa, *Int. J. Pharm.*, 2:245 (1979).
104. H. Tanigawa, T. Shimomatsuya, K. Takahashi, K. Inomata, K. Tanaka, K. Satomura, K. Hikasa, M. Hashiba, S. Muranishi, and H. Sezaki, *Cancer*, 68:741 (1987).
105. A. T. Florence and D. Whitehill, *Int. J. Pharm.*, 11:277 (1982).
106. A. T. Florence and D. Whitehill, Stability and stabilization of water-in oil-in-water multiple emulsions. In: *Macro- and Microemulsions: Theory and Applications* (D. O. Shah, ed.). American Chemical Society, Washington, DC, 1985, pp. 259–388.

107. S. S. Davis, *Chem. Ind.*, 684 (1981).
108. Y. Mine, M. Shimizu, and T. Nakashima, Colloids Surfaces B, 6:261 (1996).
109. W. J. Herbert, *Lancet*, ii:771 (1965).
110. P. J. Taylor, C. L. Miller, T. M. Pollack, F. T. Perkins, and J. Westwood, *J. Hyg.* (Cambridge), 67:485 (1969).
111. L. A. Elson, C. V. Mitcheley, A. J. Colling, and R. Schneider, *Eur. J. Clin. Biochem. Res.*, 15:87 (1970).
112. T. Takahashi, M. Hizuno, Fujita, S. Ueda, B. Nishioka, and S. Majima, *Gann*, 64:345 (1973).
113. A. F. Broden, D. R. Kavaliunas, and S. G. Frank, *Acta Pharm. Suec.*, 15:1 (1978).
114. N. N. Li, *Membrane Processing in Industry and Biomedicine*. Plenum Press, New York, 1971.
115. N. N. Li and A. L. Shrier, *Rec. Dev. Sep. Sci.*, 1:163 (1972).
116. Y. Morimoto, K. Sugibayashi, Y. Yamaguchi, and Y. Kato, *Chem. Pharm. Bull.*, 27:3188 (1979).
117. C. Chiang, G. C. Fuller, J. W. Frankenfeld, and C. T. Rhodes, *J. Pharm. Sci.*, 67:63 (1978).
118. S. W. May and N. N. Li, *Enzyme Eng.*, 2:77 (1974).
119. Y. Garti, M. Frenkel, and R. Shwartz, *J. Dispersion Sci. Technol.*, 4:237 (1983).
120. A. Vaziri and B. Warburton, *J. Microencapsulation*, 11:641 (1994).
121. F. Sebba, *J. Colloid Interface Sci.*, 40:468 (1972).
122. S. Matsumoto, Y. Kita, and D. Yonezawa, *J. Colloid Interface Sci.*, 57:353 (1976).
123. S. Magdassi, M. Frenkel, N. Garti, and R. Kasan, *J. Colloid Interface Sci.*, 97:374 (1984).
124. S. Magdassi, M. Frenkel, and N. Garti, *J. Dispersion Sci. Technol.*, 5:49 (1984).
125. A. T. Florence and D. Whitehill, *J. Colloid Interface Sci.*, 79:243 (1981).
126. D. R. Kavaliunas and S. G. Frank, *J. Colloid Interface Sci.*, 66:586 (1978).
127. S. Matsumoto, Formation and stability of w/o/w emulsions. In: *Macro- and Microemulsions* (D. O. Shah, ed.). American Chemical Society, Washington, DC, 1985, p. 430.
128. S. Friberg, L. Mandell, and M. Larsson, *J. Colloid Interface Sci.*, 29:155 (1969).
129. B. W. Barry and G. M. Saunders, *J. Colloid Interface Sci.*, 38:626 (1972).
130. T. Suzuki, H. Tsutsumi, and A. Ishida, *J. Dispersion Sci. Technol.*, 5:119 (1984).
131. D. H. Smith, G. L. Covatch, and K-H. Lim, *Langmuir*, 7:1585 (1991).
132. Y. Kita, S. Matsumoto, and D. Yonezawa, *J. Colloid Interface Sci.*, 62:87 (1977).
133. S. Matsumoto, T. Inoue, M. Kohda, and J. Ikura, *J. Colloid Interface Sci.*, 77:555 (1980).
134. S. S. Davis and A. S. Burbage, *J. Colloid Interface Sci.*, 62:361 (1977).
135. M. Tomita, Y. Abe, and T. Kondo, *J. Pharm. Sci.*, 71:332 (1982).
136. S. Magdassi and N. Garti, *Colloids Surfaces*, 12:367 (1984).
137. Y. Kita, S. Matsumoto, and D. Yonezawa, *Nippon Kagaku Kaishi*, 11 (1978).
138. S. Goto, K. Nakata, T. Miyakawa, S. Zhang, and T. Uchida, *Yakugaku Zasshi*, 101:782 (1991).
139. N. Barker and J. Hadgraft, *Int. J. Pharm.*, 8:193 (1981).
140. R. H. Guy, J. Hadgraft, I. W. Kellaway and J. M. Taylor, *Int. J. Pharm.*, 11:199 (1982).
141. A. T. Florence and D. Whitehill, *J. Pharm. Pharmacol.*, 79:243 (1981).
142. T. Yoshioka, K. Ikeuchi, M. Hashida, S. Muranishi, and H. Kezaki, *Chem. Pharm. Bull.*, 30:140B (1982).
143. T. K. Law, A. T. Florence, and T. L. Whateley, *Colloid Polymer Sci.*, 264:, 167 (1986).
144. M. L. Cole and T. L. Whately, *J. Colloid Interface Sci.*, 175:281 (1995).
145. T. K. Law, A. T. Florence, and T. L. Whateley, *J. Pharm. Pharmacol.*, (Suppl.), 36:50 (1984).
146. A. T. Florence, T. K. Law, and T. L. Whateley, *J. Colloid Interface Sci.*, 107:584 (1985).
147. Y. Kawashima, T. Hino, H. Takeuchi, T. Niwa, and K. Horibe, *Int. J. Pharm.*, 72:65 (1991).

148. I. Terrisse, M. Seiller, J. L. Grossiord, A. Magnet, and C. Lehen-Ferrenbach, *Colloids Surfaces*, 91:121 (1994).
149. Y. Sela, S. Magdassi, and W. Garti, *Colloids Surfaces*, 83:143 (1994).
150. A. Berthod, M. Rollet, and N. Farah, *J. Pharm. Sci.*, 77:216 (1988).
151. M. L. Shively and S. Myers, *Pharm. Res.*, 10:1071 (1993).
152. N. J. Morris and B. Warburton, *J. Pharm. Pharmacol.*, 34:475 (1982).
153. M. Nihant, G. Schugens, C. Grandfils. R. Jerome, and P. Teyssie, *Pharm. Res.*, 11:479 (1994).
154. H. Kunieda, C. Solans, N. Sitida, and T. L. Parra, *Colloids Surfaces*, 24:225 (1987).
155. R. Pons, L. Carrera, P. Erra, H. Kunieda, and C. Solans, *Colloids Surfaces*, 91:239 (1994).
156. R. Pons, S. C. Ravey, S. Sauvage. M. J. Stébé, P. Erra. and C. Solans, *Colloid Surfaces A*, 76:171 (1993).
157. H. Kunieda, Y. Fukui, H. Uchiyama, and C. Solans, *Langmuir*, 12:2136 (1996).
158. F. M. Menger, N. Balachander, E. van der Linden, and G. S. Hammond, *J. Amer. Chem. Soc.*, 113:51119 (1991).
159. E. Ruckenstein and K. J. Kim, *Polymer*, 31:2397 (1990).
160. G. Calderó, M. T. Garcia-Celina, C. Solans, M. Plaza, and R. Pons, *Langmuir*, 13:3185 (1997).
161. D. Stoefler and H. F. Eicke, *Physica*, 182:29 (1992).
162. P. L. Luisi, R. Scartazzini, G. Haering, and P. Schurtenberger, *Colloid Polymer Sci.*, 268:356 (1990)
163. T. Musataka, T. Imae, S. Tanaka, H. Ohki, and S. Suzuki, *Colloids Surfaces B*, 7:281 (1996).
164. I. Danielsson and B. Lindman, *Colloids Surfaces*, 3:391 (1981).
165. L. Scriven, In: *Micellization, Solubilization, and Microemulsions*, Vol. 2 (K. L. Mittal, ed.). Plenum Press, New York, 1977, p. 877.
166. M. Rosoff, In: *Progress in Surface and Membrane Science*, Vol. 12, (D. Cadenhead and F. Danielli, Eds.), Academic Press, New York, 1978, pp. 435–441.
167. K. Fontell, A. Ceglie, B. Lindman, and B. W. Niehaus, *Acta Chim. Scand. A*, 40:247 (1986).
168. P. Ekwall, *Adv. Liq. Cryst.*, 1:1 (1975).
169. M. Kahlweit, E. Lessner, and R. Strey, *J. Phys. Chem.*, 87:5032 (1983).
170. T. P. Hoar and J. H. Schulman, *Nature*, 152:102 (1943).
171. J. H. Schulman, W. Stoekenius, and L. M. Prinee, *J. Phys. Chem.*, 63:1677 (1959).
172. J. H. Schulman and J. B. Montagne, *Ann. N.Y. Acad. Sci.*, 92:366 (1961).
173. E. Ruckenstein and J. Chi, *J. Chem. Soc.*, Faraday Trans. 2, 71:1690 (1975).
174. J. T. G. Overbeek, P. L. de Bruyn, and F. Verhoeekx, Microemulsions. In: *Surfactants* (T. Tadros, ed.). Academic Press, New York, 1984, pp. 111–132.
175. J. N. Israelachvili, D. J. Mitchell, and B. W. Ninham, *J. Chem. Soc.*, Faraday Trans. 2, 77:601 (1981).
176. D. J. Mitchell and B. W. Ninham, *J. Chem. Soc.*, Faraday Trans. 2, 77:601 (1981).
177. D. Oakenfull, *J. Chem. Soc.*, Faraday Trans. 1, 76:1875 (1980).
178. J. Jouffroy, P. Levinson, and P. G. de Gennes, *J. Phys.* (Paris), 43:1241 (1982).
179. J. H. Fendler, *Membrane Mimetic Chemistry*, Wiley, New York, 1982.
180. P. L. Luisi znd C. Laake, *Trends Biotechnol.*, 4:153 (1986).
181. C. Tondre and A. Xenakis, *Colloid Polym. Sci.*, 260:232 (1982).
182. J. M. Wiencek, and S. Qutubuddin, *Colloids Surfaces*, 29:119 (1988).
183. K. M. Larsson, P. Adlercreutz, B. Mattiasson, and U. Olsson, *Biotechnol. Bioeng.*, 36:135 (1990).
184. H. Sagitani and S. Friberg, *J. Dispersion Sci. Technol.*, 1:151 (1980).
185. A. Jayakrishnan, K. Kalaiarasi, and D. O. Shah, *J. Soc. Cosmet. Chem.*, 34:335–350 (1983).

186. S. E. Friberg and G. Z. Li, *J. Soc. Cosmet. Chem.*, 34:73–81 (1983).
187. G. D. Smith, C. E. Donelan, and R. E. Garden, *J. Colloid Interface Sci.*, 60:488 (1977).
188. I. Rico and A. Lattes, *Nouv. J. Chim.*, 8:429–31 (1984).
189. M. Dvolaitzky, R. Ober, J. Billar, C. Taupin, J. Charvolin, and Y. Hendricks, *C. R. Acad. Sci. Paris*, 292:45 (1981).
190. S. Friberg and C. S. Wohn, *Colloid Polym. Sci.*, 263:156–159 (1985).
191. D. R. MacFarlane and C. A. Angell, *J. Phys. Chem.*, 86:1927 (1982).
192. T. Kemken, A. Ziegler., and B. W. Muller, *Pharm. Res.*, 9:554 (1992).
193. P. Speiser, *Prog. Colloid Polym. Sci.*, 59:48 (1976).
194. *Proc. Meran Symp.* Vol. 1. Schriftenreihe Bundes-Apothekerkammer, Gelbe Reihe, Vol. 10, 1982.
195. G. Birrenbach and P. Speiser, *J. Pharm. Sci.*, 65:1763 (1976).
196. G. Birrenbach, Thesis, ETH-Zurich, Switzerland, 1973.
197. P. Couvreur, P. Tulkens, M. Roland, A. Trouet, and P. Speiser, *FEBS Lett.*, 84:323 (1977).
198. M. Antonietti, S. Lohmann, and C. van Niels, *Macromolecules*, 25:1139 (1995}
199. V. H. Perez-Lusa, J. E. Puig, V. M. Castano, B. E. Rodriguez, M. K. Murtha, and E. W. Kaler, *Langmuir*, 6:1040 (1990).
200. J. N. Israelachvili, S. Marcelja, and R. G. Horn, *Q. Rev. Biophys.*, 13:121 (1980).
201. H. Kunieda, K. Nakamura, U. Olsson, and B. L. Lindman, *J. Phys. Chem.*, 97:9525 (1993).
202. A. D. Bangham, M. M. Standish, and J. C. Watkins, *J. Mol. Biol.*, 13:238 (1965).
203. D. W. Deamer and A. D. Bangham, *Biochim. Biophys. Acta*, 443:629 (1976).
204. M. H. Milsmann, R. A. Schwendener, and H. G. Weder, *Biochim. Biophys. Acta*, 512:147 (1970).
205. Battelle Memorial Institute, British patent application, 2 001 929A (1979).
206. F. Szoka and D. Papahadjopoulos, *Proc. Natl. Acad. Sci.*, 75:4194 (1978).
207. F. Szoka and D. Papahadjopoulos, Liposomes-Preparations and characterization. In: *Liposomes: From Physical Structure to Therapeutic Applications* (C. G. Knight, ed.). Elsevier, New York, 1981, p. 51.
208. S. Chandonnet, H. Korstvedt, and A. A. Siciliano, *Soap Cosmet. Chem. Spec.*, 61:37 (1985).
209. Papahadjopoulos, D, ed., *Liposomes and Their Use in Biology and Medicine*, New York Academy of Science, New York, 1978.
210. G. Gregoriadis and A. C. Allison, eds., *Liposomes in Biological Systems*, Wiley, New York, 1980.
211. D. D. Lasic, Liposomes from Physics to Applications. Elsevier, Amsterdam, 1993.
212. D. D. Lasic, In: *Vesicles* (M. Rosoff, ed.), Marcel Dekker, New York, 1996, pp. 461–467.
213. G, Haran, R. Cohen, L. K. Bar, and Y. Barenholz, *Biochim. Biophys. Acta*, 1151:201 (1993).
214. D. D. Lasic, *Angew. Chem. Int. Ed. Engl.*, 331:1471 (1994).
215. K. Maruyama, S. J. Kenel, and L. Huang, *Proc. Natl. Acad. Sci.* 87:5744 (1990).
216. C. R. Alving and E. A. Steck, *Trends Biochem. Sci.*, 4:N175 (1979).
217. J. T. Dingle, J. L. Gordon, B. L. Haxleman, C. G. Knight, D. P. Page-Thomas, D. C. Philips, I. H. Shaw, F. J. T. Fildes, J. E. Oliver, G. Jones, E. H. Turner, and J. S. Lowe, *Nature*, 271:372 (1978).
218. J. Khato, E. R. Priester, and S. M. Sieber, *Cancer Treat. Rep.*, 66:517 (1982).
219. V. J. Richardson, B. E. Ryman, R. F. Jewkes, K. Jeyasingh, M. N. H. Tattersall, E. S. Newlands, and S. B. Kaye, *Br. J. Cancer*, 40:35 (1975).
220. G. Poste and I. J. Fidler, Active nonspecific immunotherapy by administration of macrophase activation agents encapsulated in liposomes. In: *Optimization of Drug Delivery*, Alfred Benzon Symp. 17 (H. Bundegaard, A. Bagger Hansen, and H. Kofod, eds.). Munksgaard, Copenhagen, 1982, p. 418.

221. C. R. Alving and R. L. Richards, Immunologic aspects of liposomes. In: *The Liposomes* (M. Ostro, ed.), Marcel Dekker, New York, 1983, p. 209.
222. R. L. Juliano, D. Stamp, and N. McCullough, *Ann. N.Y. Acad. Sci.*, 308:411 (1978).
223. J. Lasch, M. Deicher, and R. Schubert, In: *Liposomes in Ophthalmology and Dermatology* (U. Pleyer, K-H. Schmidt, and H. J. Thiel, eds.), Hippokrates Verlag, Stuttgart, 1993, p. 195.
224. W. S. Oleniacz, U.S. patent 3,957,951, Lever Assignee (1976).
225. M. Mezei and V. Gulasekharam, *Life Sci.*, 26:1473 (1980).
226. M. Handjani-Vila, A. Ribier, B. Rondot, and G. Vanlerberghe, *Int. J. Cosmet. Sci.*, 1:303 (1979).
227. G. Smolin, M. Okumoto, S. Feiler, and D. Condon, *Am. J. Ophthalmol.*, 91:220 (1981).
228. H. Shaeffer, *Invest. Ophthalmol. Vision Sci.*, 22:220 (1982).
229. R. E. Stratford, D. C. Yang, M. A. Redell, V. Lee, and L. Hon, *Int. J. Pharm.*, 13:263 (1983).
230. A. A. Siciliano, *Cosmet. Toiletries*, 100:43 (1985).
231. S. Frokjher, E. L. Hjorth, and O. Worts, Stability and storage of liposomes. In: *Optimization of Drug Delivery* (H. Bundgaard, A. B. Hansen, and H. Kofod, eds.). Munksgaard, Copenhagen, 1982, p. 418.
232. Imperial Chemical Industries Ltd., Belgian Patent 866 697 (1978).
233. D. J. A. Crommelin and E. M. G. Van Bommel, *Pharm. Res.*, 1984, p. 159.
234. D. D. Lasic, P. M. Frederik, M. G. A. Stuart, Y. Barenholz, and T. J. MacIntosh, *FEBS Lett.*, 312:255 (1992).
235. G. Scherphof, J. Damen, and D. Hoekstra, Interactions of liposomes with plasma proteins and components of the immune system. In: *Liposomes: From Physical Structure to Therapeutic Applications* (C. G. Knight, ed.). Elsevier/North Holland, New York, 1981, p. 299.
236. R. A. Demel and B. DeKruijff, *Biochim. Biophys. Acta*, 457:109 (1976).
237. T. Yoshioka and A. T. Florence, *Int. J. Pharm.*, 108:117 (1994).
238. S. Cohen, M. C. Bañó, M. Chow, and R. Langer, *Biochim. Biophys. Acta*, 95:1063 (1991).
239. T. Kunitake, Y. Okahata, M. Shomomura, S. Yasunami, and T. Takarabe, *J. Am. Chem. Soc.*, 103:5401 (1981).
240. J. H. Fendler, *Membrane Mimetic Chemistry*, Wiley-Interscience, New York, 1982, pp. 158–183.
241. T. Kunitake and Y. Okahta, *Bull. Chem. Soc. Japan*, 51:1877 (1978).
242. T. Kunitake and S. Yamada, *Polym. Bull.*, 1:35 (1978).
243. Y. Okahata and T. Kunitake, *J. Am. Chem. Soc.*, 101:5231 (1979).
244. T. Kunitake and Y. Okahata, *Chem. Lett.*, 387 (1977).
245. Y. Okahata, S. Tanamichi, M. Nagai, and T. Kunitake, *J. Colloid Interface Sci.*, 82:401 (1981).
246. A. J. Baillie, A. T. Florence, L. R. Hume, G. T. Muichead, and A. Rogerson, *J. Pharm. Pharmacol.*, 37:863 (1985).
247. M. N. Azmin, A. T. Florence, R. M. Handjani-Vila, J. F. B. Stuart, G. Vanlerberghe, and J. S. Whitaker. *J. Pharm. Pharmacol.*, 37:237 (1985).
248. H. E. J. Hofland, J. A. Bouwstra, M. Ponec, H. E. Boddé, F. Spies, J. Coos Verhoef, and H. E. Junginger, *J. Control Rel.*, 16:155 (1991).
249. R. Zana, S. Yiu, and K. M. Kale, *J. Colloid Interface Sci.*, 77:456 (1980).
250. J.-H. Furhop and D. Fritsch, *Acc. Chem. Res.*, 19:130 (1986).
251. T. Kunitake, N. Nakashima, K. Takarabe, M. Nagai, A. Tsuge, and H. Yanagi, *J. Am. Chem. Soc.*, 103:5945 (1981).
252. D. S. Johnston, S. Sanghera, M. Pons, and D. Chapman, *Biochim. Biophys. Acta*, 602:57 (1980).
253. H.-H. Hub, B. Haufer, H. Koch, and H. Ringsdorf, *Angew. Chem.*, 92:692 (1980).
254. D. F. O'Brien, T. H. Whitesides, and R. T. Klingbiel, *J. Polym. Sci., Polym. Lett. Ed.*, 19:95 (1981).

255. L. Gros, H. Ringsdorf, and H. Schupp, *Angew. Chem., Int. Ed. Engl.*, 20:305 (1981).
256. Y. Okahata, *Acc. Chem. Res.*, 19:57 (1986).
257. M.-A. Guedeau-Boudeville, J. Ludovic, and J.-M. diMeglio, *Proc. Natl. Acad. Sci*, 92:9590 (1995).
258. C. Nicolau and A. Cudd, *Crit. Rev. Ther. Drug Carr. Syst.*, 6:239 (1989).
259. N. Zhu, D. Liggit, Y. Lin, and R. Debs, *Sci.*, 261:209 (1993).
260. R. Hyde et al., *Nature*, 362:250 (1993).
261. H. Gershon, R. Ghirlando, S. D. Guttman, and A. Minsky, *Biochem.*, 32:7143 (1993).
262. S. Wahlgren, A. L. Lindstrom, and S. E. Friberg, *J. Pharm. Sci.*, 73:1484 (1984).
263. R. Schmäcker and B. Strey, *J. Phys. Chem.*, 98:3908 (1994).
264. P. Strom, *J. Colloid Interface Sci.*, 154:184 (1992).
265. S. Miyazaki, C. Yakouchi, T. Nakamura, N. Hashiguchi, W. M. Hou, and M. Takada, *Chem. Pharm. Bull.*, 34:1801 (1986).
266. B. Ericsson, P. O. Erikson, J. C. Löfroth, and S. Engström, *ACS Symp. Ser.*, 469:251 (1991).
267. M. Maniar, D. Hannibal, S. Anselem, X. Xie, R. Burch, and A. J. Domb, *Pharm. Res.*, 8:S185 (1991).
268. S. Anselem, A. J. Domb, and C. R. Alving, *Vaccine Res.*, 1:383 (1992).
269. J.-H. Furhop, S. Svenson, C. Boetcher, E. Rössler, and H.-M. Vieth, *J. Am. Chem. Soc.*, 112:4307 (1990).

2

Liposomes

Danilo D. Lasic

Liposome Consultations, Newark, California

Norman Weiner

University of Michigan, Ann Arbor, Michigan

Mohammad Riaz

University of the Punjab, Lahore, Pakistan

Frank Martin

Sequus Pharmaceuticals, Inc., Menlo Park, California

I. INTRODUCTION

In the early 1960s, Alec Bangham [1] first described how membrane molecules, e.g., phospholipids, interact with water to form unique structures now recognized as liposomes. Considered a mere curiosity just some 30 years ago, liposomes are now established as a useful model membrane system, and they have demonstrated potential for delivering molecules as large as DNA to the intracellular compartment of cells. More important, liposomes have shown great potential as a drug delivery system. An assortment of molecules, including peptides and proteins, have been incorporated in liposomes, which can then be administered by different routes. Various amphiphathic molecules have been used to form the liposomes, and the method of preparation can be tailored to control their size and morphology. Drug molecules can either be encapsulated in the aqueous space or intercalated into the lipid bilayer; the exact location of a drug in the liposome depends on its physicochemical characteristics and the composition of the lipids.

Due to their high degree of biocompatibility, liposomes were initially conceived of as systems for intravenous delivery. It has since become apparent that liposomes can also be useful for delivery of drugs by other routes of administration. The formulator can use strategies to design liposomes for specific purposes, thereby improving the therapeutic index of a drug by increasing the percent of drug molecules that reach the target tissue, or, alternatively, decreasing the percent of drug molecules that reach sites of toxicity. Clinical trials now under way use liposomes to achieve a variety of therapeutic objectives including enhancing the activity and reducing toxicity of a widely used antineoplastic drug (Doxorubicin) and of an antifungal drug (Amphotericin B) delivered intravenously. Other clinical trials are evaluating the ability of liposomes to deliver intravenously immunomodulators (MTP-PE) to macrophages and imaging agents (^{111}Indium) to tumors. Recent studies in animals have reported the delivery of water-insoluble drugs into the eye, and the prolonged release of an immunomodulator (interferon) and a peptide hormone (calcitonin) from an intramuscular depot. These trials and animal studies provide evidence of the versatility of liposomes.

Drug-containing liposomes were tested in many clinical trials in humans. Following early experiments in the 1970s, which showed that there were no unexpected toxicities, there were several important clinical trials in humans using liposomes with encapsulated Amphotericin B and Doxorubicin carried out in the 1980s. Earlier in the 1990s there was the successful launch of a parenteral liposomal formulation containing Amphotericin B and more recently a sterically stabilized liposomal formulation containing the anticancer drug Doxorubicin [2]. This product, launched in 1995, is the first liposomal therapeutic parenteral approved in the United States.

A liposome is defined as a structure consisting of one or more concentric spheres of lipid bilayers separated by water or aqueous buffer compartments (Fig. 1). These spherical structures can be prepared with diameters ranging from 80 nm to 100 μm. When phospholipids are dispersed in an aqueous phase, hydration of the polar head groups of the lipid results in a heterogeneous mixture of structures, generally referred to as vesicles, most of which contain multiple lipid bilayers forming concentric spherical shells. These were the liposomes first described by Bangham, which are now referred to as multilamellar vesicles (MLVs). Sonication of these lipid dispersions results in size reduction of these liposomes to vesicles containing only a single bilayer with diameters ranging from 25–50 nm. These structures are referred to as small unilamellar

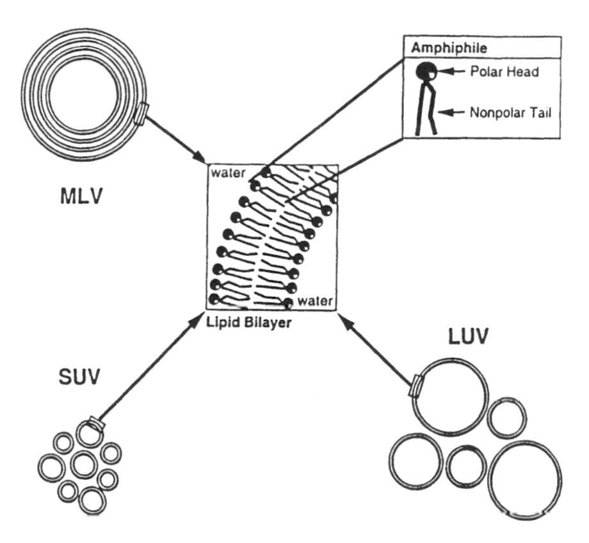

Fig. 1 Large multilamellar (MLV), large unilamellar (LUV) and small unilamellar (SUV) vesicles. From D. D. Lasic, Liposomes in Gene Delivery, CRC Press (1997).

vesicles (SUVs). Since MLVs and SUVs have certain limitations as model membrane systems and as drug delivery systems, a number of laboratories have developed single bilayer liposomes that exhibit a size range of 100–500 nm in diameter. These vesicles are referred to as large unilamellar vesicles (LUVs). The nomenclature describing liposomes can be confusing, since liposomes have been classified as a function of the number of bilayers (e.g., MLV, SUV), or as a function of the method of preparation (e.g., REV, FPV, EIV) or as a function of size (e.g., LUV, SUV). Table 1 identifies examples of frequently used nomenclatures:

Medical applications of liposomal formulations distinguish several types of liposomes based on their reactivity with surrounding media. Conventional liposomes are characterized by nonspecific reactivity. Novel liposome formulations may be either inert or can show highly specific reactivity. An important example of novel or second-generation liposomes are the so-called stealth (sterically stabilized) liposomes, which have shown reduced toxicity during anticancer therapy, and these liposomes are the basic unit for the attachment of ligands for site-specific targeting. Another exciting new area of liposome technology is their potential as vectors to complex and deliver DNA, which has shown promise in delivery of genes in gene therapy [3].

Liposomes have been used as model membranes and as carriers of drugs, DNA, ATP, enzymes, and diagnostic agents. Although there are approximately 15,000 publications dealing with liposomes, very few are centered on the pharmaceutical issues that must be addressed to bring liposomal products to the marketplace.

II. MATERIALS USED IN LIPOSOME PREPARATION

The lipids most commonly used to prepare liposomes are shown in Fig. 2.

A. Phospholipids

Glycerol-containing phospholipids are by far the most commonly used component of liposome formulations and represent more than 50% of the weight of lipid present in biological membranes. The general chemical structure of these types of lipids is exemplified by phosphatidic acid. As indicated in Fig. 3, the “backbone” of the molecule resides in the glycerol moiety. At position number 3 of the glycerol molecule the hydroxyl is esterified to phosphoric acid (hence the name glycerolphospholipids). The hydroxyls at positions 1 and 2 are usually esterified with long-chain fatty acids giving rise to the lipidic nature of these molecules. One of the remaining oxygens of phosphoric acids may be further esterified to a wide range of organic alcohols including glycerol, choline, ethanolamine, serine, and inositol. Thus, the parent compound of the

Table 1 Examples of Nomenclature Used to Describe Liposomes

Type of vesicle	Term used	Approx. size (μm)
Small, sonicated unilamellar	SUV	0.025–0.05
Large, vortexed multilamellar	MLV	0.05–10
Large unilamellar	LUV	0.1
Reverse phase evaporation	REV	0.5
French press	FPV	0.05
Ether injection	EIV	0.02

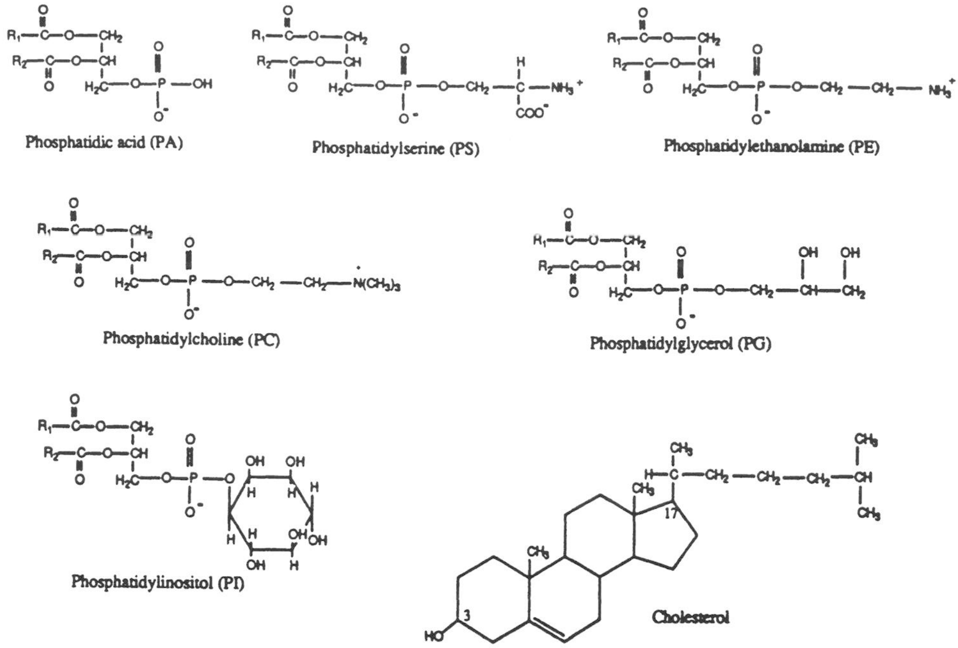

Phosphatidic acid (PA)

Phosphatidylserine (PS)

Phosphatidylethanolamine (PE)

Phosphatidylcholine (PC)

Phosphatidylglycerol (PG)

Phosphatidylinositol (PI)

Cholesterol

Fig. 2 Chemical structures of lipids commonly used to prepare liposomes. PE, PC, and cholesterol behave as neutral molecules, whereas PA, PS, PG, and PI have a net negative charge.

series is the phosphoric ester of glycerol. The phosphate moiety of phosphatidic acid carries a double negative charge only at high pH. The pH values for the two oxygens are 3 and about 7. At physiologically relevant pH values this molecule presents more than one net negative charge, but not quite two. Although phosphatidic acid occurs only in small amounts in nature, it is an important intermediate in the biosynthesis of the phosphoglycerides. The most abundant glycerol phosphatides in plants and animals are phosphatidylcholine (PC), also called lecithin, and phosphatidylethanolamine (PE), sometimes referred to as cephalin. These two phosphatides constitute the major structural component of most biological membranes. In phosphatidylserine (PS), the phosphoric acid moiety of phosphatidic acid (PA) is esterified to the hydroxyl group of the amino acid L-serine, and in phosphatidylinositol (PI) to one of the hydroxyls of the cyclic sugar alcohol inositol. In the case of phosphatidylglycerol (PG), the alcohol that is esterified to the phosphate moiety is glycerol. Although this molecule is found commonly in some bacteria, it is rare in mammalian cell membranes. However, PG is found as a natural component of the lung surfactant of humans and other higher mammals. Fatty acids are important constituents of glycerol phosphatides and are abundant as building block components of other saponifiable lipids (such as the triglycerides of fat cells). However, fatty acids occur only in trace amounts in the free (unesterified) form. Dozens of different fatty acids have been isolated, differing in the number of carbons and degrees of unsaturation. Table 2 lists the most common saturated and unsaturated fatty acids found in lipids used to form liposomes. Table 3 shows the fatty acid composition of two common phosphatidylcholines, one extracted from egg yolk and the other from soybean oil. Notice the difference in the degree of unsaturation between egg and

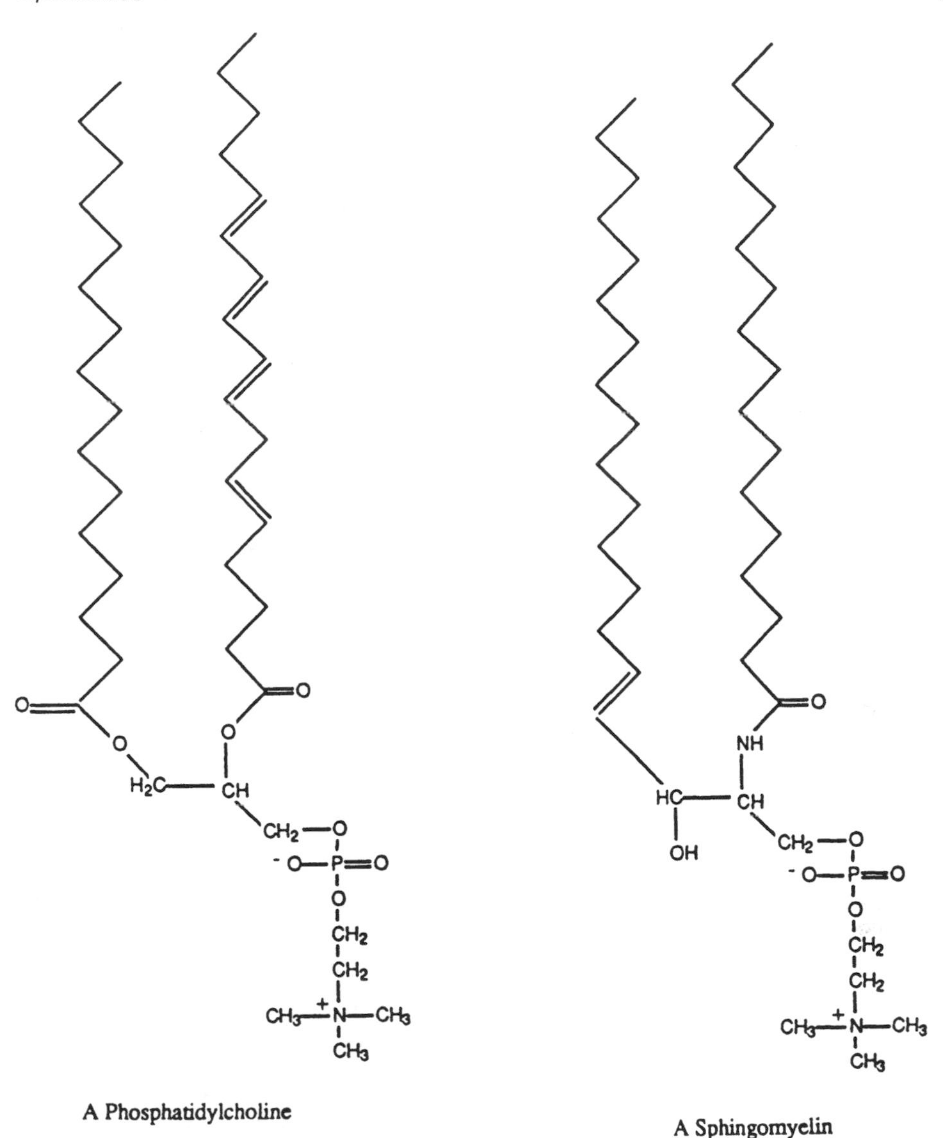

Fig. 3 Chemical structures of two representative phospholipids showing their polar head groups and nonpolar hydrocarbon chains.

soy PC. Soy PC contains a greater proportion of unsaturated bonds and is thus more susceptible to peroxidation.

B. Sphingolipids

In contrast to the phospholipids, sphingolipids contain as their structural backbone sphingosine or a related base. Such molecules are important components of both plant and animal cells, particularly in brain and nerve tissue. Sphingolipids contain three characteristic building block elements: a molecule of fatty acid, a molecule of sphingosine

Table 2 Some Common Saturated and Unsaturated Fatty Acids Found in Lipids Used to Form Liposomes

Molecular formula	Common name	Systematic name	Structural formula
		Saturated fatty acids	
$C_{16}H_{32}O_2$	Palmitic	n-hexadecanoic	$CH_3(CH_2)_{14}COOH$
$C_{18}H_{36}O_2$	Stearic	*n*-octadecanoic	$CH_3(CH_2)_{16}COOH$
		Unsaturated fatty acids	
$C_{16}H_{30}O_2$	Palmitoleic	9-hexadecenoic	$CH_3(CH_2)_5CH=CH(CH_2)_7COOH$
$C_{18}H_{34}O_2$	Oleic	*cis*-9-octadecenoic	$CH_3(CH_2)_7CH=CH(CH_2)_7COOH$
$C_{18}H_{32}O_2$	Linoleic	*cis*,*cis*-9,12-octadecadienoic	$CH_3(CH_2)_4CH=CHCH_2CH=CH(CH_2)_7COOH$

Table 3 Fatty Acid Composition of Two Common Phosphatidylcholines, One Extracted from Egg Yolk and the Other from Soybean Oil

Fatty acid composition		Egg PC	Soy PC
16:0;	Palmitic	32	12
16:1;	Palmitoleic	1.5	<0.2
18:0;	Stearic	16	2.3
18:1;	Oleic	26	10
18:2;	Linoleic	13	68
18:3;	Linolenic	<0.3	5
20:4;	Arachidonic	4.8	<0.1
22:6;	Docosahexanoic	4.0	<0.1

(or related derivative) and a head group that can vary from simple alcohols such as choline to very complex carbohydrates. The most abundant sphingolipid in higher animals is sphingomyelin, which contains either phosphorylcholine or phosphorylethanolamine esterified at the 1-hydroxy group of ceramide as its polar head group. In terms of their physical properties, sphingomyelins behave quite similarly to phospholipids; they are zwitterions at pH 7 and readily form bilayer structures in aqueous media.

C. Glycosphingolipids

Gangliosides, a second class of sphingolipid, are found mainly in the gray matter of brain tissue of higher animals and are often used as a minor component of some liposome formulations. These molecules contain complex oligosaccharides with one or more sialic acid residues in their polar head group and thus have one or more net negative charges at neutral pH. Gangliosides are included in liposomes to provide a layer of surface-charged groups.

D. Sterols

Cholesterol and its derivatives are quite often included as components of liposomal membranes. These are derivatives of the tetracyclic hydrocarbon perhydrocyclopentanophenanthrene. Cholesterol is abundant in animal tissues and is primarily localized in cell membranes. Its inclusion in liposomal membranes has three recognized effects: (1) decreasing the fluidity or microviscosity of the bilayer; (2) reducing the permeability of the membrane to water-soluble molecules; and (3) stabilizing the membrane in the presence of biological fluids such as plasma. This latter effect has proven useful in formulating liposomes for drug delivery applications that use the intravenous route of administration. Liposomes without cholesterol are known to interact rapidly with plasma proteins such as albumin, transferrin, and macroglobulins. These proteins tend to extract bulk phospholipids from liposomes, thereby depleting the outer monolayer of the vesicles leading to physical instability. Cholesterol appears to substantially reduce this type of interaction. Cholesterol has been called the "mortar" of bilayers, because, by virtue of its molecular shape and solubility properties, it fills in empty spaces among the phospholipid molecules, anchoring them more strongly into the structure. The hydroxyl group at the 3-position provides a small polar head group, and the hydrocarbon chain attached to the 17-position becomes the nonpolar end. In bilayers, cholesterol intercalates among the phospholipid molecules with its hydroxyl group oriented toward the water phase, the tetracyclic ring sandwiched between the first few carbons of the fatty acyl chains and the tail dipping fairly deep, and at a slight angle, into the hydrocarbon core of the bilayer (perhaps to about carbon 7-8 of the fatty acids).

E. Metabolic Fate of BIlayer Forming Lipids

An attribute of liposomes that translates into unusually high in vivo tolerance is the fact that the structural components of the system, phospholipids and cholesterol, are treated no differently than biological membrane lipids. In the body they are broken down by enzyme systems into natural intermediates like glycerol phosphate, fatty acids, ethanolamine, choline, and acyl-Co-A, and either metabolized further to provide energy, or enter lipid pools, which are drawn on to build new lipids, and which replace those that naturally turn over in biological membranes. Phospholipids are hydrolyzed in vivo by specific phospholipases (Fig. 4). Phospholipase A-1 specifically removes the fatty acid from the 1-position of the glycerol, and phospholipase A-2 removes the acid from the 2-carbon position by catalyzing the addition of water to the ester bond linking the two fatty acids to glycerol. Phospholipase B is a mixture of these two enzymes and can therefore selectively remove both fatty acid chains. Phospholipase C, which is found only in small amounts in cells, catalyzes the hydrolysis of the bond between phosphoric acid and glycerol, while phospholipase D cleaves off the polar alcohol head group to leave phosphatidic acid. Fatty acids liberated by the action of phospholipases enter the fatty acid pool and may be used as precursors to regenerate new phospholipids or triglycerides or be activated to acyl-Co-A by the action of coenzyme A and oxidized to CO_2 and water to yield energy via the beta-oxidation spiral as described elsewhere [4]. Glycerol phosphate is not usually broken down any further, because glycerol, like glucose, is a very good phosphate acceptor. Thus, intracellularly, the equilibrium strongly favors the phosphorylated form of glycerol. This molecule is usually recycled by the cell serving as the backbone for the formation of new phospholipids or triglycerides. Phospholipases can be used in vitro to modify natural lipids. For example, phos-

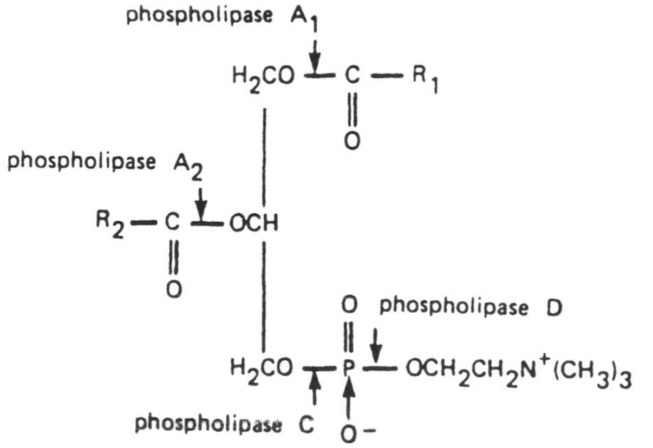

phospholipase A_1
H_2CO — C — R_1
phospholipase A_2
R_2 — C — OCH
O phospholipase D
H_2CO — P — $OCH_2CH_2N^+(CH_3)_3$
phospholipase C O^-

Fig. 4 Sites of action of phospholipases on phosphatidylcholine.

pholipase A-2 isolated from snake venom has been used to produce lysophosphatidylcholine from natural phosphatidylcholine. Phospholipase D is being used commercially to produce "semisynthetic" PS, PA, and PG from PC. The enzyme is usually extracted from cabbage and dissolved in a buffer containing an excess of the organic alcohol that one wishes to exchange for choline on PC (i.e., serine, glycerol, etc.). PC dissolved in ether is added and the two phases emulsified. The enzyme catalyzes head group exchange. The rates and yields of the conversion are dependent on the activity of the enzyme and the molar excess of the alcohol to be exchanged for choline.

The liver serves both as the chief source and chief organ for the disposal of cholesterol. A major portion of the cholesterol removed from plasma lipoproteins by the liver is excreted in the bile. Cholesterol is also excreted across the intestinal mucosa. In the lumen of the gut, cholesterol is broken down to coprostanol by the action of intestinal bacteria. About 80% of the cholesterol taken up by the liver, however, is transformed into bile acids that are excreted into the gut to aid digestion of fats. Most of the bile salts are reabsorbed along with the fats they help to emulsify. Only about 1 g of bile acid is lost in feces daily. Cholesterol also serves as the precursor of steroid hormones.

F. Synthetic Phospholipids

Generally used saturated phospholipids include dipalmitoylphosphatidylcholine (DPPC), distearoylphosphatidylcholine (DSPC), dipalmitoylphosphatidylethanolamine (DPPE), dipalmitoylphosphatidylserine (DPPS), dipalmitoylphosphatidic acid (DPPA) and dipalmitoylphosphatidylglycerol (DPPG). Several unsaturated phospholipids have also been used for preparing liposomes; these include dioleoylphosphatidylcholine (DOPC) and dioleoylphosphatidylglycerol (DOPG).

G. Polymeric Materials

Synthetic phospholipids with diactylenic group(s) in the hydrocarbon chain(s) undergo polymerization when exposed to ultraviolet light. The physical properties of the monomeric diacetylenic phospholipids were found to resemble those of naturally occurring phospholipids. Liposomes are formed using these materials, and the resultant vesicles are exposed to ultraviolet light, leading to the formation of polymerized liposomes having significantly tighter permeability barriers to drugs entrapped in their aqueous compartments [5]. A large variety of polymerizable lipids (Fig. 5), which can form vesicles, has been synthesized [6]. These include lipids containing conjugated diene, methacrylate, and thiol groups as polymerizable moieties. Several polymerizable surfactants have also been synthesized. It should be pointed out that such systems no longer can be considered "natural phospholipids" and their metabolic fates have yet to be established.

H. Polymer Bearing Lipids

In conventional liposomes, stability and repulsive interactions with macromolecules are governed mostly by repulsive electrostatic forces and weaker hydration and, in the case of large fluid membranes, undulation forces. Only recently, was it realized that repulsive forces between liposomes and macromolecules and other colloidal particles can be induced by coating liposome surfaces with appropriate polymers. Nonionic and water-

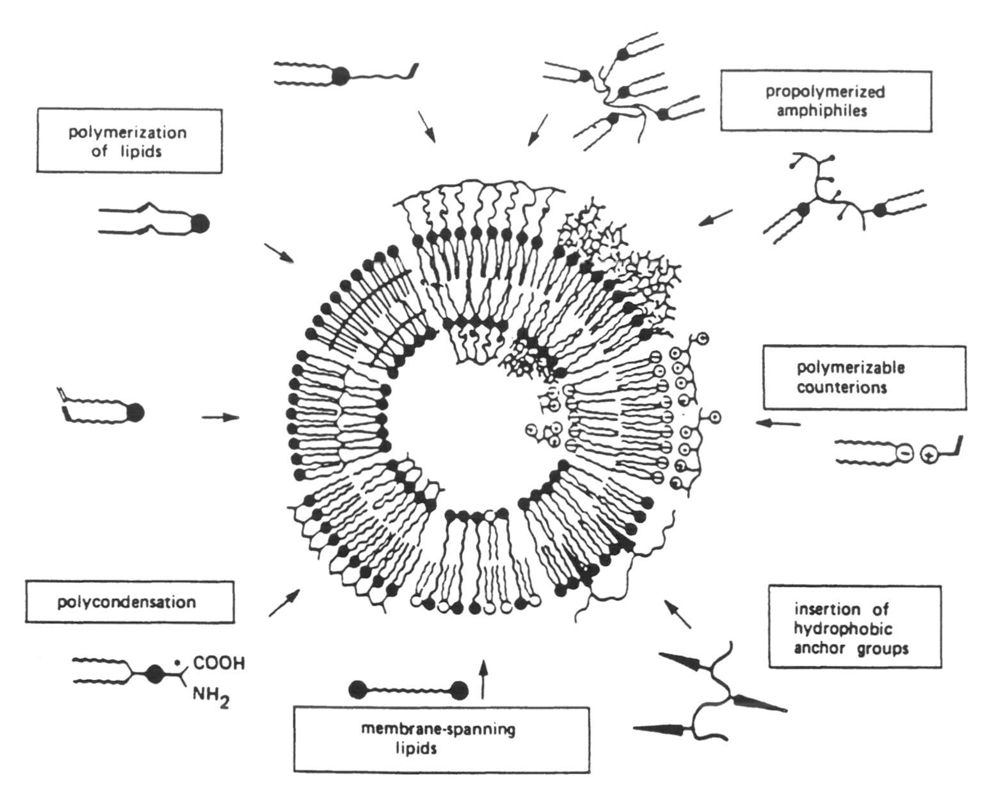

Fig. 5 Molecular architecture and function of polymerized liposomes. (From Ref. 6.)

compatible polymers, such as polyethylene oxide, polyvinyl alcohol, polyoxazolidines, and the like convey the highest stability. Adsorption of block copolymers containing alternating sequences of such hydrophilic segments with a hydrophobic part leads to liposome leakage, and, therefore, best results can be achieved by covalently attaching polymers to phospholipids. Most frequently this is diacylphosphatidylethanolamine with PEG polymer linked via a carbamate or succinate bond. The degree of polymerization varies from 15 to 120 units. Longer polymers give rise to aqueous solubility of polymer-lipids and their fast removal from membranes in nonequilibrium conditions, while shorter polymers do not offer enough repulsive pressure because van der Waals attraction is a long-range force. Figure 6 shows a diagrammatic representation of a long-circulating stealth liposome (7).

I. Cationic Lipids

The recent discovery that cationic liposomes can condense DNA and significantly improve the delivery of genes into cells caused the renewed interest in the synthesis of numerous cationic lipids. The best known are DODAB/C (dioctadecyldimethyl ammonium bromide/chloride) and DOTAP (dioleoyl trimethyl ammonium propane), which were synthesized in the 1970s and DOTMA (dioleoyl propyl trimethyl ammonium chloride, an ether analogue to DOTAP), which was introduced in 1987 [8]. Since then

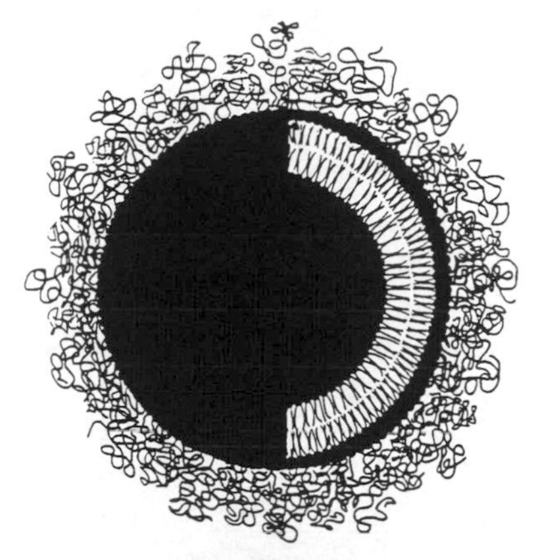

Fig. 6 Stealth and conventional liposomes. The Stealth liposomes are coated with long chains of polyethylene glycol, which prevent plasma components (opsonins) from adhering to the surface.

numerous cationic lipids have been synthesized, including various analogues of DOTMA and cationic derivatives of cholesterol.

J. Other Substances

A variety of other lipids and surfactants can be used to form liposomes. Many single-chain surfactants can form liposomes in mixtures with cholesterol. Nonionic lipids: a variety of polyglycerol and polyethoxylated mono- and di-alkyl amphiphiles have been synthesized. While they have been used primarily in cosmetic preparations, they have shown significant potential for use in topical pharmaceutical preparations [9]. A number of single- and double-chain lipids having fluorocarbon chains have recently been prepared. They show a very diverse phase behavior and can form very stable liposomes [10]. Stearylamine and dicetylphosphate have been incorporated into liposomes so as to impart either a negative or a positive surface charge to these structures. Also, a number of compounds having a single long-chain hydrocarbon and an ionic head group have been synthesized and found to be capable of forming vesicles. These include quaternary ammonium salts and dialkyl phosphates [6].

K. Lipid Selection

Many liposome-based pharmaceutical products are entering the clinical trial stage of development; several of these may reach the marketplace in a few years. The lipid component of these products must meet stringent "pharmaceutical" requirements in order to obtain regulatory approval for large-scale human testing and marketing. These include suitable purity, safety and microbial/endotoxin limits, and adequate stability. Currently available pharmaceutical-grade lipid products (such as egg and soy phosphatides) were developed primarily for the parenteral emulsions industry and, in general, they are not well suited for liposome formulations. Liposomes composed of crude egg yolk phosphatides, for example, are not stable at ambient temperatures for more than a few months. Thus developers of liposome products have been relying on specialty chemical firms to supply highly purified lipids for their raw material needs. The current cost for these high-purity lipids will need to be lowered for large-scale production and commercialization of liposome products. Although liposome-based products are within reach, and their market potential is large, successful commercialization depends in part on the willingness of lipid suppliers to differentiate their product lines in response to the needs of the rapidly emerging liposome industry. Since each liposome-based product has its unique stability, safety, and purity requirements, it is likely that a range of lipids including natural products, semisynthetics, and synthetics of varying degrees of purity will be needed. The key pharmaceutical and commercial issues that remain to be addressed by both the lipid suppliers and liposome product developers include the following:

1. Quantities, purities, and pharmaceutical attributes of lipids required for liposome products
2. Detailed specifications for each lipid, including standardized nomenclature and analytical quality control procedures
3. Introduction of the above into official compendia
4. Expansion/centralization of a database on the stability and safety of key lipids

L. Special Considerations

Various applications require different types of liposomes, as discussed in the introduction. For nonirritating topical formulations one can choose nonionic liposomes. For localized injections one can use conventional anionic liposomes. Such liposomes are also the delivery system of choice for targeting phagocytic cells of the immune system and for vaccination. For other systemic applications liposomes with covalently attached polymers provide the best delivery system. These liposomes, which are characterized by nonspecific repulsion, can be made specific by covalent attachment of antibodies on their surface, preferably at the far end of the polymer chain [11]. When nucleic acid (DNA, RNA, or short antisense single-strand oligonucleotides) complexation is desired, one chooses cationic liposomes that normally contain large fractions of neutral lipids, normally 50 mol% of dioleoylphosphatidylethanolamine, which due to its H(II) phase formation ability may improve interactions of liposome-DNA complexes with various membranes.

III. WHY LIPOSOMES ARE FORMED

As indicated in Fig. 1, lipids capable of forming liposomes (or other colloidal structures) exhibit a dual chemical nature. Their head groups are hydrophilic (water loving) and their fatty acyl chains are hydrophobic (water hating). It has been estimated that each zwitterionic head group of phosphatidylcholine has on the order of 15 molecules of water weakly bound to it, which explains its overwhelming preference for the water phase. The hydrocarbon fatty acid chains, on the other hand, vastly prefer each other's company to that of water. This phenomenon can be understood in quantitative terms by considering the critical micelle concentration (CMC) of PC in water. The CMC is defined as the concentration of the lipid in water (usually expressed as moles per liter) above which the lipid forms either micelles or bilayer structures rather than remaining in the solution as monomers. The CMC of dipalmitoylphosphatidylcholine has been measured by Smith and Tanford [12] and found to be 4.6×10^{-10} M in water. This value is in agreement with those obtained for similar amphiphiles. Clearly, this is a very small number, indicating the overwhelming preference of this molecule for a hydrophobic environment such as that found in the core of a micelle or bilayer.

From a knowledge of the total concentration of a lipid-like phosphatidylcholine and its CMC in aqueous systems, it is possible to derive the free energy for transfer of one mole of the lipid from water to micelle using the simple expression $\Delta F = -RT \ln(\mathrm{CMC})$ as described by Shinoda et al. [13]. Applying this equation, a unitary free energy of transfer from water to micelle of ≈15.3 kcal/mol is obtained in the case of dipalmitoylphosphatidylcholine and ≈13.0 kcal/mol for dimyristoylphosphatidylcholine. These results clearly point out the thermodynamic basis of bilayer assembly that has been termed the "hydrophobic effect" by Tanford and his colleagues. The large free-energy change between a water and a hydrophobic environment explains the overwhelming preference of typical lipids to assemble in bilayer structures excluding water as much as possible from the hydrophobic core in order to achieve the lowest free-energy level and hence the highest stability for the aggregate structure.

A high degree of surface activity of a given molecule does not guarantee its ability to form bilayer structures in the presence of water. The type of physical structure they attain under a given set of conditions will depend on their interactions with neighboring molecules, their interaction with water and, most important, whether the surface area of the polar head group, upon hydration, is smaller or larger than the surface area of the hydrophobic group (Fig. 7). For example, phosphatidylcholine, sphingomyelin, phosphatidylserine, phosphatidylinositol and phosphatidylglycerol have a preference for bilayer structures (liposomes). On the other hand, lysophospholipids form micelles and phosphatidylethanolamines, and negatively charged phospholipids under certain conditions (low pH and in the presence of divalent cations) form hexagonal [H(II)] structures. It should be pointed out that under proper conditions, relatively large amounts of lipids, which normally tend to form hexagonal or micellar structures, can be successfully incorporated into liposomes.

The thermodynamic analysis described above, which is not very different from the thermodynamic description of micelle formation, is only a good approximation for the explanation of liposome structure and stability. Low values of CMC may indicate mostly a solubility property and not a dynamic exchange process as observed in micelles. This thermodynamic model was improved by introducing the geometric properties of molecules resulting in a semi-empirical hard-core excluded volume packing model of amphiphile assembly. Such a model can explain different structures but fails to explain the dependence of liposome size on the experimental conditions, instability of liposomes and broad size distribution of liposomes. Also, the assumption that vesicle formation is an equilibrium thermodynamic process is in most cases false. Additionally, such thermodynamic models explain only the starting and final states and not the

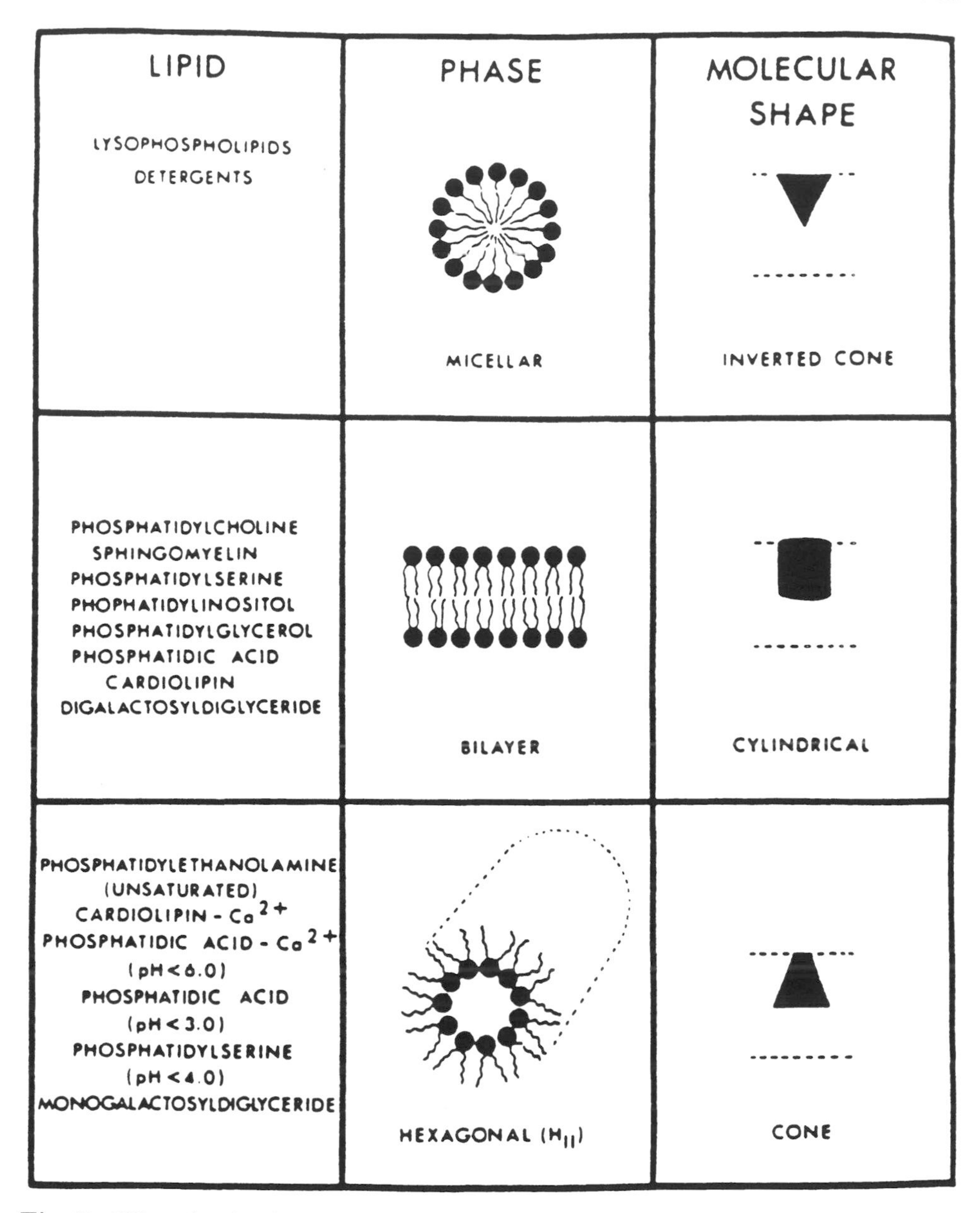

Fig. 7 Effect of molecular geometry of phase properties of lipids. (Adapted from Israelachvili and colleagues.) *Source*: P. R. Cullis and M. J. Hope, Physical properties and functional roles of lipids. In: *Biochemistry of Lipids and Membranes* (D. E. Vance and J. E. Vance, eds.), Benjamin/Cummings Inc., 1985, p. 56.

actual mechanism of liposome formation or disintegration and the intermediate structures in these processes (which in many cases, such as depletion/addition of the detergents, may not be the same due to the absence/presence of compartmentalized space and transmembrane gradients).

While such models can give a useful explanation of various liposome morphologies, they obviously have to be expanded. It is a common experience that one has to expend energy into the lipid-water mixture during vesicle preparation and that the vesicles produced are not infinitely stable systems. This all indicates that vesicles are in general not a thermodynamically stable state, but that they are rather a "kinetically trapped state," trapped in a more or less stable metastable state. This higher energy state can be easily explained by bending elasticity of bilayers. The minimal energy state of an infinite symmetric bilayer is in a flat configuration. Its bending to produce vesicles requires energy proportional to the radius of curvature and bending elasticity of the bilayer. Liposomes are of finite size, and therefore one cannot neglect edge effects of open bilayered fragments where the nonpolar interior is exposed to a polar environment. The actual minimization of these two unfavorable energy contributions results in vesicle formation. Normally the free-energy contribution from entropic terms is an order of magnitude smaller, typically excess energy due to vesicle curvature is of the order 15–50 kT, while entropy contributions are of the order of 1 kT. This approach allows one to model liposome formation [2].

First, the dynamic processes involved in vesicle formation by detergent depletion techniques were explained. Briefly, small mixed micelles of detergent-(phospho)lipid fuse upon removal of detergent, because the system is minimizing its total edge energy, which is proportional to the perimeter of disklike micelles. This growth is opposed by entropy and the reduced amount of detergent molecules that can shield the edges. These forces cause the large disklike micelles to bend and eventually self-close into closed vesicles. This eliminates the unfavorable exposure of the edges but increases bending energy of the system, and it was shown that liposomes are, in agreement with experimental observations, thermodynamically a metastable state. When disklike micelles grow, the unfavorable interactions scale with the number of molecules n, as $n^{1/2}$, while the bending energy scales as n until it reaches a constant value of $8\pi K$ (K is bending elasticity of the bilayer). Obviously, at some point the associated edge energy $E(n^{1/2}) > 8\pi K$, and the open fragment closes. Because $n^{1/2}$ is not a steep function the difference of its corresponding energy value with $8\pi K$ is below a few kT (thermal energy) for a wide range of n, which is consistent with the broad size distribution of vesicles [2]. This reasoning of an open bilayered fragment as a transition structure in the vesiculation process was used to explain some other preparation techniques as well. It is likely that all high-energy treatments of large multilamellar vesicles cause bilayer fragmentation, and fragments, after fusion, self-close into unilamellar vesicles [2].

The formation of large multilamellar liposomes from the swelling of dried lipid bilayers was explained by the difference between the area of polar heads on the outer and inner side of bilayers, which induce curvature, causes area mismatches, and results in budding off when the excess of the outer monolayer surface area over the inner one exceeds some critical value. This normally does not happen spontaneously, if no crystal defects are present. It was shown, however, that because liposomes are thermodynamically at a higher energy level than the hydrated lamellar phase in excess water, some energy input is normally required to form liposomes, which can be described as a kinetic trap. Therefore, agitating hydrating myelin figures causes rupture of long cylindrical lipid tubules, and open edges quickly reseal and form self-closed liposomes.

The consensus today is that liposomes are produced either by self-closure of smaller bilayered fragments or by fission due to a surface area difference between the two opposing monolayers, which can be induced by various gradients, across the bilayer [2]. Such gradients include pH and other ions and the asymmetric presence of molecules that can adsorb or insert into the membrane.

One can readily understand that kinetically trapped systems are advantageous over systems at thermodynamic equilibrium, since a system in equilibrium quickly adapts to the external change, such as dilution (concentration), pressure, presence of other substances, injection into body fluids, or by topical application with a resultant loss of encapsulated material. For instance, micelles or microemulsion droplets quickly disintegrate upon dilution (i.e., move to different regions in a phase diagram), whereas liposomes made from phospholipids with low values of critical micelle concentration are stable against dilution and, therefore, represent a better drug delivery system.

A. Characterization of Liposomes

1. *Factors Affecting Drug Entrapment*

The amount and location of a drug within a liposome is dependent on a number of factors. The location of drug within a liposome is based on the partition coefficient of the drug between aqueous compartments and lipid bilayers, and the maximum amount of drug that can be entrapped within a liposome is dependent on its total solubility in each phase. For example, very little 6-mercaptopurine can be encapsulated in liposomes because this drug has limited solubility in both polar and nonpolar solvents. The total amount of liposomal lipid used and the internal volume of the liposome will affect the total amount of nonpolar and polar drug, respectively, that can be loaded into a liposome. Efficient capture will depend on the use of drugs at concentrations that do not exceed the saturation limit of the drug in the aqueous compartment (for polar drugs) or the lipid bilayers (for nonpolar drugs). The method of preparation can also affect drug location and overall trapping efficiency. Figure 8 diagrammatically represents the various sites in the liposome that are available for drug entrapment.

Incorporation of drugs that have intermediate partition coefficients (significant solubility in both the aqueous phase and the bilayer) may be undesirable. If liposomes are prepared by mixing such a drug with the lipids, the drug will eventually partition to an extent depending on the partition coefficient of the drug and the phase volume ratio of water to bilayer. Also, the rate of partitioning will be a function of its diffusivity in each phase. Release rates (a measure of instability) are highest when the drug has an intermediate partition coefficient. Bilayer/aqueous compartment partition coefficients are usually estimated by determining their organic solvent/water (e.g., octanol/water) partition coefficients. They can also be determined precisely by a method described by Bakouche and Gerlier [14], which is based on the physical separation of the aqueous and bilayer phases by ultracentrifugation after mechanical (ultrasonics at low temperatures) disruption of the liposomes followed by analysis of each phase for drug.

2. *Internal Volume and Encapsulation Efficiency*

Internal volume and encapsulation efficiency are two parameters used to describe entrapment of water-soluble drugs in the aqueous compartments of liposomes. The internal or trapped or capture volume is expressed as aqueous entrapped volume per unit quantity of lipid (μL/μmol or μL/mg). It is determined by entrapping a water-soluble marker such as 6-carboxyfluorescein, ^{14}C or ^{3}H-glucose or sucrose and then lysing the liposomes by the use of a detergent such as Triton X-100. Determination of the amount

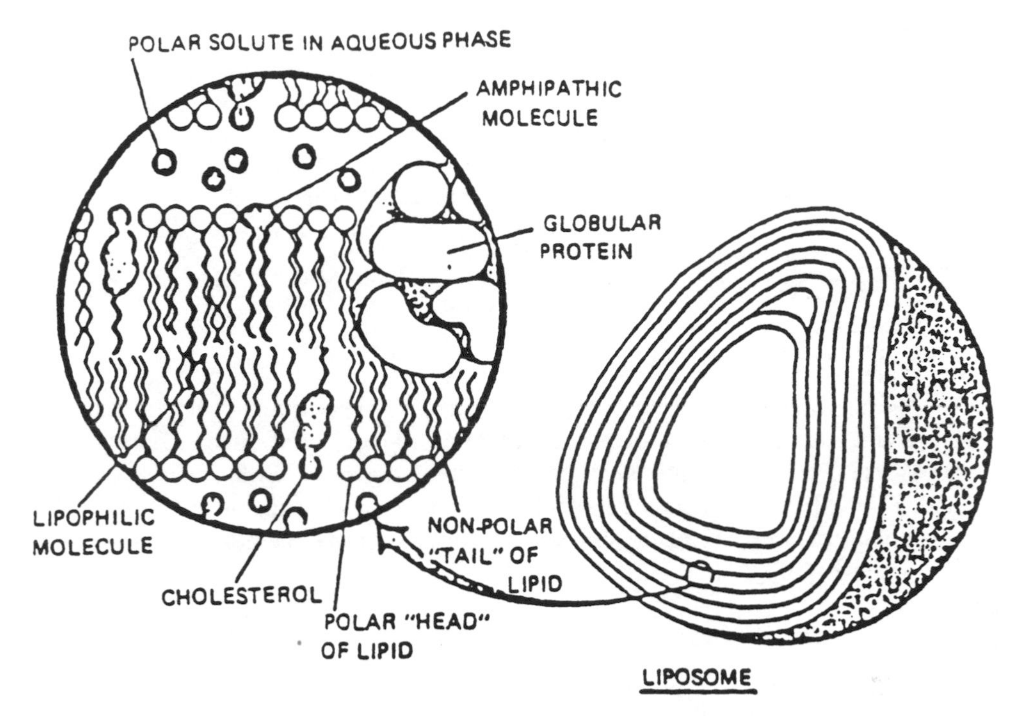

Fig. 8 various sites in the liposome available for drug entrapment. *Source*: J. Fendler in (G. Gregoriadis and A. C. Allison eds.), *Liposomes in Biological Systems*, John Wiley & Sons, New York, 1980, p. 89.

of marker that was trapped enables one to back-calculate the volume of entrapped water. The encapsulation efficiency describes the percent of the aqueous phase (and hence the percent of water-soluble drug) that becomes entrapped during liposome preparation. The remaining drug remains outside of the liposome and is therefore "wasted." Encapsulation efficiency is usually expressed as percentage entrapment per milligrams of lipid.

The internal or trapped volume and encapsulation efficiency greatly depend on liposomal content, lipid concentration, method of preparation, and the drug used. Some typical values are:

Liposome type	Internal volume (μL/μmol)	Entrapment efficiency (%/mg)
SUV	< 0.5	< 1
MLV	> 4	5–15
REV	> 10	35–65

Incorporation of charged lipids into bilayers increases the volume of the aqueous compartments by separating adjacent bilayers due to charge repulsion, resulting in increases in trapped volume. It should be pointed out that for hydrophobic drugs, entrapment efficiency usually approaches 100% almost irrespective of liposomal type and composition.

This high encapsulation efficiency, however, is normally observed only in the test tube, while upon application or simple dilution, the majority of the drug is quickly lost from the liposomes. The same is true for water-soluble drugs with good membrane permeability such as antibiotics and many anticancer agents. Some drugs, which are weak acids or weak bases, can be loaded into liposomes by the use of a pH gradient [15]. Recently ammonium salt gradient was introduced [16], which, in addition, can cause the precipitation of the drug in the liposome interior and thus greatly increase the stability of the encapsulation [17].

3. *Lamellarity*

The average number of bilayers present in liposomes can be found by freeze-fracture electron microscopy and ^{31}P-NMR. In the latter technique, the signals are recorded before and after the addition of nonpermeable broadening agent such as Mn^{2+}. Manganese ions interact with the outer leaflet of the outermost bilayer. Thus, a 50% reduction in NMR signal means that the liposome preparation is unilamellar, and a 25% reduction in the intensity of the original NMR signal means there are two bilayers in the liposomes [18].

4. *Size and Size Distribution*

The average size and size distribution of liposomes are important parameters with respect to physical properties and biological fate of the liposomes and their entrapped substances. There are a number of methods used to determine this parameter, but the most commonly used methods are the following.

a. Light Scattering. A variety of techniques are available to size liposomes based on light scattering. The popularity of this method depends on its ease of operation and the speed by which one can obtain data. The newer instruments are based on dynamic laser light scattering.

If the liposomes to be analyzed were monodisperse, light scattering would be the method of choice; unfortunately, most preparations are heterogeneous, and they require an accurate estimation of their size-frequency distributions. Light scattering methods rely on algorithms to determine particle size distributions, and the results obtained can be very misleading. Some complex algorithms have been developed in an attempt to deal with this problem. Furthermore, such methods cannot distinguish between a large particle and a flocculated mass of smaller particles. Most important, it may be necessary to remove any micron-sized particles that are present in the same prior to analysis.

The difficulty in interpreting particle size data can be demonstrated by taking a simple example of a dispersion comprised of 97% unilamellar vesicles with a radius of 15 nm and 3% multilamellar vesicles with a radius three times greater (45 nm):

	15 nm Particles		45 nm Particles
Percent particles	97		3
Percent surface area	78		22
Percent total volume	54		46
True statistical average radius		15.9 nm	
Instrumental average radius		25.3 nm	

Thus, 3% of the particles comprise almost one-half the volume of liposomes. Of course, the same problem of data analysis occurs with other disperse systems such as emulsions and suspensions.

b. Light Microscopy. This method can be used to examine the gross size distribution of large vesicle preparations such as MLVs. The inclusion of a fluorescent probe in the bilayer permits examination of liposomes under a fluorescent microscope and is a very convenient method to obtain an estimate of at least the upper end of the size distribution.

c. Negative Stain Electron Microscopy. This method, using either molybdate or phosphotungstate as a stain, is the method of choice for size distribution analysis of any size below 5 μm. It should be used to validate light scattering data that will ultimately be used for quality assurance. For accurate statistical evaluation (±5%), one should count at least 400 particles and not rely on a single specimen for counting.

d. Freeze Fracture Electron Microscopy. This method is especially useful for observing the morphological structure of liposomes. Since the fracture plane passes through vesicles that are randomly positioned in the frozen section, resulting in nonmidplane fractures, the observed profile diameter depends on the distance of the vesicle center from the plane of the fracture. Mathematical methods have been devised to correct for this effect. Fig. 9 shows micrographs for liposomes after 10 passes through a 0.1 μm polycarbonate filter.

For all of the microscopy procedures used, one should always be on the lookout for aggregated particles or flocs.

e. Cryoelectron Microscopy. This is a relatively new technique that allows direct observation of quickly frozen samples without any staining and is, therefore, the least prone to artifacts. Numerous tests have shown that very quick freezing can preserve the structure, while it may give rise to unreal size distributions due to the fact that larger particles are excluded from the thin (0.2–0.4 μm) film of ice on the microscopic grid [2].

f. Gel Chromatography. Since the introduction of large pore size gel (Sephacryl S 1000), an easy and quantitative determination of liposome size distribution is possible. In contrast to all other techniques, this method gives a true (i.e., "fit-independent") distribution according to their true hydrodynamic radius for liposomes smaller than 0.3–0.4 μm [2].

5. *Application of Double Layer Theory to Liposomes*

Once assembled, liposomes behave in much the same way as other charged colloidal particles suspended in water or electrolyte solutions. Under conditions where the charge on each particle is weak, the electrostatic repulsive force among the particles is also weak, increasing the opportunity for close approach. Some neutral particles tend either to flocculate or aggregate and sediment from suspension for this reason. Similarly, two populations of liposomes bearing opposite electric charges will aggregate at a rate that is a function of the electrostatic attractive forces among the particles. Particles bearing net negative charges may be induced to aggregate strongly in the presence of di- or trivalent cations. For example, calcium in the 1–2 mM range will induce liposomes containing more than 50 mol% PS to aggregate. These phenomena have dramatic effects on the physical stability of liposomes and lead to fusion of liposomes with one another resulting in increases in their overall size. Like aggregation, particle size

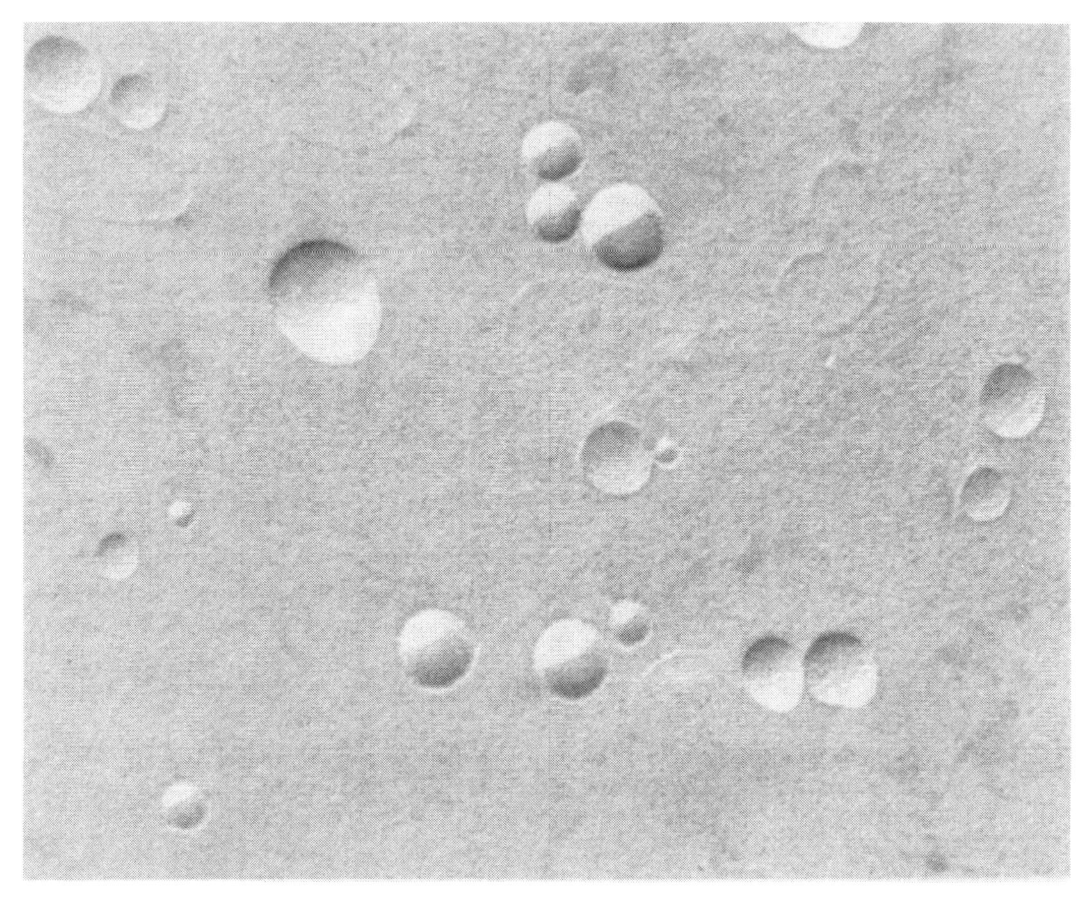

Fig. 9 Freeze fracture electron micrograph of EL:mixed phosphoinositides (4:1) liposomes after 10 passes through a 0.1 μm polycarbonate filter.

growth, particularly during storage, would be undesirable in most products. Fortunately the tendency of liposomes to aggregate and fuse can be controlled by the inclusion of small amounts of negatively charged lipids such as PS or PG or positively charged amphiphiles such as stearylamine in the formulation. Knowing the number and the sign of charged groups added and the valency and concentration of electrolytes in the medium, the magnitude of the electrostatic forces generated by these charged groups can be closely approximated by using the double layer theory. These results can then be correlated with physical stability of liposomes and used to guide formulation efforts. The amount of charged component and ionic conditions in a particular liposome dosage form can be adjusted to produce a high-enough zeta potential to inhibit close approach of vesicles and prevent their aggregation. In practice it is usually necessary to determine empirically the magnitude of the zeta potential required to prevent aggregation in a particular system. However, once this has been done, it is possible to use the zeta potential as a quality control check to insure that each batch of liposomes contains sufficient charged groups to avoid aggregation during storage.

6. *Phase Behavior of Liposomes*

An important feature of membrane lipids is the existence of a temperature-dependent reversible phase transition, where the hydrocarbon chains of the phospholipid undergo a transformation from an ordered (gel) state to a more disordered fluid (liquid crystalline) state. These changes have been documented by freeze-fracture electron microscopy but are most easily demonstrated by differential scanning calorimetry. Figure 10 illustrates the phase transition region for a typical phospholipid.

The physical state of the bilayer profoundly affects the permeability, leakage rates, and overall stability of the liposomes. The phase transition temperature *Tm* is a function of the phospholipid content of the bilayer (Table 4).

By proper admixture of bilayer-forming materials, one may design liposomes to "melt" at any reasonable temperature. This strategy has been used to deliver methotrexate to solid tumors, which are heated to the phase transition temperature of the custom-designed liposomal phospholipids. The phase transition temperature can be altered by using phospholipid mixtures or by adding sterols such as cholesterol. The *Tm*-value can give important clues as to liposomal stability and permeability and as to whether a drug is entrapped in the bilayer or the aqueous compartment.

B. Liposome Preparation Methods

1. *Multilamellar Vesicles*

Multilamellar vesicles are by far the most widely studied type of liposome and, as pointed out by Alec Bangham in 1964, exceptionally simple to make. In general a mix-

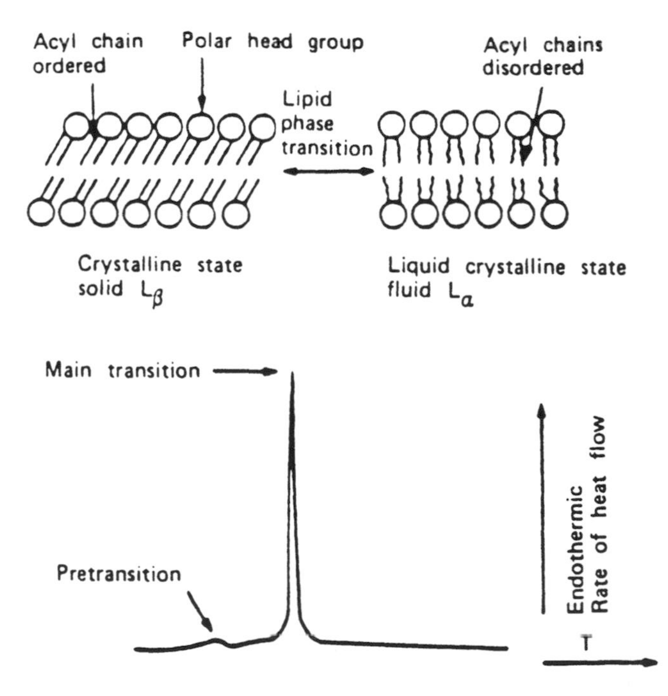

Fig. 10 Phospholipid gel to liquid crystalline phase transition. (Source: P. R. Cullis and M. J. Hope, Physical properties and functional roles of lipids in membranes. In: *Biochemistry of Lipids and Membranes* (D. E. Vance and J. E. Vance, eds.), Benjamin/Cummings Publishing Co., California, (1985). p. 43.

Table 4 Phase Transition Temperatures of Some Synthetic Phospholipids Used to Prepare Liposomes

Lipid	Charge	*Tm* (°C)
Dilauroyl phosphatidylcholine	0	0
Dimyristoyl phosphatidylcholine	0	23
Dipalmitoyl phosphatidylcholine	0	41
Dimyristoyl phosphatidylethanolamine	0	48
Distearoyl phosphatidylcholine	0	58
Dipalmitoyl phosphatidylethanolamine	0	60
Dioleoyl phosphatidylglycerol	-1	-18
Dilauroyl phosphatidylglycerol	-1	4
Dimyristoyl phosphatidylglycerol	-1	23
Dipalmitoyl phosphatidylglycerol	-1	41
Distearoyl phosphatidylglycerol	-1	55

ture of lipids is deposited as a thin film on the bottom of a round-bottom flask by rotary evaporation under reduced pressure. MLVs form spontaneously when an excess volume of aqueous buffer is added to the dry lipid. However, in many cases, MLVs have not been rigorously characterized with respect to size, polydispersity, number of lamallae, encapsulated volume, and stability. Due to their ease of production, many investigators have simply made a preparation of MLVs for use in both in vitro and in vivo experiments without taking the time to fully characterize them. This has led to a great deal of confusion in the interpretation of experimental results, because, as will be explained below, minor changes in the method of preparation can lead to major differences in the behavior of liposomes.

2. *Slow versus Fast Hydration, Thickness of the Lipid Film*

The time allowed for hydration and conditions of agitation are critical in determining the amount of the aqueous buffer (or drug solution) entrapped within the internal compartments of the MLV. For example, Szoka and Papahadjopoulos [19], reported that a similar lipid concentration can encapsulate 50% more aqueous buffer per mole of lipid when hydrated for 20 h with gentle shaking, compared to a hydration period of 2 h with vigorous shaking, despite the fact that the two preparations exhibit a roughly similar particle size distribution. If hydration time is reduced to a few minutes with vortexing, a suspension will exhibit a still lower capture volume and a smaller mean diameter. Bangham [1] showed that the hydration and entrapping process is most efficient when the film of dry lipid is kept thin. This means that different-sized round-bottom flasks should be used for different quantities of lipid. Glass beads have been used by some investigators to increase the surface area available for film deposition. Thus the hydration time, method of suspension of the lipids, and the thickness of the film can result in markedly different preparations of MLVs, in spite of identical lipid concentrations and compositions, and volume of the suspending aqueous phase.

3. *Effect of Charged Lipids*

The presence of negatively charged lipids, such PS, PA, PI or PG, or positively charged detergents such as stearylamine, will tend to increase the interlamellar distance between successive bilayers in the MLV structure and thus lead to a greater overall entrapped volume. This is particularly true in low ionic strength buffers or nonelectrolytes (such as sucrose), since the electrostatic repulsive forces, which give rise to the effect, are greater under these conditions. Generally, about a 10–20 mol percent of a charged species is used, although it is possible to produce MLVs from a singly charged lipid such as PS. The presence of charged lipids also reduces the likelihood of aggregation following the formation of MLVs.

4. *Hydration in the Presence of Solvent*

MLVs with high entrapment of solutes can be produced by hydrating the lipid in the presence of organic solvents. A method introduced by Papahadjopoulos and Watkins [20] begins with a two-phase system consisting of equal volumes of petroleum ether containing bilayer-forming lipids and of an aqueous phase. The contents of the tube are emulsified by vigorous vortexing, and the ether is removed by passing a stream of nitrogen gas over the mixture. As the ether is removed in the carrier gas, MLVs form in the aqueous phase. A similar method was reported by Gruner et al. [21] except that diethyl ether was used as the solvent, sonication was used in place of vortexing, and the aqueous phase was reduced to a relatively small proportion. Typically, the lipids are dissolved in about 5 mL ether, and about 0.3 mL of the aqueous phase to be entrapped is added. The two phases are emulsified by sonication while a gentle stream of nitrogen gas is passed over the mixture. The resulting MLV preparation encapsulates up to 40% of the solvent throughout the hydration step, and the concentration of solute molecules is in equilibrium across all the bilayers, a feature that is claimed to translate into enhanced stability to leakage.

5. *MLVs Formed by Freeze Drying SUV Dispersions*

A simple method for preparing MLVs with high entrapment efficiency was developed by Ohsawa et al. [22] and Kirby and Gregoriadis [23]. The aqueous phase containing the molecules to be encapsulated is mixed with a preformed suspension of SUVs, and the mixture is freeze-dried by conventional means. Large MLVs are formed when the dry lipid is rehydrated, usually with a small volume of distilled water. Encapsulation efficiencies up to 40% have been reported for this method.

6. *Small Unilamellar Vesicles*

The classical methods of dispersing phospholipids in water to form optically clear suspensions with a particle weight of about 2×10^6 daltons involve various mechanical means and began with the sonication method reported in the mid 1960s [24], followed by refinements introduced by Hamilton and Guo [25] and Barenholz et al. [26] in the mid-1970s, who used a high-pressure device to produce the same effect in larger volumes. These types of SUV dispersions have been rigorously characterized by Huang [27] and others and shown to consist of rather uniform closed bilayer vesicles of about 25–50 nm diameter. Solvent-injection methods have also been devised to produce SUVs. These typically involve the slow injection of a lipid solution in either ethanol or ether into warm water containing a drug or other marker to be entrapped. All of these methods are discussed in greater detail below.

7. Sonicated SUVs

The preparation of sonicated SUVs has been reviewed in detail by Bangham [28]. Briefly, the usual MLV preparation is subsequently sonicated either with a bath-type sonicator or a probe sonicator under an inert atmosphere (usually nitrogen or argon). Although probe sonication leads to more rapid size reduction of the MLVs, degradation of lipids, metal particle shedding from the probe tip, and aerosol generation can present problems. Bath-type sonicators also have disadvantages (such as the need to pay greater attention to position of the tube and water level in the bath), but temperature can be accurately regulated. Also, the tube containing the specimen is sealed allowing for aseptic operations and little likelihood of personnel exposure to aerosols.

8. French Pressure Cell

Dispersions of MLVs can be converted to SUVs by passage through a small orifice under high pressure. A French pressure cell was used by Hamilton et al. [29] for this purpose. MLV dispersions are placed in the French press and extruded at about 20,000 psi at 4°C. One pass through the cell produces a heterogeneous population of vesicles ranging from several microns in diameter to SUV size. Multiple extrusions result in a progressive decrease in the mean particle diameter. Following about four to five passes, approximately 95% of the vesicles are converted to SUVs as judged by size exclusion chromatography. The resulting vesicles are somewhat larger than sonicated SUVs ranging in size from 30 to 50 nm. The method is simple, reproducible, and nondestructive. However, temperature control is difficult (the pressure cell must be allowed to cool between extrusions or the temperature rise may damage the lipids), and the working volumes are relatively small (about 50 mL maximum).

9. Solvent Injection Method

a. Ether Infusion. A method introduced by Deamer and Bangham in 1976 [30] provides a means of making SUVs by slowly introducing a solution of lipids dissolved in diethyl ether (or ether/methanol mixtures) into warm water. Typically, the lipid mixture is injected into an aqueous solution of the material to be encapsulated (using a syringe-type infusion pump) at 55–65°C or under reduced pressure. Subsequent removal of residual ether under vacuum leads to the formation of single-layer vesicles. Depending on the condition used, the diameters of the resulting vesicles ranges from 50–200 nm. The usual lipid concentration is about 2 mg/mL ether, and about 2 mL of this solution is infused into 4 mL of the aqueous phase at a rate of 0.2 mL/min at 50–60°C.

b. Ethanol Injection. An alternative method for producing SUVs that avoids both sonication and exposure to high pressure is the ethanol-injection technique described by Batzri and Korn [31]. Lipids dissolved in ethanol are rapidly injected into a vast excess of buffer solution forming SUVs spontaneously. The procedure is simple and rapid, and avoids exposure to harsh conditions of both lipids and the material to be entrapped. Unfortunately, the method is restricted to the production of relatively dilute SUV suspensions. The final concentration of ethanol cannot exceed about 10% by volume, or the SUVs will not form. Removal of residual ethanol can also present a problem, since ethanol forms an azeotrope with water, which is difficult to remove under vacuum or by distillation. Various available ultrafiltration apparatus may be used to both concentrate the suspension and remove ethanol, but, these procedures tend to be slow and expensive to scale up. Another limitation of the method is related to the susceptibility of various biologically active macromolecules to inactivation in the presence of even low amounts of ethanol.

10. Large Unilamellar Vesicles and Intermediate-Sized Unilamellar Vesicles

Large unilamellar vesicles provide a number of important advantages as compared to MLVs, including high encapsulation of water-soluble drugs, economy of lipid, and reproducible drug release rates. However, LUVs are perhaps the most difficult type of liposome to produce. "Large" in the context of liposomes usually means any structure larger than 100 nm; thus large unilamellar vesicles refer to vesicles bounded by a single bilayer membrane that are above 100 nm in diameter. Some author have referred to liposomes between 50 and 100 nm as "large," but these would be more approximately called intermediate-sized. Two primary methods are used to produce LUVs, one involving detergent dialysis, the other a sophisticated reverse emulsification technique. Intermediate-sized single-layered vesicles can be generated from MLV dispersions by sequential extrusion through small pore size polycarbonate membranes under high pressure. A number of other techniques for producing LUVs have been reported including freeze thawing, slow swelling in nonelectrolytes, dehydration followed by rehydration, and dilution or dialysis of lipids in the presence of chaotropic ions. Each of these methods is reviewed below.

11. LUVs Formed by Detergent Removal

An essentially different approach to produce liposomes is dependent on the removal of detergent molecules from aqueous dispersions of phospholipid/detergent mixed micelles. As the detergent is removed, the micelles become progressively richer in phospholipid and finally coalesce to form closed single-bilayer vesicles. Three methods of detergent removal appropriate for this purpose have been described in the literature and are treated separately below.

a. Dialysis. Kagawa and Racker [32] were the first to introduce the dialysis method for lipid vesicle preparation. Although these authors were primarily interested in reconstituting biological membranes solubilized with detergents, their method is applicable to the formation of liposomes as well. Detergents commonly used for this purpose exhibit a reasonably high critical micelle concentration (on the order of 10–20 mM) in order to facilitate their removal and include the bile salts sodium cholate and sodium deoxycholate and synthetic detergents such as octylglucoside. The treatment of egg PC with a 2:1 molar ratio of sodium cholate followed by dialysis results in the formation of vesicles in the 100 nm diameter range within a few hours. Another modification of the cholate removal technique is one in which the rate of efflux of the detergent from the mixture is controlled. This procedure, described in detail by Milsmann et al. [33], uses a phospholipid:detergent ratio of 0.625 and rapidly removes the detergent in a flow-through dialysis cell. The procedure forms a homogeneous population of single-layered vesicles with mean diameters of 50–100 nm.

b. Column Chromatography. The formation of 100 nm single-layered phospholipid vesicles during removal of deoxycholate by column chromatography has been reported by Enoch and Strittmatter [34]. The method involves the treatment of phospholipid, in the form of either small sonicated vesicles or a dry lipid film, at a molar ratio of deoxycholate to phospholipid of 1:2. Subsequent removal of the detergent during passage of the dispersion over a Sephadex G-25 column results in the formation of uniform 100 nm vesicles that are readily separated from small sonicated vesicles.

c. Bio-Beads™. Another promising method for forming reconstituted membranes reported by Gerritsen et al. [35] may also be applicable to LUV preparation. The system involves the removal of a nonionic detergent, Triton X-100, from detergent/phospholipid mixtures. This method is based on the ability of Bio-beads SM-2 to adsorb Triton X-100 selectively and rapidly. The dried lipid is suspended in 0.5–1.0% Triton X-100, and washed Bio-beads are added directly to the solution (about 0.3 g wet Bio-beads per mL of dispersion) and rocked for about 2 h at 4°C. The beads are removed by filtration. The final particle size is determined by the conditions used including lipid composition, buffer composition, temperature, and, most critically, the amount and activity of the beads themselves.

12. Reverse Phase Evaporation Technique

LUVs can also be prepared by forming a water-in-oil emulsion of phospholipids and buffer in excess organic phase followed by removal of the organic phase under reduced pressure (the so called "Reverse Phase Evaporation," or REV, method). The two phases are usually emulsified by sonication, but other mechanical means have also been used. Removal of the organic solvent under vacuum causes the phospholipid-coated droplets of water to coalesce and eventually form a viscous gel. Removal of the final traces of solvent results in the collapse of the gel into a smooth suspension of LUVs. With some lipid compositions the transition from emulsion to LUV suspension is so rapid that the intermediate gel phase appears not to form. The method, which was pioneered by Szoka and Papahadjopoulos in 1978 [36], has been used extensively for applications that require high encapsulation of a water-soluble drug. Entrapment efficiencies up to 65% can be obtained with this method. The phospholipids are first dissolved in an organic solvent such as ethylether, isopropylether or mixtures of two solvents, such as isopropylether and chloroform. The emulsification is most easily accomplished when the density of the organic phase matches that of the buffer (i.e., about 1). For this reason, ether (density of about 0.7) is often mixed with a solvent of higher density such as trichlorotrifluoroethane (density of 1.4), to produce a solvent system with a density close to water. The aqueous phase containing the material to be entrapped is added directly to the phospholipid-solvent mixture. The ratio of aqueous phase to organic phase is usually about 1:3 for ether and 1:6 for isopropylether-chloroform mixtures. Preparations using even greater proportions of organic phase have been reported. The two phases are emulsified by sonication for a few minutes, and the organic phase is removed slowly under a partial vacuum produced by a water aspirator on a rotary evaporator at 20–30°C. The vacuum is usually maintained at about 500 microns for the first few minutes (using a nitrogen gas bleed to lower the vacuum and a gauge to measure the vacuum) and then raised cautiously to fill the aspirator vacuum to prevent the ether from evaporating too quickly. A typical preparation contains 60 μmol lipid dissolved in 3 mL ether and 1 mL aqueous phase contained in a sealed screw cap tube. The mixture is sonicated in a bath-type sonicator for about 5 min or until a homogeneous emulsion is formed. For a quick check to determine if emulsification is complete, one can interrupt sonication and allow the tube to stand for about a minute. If a clear layer of ether is observed over the aqueous phase, sonication should be continued for an additional period. Maximal encapsulation (65%) is obtained when the ionic strength of the aqueous phase is low. The method has been used to encapsulate both small and large molecules. Biologically active macromolecules, such as RNA and various enzymes, have been encapsulated without loss of activity. The principal disadvantage of the method is the exposure of the material to be encapsulated to organic solvents and mechanical agitation, conditions that lead to the denaturation of some proteins or breakage of DNA strands.

13. Formation of Intermediate-Sized Unilamellar Vesicles by High-Pressure Extrusion

As mentioned above, MLV suspensions rich in acidic lipids, such as PS or PG, tend to have large interbilayer distances and large internal aqueous cores due to electrostatic repulsive forces among the bilayers. Hope et al. [37] showed that as MLVs are repeatedly extruded through very small pore diameter polycarbonate membranes (0.8–1.0 micron) under high pressure (up to 250 psi) their average diameters becomes progressively smaller reaching a minimum of 60–80 nm after about five to 10 passes. Moreover, as the average size is reduced, the vesicles become more and more single layered. The mechanism at work during such high-pressure extrusion appears to be much like the peeling of an onion. As the MLVs are forced through the small pores, successive layers are "peeled" off until only one remains. For this method to generate truly single-layered vesicles, however, the aqueous core of the starting MLV must be greater than about 70 nm in diameter. Although this appears to be the case for vesicles composed predominantly of acidic lipids, neutral vesicles or vesicles with only a few mole percent acidic lipids are not likely to convert to true single lamellar vesicles using this technique, because the diameter of the inner most bilayer is probably significantly less than 70 nm.

14. Miscellaneous Methods

a. Slow Swelling in Nonelectrolyte Solutions. In 1969, Reeves and Dowben [38] reported a method for producing very large (up to several 10s of microns) single-layered liposomes by allowing a thinly spread layer of hydrated phospholipids to slowly swell in distilled water or a nonelectrolyte solution. Typically, a mixture of lipids in ether or chloroform is deposited as a thin film on the bottom of a flat-bottomed beaker. The lipid is slowly hydrated by passing nitrogen gas saturated with water vapor over the film for several hours. When the film has completely hydrated, it will become opaque in appearance. Following hydration, distilled water or a nonelectrolyte solution (e.g., sucrose) is carefully layered over the film, and the beaker is placed in a 37°C water bath for several more hours. During this period very large single-walled vesicles are formed by a mechanism that begins with single bilayers swelling and budding from the film, pinching off, and eluting into the aqueous medium. The yield of single-layered vesicles is high if conditions are right, but the main disadvantage of the technique is its sensitivity to any kind of mechanical agitation during vesicle formation. Also, since a very thin film is required and swelling times are long, this method would be difficult to scale up.

b. Removal of Chaotropic Ions. Oku and MacDonald [39] developed a method of forming giant single-lamellar vesicles with diameters in the range of 10–20 microns by removal of sodium trichloroacetate by dialysis or dilution from a solution containing egg phospholipids and molar concentrations of sodium trichloroacetate. The yield of giant vesicles was critically dependent on the starting concentration of the chaotropic ion and temperature. Inclusion of a freeze-thaw step reduced the required concentration of trichloroacetate to about 0.1 M. The giant liposomes apparently were formed from concentrations of the ion that induced the transformation of phospholipids from the lamellar phase to the micellar phase. Other chaotropic ions were also shown to be effective, including urea or guanidine-HCl.

c. *Freeze/Thaw.* A method for the reconstitution of membrane proteins based on rapid freezing of sonicated phospholipid mixtures followed by thawing and brief sonication was originally described by Kasahara and Hinkle [40]. In 1981, Pick [41] reported that vesicles formed by this simple procedure exhibited specific trapping volumes of up to 10 μL per μmol lipid with encapsulation efficiencies of 20–30%. Formation of large liposomes by this technique probably results from the fusion of small vesicles during freezing and/or thawing of the suspension of small vesicles. This type of fusion is strongly inhibited by increasing the ionic strength of the medium, e.g., adding sucrose, and by increasing the lipid concentration. For an unexplained reason, pure phosphatidylcholine vesicles do not appear to be good candidates for this type of fusion induced growth, however. Ohu and MacDonald [42] have shown that freeze/thawing of SUVs prepared in high concentrations of alkali metal chlorides also results in the formation of giant single-layered liposomes. The method involves the formation of fully hydrated small vesicles in dilute buffer by sonication followed by freeze/thawing in the presence of high concentrations of the electrolyte of interest in order to induce equilibration of the electrolyte across the bilayer membranes of the small vesicles. In the final step of the process, the electrolyte concentration is reduced by dialysis against dilute buffer. This results in the influx of water into the small vesicles (driven by the osmotic imbalance) causing them to swell and fuse into giant vesicles. The method is rather involved and not easily scaled up.

d. *Dehydration/Rehydration of SUVs.* Large unilamellar and oligolamellar vesicles with high entrapment efficiencies have been formed by a clever method reported by Shew and Deamer [43]. In this method, sonicated vesicles are mixed in an aqueous solution with the solute desired to be encapsulated, and the mixture is dried under a stream of nitrogen. As the sample is dehydrated, the small vesicles fuse to form a multilamellar film that effectively sandwiches the solute molecules between successive layers. Upon rehydration, large vesicles are produced that have encapsulated a significant proportion of the solute. The optimal mass ratio of lipid to solute was reported to be approximately 1:2 to 1:3. This method has been applied in a number of settings, since it depends only on controlled drying and rehydration processes and does not require extensive use of organic solvents, detergents, or dialysis systems.

C. Methods for Controlling the Particle Size and Size Distribution of Liposomes

In most studies, using liposomes as drug carriers, particle size has not been rigorously controlled. In studies on tissue distribution reported to date, for example, various investigations have used either the initial liposome preparation containing a wide distribution of sizes (ranging from 0.2 to 10s of microns) or sonicated vesicles that, although exhibiting a narrow size distribution, are quite small and thus have a limited capacity to carry drugs. Judging from the few studies using controlled particle size, it is clear that vesicle size can have dramatic effects on the in vivo behavior of liposomes. Therefore, before liposome drug carrier systems can be taken seriously for pharmaceutical applications, their size will have to be controlled within reasonable limits. Three possible approaches have been explored for controlling the particle size distribution of liposome preparations: (1) fractionation of the size of interest from a heterogeneous population; (2) homogenization of a polydisperse dispersion to yield a population of smaller vesicles with a narrow size distribution; and (3) extrusion of a heterogeneous preparation through capillary pore membranes of known pore diameter to yield an average size that approximates the pore diameter.

1. Fractionation

Two methods have enjoyed widespread use for fractionating liposomes of the desired size from a heterogeneously sized population: centrifugation and size exclusion chromatography. Both can be used to enrich the product with the desired particle size but are limited in terms of the volumes that can be easily handled.

a. Centrifugation. Liposome sediment in a centrifugal field at a rate that is dependent on their size and density. Large liposomes composed of neutral lipids such as PC can easily be pelleted at fairly low *g* forces in a conventional centrifuge. Under proper conditions the smaller liposomes will remain in the supernatant. This method is useful for making gross cuts between small and larger liposomes but not for generating narrow particle size distributions. Also, the volumes that can be handled are limited by the volume capacity of the centrifuge. However, zonal rotors or continuous-flow centrifuges may be adaptable to this application. Another disadvantage to centrifugation is that liposomes smaller than about 0.5 micron tend to require high *g* forces and long spinning times in order to achieve effective separation from slightly larger particles in the 0.1–0.2 micron range. Also the capacity of the ultracentrifuges normally used for this purpose is limited to a few hundred mL per run.

b. Size Exclusion Chromatography. Column chromatography has been used for many years as an analytical method to assess the particle size of liposomes. Preparative scale chromatography has also been applied to produce liposomes of fairly homogeneous sizes. This method is particularly useful for separating SUVs from larger structures. Typically, a column of Sepharose 4B is equilibrated with a buffer of the same osmolarity as the medium in which the vesicles were prepared, and an aliquot of the liposomes is applied to the column. The column is eluted with the same buffer and fractions are collected. Large liposomes appear in the void volume, whereas SUVs elute with the included volume. Larger pore size chromatographic media, such as Sephacryl S1000, have been used in a similar fashion to fractionate populations of larger particles. In general, however, such chromatographic separations are quite limited in terms of volumes, and throughput must be carried out in batches, resulting in significant dilution of the product.

2. Homogenization

In those cases in which a fairly small particle size is desirable, homogenization has proved a useful approach. In much the same way as milk is homogenized, the average particle size and polydispersity of vesicle dispersions can be reduced by passage through a high-pressure homogenizer. One such device marketed by Microfluidics Corp., Newton, MA, under the trade name Microfluidizer™ (Fig. 11) has been shown by Mayhew et al. [44] to generate vesicles in the 50–200 nm size range. Such homogenizers are

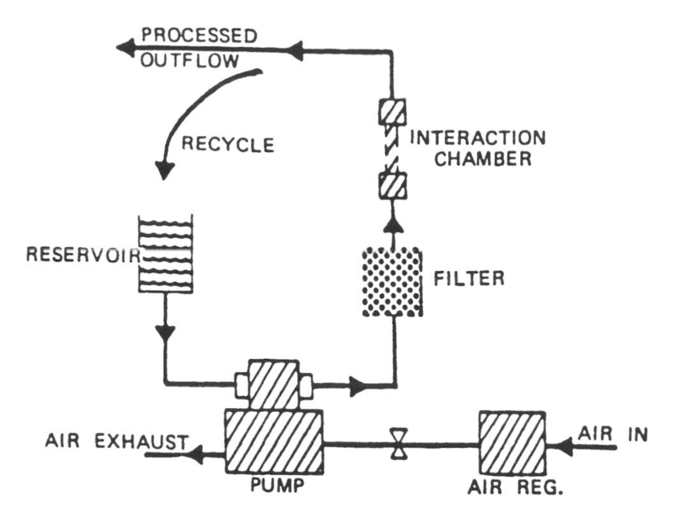

Fig. 11 The Microfluidizer. (From Ref. 44.)

amenable to scale up, and throughput rates are high. As with other high-pressure devices, however, heat regulation can sometimes present problems, and the shear forces developed within the reaction chamber can lead to partial degradation of the lipids. Another disadvantage relates to the empirical observation that conditions designed to produce approximately 200 nm particles often results in a bimodal distribution, with the bulk of the vesicles in the desired size range contaminated by a significant proportion of very small vesicles (less than 50 nm).

3. *Capillary Pore Membrane Extrusion*

A technique that has gained widespread acceptance for the production of liposomes of defined size and narrow size distribution, introduced by Olson et al. [45] in 1979, involves the extrusion of a heterogeneous population of fairly large liposomes through polycarbonate membranes under moderate pressures (100–250 psi). Such membranes have uniform straight-through capillary pores of defined size, and polycarbonate does not bind liposomes containing charged species. This simple technique can reduce a heterogeneous population of MLVs or REVs to a more homogeneous suspension of vesicles exhibiting a mean particle size that approaches that of the pores through which they were extruded. MLVs with a mean diameter of 260 nm can be obtained following a single extrusion through 200 nm pore size polycarbonate membranes; 75% of the encapsulated volume resides in vesicles between 170 and 370 nm (as measured by negative stain electron microscopy). Upon additional extrusions through the same pore size membrane, the average size is reduced further, finally approaching about 190 nm with greater than 85% of the particles in the 170–210 nm range. Compared to SUV preparations this still represents a rather broad distribution of vesicle sizes, but compared to the original MLV population, which ranges in size from about 500 nm to several microns, it represents a considerable reduction of both average particle size and polydispersity. In practice, it is sometimes preferable to extrude sequentially through membranes of decreasing pore diameter. For example, a concentrated dispersion of MLVs may be difficult to extrude directly through a 200 nm pore size membrane under normal operating pressures (about 90 psi). It is advisable to begin the process by extrusion through a 0.8, 0.6, 0.4, and finally 0.2 micron pore size. Alternatively, it is possible use higher pressures to extrude concentrated dispersions through the smaller pore size membranes directly. A special high-pressure filter holder is required, however, since operating pressures may reach 250 psi. One such device is available commercially under the trade name LUVET™, which can accommodate up to 10 mL and is equipped with a recirculation mechanism, which permits multiple extrusion with little difficulty.

IV. TYPICAL LIPOSOME FORMULATIONS

The lipid composition of liposomes is as important as their size and type in defining their pharmaceutical and therapeutic properties. For mixed-lipid vesicles, compositions can be described in terms of mole fraction, mole ratio, or weight ratio of the bilayer forming materials, one of which may be a drug incorporated into the bilayer structure. A typical formula follows:

Ingredient	mol %
Egg lecithin	45
Phosphatidylserine	9.9
Cholesterol	45
α-Tocopherol	0.1

Lecithin is the most frequently used lipid for liposome formulations. Whereas natural lecithins are relatively inexpensive, they suffer problems of potential chemical instability. Saturated synthetic phospholipids, on the other hand, may be prohibitively expensive for most applications. A compromise solution may reside in the carefully controlled partial hydrogenation of phospholipids derived from natural sources. Generally, the mol% of lecithin used in liposomes for pharmaceutical purposes range from 50 to 100%. Sphingomyelin is sometimes substituted for lecithin as the principal component. It produces more rigid membranes, which are thought to have longer half-lives in the in vivo circulation. Phosphatidylserine is an example of a lipid used to impart a charge to the liposome. Ten percent is quite sufficient to reduce flocculation over the shelf-life of the product. Higher percentages of charged lipid are sometimes used for purposes of directing liposomes to biological targets. Cholesterol is added to "toughen" liposomes in that it reduces leakage of entrapped drug on the shelf as well as in a number of biological environments. For example, the incorporation of an amount of cholesterol equivalent to the amount of lecithin used dramatically reduces the rate of bilayer lipid loss to blood proteins and lipoproteins as well as their uptake by cells of the reticuloendothelial system. Generally, the mol% cholesterol used in liposomes for pharmaceutical purposes ranges from 25 to 50%. α-Tocopherol and BHT are examples of effective antioxidants used to prevent oxidation of unsaturated phospholipids.

Two kits are commonly available from Sigma Chemicals for the formation of liposomes:

Positive Liposomes Kit		Negative Liposomes Kit	
egg lecithin	63 μmoles	egg lecithin	63 μmoles
stearylamine	18 μmoles	dicetylphosphate	18 μmoles
cholesterol	9 μmoles	cholesterol	9 μmoles

V. STABILITY OF LIPOSOMES

The stability of any pharmaceutical product is usually defined as the capacity of the formulation to remain within defined limits for a predetermined period of time (shelf-life of the product). The first step in designing any type of stability testing program is to specify these limits by establishing parameters defined in terms of chemical stability, physical stability, and microbial stability. Next, methods must be established to evaluate each of these parameters. One must treat liposomal drug delivery systems in the same way as the more traditional pharmaceutical dosage forms are treated with respect to the establishment of clearly defined protocols for their characterization, manufacture, stability testing, and efficacy.

A. Physical Stability

Stability of liposomes can be described by classical models from colloid science. Colloidal systems can be stabilized electrostatically, sterically, or electrosterically. In addition to normal colloids, self-assembling colloids can undergo other changes, such as fusion or phase change after aggregation. A liposome dispersion in a test tube exhibits a given physical and chemical stability. Generally, the former deals with the preservation of liposome structure while the latter one with the chemical structure of molecules. Therefore, physical stability means the preservation of liposome size distribution and the amount of the material encapsulated. Obviously this depends on mechanical properties of liposome membranes, their thermodynamics, and colloidal properties of the system.

B. Chemical Stability

Chemically, phospholipids are susceptible to hydrolysis. Additionally, phospholipids containing unsaturated fatty acids can undergo oxidative reactions. Much of the data on liposomes that have appeared in the literature can be considered suspect due to the use of phospholipids containing significant amounts of oxidation and hydrolysis products. These reaction production can cause dramatic changes in the permeability properties of liposomes. Preparative procedures (e.g., sonication, homogenation) or storage conditions (e.g., exposure to different pH values) can affect the decomposition rate of the liposomal lipids.

1. Lipid Peroxidation

Most of the phospholipid liposomal dispersions used contain unsaturated acyl chains as part of the molecular structure. These chains are vulnerable to oxidative degradation (lipid peroxidation). The oxidation reactions can occur during preparation, storage, or actual use. Oxidative deterioration of lipids is a complex process involving free-radical generation and results in the formation of cyclic peroxides and hydroperoxides.

Most of the procedures used to measure lipid peroxidation are nonspecific and are either based on the disappearance of unsaturated fatty acids (determined by lipid extraction techniques followed by GLC analysis) or the appearance of conjugated dienes. The latter technique is now widely used, since oxidation is accompanied by increased ultraviolet absorption in the 230–260 nm range. If unsaturated phospholipids are used to prepare liposomes, and no special precautions are used to minimize oxidation, the reaction will occur readily. Oxidation of the phospholipids may be minimized by a number of methods:

1. Minimum use of unsaturated phospholipids (if appropriate)
2. Use of argon or nitrogen to minimize exposure to O_2
3. Use of light-resistant containers
4. Removal of heavy metals (EDTA)
5. Use of antioxidants such as α-Tocopherol or BHT

2. *Lipid Hydrolysis*

The most important degradation product resulting from lecithin hydrolysis is lyso-lecithin (lyso-PC), which results from hydrolysis of the ester bond at the C_2 position of the glycerol moiety. Many workers choose the formation of lyso-PC as a standard measure for the chemical stability of phospholipids, since the presence of lyso-PC in lipid bilayers greatly enhances the permeability of liposomes. It is, therefore, extremely important that the formation of lyso-PC be kept to a minimum during storage. Lyso-PC is usually analyzed by phospholipid extraction followed by separation of PC and lyso-PC by TLC. The spots are then usually scraped and assayed for total phosphorus content.

Although factors such as sonication could affect the degree of lyso-PC formation, probably the single most important method of minimizing this problem is by the proper sourcing of the phospholipids to be used. They should be essentially free of any lyso-PC to start with and, of course, be free of any lipases.

2. *Miscellaneous Chemical Stability Concerns*

One must not ignore the fact that the other bilayer lipids, which may be present, can also decompose. For example, cholesterol, in aqueous dispersion, has been shown to oxidize rapidly when unprotected. Finally, the drug itself must be considered. The stability profile of the "free" drug may be quite different from its profile in the encapsulated state. In fact, a number of strategies have been developed that are based on protecting drugs from biological environments by encapsulating them in liposomes. Examples include the protection of insulin from proteolytic enzymes of the gastrointestinal tract and the prolongation of ester hydrolysis of prodrugs (e.g., cortisone hexadecanoate) after intramuscular administration.

VI. STABILITY TESTING (GENERAL CONSIDERATIONS)

Stability testing of liquid disperse systems is one of the most difficult problems faced by formulation chemists. The scientist is often asked to predict the shelf-life of a product or choose between experimental formulations based on estimates of how well they will hold up with time. There are no standardized tests available to determine physical stability, and quite often there is no certainty of what type of stability is being investigated. The first order of priority for solving stability problems of disperse systems is to define clearly the type or types of stability of concern. Categorizing stability as either physical or chemical is not sufficient. The various groups that are concerned with the product (produce development, production, analysys, marketing, etc.) must have a clear and precise reference frame of stability.

An understanding of the factors that lead to stability problems can help determine which methods of testing are most likely to yield information applicable to the estimation of the product's shelf-life. Stability tests commonly stress the system to limits beyond those that the product will ever encounter. Typical examples of stress tests include exposure of the product to high temperatures and large gravitational forces. It is important to understand whether these tests are being performed because the product is expected to encounter these conditions or because, even though these conditions will never be approached, the results will help predict shelf-life at more moderate conditions.

High-temperature testing (> 25°C) is almost universally used for heterogeneous products. Various laboratories store their products at temperatures ranging from 4°C (refrigerator temperature) to 50°C (or perhaps even higher). The temperatures used in heat-cool cycling are also quite varied, often without regard for the nature of the product. What will the increase in temperature likely do to the properties of the systems under study?

For liposomes, elevated temperatures may dramatically alter the nature of the interfacial film, especially if the phase transition temperature is reached. If one expects the product to be exposed to a temperature of 45°C for extended period of time or for short durations, (shipping and warehouse storage), studies at 45–50°C, (long-term and heat-cool cycling), are quite justified. A study of a product at these temperatures determines (1) How the product is holding up at this elevated temperature; and (2) whether the damage is reversible or irreversible when the product is brought back to room temperature. If temperatures higher than the system will ever encounter are used, even in short-term heat-cool cycling, there is a risk of irreversibly damaging the bilayers so that when it is brought back to room temperature, the membrane cannot heal.

If a liposomal dispersion is partially frozen and then thawed, ice crystals nucleate and grow at the expense of water. The liposomes may then be pressed together against the ice crystals under great pressure. If the crystal grows to a size greater than the void spaces, instability is more likely. That is why a slower rate of cooling, resulting in larger ice crystals, produces greater instability. Polymers may retard ice crystal growth.

Van Bommel and Crommelin [46] showed that even one freeze-thaw cycle causes almost complete rapid leakage of carboxyfluorescein from liposomes (REVs) prepared from unsaturated phospholipids (even when cholesterol is added). However, liposomes composed of distearoylphosphatidylcholine, dipalmitoylphosphatidylglycerol, and cholesterol show slightly better freeze-thaw stability.

Stability testing protocols should be developed for liposomal products on a case-by-case basis. A typical protocol for a product that would be shipped in vehicles not equipped with climate control and stored in warehouses for prolonged periods under similar conditions, might include testing under the following conditions:

1. One month at 45°C
2. One month at 4°C
3. Six months at 37°C
4. 12–24 months at room temperature
5. 12–24 months at various light intensities
6. Two to three "freeze-thaw" cycles (–20°C ↔ 25°C)
7. Six to eight "heat-cool" cycles (5°C ↔ 45°C, 48 hours at each temperature)

8. 24–48 h on a reciprocating shaker at 60 cycles/min (estimates transportation conditions).

One should be certain that studies are performed using all types and sizes of containers. Under each of the test conditions, the following data can be collected:

1. Visual and microscopic observations, e.g., flocculation
2. Particle size profiles
3. Rheological profiles
4. Chemical stability
5. Extent of leakage

VII. FREEZE-DRYING (LYOPHILIZATION)

Freeze-drying involves the removal of water from products in the frozen state at extremely low pressures. The process is generally used to dry products that are thermolabile and would be destroyed by heat-drying.

Lyophilization has great potential as a method to solve long-term stability problems of liposomes. Intuitively, one would suspect that liposomes containing drugs entrapped in their bilayers would be better candidates for lyophilization than liposomes containing drugs entrapped in their aqueous compartments, since the lyophilization procedure would be expected to cause some bilayer disruption and subsequent leakage.

Various studies have shown that water-soluble markers, such as carboxyfluorescein, do not survive freeze-drying in that even under the best of circumstances (use of saturated lipids and incorporation of cryoprotectants), a significant portion of the marker is lost on reconstitution. On the other hand, liposomes can retain > 90% of lipid-soluble drugs, such as Doxorubicin, on reconstitution. The amount retained depends on the use of cryoprotectants, lipid composition, liposome type, and loading dose.

If the leaked-out drug is removed and the preparation frozen for a second time, essentially 100% of the drug is recoverable on reconstitution. This indicates that the original loss represents the portion of the drug residing in the aqueous compartment. Thus, when formulating, one must ensure that essentially all the drug is placed in the bilayer or accept a certain percentage of loss to the external medium.

Recently, it was found that trehalose, a carbohydrate commonly found at high concentrations in organisms capable of surviving dehydration, is an excellent cryoprotectant for liposomes. It may work by stabilizing the bilayers, especially at their phase transition temperatures, during both freezing and thawing [47].

VIII. STABILITY OF LIPOSOMES IN BIOLOGICAL FLUIDS

The ultimate efficacy of a liposomal dosage form will be judged by the ability of the formulator to reliably control the amount of free drug that reaches the site of action over a given period of time. Generally, the exact "site of action" or receptor site at the molecular level is not known and one relies on attaining reproducible blood levels of the drug. With traditional nonparenteral dosage forms, only the free drug is absorbed, and once the drug is in the bloodstream, it has no memory about where it came from. Thus, the only method available to control the pharmacokinetics of a drug is to adjust the amount of drug that enters the blood as a function of time.

Parenteral, especially intravenous, administration of liposomally encapsulated drugs presents the formulator with additional means to control the pharmacokinetics of the drug. Factors that affect the pharmacokinetics of parenteral liposome administration include

1. Concentration of free drug in blood
2. Concentration of liposomes and their entrapped drug in blood
3. Leakage rate of drug from the liposome in the blood
4. Disposition of the interact drug-carrying liposomes in the blood

In order to reliably control the pharmacokinetics of these complex systems, one must be able to separate the

1. Stability (leakage rate) of drug from the liposome in the blood.
2. Disposition of the intact drug-carrying liposome in the blood. The pharmacokinetics of intact liposomes is beyond the scope of this chapter and has been thoroughly reviewed elsewhere [e.g., 48].

A. Liposome Stability in Blood and Plasma

The inability of liposomes to retain entrapped substances, when incubated with blood or plasma, has been known for about a decade. The fact that high molecular weight substances, such as inulin and even albumin, leak out on incubation with plasma suggests that more than superficial damage is being done to the liposomes even though their gross morphology appears unchanged. The instability of liposomes in plasma appears to be the result of the transfer of bilayer lipids to albumin and high-density lipoproteins (HDLs). Additionally, some of the protein is transferred from the lipoprotein to the liposome. Both lecithin and cholesterol also exchange with the membranes of red blood cells. Liposomes are most susceptible to HDL attack at their gel to liquid crystalline phase transition temperature. It is, therefore, worthwhile to determine by differential scanning calorimetry whether the formulation has a phase transition temperature close to 37°C.

The susceptibility of liposomal phospholipid to lipoprotein and phospholipase attack is strongly dependent on liposome size and type. Generally, MLVs are most stable, since only a portion of the phospholipid is exposed to attack, and SUVs are the least stable because of the stresses imposed by their curvature. Liposomes prepared with higher chain length phospholipids are most stable both in buffer and in plasma. Incorporation of charged lipid into the bilayer decreases stability in plasma even when cholesterol is included to bring the liposomes to the gel state. Cholesterol and sphingomyelin are generally very effective in reducing the instability of liposomes in contact with plasma. It is believed that the primary reason for this effect is not the increased bilayer tightness produced by cholesterol but the prevention of transfer of phospholipid to the plasma lipoprotein and red blood cell membrane. Table 5 shows that liposomal stability in plasma increases as the ratio of cholesterol in the liposome increases.

B. Stealth® Liposomes

One of the main disappointments of drug delivery using conventional liposomes was the realization that neither mechanical nor electrostatic stabilization can increase liposome stability in biological systems, such as in blood circulation. This is a very

Table 5 Release of Solutes From SUVs in the Presence of Plasma

Liposomes	Sucrose	Inulin	PVP
PC	80.6 ± 10.4	68.9 ± 6.9	26.9 ± 3.7
PC-CH (7:2)	42.2 ± 2.8	31.1 ± 7.2	26.1 ± 3.0
PC-CH (7:7)	4.1 ± 2.1	7.7 ± 0.9	6.6

Source: C. Kirby and G. Gregoriadis, *Biochem. J.*, 199:251–254 (1981).

"liposomicidal" environment, because the body protects itself with an elaborate immune system, and liposomes, if not quickly degraded by lipoproteins, are rapidly recognized as foreign particles and quickly taken up by the phagocytic cells of the body's immune system, which are located mostly in the liver and spleen.

A major breakthrough in liposome application was the realization that external steric stabilization can increase liposome stability in a biological environment. It was discovered that covering liposomes with hydrophilic, nonionic polymers greatly increased their stability in blood circulation [3, 49]. They must be inert, well solvated, and compatible with the solvent, and have polarizability close to that of water. Steric stabilization can be induced by the surface attachment of various natural or synthetic polymers, either by adsorption, hydrophobic insertion, electrostatic binding, or, preferably, by grafting via covalent bond. Nonionic, water-compatible, flexible, and well-hydrated polymers are preferred. The repulsion between surfaces with attached polymers was shown to be dependent on the grafting density and degree of polymerization. Normally, liposome bilayers contain 5 mol% of lipid with covalently attached polyoxyethylene glycol with molecular weight of 2,000 Da. Neither introduction of a larger amount of PEG-lipid nor longer polymer chains resulted in an improved stability in circulation. This is due to increased lateral pressure of the polymer above the surface and increased aqueous solubility of these large molecules. Normally, distearoyl chains are used to increase the anchoring effect.

Qualitatively one can explain the enhanced stability of such sterically stabilized liposomes in liposomicidal environments by their ability to prevent adsorption of various blood components and their close approach [2]. It was proposed that the major mechanism of liposome uptake and disintegration in plasma is reaction with proteins of the immune system, which adsorb onto foreign colloidal particles and tag them for subsequent macrophage uptake [2,49]. It is readily assumed that in the presence of surface-attached polymer, the adsorption of immunoglobulins or proteins of the complement cascade onto liposomes is reduced, and that lipid exchange interactions, which deplete liposome lipids, are minimized.

C. Liposome Stability in the Gastrointestinal Tract

Although many papers have been published on the oral administration of liposomally encapsulated drugs, especially insulin, very little effort has been made to critically assess the stability of liposomes in the environment of the gastrointestinal tract. Rowland and Woodley [50] have shown that most of the liposomal formulations that have been used are quite unstable to the gastrointestinal environment (low pH, bile, and/or phospholipase). Even distearoylphosphatidylcholine/cholesterol liposomes are very unstable in the gastrointestinal tract, and liposomally encapsulated and free drug give about the same pharmacokinetics when administered by the oral route to rats (Fig. 12).

IX. MEDICAL APPLICATIONS

Due to their biocompatibility, biodegradability, and colloidal properties, liposomes are one of the most studied drug delivery systems. They can act as a disperser for drugs that are difficult to solubilize, a sustained release system for microencapsulated agents, penetration enhancers, and site-specific delivery vehicles. In addition, in many cases the toxicity of the encapsulated drug is reduced because liposomes do not accumulate in some organs, such as the heart and kidneys. [51]

While drug-laden liposomes can be applied via all administration routes, including parenteral, topical, oral, and pulmonary, systemic applications are the most widely used. The major problem is the rapid leakage of the encapsulated drugs and quick uptake of liposomes by the cells of the reticuloendothlial system (RES) [7,49]. This natural fate of liposomes can be used to target these cells resulting in major improvements of the therapeutic index of drugs in the treatment of parasitic infections of these cells [52–54]. Recently, sterically stabilized liposomes, which can evade rapid clearance by the cells of RES, were developed, and improvements in anticancer therapy in animal models [7,55–58] as well as in human patients were reports [7,59,60].

The therapeutic promise of liposomes as a drug delivery system is fast becoming a reality. One must bear in mind that only in the last 15 years or so have real advances been made in translating the progress from university laboratories into pharmaceutically acceptable dosage forms. Pharmaceutical scientists collaborating with process engineers have been able to produce large volumes of sterile, pyrogen-free liposomes with acceptable shelf-lives. With current emphasis on increasing therapeutic indices of drugs,

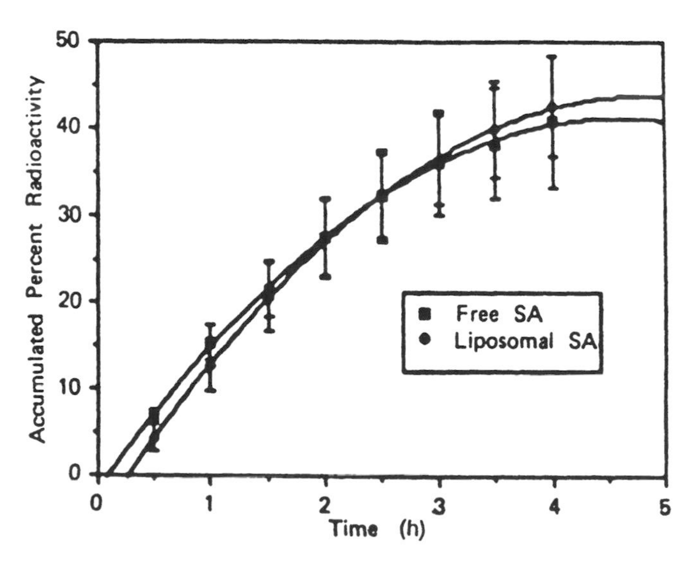

Fig. 12 Urinary excretion profiles after oral administration of free and liposomally encapsulated (DSPC:CH; 2:1 multilamellar vesicles) salicylic acid to fasted rats. (Source: N. Weiner and C. M. Chiang, Gastrointestinal uptake of liposomes. In: *Liposomes as Drug Carriers*, (G. Gregoriadis, ed.), 1988, p. 606.

it appears quite likely that these biocompatible, biodegradable vehicles will continue to receive increased attention from the pharmaceutical industry.

As of this printing, more than 10 companies plan to or have applied to the Food and Drug Administration for approval to test approximately 20 liposomally entrapped drug products. These dosage forms include anticancer and antifungal agents as well as drugs to combat arthritis, glaucoma, and dry eye.

Within a short period of time one might expect to see a broad range of liposomal products in various stages of clinical testing. The most promising appear to be liposomal products specifically formulated to facilitate the following:

1. Site-Specific Delivery. Particular emphasis has been placed on disease states involving the RES. Examples include antimonial compounds for parasitic disease, immunomodulation using macrophage activating agents, and antiviral treatment using ribavirin.
2. Site-Avoidance Delivery. The most promising examples are liposomal Doxorubicin (reduced cardiotoxicity) and liposomal Amphotericin B (reduced nephrotoxicity).
3. Sustained or Controlled Release. Examples include inhalation of bronchodilators, ocular delivery of antibiotics, intramuscular delivery of peptides, and topical delivery of a variety of drugs.
4. Passive targeting. Small particles can extravasate at the sites where the vascular system is leaky. This often happens in tumors and sites of trauma, either due to badly formed blood vessels in fast-growing tumors or due to the healing process itself. The accumulation in these sites is proportional to blood circulation times, and long-circulating liposomes were shown to accumulate in tumors [7,61].

   Still on a laboratory scale, are attempts of active targeting of (drug-laden) liposomes with attached ligands. Chemistry of attachment of various ligands to liposome surface or to the terminus of surface-grafted polymers was developed by Hansen et al. [62].
5. Gene therapy. Conventional liposomes have also been tried as delivery systems to deliver DNA into cells [63–65]. The rationale was the ability of liposomes to enhance intracellular accumulation, i.e., facilitate transfer of these large and heavily charged molecules across rather impermeable cell membranes. The procedures were rather cumbersome and inefficient, especially in the encapsulation of larger fragments. Because the technique was applicable in practice only in vitro or ex vivo, and required addition of fusogenic agents, its use rapidly declined with the emergence of electroporation [66]. Improvements achieved by using pH-sensitive liposomes did not change the trend [67]. Another candidate for DNA delivery is the virosome, a liposome containing fusogenic viral proteins [68]. The problem with such structures is effective DNA encapsulation. Even with a possible addition of condensing agents DNA particles are still rather large and not easy to encapsulate into relatively small liposomes and virosomes. Because these structures contain viral or bacterial proteins, immunogenicity may be a problem as well.

The use of liposomes in transfection to deliver a gene that encodes a particular protein into appropriate cells, with the aim of inducing the production of the encoded protein by the targeted cell, was revived with the emergence of cationic liposome-nucleic acid complexes. These colloidal particles that are composed of genes and liposomes/lipids are called genosomes. In an analogy to complexation of DNA with cations, and later cationic polyelectrolytes, it was shown that cationic surfactants, such as micelles and liposomes can condense and compact DNA [69,70]. Soon it was shown that such complexes can be several orders of magnitude more efficient in transfection than other approaches [8,71,72]. This complexation system can also be used for the delivery of RNA [73], single-stranded segments, and antisense oligonucleotides [74]. Obviously, it can be used to deliver ribozymes, some proteins, and other negatively charged molecules and macromolecues into various cells as well.

The major breakthrough was the observation that gene transfection using genosomes can also be achieved in vivo [75]. This raises the promise for the treatment of many diseases. Liposomes seem to be favored with respect to viral vectors because of the larger carrying capacity, apparent lack of immunogenicity, and for safety reasons. Efficient transfection can be achieved either by using supercoiled or linear plasmids. Their size varies but the range is normally between half and several tens of kilobase pairs.

Following difficulties to transfect cells with "DNA-micelle" complexes, bilayer forming diacyl cationic lipids are primarily used for transfection. The first two important lipids used and synthesized were dioctadecyl dimethyl ammonium bromide (DODAB) [76], and the 1,2-diacyl-3-trimethylammonium propane (DOTAP) family [77], respectively. Both are quaternary ammonium salts. An analogue of DOTAP is 2,3-bis(oleoyl)propyl trimethyl ammonium chloride (DOTMA), in which fatty acids are attached to the propyl backbone via ether, instead of ester bonds. Following the success of these lipids in gene transfection, many other lipids have been synthesized. These include variations of these structures as well as attachment of polyelectrolytes (polylysine) and polycations (spermine, spermidine) onto the diacylated backbone or the sterol group. The best known examples are 2,3 dioleoyloxy-N-[sperminecarboxamino)ethyl]-N,N-dimethyl-1-propanaminium triflurocetate (DOSPA) [78,79] and polylysine-lipid [80]. An alternative approach is to impose a positive charge on cholesterol, and a series of such molecules was synthesized [72]. Natural zwitterionic lipids can be rendered cationic by reacting and thus eliminating the negative charge on the phosphate group, and the zwitterionic and positively charged amino acids can be (di) acylated to form positively charged acyl esters or (di)acylated basic amino acids. Additional possibilities include using other charges (e.g., oxonium and phosphonium cations, metalo-lipids) as well as delocalized charges on heterocyclic polar heads. Novel approaches will very likely also exploit longer polyelectrolytes [81], lipopolyelectrolytes, and other polymers, such as dendrimers [82] and block-copolymers.

Cationic liposomes are mainly composed of binary mixtures of cationic lipid and a neutral component. Work in the early 1990s showed that the nature of cationic lipid, i.e., length and saturation of fatty acids; nature of chemical bonds between various parts of the molecule; the spacer length between the charge and the hydrophobic part of the molecule; presence and nature of the backbone; nature of the charge and its pK value; charge density and number of charges per molecule; and hydroxylations, ethylhydroxylations, methylations, etc. of polar head, can influence the transfection efficacy.

As a general rule, these lipids are mixed with DOPE, normally at equimolar ratios. Only rarely are other neutral lipids, such as lecithin or cholesterol, or no lipids used. Namely, for some cell types, it is known that pure cationic lipid liposomes, most notably DOTAP, transfect better than their mixtures with DOPE. In some cases, cholesterol was found to be more effective than DOPE [83]. All this shows that there may

Table 6 Composition of Several Commercially Available Cationic Liposomes as Transfection Kits

Name	Composition (w/w)	conc. (mg/mL)	conc. (mM)
Lipofectin	DOTMA:DOPE (1:1)	1	1.45
Lipofectamine	DOSPA:DOPE (3:1)	2	2.04
Lipofectace	DODAB:DOPE (1:2.5)	1	1.41
DOTAP	DOTAP	1	1.36
CellFectin	TMTPSp:DOPE (1:1.5)	1	1.12
Transfectam	DOGS	1	1.11
TFX-50	TDA:DOPE (1:1)	—	2.1
DC-Chol	DC-Chol:DOPE (3:2)	—	2.0

be no general mechanism of transfection. Table 6 shows the composition of several commercially available cationic liposomes as transfection kits [84].

X. NONMEDICAL APPLICATIONS OF LIPOSOMES

Liposomes make very useful models for studying biomembranes and membrane proteins. In addition to applications in basic sciences and medical applications, their properties can be exploited for various other uses. Because hydrophobic substances can be dispersed in a hydrophilic medium they can be used in the food industry. The best known example is microencapsulated enzymes in cheese ripening.

Liposomes can be used as a signal carrier in diagnostics. Normally ligands contain one marker group (fluorescent or radioactive). If ligands are conjugated to liposomes, which can contain hundreds of markers, the signal can be accordingly amplified.

Other uses involve the use of liposomes in the coating industry and ecology. It was shown that bioreclamation can be improved if nutrients and bacteria are dispersed in/with liposomes.

Liposomes, however, have achieved the largest impact in cosmetics. Hundreds of products exist, ranging from skin creams, to after-shaves, to body lotions, to sunscreens. Formulations exist from simple liposome kits with which customers can mix in their own ingredients to sophisticated formulations with encapsulated enzymes and antibiotics. While real benefits are debated, we can simply state that liposomes offer a biocompatible, natural, and water-based system to solubilize hydrophobic ingredients in what seems more appropriate than the use of detergents, oils, or alcohols in nonliposomal formulations [85].

There is little doubt that with improved understanding of liposome stability and interaction characteristics, many other successful applications will follow.

REFERENCES

1. A. D. Bangham, *Methods Membr. Biol.*, 1:1–68 (1974).
2. D. D. Lasic, *Liposomes: From Physics to Applications*, Elsevier, Amsterdam, 1993.
3. D. D. Lasic, D. Papahadjopoulos, *Sci.*, 267:135–136 (1995).

4. M. Florkin, and E. H. Stotz, eds., *Lipid Metabolism, Vol. 18 of Comprehensive Biochemistry*, American Elsevier, New York, 1967.
5. S. L. Regen, Polymerized liposomes. In: *Liposomes from Biophysics to Therapeutics*, (M. J. Ostro, ed.), Marcel Dekker, Inc., New York, Chap. 3, pp. 73–108 (1987).
6. H. Ringsdorf, B. Schlarb, and J. Venzmer, *Angew Chem. Int. Ed. Engl.*, 27:113–158 (1988).
7. D. D. Lasic and F. J. Martin, (eds.) *Stealth Liposomes*, CRC Press, Boca Raton, Fla., 1995.
8. E. Felgner, *Proc. Natl. Acad. Sci.*, 84:7413–7417 (1987).
9. N. Weiner, S. Niemec, C. Ramachandaran, Z. Hu, and K. Egbaria, *J. Drug Targeting*, 2:405–410 (1994).
10. J. Riess, In: *Nonmedical Applications of Liposomes* Vol. III, (D. D. Lasic and Y. Barenholz, eds.), CRC Press, Boca Raton, Fla. 1996.
11. S. Zalipsky, *Bioconj. Chem.*, 6:150–165 (1995).
12. R. Smith and C. Tanford, *J. Mol. Biol.*, 67:75–83 (1972).
13. K. Shinoda, T. Nakagaawa, B. Tamamushi, and T. Isemura, *Colloidal Surfactants*, Academic Press, New York (1963).
14. O. Bakouche and D. Gerlier, *Anal. Biochem.*, 130, 1983 (1983).
15. D. W. Deamer, *Biochim. Biophys. Acta*, 274:323–331 (1972).
16. G. Haran, R. Cohen, L. Bar, and Y. Barenholz, *Biochim. Biophys. Acta*, 1151:201–214 (1993).
17. D. D. Lasic, P. M. Frederik, M. C. A. Sutart, Y. Barenholz, and J. MacIntosh, *FEBS Lett.*, 1992.
18. M. J. Hope, M. B. Bally, G. Webb, and P. R. Cullis, *Biochim. Biophys. Acta*, 812:55 (1985).
19. F. Szoka and D. Papahadjopoulos, *Biochem.*, 75:4194–4198 (1978).
20. D. Papahadjopoulos and J. C. Watkins, *Biochim. Biophys. Acta*, 135:639–652 (1967).
21. S. M. Gruner, R. P. Lenk, S. Janoff and M. J. Ostro, *Biochem.*, 24:2833–2842 (1985).
22. T. Ohsawa, H. Miura, and K. Harada, *Chem. Pharm. Bull.*, 32:2442–2445 (1984).
23. C. J. Kirby and G. Gregoriadis, A simple procedure for preparing liposomes capable of high encapsulation efficiency under mild conditions. In: *Liposome Technology*, Vol. 1, CRC Press, Boca Raton, FL, 1984.
24. L. Saunders, *J. Pharm. Pharmacol.*, 14:567–572 (1962).
25. R. L. Hamilton and L. Guo, French pressure cell liposomes: Preparation, properties and potential. In: *Liposome Technology*, Vol 1, CRC Press, Boca Raton, FL, 1984.
26. Y. Barenholz, S. Amselem, B. J. Litman, J. Goll, T. E. Thompson, and F. D. Carson, *Biochem.*, 16:2806–2810 (1977).
27. C. H. Huang, *Biochem.*, 8:344–352 (1969).
28. A. D. Bangham, *Methods Membr. Biol.*, 1:1–68 (1974).
29. R. Hamilton, J. Goerke, and L. Guo, *J. Lipid Res.*, 21:981–992 (1980).
30. D. Deamer and A. D. Bangham, *Biochim. Biophys. Acta*, 443:629–634 (1976).
31. S. Batzri and E. D. Korn, *Biochim. Biophys. Acta*, 298:1015–1019 (1973).
32. Y. Kagawa and E. Raker, *J. Biol. Chem.*, 246:5477–5487 (1971).
33. M. H. W. Milsmann, R. A. Schwendner, and H. Weber, *Biochim. Biophys. Acta*, 512:147–155 (1978).
34. H. G. Enoch and P. Strittmatter, *Biochem.*, 76:145–149 (1979).
35. W. J. Gerritsen, A. J. Verkleij, R. F. A. Zwall, and L. L. Van Deenan, *Eur. J. Biochem.*, 75:4194–4198 (1978).
36. F. Szoka and D. Papahadjopoulos, *Biochem.*, 75:4194–4198 (1978).
37. M. J. Hope, M. B. Bally, G. Webb, and P. Cullis, *Biochim. Biophys. Acta*, 812:55–65 (1985).
38. J. P. Reeves and R. M. Dowben, *J. Cell Physiol.*, 73:49–60 (1969).
39. N. Oku and R. C. MacDonald, *Biochim. Biophys. Acta*, 734:54–61 (1983).
40. A. Kasahara, and A. Hinkle, *J. Biol. Chem.*, 252:7384–7390 (1977).

41. U. Pick, *Arch. Biochem. Biophys.*, 212:186-194 (1981).
42. N. Oku and R. C. MacDonald, *Biochem.*, 22:855-863 (1983).
43. R. L. Shew and D. Deamer, *Biochim. Biophys. Acta*, 816:1-8 (1985).
44. E. Mayhew, G. T. Nikolopoulos, J. J. King, and A. A. Siciliano, *Pharm. Manufac.*, 2:18-22 (1985).
45. F. Olson, C. A. Hunt, F. Szoka, W. J. Vail, and D. Papahadjopoulos, *Biochim. Biophys. Acta*, 557:9-23 (1979).
46. E. M. G. van Bommel and D. J. A. Crommelin, *Int. J. Pharm.*, 22:299-310 (1984).
47. J. H. Crowe and L. M. Crowe, *Biochim. Biophys. Acta*, 939:327-334 (1988).
48. G. Poste, *Biol. Cell*, 47:19-38 (1983).
49. D. Papahadjopoulos, *Proc. Natl. Acad. Sci.*, 88:11460-11464 (1991).
50. R. N. Rowland and J. F. Woodley, *Biochim. Biophys. Acta*, 620:400-409 (1980).
51. G. Gregoriadis, *J. Antimicrobiol. Therapy*, 28:39-48 (1991).
52. R. R. C. New, M. D. Chance, W. Thomas, and W. Peter, *Nature*, 272:55-56 (1978).
53. P. M. Peters, C. W. Huiskamp, W. Eling, and D. J. A. Crommelin, *Parasitology*, 98:397-386 (1983).
54. J. P. Adler-Moore and R. T. Proffitt, *J. Liposome Res.*, 3:429-450 (1993).
55. E. Mayhew, D. D. Lasic, S. Babbar, and F. J. Martin, *Int. J. Cancer*, 51:302-309 (1992).
56. J. Vaage, E. Mayhew, D. D. Lasic, and F. J. Martin, *Int. J. Cancer*, 51:942-947 (1992).
57. T. M. Allen, T. Mehra, C. Hansen, and Y. C. Chin, *Cancer Res.*, 52:2431-2436 (1992).
58. S. S. Williams, T. R. Alsoco, E. Mayhew, D. D. Lasic, F. J. Martin, and R. B. Bankert, *Cancer Res.*, 53:3954-3958 (1993).
59. A. Gabizon, et al., *Cancer Res.* 54:987-992 (1994).
60. B. Uziely, et al., *J. Clin. Oncology*, 13:1777-1785 (1995).
61. D. Needham, K. Hristova, T. J. MacIntosh, D. Dewhirst, N. Wu, and D. D. Lasic, *J. Liposome Res.*, 2:411-439 (1992).
62. C. B. Hansen, G. Y. Kao, S. Zalipsky, and T. M. Allen, *Biochim. Biophys. Acta*, 1239:133-144 (1995).
63. R. P. Fraley, S. Subramani, P. Berg, and D. Papahadjopoulos, *J. Biol. Chem.*, 255:10431-10435 (1980).
64. R. P. Fraley, D. Papahadjopoulos, *Curr. Top. Microbiol. Immunol.*, 96:171-187, (1982).
65. C. Nicolau and A. Cudd, *Crit. Rev. Ther. Drug Carrier Sys.*, 6:239-271 (1989).
66. D. D. Lasic, N. S. Templeton, *Adv. Drug. Del. Rev.* 20:221-260, (1996).
67. J. Y. Legendre and F. C. Szoka, *Proc. Natl. Acad. Sci.*, 90:893-897 (1993).
68. K. L. Brigham and H. Schreier, *J. Liposome Res.*, 3:31-49, (1993).
69. J. P. Behr, *Tetrahedron Lett.*, 27:5861-5864, (1986).
70. A. Minsky, R. Ghirlando, and H. Gershon, In: *Liposomes: from Gene Therapy to Diagnostics and Ecology* (D. D. Lasic and Y. Barenholz, eds.) CRC Press, Boca Raton, FL.
71. N. Ballas, N. Zakai, I. Sela, and A. Loyter, *Biochim. Biophy. Acta*, 939:8-18 (1988).
72. X. Gao and L. Huang, *Biochem. Biophys. Res. Comm.*, 179:280-285, (1991).
73. R. W. Malone, P. Felgner, and I. M. Verman, *Proc. Natl. Acad. Sci.*, 88:6077-6081, (1989).
74. M. Chiang, H. Chan, M. A. Zounes, S. Freier, W. Loma, and C. F. Bennett, *J. Biol. Chem.*, 266:1862-1871, (1991).
75. N. Zhu, D. Liggitt, Y. Liu, and R. Debs, *Sci.*, 261:209-211 (1993).
76. K. Kunitake and Y. Okahata, *J. Amer. Chem. Soc.*, 99:3860-3861 (1977).
77. H. Eibl and P. Wooley, *Biophys Chem.*, 10:261-271, (1979).
78. J. P. Behr, B. Demeneix, J. P. Loeffler, and J. P. Mutul, *Proc. Natl. Acad. Sci.*, 86:6982-6986 (1989).
79. J. S. Remy, A. Kichler, V. Mordovinov, F. Schuber, and J. P. Behr, *Proc. Natl. Acad. Sci.*, 92:1744-1748 (1995).

80. X. Zhou and L. Huang, *Biochim. Biophys. Acta*, 1189:195–203 (1994).
81. O. Boussif, F. Lezoualch, M. A. Zanta, M. Merggny, D. Schermann, B. Demeneix, and J. P. Behr, *Proc. Natl. Acad. Sci.*, 92.
82. R. F. Service, *Sci.*, 268:458–459 (1995).
83. M. J. Bennett, R. W. Malone, and M. H. Nantz, *Tetrahed. Lett.*, 36:2207–2210 (1995).
84. D. D. Lasic, *Liposomes in Gene Delivery*, CRC Press, Boca Raton, FL, 1997.
85. D. D. Lasic and Y. Barenholz, eds., *Handbook on Nonmedical Applications of Liposomes*, Vols. 1–4, CRC Press, Boca Raton, FL, 1996.

3

Polymeric Dispersions as Drug Carriers

Roland Bodmeier

Freie Universität Berlin, Berlin, Germany

Philippe Maincent

University of Nancy, Nancy, France

I. INTRODUCTION

During the last two decades, considerable attention has been given to the development of novel drug delivery systems. The rationale for controlled drug delivery is to alter the pharmacokinetics and pharmacodynamics of drug substances in order to improve the therapeutic efficacy and safety through the use of novel drug delivery systems. Besides more traditional matrix or reservoir drug delivery systems, colloidal drug delivery systems have gained in popularity. The major colloidal drug delivery systems investigated include liposomes and polymeric nanoparticles [1]. These systems have been investigated primarily for site-specific drug delivery, for controlled drug delivery, and also for the enhancement of the dissolution rate/bioavailability of poorly water-soluble drugs. The primary routes of administration under investigation are parenteral routes; however, other routes such as the oral, ocular, or topical routes are also being investigated.

While numerous publications have appeared on liposomes from international research teams, the research on polymeric nanoparticles has been primarily performed by a few research groups in Europe (Speiser et al., Couvreur et al., Davis et al., Gurny et al., Kreuter et al.). Nanoparticles are being investigated as an alternative colloidal drug delivery system that could potentially avoid some of the technical problems observed with liposomes. When compared to the fluid or semifluid liposome systems, the polymeric nanoparticle systems could be classified as solid particles.

Nanoparticles are colloidal particles with a size smaller than 1 mm. The active compound can be present in various physical states; it can be dissolved in the polymeric matrix, it can be encapsulated, or it can be adsorbed or attached to the surface of the colloidal carrier. The term nanoparticles encompasses both nanocapsules and nanospheres (Fig. 1). Nanocapsules have a core-shell structure (a reservoir system), while nanospheres represent a matrix-system. It is often difficult to distinguish between the different structures and morphologies of the particles; the term nanoparticles is thus used as a general term.

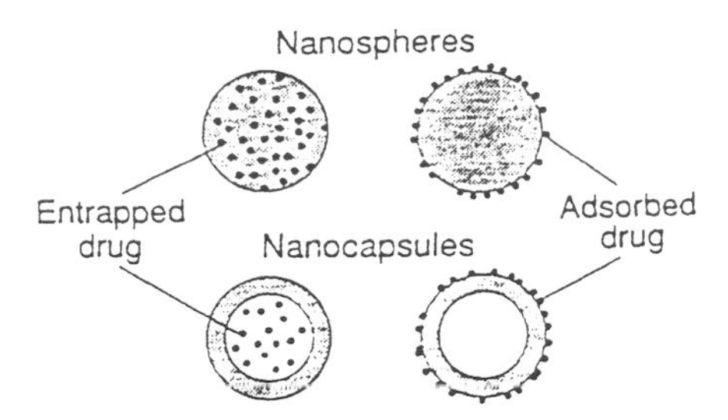

Fig. 1 Various types of drug-loaded nanoparticles. (Reproduced with permission from Ref. 63.)

This chapter presents an overview of the different methods used to prepare polymeric nanoparticles and the physico-chemical methods used to characterize the nanoparticles. In vivo results with nanoparticles are not reviewed. Several excellent reviews exist on this subject [2–4].

II. PREPARATION OF NANOPARTICLES

The techniques used to prepare nanoparticles are generally classified in two groups. In the first group, nanoparticles are formed from preformed polymers. The polymers include both water-insoluble and -soluble polymers of synthetic, semisynthetic, or natural origin. Alternatively, nanoparticles are not prepared from preformed polymers but through various polymerization reactions of lipophilic or hydrophilic monomers.

The techniques used for the preparation of nanoparticles based on water-insoluble polymers are primarily derived from methods developed for the preparation of aqueous colloidal polymer dispersions, which are used in the coating of solid dosage forms [5]. Latexes are obtained by emulsion polymerization (e.g., acrylic latexes—Eudragits®) and pseudolatexes through the emulsification of polymer solutions or melts (e.g., ethyl cellulose—Aquacoat®, Surelease®). The pharmaceutical industry adopted the colloidal polymer dispersions from other industrial applications including paints, varnishes, adhesives, and paper coatings. The aqueous colloidal polymer dispersions have been developed to avoid problems associated with the use of organic polymer solutions during the coating of solid dosage forms. The polymer dispersions allow the formation of water-insoluble coatings through the coalescence of the colloidal polymer particles into a homogeneous film from a completely aqueous coating medium. Problems associated with organic solvents, such as industrial and environmental hazards, pollution, and increased costs could be avoided.

Drug-containing colloidal polymer particles (nanoparticles) can be obtained through the incorporation of a drug substance during or after the preparation of the polymer dispersions. As with microencapsulation methods, there is not one universal technique to prepare nanoparticles. The choice of a particular preparation method and a suitable polymer depends on the physico-chemical properties of the drug substance, the desired release characteristics, the therapeutic goal, the route of administration, the biodegradability/biocompatibility of the carrier material, and regulatory considerations. From a technological point of view, the successful selection of a preparation method is determined by the ability to achieve high drug loadings, high encapsulation efficiencies, and high product yields, and the potential for easy scale-up. For example, methods with high encapsulation efficiencies but with only low drug loading capacity are limited to very potent drugs. Methods that result in nanoparticles with high drug loadings and high encapsulation efficiencies are preferred. The term nanoparticles often actually describes the suspended system of the nanoparticles in an aqueous phase, which is the colloidal polymer dispersion. Besides high drug loadings and encapsulation efficiencies, a preparation method should also be able to yield polymer dispersions with a high nanoparticle content, which also directly relates to the size of the dose that can be reasonably administered.

The techniques for preparing nanoparticles from preformed polymers are presented first, followed by techniques to prepare nanoparticles by polymerization of various monomers.

III. NANOPARTICLES PREPARED FROM PREFORMED POLYMERS

Nanoparticles have been prepared from a variety of both water-soluble and water-insoluble polymers of synthetic, semisynthetic, and natural origin. The use of preformed polymers has several advantages when compared to use of nanoparticles prepared through the polymerization of monomers. These include the use of polymers with well-characterized physico-chemical properties, established safety and approval standards, the absence of residual monomers or polymerization reagents (e.g., initiators or catalysts) and the lack of possible reactions between drugs and monomers. In addition, it is not possible to obtain nanoparticles by emulsion polymerization for polymers where suitable monomers are not available. These polymers include, in particular, polyesters such as polylactic acid (PLA) and polylactic/glycolic acid copolymers, cellulose derivatives such as ethyl cellulose and cellulose acetate, and natural polymers such as gelatin or sodium alginate.

Most techniques for preparing nanoparticles from preformed polymers are derived from conventional emulsification technologies, which have been used to prepare microparticles. The nanoparticles can be formed either by dispersing the polymer-drug phase into an external phase or by phase inversion techniques, whereby the external phase is first added to the internal phase containing the drug and the polymer. These techniques also allow the use of mixtures of drugs or polymers. The preparation techniques using polymers are classified based on the solubility of the polymer.

A. Nanoparticles Prepared from Water-Insoluble Polymers

1. Solvent Evaporation Method—Water-Immiscible Organic Solvents

The solvent evaporation method is a technique widely used for preparing biodegradable and nondegradable microspheres [6–8]. In this method, the drug and polymer are dispersed/dissolved or dissolved in a volatile, water-immiscible organic solvent (e.g., methylene chloride, chloroform, ethyl acetate). This solution or dispersion is then emulsified in an external aqueous phase containing an emulsifying agent by using conventional emulsification equipment to form an oil-in-water (O/W) emulsion. Microspheres with a size range generally between 5 and 250 μm are obtained after solvent evaporation. To prepare nanoparticles and not microspheres, the O/W emulsion is homogenized under high shear with appropriate homogenization equipment (e.g., microfluidizer, sonication) prior to the precipitation of the polymer to further reduce the particle size of the internal organic phase into the colloidal size range (Fig. 2) [9–12]. Drug-polymer particles in the nanometer size range are formed after solvent diffusion

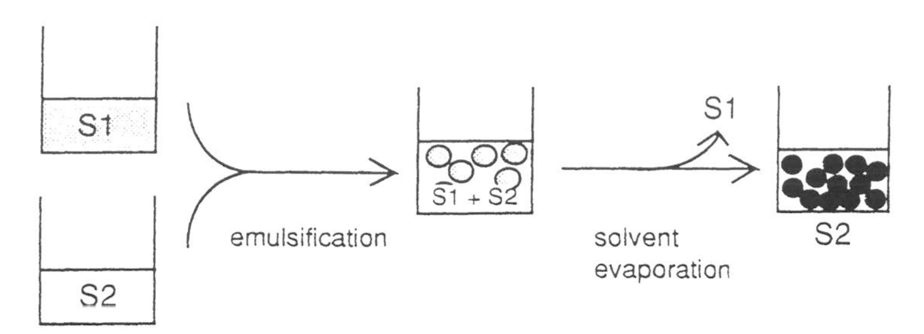

Fig. 2 Preparation of nanoparticles by solvent evaporation. S1: organic solution of polymer. S2: nonsolvent of polymer. (Reproduced with permission from Ref. 61.)

into the aqueous phase and evaporation at the water/air interface causing polymer and drug precipitation. The solvent removal process can be accelerated through heating and the application of a vacuum. The polymers used to prepare the nanoparticles include biodegradable polymers such as PLA and its copolymers and cellulose derivatives such as ethyl cellulose or cellulose acetate phthalate. In principle, any water-insoluble polymer being soluble in a water-immiscible organic solvent could be converted into a colloidal polymer dispersion.

The high-pressure emulsification-solvent evaporation method is limited to water-insoluble drugs. Unless specific binding exists, water-soluble drugs cannot be encapsulated because of complete partitioning into the external aqueous phase during emulsification. One problem with this method, especially at higher drug loadings, is the appearance of drug crystals in the external aqueous phase. Drug crystallization in the external phase can be explained with the change in drug solubility in the aqueous phase during nanoparticle preparation [12]. The solubility of the drug in the external aqueous phase changes during the preparation of the nanoparticles. It first increases because of organic solvent diffusing from the internal organic phase into the external aqueous phase. After solvent evaporation, the drug solubility decreases and the amount of drug present in the aqueous phase exceeds the drug solubility, resulting in drug precipitation. The drug crystallization depends on the polymer, organic solvent, and type and concentration of surfactant used. The rate of solvent diffusion into the aqueous phase influences the rate of polymer precipitation. It is related to the water solubility and rate of evaporation of the organic solvent. A faster polymer precipitation results in a decrease in drug crystallization in the external phase. Another important factor for high drug entrapment (low crystallization in the external aqueous phase) is the drug-polymer compatibility. Drug crystallization in the aqueous phase occurs at higher drug loadings with polymers having a higher affinity for the drug. Drug used in excess of the solubility of the drug in the polymer precipitates in the external phase. In contrast to microparticles, where the drug can be either dissolved or dispersed in the polymeric matrix, the drug is only dissolved in the polymeric nanoparticle matrix. Increasing the concentration of sodium lauryl sulfate in the aqueous phase also enhances drug crystallization. The surfactant not only solubilizes the drug but also solubilizes the organic solvent during the nanoparticle preparation. This results in higher amounts of drug-crystallizing in the aqueous phase after solvent evaporation.

Various processing and formulation variables influence the particle size of the resulting nanoparticles. The particle size of the solidified nanoparticles is determined by the size of the emulsified polymer-drug-solvent droplets, and hence depends on the homogenization equipment. Nanoparticles have been prepared with conventional laboratory homogenizers [9], by ultrasonication [10,11,13], and by microfluidization [12]. Microfluidizer processing is a patented mixing technology available for the preparation of dispersed systems on both a laboratory- and production-size scale. In this process, the liquid (e.g., O/W-emulsion) is pumped through microchannels to an impingement area at high operating pressures. Cavitation and the accompanying shear and impact are responsible for the particle size reduction within the "interaction chamber." Other pharmaceutical applications of microfluidization include the preparation of liposomes, microemulsions, and parenteral nutrition emulsions.

With the microfluidizer, the particle size decreases with increasing operating pressure and increasing number of cycles [12]. The particle size of the nanoparticles is also affected by the type of surfactant and surfactant concentration, the viscosity of the organic polymer solution, and the phase ratio of the internal to the aqueous phase [12,14]. As expected, the particle size decreases with decreasing viscosity of the organic phase (lower polymer concentration) and increasing poly(vinyl alcohol) concentration in the aqueous phase (up to a optimum concentration).

Besides high-shear emulsification techniques, surfactants, which reduce the interfacial tension, are necessary in order to obtain particles in the submicron range. The surfactants or polymeric stabilizers used include, among others, various polysorbates, sodium lauryl sulfate, poly(vinyl alcohol), and gelatin. For nanoparticles to be administered by injection, purification of the nanoparticles from surfactants may be necessary. The emulsifiers can be removed by dialysis or by separating the nanoparticles from the aqueous phase by ultracentrifugation followed by washing steps with surfactant-free water. This, however, may lead to destabilization of the nanoparticle suspension. Nanoparticles of the PLA have been prepared using albumin as a colloidal stabilizer, resulting in a fully biodegradable system [15].

Nanoparticles can also be formed by phase inversion. In this method, an aqueous phase is dispersed into the drug-containing organic polymer solution in order to form a water-in-oil (W/O) emulsion. Phase inversion occurs upon further addition of water. Various emulsifier combinations can be used. One popular method, which has been used to prepare a commercial aqueous ethyl cellulose dispersion (Surelease®) for coating purposes, is based on the in situ formation of the emulsifying agent [5]. Fatty acids, such as oleic or stearic acid, are added to the polymer melt. The aqueous phase containing an alkaline agent is added to the organic polymer phase, resulting in the ionization of the fatty acid at the interface.

2. *Solvent Evaporation Method—Water-Miscible Organic Solvents*

Replacing the organic solvents (e.g., methylene chloride) commonly used in the solvent evaporation method with less toxic solvents such as ethanol or acetone would be highly desirable from a toxicological point of view. However, the addition of solutions of polymers, such as PLA or ethyl cellulose, in these water-miscible solvents to an external aqueous phase would not result in nanoparticles under the same experimental conditions. Because of the complete miscibility of the organic solvent and water, large precipitates or agglomerates would form at the commonly used polymer concentrations. Various modified solvent evaporation methods, as described below, have been developed for water-miscible solvents. These methods are based either on the use of much lower polymer solution concentrations or on the use of polymers with stabilizing functional groups, or on a reduced miscibility of the solvent-water system based on a "salting-out mechanism."

Liposomes, another colloidal delivery system, have been prepared with water-miscible solvents by a so-called solvent injection method [16]. The injection of a solution of the lipids in ethanol into an aqueous phase results in small unilamellar vesicles, whereas the injection of solutions of the lipid in water-immiscible solvents results in large unilamellar vesicles (17). The preparation of polymeric nanoparticles with water-miscible solvents is derived from this technology (Fig. 3a). The methodology is similar to the solvent evaporation method described above, with only the water-immiscible organic solvent being replaced with a water-miscible solvent such as acetone or ethanol [18]. The drug-polymer solution is then poured into an external aqueous phase containing the emulsifying agent, to form droplets spontaneously. Nanospheres (a matrix-system) are formed after solvent evaporation. As mentioned above, only very dilute solutions of most polymers, in particular polymers without stabilizing functional groups can be used because of polymer precipitation at higher concentrations.

Besides using the traditional solvent evaporation method, stealth nanoparticles have also been prepared with acetone by using poly(*d,l*-lactide)-methoxy (polyethylene glycol), a proprietary, more hydrophilic PLA-derivative [19]. In analogy to stealth liposomes, the circulation time of the nanoparticles could be increased because of the more hydrophilic character of the nanoparticles. Oligonucleotides have been entrapped within these stealth nanoparticles through the complexation with oligopeptides. The oligonucle-

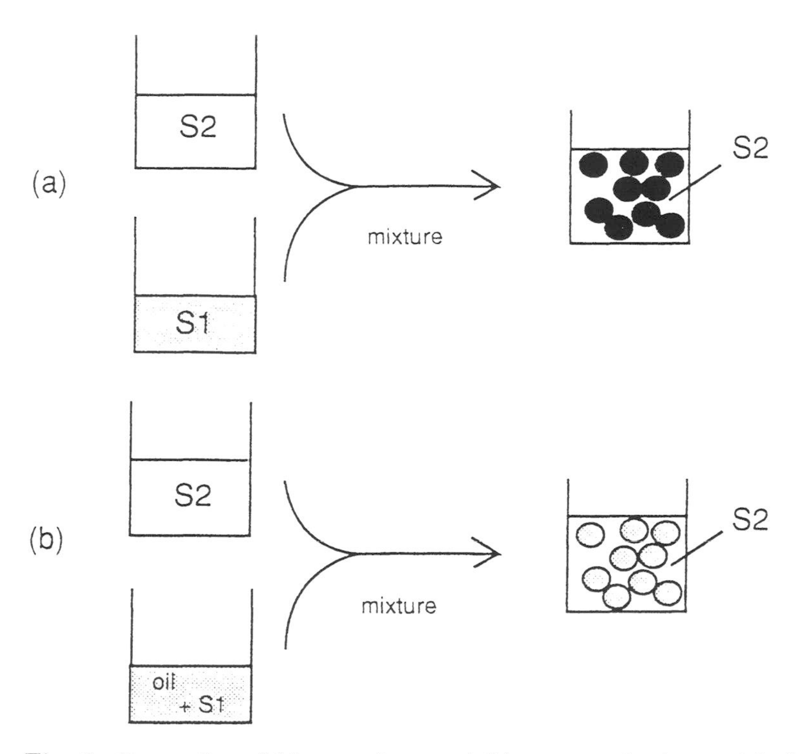

Fig. 3 Preparation of (a) nanospheres and (b) nanocapsules by precipitation using polymer solvent/nonsolvent systems. S1: organic solution of polymer. 2S: nonsolvent of polymer. (Reproduced with permission from Ref. 61.)

otide/oligopeptide coprecipitates with the polymer after addition of an acetonic solution to water. However, it has yet to be proven that the complex is truly encapsulated within the nanoparticles or if it just coexists in submicron size beside the nanoparticles.

Drug-containing acrylic nanoparticles can be formed spontaneously at high polymer solution concentrations (up to 40%) with pharmaceutically acceptable organic solvents, without the use of surfactants or sophisticated equipment, with polymers containing stabilizing functional groups [20]. The polymers used were acrylic copolymers with quaternary ammonium groups, Eudragit® RS and RL 100. Aqueous latexes of these polymers are used extensively in the coating of solid dosage forms and have been prepared by emulsification of the solid polymer [21], or of solutions of the polymers in acetone [22], without the use of emulsifying agents. In contrast to the conventional solvent evaporation method, which uses primarily toxic, water-immisible solvents such as methylene chloride, water-miscible solvents such as acetone or ethanol are used with these acrylic polymers. The drug-polymer solution is simply added to an aqueous phase and the nanoparticles form spontaneously without high shear. Eudragit® RS and RL 100 are cationic polymers, being poly (ethyl acrylate-methyl methacrylate-trimethyl ammonio ethyl methacrylate chloride) copolymers with ratios of 1:2:0.1 and 1:2:0.2. The nanoparticles are stabilized by the quaternary ammonium groups present in the polymer. The average particle size of nanoparticles prepared by microfluidization with the water-immiscible solvent, methylene chloride, is larger when compared to the size of spontaneously formed nanoparticles. As expected, the particle size increases with increasing polymer concentration because of increased solution viscosity. Commercially available aqueous colloidal polymer dispersions such as acrylic or cellulosic latexes or pseudolatexes have been converted into redispersible powders to reduce the bulk volume, to allow more flexible compounding, and to improve stability of polymers that hydrolyze in aqueous media during storage. Freeze- or spray-dried Eudragit® RS or RL nanoparticles can be easily resuspended because of the quaternary ammonium groups. The particle size closely matches the original particle size.

Besides the parenteral route, potential application of colloidal polymer dispersions include the oral delivery of bad-tasting or irritating drugs, or topical administration in the form of brush-on nanosuspensions or free films for transdermal systems. Upon drying of nanosuspensions on the skin, films or individual nanoparticles could be formed depending on the minimum film formation temperature of the polymer-plasticizer-drug combination. Drug-containing films of water-insoluble polymers are generally prepared by casting and drying an organic drug-polymer solution or suspensions. To circumvent problems associated with the use of organic solvents, polymeric films have been prepared from aqueous polymer dispersions. Water-soluble drugs can be dissolved in the colloidal polymer dispersion prior to film casting. Several alternatives are available for the incorporation of water-insoluble drugs into colloidal polymer dispersions. Liquid lipophilic drugs have been emulsified directly into latices prior to the preparation of films for transdermal applications [23]. In a previous study, propranolol has been dispersed into an ethyl cellulose pseudolatex. Drug settling during the drying process results in a nonhomogenous drug distribution in the polymer film. A more homogenous distribution is obtained by dissolving the drug in a water-insoluble plasticizer, dibutyl sebacate, and emulsifying the drug-plasticizer solution into the latex prior to film casting [24]. Lidocaine has been incorporated into an ethyl cellulose pseudolatex during the preparation of the colloidal dispersion by a sonication-solvent evaporation method.

The drug is added to the organic polymer solution prior to the emulsification step. The pseudolatex forms a clear, flexible film on the skin with local anaesthetic activity [25].

Niwa et al. used a mixed-solvent system of methylene chloride and acetone to prepare PLGA-nanoparticles [26]. The addition of the water-miscible solvent, acetone, results in nanoparticles in the submicron range; this is not possible with only the water-immiscible organic solvent, under the experimental conditions. The addition of acetone decreases the interfacial tension between the organic and the aqueous phase and, in addition, results in the perturbation of the droplet interface because of the rapid diffusion of acetone in the aqueous phase. This technique is named "spontaneous emulsification solvent diffusion method."

Nanocapsules of biodegradable polymers, such as PLA and PLA-copolymers or poly (e-caprolactone), have been prepared by an "interfacial polymer deposition mechanism" [27–33]. An additional component, a water-immiscible oil, is added to the drug-polymer-solvent mixture (Fig. 3b). A solution of the polymer, the drug, and a water-immiscible oil in a water-miscible solvent such as acetone is added to an external aqueous phase. The water-miscible organic solvent diffuses rapidly into the aqueous phase. The polymer precipitates at the oil/water interphase, surrounding the drug-containing oil core. It is hypothesized that the polymer acts as a surfactant stabilizing an O/W emulsion. This technique results in nanocapsules because of a core-shell structure. The drug must have a high solubility in the oil-solvent mixture in order to obtain nanoparticles with high drug loadings. When compared to nanoparticles prepared with water-immiscible organic solvents, the disadvantage of this interfacial deposition method is the use of only very diluted polymer solutions. A polymer precipitate, and not individual nanoparticles are formed at higher polymer concentrations because of the water-miscibility of the organic solvent. To obtain more concentrated dispersions, the polymer dispersion can be concentrated under reduced pressure or be converted into a redispersible powder by spray- or freeze-drying techniques.

An interesting modification of the solvent evaporation method is the "salting-out procedure" (Fig. 4) [34–39]. The nanoparticles are prepared by adding an aqueous phase saturated with an electrolyte or nonelectrolyte to a solution of the polymer and drug in a water-miscible organic solvent under agitation until an O/W-emulsion forms. Poly-(vinyl alcohol) has been added to the aqueous phase to act as a viscosity-enhancing and emulsifying agent. Saturating the aqueous phase reduces the miscibility of acetone and water by a salting-out process and allows the formation of an O/W emulsion from the otherwise miscible phases. After the formation of the internal organic phase droplets, water is added to allow diffusion of the organic solvent into the external phase and precipitation of the polymer resulting in nanoparticle formation. The solubility/miscibility of acetone with the external aqueous phase is thereby gradually increased up to complete miscibility by diluting the initially saturated aqueous phase with water. The salt or nonelectrolyte can be removed by centrifugation and subsequent washing steps from the polymeric nanoparticles. Various polymers including cellulose acetate phthalate, methacrylic acid copolymers, ethyl cellulose, and PLA have been used. Cellulose acetate phthalate nanoparticles have been investigated as an in-situ gelling ophthalmic drug delivery system [35]. The nanoparticles are insoluble in water; however, the enteric polymer dissolves/gels at the pH of the tear fluids. Cellulose acetate phthalate, an ester, is not stable and hydrolyzes in aqueous media. The polymer suspension could be converted in a redispersible nanoparticles powder, which, however, may not be a practical approach for ophthalmic drug delivery. Enteric polymers based on acrylates

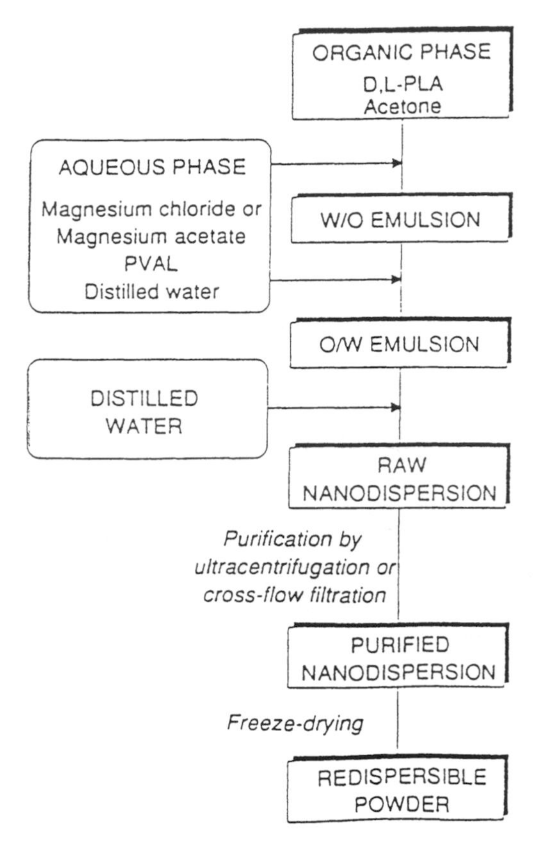

Fig. 4 Preparation of nanoparticles by the salting-out method. (Reproduced with permission from Ref. 39.)

(e.g., Eudragit® L) are stable in aqueous colloidal dispersion and could be used instead of cellulose acetate phthalate. The major advantages of the salting-out method are the avoidance of chlorinated solvents and surfactants commonly used with the conventional solvent evaporation method. Converting the polymer dispersion into a redispersible powder without surfactants could be a problem with nanoparticles prepared with polymers without surface-active functional groups.

When compared to nanoparticles prepared by emulsion polymerization techniques, the particle size of nanoparticles obtained by solvent evaporation methods is generally larger and the particle size distribution is not monodisperse. The polymer dispersions prepared by emulsification techniques are, therefore, generally physically less stable than nanoparticles prepared by emulsion polymerization techniques. The use of high-shear homogenization or sonication techniques is energy consuming and could potentially result in the degradation of the drug or the polymeric carrier. With nanoparticles intended for parenteral use, either biocompatible/biodegradable surfactants have to be used or the surfactant has to be removed by techniques such as dialysis. This, however, may result in physical stability problems of the polymer dispersion. Besides physical stability, the chemical stability of not only the drug but also of the polymeric carrier in the aqueous medium has to be investigated. In particular a biodegradable polyester, such as PLA, is prone to hydrolysis [40]. With PLA nanoparticles, the conversion of the polymer dispersion to a redispersible powder by freeze-drying is advised. In addition, with nanoparticles prepared by the solvent evaporation method, the issue of residual organic solvents must be addressed.

B. Nanoparticles Prepared from Hydrophilic Polymers

When compared to the synthetic, water-insoluble polymers described above, the hydrophilic polymers described below are primarily of natural origin. The nanoparticles are prepared either by a W/O-emulsification method or by aqueous phase separation techniques (e.g., coacervation).

1. *W/O-Emulsification Methods*

Nanoparticles of hydrophilic polymers (e.g., albumin, chitosan, gelatin, or carbohydrates) can be prepared by W/O-emulsification techniques (Fig. 5). This process has been developed for the preparation of albumin microspheres [41–44]; however, as with microspheres prepared by the solvent evaporation method, the use of high-shear homogenization equipment or ultrasonication allowed the formation of emulsions in the nanometer size range. Basically, an aqueous polymer solution is emulsified into an external, water-immiscible phase, such as an oil or organic solvent followed by homogenization. Upon water removal, the polymer droplets solidify. Upon contact with wa-

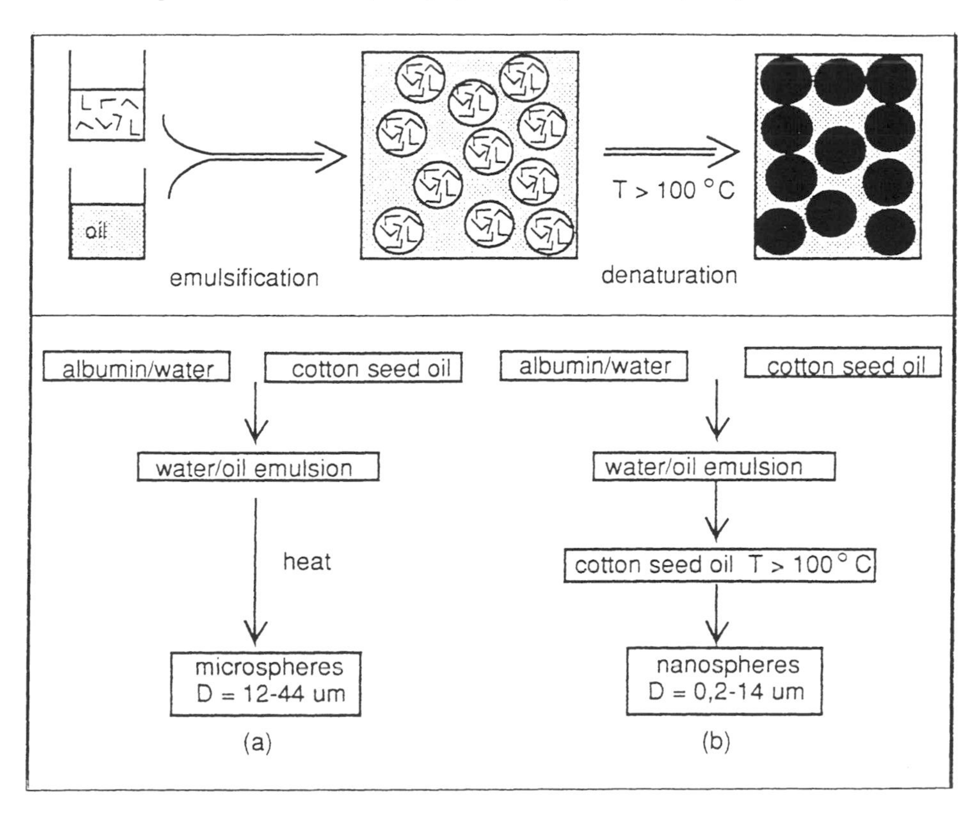

Fig. 5 Preparation of (a) microspheres and (b) nanospheres by thermal denaturation of albumin. (Reproduced with permission from Ref. 61.)

ter, the resulting nanoparticles would completely dissolve because of their high water solubility. Insoluble nanoparticles can be obtained by further hardening/insolubilizing the polymer through chemical cross-linking with aldehydes or other cross-linking agents or through denaturation at elevated temperatures.

To obtain high encapsulation efficiencies, the drug must be insoluble in the external phase. Hydrophilic polymer nanoparticles prepared by W/O-emulsification techniques are, therefore, limited to water-soluble drugs. The drug can be added prior to the emulsification step into the aqueous polymer solution or it can be adsorbed onto the nanoparticles after separation and purification. Water-soluble drugs can be entrapped and bound to the polymer, for example through an ion-exchange mechanism.

Albumin nanoparticles have been prepared by emulsification of an aqueous albumin solution (up to 50% w/w polymer content) in oil followed by either heat denaturation or chemical cross-linking of the protein. In the heat denaturation method, the aqueous albumin solution is emulsified in cottonseed oil, homogenized, and then poured into heated oil to denature the protein. The denaturation temperature and time affect the degradation rate of the particles. The nanoparticles could also be prepared at room temperature through chemical cross-linking with aldehydes (e.g., glutaraldehyde, butanedione). The albumin nanoparticles are then separated and washed with organic solvents to remove the adhering oil. Only relatively small amounts of aqueous albumin solution can be emulsified into large amounts of oil, because significantly longer drying times and possible agglomeration result at higher ratios of internal to external phase. The effect of various process and formulation variables on particle size and drug release has been studied in detail by Gallo et al. [45] and Gupta et al. [46].

Gelatin nanoparticles have been prepared by emulsifying a concentrated gelatin solution (30%) into hydrogenated castor oil containing proper emulsifying agents above the gelation temperature of the gelatin solution [47]. The W/O emulsion is then cooled in order to gel the aqueous gelatin droplets. It is then diluted with acetone to probably dehydrate the gelatin nanoparticles and to ease the removal of the oil phase by filtration through a membrane filter with a pore size of 50 nm. The nanoparticles are washed with acetone and insolubilized/hardened with a formaldehyde solution. The nanoparticles have an average diameter of less than 300 nm. The degradation of the nanoparticles and hence the drug release are dependent on the degree of cross-linking with glutaraldehyde.

Besides emulsifying into an oil phase, the aqueous polymer solution can also be emulsified into organic solvents such as chloroform. Gelatin nanoparticles have been prepared with a mixture of chloroform/toluene as the external phase [48]. The use of organic solvents is, however, undesirable.

Chitosan is one of the few naturally derived polysaccharides carrying basic functional groups. It is obtained through the deacetylation of chitin. Since it is only soluble in acidic media, chitosan micro- and nanoparticles are prepared by emulsifying aqueous solutions of chitosan in acetic acid into an external oil phase [49]. The cross-linking agent can be added to the aqueous chitosan solution prior to emulsification. The nanoparticles are obtained after solidification of the internal aqueous phase through the removal of water at elevated temperature and reduced pressure. Colloidal magnetite particles can be added to the aqueous phase prior to emulsification to produce externally guidable magnetic nanoparticles. Magnetic albumin nano- and microparticles have been prepared previously in a similar manner [50,51].

The polysaccharides, chitosan and sodium alginate, form gels with counterions such as tripolyphosphate or calcium chloride. Nanoparticles can be prepared by emulsifying an aqueous polysaccharide solution into the oil phase followed by emulsification of an aqueous solution of the counterion solution.

The drawbacks of the W/O-emulsification method include the use of large amounts of oils as the external phase, which must be removed by washing with organic solvents; heat stability problems of drugs; possible interactions of the cross-linking agent with the drug; and, as with all nanoparticles prepared by emulsification techniques, a fairly broad particle size distribution.

2. *Aqueous Phase Separation Techniques*

Phase separation (e.g., precipitation or coacervation) of water-soluble macromolecules such as gelatin or albumin can be induced through pH-changes or the addition of desolvating agents such as salts or water-miscible organic solvents (e.g., ethanol, isopropanol). The preparation of gelatin microcapsules by simple or complex coacervation is well known.

Gelatin and albumin nanoparticles have been prepared through desolvation of the dissolved macromolecules by either salts (e.g., sodium sulfate or ammonium sulfate) or ethanol [52–55]. This is, in principle, similar to a simple coacervation method. Bulk precipitation of the polymer must be avoided. The system generally has to be processed with a homogenizer to minimize particle-particle association. The particles can then be insolubilized through cross-linking with an optimum amount of aldehydes. Higher concentrations of cross-linking agent result in larger agglomerates. The coacervation and hardening process must be optimized to avoid the formation of larger agglomerates. Turbidity measurements are used to follow the phase separation/desolvation process and to establish three component phase diagrams. The hardening reaction can be terminated through the addition of sodium metabisulphite [55].

These phase separation methods avoid the use of oils as the external phase. However, the nanoparticle suspension must be purified from the desolvating and cross-linking agents. In addition, drug loadings may be low, unless specific binding of the drug to the carrier molecules or covalent linking occurs. To load the drug in the nanoparticles, the drug can either be dissolved in the aqueous phase before nanoparticle formation or it can be added to the cross-linked, but empty, nanoparticles.

Gelatin nanocapsules containing triamcinolone acetonide have been prepared by a O/W-emulsification method, which, in principle, is similar to a simple coacervation technique [56]. A solution of the drug in chloroform is emulsified into the gelatin solution under sonication. A solution of sodium sulfate is added to induce dehydration of gelatin and coacervation around the chloroform droplets. The capsule walls are hardened with glutaraldehyde. The capsular structure of the nanoparticles is confirmed by transmission electron microscopy. Holes in the shell are the result of the removal of chloroform. They can be avoided by restricting the applied pressure during the freeze-drying process.

Alginate nanoparticles have been prepared by the addition of an optimized amount of calcium chloride to a sodium alginate solution [57,58]. This so-called microgel can be cross-linked with the oppositely charged polymer, poly-*l*-lysine, to form nanospheres. This technique uses the well-known gelification phenomenon of sodium alginate with Ca-ions, however, at much lower concentrations than normally used for gel formation.

Again, since only one phase is used, high drug loadings are difficult to obtain unless specific binding of the drug to the anionic polymeric carrier occurs. More than 50 mg doxorubicin, a cationic drug, can be loaded onto 100 mg sodium alginate. The drug can probably be released by an ion-exchange mechanism, which generally occurs quite rapidly.

IV. NANOPARTICLES PREPARED THROUGH POLYMERIZATION OF MONOMERS

The methods used to prepare nanoparticles by polymerization are derived from methods developed to prepare latexes. The monomers, which are primarily of acrylic origin, can be either dissolved or emulsified into the continuous phase, with the resulting polymer being insoluble in either case. The preparation of nanoparticles by polymerization reactions is generally classified according to the resulting polymer or the resulting particle structure. In this chapter, the preparation methods are divided into methods resulting in either nanospheres or nanocapsules.

A. Nanospheres

1. *Poly(alkyl methacrylate) Nanoparticles*

These nanoparticles are prepared by polymerization of alkyl methacrylate monomers. The most frequently used monomer, methyl methacrylate, has a water solubility of approximately 1.5%. After dissolving it in the aqueous phase, the polymerization can be initiated either chemically by the addition of initiators such as ammonium or potassium peroxodisulfate at elevated temperatures [59] or by high-energy radiation [60]. The last method has the advantage that no additional agents are needed. This simplifies the purification procedures. Nucleation of the polymerization is initiated directly in the monomer solution. Above a certain molecular weight, the oligomers precipitate to form aggregates that are stabilized by surfactant molecules. The nanoparticles are formed through the growth of the aggregates [61].

The molecular weight and the particle size of the nanoparticles increase with increasing monomer concentration and decrease slightly with increasing temperature and with increasing initiator concentration [59,62]. Above a certain temperature or initiator concentration, the number of nucleating radicals remains constant, therefore resulting in a constant number of particles. Additional monomer, however, will increase the molecular weight of the resulting particles. Although poly(methyl methacrylate) (PMMA) nanoparticles are generally produced without surfactants, a more homogeneous particle size distribution can be obtained with the addition of hydrophilic macromolecules to the aqueous phase [2]. These molecules may also be bound to the surface of the nanoparticles.

The active substance can be added to the aqueous phase either prior, during, or after the polymerization reaction or it can be chemically linked to the polymer particles. An overview of the incorporation modes and the entrapment efficiencies is described by Allemann et al. [63]. Hydrophilic drugs are predominantly added/adsorbed to the already formed nanoparticles, whereas lipophilic drugs can be dissolved in the monomer or in an organic solvent miscible with the monomer but immiscible with the aqueous phase. When adding drugs to the already formed nanoparticles, an important factor is the compatibility of the drug with the polymer dispersion. Potential interactions between the drug and polymer or surfactant can result in flocculation of the polymer suspension.

Methyl methacrylate has been copolymerized with various other monomers, such as hydroxypropyl methacrylate or methacrylic acid, with the intention of preparing nanoparticles with different hydrophilicities in order to modify their circulation time and body distribution [64].

Thermal- and pH-sensitive poly(*N*-isopropylacrylamide-co-methacrylic acid) nanospheres have been prepared by a dispersion polymerization process [65]. The monomer mixture is dissolved in water. Potassium persulfate is used as an initiator, sodium lauryl sulfate as a stabilizer, and *N,N'*-methylenebisacrylamide as the cross-linking agent. The volume phase transition of the nanoparticles (swelling, shrinking), as measured by transmittance, is pH and temperature sensitive. It can be controlled by the ratio of the two monomers.

The research on PMMA nanoparticles has declined in recent years because of their lack of acceptable biodegradability. More rapidly degrading polymers, such as polyesters or poly(cyano acrylates), are required for intravenous drug delivery. PMMA-nanoparticles are primarily used for basic body distribution studies in animals in order to determine the fate of the nanoparticles within the body and the distribution in various organs as a function of time. This allows conclusions to be drawn with respect to site-specific drug delivery. In addition, they have also been used as adjuvants for vaccines [66]. The antigens can be added either prior to the polymerization reaction, or, in order to avoid potential destruction of the antigen, to the already-polymerized nanoparticles. The adjuvant effect increases with decreasing particle size and increasing hydrophobicity of the nanoparticles.

2. *Poly(alkyl cyanoacrylate) Nanoparticles*

In recent years, the interest in polymeric nanoparticles has shifted from poly(alkyl methacrylate) to poly(alkyl cyanoacrylate) nanoparticles because of their biodegradability and a simpler polymerization procedure [67]. They are prepared by emulsion polymerization (Fig. 6).

Emulsion polymerization is a frequently used method for the preparation of nanoparticles from monomers [2]. In this method, a monomer is emulsified in an immiscible external aqueous phase containing a surfactant. Above the critical micelle concentration, micelles form and are able to solubilize the monomer molecules. In addition to micelle formation, the surfactants also adsorb on the monomer emulsion droplets and stabilize the emulsion and then the polymeric nanoparticles. The polymerization reaction can be initiated within the micelles, or, with more soluble monomers, also in the continuous phase. In the first case, the monomer molecules diffuse from the emulsion droplets through the aqueous phase to the micelles. The solubilized monomer molecules within the micelles are then polymerized to form the polymer dispersion. Polymerization and chain growth are maintained by further monomer molecules diffusing to the growing polymer. The emulsified monomer droplets act as a reservoir for the monomer. After the polymerization reaction, the emulsifier molecules stabilize the colloidal polymer dispersion against physical instability (coagulation, flocculation, coalescence). Later it was found that emulsion polymerization could be carried out without any emulsifier molecules being present. It was concluded that the initiation of the polymerization occurred in the solvent phase. The polymerization is, therefore,

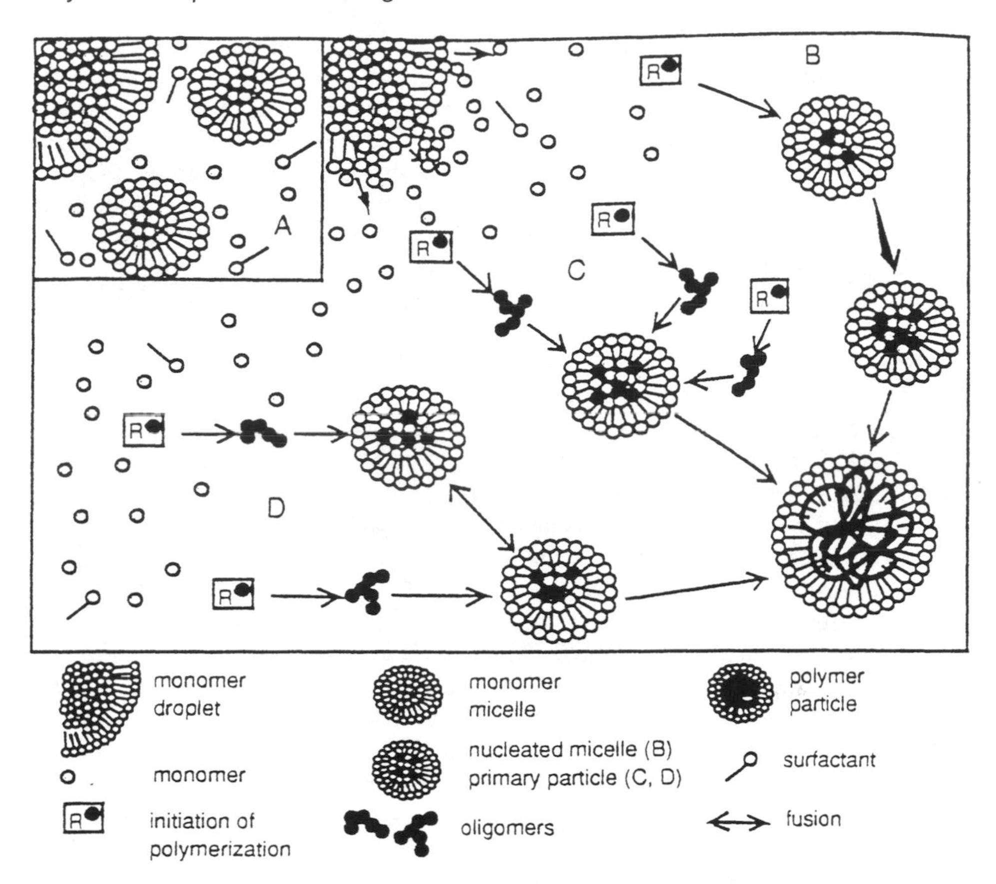

Fig. 6 Mechanism of emulsion polymerization. A: structure of initial emulsion. B, C, D: emulsion polymerization. B: initiation by a micellar nucleation mechanism. C, D: initiation by a homogenous nucleation mechanism associated with the growth of primary particles. C: by oligoradical capture. D: by flocculation. (Reproduced with permission from Ref. 61.)

initiated with dissolved monomer molecules. Initially, the growing polymer molecules are still dissolved in the external phase. After reaching a critical molecular weight, the molecules become insoluble, and phase separation and nanoparticle formation occurs. A nanoparticle, therefore, consists of a large number of individual polymer molecules.

In general, alkylcyanoacrylate monomers have a lower water solubility than alkyl methacrylate monomers. The nanoparticles are, therefore, prepared by emulsion polymerization, whereby the water-insoluble monomers are emulsified into the external phase under agitation. The monomer droplets are usually stabilized through the addition of surfactants or steric stabilizers to the aqueous phase (e.g., 1% w/w dextran 70, poloxamers). The nanoparticles are formed through an anionic polymerization reaction, initiated by the OH-ions present in water [2,3]. This reaction proceeds very rapidly. To decrease the polymerization rate, the pH of the aqueous phase is adjusted to acidic values (pH 3.5) with diluted hydrochloric or citric acid. The pH-value of the aqueous phase influences the particle size and the molecular weight of the nanoparticles; the lower the pH, the smaller the particles and the lower the molecular weight [68]. Bubbling sulfur dioxide through the monomer shortly before the polymerization reaction produces even smaller particles [69]. The concentration of sulfur dioxide is varied through the bubbling time. Increasing the concentration of sulfur dioxide from 0 to 3% based on the monomer results in a decrease in nanoparticle size from 160 nm to 18 nm. Zeta potential measurements show a significant increase in negative charge in the presence of sulfur dioxide. The higher net charge also explains the better physical stability of the polymer dispersion. In addition, the nanoparticles can be prepared without the stabilizer, dextran, which can potentially cause anaphylactoid reactions.

To enable transendothelial passage, nanoparticles smaller than 50 nm have been prepared by optimizing the surfactant concentration (Pluronic® F-68, investigate range, 0.2–10%) and the pH of the aqueous phase [70]. The particle size decreases with an increase from 0.2 to 3% Pluronic® F-68, and then remains fairly unchanged at higher surfactant concentrations. Smaller nanoparticles are obtained at lower HCl concentrations, which is consistent with the anionic polymerization mechanism, and lower monomer concentrations. The addition of cosolvents, such as acetone or methanol, to the aqueous phase in order to solubilize poorly soluble drugs increases the nanoparticle size dramatically, for example to more than 300 nm at 30% methanol. Both hydrophilic (ampicillin) and lipophilic (dexamethasone) drugs can be adsorbed efficiently.

Stabilizers, which are added to stabilize the monomer emulsion and also the nanoparticle suspension, can significantly influence the particle size and the molecular weight of the nanoparticles [71,72]. The molecular weight varies between 15,300 for Pluronic® F-68 and 4,177 for Triton® X100 [68], and increases with increasing surfactant concentration. A higher molecular weight can also be the result of the incorporation of the stabilizers (dextrans or poloxamers) into the polymer chain [72]. The particle size of the nanoparticles can also be affected by the monomer concentration and the stirring speed. Interestingly, the particle size grows with increasing stirring rate [71]. This is attributed to the higher kinetic energy overcoming the interfacial energy barrier of the particles, thus resulting in coalescence. Colloidal polymer dispersions may be converted into a redispersible powder by freeze-drying. Freeze-drying and redispersion of the nanoparticles in water result in similar particle size distribution and drug loading when compared to the original polymer dispersions [73].

Among colloidal polymeric drug carriers, alkyl cyanoacrylate nanoparticles have been most extensively investigated, both in vitro and in vivo. A summary of the drugs used and the incorporation method and encapsulation efficiency is given in Allemann et al. [63]. As with the other polymerization methods, the drugs can be incorporated prior or after the polymerization reaction. The encapsulation efficiency has been shown to be higher with monomers having longer side chains and by adding the drug prior to the polymerization reaction [74–76]. Lipophilic drugs can be incorporated by dissolving them in the monomer or in an organic solvent to be added to the monomer. The encapsulation efficiency depends on the partitioning behavior of the drug between the monomer/polymer and the aqueous phase. It generally decreases with increasing theoretical drug loading (the amount of drug added to the polymerization medium). Ionizable hydrophilic compounds can be entrapped within cyanoacrylate nanoparticles through the formation of ion pairs with oppositely charged counterions. Vidarabine has been incorporated in the presence of the anionic surfactant, dioctylsulfosuccinate [77] and antisense oligonucleotides have been complexed with hydrophobic cations such as quaternary ammonium salts. The oligonucleotides are added to the already polymerized nanoparticle suspension [78]. Only 1–2% of oligonucleotides are adsorbed in the absence of the counterions, whereas complete adsorption can be achieved in the presence of the counterions. It is suggested that the formed ion pairs are adsorbed on the nanoparticles by hydrophobic interactions between the hydrophobic moiety of the cation and the alkyl chains of the poly(cyanoacrylate) polymer. The extent of the adsorption increases with increasing oligonucleotide chain length and depends on the hydrophobic character of the counterions.

Harmia et al. have investigated the sorption of pilocarpine onto nanoparticles as a function of polymer type, counterion, added electrolyte, and surfactants [79]. The adsorption follows Langmuir and Freundlich isotherms and is enhanced by using the less soluble drug salt. Nanoparticles with a high specific surface area are prepared from less hydrophobic polymers and an increasing electrolyte concentration.

The time at which the peptide, a growth-hormone–releasing factor, is added after the initiation of the polymerization reaction is critical [80]. If it is added shortly after the polymerization is started, it is highly associated with the polymer; however, it also becomes covalently bound to the polymer. With regard to the drug release mechanism, the peptide is released by erosion of the polymeric matrix rather than by passive diffusion through the matrix. In the absence of esterases, no drug is released.

High encapsulation efficiencies at high drug loadings, as is frequently obtained with microparticles, are more difficult to achieve with colloidal drug carriers. The solids content of the polymer dispersion is often quite dilute ($< 1\%$ w/w).

The biodegradability of cyanoacrylate nanoparticles can be controlled by using different monomers. The rate of biodegradation is inversely proportional to the length of the side chain length [81]. For this reason, butyl, isobutyl, or isohexyl cyanoacrylate monomers are preferably used today rather than methyl and ethyl cyanoacrylate monomers. The degradation of the nanoparticles has been described to be controlled by surface erosion [82]. The side chains are enzymatically hydrolyzed. The backbone is not degraded, but gradually becomes more hydrophilic and ultimately water soluble. It has been shown that the drug release from nanospheres is controlled by the erosion of the polymer [80,83]. In addition, the toxicity of the monomers decreases with increasing chain length. The degradation of the nanoparticles will also be affected by the molecular weight of the polymer and its particle size. The pH of the degradation medium also affects the degradation. Poly(ethyl cyanoacrylate) nanoparticles do not show significant degradation at pH 1.2, but degrade more rapidly with an increase in the pH of the medium [84]. This may have implications, in particular for the oral application of nanoparticles due to the pH gradient in the gastrointestinal tract. Nanoparticles from copolymers can be prepared by merely mixing different monomers [85].

The polymerization reaction is an anionic process; basic drugs can, therefore, act as polymerization starters [77]. They can be incorporated into the polymer chain. If the drug-monomer interaction is unwanted, the drug can be added to the finished nanoparticles after polymerization.

To alter the natural body distribution of the nanoparticles (passive targeting) and their rapid clearance through the reticuloendothelial system, colloidal magnetites have been incorporated into polycyanoacrylate [86] or albumin nanoparticles [50,51]. After intravenous administration, the nanoparticles can then be trapped in specific locations with an external magnetic field. The body distribution can also be altered or prolonged through the coating of the nanoparticles with, for example, hydrophilic surfactants (active targeting). In analogy to stealth liposomes or PLA nanoparticles, PEG has been adsorbed to poly(alkyl cyanoacrylate) nanoparticles during the nanoparticle formation [87].

3. Polyacrylamide Nanoparticles

Nanoparticles can be prepared from water-soluble monomers emulsified in an external organic phase [88–90]. For example, polyacrylamide nanoparticles have been prepared by solubilizing acrylamide and the cross-linking agent, *N,N'* bisacrylamide with surfactants in an organic phase (W/O microemulsion). The polymerization reaction can be initiated within the inverse micelles chemically by using *N,N,N',N'*-tetramethylethylene-diamine and potassium peroxodisulfate as initiators or through gamma-, ultraviolet-, or light-irradiation. The drugs are generally added prior to the polymerization reaction. This method is primarily suitable for water-soluble drugs being insoluble in the external phase. The encapsulation efficieny depends on the distribution of the drug between the external phase and the internal monomer/nanoparticle phase.

The drawback of the method of preparing polyacrylamide nanoparticles is the use of toxic monomers, organic solvents (e.g., hexane, chloroform), and surfactants necessary to stabilize the microemulsion. Purification techniques are required to remove these toxic components and, if used, the initiators. Because of these factors and the nondegradability of the particles, interest in this technique is decreasing.

B. Nanocapsules

In anology to nanospheres prepared from preformed polymers by the interfacial polymer deposition mechanism, nanospheres from cyanoacrylate monomers can be prepared based on a similar principle [91–93]. In the interfacial polymerization technique, the alkyl cyanoacrylate monomer (e.g., isobutyl cyanoacrylate) and the drug are dissolved in a water-miscible solvent (e.g., ethanol) and an oil (e.g., Miglyol®, ethyl oleate, phospholipids, or benzyl benzoate). This organic phase is then added slowly to an external aqueous phase (+ stabilizer) under agitation. The organic solvent diffuses rapidly in the aqueous phase, and the monomer polymerizes at the droplet/water interface by the anionic polymerization mechanism described above. Because of the oil, polymerization occurs only at the interface and not throughout the droplet. Interestingly, bulk polymerization of the monomer in the oil droplets, resulting in a matrix and not a reservoir system, occurs in the absence of ethanol. This matrix system has the oil dispersed throughout the polymeric matrix. The water-miscible cosolvent, ethanol, is needed for the formation of the nanocapsules (core-shell structure). The monomer is transported to the interface by the diffusion of ethanol from the oil to the external aqueous phase. The cosolvent must therefore be miscible with the oil and water and also be a solvent for the monomer. The formation of the nanocapsules is a spontaneous process, the particle size of the resulting nanocapsules is independent of the agitation rate. A more detailed description of the formation mechanism of the nanospheres can be found in Puisievy et al. [3]. The process usually results in the formation of nanocapsules as well as nanospheres, which have a smaller size than the nanocapsules but a similar molecular weight. The different types of nanoparticles can be separated by ultracentrifugation according to differences in density and are then identified by transmission electron microscopy [3]. Nanocapsules with the oil core have a lower density. The drug release from the nanocapsules is slower when compared to the release from nanospheres of similar particle size.

The ethanol is removed under vacuum, which can also be used to further concentrate the polymer dispersion. After solvent removal, the nanoparticles are filtered through a sintered glass filter to remove larger aggregates of polymer or free drug crystallized in the aqueous phase. The resulting nanocapsules have a mean diameter between 200 and 300 nm and have an oily core surrounded by a polymeric shell. A polymerization inhibitor (sulfur dioxide) has to be added with the less toxic isohexyl cyanoacrylate, since the monomer reacts immediately in the presence of ethanol. The interfacial polymerization technique allows the encapsulation of oil-soluble drugs, which is more difficult with nanospheres prepared by emulsion polymerization technique. The addition of the oil affects the partition coefficient nanoparticle/aqueous phase and therefore the drug loading. The particle size is determined by the amount of oil used; lower concentrations of oil result in smaller particles [94]. In general, because of the oil, the particle size of nanocapsules is larger than comparable nanospheres. Surfactants are not needed for the preparation of the nanocapsules, but they are needed for stabilization of the colloidal dispersion during storage against aggregation into a nonredispersible cake. With this technique, the amount of unreacted monomer within the nanocapsules can be potentially higher when compared to nanospheres prepared by emulsion polymerization.

A nonaqueous interfacial polymerization technique, in which the monomer is not dissolved in the internal phase, as described above, but in an external organic phase has been developed by Krause et al. [95]. The drug is dissolved in alkanized methanol; this solution is then emulsified under sonication into isooctane containing a nonionic, low HLB-sorbitan ester surfactant. Methyl cyanoacrylate is added to the external phase and polymerizes at the methanol/isooctane interface. The drug has to be insoluble in the external isooctane phase in order to get high drug loadings and encapsulation efficiencies. Besides nanocapsules, nanospheres are also formed.

Water-soluble drugs can be entrapped within nanocapsules by dissolving the drug in water, followed by emulsification into an external organic phase (e.g., chloroform, cyclohexane) to form a W/O emulsion [96,97]. The monomer is then added to the organic phase to induce polymerization. Since the monomer is added to the external phase, the residual monomer concentration in the nanocapsules is probably lower than with the interfacial polymerization method described above, whereby the monomer is added to the internal phase. However, large amounts of organic solvents used as external phase have to be removed. Similarly, cyanoacrylate nanoparticles are prepared from a W/O microemulsion by using isopropyl myristate as the external phase [98]. The drug, which has been encapsulated with an external lipophilic phase, may be rapidly released upon contact with aqueous media. With the last two methods, the core consists of drug dissolved in either methanol or water, which, when compared with oil, can be easily removed during a drying process (freeze- or spray-drying). This can result, however, in the collapse of the particles.

Submicron nanocapsules were prepared by the formation of a thin polymeric shell on nm-sized particles by aqueous emulsion polymerization techniques. In principle, seed colloidal polymer particles, which were first polymerized, were used at high enough concentrations to prevent the formation of separate nanoparticles during the second polymerization reaction (formation of the shell) [99]. This appears to be an interesting technique for coating drug-containing nanoparticles formed in a first polymerization reaction with a polymer formed in a second polymerization reaction.

Nanocapsules can possibly also be formed with two miscible monomers (or monomers soluble in the same organic solvent) that separate during the polymerization reaction in a core-shell structure with the respective polymers.

V. CHARACTERIZATION OF NANOPARTICLES

Formulation and process parameters of the previously described preparation techniques strongly affect the performance of the colloidal polymer dispersions. As with all drug delivery systems, a precise characterization of the physico-chemical parameters is important in order to develop a drug product of high quality [1,4,100].

The parameters generally used to characterize nanoparticles include the particle size, morphology, and surface charge; the drug loading and encapsulation efficiency; the drug release profile; the physico-chemical state of the drug and polymer; molecular weight of the polymeric carriers; and in vitro biodegradation. Although less popular, the following techniques have been used to obtain further information, in particular on the surface properties of the nanoparticles: dye adsorption, hydrophobic interaction chromatography, aqueous two-phase partitioning, measurement of the dielectric constant, electron spectroscopy for chemical analysis (ESCA), secondary ion mass spectrometry (SIMS), small-angle x-ray scattering (SAXS), advanced microscopic techniques (atomic force and scanning tunnelling microscope). The major techniques are summarized in Table 1.

A. Particle Size, Morphology, and Charge

Nanoparticles are generally characterized by their average particle diameter, their size distribution, and charge. These parameters affect, for example, the physical stability and the in vivo distribution of the nanoparticles. The first two parameters are usually determined by photon correlation spectroscopy (PCS). Although it is now well-known that the in vivo fate of nanoparticles is primarily a function of their size, the shape of

Table 1 Major Techniques Used for Physico-Chemical Characterization of Nanoparticles

Parameter	Techniques
Particle size	Photon correlation spectroscopy, scanning electron microscopy, transmission electron microscopy, small-angle x-ray scattering
Morphology	Scanning electron microscopy, transmission electron microscopy, electron spectroscopy for chemical analysis, secondary ion mass spectrometry, small-angle x-ray scattering, atomic force microscopy, scanning tunnelling microscopy
Surface characteristics	Charge: laser Doppler anemometry, electrophoresis, dielectricity Hydrophobicity: hydrophobic interaction chromatography, contact angle
Drug loading	Classical analytical methods (UV, HPLC) after ultracentrifugation, ultrafiltration, gel filtration, or centrifugal ultrafiltration
Drug release	Dialysis, direct dilution + ultracentrifugation, ultrafiltration or centrifugal ultrafiltration
Physical state of drug and polymer	Differential thermal analysis, differential scanning calorimetry, x-ray diffraction
Molecular weight	Gel permeation chromatography

the nanoparticles may also determine their toxicity [4]. Electron microscopy techniques are very useful in ascertaining the overall shape of polymeric carriers. The surface charge of the nanoparticles affects the physical stability and redispersibility of the polymer dispersion as well as their in vivo performance. Coating nanoparticles with nonionic surfactants has a dramatic influence on their surface charge as well as on their hydrophilic/lipophilic surface properties and, therefore, on their opsonization mechanism and their fate upon parenteral injection [4,101–103].

1. *Photon Correlation Spectroscopy*

PCS (also referred to as QELS: quasi-elastic light scattering), a technique based on dynamic laser light scattering, allows the determination of the particle size of nanoparticles in the range of 5 nm to approximately 5 μm. It is an established technique, which is reproducible, rapid, accurate, and nondestructive. The PCS equipment consists of a laser source, a temperature-controlled sample cell and a photomultiplier for detection of the light scattered at a specific angle. More details about the theoretical aspects of PCS can be found in [4]. Nanoparticles are usually polydisperse in size, and PCS will give an average value of the particle size. Another parameter, the polydispersity index, is necessary to characterize the particle size distribution of the polymer dispersion. For a theoretically monodisperse colloidal suspension, the polydispersity index is zero. However, polymer dispersions with a polydispersity index between 0.03 and 0.06 are still considered monodisperse. A narrow size distribution, corresponding to a polydispersity index between 0.1 and 0.2, is generally found with colloidal drug carriers. Larger polydispersity indexes indicate a broad size distribution. Monodisperse polystyrene standards with different particle sizes are generally used to calibrate the PCS equipment.

Prior to determination of particle size, the samples have to be diluted to get the proper light scattering [104]. The dilution could cause changes in the particle size because of swelling or desorption of surfactants from the nanoparticle surface. To compare results, it is recommended to use similar experimental conditions.

For samples with broad particle size distributions, differences in particle size can be found by comparison with results obtained by electron microscopy. McCracken and Sammons [105] obtained similar mean diameters of liposomes using three different techniques (electron microscopy, PCS, ultracentrifugation), but the PCS technique displayed a broader distribution attributed to the presence of larger liposomes. Gallardo et al. (92) obtained similar results by PCS and transmission electron microscopy.

2. *Electron Microscopy Techniques*

Microscopic techniques allow determination of both particle size and distribution and observation of particle shape. In addition, other parameters, such as morphology or surface roughness of the nanoparticles, can be observed. Microscopic observations are very useful for the detection of free, nonentrapped drug crystals within the polymer dispersion [100]. The internal structure of polymeric nanoparticles can be studied through freeze fracture techniques [31]. However, this latter technique is more difficult and cannot be carried out on a routine basis. With colloidal polymer dispersions, scanning and transmission electron microscopy (SEM and TEM) have been used.

For SEM studies, the polymeric dispersions have to be first converted into a dry powder, which is then mounted on a stage followed by coating with a conductive metal. The specimen is scanned with a focused fine beam of electrons. The information on the surface characteristics is obtained from secondary electrons emitted from the specimen surface. The nanoparticles must be able to withstand a strong vacuum, and the electron beam can result in the "burning" of the polymer. SEM pictures of polymeric drug carriers generally reveal smooth and spherical particles. The mean size obtained by SEM is generally comparable with results obtained by PCS [92]. Drying of the polymeric dispersion during sample preparation can result in the precipitation of drug and the appearance of drug crystals with possible misinterpretations. In addition, other excipients in the aqueous phase (e.g., surfactants, viscosity enhancing agents, electrolytes) may lead to artifacts.

Although the principle of TEM is different from SEM (TEM uses electrons transmitted through thin specimens), the same type of information as with SEM can be obtained by this technique. The colloidal polymer dispersions are usually placed on a collodion support on copper grids and are negatively stained with phosphotungstic acid or derivatives. This technique, alone or in combination with freeze fracture procedures, is useful for revealing structural differences between nanospheres and nanocapsules. Nanospheres display a continuous or porous matrix [106], whereas nanocapsules have a core-shell structure with a very thin polymeric wall of a few nanometers [31]. Fessi et al. were able to determine the thickness of the polymeric wall of PLA nanocapsules to be approximately 10 nm with TEM (31). Nanocapsules were larger when compared to nanospheres (92).

Although TEM is an excellent technique, it is time-consuming and difficult to carry out on a routine basis.

3. *Surface Charge: Laser Doppler Anemometry*

Laser Doppler anemometry provides useful information on the surface charge of colloidal particles through the measurement of the zeta potential. The zeta potential is an indirect measure of the surface charge: it measures the potential difference between the outer Helmoltz plane and the surface of shear. The technique is based on the velocity of the nanoparticles within an applied electric field. The velocity of the particles is evaluated by the shift caused in the interference fringe, which is produced by the intersection of two laser beams. This electrophoretic mobility is then transformed in zeta potential values with the help of the Smoluchowsky equation [4]. Data on ionic strength are often missing in the literature. The ionic strength should be kept constant in order to compare results from different experiments. It is also well-known that the pH influences zeta potential measurements, particularly for particles having ionizable functional groups. It is very important that ionic strength and pH are kept constant in order to correlate zeta potential measurements.

Most colloidal particles have negative zeta potential values ranging from about −100 to −5 mV. When polymeric carriers are prepared with specific excipients, for example cationic diethylaminoethyl-Dextran, then zeta potential measurements may result in positive values [94]. The use of nonionic surface active agents in the preparation of nanoparticles may lead to zeta potential values close to neutrality. The surfactants are useful for the preparation of long circulating nanoparticles, since strongly negative charged particles are rapidly removed from the blood circulation by the mononuclear phagocyte system [4,107]. However, high surface charges are also responsible for the physical stability of the colloidal polymer dispersions, since the repulsive forces prevent aggregation upon aging. High potential values (< -30 mV) should be achieved in order to ensure a high-energy barrier [108].

B. Drug Loading

The drug loading of the nanoparticles is generally defined as the amount of drug bound per mass of polymer (usually moles of drug per mg polymer or mg drug per mg polymer); it could also be given on a percentage basis based on the polymer. The encapsulation efficiency refers to the ratio of the amount of drug encapsulated/adsorbed to the total (theoretical) amount of drug used. With regard to the final drug delivery system—the dispersion of the nanoparticles in an aqueous phase—it is also interesting to express the drug loading as the amount of drug per mL of polymer dispersion.

The drug loading of the dispersion can be fairly easily determined by extracting a certain amount of polymer dispersion and quantitating the amount of drug. This drug loading value, however, includes the drug entrapped within the nanoparticles and also free, nonentrapped drug in the external phase. To determine the loading of the nanoparticles, the nanoparticles have to be separated from the aqueous phase prior to their analysis for drug content. Because of the colloidal size of the polymer phase, colloidal polymer dispersions are physically stable and more difficult to separate when compared to a multiparticulate system in the μm-size range, for example microspheres. The techniques used to determine the drug loading of the nanoparticles are based on the separation of the dispersed nanoparticles from the continuous aqueous phase. The two phases should be separated as rapidly as possible, since there is an equilibrium between adsorption and desorption. A long separation process may lead to drug loss of the adsorbed (weakly bound) drug. It is also important to characterize the influence of temperature on the adsorption/incorporation process, since nanoparticles are often prepared at room temperature, centrifugated at 4°C and administered in vivo at 37°C. The partitioning of the drug can be temperature sensitive, and the drug loading values can be affected by changing temperature conditions.

The following techniques are used routinely in many laboratories to determine the drug loading.

1. *Ultracentrifugation*

The most popular separation technique for polymer dispersions is probably ultracentrifugation. In this method, the nanoparticles are separated from the aqueous phase at high centrifugation speeds (from 40,000 to 100,000 g for 0.5 to 1.5 h). Due to the strong centrifugal forces, most colloids settle during the process. This is the case for nanospheres. Nanocapsules, depending on the density of the incorporated oil (< 1 for olive oil or Miglyol® 812), float on the top of the aqueous phase [33]. After careful separation of both phases, the unencapsulated drug can be assayed in the aqueous phase by standard analytical techniques (e.g., UV, HPLC).

It is generally assumed that the amount of drug encapsulated within the nanoparticles is equal to the difference between the total amount of drug used and the amount of free drug in the aqueous phase. The total amount of drug used can either be calculated based on the initial amount of drug added to the preparation medium or obtained by completely dissolving the drug-containing polymer dispersion in an appropriate solvent such as, for example, acetone for poly(e-caprolactone) nanoparticles. The latter method can potentially lead to misleading results of the encapsulation efficiency because of the possibility of drug adsorption to glassware, etc. It is recommended that the amount of drug in the supernatant as well as in the nanoparticle sediment be determined. This mass balance study will reveal any degradation or sorption process by comparing the actual total amount of drug present in the polymer dispersion with the theoretical amount added. The amount of drug in the nanoparticle sediment can be determined by extraction or dissolution/drug analysis. Due to the potential difficulty of redisolving or extracting the drug, a simpler method to quantitate the drug loading in the nanoparticle sediment is the use of labeled (generally ^{14}C) compounds.

It should be noted that unencapsulated drug can not only be dissolved in the aqueous phase but can also be present in the form of drug crystals. Drug crystals in the aqueous phase have been observed in polymer dispersions prepared by the solvent evaporation method, especially at high theoretical drug loadings [12]. The drug diffuses in the external aqueous phase and precipitates after solvent removal. Ultracentrifugation does not result in the separation of the drug crystals from the nanoparticles: the drug crystals can also settle and be present in the nanoparticle sediment. Microscopic methods (e.g., TEM or observation under polarized light) are useful for detecting the presence of drug crystals [100]. A possible alternative to separating the drug crystals from the polymer dispersion can be an initial centrifugation step at lower centrifugation speeds, during which the nanoparticles remain suspended and the drug crystals settle. The polymeric dispersion in the supernatant can then be separated into the nanoparticles and the aqueous phase by ultracentrifugation after removal from the drug crystals.

Another potential problem associated with ultracentrifugation is the influence of temperature on the drug distribution between the nanoparticles and the external phase. Nanoparticles are generally centrifuged at low temperatures (below 10°C), while they are prepared and stored at room temperature. The solubility of the drug in the external or polymer phases or drug sorption/desorption processes can be temperature dependent. It should be verified that the system behaves similarly at different temperatures. In addition, ultracentrifugation is a destructive method, since it is difficult to resuspend the nanoparticle sediment into the original polymer dispersion.

2. *Ultrafiltration*

In this technique, free (nonencapsulated) drug is determined in the clear supernatant following the separation of the nanoparticles by ultrafiltration with membrane filters having a pore size between 10 and 20 nm. Centrifugal ultrafiltration and ultrafiltration at low pressure are primarily used.

In the centrifugal ultrafiltration technique, the use of filters such as Ultrafree®-MC (Millipore®) or Ultrasart® 10 (Sartorius) allows the separation of the colloidal particles from the aqueous medium with the help of classical centrifugation techniques. The samples are placed in tubes and during centrifugation (12,000 g for 30 min) the aqueous phase is forced through the filter because of centrifugal forces [109]. These filters allow the use of small volumes of polymer dispersions, which are very useful for potent or expensive agents. However, the small sample volume may also result in volumes too little for analytical assays.

Ultrafiltration at low pressure, which is used for both nanoparticles and emulsions, requires a larger volume (40 mL) of the nanoparticle suspension. The polymer dispersion is placed in a stirred vessel under positive nitrogen pressure (less than 0.5 bar). The use of Y. M. 10 Amicon ultrafiltrate membranes allows samples of about 2 mL to be collected continuously until 20% of the total volume has been filtered [108]. This method could also be used to determine the drug release from nanoparticles, if the polymer dispersion is added to the dissolution medium before the pressure is applied. Ultrafiltration at low pressure is faster than the classical ultracentrifugation technique.

For both techniques, the potential risk of drug sorption to the filter should be addressed; it is often necessary to saturate the filter membrane with drug [110]. In addition, it is difficult to recover the particles from the filter membrane, which could also be clogged.

3. Gel Filtration

The separation of the nanoparticles from the unencapsulated drug is based on the differences in size between the polymeric particles and the free drug. The larger nanoparticles are eluted rapidly while the drug is retained longer and, therefore, eluted later [111]. The appearance of the nanoparticles in the eluent can be observed visually or by means of a turbidimetric method. Once the separation has been performed, the drug concentrations in the aqueous samples can be determined. The eluted samples containing the nanoparticles can be assayed for the actual drug loading after dissolution/extraction with appropriate solvents.

One critical variable of the gel filtration technique is the choice of a suitable gel that is able to separate the two phases. The use of Sephadex® and Sepharose® has been reported for poly(butyl cyanoacrylate) [111] or poly(lactide) nanoparticles [112], respectively. With gel filtration, there is a risk of underestimating the actual drug loading. Gel filtration is based on the continuous flow of solvent through the column and the sample. This can result in the extraction (release) of drug from the nanoparticles upon contact with the elution medium, resulting in lower than actual drug loadings. This problem can be minimized by reducing the elution time for the nanoparticles through the coupling of the column to a middle-pressure chromatographic pump. Using such equipment, Beck et al. reported the separation of loaded nanoparticles from the free drug in about 35 min [110].

C. Drug Release Studies

Depending on the desired use, nanoparticles should be able to release the drug rapidly or in a controlled fashion. The release profile may indeed be very different if one wishes either a slow release of the drug in the blood like with stealth nanoparticles or a fast release in the vicinity of the targeted area.

The selection and optimization of drug release studies are a problem with colloidal polymer particles. While the dissolution medium can be easily separated from most polymeric drug delivery systems (e.g., tablets, capsules, microspheres), this is not the case with colloidal polymer dispersions. It is generally necessary to separate the release medium from the nanoparticles by either a membrane or ultracentrifugation.

1. Dialysis Techniques

Dialysis techniques are very popular in order to study the release of drugs from colloidal polymer dispersions. They can be divided into two major methods, namely the classical dialysis technique and the bulk-equilibrium reverse dialysis technique.

In the classical dialysis technique, the nanoparticle dispersion is placed in the dialysis bag, which is then hermetically sealed and dropped into the dissolution medium. The released drug diffuses through the dialysis membrane and is then assayed in the sampled dissolution medium. The release studies are generally carried out in an aqueous environment. Sink conditions must be maintained. Depending on the solubility of the drug, cosolvents (e.g., ethanol, PEG) may be used as well as nonaqueous media. Hubert et al. used PEG 600 to study the release of darodipine [113]. Serum as well as human gastric juice have also been used to mimic in vivo conditions [113].

Potential drawbacks of the dialysis method were reported by Washington [114]. The colloidal polymer dispersions are not significantly diluted by the dissolution medium inside the bag, which can result in sink problems. The release rate of the drug and its appearance in the dissolution medium is governed by the partition coefficient of the drug between the polymeric phase and the aqueous environment within the dialysis bag, and by the diffusion of the drug across the membrane. Certain drugs, in particular lipophilic drugs, can have a high affinity for the dialysis membrane. The drug release can then be controlled by the membrane and not by the colloidal drug carrier. This technique allows the comparison of different formulations and is useful for quality control.

The second method has been reported by Levy and Benita and is referred to as the bulk-equilibrium reverse dialysis technique [115]. In this method, the colloidal polymer dispersion is diluted with the release medium so that sink conditions can be maintained. Small dialysis bags containing 1 mL of buffer are then added to the nanoparticle-containing dissolution medium. The released drug diffuses into the bag. The dialysis bags are removed at predetermined time intervals, the amount of drug released is determined, and the release curve can be drawn. The main drawbacks of this technique are that it is not sensitive enough to characterize rapid release from nanoparticles and the potential drug binding to the membrane.

2. *Other Methods*

Because of the potential problems with the membrane methods described above, methods without membranes are preferable under certain conditions. The nanoparticle dispersion is added to the dissolution medium, and samples of the diluted dispersion are drawn at specified time intervals and analyzed for the amount of drug released. The amount of drug release can only be determined after separation of the nanoparticles from the aqueous phase, which obviously should occur as rapidly as possible to prevent further drug release in the dissolution medium after sampling.

The dissolution medium and the dilution factor (reflecting the amount of dissolution medium added) have to be carefully selected. There is a large variety of dissolution media available, ranging from water to plasma or serum. The choice of the dilution factor is important since it should reflect the potential therapeutic use of the polymeric dispersion. The requirements of the dilution factor depend on the potential target, the volume of colloidal polymer dispersion administered, and, eventually, the animal model. The dilution factor needs to be different, for example, for nanoparticles that are intended for either intravenous or ocular delivery. Dilution factors of at least 100 should be used if the polymeric dispersion is intended to be administered intravenously. On the contrary, a low dilution factor (<1) should be used if the dispersion is intended for ocular delivery.

To determine the amount of drug released, the nanoparticles have to be separated rapidly from the dissolution medium. The same techniques (e.g., ultracentrifugation, ultrafiltration, or gel chromatography) already described for the determination of the drug loading can also be used here. The separation needs to be immediate to avoid further drug release. Ultracentrifugation and ultrafiltration techniques appear to be more appropriate than a gel chromatography method. However, the ultracentrifugation time (usually more than 30 min) and the risk of filter clogging have to be taken into account.

A new ultrafiltration technique at low pressure, which is able to discriminate between the drug release of different colloidal formulations, was proposed by Magenheim et al. [116]. The method is based on the direct dilution of the nanoparticles with the release medium. The amount of drug released from the carrier is forced by pressure through the membrane. The filtrate is sampled and assayed at set time intervals. A different release behavior between submicron emulsions and nanoparticles is observed. The drug model, miconazole, is immediately released under sink conditions from microemulsions while sustained release is obtained with nanoparticles containing indomethacin.

With regard to the nonstatic conditions observed in vivo, continuous flow methods are a viable option. The colloidal polymer dispersion is added into an ultrafiltration cell and the dissolution medium is continuously forced through the ultrafiltration membrane and analyzed. However, there is a risk of clogging the filter with this technique [114].

In conclusion, there are several methods available to study drug release from nanoparticle suspensions. They all have advantages and disadvantages. Rather than obtaining "absolute" drug release parameters, the different techniques offer a broad range of comparisons of formulations and can also be used as a tool to check the reproducibility of preparations. There is a need for a better correlation between in vivo and in vitro results in recommending one method over another.

D. Characterization of the Physico-Chemical State of Drug and Polymer in Nanoparticles

It is of primary importance to determine the physical state of both the drug and the polymer, since these states influence the in vitro and in vivo release characteristics as well as the aging behavior of the polymer dispersions. A drug may be present either as a solid solution or a solid dispersion in an amorphous or crystalline polymer. Other combinations with semicrystalline polymers can also occur.

To determine the physico-chemical state of the drug and the polymer, the samples are mainly analyzed in solid form after drying the polymer dispersions. Basic techniques include thermal analysis or x-ray diffraction. In colloidal polymer dispersions, there is an equilibrium between the adsorbed/entrapped drug and the drug dissolved or even dispersed in the external aqueous phase. During the drying process, free drug in the aqueous phase may precipitate either as discrete separate particles or directly on the carriers surface. Care should therefore be taken to discriminate between the drug entrapped in or adsorbed to the polymer and precipitated drug. One way to overcome the problem of dissolved drug precipitating during the drying process is to separate the nanoparticles from the aqueous phase prior to the analysis of the nanoparticle sediment. However, solid drug crystals originally present in the aqueous medium can also be present in the sediment. They have to be separated from the nanoparticles.

1. Thermal Analysis

Thermal analysis covers a group of techniques that measure a physical property of a substance as a function of temperature. It includes, among others, differential thermal analysis (DTA) and differential scanning calorimetry (DSC)—DTA measures the temperature differences between a reference and the sample, whereas DSC measure the quantity of heat necessary to maintain the sample and the reference at the same temperature. Crystalline transitions, fusion, evaporation, and sublimation are easily quantified on DSC curves. Because of small samples (a few mg) and rapid experimentation, thermal analysis has found widespread applications in both research laboratories and routine quality control.

With colloidal polymer dispersions, the drug within the polymeric matrix is probably molecularly dispersed. With the solvent evaporation method, drug in excess of the solubility of the drug in the polymer would precipitate in the external aqueous phase. It would be unusual to observe melting endotherms of the drug. This could be interpreted as the drug being in the amorphous but dispersed state or being dissolved in the polymer matrix, with the latter being the probable conclusion.

The thermal methods are destructive and could potentially yield wrong results. Drugs, which may be present in a crystalline state in the polymer, can dissolve in a polymer that melts at a lower temperature. The melting transition of a drug would not be visible and the wrong conclusion of the drug's being soluble in the polymeric system could be drawn. In addition, analyzing the thermal events of multicomponent mixtures like nanoparticles (e.g., drug, polymer, surfactants) can be complicated.

2. *X-Ray Diffraction*

When a monochromatic x-ray beam is focused on a crystal, the scattered x-rays from the regularly placed atoms interfere with each other, giving strong diffraction signals in particular directions, since the interatomic distances are of the same order as the x-ray wavelength. The directions of the diffracted beams are related to the shape and dimensions of the unit cell of the crystalline lattice. The diffraction intensity depends on the disposition of the atoms within the unit cell. This technique allows amorphous and crystalline materials to be differentiated. Crystalline materials display many diffraction bands, whereas amorphous compounds present a more or less regular baseline. The determination of the crystalline nature of the polymer is important, since crystalline polymers degrade much slower than amorphous ones. In the same polymer, amorphous as well as crystalline domains can be found and the degradation process, both in vivo and in vitro, can be different for the two domains [117].

There are only a few reports on the x-ray diffraction pattern of nanoparticles in the literature. Polyacrylic nanoparticles [118] and poly(butyl cyanoacrylate) nanoparticles [119] are amorphous and show no sign of crystallinity. The diffraction pattern of poly(butyl cyanoacrylate) nanoparticles containing pilocarpine is different from those of empty nanoparticles and of the drug. Some distinctive pilocarpine diffraction peaks disappear or decrease in intensity. This is in favor of a solid dispersion of the drug in the polymer matrix. Maincent et al. (unpublished data) confirmed by x-ray diffraction and DSC that poly(e-caprolactone) kept its semicrystalline characteristics in poly(e-caprolactone) nanoparticles prepared by the precipitation method.

E. Molecular Weight of the Polymeric Carrier

Almost all synthetic polymers are heterodisperse with respect to their molecular weight. They are generally described by an average molecular weight and by a molecular weight distribution. The molecular weight is a very important parameter since it will influence physico-chemical properties such as the diffusion rate of a drug or the degradation of the polymer [120].

There are different ways to express the molecular weight: number average molecular weight (Mn) and weight average molecular weight (Mw) are the most popular. The ratio Mw/Mn is a measure of the polydispersity of a polymer.

Size exclusion chromatography (SEC), also named gel permeation chromatography (GPC), is the standard technique for obtaining the molecular weight of polymers that are used as drug carriers. The separation is determined by a size exclusion mechanism in which the pore volume in the gel particles accessible to a given molecule is determined by the pore size distribution and solute size. The polymer molecules are fractionated according to their molecular weight and quantitated, generally with a RI-detector. Monodisperse polystyrene standards with a range of molecular weights are used to calibrate the column in the case of water-insoluble polymers. The molecular weight values of the polymers will, therefore, not have absolute values but relative ones with respect to the polystyrene standards. With hydrophilic polymers, poly(ethylene oxide) standards are used. The determination of the molecular weight is important with regard to the characterization of the nanoparticles. For example, the molecular weight of the polymer in nanoparticles prepared from preformed polymers by the solvent evaporation method may be different from the original molecular weight. Low molecular weight fractions can leach into the aqueous phase, or high shear during the preparation of the nanoparticles may result in the degradation of the polymer. The molecular weight of the polymer obtained after polymerization of monomers is an important characteristic. Molecular weight analysis is essential for following the in vitro and in vivo degradation of the nanoparticles. Potential degradation of the polymeric nanoparticles during storage may necessitate the conversion of the polymer dispersion in a redispersible powder. Polyesters such as polylactides or cellulose esters are prone to hydrolysis. The effect of gamma-irradiation, a sterilization method, on molecular weight has also been reported [121].

1. In Vitro Degradation of Nanoparticles

The goal of the degradation studies is to obtain useful information about the chemical stability of the polymer and particularly on the influence of the biodegradation of the nanoparticles on the drug release. These studies can be carried out at different temperatures (for example -20°C, 4°C, 25°C, 37°C, 40°C) of the polymer dispersion to obtain results on the chemical stability of the polymer in an aqueous environment. Chemical instability can also result in physical stability problems of the dispersion. The in vitro degradation can be followed with the previously described methods. The quantification of the in vitro degradation is related (not limited) to changes in polymer molecular weight, in particle size, in surface potential, and in overall shape. For example, poly(butyl cyanoacrylate) nanoparticles have been incubated in buffers of various pH values (3.0, 6.0, 10.0) at room temperature and at 50°C. After 9 days at room temperature, complete degradation was reported at pH 10.0, whereas only slight degradation was observed at pH 3 and 6 [122]. Using SEC, Chouinard et al. showed that the molecular weight of poly(alkyl cyanoacrylate) nanocapsules is dependent on the level of SO_2 present in the polymerization medium: the higher the SO_2 level, the smaller the molecular weight (93). By following changes in molecular weights and in pH, Lemoine et al. [40] showed that the stability of nanoparticles in vitro is influenced by (1) the type of polymer (PCL and PLA > PLGA); (2) the temperature (–16°C > 4°C > RT

>> 37°C); (3) the presence of water (freeze dried > water environment); and (4) the pH (pH 7.4 buffer > pH 6, nonbuffered).

Enzymes can be added to the aqueous phase with the goal of obtaining data, which can be correlated to the in vivo behavior. A faster drug release has been observed in the presence of enzymes [123]. In addition, the protective effect of the nanoparticles for enzyme-labile drugs has been shown [124]. Calcitonin and insulin can be protected against intestinal proteases when encapsulated within poly(isobutyl cyanoacrylate) nanoparticles.

F. Miscellaneous Techniques

Among the less popular techniques, principally because of the cost of equipment, are ESCA (also called XPS for x-ray photoelectron spectroscopy), SIMS, dielectric constant measurements, small-angle x-ray scattering experiments, field flow fractionation, atomic force microscopy (AFM), and scanning tunnelling microscopy (STM).

1. Electron Spectroscopy for Chemical Analysis

One of the most powerful tools for the nondestructive determination of details of surface structure, bonding, and reactivity is ESCA. The technique is based on the irradiation of a surface with a beam of "soft" x-rays, resulting in the emission of photoelectrons. Only those having a sufficient kinetic energy (emitted from the sample's immediate surface) can pass through the studied material and be detected. The binding energy of the emitted electrons is recorded: it depends on the atomic environment of the irradiated atoms. Each polymer has a specific fingerprint. Sampling depth, in the range of typically 1–10 nm [125], depends on electron kinetic energy, nature of the material, and angle of the analyzer with respect to the surface. Compilations of binding energies are available [126], enabling detailed interpretation of XPS data. This technique is used to confirm the presence of PEG on nanospheres prepared from block PEG-PLGA polymers [101]. Das et al. (127) confirmed with ESCA the continuity of the polymer surface and attested the elemental purity of the polymer in poly(isobutyl cyanoacrylate) nanoparticles.

2. Secondary Ion Mass Spectrometry

In the SIMS technique, the sample is bombarded by a primary ion beam. About 5% of the material spattered from the surface consists of positively and negatively charged atoms (elemental ions) or molecular fragments (cluster ions), and these are mass-analyzed to produce positive and negative secondary ion mass spectra. Vacuum conditions are more stringent than for XPS. Different classes of polymers are therefore readily distinguishable, and individual members of a class can usually be identified by careful spectral interpretation. Copolymer identification is also possible. The advantages of SIMS are high absolute sensitivity and low information depth, molecular specificity, and imaging potential. The main disadvantage is that the quantification of data is not straightforward.

XPS and SIMS complement each other very well and polymer surface characterization may benefit from their combined application.

3. Dielectricity Measurements

Dielectricity measurements are based on the high-frequency measurement of the dielectric constant and/or the permittivity of colloidal suspensions [128]. The method consists of introducing the sample (colloidal suspension) into a capacitive cell. The cell is submitted to a high-frequency field (1–100 MHz) and a measure of Z, the impedance, is obtained. From the complex impedance of the cell, with the help of a physical model, Z is converted into a complex permittivity $\hat{e}^*$ according to the following equation: $\hat{e}^* = \hat{e}' - j\hat{e}''$, where $\hat{e}'$ and $\hat{e}''$ are the dielectric constant and dielectric permittivity, respectively. In the area of colloidal drug carriers, the two latter parameters are used to monitor the adsorption of components onto surfaces. This technique has been used to determine the adsorption isotherm plateau of propranolol hydrochloride onto poly(isobutylcyanoacrylate) nanoparticles [129].

One note of caution should be added, since the presence of ionized species with the ability to form ion pairs is necessary for a dielectric response. This is the case when poly(isobutyl cyanoacrylate) nanoparticles are prepared with dextran sulfate as the stabilizing agent and propranolol hydrochloride as the drug. Under these conditions, the technique is very easy and reproducible and allows the construction of adsorption isotherms in a much quicker time than the classical depletion methods [130].

4. *Small-Angle X-Ray Scattering Experiments*

X-ray small-angle analysis is a technique for studying structural features of colloidal size. Any scattering process is characterized by a reciprocity law, which gives an inverse relationship between particle size and scattering angle. Colloidal dimensions are enormously large compared to x-ray wavelength, which makes the angular range of observable scattering correspondingly small. Small-angle x-ray scattering is always observed, and only observed, when electron-density inhomogeneities of colloidal size exist in the sample. The problem of SAXS consists in deducing size, shape, and mass from the scattering curve. One has to find a model particle, which is equivalent in scattering with the particle in suspension [131]. However, SAXS allows a very precise characterization of nanoparticles including shape, diameter, surface structure, and polymer molecular weight. This technique can be very useful in the field of colloidal carriers, since it is possible to obtain simultaneously various parameters. Different subpopulations as well as different hydrophobic areas are present on the surface of acrylic acid copolymer nanoparticles [132].

5. *Surface Hydrophobicity*

Nanoparticles with hydrophobic surface properties are more rapidly cleared than more hydrophilic nanoparticles. The adsorption of opsonins, and, therefore, the recognition by the mononuclear phagocyte system depends on (among other parameters such as charge, shape of the particles, etc.) surface hydrophobicity. In addition, surface hydrophobicity has a direct influence on the binding of both drugs and surfactants to the surface of nanoparticles [4].

This dye adsorption technique is one of the first methods used to measure the hydrophobicity of nanoparticles [133]. A lipophilic compound, Rose Bengal, is adsorbed at different concentrations on poly(butyl cyanoacrylate) nanoparticles. From the adsorption isotherms, a binding constant affinity can be calculated, which may allow comparisons to be drawn between different polymeric nanoparticles. However, this technique does not give any information on potential subpopulations differing in surface hydrophobicity [134]. An additional problem with this technique is the fast and easy sorption of Rose Bengal on materials used either for the preparation or the centrifugation of nanoparticles.

Hydrophobic interaction chromatography allows the separation of nanoparticles or solutes on the basis of their different hydrophobic interaction with a hydrophobic column material [135]. Basically, nanoparticles are forced through a hydrophobic alkyl-agarose column: hydrophilic nanoparticles pass the column without interactions and are eluted rapidly. The retention time of hydrophobic particles depends on the degree of hydrophobicity. Very hydrophobic nanoparticles may not be eluted at all. Nanoparticles coated with different coating materials (Poloxamers® and Pluronics® with different hydrophilic/hydrophobic properties), can be distinguished with hydrophobic interaction chromatography [4,135]. A miniaturized version of the technique has been proposed, which allows the characterization of the surface properties within minutes [134].

The aqueous two-phase partitioning technique is based on the partitioning of colloidal suspensions in a two-phase system. The two-phase system is based on the mixing of dextran and polyethylene glycol solutions at different concentrations. Above certain concentrations, two phase regions are obtained. However, it is not possible to detect differences in the partitioning of hydrophobic nanoparticles. The change in the partitioning behavior after coating the nanoparticles with different surface active agents or after drug incorporation reflects the extent of the changes in surface properties [4].

6. *Advanced Microscopic Techniques*

There is an increasing interest in using scanning probe microscopes such as the AFM and the STM. In these techniques, a small probe is brought into close proximity to the object and by guiding the probe over the surface, a three-dimensional relief of the object is obtained that reflects the nature of the local interaction between the probe and the sample. In the case of the STM, the probe is guided such that a given tunnelling current between the probe and the sample remains constant. In the case of the AFM, force fields between the probe and the sample are used to guide the probe over the surface [136]. Both techniques are carried out under ambient conditions.

These techniques allow the observation of nanoparticles without any, potentially critical, sample preparation [137]. The spherical geometry and a high degree of monodispersity of nanospheres can be confirmed. Greff et al. characterized spherical polymeric nanospheres with atomic force microscopy [101].

G. Sterility of Colloidal Polymer Dispersions

Most colloidal drug carriers are investigated in a parenteral drug delivery system. Due to technical difficulties, colloidal polymer dispersions used in research are rarely sterilized, even in the case of intravenous administration, since the objective of the initial in vivo studies is to show a potentially interesting pharmacological effect. With regard to the final marketable product, sterility of the colloidal polymer dispersion is of utmost importance. Sterile liposome formulations are already on the market; they are marketed either as a suspension in an aqueous medium or as a freeze-dried powder.

Filtrative sterilization cannot be applied to colloidal drug carriers. The pore size of the filters (0.2 μm) has to be larger than the nanoparticles, which is rarely the case. The most popular sterilization method is heat sterilization (autoclaving at 121°C for at least 15 min). With colloidal polymer dispersion, heat sterilization can lead to the agglomeration of particles or to an increase in mean size [106], which can totally change the in vivo behavior of the vectors. Heat sterilization is only applicable for polymers that do not melt during heat sterilization. Rolland [138] obtained sterile poly(acrylate) nanoparticles for human use by heat sterilization (autoclaving at 121°C for 20 min). The sterilization process induces only a slight increase in the mean diameter of the particles. The same observation was made by Al Khouri-Fallouh et al. (91) after autoclaving of poly(isobutyl cyanoacrylate) nanocapsules. On the other hand, nanoparticles made with poly(e-caprolactone) or poly(lactide) can not be heat sterilized because of their low transition temperatures. Even if the polymer does not change physically or chemically during the sterilization, the drug can probably leach into the external phase during sterilization at elevated temperatures and precipitate upon cooling of the dispersion.

Other techniques, such as gamma sterilization or aseptic procedures, can be chosen. Gamma sterilization techniques have been applied to the sterilization of biodegradable microspheres or implants. A decrease in the molecular weight of the polymer is usually observed after gamma irradiation [121]. Similar observations have been presented by Lemoine et al. [40], who reported a decrease of 28% in the molecular weight of PLGA nanoparticles after a 21,000 Gy irradiation dose. A decrease of the polymer molecular weight may totally change the in vitro, and consequently the in vivo, release profile of the incorporated drugs. After a 2.5 Mrad dose, Alleman et al. reported the dramatic increase of the dissolution rate of savopexine from poly(DL-lactic acid) nanoparticles [38]. This is apparently due to a dramatic decrease in the number-average molecular weight at sterilization doses above 0.1 Mrad. The degradation of the polymer during sterilization and its influence on important product characteristics, such as drug release and biodegradation, has yet to be characterized in detail.

Aseptic procedures have also been applied in order to produce sterile nanoparticles. Doxorubicin poly(isobutyl cyanoacrylate) nanoparticles have been prepared for clinical trials by an aseptic method [73]. The polymerization process is carried out in an aseptic room under laminar flow with dry-heat sterilized material. The aqueous acidic medium containing doxorubicin is sterilized by filtration before the polymerization process. It is also possible to obtain sterile nanospheres or nanocapsules with the precipitation method described by Fessi et al. [31]. The two phases may be mixed in a sterile chamber in a continuous process: the two phases are mixed in a constant phase volume ratio with continuous removal of the formed colloidal suspension. The solvent is continually eliminated, as well as part or all of the water, in order to obtain the desired concentration of nanospheres or a dry product. If the mixing chamber is kept sterile, the two phases can be sterilized by filtrative sterilization and it is, therefore, possible to obtain sterile nanoparticles according to this continuous process.

VI. CONCLUSIONS

A variety of methods for the preparation and characterization of polymeric nanoparticles were briefly reviewed in this chapter. Only a combination of these techniques will result in a well-characterized product with optimal physico-chemical and biological properties.

While drug-free colloidal polymer particles (latex or pseudolatex) are widely used in the pharmaceutical industry in the coating of solid dosage forms, drug-loaded nanoparticles still circulate primarily in the academic environment, trying to escape to intermediate targets (pharmaceutical industry). In the future, polymeric nanoparticles may represent an alternative to other colloidal carriers (e.g., liposomes), few of which have reached their final target not too long ago.

REFERENCES

1. J. Kreuter, (ed.), *Colloidal Drug Delivery Systems*, Marcel Dekker, New York, 1994.
2. J. Kreuter, Nanoparticles. In *Colloidal Drug Delivery Systems*, (J. Kreuter, ed.) Marcel Dekker, New York, 1994, pp 219–342.
3. F. Puisieux, G. Barratt, G. Couarraze, P. Couvreur, J. P. Devissaguet, C. Dubernet, E. Fattal, H. Fessi, and C. Vauthier, Polymeric micro- and nanoparticles as drug carriers. In: *Polymeric Biomaterials*, (S. Dumitriu, ed.), Marcel Dekker, New York, 1994, pp. 749–794.
4. R. H. Müller (ed.), *Colloidal Carriers for Controlled Drug Delivery and Targeting: Modification, Characterization and In Vivo Distribution*, Wiss. Verl.-Ges., Stuttgart, Germany, 1990.
5. J. W. McGinity (ed.), *Aqueous Polymeric Coating for Pharmaceutical Applications*, Marcel Dekker, New York, 1989.
6. L. R. Beck, D. R. Cowsar, D. H. Lewis, R. J. Cosgrove, C. T. Riddle, S. L. Lowry, and T. A. Epperly, A new long-acting injectable microcapsule system for the administration of progesterone, *Fertility and Sterility*, 31:545-551 (1979).
7. S. Benita, J. P. Benoit, F. Puisieux, and C. Thies, Characterization of drug-loaded poly(d,l-lactide) microspheres. *J. Pharm. Sci.*, 73:1721–1724 (1984).
8. R. Bodmeier, H. Chen, P. Tyle, and P. Jarosz, Pseudoephedrine HCl microspheres formulated into an oral suspension dosage form, *J. Control. Rel.*, 15:65–77 (1991).
9. R. Gurny, N. A. Peppas, D. D. Harrington, and G. S. Banker, Development of biodegradable and injectable latices for controlled release of potent drugs, *Drug. Dev. Ind. Pharm.*, 7:1–25 (1981).
10. H. J. Krause, A. Schwarz, and P. Rohdewald, Polylactic acid nanoparticles, a colloidal drug delivery system for lipophilic drugs, *Int. J. Pharm.*, 27:145–155 (1985).
11. F. Koosha, R. H. Müller, S. S. Davis, and M. C. Davies, The surface chemical structure of poly(β-hydroxybutyrate) microparticles produced by solvent evaporation process, *J. Control. Rel.*, 9:149 (1989).
12. R. Bodmeier, and H. Chen, Indomethacin polymeric nanosuspensions prepared by microfluidization, *J. Control. Rel.*, 12:223-233 (1990).
13. F. Koosha, R. H. Müller, and C. Washington, Production of polyhydroxybutyrate (PHB) nanoparticles for drug targeting, *J. Pharm. Pharmacol.*, 39:136P (1987).
14. P. D. Scholes, A. G. A. Coombes, L. Illum, S. S. Davis, M. Vert, and M. C. Davies, The preparation of sub-200 nm poly(lactide-co-glycolide) microspheres for site-specific drug delivery, *J. Control. Rel.*, 25:145–153 (1993).
15. T. Verrecchia, G. Spenlehauer, D. V. Bazile, A. Murry-Brelier, Y. Archimbaud, and M. Veillard, Non-stealth (poly(lactic acid/albumin)) and stealth (poly(lactic acid-polyethylene glycol)) nanoparticles as injectable drug carriers, *J. Control. Rel.*, 36:49–61 (1995).
16. S. Batzri, and E. D. Korn, Single bilayer liposomes prepared without sonication, *Biochimica et Biophysica Acta*, 443:629–634 (1973).
17. D. Deamer, and A. D. Bangham, Large volume liposomes by an ether vaporization method, *Biochimica et Biophysica Acta*, 298:1015–1019 (1976).
18. H. Fessi, J. P. Devissaguet, F. Puisieux, and C. Thies, Procede de preparation de systemes colloidaux dispersibles d'une substance, sous forme de nanoparticules, French Patent 2,608,988, (1986).
19. C. Emile, D. Bazile, F. Herman, C. Helene, and M. Veillard, Encapsulation of oligonucleotides in stealth Me.PEG-PLA50 nanoparticles by complexation with structured oligopeptides. In: *Proc. 7th Int. Pharmaceutical Technol. Conf.*, Budapest, Hungary, pp 461 462 (1995).
20. R. Bodmeier, H. Chen, P. Tyle, and P. Jarosz, Spontaneous formation of drug-containing acrylic nanoparticles, *J. Microencapsulation*, 8(2):161–170 (1991).

21. K. O. R. Lehmann, Chemistry and application properties of polymethacrylate coating systems, In: *Aqueous Polymeric Coating for Pharmaceutical Applications*, (J. W. McGinity, ed.) Marcel Dekker, New York, pp. 153–245.
22. R. K. Chang, J. C. Price, and C. Hsiao, Preparation and preliminary evaluation of Eudragit RL and RS pseudolatices for controlled drug release, *Drug Dev. Ind. Pharm.*, 15:361–372 (1989).
23. R. Lichtenberger, K. Wendel, and H. P. Merkle, Polymer films from aqueous latex dispersions as carriers for transdermal delivery of lipophilic drugs. In: *Proc. 15th Int. Symp. on Controlled Release Bioactive Materials*, Basel, Switzerland, pp. 147–148 (1988).
24. R. Bodmeier, and O. Paeratakul, Drug release from polymeric films and laminates prepared from aqueous latexes, *Proc. 5th Int. Pharmaceutical Technol. Conf.*, Paris, France, pp. 61–68 (1989).
25. S. Büyükyaylaci, Y. M. Joshi, G. E. Peck, and G. S. Banker, Polymeric dispersions as a new topical drug delivery system. In: *Recent Advances in Drug Delivery Systems*, Plenum Press, New York, pp. 291–306 (1984).
26. T. Niwa, H. Takeuchi, T. Hino, N. Kunou, and Y. Kawashima, Preparations of biodegradable nanospheres of water-soluble and insoluble drugs with d,l-lactide/glycolide copolymer by a novel spontaneous emulsification solvent diffusion method, and the drug release behavior, *J. Control. Rel.*, 25:89–98 (1993).
27. H. Fessi, F. Puisieux, and J. P. Devissaguet, Procede de preparation de systemes colloidaux dispersibles d'une substance, sous forme de nanocapsules, European patent 274 961 (1987).
28. N. Ammoury, H. Fessi, J. P. Devissaguet, F. Puisieux, and S. Benita, Physicochemical characterization of polymeric nanocapsules and in vitro release evaluation, *S.T.P. Pharma Sci.*, 5:647–651 (1989).
29. N. Ammoury, H. Fessi, J. P. Devissaguet, M. Allix, M. Plotkine, and R. G. Boulu, Effect on cerebral blood flow of orally administered indomethacin-loaded poly(isobutylcyanoacrylate) and poly(d,l-lactide) nanocapsules, *J. Pharm. Pharmacol.*, 42:558–561 (1990).
30. N. Ammoury, H. Fessi, J. P. Devissaguet, M. Dubrasquet, and S. Benita, Jejunal absorption, pharmacological activity, and pharmacokinetic evaluation of indomethacin-loaded poly(d,l-lactide) and poly(isobutyl-cyanoacrylate) nanocapsules in rats, *Pharm. Res.*, 8:101–105 (1991).
31. H. Fessi, F. Puisieux, J. P. Devissaguet, N. Ammoury, and S. Benita, Nanocapsule formation by interfacial polymer deposition following solvent displacement, *Int. J. Pharm.*, 55:R1–R4 (1989).
32. N. Ammoury, H. Fessi, J. P. Devissaguet, F. Puisieux, and S. Benita, In vitro release kinetic pattern of indomethacin from poly(d,l-lactide) nanocapsules, *J. Pharm. Sci.*, 79(9):763–767 (1990).
33. L. Marchal-Heussler, H. Fessi, J. P. Devissaguet, M. Hoffman, and P. Maincent, Colloidal drug delivery systems for the eye. A comparison of the efficacy of three different polymers: Polyisobutylcyanoacrylate, polylactic-co-glycolic acid, poly-epsiloncaprolacton, *S.T.P. Pharma Sci.*, 2:98–104 (1992).
34. C. Bindschaedler, R. Gurny, and E. Doelker, Process for preparing a powder of water-insoluble polymer, which can be redispersed in a liquid phase, the resulting powder and utilization therof, Swiss Patent 1497/88 (1988).
35. H. Ibrahim, C. Bindschaedler, E. Doelker, P. Buri, and R. Gurny, Concept and development of opthalmic pseudo-latexes triggered by pH, *Int. J. Pharm.*, 77:211–219 (1991).
36. H. Ibrahim, C. Bindschaedler, E. Doelker, P. Buri, and R. Gurny, Aqueous nanodispersions prepared by a salting-out process, *Int. J. Pharm.*, 87:239–246 (1992).
37. E. Allemann, R. Gurny, and E. Doelker, Preparation of aqueous polymeric nanodispersions by a reversible salting-out process, influence of process parameters on particle size, *Int. J. Pharm.*, 87:247–253 (1992).

38. E. Allemann, J.-C. Leroux, R. Gurny, and E. Doelker, In vitro extended-release properties of drug loaded poly(dl-lactic acid) nanoparticles produced by a salting-out procedure, *Pharm. Res.*, 10(12):1732–1737 (1993).
39. E. Allemann, E. Doelker, and R. Gurny, Drug loaded poly(lactic acid) nanoparticles produced by a reversible salting-out process: purification of an injectable dosage form, *Eur. J. Pharm. Biopharm.*, 39:13–18 (1992).
40. D. Lemoine, C. Francois, V. Berlage, F. Kedzierewicz, P. Maincent, M. Hoffman, and V. Preat, Stability study of poly(d,l lactide), poly(d,l-lactice-co-glycolide) and poly(ε-caprolactone) nanoparticles. *Proc. 7th Int. Pharmaceutical Technol. Conf.*, Budapest, Hungary, pp. 481–482 (1995).
41. B. Ekman, and I. Sjöholm, Incorporation of macromolecules in microparticles: Preparation and characteristics, *Biochem.* 15:5115 (1976).
42. A. F. Yapel, Jr., Albumin microspheres: Heat and chemical stabilization. In: *Methods in Enzymology: Part A, Drug and Enzyme Targeting*, (K. J. Widder and R. Green, eds.), Academic Press, Orlando, 1985, pp 3–18.
43. W. E. Longo, and E. P. Goldberg, Hydrophilic albumin microspheres. In: *Methods in Enzymology: Part A, Drug and Enzyme Targeting*, (K. J. Widder and R. Green, eds.), Academic Press, Orlando, 1985, pp. 18–26.
44. E. Tomlinson, and J. J. Burger, Incorporation of water-soluble drugs in albumin microspheres. In: *Methods in Enzymology: Part A, Drug and Enzyme Targeting*, (K. J. Widder and R. Green, eds.), Academic Press, Orlando, 27–43.
45. J. M. Gallo, C. T. Hung, and D. G. Perrier, Analysis of albumin microsphere preparation, *Int. J. Pharm.*, 22:63–74 (1984).
46. P. K. Gupta, C. T. Hung, and D. G. Perrier, Albumin microspheres II. Effect of stabilisation temperature on the release of adriamycin, *Int. J. Pharm.*, 33:147 (1986).
47. T. Yoshioka, M. Hashida, S. Muranishi, and H. Sezaki, Specific delivery of mitomycin C to the liver, spleen, and lung: Nano- and microspherical carriers of gelatin, *Int. J. Pharm.*, 8:131–141 (1981).
48. Y. Tabata and Y. Ikada, Synthesis of gelatin microspheres containing interferon, *Pharm. Res.*, 6:422 (1989).
49. E. E. Hassan, R. C. Parish, and J. M. Gallo, Optimized formulation of magnetic microspheres containing the anticancer agent, oxantrazole, *Pharm. Res.*, 9:390 (1992).
50. K. Widder, G. Flouret, and A. Senyei, Magnetic microspheres: Synthesis of a novel parenteral drug carrier, *J. Pharm. Sci.*, 68:79–82 (1979).
51. K. Widder, R. M. Morris, R. G. Poore, P. H. Howards, and A. Senyei, Selective targeting of magnetic albumin microspheres containing low-dose doxorubicin: Total remission in Yoshida Sarcoma-bearing rats, *Eur. J. Cancer and Clinical Oncology*, 19:141–147 (1983).
52. J. J. Marty, R. C. Oppenheim, and P. Speiser, Nanoparticles—a new colloidal drug delivery system, *Pharm. Acta Helv.*, 53:17–22 (1978).
53. M. El-Samaligy and P. Rohdewald, Triamcinolone diacetate nanoparticles, a sustained release drug delivery system suitable for parenteral administration, *Pharm. Acta Helv.*, 57:201 (1982).
54. M. El-Samaligy, and P. Rohdewald, Reconstituted collagen nanoparticles, a novel drug carrier delivery system, *J. Pharm. Pharmacol.*, 35:537–539 (1983).
55. R. C. Oppenheim, Solid colloidal drug delivery systems: Nanoparticles, *Int. J. Pharm.*, 8:217–234.
56. H. J. Krause, and P. Rohdewald, Preparation of gelatin nanocapsules and their pharmaceutical characterization, *Pharm. Res.*, 2:239–243 (1985).
57. M. Rajanorivony, C. Vauthier, G. Couarraze, F. Puisieux, and P. Couvreur, Development of a new drug carrier made from alginate, *J. Pharm. Sci.*, 82(9):912–917 (1993).
58. J. E. Diederichs, C. Vauthier, H. Alphandary, P. Couvreur, and R. H. Müller, Forma-

tion process and determination of microviscosity of alginate nanoparticles. In: *Proc. 21st Int. Symp. on Controlled Release of Bioactive Materials*, Nice, France, pp. 511–512 (1994).
59. U. E. Berg, J. Kreuter, P. P. Speiser, and M. Soliva, Herstellung und in vitro-Prüfung von polymeren Adjuvantien für Impfstoffe, *Pharm. Ind.*, 48:75 (1986).
60. J. Kreuter, and H. J. Zehnder, The use of ^{60}Co-γ-irradiation for the production of vaccines, *Radiat. Effects.*, 35:161 (1978).
61. C. Vauthier-Holtzscherer, S. Benabbou, G. Spenlehauer, M. Veillard, and P. Couvreur, Methodology for the preparation of ultra-dispersed polymer systems, *S.T.P. Pharma Sci.*, 1(2):109–116 (1991).
62. J. Kreuter, Evaluation of nanoparticles as drug-delivery systems. I. Preparation methods, *Pharm. Acta Helv.*, 58:196 (1983).
63. E. Allemann, R. Gurny, and E. Doelker, Drug-loaded nanoparticles-preparation methods and drug targeting issues, *Eur. J. Pharm. Biopharm.*, 39(5):173–191.
64. A. Rolland, D. Gibassier, P. Sado, and R. Le Verge Methodologie de preparation de vecteurs nanoparticulaires a base de polymeres acryliques, *J. Pharm. Belg.*, 41:83–93 (1986).
65. X. Y. Wu, and P. L. Lee, Preparation and characterization of thermal- and pH-sensitive nanospheres, *Pharm. Res.*, 10(10):1544–1547 (1993).
66. J. Kreuter, Possibilities of using nanoparticles as carriers for drugs and vaccines, *J. Microencapsulation*, 5(2):115–127 (1988).
67. P. Couvreur, and C. Vauthier, Polyalkylcyanoacrylate nanoparticles as drug carrier: present state and perspectives, *J. Control. Rel.*, 17:187–198 (1991).
68. G. Puglisi, G. Giammona, M. Fresta, B. Carlisi, N. Micali, and A. Villari, Evaluation of polyalkylcyanoacrylate nanoparticles as a potential drug carrier: preparation, morphological characterization and loading capacity, *J. Microencapsulation*. 10(3):353–366 (1993).
69. V. Lenaerts, P. Raymond, J. Juhasz, M. A. Simard, and C. Jolicoeur, New method for the preparation of cyanoacrylic nanoparticles with improved colloidal properties, *J. Pharm. Sci.*, 78(12):1051–1052 (1989).
70. B. Seijo, E. Fattal, L. Roblot-Treupel, and P. Couvreur, Design of nanoparticles of less than 50 nm diameter: preparation, characterization and drug loading, *Int. J. Pharm.* 62:1–7 (1990).
71. S. J. Douglas, L. Illum, and S. S. Davis, Particle size and size distribution of poly(butyl-2-cyanoacrylate) nanoparticles. I. Influence of physicochemical factors, *J. Colloid Interface Sci.*, 101:149 (1985).
72. S. J. Douglas, L. Illum, and S. S. Davis, Particle size and size distribution of poly(butyl-2-cyanoacrylate) nanoparticles II. Influence of the stabilizers, *J. Colloid Interface Sci.*, 103:154 (1985).
73. C. Verdun, P. Couvreur, H. Vranckx, V. Lenaerts, and M. Roland, Development of a nanoparticle controlled-release formulation for human use, *J. Control. Rel.*, 3:205–210 (1986).
74. N. Bapat, and M. Boroujerdi, Uptake capacity and adsorption isotherms of doxorubicin on polymeric nanoparticles: effect of methods of preparation, *Drug Dev. Ind. Pharm.*, 18:65–77 (1992).
75. M. J. Alonso, C. Losa, P. Calvo, and J.-L. Vila Jato, Approaches to improve the association of amikacin sulphate to poly(alkylcyanoacrylate) nanoparticles, *Int. J. Pharm.*, 68:69–76 (1991).
76. P. Couvreur, B. Kante, M. Roland, and P. Speiser, Adsorption of antineoplastic drugs to polyalkylcyanoacrylate nanoparticles and their release in calf serum, *J. Pharm. Sci.*, 68:1521–1524 (1979).
77. V. Guise, J. Y. Drouin, J. Benoit, J. Mahuteau, P. Dumont, and P. Couvreur, Vidarabine-loaded nanoparticles: A physicochemical study, *Pharm. Res.*, 7:736–741 (1990).

78. C. Chavany, T. L. Doan, P. Couvreur, F. Puisieux, and C. Helene, Polyalkylcyanoacrylate nanoparticles as polymeric carriers for antisense oligonucleotides, *Pharm. Res.*, 9(4):441-449 (1992).
79. T. Harmia, P. Speiser, and J. Kreuter, Optimization of pilocarpine loading onto nanoparticles by sorption procedures, *Int. J. Pharm.*, 33:45-54 (1986).
80. J. L. Grangier, M. Puygrenier, J. C. Gautier, and P. Couvreur, Nanoparticles as carriers for growth hormone releasing factor, *J. Control. Rel.*, 15:3-13 (1991).
81. F. Leonard, R. K. Kulkarni, G. Brandes, J. Nelson, and J. J. Mameron, Synthesis and degradation of poly(alkylcyanoacrylates), *J. Appl. Polym. Sci.*, 10:259-272 (1966).
82. R, Müller, C. Lherm, J. Herbort, and P. Couvreur, In vitro model for the degradation of alkylcyanoacrylate nanoparticles, *Biomaterials*, 11:590-595 (1990).
83. V. Lenaerts, P. Couvreur, D. Christiaens-Leyh, E. Joiris, M. Roland, B. Rollman, and P. Speiser, Degradation of polyisobutyl-cyanoacrylate nanoparticles, *Biomaterials*, 5:65-68 (1984).
84. A. Piskin, A. Tuncel, A. Denizli, E. B. Denkbas, H. Ayhan, H. Cicek, and K. T. Xu, Nondegradable and biodegradable polymeric particles, in Diagnostic Biosensor Polymers. (A. M. Usmani, and N. Akmal, eds.), American Chemical Society, Washington, DC, pp. 222-237 (1994).
85. B. Kante, P. Couvreur, G. Dubois-Crack, C. De Meester, P. Guiot, M. Roland, M. Mercier, and P. Speiser, Toxicity of polyalkylcyanoacrylate nanoparticles I: Free nanoparticles, *J. Pharm. Sci.*, 7:786-790 (1982).
86. A. Ibrahim, P. Couvreur, M. Roland, and P. Speiser, New magnetic drug carrier, *J. Pharm. Pharmacol.*, 35:59-61 (1982).
87. C. Vauthier, M. I. Popa, F. Puisieux, and P. Couvreur, Evaluation of potentiality to graft PEG to poly(alkylcyanoacrylate) nanoparticles in the course of the formation of the nanoparticles in absence of surfactants. *Proc. 22nd Int. Symp. on Controlled Release of Bioactive Materials*, Seattle, WA 592-593 (1995).
88. G. Birrenbach, and P. P. Speiser, Polymerized micelles and their use as adjuvants in immunology, *J. Pharm. Sci.*, 65:1763-1766 (1976).
89. B. Ekman, and I. Sjöholm, Improved stability of proteins immobilized in microparticles prepared by a modified emulsion polymerization technique, *J. Pharm. Sci.*, 67:693 (1978).
90. P. Edman, B. Ekman, and I. Sjöholm, Immobilization of proteins in microspheres of biodegradable polyacryldextran, *J. Pharm. Sci.*, 69:838 (1980).
91. N. Al-Khouri-Fallouh, L. Roblot-Treupel, H. Fessi, J. P. Devissaguet, and F. Puisieux, Development of a new process for the manufacture of polyisobutylcyanoacrylate nanocapsules, *Int. J. Pharm.*, 28:125-132 (1986).
92. M. Gallardo, G. Çouarraze, B. Denizot, L. Treupel, P. Couvreur, and F. Puisieux, Study of the mechanisms of formation of nanoparticles and nanocapsules of polyisobutyl-2-cyanoacrylate, *Int. J. Pharm.*, 100, 55-64 (1993).
93. F. Chouinard, S. Buczkowski, and V. Lenaerts, Poly(alkylcyanoacrylate) nanoparticles: Physicochemical characterization and mechanism of formation, *Pharm. Res.*, 11(6):869-874 (1994).
94. F. Chouinard, F. W. K. Kan, J.-C. Leroux, C. Foucher, and V. Lenaerts, Preparation and purification of polyisohexylcyanoacrylate nanocapsules, *Int. J. Pharm.*, 72:211 (1991).
95. H. J. Krause, A. Schwartz, and P. Rohdewald, Interfacial polymerization, a useful method for the preparation of polymethylcyanoacrylate nanoparticles, *Drug. Dev. Ind. Pharm.*, 12:527-552 (1986).
96. M. S. El Samaligy, P. Rohdewald, and H. A. Mahmoud, Polyalkylcyanoacrylate nanocapsules, *J. Pharm. Pharmacol.*, 38:216-218 (1986).
97. M. R. Gasco, S. Morel, M. Trotta, and I. Viano, Doxorubicine englobed in polybutylcyanoacrylate nanocapsules: behavior in vitro and in vivo, *Pharm. Acta Helv.*, 66:47-49 (1991).

98. M. R. Gasco, and M. Trotta, Nanoparticles from microemulsions, *Int. J. Pharm.*, 29:267–268 (1986).
99. W.-D. Hergeth, U.-J. Steinau, H.-J. Bittrich, K. Schmutzler, and S. Wartewig, Submicron particles with thin polymer shells, *Progr. Colloid Polym. Sci.*, 85:82–90 (1991).
100. B. Magenheim and S. Benita, Nanoparticle characterization: a comprehensive physicochemical approach, *S.T.P. Pharma Sci.*, 1(4):221–241 (1991).
101. R. Greff, Y. Minamitake, M. T. Perrachia, V. Trubetskoy, V. Torchilin, and R. Langer, Biodegradable long circulating polymeric nanospheres, *Sci.*, 263:1600–1603 (1994).
102. S. Stolnik, S. E. Dunn, M. C. Garnett, M. C. Davies, A. G. A. Coombes, D. C. Taylor, M. P. Irving, S. C. Purkiss, T. F. Tadros, S. S. Davis, and L. Illum, Surface modification of poly(lactide-co-glycolide) nanospheres by biodegradable poly(lactide)-poly(ethylene glycol) copolymers, *Pharm. Res.*, 11(12):1800–1808 (1994).
103. D. Bazilc, C. Prud'Homme, M. T. Bassoulet, M. Marlard, G. Spenlehauer, and M. Veillard, Stealth Me.PEG-PLA nanoparticles avoid uptake by the mononuclear phagocytes system, *J. Pharm. Sci.*, 84(4):493–498 (1995).
104. *Particle sizing by light scattering. Application manual 0065*. Malvern Instruments, Orsay, France, 1993.
105. M. S. McCracken and M. C. Sammons, Sizing of a vesicle drug formulation by quasi-elastic light scattering and comparison with electron microscopy and ultracentrifugation, *J. Pharm. Sci.*, 76(1):56–59 (1987).
106. J. M. Rollot, P. Couvreur, L. Roblot-Treupel, and F. Puisieux, Physicochemical and morphological characterization of polyisobutylcyanoacrylate nanocapsules, *J. Pharm. Sci.*, 75(4):361–364 (1986).
107. S. E. Dunn, A. Brindley, S. S. Davis, M. C. Davies, and L. Illum, Polystyrene-poly (ethylene glycol) (PS-PEG2000) particles as model systems for site specific drug delivery. 2. The effect of PEG surface density on the in vitro cell interaction and in vivo biodistribution, *Pharm. Res.*, 11(7):1016–1022 (1994).
108. S. Benita and M. Y. Levy, Submicron emulsions as colloidal drug carriers for intravenous administration: comprehensive physicochemical characterization, *J. Pharm. Sci.*, 82(11):1069–1079 (1993).
109. C. Losa, L. Marchal-Heussler, F. Orallo, J. L. Vila Jato, and M. J. Alonso, Design of new formulations for topical ocular administration: polymeric nanocapsules containing metipranolol, *Pharm. Res.*, 10(1):80–87 (1993).
110. P. Beck, J. Kreuter, R. Reszka, and I. Fichtner, Influence of polybutylcyanoacrylate nanoparticles and liposomes on the efficacy and toxicity of the anticancer drug mitoxantrone in murine tumour models, *J. Microencapsulation*, 10(1):101–114 (1993).
111. P. Beck, D. Scherer, and J. Kreuter, Separation of drug-loaded nanoparticles from free drug by gel filtration, *J. Microencapsulation*, 7(4):491–496 (1990).
112. T. Verrecchia, P. Huve, D. Bazile, M. Veillard, G. Spenlehauer, and P. Couvreur, Adsorption/desorption of human serum albumin at the surface of poly(lactic acid) nanoparticles prepared by a solvent evaporation process, *J. Biomed. Mat. Res.*, 27:1019–1028 (1993).
113. B. Hubert, J. Atkinson, M. Guerret, M. Hoffman, J. P. Devissaguet, and P. Maincent, The preparation and acute antihypertensive effects of a nanocapsular form of darodipine, a dihydropyridine calcium entry blocker, *Pharm. Res.*, 8(6):734–738 (1991).
114. C. Washington, Drug release from microdisperse systems. A critical review, *Int. J. Pharm.*, 58:1–12 (1990).
115. M. Y. Levy and S. Benita, Drug release from submicron o/w emulsion: A new in vitro kinetic evaluation method, *Int. J. Pharm.*, 66:29–37 (1990).
116. B. Magenheim, M. Y. Levy, and S. Benita, A new in vitro technique for the evaluation of drug release profile from colloidal carriers—ultrafiltration technique at low pressure, *Int. J. Pharm.*, 94:115–123 (1993).

117. S. Li and M. Vert, Morphological changes resulting from the hydrolytic degradation of stereocopolymers derives from L- and DL-lactides, *Macromolecules*, 27:3107–3110 (1994).
118. J. Kreuter, Physicochemical characterization of polyacrylic nanoparticles, *Int. J. Pharm.*, 14:43–58 (1983).
119. T. Harmia, P. Speiser, and J. Kreuter, A solid colloidal drug delivery system for the eye: encapsulation of pilocarpin in nanoparticles, *J. Microencapsulation*, 3(1):3–12 (1986).
120. T. G. Park, Degradation of poly(D-L-lactic acid) microspheres: Effect of molecular weight, *J. Control. Release*, 30:161–173 (1994).
121. A. G. Hausberg, R. A. Kenley, and P. P. DeLuca, Gamma irradiation effects on molecular weight and in vitro degradation of poly(D,L-lactide-co-glycolide) microparticles, *Pharm. Res.*, 12(6):851–856 (1995).
122. M. Stein and E. Hamacher, Degradation of polyisobutyl 2-cyanoacrylate microparticles, *Int. J. Pharm.*, 80:R11–R13 (1992).
123. D. Scherer, J. R. Robinson, and J. Kreuter, Influence of enzymes on the stability of polybutylcyanoacrylate nanoparticles, *Int. J. Pharm.*, 101:165–168 (1994).
124. P. J. Lowe and C. S. Temple, Calcitonin and insulin in isobutylcyanoacrylate nanocapsules: Protection against proteases and effect on intestinal absorption in rats, *J. Pharm. Pharmacol.*, 46:547–552 (1994).
125. J. A. Hayward, A. Durrani, Y. Lu, C. Clayton, and D. Chapman, Biomembranes as models for polymer surfaces. IV. ESCA analysis of phosphorylcholine surface covalently bound to hydroxylate substrates, *Biomaterials*, 7:252–258 (1986).
126. D. T. Clark and H. R. Thomas, Applications of ESCA to polymer chemistry. XVII. Systematic investigation of the core levels of simple homopolymers, *J. Polym. Sci.: Polymer Chem. Ed.*, 16:791–820 (1986).
127. S. K. Das, I. G. Tucker, D. J. T. Hill, and N. Ganguly, Evaluation of poly(isobutylcyanaoacrylate) nanoparticles for mucoadhesive ocular drug delivery. I. Effect of formulation variables on physico-chemical characteristics of nanoparticles, *Pharm. Res.*, 12(4):534–540 (1995).
128. E. Benoit, P. Maincent, and J. Bessière, Applicability of dielectric measurements to the adsorption of drugs onto nanoparticles, *Int. J. Pharm.*, 84:283–286 (1992).
129. E. Benoit, O. Prot, P. Maincent, and J. Bessière, Adsorption of beta-blockers onto polyisobutylcyanoacrylate nanoparticles measured by depletion and dielectric methods, *Pharm. Res.*, 11(4):585–588 (1994).
130. O. Prot, E. Benoit, P. Maincent, and J. Bessière, Adsorption of an ionic drug onto poly(isobutylcyanoacrylate) nanoparticles: adsorption model and dielectric interpretation, *J. Colloid Interf. Sci.*, 184:251–258, (1996).
131. O. Kratky, A survey. In: *Small Angle X-ray Scattering* (O. Glatter and O. Kratky eds.), Academic Press, London, pp 3–13.
132. J. J. Müller, G. Lubowski, R. Kröber, G. Damaschun, and M. Dittgen, Acrylic acid copolymer nanoparticles for drug delivery: structural characterization of nanoparticles by small-angle x-ray scattering, *Colloid Polym. Sci.*, 272:755–769 (1994).
133. S. J. Douglas, L. Illum, and S. S. Davis, Poly(butyl 2-cyanoacrylate) nanoparticles with differing surface charges, *J. Control. Release*, 3:15–23 (1986).
134. K. H. Wallis and R. H. Müller, Determination of the surface hydrophobicity of colloidal dispersions by mini-hydrophobic intercation chromatography, *Pharm. Ind.*, 55(12):1124–1128 (1993).
135. H. Carstensen, B. W. Müller, and R. H. Müller, Adsorption of ethoxylated surfactants on nanoparticles. I. Characterization by hydrophobic interaction chromatography, *Int. J. Pharm.*, 67:29–37 (1991).
136. M. Radmacher, R. W. Tillmann, M. Fritz, and H. E. Gaub, From molecules to cells: Imaging soft samples with the atomic force microscope, *Sci.*, 257:1900–1905 (1992).

137. M. Skiba, F. Puisieux, D. Duchene, and D. Wouesidjewe, Direct imaging of modified β-cyclodextrin nanospheres by photon scanning tunnelling and scanning force microscopy, *Int. J. Pharm.*, 120:1–11 (1995).
138. A. Rolland, Clinical pharmacokinetics of doxorubicin in hepatoma patients after a single intravenous injection of free or nanoparticle-bound anthracycline, *Int. J. Pharm.*, 54:113–121 (1989).

4

Aqueous Polymeric Dispersions as Film Formers

Jean Wang and Isaac Ghebre-Sellassie

Parke-Davis Pharmaceutical Research, Warner-Lambert Company, Morris Plains, New Jersey

I. INTRODUCTION

Although the words *film* and *coating* have been referred to interchangeably in many industries [1], such as the paint, photography, and pharmaceutical industries, attempts have also been made at times to distinguish between the two terms. Film refers to thin sheets of materials of diverse origins, while a coating represents the formation of a continuous film over a surface. Materials used for film formation differ in physicochemical characteristics depending on the end use of the coated product. For instance, the chemical composition of film formers commonly used in the pharmaceutical industry ranges from small molecules, such as sugars, to complex polymeric materials. Over the years, pharmaceutical coating processes have evolved from operations that were based on an art practiced by few skilled operators to a discipline governed by well-controlled engineering principles. At the same time, significant advances in coating equipment were made and opened up a new frontier in coating technology, which made it possible to easily and efficiently handle water-based coating formulations. These advances then allowed the development of aqueous dispersions consisting of water-insoluble polymers as film formers, which are the subject of this chapter.

A. Historical Overview

Although pharmaceutical coatings have over time undergone evolutionary changes especially in the areas of coating materials and processing technologies, progress has been slow, particularly in the types of materials that could be used as film formers due to the stringent toxicological requirements the pharmaceutical industry requires. Several conventional coating technologies that are widely used in the pharmaceutical industry are briefly described below.

B. Sugar Coatings

Sugar coating has a long history in the pharmaceutical industry. It uses sugar solutions or syrups as coating solutions that are sprayed onto substrates, such as tablets, pellets, and granules. Sugar coating has been primarily used for taste masking and for aesthetic purposes. The most serious drawback of sugar coating is the low molecular weight of the film forming materials, specifically glucose and/or sucrose. The formed films are weak and have low mechanical strength. Consequently, the coating levels that are applied onto substrates are generally high, contributing to as much as 30–70% of the total weight of the product, thereby resulting in long processing times. Another disadvantage of the sugar coating process is that the coating solutions are susceptible to bacterial and fungal growth, and they must be carefully monitored. In spite of these potential challenges, sugar coatings have been used successfully to coat numerous products and are still popular in certain quarters.

C. Polymeric-Solution-Based Coatings

In response to a growing demand for more efficient coating processes, coating equipment manufacturers started to introduce equipment with high drying efficiency. Simultaneously, pharmaceutical scientists identified new polymeric materials that are suitable for coating pharmaceutical dosage forms. As a result, scientists at Abbot Laboratories in the early 1950s developed the first film-coated tablets that used a polymer as a film former [2]. Shortly thereafter, other companies followed suit. In all cases, the initial objective was to satisfy the functional requirement of sugar coating. Later on, however, the objective was expanded to include the development of modified release coatings. The coatings were competed almost exclusively of cellulose derivatives, including methyl cellulose, ethyl cellulose, and hydroxypropyl methyl cellulose, although acrylate derivatives have also been used to some extent [3,4]. All of them were dissolved in organic solvents to form coating solutions. These coatings generally contribute less than 10% of the total weight of finished products and require short processing times. As a result, use of organic-solution-based polymers opened a new era in pharmaceutical coating. Despite significant advantages, organic-solvent-based film coating processes still have some obvious shortcomings. The high cost of organic solvents and the need for expensive solvent recovery systems coupled with the pollution and safety hazards associated with the use of organic solvents have limited the extensive use of organic-solvent-based film coatings in pharmaceutical dosage forms [5]. As a result, water-based coating systems were reintroduced in the early 1970s and gradually replaced organic-solvent-based solutions in many applications.

Initially, the focus was to use water-soluble polymeric materials, such as hydroxypropyl methyl cellulose, gelatin, polyvinylpyrrolidone, methyl cellulose, and hydroxypropyl cellulose, to mask taste and improve the appearance of immediate-release dosage forms. While these coating systems provide many of the desired attributes, they also exhibit significant shortcomings. Due to their high water solubility, these polymers could not be used as modified-release coatings. In addition, relatively high temperatures are required to drive off the water during the coating operation. Prolonged exposure to elevated temperature could detrimentally affect water-sensitive and/or heat-labile drugs, and hence restricts the use of aqueous-solution-based coating formulations. These limitations eventually led to the development of a third generation of pharmaceutical coating systems, i.e., aqueous polymeric dispersions.

D. Dispersion-Based Film Coatings

Aqueous polymeric dispersions, latices, or pseudolatices, are all colloidal systems in which high molecular weight polymers are homogeneously dispersed in submicron sizes with the aid of surfactant(s) and other stabilizing agents. Latex technology was initially developed to manufacture synthetic rubber but was later used for other applications including surface coatings. The polymeric coatings are classified as thermoplastic or thermoset polymers based on the type of polymers used. A thermoplastic coating contains at least one un–cross-linked high molecular weight polymer that provides the required mechanical strength to a film. Thermosetting coatings contain reactive low molecular weight polymers or oligomers that can polymerize after application to the substrates under the influence of heat or radiation [6]. The resulting thermosetting coatings frequently contain residual reactive materials that might be toxic or bio-incompatible, and, therefore, not suitable for pharmaceutical applications. In contrast, thermoplastic coatings are derived from pre-existing or synthetic polymers that have well-established safety profiles and hence satisfy all regulatory requirements on excipient safeties. Therefore, the term aqueous polymeric dispersions refers exclusively to the dispersions of thermoplastic polymers in water.

II. MANUFACTURING PROCESSES OF AQUEOUS POLYMERIC DISPERSIONS

Aqueous polymeric dispersions that are widely used by the pharmaceutical industry are manufactured using different processes and different starting materials. They are classified into latices and pseudolatices, depending on the starting materials used during manufacture, and are described below.

A. Latices

Latices are aqueous polymeric dispersions produced by emulsion polymerization. The homogeneous and heterogeneous nucleation models have been proposed to describe such processes. In this chapter, only the homogeneous nucleation model is introduced since it describes the process of emulsion polymerization involving water-soluble monomers or situations in which the concentration of the surfactant is below its critical micelle concentration (CMC) [7–9]. This model is applicable because acrylic latices that are commonly used in the pharmaceutical industry are derived from the highly water soluble acrylic acid monomers.

According to the homogeneous nucleation model, emulsion polymerization is divided into three phases. During the first phase, initiators generate free radicals in the continuous water phase in the presence of monomers and surfactants to start the polymerization reaction. Polymerization proceeds very rapidly in the continuous aqueous phase through the formation of oligomer radicals. These polymers are insoluble and, therefore, precipitate from solution to form primary particles. These primary particles are dispersed with the help of surfactants to form polymer droplets. During this phase, the number of polymer particles increases as the propagation reaction proceeds. The second phase starts when the polymer particles are stabilized and continues until the dispersed monomer phase disappears. Since monomers and intermediate oligomer radicals generated in the process can be captured by polymer particles, polymerization continues inside the polymer particles, which swells the polymer particles and increases their size. During the third phase, the monomer concentration within the polymer particles is dramatically depleted and polymerization rates approach zero order. During polymerization, the size of the polymers increases until the glass-transition temperature of the polymer is higher than the reaction temperature at which point the reaction is terminated due to lack of mobility.

As a direct outcome of the manufacturing process, latices may contain not only polymers, but also surfactants, as well as traces of initiators and monomers that must be eliminated as completely as possible from the latices used in pharmaceutical applications. Water-soluble monomers are removed from the polymer particles simply by washing the products or by distilling, since initiators and monomers are more volatile than the formed polymers.

B. Pseudolatices

Although emulsion polymerization produces aqueous polymeric dispersions that are suitable for pharmaceutical applications, its use is limited by the potential presence of residual monomers and initiators in the final product. In the mid-1970s, scientists at the Center for Surface Coating at Lehigh University developed an alternative manufacturing method for polymeric dispersions that mechanically converted pre-existing water-insoluble polymers into colloidal polymeric dispersions. These polymeric dispersions are collectively referred to as pseudolatices. Later on, this technique was applied by scientists at Purdue University to manufacture aqueous polymeric dispersions of ethyl cellulose, an FDA approved polymer [3]. Since then, additional pseudolatices manufactured using other methods were developed and are being used for pharmaceutical applications. The three most common methods used to manufacture pseudolatices are emulsion-solvent evaporation (emulsion hardening), phase inversion, and solvent change.

1. Emulsion-Solvent Evaporation

During the manufacture of a pseudolatex by the emulsion-solvent evaporation method [10], the polymer is first dissolved in a water-immiscible organic solvent and then mixed with an aqueous phase that contains a surfactant and/or a stabilizer. The mixture is then emulsified to form an O/W type of emulsion (i.e., polymer as an internal phase). The crude emulsion is then subjected to ultrasonication or homogenization to generate a fine emulsion containing submicron droplets. Finally, the organic solvent is removed at elevated temperature and/or reduced pressure to form the pseudolatex. The process generates a polymer dispersion in which the polymer particles are so small that they cannot be observed by a conventional optical microscope. Aquacoat® (FMC Inc., Philadelphia) is prepared using this procedure.

2. Phase Inversion

Phase inversion [11] is a process in which the polymer and plasticizer first undergo hot melting or solvent gelation and are subsequently combined with long-chain fatty acids in an extruder to form a homogeneous mixture. A diluted alkali solution is slowly added to the mixture under vigorous agitation to form initially a W/O type of emulsion (i.e., water in polymer emulsion). As more aqueous alkali is added, phase inversion occurs, and a dispersion of polymer in water (i.e., O/W emulsion) is produced. Surelease® (Colorcon, West Point, PA) aqueous dispersion is manufactured using the phase inversion.

3. Solvent Change

Polymers with ionizable groups can be formulated into aqueous dispersions without the addition of emulsifiers/stabilizers by a method known as solvent change [12]. In practice, ionic polymers are dissolved in a water-miscible organic solvent or solvent mixtures to produce a polymer solution. This solution is then added to deionized water or the water is added to the solution with mild agitation. The organic solvents are then removed under vacuum to generate a pseudolatex. Absence of the emulsifier in the dispersions gives the pseudolatex system some unique properties, including good stability to heat and mechanical shearing, and good compatibility with organic solvents. Eudragit® RL 30D and RS 30D (Röhm Pharma, Weiterstadt, Germany) are prepared using this procedure.

III. STABILITY OF AQUEOUS POLYMERIC DISPERSIONS

Aqueous polymeric dispersions are thermodynamically unstable systems due to the large surface area of the dispersed particles and have a tendency to settle. The major forces that determine the stability of dispersed particles are (1) gravity of the dispersed polymer particles; (2) repulsive forces due to the same charges on the surfaces of the polymer particles; (3) entropic repulsion forces that are generated as a result of the steric hindrance of the solvent on the outside layer of the polymer particles; and (4) attractive forces due to London and van der Waals forces. Repulsion forces repel particles from each other and stabilize the dispersions, whereas attractive forces cause the polymer particles to agglomerate and destabilize the system. The stability of the system is therefore determined by the balance among these three forces. Aqueous polymeric dispersions are also susceptible to disturbance from a number of external factors such as electrolytes, surface active agents, mechanical agitation, thermal affect (freeze-thaw, heating), and organic solvents, and thus should not be subjected to these conditions during storage or use.

A. Effect of Electrolytes

The electrostatic charge that exists on the particle-liquid interface in dispersions can be introduced by various means, including the adsorption of ionic substances on the surfaces of the particles. The electric double layer, which is the region that extends from the surface of the particle into the bulk solution with zero surface potential, controls the stability of the dispersions against coagulation. Any condition that destroys the electrical double layer, such as the addition of electrolytes, destabilizes the polymeric dispersion and can lead to coagulation in which polymers precipitate from the aqueous phase. The minimum amount of electrolyte that is needed to bring about coagulation of a dispersion is called the critical coagulation concentration (CCC) and is directly related to the valence of the counter-ion of the coagulating electrolyte [13]. An aqueous polymeric dispersion is in part stabilized by the electrostatic repulsions derived from the surface charges of the dispersed particles. The isoelectric point is the pH at which the particles have zero charge. Below this pH, the particles carry positive charges, whereas above this pH they exhibit negative charges [14].

B. Effect of Particle Sizes

Reduction of the surface area of the latex particles decreases the system energy. Both theory and practice have proven that large particles yield better stability of latex systems as long as they remain in a colloidal state. Instability of the latex occurs when small and larger particles co-exist in the system due to small particle-interactions. In addition, coagulation tends to occur with latices that exhibit wide particle size distribution [15,16], presumably due to the Ostwald effect, in which smaller particles are dissolved or absorbed into large particles.

C. Effect of Surface Active Agents

A nonionic surfactant can be adsorbed onto the particle surfaces and stabilize the latex by forming a steric barrier. Optimal stability is achieved when the entire surface of the polymer particles is completely covered by the surfactant.

D. Effect of Mechanical Agitation

Agitation provides kinetic energy, thereby increasing collisions among the dispersed polymeric latex particles. This process counteracts repulsive forces that keep the particles apart and decreases the stability of the aqueous polymeric dispersions.

E. Thermal Effect

Elevated temperatures increase frequency of collisions among polymer particles and destabilize the dispersions. Reducing the temperature can stabilize the system, although temperatures below the freezing point force the dispersed particles closer together and may destabilize the system [17].

F. Effect of Viscosity

Two factors can contribute to a viscosity increase of an aqueous polymeric dispersion, i.e., addition of thickening agent and concentration of polymers in the dispersion. An increase in viscosity due to the addition of a thickening agent prevents settlement of polymer particles and hence stabilizes the system. However, viscosity also increases linearly with an increase in the concentration of polymer in the dispersion. At high levels of polymer concentration, the interparticle distance in the dispersion is short, and particle-particle interactions become significant, leading to a rise in viscosity [18]. Viscosity is also impacted critically by particle size distribution. For instance, large particles usually decrease the viscosity of dispersion and improve stability.

IV. CHARACTERIZATION OF AQUEOUS POLYMERIC DISPERSIONS

There are two important parameters that characterize a polymeric dispersions, particle size distribution and viscosity, which together determine the stability of the dispersion and the quality of films formed.

A. Particle Size Distributions

As indicated earlier, the physical stability of latex or pseudolatex dispersions depends in part on the particle size and size distribution of the dispersed particles. Particle sizes and their distribution can be conveniently monitored by using a microscope. Particle sizes as low as 1 μm can be measured using an optical microscope, while smaller particles are measured using an electron microscope. Measurement with an electron microscope, however, requires special precautions, since the particles tend to soften when exposed to the electron beam. Polymer particles that have low glass-transition temperatures have the tendency to soften under heat generated by the electron beam, especially if the measurement is carried out above the glass transition of the polymer. This problem can be overcome by carrying out the measurement at a lower temperature using an electron microscope fitted with a cold stage attachment [19]. Other techniques that have been reported include turbidity measurement and dynamic chromatography [20]. Turbidity measurement uses a curve-fitting technique to estimate the particle sizes of a dispersion system based on limited standards of known mean particle sizes [21,22]. This technique is not expected to give an accurate measurement of particle size distribution.

B. Viscosity

An important bulk property of aqueous polymeric dispersions is viscosity. Viscosity is also a valuable parameter that needs to be carefully monitored during manufacturing, handling, and coating applications of latices and pseudolatices. Viscosity measurements are usually carried out using a rotational viscometer.

Although clarity and conductivity of dispersed systems are important properties that are determined routinely in other industries, their application in pharmaceutical coatings is very limited and, therefore, is not described here.

V. LATICES USED IN PHARMACEUTICAL COATINGS

A. Eudragit L 30D and Eudragit NE 30D

The first aqueous polymeric dispersions developed for use in the pharmaceutical industry were based on a group of acrylic polymers, collectively known commercially as Eudragit polymers. These aqueous polymeric dispersions, which are prepared by emulsion polymerization, are marketed as Eudragit L 30D and Eudragit NE 30D and both contain approximately 30% (w/w) solids. The film former in Eudragit L 30D is a copolymer of methacrylic acid and methacrylic acid methyl ester with the ratio of acid to ester of 1:1 (Fig. 1A). The dispersion also contains sodium lauryl sulfate and polysorbate 80 as emulsifiers. Films prepared from this polymeric dispersion are insoluble in gastric fluid but are readily soluble in aqueous media of pH 5 or higher. Hence Eudragit L 30D has been widely used in enteric coating formulations. The dissolution rate of L 30D films at pH 7 is approximately 120 mg/min per gram [23]. Since the glass-transition temperature of Eudragit L 30D polymer is high ($> 60°C$), the dispersion is mixed with plasticizers before coating. Recommended plasticizers include polyethylene glycol, citric acid esters, and dibutylphthalate.

The film former in Eudragit NE 30D is a neutral copolymer based on ethyl and methyl methacrylic acid esters (Fig. 1B). The dispersion also contains the nonionic surfactant, nonoxynol 100, as an emulsifier. A Eudragit NE 30D aqueous dispersion produces insoluble film, swellable in solutions within the physiological pH, and hence is suitable for development of controlled-release products. This dispersion does not need to be plasticized due to the low glass-transition temperature of the polymer ($< 8°C$). However, it does require antiadherent such as talc, to minimize tackiness during the coating operation and the storage of the coated products. The films prepared from Eudragit NE 30D are not soluble, but are swellable in the entire gastro-intestinal region. Compositions of Eudragit products are summarized in Table 1.

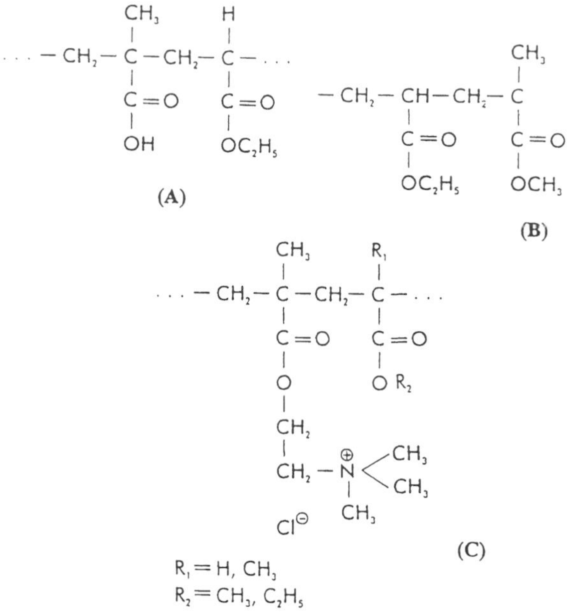

(A)

(B)

(C)

$R_1 = H, CH_3$
$R_2 = CH_3, C_2H_5$

Fig. 1 Structures of Eudragit: (A) L 30D; (b) NE 30D; and (C) RL/RS 30D.

B. Eudragit RL 30D and Eudragit RS 30D

The second group of acrylic-based aqueous polymeric dispersions is formulated from Eudragit RL 30D and Eudragit RS 30D. The film-forming polymers in these dispersions are composed of acrylic acid and methylacrylic acid esters that have low content of quaternary ammonium groups (Fig. 1C). The cationic density on the polymeric backbone is 1 per 20 repeating units for RL and 1 per 40 repeating units for RS. Films of these polymers are not soluble, but are swellable at physiological pH, and hence are suitable for sustained-release coatings. The small number of quaternary ammonium groups in these polymers does not change the hydrophobicity of the resulting films. Both Eudragit RL 30D and RS 30D contain sorbic acid as a preservative. The water vapor transmission rates for unplasticized Eudragit RL 30D and Eudragit RS 30D films are approximately 390 and 170 g/m^2 · day, respectively [24]. The dispersions are often used in combination with other water-soluble additives to form films that have the desired drug permeabilities. Due to a high glass-transition temperature of the polymers (~ 90°C) [25], Eudragit RL/RS 30D dispersions are plasticized prior to the coating op-

Table 1 Composition of Eudragit Products

	L 30D	NE 30D	RS/RL 30D
Polymer	30%	30%	30%
Emulsifier	Sodium lauryl sulfate, 0.21%	polysorbate 80, 0.69%	Nonoxynol 100
Preservatives	None	None	0.25% sorbic acid
pH	2–3	Neutral	Neutral

erations. Recommended plasticizers are polyethylene glycol, triacetin, dibutylphthalate, and citric acid esters.

C. Aquacoat

The first aqueous polymeric dispersion that was manufactured from a pre-existing thermoplastic polymer was Aquacoat. Aquacoat is an ethylcellulose aqueous dispersion produced by emulsion/solvent evaporation. The structure of cellulose is shown in Fig. 2 and the composition of the dispersion is given in Table 2 [26].

In addition to ethylcellulose, the film-forming polymer, Aquacoat contains sodium lauryl sulfate as an emulsifier and cetyl alcohol as a stabilizer. The size of the polymer particle ranges from 0.05 to 0.3 μm with an average diameter of 0.2 μm. Because ethylcellulose films are insoluble at physiological pH, Aquacoat is mainly used to sustain the dissolution or release of drugs from various dosage forms. However, water-soluble polymers and other excipients can also be incorporated into an Aquacoat formulation to form gastric-soluble coatings that can be used for immediate release products and for taste masking. Aquacoat does not contain a plasticizer and needs to be plasticized before use to ensure the formation of an appropriate film. Both water-insoluble and -soluble plasticizers can be used. Water-insoluble plasticizers tend to give better film properties, particularly when used for controlled-release purposes due to their high solubilities in the ethylcellulose polymer. In contrast, water-soluble plasticizers tend to partition in the water phase of the pseudolatex formulation and hence may not always efficiently interact with the polymer. Suitable plasticizers include dibutyl sebacate, diethyl phthalate, triethyl citrate, tributyl citrate, and triacetin. The plasticizer must be added to the Aquacoat dispersion, preferably following dilution with water, under moderate shear for 20–30 min before other additives are incorporated. Colorants that could be used in Aquacoat formulations include alcohol-based or propylene-glycol-based color dispersions, milled aluminum lakes, or opacifiers (titanium dioxide). The colorants should first be properly mixed with water or water-soluble polymer solutions under agitation to form suspensions that are added slowly into a plasticized Aquacoat formulation. Direct addition of alcohol-based colorants into an Aquacoat formulation causes the dispersion to coagulate. Since colorants tend to increase the water permeability of films, they are applied as part of an overcoat formulation over the sustaining coat of

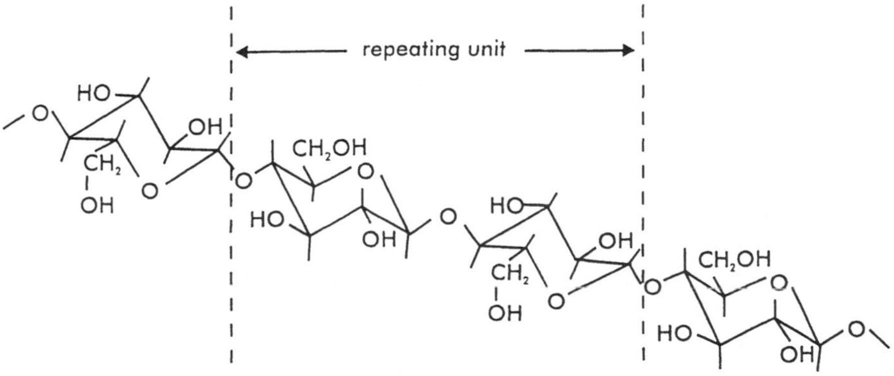

Fig. 2 Structure of cellulose polymer. Ethyl cellulose is the polymer in which -OH groups are substituted (partially or completely) with ethyl groups.

Table 2 Composition of Aquacoat Aqueous Dispersion

Materials	Weight fraction
Ethylcellulose	24.5–29.5%
Sodium lauryl sulfate	0.9–1.7%
Cetyl alcohol	1.7–3.3%
Total solid	29–32%
pH	4.0–7.0
Viscosity	150 cps

controlled release product. In other words, the inner layer, which contacts the core, is the rate-controlling Aquacoat film, and the outer layer carries the colorant.

Aquacoat films, prepared under normal processing conditions, undergo gradual post-coating coalescence. This is believed to last at room temperature for approximately three or more weeks before the film properties stabilize. Curing the film at elevated temperatures (40–75°C) accelerates the coalescence process. Consequently, Aquacoat-coated products are cured after the coating process is completed either in the same coating equipment or in a temperature-controlled oven.

D. Surelease

Another ethylcellulose aqueous dispersion that has had wide pharmaceutical application is Surelease. Surelease is manufactured by phase inversion and is a fully plasticized aqueous ethylcellulose dispersion that does not need additional plasticization before its application. The aqueous dispersion is simply diluted with water and used as is unless other additives are needed to modify the drug release characteristics of the films. Surelease is available in two forms, E-7050 and E-7-19010, as shown in Table 3. Formulation E-7050 contains dibutyl sebacate as a plasticizer while E-7-19010, developed to satisfy international pharmaceutical regulatory requirements, contains a fractionated coconut oil as plasticizer. The dispersions also contain oleic acid, which ionizes under basic conditions and imparts charges onto the polymer particles to help stabilize the pseudolatex. Ammonium hydroxide is used to bring the dispersion to pH 11.

Table 3 Composition of Surelease Aqueous Polymeric Dispersions

E-7050[a]	E-7-19010[a]
Ethylcellulose	ethylcellulose
—	Fractionated coconut oil
Dibutyl sebacate	—
Oleic acid	Oleic acid
Ammonium hydroxide	Ammonium hydroxide

[a]Total solid, 24–26%; Final pH 9.5–11.5.

E. Other Aqueous Dispersions

Aqueous dispersions containing silicone elastomers were also developed by Dow Corning Corp. A number of articles that describe the use of such dispersions for possible pharmaceutical applications have been published [27,28]. Interest is based on the unique properties of silicone. The basic structure of silicone polymers is polydimethylsiloxane (PDMS). Pure PDMS polymers have well-established safety profiles and are generally considered biocompatible. In addition, PDMS and its derivatives have been used for a long time in coating and adhesion industries. Silicone elastomer latices used by these industries are manufactured by emulsion polymerization followed by cross-linking of silicone polymers [29]. In addition to the ethylcellulose polymers, several other cellulose ester polymers, such as cellulose acetate, cellulose triacetate, and cellulose acetate butyrate, are currently being investigated by FMC Corp. to see if they are suitable for the development of aqueous polymeric dispersions.

VI. MECHANISM OF FILM FORMATION

The mechanism of film formation from aqueous polymeric dispersions has been thoroughly studied and is well understood. Irrespective of the manufacturing technique, aqueous polymeric dispersions may contain, besides the film-forming polymers, emulsifiers, stabilizers, and/or plasticizers. Moreover, additional ingredients may be incorporated in the final formulation to modify the properties of the films. Various theories have been proposed to explain the mechanism of film formation from formulations containing aqueous polymeric dispersions, which are described below.

A. Viscous Flow Theory

One of the early theories that describes film formation from latex formulation (i.e., latex particles) was the viscous flow theory, which is based on the so called sintering of powdered particles [30]. Sintering is defined as the bonding of two or more particles by the application of heat at a temperature below the melting point of any component of the system. The sequence of events that occur during the sintering of powdered particles include viscous flow, evaporation-condensation, volume diffusion, and surface diffusion.

The cross section of two sintered spherical particles is shown in Fig. 3. The radius of the latex particles is r and the half contact angle is θ. The change in surface area that occurs as latex particles coalesce generates sufficient energy to induce viscous flow [31]. When dealing with a single sphere, two forces are considered: (1) sur-

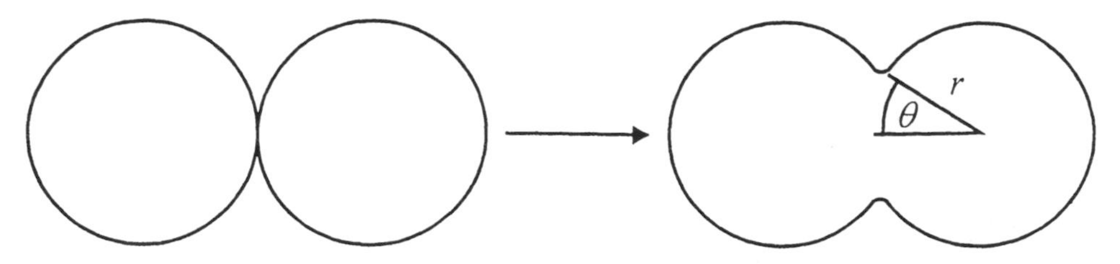

Fig. 3 Fusion of two contacting particles through viscous flow. (From Ref. 31.)

face tension, which is generated on the surface and tends to shrink the sphere to reduce the surface area; and (2) shearing stress, which is generated from inside the radius and tends to expand it. At equilibrium, the two forces are equal and the shape of the sphere remains unchanged. Therefore, the surface free energy is equal to the shear energy as described in Eqs. (1) and (2). That is

$$\pi r^2 P = 2\pi r \gamma \tag{1}$$

$$P = \frac{2\gamma}{r} \tag{2}$$

where r is the radius of a sphere; P is shear stress; and γ is surface tension. Equation (2) demonstrates that the shearing stress is proportional to the surface tension and inversely proportional to the radius of a sphere.

If there is sufficient shearing stress, the two latex particles are brought in close contact and eventually coalesce. The degree of coalescence between two spheres is measured by Frenkel's equation, which describes the process as purely viscous flow

$$\theta^2 = \frac{3\gamma t}{2\pi\eta r} \tag{3}$$

where γ is the surface tension; η is the viscosity coefficient; r is the radius of the sphere; and θ is the contact angle.

When a coating dispersion is applied on a surface, the viscosity of the dispersion increases progressively due to water evaporation. This process continues until the dispersed polymer particles touch each other. As a result, the void between polymer particles decreases, which in turn exerts a shearing stress on the surrounding polymer particles as described in Eq. (2). The shearing stress continues to increase as the size of the void decreases. Alternatively, the void between two polymer particles can also be defined by the contact angle, which is related to the physical properties of the film and is suitable for the measurement of latex sintering. As the contact angle increases, the voids decrease, and the tensile strength increases. Equation (3) illustrates that a decrease in either r, the radius of the sphere, or η, the viscosity coefficient, leads to an increase in θ, the contact angle. The radius of the sphere r is directly related to the particle size of the latex spheres, and θ is a function of the properties of latex particle and the processing temperature. Therefore, a decrease in the particle size of the latex particles enhances sintering or coalescence. Addition of the plasticizer or an increase in processing temperature produces better fused films with high tensile strength. Since processing temperature is determined by many other factors, such as equipment and facility limitations, as well as thermal sensitivity of the active ingredients, incorporated of a plasticizer into the aqueous polymeric dispersion is usually considered first.

In short, the viscous flow theory states that sintering of latex particles takes place as a result of the shearing stress generated by the surface tension of the particles, a process described by Eq. 3. In addition, more cohesive films can be formed by using coating dispersions that consist of small polymer particles, exhibit low viscosity, contain plasticizers, and are processed at elevated temperature. This is due to enhanced sintering or coalescence of the latex particles.

B. Capillary Force Theory

Although the viscous flow theory adequately addresses the coalescence process, it does not fully account for the following events that occur during film formation [32]:

1. The formation of films from many latices and pseudolatices proceeds concurrently with water evaporation and ceases when the evaporation of water is complete. These events imply that the polymer/water interfacial tension, rather than the polymer surface tension, provides the shearing stress for sintering.
2. The rate of water removal determines the rate of coalescence of the polymer particles and can be altered by changing the capillary pressure or relative humidity surrounding the substrate.
3. Porous films containing some uncoalesced particles may be formed if the temperature is kept below a certain critical value during water evaporation.
4. Partially cross-linked latex polymers can form continuous films, even though the mutual interpenetration of the latex particles is hindered by cross-links.

As a result, the capillary force theory was developed to address the deficiencies of the viscous flow theory. Film formation is a thermodynamically favored process, since it is accompanied by a decrease in free energy associated with the reduction in total surface energy of the dispersed particles. The forces that contribute to film formation are *Fs*, the force produced by the negative curvature of the polymer surface when two particles come into contact with each other; *Fc*, the capillary force generated from the water-polymer interfacial tension that exists in the interstitial capillary system during the water evaporation phase; *Fv*, van der Walls attraction forces that are formed between adjacent spheres; and *Fg*, gravitational forces that induce settling in a dispersion. These forces are opposed by other forces that resist polymer sphere deformation, and include the surface tension of the sphere F_G, which maintains the shape of the sphere; and the coulombic repulsion forces *Fe*, which exist between two charged spheres. Therefore, for coalescence to occur, the following relationship must exist not only initially but throughout the course of the coalescence process:

$$\mathrm{Fs} + \mathrm{Fc} + \mathrm{Fv} + \mathrm{Fg} > \mathrm{F_G} + \mathrm{Fe} \quad (4)$$

The capillary force generated from interfacial tension is low, especially when the dispersed polymer particles contain surfactants [33,34]. In general, however, film formation takes place when the capillary force, *Fc* overcomes the resistance to deformation by the dispersed particles F_G, i.e.

$$\mathrm{Fc} > \mathrm{F_G} \quad (5)$$

The capillary force theory is illustrated in Fig. 4, where *r* represents the radius of a polymer particle and *R* represents the radius of a water droplet. When a polymeric-dispersion-based coating formulation is applied onto a surface, the polymeric particles are initially separated by water. As the water evaporates, the particles come closer together until the capillary pressure increases and the capillary pressure reaches a maximum when the water level in the interstitial capillary space approaches minimum. This situation is depicted in Fig. 5 and indicates that the maximum capillary pressure is reached when the half contact angle is 30°.

If the polymer is sufficiently rigid to resist deformation while the capillary pressure lasts, water evaporation continues without concurrent particle coalescence. On the

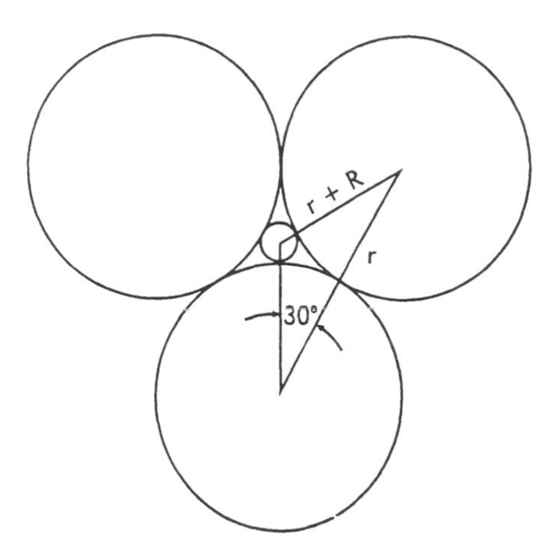

Fig. 4 Contracting forces for polymer particles coalescence, as a result of capillary pressure generated from water-polymer interfaces. (From. Ref. 32.)

other hand, if the force is sufficient to induce particle deformation, coalescence occurs. The contact area between water and a polymer particle decreases as water evaporates, which results in a decrease in the contact angle. Thus, as the curvature of the water surface increases, there is an increase in the capillary force with an initial increase in the capillary pressure, which is followed by a decrease in the capillary forces and the capillary pressure as the water evaporates. In summary, as coalescence proceeds, both the geometry of the deformed particles and contact angle between particles change, and result in a change in the forces required for continuous coalescence.

There are two extreme types of deformation of polymer particles: purely elastic deformation and purely viscous deformation. During elastic deformation, the force is solely dependent on the properties of the particles. If at any point during the process the capillary force becomes insufficient to induce complete deformation, the partially

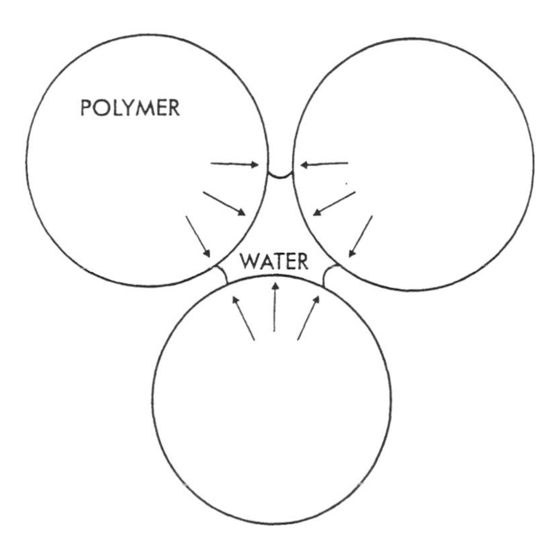

Fig. 5 How maximum capillary pressure is generated. (From Ref. 32.)

deformed particles rebound back to their original shape. During pure viscous flow, the polymer particles deform in proportion to the applied force. If the polymer flow is sufficiently rapid to move into the void space that existed before water is removed, then coalescence can occur as the water evaporates during the drying phase. Coalescence continues to proceed even after the complete evaporation of water due to the forces created by surface tension and gravity. The necessary condition for film formation is described by the following equation:

$$G < \frac{35\gamma}{r} \tag{6}$$

where G is the shearing modulus; γ is the surface tension; and r is the radius of the polymer particles.

C. Autoadhesion Theory

When an aqueous polymeric dispersion is applied over a surface, the milky dispersion is gradually transformed into a transparent film, a sign usually viewed as complete coalescence. Upon aging, the physical properties of the film generally improve. Electron microscopic inspection of fresh films showed that contours of individual particles were still present even though the film was clear and transparent. These contours, however, gradually disappeared during storage, a phenomenon that cannot be explained by the capillary pressure theory, particularly when water is absent from the system. However, another theory, the autoadhesion theory proposed by Voyutskii [35], can explain the above observations well. According to this theory, simple physical contact between two polymer particles cannot form a stable and continuous film. Conglomeration or autoadhesion of polymer particles is caused by mutual interdiffusion of the free polymer chain ends when the terminal segment of polymer chains diffuse from one globule (polymer coil) to another to form a cohesive film caused by extensive chain entanglements. Consistent with this theory, Chainey et al. also attributed the decrease in gas permeability of films, an indirect measure of the formation of continuous films, to the gradual loss of mobility of polymer chain ends [36].

According to thermodynamic principles

$$\Delta F = \Delta U - T\,\Delta S \tag{7}$$

where F is the free energy; U is the internal energy; S is entropy; and T is the temperature of the system. Considering that autoadhesion generally requires some length of time for polymer interdiffusion to take place and that the rate of autoadhesion can be significantly increased by an increase in the contact time between two polymer particles, formation of cohesive films depends mainly on the contribution of the entropy factor. Vanderhoff et al. suspended polymer spheres in a water droplet to demonstrate the coalescence process [37]. When water evaporates during the drying process, the polymer particles are brought into close proximity until the stabilizing layers on the particles resist further closer contact. As water continues to evaporate, the force created by the surface tension of water pushes the particles together until the stabilizing layers are ruptured, resulting polymer-polymer contact. At this point, forces derived from the polymer/water interfacial tension exert additional pressure on the particles. Pressure-induced fusion of polymeric particles, as required by the autoadhesion theory, is essentially independent of latex particle sizes. It asserts that particle size has no effect on either the minimum film formation temperature (MFT) of formulations based on aqueous polymeric dispersions or on the overall process of film formation. However, other investigators subsequently discovered that MFT increases with an increase in the size of the dispersed particles [38], and their findings have led to the development of additional models.

D. Viscoelastic Deformation Theory

The key assumption behind the models cited above is that polymer deformation is an elastic phenomenon, i.e., polymer particles will rebound back after release of the applied force. Actually, polymers are viscoelastic materials that respond to an applied force by a combination of viscous flow and elastic deformation. The contact area of viscoelastic materials varies with time, and consequently analysis of the deformation process is complicated. Lee and Radok [39] derived a model based on the Maxwell element by assuming a creep deformation during latex film formation as explained by

$$\frac{1}{J_c(t')} \leq \frac{34\sigma}{R} \tag{8}$$

where $Jc(t')$ is the time-dependent creep compliance of the polymer when particles are in contact with each other; R is the radius of the particle; and σ is the polymer-water interfacial tension. In a linear viscoelastic region, the shear modulus $G(t)$ is equal to $1/Jc(t')$, a relationship that is in agreement with Brown's capillary model. Eckersley and Rudin [40] had experimentally verified such a prediction by comparing results from an electron microscopic examination of the coalescence of a synthetic latex emulsion with the model-predicted particle deformation.

The mechanism of film formation from aqueous polymeric dispersions can be divided into two stages [41]. In the first stage, an irreversible contact of polymer particles occurs as a result of interfacial pressure or capillary pressure. The driving force behind this phase of film formation is the rate of water evaporation. In this stage, interfacial tension along with the capillary forces cause the coalescence of latex particles. The second stage is characterized by autoadhesion or mutual interdiffusion of polymer chain ends to form a continuous cohesive film. The degree of film formation is a function of the concentration of polymer particles and the drying temperature above its MFT.

VII. CHARACTERIZATION OF POLYMERIC FILMS

Over the years, the paint industry found it necessary to develop a series of testing procedures to evaluate coatings and correlate their film characteristics with their coating performance. Some of the tests, which were later adopted by the American Society for Testing and Materials (ASTM), are still widely used today by various industries, including pharmaceutical companies [42]. These tests are used for quality control purposes or to predict performance of the materials. While the tests can never completely predict actual performance, quality control tests are used to evaluate coatings. Generally, coatings must have good adhesion to their substrates, should have an attractive appearance and be glossy, should provide complete coverage of the substrate, should be abrasion resistant, and should have exterior durability (no peeling or cracking during storage), to mention but a few. These properties are measured indirectly using the tests recommended by ASTM. Tests include tensile strength and modules, tensile strength-to-break, elongation-to-break, work-to-break, loss and storage modules, and loss tangent (tan δ). Parameters obtained from these measurements are, to a large degree, related to the coating structure and are correlated with its actual performance [43]. In addition, pharmaceutical coatings must meet stringent drug safety standards, be biocompatible, be permeable to water and drugs, and exhibit nonreactivity toward any other excipient in the dosage forms.

A. Free Film Characterization

The ultimate goal for the evaluation of free films is to determine the physical and mechanical properties of their coatings and predict their performance in a relatively simple and efficient way. Therefore, testing should be properly designed so that meaningful conclusions can be made. Measurements of film properties are expressed according to the ASTM procedures in which a film is defined as a sheet of material having uniform thickness not greater than 1.0 mm. Some of the commonly used techniques that are used to characterize free films are described below.

1. Thermal Analysis

The polymeric components of aqueous dispersions that are suitable for pharmaceutical applications are viscoelastomers; i.e., components of these polymers exhibit both elastic and viscous behavior. Viscoeslasticity also refers to both the time and temperature dependence of mechanical properties. Figure 6 depicts the five regions of viscoelastic behavior of polymers as a function of temperature. Thermal analysis, which provides a measure of the mechanical properties of polymeric films as a function of tempera-

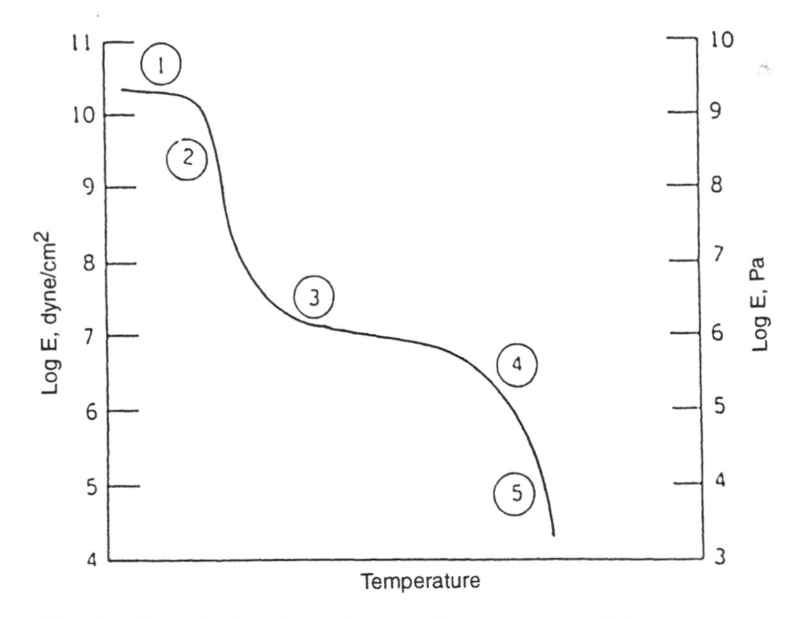

Fig. 6 Viscoelasticity behavior of a linear, un-cross-linked, amorphous polymer: region 1, glassy region; region 2, glass transition region; region 3, rubbery plateau region; region 4, rubbery flow region; region 5, liquid flow region. (From Ref. 46.)

ture, is ideal for the characterization of films derived from aqueous polymeric dispersions. It is particularly useful for the measurement of the glass-transition temperature T_g, a critical property of amorphous polymers. Glass-transition temperature is defined as the maximum rate of decline of the modulus at the glass-transition region, as shown in Fig. 6.

Thermal analysis can be based either on energy change or polymeric segmental motion that occurs during phase transition, both of which are described below.

2. Analysis Based on Energy Change Associated with Phase Transition

Differential scanning calorimetry (DSC) and thermal differential analysis (TDA) are used to determine the energy change that occurs during phase transition. Differential scanning calorimetry and TDA operate under the same principle except that DSC records the energy required to establish the zero temperature difference between the sample and the reference whereas TDA records the difference in temperature between the sample and the reference. Though widely used, DSC and TDA may not be sufficiently sensitive to determine the glass-transition temperature T_g of films prepared from aqueous polymeric dispersions because both techniques measure secondary indications associated with T_g instead of the transition itself.

3. Analysis Based on Segmental Motion of Polymer Chains

The most widely used technique to determine segmental motion of polymeric chains in films are TMA and DMTA. Thermal mechanical analysis (TMA) records T_g as a temperature at which materials become soft, indicating transition from a glassy region to a rubbery state. However, this measurement is not sufficiently sensitive for most pharmaceutical coatings since the glass-transition region is not sharp. What is actually measured is the softening temperature. Dynamic mechanical thermal analysis (DMTA) is perhaps the most reliable method for determining T_g. Commonly used dynamic mechanical instruments measure the deformation of a polymeric material in response to vibrational forces. As indicated earlier, the majority of polymer coatings are viscoelastic materials that have both elastic and viscous properties. The elastic portion of the polymer can absorb mechanical energy and its deformation under force can be restored upon the removal of the force. The viscous portion of the polymer dissipates energy as heat. In a DMTA test, a torsional strain with certain frequency is applied on the polymer film to cause deformation. The resulting stress is monitored as a function of temperature. Dynamic mechanical thermal analysis measurement generates a profile of the modulus of a polymer film as a function of temperature under certain frequency. The temperature that corresponds to the steepest decline of the modulus represent T_g [44].

As described previously, viscoelasticity is a time- and temperature-dependent phenomenon. The rate of applying heat is, therefore, critical to the measurement of T_g. For example, T_g increases as the heating rate increases, which makes the technique sensitive to operating conditions. The absolute T_g is, therefore, virtually impossible to measure experimentally. Thus, it is critically important to specify the conditions under which T_g is measured in order to make meaningful comparisons.

B. Mechanical/Dynamic Tests

Mechanical testing is the most frequently used technique to evaluate pharmaceutical coatings. It measures the strength or toughness of films and their deformation characteristics. A brief description of two types of mechanical tests is given below.

1. Stress-Strain Tests

A mechanical test that has been extensively used to characterize films is the stress-strain test [45]. Strain ε is deformation of a material caused by the application of an external force. It is obtained by dividing elongation of film by the original length of the film

$$\varepsilon = \frac{\Delta L}{L_0} \tag{9}$$

Stress or tensile stress, σ is the resistance to deformation developed within a material after it has been subjected to an external force. Its unit is "force per unit area" and is determined by dividing the maximum force before the break by the original minimum cross-sectional area of the film.

Young's modules E is a measure of the stiffness of the film produced from an elastic material and is expressed as the ratio of a change in stress to a change in strain within the elastic region of the material

$$E = \frac{\sigma}{\varepsilon} \tag{10}$$

In a stress-strain test, a sample is deformed at a constant rate and the built-up stress is measured over time. A typical stress-strain curve is presented in Fig. 7 [46]. Parameters that can be derived from such a curve are percent elongation at break (the ratio of maximum elongation L to the original length L_0), and the yield point (the point beyond which elongation is not completely recoverable and permanent deformation takes place).

If a shearing or a twisting motion is applied instead of elongation, the measure of stiffness is shear modulus G. The shear modulus is calculated as a ratio of shear force f and shear strain s and can be determined using Maxwell and Kelvin elements

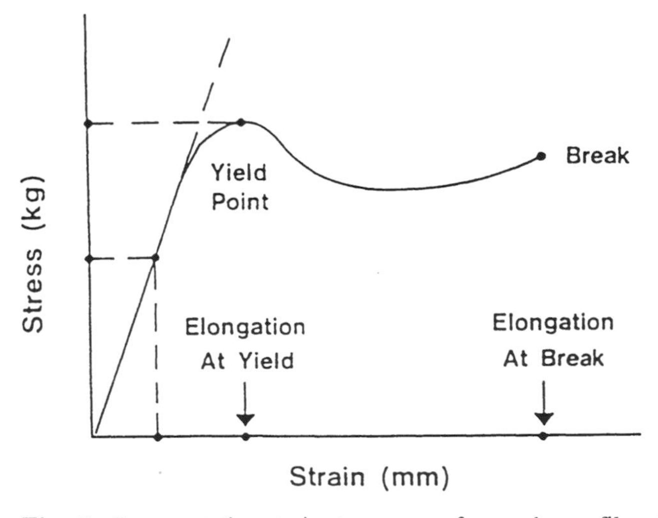

Fig. 7 Representative strain-stress curve for a polymer film. (From Ref. 58.)

$$G = \frac{f}{s} \tag{11}$$

2. *Stress Relaxation and Creep Tests*

In a stress relaxation experiment, the sample is rapidly stretched to a required length, and the stress that is built up in the sample is recorded as a function of time. The length of the sample remains constant, as does the temperature. Creep experiments are conducted in an inverse manner. A constant stress is applied to a sample, and the dimensions of the samples are recorded as a function of time [47].

A stress relaxation test determines the toughness of the material as a function of time, i.e., modulus $E(t)$, whereas a creep test measures the softness of material, i.e., the compliance $J(t)$. Under conditions far removed from phase transition, compliance is approximately equal to the reciprocal of modulus, i.e., $J = 1/E$.

Mathematical analysis of the relaxation and creep phenomena can be carried out using spring and dashpot elements. A spring can be stretched instantly under stress and then held under the same stress indefinitely. In a dashpot, the plunger moves under stress through a viscous fluid at a rate proportional to the stress; removal of the stress does not lead to recovery. Data obtained from springs and dashpots can be combined to develop mathematical models of viscoelastic behavior of polymers. Maxwell and Kelvin elements, as illustrated in Fig. 8, are two of the simple spring and dashpot arrangements. The spring and dashpot are configured in series in the Maxwell element, while they are arranged in parallel in the Kelvin element. Creep behavior of the Maxwell and Kelvin elements are shown in Fig. 9.

C. Permeability Tests

1. *Solute Diffusion Test*

The rate of diffusion of a solute through a free film can provide a good measure of the inherent properties of the film, especially as related to the diffusion rate of the drug through its coating. Based on the mass transport principle, diffusion is a process in which molecules move from a position of high concentration to a position of lower concentration by random motion. For diffusion to proceed through films or other condensed

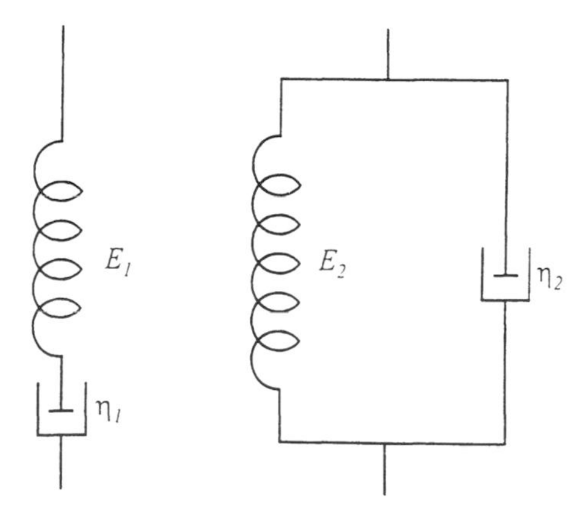

Fig. 8 Maxwell and Kelvin Elements. (From Ref. 47.)

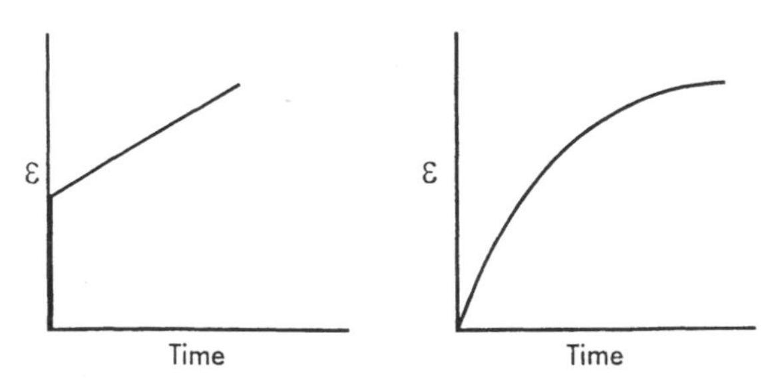

Fig. 9 Creep behavior of Maxwell and Kelvin elements. The Maxwell element exhibits viscous flow during deformation, whereas the Kelvin element reaches an asymptotic limit to deformation. (From Ref. 47.)

structures, the permeating materials must have sufficient kinetic energy to move to a new position, and the pores in the film should have a size equivalent to or greater than the size of the diffusive material or the drug. The total porous space in a diffusion medium is known as the free volume. Any change that causes a decrease of polymer flexibility and a reduction of the free volume will decrease the diffusivity of the permeating molecules. Another factor that determines permeation is the hydrophilicity or hydrophobicity of the polymer that makes up the film.

Permeability of a film can be determined using an equation derived from Fick's Law [48,49]

$$J = -D\frac{dC}{dx} \tag{12}$$

where J is the mass flux of solute across a unit area in a unit time; dC/dx is the concentration gradient of solute across a film over a small distance; and D is the diffusion coefficient of the solute in medium (film). Here, D is directly related to the free volume of the film and size of solute molecules.

If the concentrations of solute on either side of the film in the donor and receiver compartment are maintained constant and the concentration gradient of solute across the film becomes constant, the steady-state flux is given by

$$J = \frac{KD}{h}\frac{\phi}{\tau}(C_d - C_r) \tag{13}$$

where K is the film/liquid partition coefficient; h is the film thickness; ϕ is the nonfilled polymer free volume; τ is the film tortosity; and C_d and C_r are the concentrations of solute in the donor and receiver chambers, respectively.

2. *Water Vapor Transmission Test*

Another parameter that is used to characterize free films is water vapor transmission (WVT). Such studies can be used to evaluate film integrity and the permeability of the film to water vapor. Water vapor permeability, as defined by ASTM, is "the rate of water vapor transmission through unit area of flat material of unit thickness induced by unit vapor pressure difference between two specific surfaces, under specific temperature and humidity [50]. The WVT properties of pharmaceutical coatings are critical, especially when the films are used as protective coatings or to coat hygroscopic or moisture-sensitive dosage forms. The WVT of films can be determined using [51]

$$\frac{dQ}{Adt} = \frac{P}{h}(p_{\text{water}_i} - p_{\text{water}_o}) \tag{14}$$

where dQ/dt is the mass of water vapor crossing the film in a unit time (g/hr); A is the area of the film where transmission of water vapor (cm^2) occurs; P is the water vapor transmission constant (g/cm·hr·mmHg); h is the film thickness (cm); and $(\rho_{\text{water}_i} - \rho_{\text{water}_o})$ is the difference of partial water vapor pressure across the film (mmHg). The water vapor transmission constant is usually used to compare the permeability of different films to water vapor under different conditions [52,53]. Standard devices that are used to determine WVT are described in ASTM publications [54]. However, WVT can also be determined using other devices as long as the necessary conditions expressed by Eq. (14) are satisfied.

Two methods that are routinely used to measure WVT are the desiccant method and the water method. In the desiccant method, the film is sealed to an open mouth of a test dish containing a desiccant. The assembly is then placed under controlled environmental conditions, i.e., controlled temperature and humidity. The mass transport of water vapor through the film into the desiccant is then determined by periodic weighing. In the water method, a dish containing distilled water is sealed by a test film. The assembly is then placed under controlled temperature and humidity. The whole assembly is periodically weighed to determine weight loss due to water vapor transmission through the film.

3. *Oxygen Permeation*

Although limited studies regarding oxygen permeation through pharmaceutical coatings have been reported, it is likely that oxygen plays a significant role in the stability of drugs susceptible to oxidative degradation. Permeation of oxygen is based on the same principle as that of WVT. However, it is not as straightforward and may be difficult to precisely quantify.

D. Characterization of Film Coatings

1. *Appearance Tests*

The simplest method to evaluate coated surfaces is by appearance. Appearance tests comprise physical tests that measure opacity, roughness, and gloss. Coatings are also visually inspected for defects.

2. *Adhesion*

Adhesion of a film to the surface of a tablet is essential to achieve the desired level of coating without any loss of coating material from the tablets due to peeling off of the film during the coating process. Scientists have, therefore, developed a peel test to evaluate the forces of adhesion between the film and tablet surface. Some investigators used a modified tensile test [55]; other measured the force required to remove a film at a 90° angle from the surface of a tablet [56]. This was accomplished by first pressing a piece of double-sided adhesive tape onto the film-coated tablet so that the tablet can be glued onto a stationary surface. Then, the area of film under test is cut off with a sharp knife and removed from the rest of the coatings. Adhesion is reported as the force required to remove the coating from unit area of the tablet surface.

3. Indentation

Indentation is another method used to characterize coatings. It is a measure of the resistance of deformation (hardness). The procedure involves the application of a known load on an indenter tip of known surface area that, in turn, penetrates the test coating [57,58]. Since the depth of penetration can be time dependent, the testing period should be specified, and is usually 30 sec. The hardness if calculated as the load applied to the indenter divided by the area of the indenter tip.

4. Puncture and Shear

Pharmaceutical coatings [59] can also be directly examined using puncture or shear tests. The puncture test involves the measurement of resistance of the coating to deformation using a puncturing probe. Stress is then measured as a function of probe displacement. Properties, such as elongation to puncture and puncture strength, are calculated. The shear test measures resistance of the coating to shear, and shear strength is then calculated.

VIII. EFFECT OF FORMULATION AND PROCESSING FACTORS

The properties of films or coatings are influenced by both formulation components and processing conditions, and consequently a change in either of these is usually accompanied by significant changes in the properties of the coatings. The effects of some of the critical formulation and process parameters on film characteristics are discussed below.

A. Formulation Variables

While the polymeric component of aqueous dispersions mainly dictates the physical properties of the coatings, various formulation parameters also play a significant role during and subsequent to film formation.

1. Solid Content

Polymer concentrations in an aqueous polymeric coating dispersion are mainly determined by the type of polymers dispersed in the aqueous dispersion and the physicochemical properties of the active substance present in the core. In general, the solid content should be as high as possible to reduce the coating time as long as the quality of the coating is not affected. However, in certain circumstances, solid content could alter the drug release profile even though the weight gain due to the coating itself remains constant and needs to be optimized. For instance, when theophylline pellets were coated with an aqueous formulation of an ethylcellulose dispersion at 35°C for 3% weight gain using formulations containing different polymer contents, the release of drug was different. A formulation containing 5% of ethylcellulose polymer produced a lower rate and extent of drug release during in vitro dissolution tests than those formulations containing 10% and 15% polymer. A similar observation was also made with Eudragit RL/RS 30D coatings [60]. Such a finding indicates that, at lower concentrations, more layers of coating, and hence more continuous films, are formed for a given coating thickness. Such a relationship between solid content and coating performance is also dependent on the coating conditions (i.e., coating temperature, spraying rate, bed load, etc.) and can be potentially minimized by properly optimizing the coating conditions.

2. *Plasticizers*

Plasticizers are critical formulation variables that facilitate proper film formation and hence have an impact on the quality of the resulting film [61,62]. Commonly used plasticizers include triacetin, triethyl citrate, polyethylene glycol 200 to 8000, glycerol monostearate, tributyl citrate, tributyrin, acetyltriethyl citrate, dibutyl sebacate, diethylsebacate, acetyl tributyl citrate, and dibutyl phthalate. The amount of plasticizer required in a formulation can be calculated using an equation derived by Couchman [63]

$$\frac{1}{T_g} = \frac{w_1}{T_1} + \frac{w_2}{T_2} \tag{15}$$

where w_1 and w_2 are the weight fractions of the polymer and plasticizer in formulation; T_1 and T_2 are the corresponding glass-transition temperatures (°K) of polymer and plasticizer, respectively; and T_g is the glass-transition temperature of the resulting coating. If the glass-transition temperatures of the polymer and plasticizer are known, the amount of plasticizer required to plasticize the polymer to attain a given transition temperature can be calculated. When the calculated amount of plasticizer from Eq. (15) is incorporated into a coating, the resulting film will have a designated glass-transition temperature. Significant deviation from the theoretical value will either cause incomplete film formation due to insufficient plasticizing or a tacky film due to over plasticizing.

The hydrophilicity of a plasticizer can also have a profound effect on the resulting film properties due to differences in partitioning between the polymer and water. For example, triethyl citrate, a relatively highly water soluble plasticizer, has a much greater effect on reducing film puncture strength and the increase in the elongation of films prepared from Eudragit RS 30D formulations than the water-insoluble plasticizer, acetyl tributyl citrate [64]. However, the presence of either plasticizer in the films did not have any detectable effect on the moisture uptake of the coated dosage forms.

3. *Emulsifiers and Stabilizers*

Emulsifiers or stabilizers are added into aqueous polymeric dispersions to stabilize the dispersed polymeric particles. They also exert a strong influence on the colloidal properties of the dispersions and the corresponding films.

Aqueous polymeric dispersions may contain stabilizer(s) to enhance the physical stability of the dispersion during storage or transportation. Stabilizers remain in the film either as an independent network or dissolve in the polymer during the water evaporation phase. In the latter case, the stabilizer can soften or swell the network and enhance film formation [65]. In some systems, these stabilizers are squeezed out of the film, i.e., form extrusions or blisters on the surface of film due to poor miscibility between polymer and the stabilizers. Eventually, the stabilizers may slowly evaporate or sublime, spread over the surface of film, or get oxidized to produce different molecules that have better solubility in the polymer. In contrast, emulsifiers may reside at the interface between the particles, form small islands on their own, or diffuse into the polymer particles [66].

4. Water-Soluble Additives

Hydrophilic polymers, such as PEG and PVP, have a dual effect on film properties and drug permeation. They may function as plasticizers and hence lower the T_g of the film and promote coalescence of the polymeric particles. During dissolution, these polymers tend to leach out from the film and dissolve in the releasing medium, leaving behind pores through which drug molecules can diffuse easily. The size of the pores and thus the rate of the drug release depends on the molecular weight of the additives. For instance, a dissolution test in 0.06N HCl involving PEG containing Eudragit RS 30D film coatings indicated that the drug release rate can be enhanced by an increase in the molecular weight of PEG from 200 to 8000 Daltons [67].

5. Inorganic Additives

High-density inorganic substances, such as titanium dioxide, silica, talc, kaolin, etc., are frequently incorporated into aqueous polymeric dispersion formulations to enhance film strength, to reduce tackiness, and to improve the appearance of coatings. Selection of an inorganic additive for a coating formulation depends on chemical compatibility of the additive with other excipients in the formulation, on the intended purpose, and on viscosity limitations. There is a maximum amount of inorganic additive that a coating formulation can tolerate. This amount is known as the critical pigment volume concentration (CPVC), where pigment refers to inorganic additives and solid colorants. At the CPVC, a characteristic phase transition takes place. When the pigment volume concentration is below the CPVC, the coating is a continuous film in which the pigment particles are randomly distributed. Above the CPVC, voids are formed in the film due to an insufficient amount of polymer. Consequently, the film becomes locally discontinuous and loses its mechanical strength [68].

6. Secondary Polymeric Dispersions

Sometimes, aqueous polymeric dispersions may be added to the primary dispersion to modify the drug release characteristics, provided that the primary and secondary dispersions are compatible. For instance, Eudragit RS and RL 30D are frequently used in combination in order to achieve a desired film permeability.

B. Effect of Process Variables

The selection of the appropriate coating equipment and optimum processing conditions are essential factors that significantly influence the formation of continuous and reliable coatings. Processing parameters need to be optimized to achieve the complete coalescence of the polymer particles during or subsequent to the coating operation. Critical parameters that have a severe impact on the formation of optimum films include product temperature, inlet air humidity and temperature, air flow, spraying rate, and post-coating curing. While each of these parameters individually influences the properties of coatings, it is the interplay between the various parameters that collectively determines the quality of the final film.

Drying temperature is a critical process variable that dictates the physical nature of the films. Selection of the drying temperature should be based on the properties of the dispersed polymer. In general, product temperature is kept above the glass-transition temperature of the film to promote mobility of the polymer chains, which in turn allows the interpenetration of the terminal polymer segments. If the product temperature is kept below the glass transition temperature, it tends to slow down or even retard polymer coalescence and leads to prolonged drying time and/or formation of incomplete films. Product temperatures that are significantly above the T_g of the film may cause tackiness and increase agglomeration of the coated particles. A product temperature that is too high may also lead to premature water evaporation, thereby leading to the formation of discontinuous or porous films. The effect of drying temperature on the structural makeup of the coatings can be evaluated through dissolution testing of the coated products [69]. For example, diphenhydramine hydrochloride pellets were first coated with Aquacoat sustained-release coating formulation to a 15% weight increase. An additional coating of 1% of water-soluble polymer was then applied as an overcoat to reduce tackiness during high-temperature curing. The coated pellets were tested for drug release in distilled water using the USP II apparatus. As shown in Fig. 10, the release rate of diphenhydramine hydrochloride decreased when the coating temperature was increased from 22 to 35°C. A further increase in the coating temperature from 40 to 50°C led to a faster release. Thus, the optimum product temperature for the formulation studied was between 35 and 40°C, where the conditions for the formation of a continuous film were apparently satisfied. The fast release at low temperatures is attributed to poor coalescence of the film due to the hardening of the polymeric particles, whereas the fast release at high temperatures is believed to be due to fast evaporation of water and the subsequent formation of discontinuous films.

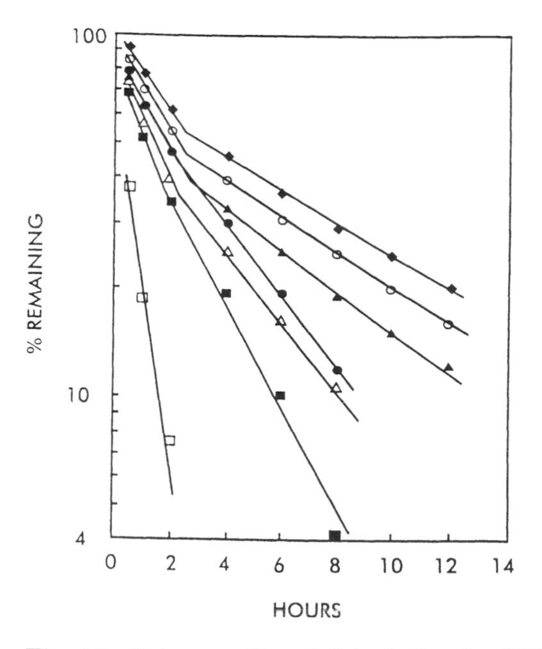

Fig. 10 Release profiles of diphenhydramine·HCl from pellets coated with Aquacoat at 22°C (□), 25°C (Δ), 30°C (○), 35°C (◆), 45°C (●), and 50°C (■). (From Ref. 69.)

Post-coating curing also has a marked effect on coating integrity as well as on the drug release characteristics. If film formation is incomplete during the coating process due to low coating temperatures or shorter coating times, further coalescence of the polymeric particles could potentially occur if the coating product is cured at elevated temperatures [70]. Curing can be carried out either immediately following the coating process in the same processing equipment or in ovens. This is consistent with the findings related to a gradual decrease in water vapor permeation rate through cellulose acetate films as a function of time and temperature [71]. The effects of coating curing temperatures on drug release from Surelease-coated diphenhydramine·HCl pellets have been investigated [72]. Diphenhydramine hydrochloride pellets were coated at 30–55°C in a fluidized bed with Surelease formulation to a final weight gain of approximately 15%. The coated pellets were tested in deaerated deionized water at 37°C using a USP II apparatus. As shown in Fig. 11, an increase in the curing temperature resulted in a significant reduction in both the rate and extent of drug release for pellets coated at a product temperature of 30°C. However, curing had less effect on drug release profiles when the coating was conducted at a higher temperature (Fig. 12). These observations demonstrate that the need for post-coating curing depends on the coating conditions.

IX. CONCLUSION

Aqueous polymeric dispersions have been and will continue to be extensively used in the pharmaceutical coating industry for manufacture of therapeutic and health care products. As discussed in this chapter, the quality and functionality of a coating are dependent on the properties of polymeric materials, formulation compositions, and process-

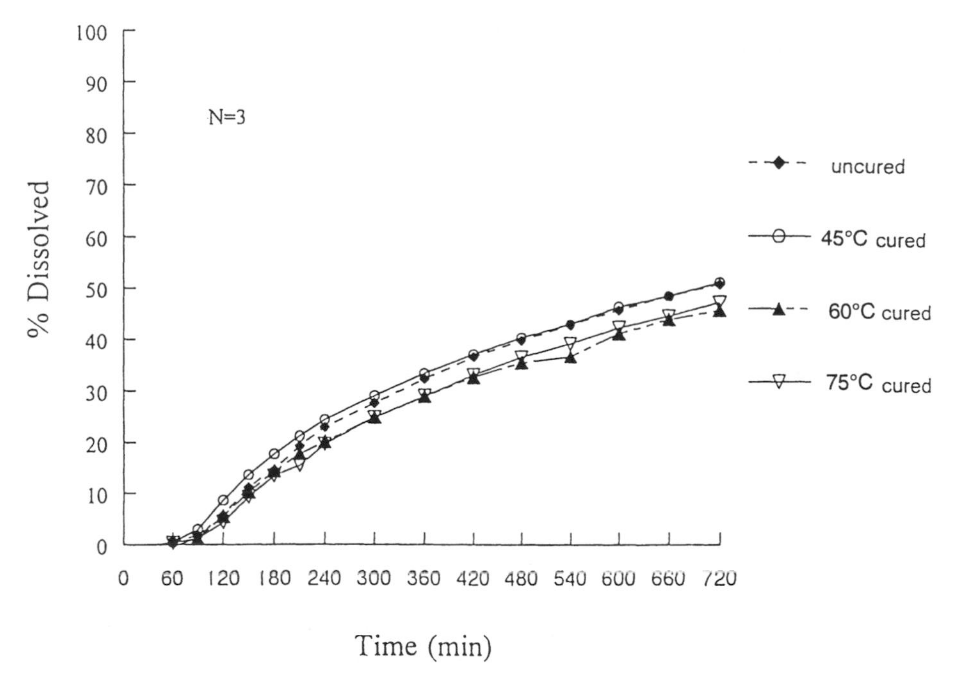

Fig. 11 Release profiles from diphenhydramine·HCl pellets coated with Surelease E-7-19010 at 30°C and then cured at 45°C (○), 60°C (▲), and 75°C (∇); no curing (◆). (From Ref. 72.)

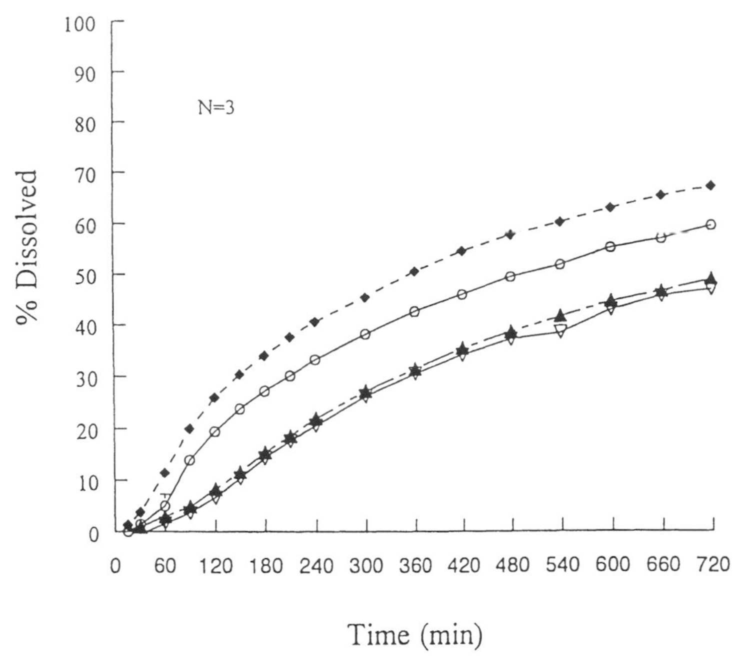

Fig. 12 Release profiles from diphenhydramine·HCl pellets coated with Surelease E-7-19010 at 55°C and then cured at 45°C (○), 60°C (▲), and 75°C (∇); no curing (◆). (From Ref. 72.)

ing conditions. The goal of any coating application is to generate a complete coalesced film in order to obtain a high-quality product and maintain its consistent performance over time. Film formation from latex and pseudolatex systems involves four different mechanisms. There is no single theory that alone can describe the film formation process. Based on the composition of the polymeric dispersions and the conditions encountered during the film formation process, one theory may be more relevant than another. Even for the same formulation, film formation may be governed by different mechanisms during different stages of the coating process. For instance, the autoadhesion theory may describe the predominant mechanism during the gradual coalescence of polymer particles following complete removal of water from the system. However, in the early stage of film formation, other mechanisms, such as viscous flow and capillary pressure theories, may be more pertinent for describing polymer particle deformations. Parameters, such as average polymer molecular weight; molecular weight distribution; polymer particle size and distribution; and the presence of stabilizers, plasticizers, and other ingredients, play very important roles. The manufacturing process of the dispersions can also influence the properties of the ingredients in the dispersion and in turn determine the properties of the film. In practice, film characteristics related to mechanical, thermal, and dynamic properties can be measured using DSC, TDA, TMA, DMTA, stress-strain tests, creep compliance tests, substrate adhesion tests, and hardness and indentation tests. A full understanding of film formation mechanisms and formulation and processing factors that influence film formation can lead to a rational design of a coating process that can produce optimum coatings and high-quality products.

X. EXAMPLES

A. Formula 1: Control Release Coating with Eudragit® NE 30D

Pigment suspension		Dispersion	
Water	700	Water	333
Talc	160	Eudragit NE 30D	667
Titanium dioxide	70		
Red-brown pigment R26495	30		
Hydroxypropylmethylcellulose	10		
Polyethylene glycol 6000	30		

The pigment suspension is poured into the Eudragit NE 30D suspension with stirring shortly before the coating process. The mixture is then filtered through a 0.1 mesh sieve and is continuously stirred during coating. It is recommended that this coating dispersion be used within 48 hours [73].

B. Formula 2: Colorless Enteric Coatings

	Parts by weight
Eudragit L 30D	1000
Plasticizer	30
Talc	70
Antifoam emulsion	2
Water	898
Solid content	20%

This is a recommended master formula for enteric contains (Röhm Pharma, Technique Sheet, Eudragit L 30D, p. 4).

C. Formula 3: Aqueous Controlled Release Coatings of Pseudoephedrine Beads Using Surelease

		Weight (g)
Drug layer suspension	Pseudoephedrine	3000
	Opadry (YS-1-7065)	300
	Micronized talc	300
	Deionized water	3600
Seal coat	Opadry (YS-1-7472)	40
	Deionized water	360
Sustained release coat	Surelease (E-7-7050)	960
	Deionized water	640

Drug layer suspension is applied to 1.0 kg of 25–30 mesh nonparels, and then the remaining coating solutions are applied [Colorcon Technique Data, Surelease PT-37 (10-92)].

D. Formula 4: Coating of Multiple Vitamin Tablets

		Weight (g)
Coating	Aquacoat® 30%	883.3
	HPMC E-5 (16% solution)	1656.3
	Dibutyl sebacate (20% bass on Aquacoat solids)	53.0
	Tween 80 (0.5% based on total suspension)	25.0
	Color	400.0
Gloss solution	HPMC E-5 (16% solution)	250.0
	PEG 400	20.0
	Water	1730.0

The coating suspension is prepared via the following steps: (1) add DBS to Aquacoat and mix for 30 min; (2) mix HPMC solution, Tween 80, and colorant for 10 min; (3) mix (1) and (2) for 10 min, and adjust the final volume with water and mix for 5 min. The gloss solution is prepared as follows: (1) mix all ingredients and stir for 10 min, (2) adjust final volume with water and mix for 5 min (Aquacoat technique sheet on coatings using 24 in. Accela-Cota for preparation of 10 kg batch, FMC, p. 9).

E. Example 5: Effect of Coating Composition

One investigation showed that as the amount of kaolin in a formulation containing Eudragit NE 30D increased, the percentage of drug release as a function of time is correspondingly increased because of an increase in the film porosity [74].

F. Example 6: Effect of Secondary Film Forming Polymer

Theophylline pellets with 75% of drug load and size ranging from 20–40 mesh were coated with an Aquacoat dispersion that contained HPMC in order to achieve target release characteristics. The coating formulations were plasticized with 24% of dibutyl sebacate, and coating operations were conducted in Glatt WSG-5 unit with a Wurster insert. While the coating level was held at 13% for all the formulations, HPMC content was varied. The target release was obtained with 7–10% of HPMC in the coating formulation [75].

REFERENCES

1. *McGraw-Hill Dictionary of Scientific and Technical Terms*, 4th Ed. McGraw-Hill Book Co., 1994, pp. 394.
2. J. A. Seitz, S. P. Mehta, and J. L. Yeager, Chapter 12: Tablet Coating, In: *The Theory and Practice of Industrial Pharmacy*, 3rd Ed., (L. Lachman, H. A. Lieberman, J. L. Kanig, eds.). Lea & Febiger, Philadelphia, 1986, pp. 346.
3. G. S. Banker and G. E. Peck, The new water-based colloidal dispersions, *Pharm. Technol.*, 5:55–61 (1981).
4. S. C. Porter and J. E. Hogan, Tablet film-coating, *Pharmacy Int.*, May:122–127 (1984).
5. J. E. Hogan, Aqueous versus organic solvent film coating, *Int. J. Pharm. Tech. & Prod. Mfr.*, 3(1):17–20 (1982).
6. Coatings, *Kirk-Othmer Encyclopedia of Chemical Technology*, 4th Ed., John Wiley & Sons, Vol. 6, 1993, pp. 670.

7. W. D. Harkins, A general theory of the mechanism of emulsion polymerization, *J. Am. Chem. Soc.*, 69:1428–1444 (1947).
8. G.Odian, Chapter 4: Emulsion polymerization, In: *Principles of Polymerization*, 2nd Ed., 1981, pp. 319–325.
9. R. M. Fitch and C. H. Tsai, Particle formation in nonmicellar systems, *J. Pharm. Sci.*, Part B, 8(10):703–710 (1970).
10. G. W. Burton and C. P. O'Farrell, Preparation of artificial latexes, *Rubber Chem. Technol.*, 49(2):394–397, (1976).
11. R. K. Chang, C. H. Hsiao, and J. R. Robinson, A review of aqueous coating techniques and preliminary data on release from a theophylline product, *Pharm. Technol.*, 11(3):56–67 (1987).
12. D. Dieterich, W. Keberle, and R. Wuest, Aqueous dispersions of polyurethane ionomers, *J. Oil Col. Chem. Assoc.*, 53(5):363–379 (1970).
13. E. J. Verwey and J. Th. G. Overbeek, Theory of stability of lyophobic colloids, Elsevier, New York, 1984.
14. J. W. Vanderhoff and G. S. Carl, Measure of the degree of dispersion of aqueous silicone carbide dispersions, *Polym. Mater. Sci. And Eng., Proc. ACS Div. of Polymeric Materials: Sci. and Eng.*, 52:404–408 (1985).
15. J. N. Shaw and R. H. Ottewill, Relaxation effects in the electrophoresis of dispersed systems, *Nature*, 208(5011):681–2 (1965).
16. A. Kitahara and H. Ushiyama, Flocculation in mixed latices, *J. Colloid Interface Sci.*, 43(1):73–77 (1973).
17. W. G. Barb and W. Mikucki, The coagulation of polymer latices by freezing and thawing, *J. Polym. Sci.*, 37:499–514 (1959).
18. I. M. Kreiger, Rheology of monodisperse latices, *Adv. Colloid Interface Sci.*, 3(2):111–136 (1972).
19. E. Vanzo, R. H. Marchessault, and V. Stannett, Synthetic high polymers, *J. Colloid Sci.*, 19(6):578–583 (1964).
20. H. Smith, *J. Colloid Interface Sci.*, 48:147–151 (1974).
21. L. D. Maxim, A. Klein, M. E. Mayer, and C. H. Kuist, Particle size distribution by turbidity, *J. Ploym. Sci.*, Part C, 27:195–205 (1969).
22. R. R. Irani and C. F. Callis, *Particle Measurement Interpretation and Applications*, John Wiley & Sons, Inc., New York, 1963.
23. Technical Sheets of Eudragit® L 30D, Röhm Pharma, Weiterstadt, Germany, 1995.
24. Technical Sheets of Eudragit® RL/RS 30D, Röhm Pharma, Weiterstadt, Germany, 1995.
25. Y. Fukumori, Y. Yamaoka, H. Ichikawa, Y. Takeuchi, T. Fukuda, and Y. Osako, Coating of pharmaceutical powders by fluidized bed process. IV. 1) Softening temperatures of acrylic copolymers and its relation to film-formation in aqueous coating, *Chem. Pharm. Bull.*, 36(12):427–493 (1988).
26. FMC Manual, Specifications and method of assay: Aquacoat®, aqueous polymeric dispersion, 1995, p. 3.
27. L. C. Li and G. E. Peck, Water based silicone elastomer control release tablet film coating II: Formulation considerations and coating evaluation, *Drug Dev. Ind. Pharm.*, 15(4):499–531 (1989).
28. L. C. Li and G. E. Peck, Water based silicone elastomer control release tablet film coating IV: Process evaluation, *Drug Dev. Ind. Pharm.*, 16(3):499–531 (1989).
29. D. T. Liles and H. V. Lefler III, Silicone rubber latices in water based coatings, *Modern Paint and Coatings*, 81(8):42–48 (1991).
30. G. C. Kúczynski, *J. Applied Phys.*, 20:1160 (1949).
31. R. E. Dillon, L. A. Matheson, and E. B. Bradford, Sintering of synthetic latex particles, *J. Colloid. Sci.*, 6:108–117 (1951).

32. G. L. Brown, Formation of films from polymer dispersions, *J. Polym. Sci.*, XXII:423–434 (1956).
33. S. S. Kistler, Coherent expanded aerogels and jellies, *Nature*, 127:741 (1930).
34. S. S. Kistler, Coherent expandable aerogels, *J. Phys. Chem.*, 36:52–64 (1932).
35. S. S. Voyutskii, Concerning mechanism of film formation from high polymer dispersions, *J. Polym. Sci.*, XXXII(125):528–529 (1958).
36. M. Chainey, M. C. Wilkinson, and J. Hearn, Permeation through homopolymer latex films, *J. Polym. Sci., Polymer Chem. Ed.*, 23:947 (1985).
37. J. W. Vanderhoff, H. L. Tarkowski, M. C. Jenkins, and E. B. Bradford, Theoretical considerations of interfacial forces involved in the coalescence of latex particles, *J. Macromol. Chem.*, 1:361–397 (1966).
38. H. D. Cogan and D. F. Clarke, New developments in chemicals for emulsion paints, *Offic. Dig., Federation Paint & Varnish Production Clubs*, 28:764–772 (1956).
39. E. H. Lee and J. R. M. Radok, The contact problem for viscoelastic bodies, *J. Appl. Mech.*, 439 (1960).
40. S. T. Eckersley, and A. Rudin, Mechanism of film formation from polymer latexes, *J. Coat. Tech.*, 62(780):89–100 (1990).
41. W. A. Henson, D. A. Taber, and E. B. Bradford, Mechanism of film formation of latex paint, *Ind. Eng. Chem.*, 45(4):735–739 (1953).
42. *Annual Book of Standards*, Vols. 06.01, 06.02, 06.03, ASTM, Philadelphia.
43. Coatings, *Kirk-Othmer Encyclopedia of Chemical Technology*, 4th Ed., John Wiely & Sons, New York, Vol. 6, 1993, pp. 669–746.
44. M. Takayuki, *Dynamic Mechanical Analysis of Polymeric Materials*, Elsevier Scientific Publishing Co., New York, 1978, pp. 1–12.
45. Film coating and film-forming materials evaluation, *Encyclopedia of Pharmaceutical Technology*, Marcel Dekker, New York, 1992, 6:1–12.
46. L. H. Sperling, *Introduction to Physical Polymer Science*, 2nd Ed., John Wiley & Sons, New York, 1992, pp. 310.
47. L. H. Sperling, *Introduction to Physical Polymer Science*, 2nd Ed., John Wiley & Sons, New York, 1992, pp. 458–462.
48. J. Crank, *Diffusion in Polymers*, Academic Press, New York, 1968.
49. J. Crank, *The Mathematics of Diffusion*, 2nd Ed., Oxford University Press, New York, 1975.
50. Standard test method for water vapor transmission of materials, Designation: E 96-93, *Annual Book of ASTM Standards*, ASTM, Philadelphia, 1993, p. 472.
51. Z. Y. Wang, Development and critical evaluation of solvent-free photocurable coatings for pharmaceutical application, dissertation, University of Connecticut, 1994, pp.100–102.
52. O. L. Sprockel, W. Prapaitrakul, and P. Shivanand, Permeability of cellulose polymers: Water vapor transmission rates, *J. Pharm. Pharmacol.*, 42:152–7 (1990).
53. H. N. Joshi, M. A. Kral, and E. M. Topp, Microwave drying of aqueous tablet film coatings: A study on free films, *Int. J. Pharm.*, 51:151–160 (1989).
54. Standard test method for water vapor transmission of materials, Designation: E 96-93, *Annual Book of ASTM Standards*, ASTM, Philadelphia, 1993, pp. 472–477.
55. J. A. Wood and S. W. Harder, Adhesion of film coatings to the surfaces of compressed tablets, *Can. J. Pharm. Sci.*, 5(1):18–23 (1970).
56. D. G. Fisher and R. C. Rowe, Adhesion of film coatings to tablet surface-instrumentation and preliminary evaluation, *J. Pharm. Pharmcol.*, 28:886–889 (1976).
57. M. E. Aulton, Assessment of the mechanical properties of film coating materials, *Int. J. Pharm. Tech. Prod. Mfr.*, 3:9–16 (1982).
58. Film coatings and film-forming materials: Evaluation, *Encyclopedia of Pharmaceutical Technology*, Vol. 6, Marcel Dekker, New York, 1992, pp. 1–28.
59. G. W. Radebaugh, J. L. Murtha, T. N. Julian, and J. N. Bondi, Methods for evaluating the

ouncture and shear properties of pharmaceutical polymeric films, *Int. J. Pharm.*, 45:39–46 (1988).
60. A. Laicher, C. A. Lorck, P. C. Grunenberg, H. Klemm, and F. Stainslaus, Aqueous coatings of pellets to sustained-release dosage forms in a fluid-bed coater, *Drugs Made In Germany*, 38(1):23–26 (1995).
61. Y. Fukumori, Y. Yamaoka, H. Ichikawa, Y. Takeuchi, T. Fukuda, and Y. Osako, Coating of pharmaceutical powders by fluidized bed process. IV. 1) Softening temperature of acrylic copolymers and its relation to film-formation in aqueous coating, *Chem. Pharm. Bull.*, 36(12):427–493 (1988).
62. I. Uma, W. H. Hong, N. Das, and I. Ghebre-Sellassie, Comparative evaluation of three organic solvent and dispersion based ethylcellulose coating formulations, *Pharm. Technol.*, 14(9):68–74 (1990).
63. L. H. Sperling, *Introduction to Physical Polymer Science*, 2nd Ed., John Wiley & Sons, New York, 1992, pp. 335–339.
64. R. Bodmeier, and O. Paweatakul, Dry and wet strength of polymeric films prepared from an aqueous colloidal polymer dispersion, Eudragit® RS30D, *Int. J. Pharm.*, 96:129–138 (1993).
65. S. S. Voyutskii and Y. L. Margolina, The nature of self-adhesion (TACK) of polymers. I, *Rubber Chem. and Technol.*, 30:531–543 (1957).
66. O. L. Wheeler, H. L. Jaffe, and N. A. Wellman, Mechanism of film formation of polyvinyl acetate emulsions, Rubber Chem. and Technol., 26:1239–1241 (1954).
67. N. A. Muhammad, W. Boisvert, M. R. Harris, and J. Weiss, Modifying the release properties of Eudragit® L30D, *Drug Development and Industrial Pharmacy*, 17(8):2497–2509 (1991).
68. G. P. Birrwagen, Critical pigment volume concentration (CPVC) as a transition point in the properties of coatings, *J. Coatings Technol.*, 64(806):71–75 (1992).
69. S. T. Yang, and I. Ghebre-Sellassie, The effect of product bed temperature on the microstructure of Aquacoat®-based controlled-release coatings, *Int. J. Pharm.*, 60:109–124 (1990).
70. M. Harris, I. Ghebre-Sellassie, and R. Nesbitt, A water-based coating process for sustained release, *Pharm. Technol.*, 10(9):102–107 (1986).
71. C. M. Sinko, A. F. Yee, and G. L. Amidon, Prediction of physical aging in controlled-release coatings: The application of the relaxation coupling model to glass cellulose acetate, *Pharm. Res.*, 8(6):698–705 (1991).
72. J. Wang and I. Ghebre-Sellassie, unpublished results.
73. Technique Application Pamphlet, Eudragit® E30D, Röhm Pharma, GMBH, Weiterstadt, Germany.

5

Biodegradable Nanoparticles of Poly(lactic acid) and Poly(lactic-*co*-glycolic acid) for Parenteral Administration

Eric Allémann and Robert Gurny

University of Geneva, Geneva, Switzerland

Jean-Christophe Leroux

University of Montreal, Montreal, Quebec, Canada

I. INTRODUCTION

Following parenteral administration of drugs, an active substance distributes throughout the body as a function of its physico-chemical properties and molecular structure. The amount of drug reaching the site of action is often only a small fraction of the administered dose. Accumulation at nontarget sites may lead to adverse reactions and undesirable side effects [1]. A way of modifying the original biodistribution of substances is to entrap them in submicroscopic drug carriers. Among such carriers, liposomes (see chapter 2 on liposomes in this volume), polymeric nanoparticles [2], solid lipid nanoparticles [3], and pharmacosomes [4] have been studied. Twenty-five years ago, liposomal formulations were facing stability problems as well as poor drug loading capacities. Therefore, nanoparticles were at that time seen as a promising alternative for targeting substances. Nanoparticles are solid or semisolid colloidal particles ranging in size from 10 to 1000 nm [5]. They consist of macromolecular materials and can be used as drug carriers. Polymeric nanoparticles is a collective name for nanospheres and nanocapsules [2] (Fig. 1). Nanospheres have a matrix-type structure. Drugs may be adsorbed at their surface, entrapped in the particle or dissolved in it. Nanocapsules have a polymeric shell and an inner liquid core. In this case, the active substances are usually dissolved in the inner core, but may also be adsorbed at their surface. Nanoparticles can be prepared by polymerization techniques or by dispersion of preformed polymers. For example, preparation of nanoparticles by polymerization of cyanoacrylates has been extensively studied (see chapter 3 in this volume).

The purpose of the present chapter is to review the techniques available for the preparation of poly(lactic acid) and poly(lactic-*co*-glycolic acid) nanoparticles and to

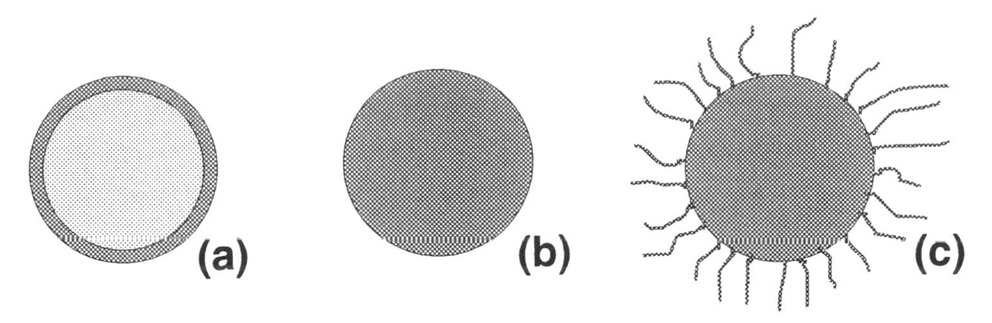

Fig. 1 Various types of nanoparticles: (a) nanocapsules; (b) nanospheres; (c) coated nanospheres.

discuss their properties and applications. These preformed polymers have the advantage of being well characterized and biodegradable, and they are already commercially used for microparticulate drug delivery systems (e.g., Decapeptyl®, Parlodel® L.A., Enantone® L.P.). After intravenous injection, surface unmodified (or plain) nanoparticles are recognized as exogenous by the cells of the mononuclear phagocyte system, mainly the Kupffer cells of the liver and the macrophages of the spleen. The ensuing massive uptake can be of interest for the treatment of diseases mainly affecting the liver (e.g., salmonella, leishmania), but for treating diseases located elsewhere, this selective uptake represents a major drawback. To solve this problem, researchers have proposed a variety of methods for modifying the surface of the particles, in order to slow down the phagocytosis of drug-loaded nanoparticles (Fig. 1). These methods mainly involve the use of hydrophilic polymers or surfactants, such as polyethylene glycol (PEG) and poloxamers, and are presented later in this chapter.

Most of the poly(α-hydroxycarboxylic acid) nanoparticles are intended to be injected, and therefore pharmaceutical aspects such as purification and sterilization are important. These topics are discussed in this chapter as well as the possibility of freeze-drying the particles to enhance their physical stability.

II. POLYMERS USED FOR PARTICLE PREPARATION

Poly(lactic acid) (PLA) and poly(lactic-*co*-glycolic acid) (PLGA) are the most common polymers used for the preparation of poly(α-hydroxycarboxylic acid) nanoparticles. Block copolymers of poly(α-hydroxycarboxylic acids with polyethylene glycol are used to modify the surface characteristics of the particles. These polymers are presented in this section.

A. Poly(α-hydroxycarboxylic acid)

Glycolic and lactic acid are the common names of α-hydroxyacetic and α-hydroxypropionic acid, respectively. These acids may be polymerized directly from the monomer. However, a practical limit in molecular weight exists for polymers prepared by this method. Generally, the polymers or the copolymers are synthesized using cyclic dimers (lactoncs) as thc starting material. Polymerization through ring opening of dimers is achieved in the presence of a catalyst [6]. For intermediate molecular weights (10,000–40,000 daltons) acids are convenient catalysts, whereas for higher molecular weight polymers, organometallic compounds are required (e.g. stannous octanoate). L-PLA is

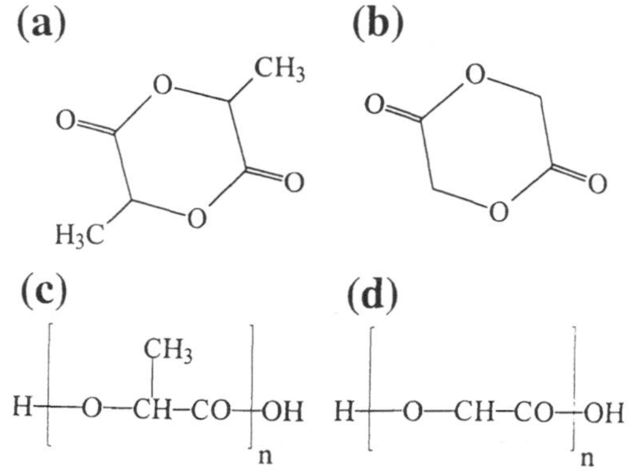

(a) **(b)**

(c) **(d)**

H–[O–CH(CH_3)–CO]$_n$–OH H–[O–CH–CO]$_n$–OH

Fig. 2 Chemical structures of (a) lactide; (b) glycolide; (c) poly(lactic acid); and (d) poly(glycolic acid).

produced from L-lactide, D,L-PLA is prepared from racemic D,L-lactide, PLGA is produced from a mixture of D,L-lactide and glycolide. PGA is produced from glycolide (Fig. 2).

The physical characteristics of the polymers depend on their composition and molecular weight. The melting behavior of poly(α-hydroxycarboxylic acids) is largely influenced by their chemical structure (Table 1). The homopolymers L-PLA and PGA and copolymers with a high percentage of L-lactide and glycolide are semicrystalline plastics. Therefore, in the differential scanning calorimetry diagram, an endothermic melting peak is found, whereas a glass transition point can be observed with other polymers of this series. Solubility characteristics depend also on the polymer composition. They are presented in Table 2.

Under physiological conditions, poly(α-hydroxycarboxylic acids) are hydrolytically degraded into a α-hydroxycarboxylic acids (e.g., glycolic acid, lactic acid) and eliminated through the Krebs cycle, primarily as carbon dioxide, and in urine [6]. The degradation rate depends on the composition of the polymer, its molecular weight, and the size of the particles, and the environmental conditions (e.g., pH) [7,8]. Degradation rates of poly(α-hydroxycarboxylic acids) have been investigated in rats [9]. For pellets, the half-life of the homopolymers and copolymers decreased from 5 months for 100% PGA to 1 week with a 50:50 PGA:PLA copolymer and increased to 6.1 months for 100% PLA. These values should be taken with caution, since they concern formulations with sizes in the millimeter range. For nanoparticles, which have a significantly smaller size and a much larger specific surface area, the half-lives are certainly shorter. Depending on the intended application, one might choose a PLGA copolymer rather

Table 1 Thermal Characteristics of Poly(α-hydroxycarboxylic acids)

Polymer	Crystallinity	Glass transition range	Melting point (°C)
PGA	Semicrystalline	20–35	222–228
D,L-PLA	Amorphous	50–65	—
L-PLA	Semicrystalline	30–65	170–185
PLGA	Amorphous	40–60	—

Table 2 Solubility of Poly(α-hydroxycarboxylic acids)

Polymer	Dichloromethane chloroform	Acetone ethylacetate benzyl alcohol	Tetrahydrofuran dimethyl-formamide	Hexafluoro-isopropanol
PGA	−	−	−	+
D,L-PLA	+	+	+	+
L PLA	+	−	−	+
PLGA	+	+	+	+

than PLA, to avoid the risk of polymer bioaccumulation. The differences in physico-chemical characteristics also influence the release properties of drug-loaded nanoparticles. For instance, it was demonstrated that indomethacin is released faster from PLGA than from PLA nanoparticles. This is ascribed to the more hydrophilic and more permeable nature of the copolymer [10].

B. Block Copolymers

Due to the extensive phagocytosis of nanoparticles after intravenous injections, several authors have proposed modifying the physico-chemical characteristics of the poly(α-hydroxycarboxylic acids) by linking chains of ethylene oxide to one end of the polymer chain. For example, Stolnik et al. [11] used a copolymer of PLA and PEG to coat nanoparticles of PLA. Gref et al. [12] used a copolymer of PLGA and PEG as a starting material for the preparation of their particles. PLA-PEG copolymers have been synthesized by ring opening polymerization of the D,L-lactide, or D,L-lactide plus glycolide, in the presence of the methoxypolyethylene glycol and stannous octanoate as a catalyst. PLGA-PEG copolymers have been synthesized by direct reaction of the terminal amine end group from monoamine monomethyoxy polyethylene glycol with PLGA (carrying a reactive carboxyl end group). PLGA-PEG can also be synthesized by the same reaction as for PLA-PEG [12].

III. PREPARATION METHODS

Numerous methods are available for the preparation of poly(α-hydroxycarboxylic acid) nanoparticles. For nanospheres, certain techniques involve the emulsification of a polymer solution (emulsification-evaporation, salting-out, emulsification-diffusion) while others involve a direct precipitation of the polymer solution. For nanocapsules, only one method is currently available. It is based on the interfacial deposition (or desolvation) of a polymer solution. The different steps of each procedure are described in this section. Due to the problem of extensive phagocytosis of plain particles, emphasis is given to the possibilities of preparing particles coated with hydrophilic polymers

A. Nanospheres

1. Emulsion-Evaporation Procedure

In 1977, Gurny et al. [13,14] first reported on the preparation of PLA nanospheres for parenteral drug delivery. The particles were prepared by the emulsification-evaporation method. In this technique, based on a patent of Vanderhoff et al. [15], the polymer is dissolved generally in a chlorinated solvent (e.g., CH_2Cl_2, $CHCl_3$) and emulsi-

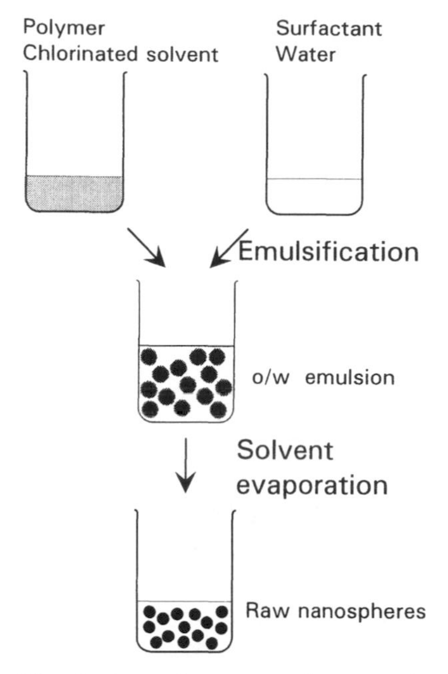

Fig. 3 Emulsion-evaporation procedure.

fied in an aqueous phase containing a surfactant [16,17]. The most common surfactants used for this type of preparation are polysorbates, poloxamers, and sodium dodecylsulfate. Emulsification can be achieved by mechanical stirring, sonication [18], or microfluidization (high-pressure homogenization through narrow channels) [19]. The organic solvent is then removed under reduced pressure (Fig. 3). Under these conditions, the organic solvent diffuses into the aqueous phase and progressively evaporates. A recent publication [12] reports on the possibility of preparing biodegradable nanospheres of poly(α-hydroxy acids) having a hydrophilic surface of PEG. To this end, the authors used a copolymer of PLA and PEG as the starting material. During the evaporation of the solvent, the PLA ends tend to position themselves in the inner part of the forming particles, whereas the PEG ends tend to position themselves on the surface of the particles (Fig. 1c).

As can be seen in Table 3, the preparation of PLA-PEG nanospheres does not necessarily require any surfactant. During the emulsification, the PEG end of the copolymer rearranges from the droplets' inner core to the water interface, while the PLA end remains inside the droplets. Upon solvent removal, the nanosphere cores solidify. As a result of the amphiphilic nature of the diblock polymers, the use of (potentially toxic) surfactants can be avoided. The main disadvantage of this technique is that chlorinated solvents are required to prepare the emulsions. Usually, they are removed by evaporation under reduced pressure. Determination of the residues of these toxic solvents is, however, required and can be achieved by headspace gas chromatography analysis [21].

2. *Salting-Out Procedure*

The salting-out technique was introduced and patented by Bindschaedler et al. and Ibrahim et al. [22–24] in 1988. Compared to the emulsification-evaporation technique,

Table 3 Formulations of Nanospheres Prepared by Emulsion-Evaporation Procedure

	Example 1		Example 2	
Organic phase				
Polymer	PLGA 50:50	100 mg	PLA-PEG 5000	25 mg
Drug	—		Lidocaine	0–25 mg
Solvent	CH_2Cl_2	10 mL	Ethyl acetate	2 mL
Aqueous phase				
Surfactant	Polysorbate 80	0.5–3.0 g	—	
	Water	100 mL	Water	30 mL
Emulsification	Microfluidization		Vortexing and sonication	
Particle size	75–300 nm		~140 nm	

Source: Example 1, Ref. 20; Example 2, Ref. 12.

the main advantage of this method is that the use of potentially toxic solvents is avoided. Here, only acetone is used and it can be easily removed in a final step by cross-flow filtration (see purification procedures).

The preparation method consists in adding, under mechanical stirring, an electrolyte-saturated solution containing a hydrocolloid, generally poly(vinyl alcohol), as a stabilizing and viscosity-increasing agent to an acetone solution of polymer. Poly(vinyl alcohol) (PVAL) has the advantage of being compatible with several electrolytes [25]. The saturated aqueous solution prevents acetone from diffusing into the water by a salting-out process. After the preparation of an oil-in-water emulsion, sufficient water or an aqueous solution of PEG is added to allow complete diffusion of acetone into the aqueous phase, thus inducing the formation of nanospheres (Fig. 4) [26–29].

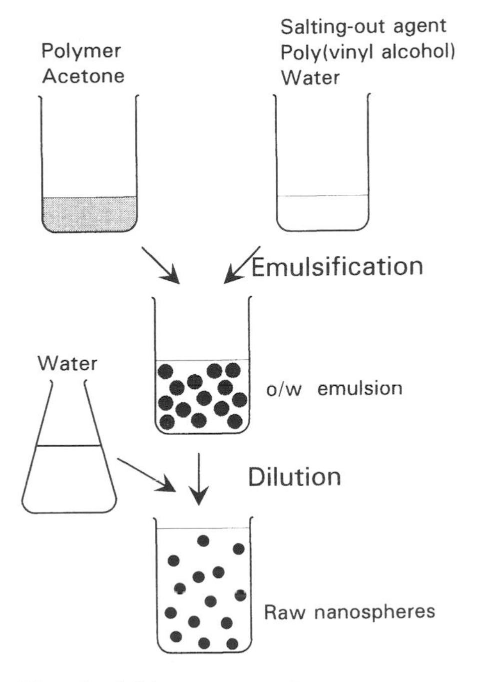

Fig. 4 Salting-out procedure.

Several important factors influence the size of the nanoparticles. By increasing the PVAL concentration in the external phase of the emulsion, a decrease in particle size is observed. This agent appears to have a steric stabilizing action on the dispersed droplets of the internal phase. Therefore, addition of a surfactant is not required. An increase in stirring rate up to 6,000 rpm (with an air-driven propeller) allows a significant reduction in particle size. To obtain small nanospheres, it is also important to work with concentrated solutions of polymer (15–25% w/v).

To obtain high drug loadings with this method, high solubility of the drug in acetone is required. Then, the salting-out agent has to be chosen depending on the pKa of the drug. The entrapment efficacy can be improved by using the appropriate salting-out agent. For example, for a neuroleptic compound (savoxepine CGP 19486) (pKa 8.3) using magnesium chloride rather than magnesium acetate has a significant effect. Due to the presence of the basic acetate salt, the pH of the aqueous phase rises from 6 to 8, thus preventing solubilization of the drug in the aqueous phase. With this improved method, entrapment efficiencies up to 95% were observed with a drug loading of 17% and an overall batch weight yield of 93% [30]. This method allows the preparation of batches of up to 30 g of nanospheres in lab-scale reactors.

Two typical formulas are presented in Table 4. In example 1 a lipophilic compound (^{14}C-CGP 19486) has been successfully incorporated. These nanoparticles have been administered to rats. Stable plasma levels of radioactivity in plasma were observed for 7 days after intramuscular injections. In example 2, PEG 5% was used as the dilution phase. This change in the preparation procedure allows modification of the hydrophobic/hydrophilic character of the surface of the particles. Compared to plain nanoparticles, this hydrophilic coating significantly reduced phagocytosis by the cells of the mononuclear phagocyte system (MPS), and therefore extended circulation times in mice after intravenous injections were observed ($AUC_{0\text{-}168h}$: 227 mg h g^{-1}, uncoated: 57 mg h g^{-1}).

Examples presented in Table 4 concern nanoparticles with a large particle size (about 750–1000 nm), but the salting-out technique also permits the preparation of particles with a mean size of 200 nm [26].

Table 4 Formulations of Nanospheres Prepared by Salting-Out Procedure

	Example 1		Example 2	
Organic phase				
Polymer	D,L-PLA	1.77 g	D,L-PLA	0.5 g
Drug	^{14}C-CGP 19486	177 mg	Hexadecafluoro phthalocyanine	
Solvent	Acetone	10 g	Acetone	10 g
Aqueous phase				
Electrolyte	Magnesium acetate $4H_2O$	8.75 g	Magnesium acetate $4H_2O$	8.75 g
Stabilizing agent	PVAL	2 g	PVAL	2.75 g
	Water	14.25 g	Water	13.5 g
Dilution phase				
Water	Water	25 g	PEG 20,000 5%	25 g
Particle size	736 nm		988 nm	

Source: Example 1, Ref. 30; Example 2, Ref. 28.

3. *Emulsification-Diffusion Procedure*

This method is derived from the salting-out procedure. It was proposed by Leroux et al. [31] to overcome the problem of using large amounts of salts in the aqueous phase, because the electrolytes sometimes cause problems of compatibility with some bioactive compounds. Here, an aqueous gel of a stabilizing hydrocolloid (PVAL or gelatin) is added to a solution of polymer dissolved in benzyl alcohol under mechanical stirring. Due to the partial miscibility of benzyl alcohol with water, a water-in-oil emulsion is first obtained. The emulsion undergoes a phase inversion upon complete addition of the aqueous gel. Since benzyl alcohol is miscible at a ratio of 1:25 (w/v) with water, a large amount of water is subsequently added to the emulsion in order to allow the complete diffusion of the organic solvent into the water, thus leading to the precipitation of the polymer as nanospheres. The overall process takes less than 10 min. A typical formula is presented in Table 5. After the preparation the reagents (benzyl alcohol and PVAL) must be removed.

With this technique, it is possible to produce particles as small as 70 nm, by increasing the PVAL concentration up to 27.5% (w/w) in the aqueous phase. This technique is very attractive since it was previously found that targeting of carriers to nonphagocytic cells is limited to carrier sizes below 150 nm [32]. The nanoparticle size can also be adjusted by varying the percentage of polymer in the internal phase. In contrast to the salting-out procedure [26], the increase in polymer concentration in the internal phase raises the nanoparticle mean size. Increasing the PLA concentration from 0.5 to 14.3% changes the organic phase density from 1.042 to 1.061. Since the external phase has a density of 1.040, the difference between the density of the two phases tends to increase with increasing PLA concentrations. According to Stokes' law [33], the internal and external phase should have approximately the same density to ensure optimal stability of the dispersion, and therefore the nanoparticle mean size increase could be partly attributed to density variations of the internal phase. On the other hand, when nanoparticles are produced using acetone as the organic solvent (salting-out procedure), the nanoparticle size decreases with increasing polymer concentrations. In this case, increasing the polymer concentration from 10 to 20% decreases the difference of density between the internal and the external phases from 0.021 to 0.002. It should

Table 5 Formulation of Nanospheres Prepared by Emulsification-Diffusion Procedure

	Example 1	
Organic phase		
Polymer	D,L-PLA	3 g
Drug	Chlorambucil	0.45 g
Solvent	Benzyl alcohol	21 g
Aqueous phase		
Stabilizing agent	PVAL (Mowiol® 4-88)[a]	
	Water	40 g
Dilution phase		
Water	Water	660 g
Particle size	295 nm	

[a]Hoechst, Frankfurt an Main, Germany.
Source: Ref. 31.

be stressed that reference to Stokes' law is simply used to help in understanding the above-mentioned phenomenon, even though this approach cannot be strictly applied to these extremely small emulsion droplets, due to the effect of Brownian motion. Other advantages of this method are that PVAL can be replaced by other stabilizing colloids, (e.g., gelatin), and due to the absence of highly concentrated solutions of salts, the aqueous phase can be buffered, which in some cases may enhance drug entrapment.

Investigations with the cytostatic drug chlorambucil as an active compound have shown that this technique allows a good overall weight yield (91%) and relatively high drug loading (8.5%).

4. *Precipitation Procedure I*

This method was first described and patented by Fessi et al. [34,35]. In this type of preparation, the polymer, generally D,L-PLA, is dissolved in a water-miscible solvent (e.g., acetone). The solution is then poured under magnetic stirring into a nonsolvent (usually water containing a surfactant), which leads to the precipitation of nanospheres (Fig. 5). Even though the patent gives examples of drug loading, it appears very difficult to efficiently load such precipitated nanospheres with an active compound. In fact, it is difficult to establish a polymer/drug/solvent/nonsolvent system that allows nanosphere precipitation while simultaneously avoiding extensive diffusion of the drug along with the solvent. Nevertheless, the preparation of small amounts of unloaded nanospheres is possible (Table 6). In this case, addition of surfactant is not required. Preparation of large-scale batches of poly(α-hydroxy carboxylic acid) nanospheres has never been reported with this technique. However, the same technique has been used for the preparation of larger batches of methacrylic acid esters pseudolatexes [36]; but in this case, a surfactant is required (e.g., sodium dodecyl sulfate).

5. *Precipitation Procedure II*

An alternative method for producing nanospheres that avoids the use of chlorinated solvents is the precipitation-solvent evaporation technique described by Stolnik et al. [11,40]. The polymer is dissolved in acetone. A nonsolvent mixture of water and ethanol is added dropwise through a needle into the polymer solution, stirred by a magnetic stirrer, until turbidity indicative of polymer precipitation is visually observed. The sus-

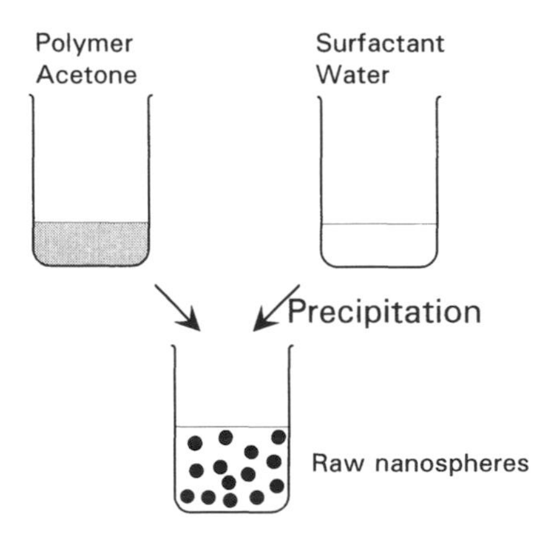

Fig. 5 Precipitation procedure.

Table 6 Formulations of Nanospheres Prepared by Precipitation Procedure (First Method)

	Example 1		Example 2	
Organic phase				
Polymer	D,L-PLA	12.9 mg	PLA-PEG copolymer	80 mg
Solvent	Acetone	750 μL	Acetone	5 mL
Aqueous phase				
	0.13 M phosphate buffer	5 mL	Water	5 mL
Mean particle size	103 ± 45 nm		205 nm	

Source: Example 1, Ref. 37; Example 2, Refs. 38 and 39.

pension of these preformed nanospheres is then added to an aqueous solution of PLA-PEG copolymer or poloxamine in order to coat the particles with hydrophilic molecules. Subsequently, the suspension is agitated at ambient temperature to allow evaporation of solvents. Complete removal of ethanol may represent a problem, since it forms an azeotrope with water. However, low concentrations of ethanol are tolerated in injectable products. Purification of the particles on Sepharose 4B gel (Pharmacia, Uppsala, Sweden) has been suggested as an alternative [40]. Table 7 presents a typical formula using this method. Average size of particles prepared by this method varies from 140 to 1220 nm with a narrow size distribution.

B. Nanocapsules

Nanocapsules having a liquid core and a polymeric shell can be prepared with a method proposed by Fessi et al. and Fawaz et al. [41–43]. In this method, the polymer is dissolved in a water-miscible solvent (generally acetone). A lipophilic drug is then added to this solution along with a phospholipid mixture (e.g., Epikuron®, Lucas Meyer, Hamburg, Germany) and benzyl benzoate. Both the phospholipid mixture and the benzyl benzoate are insoluble in water. This solution is poured into water containing a surfactant (poloxamer 188) under mild magnetic agitation. Due to the miscibility of water

Table 7 Formulation of Nanospheres Prepared by Precipitation Procedure (Second Method)

	Example 1	
Organic phase		
Polymer	PLGA 75:25	200 mg
Drug	—	
Solvent	Acetone	10 mL
Nonsolvent phase		
	Water:ethanol 1:1[a]	
Coating phase		
	PLA-PEG	150 mg
	Water	15 mL
Particle size	161 ± 3.7 nm	

[a]Sufficient amount to precipitate the particles.
Source: Ref. 11.

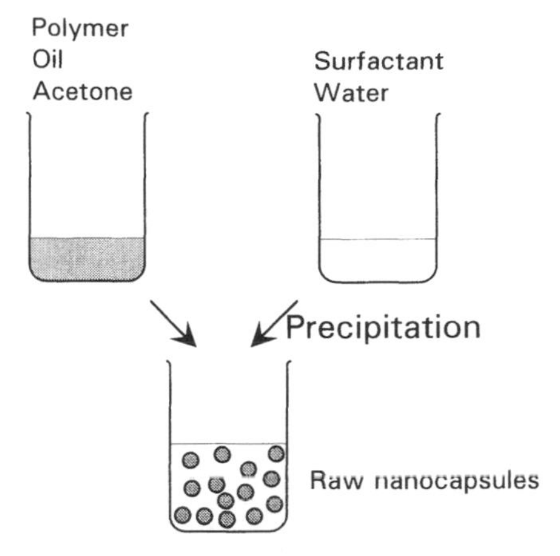

Fig. 6 Interfacial deposition procedure for nanocapsules.

and acetone, the latter diffuses immediately into the aqueous phase, thus leading to the desolvation of the polymer at the interface. Consequently, nanocapsules containing the drug in the inner core are formed (Fig. 6). After complete formation of the capsules, the acetone is removed, generally under reduced pressure. Several compounds such as indomethacin (Table 8), the immunomodulating agent muramyl dipeptide, clofibride [44], and diclofenac [45] have been incorporated in such capsules. This method can be applied to lipophilic drugs having good solubility in organic-oily phases.

IV. PURIFICATION

Since most nanoparticles of PLA and PLGA are designed for parenteral administration, a purification step of the raw nanoparticulate suspensions must be considered for all types of particles. Depending on the method of preparation, the following potentially toxic and/or undesirable preparation additives can be present in the raw suspensions: organic solvents, surfactants, stabilizers, electrolytes, and large polymer aggregates. Purification is also needed to separate free drug from the drug bound to the particles

Table 8 Formulation of Nanocapsules

	Example 1	
Organic phase		
Drug	Indomethacin	100 mg
Polymer	D,L-PLA	625 mg
Additives	Benzyl benzoate	2.5 mL
	Epikuron® 170[a]	625 mg
Solvent	Acetone	125 mL
Aqueous phase	Poloxamer 188[b]	625 mg
	Water	250 mL
Mean particle size	220 ± 20 nm	

[a]Lucas Meyer, Hamburg, Germany.
[b]Pluronic®, BASF, Ludwigshafen, Germany.
Source: Ref. 43.

in order to avoid a burst effect at the time of injection. Polymer aggregates in the micrometer range are usually easily discarded by filtration through sintered glass filters. Organic solvents (especially chlorinated solvents) must absolutely be eliminated, to be below pharmacopoeia limits. Most of the surfactants are also undesirable, and up to now, the U.S. Food and Drug Administration has only accepted the presence of poloxamer 188 and of polysorbate 20, 40, and 80 in injectable dosage forms. The purification methods available are ultracentrifugation, centrifugal ultrafiltration, cross-flow filtration, gel filtration, and dialysis.

A. Ultracentrifugation

This method is widely used for separating free from bound drug during in vitro release studies, but it is also used for purifying raw nanoparticulate suspensions (Fig. 7a). After ultracentrifuging, the supernatant is discarded, and the particles are resuspended in an appropriate solvent (generally water). This procedure is repeated four or five times to quantitatively remove the undesired preparation additives. For example, Krause et al. [18] centrifuged L-PLA nanospheres for 30 min at $19{,}500 \times g$. A fine adjustment of the relative centrifugal field must be achieved in order to avoid partial aggregation of the particles [27]. Another possibility to avoid aggregation is to ultracentrifuge the particles on sucrose gradients [46]. It should also be mentioned that ultracentrifugation is a time-consuming process only suitable for laboratory-scale batches. Upscaling of this method is not currently possible.

B. Centrifugal Ultrafiltration

This technique is also appropriate for the purification of small amounts of nanoparticles. It requires a device (Ultrafree®, Millipore, Bedford, MD; or Centricon®, Amicon®, W. R. Grace & Co., Amicon Div., Danvers, MD) that is mainly based on a centrifuge tube separated from an enclosed tube by an ultrafiltration membrane. During centrifugation, this device allows separation of nanoparticles from the dispersion medium [44].

C. Cross-Flow Filtration

To overcome the problem of scale-up, a cross-flow filtration (CFF) procedure has been developed [27]. As opposed to dead end filtration (Fig. 7b), where the flow is perpendicular to the filter, in CFF (Fig. 6c), the fluid to be purified is directed tangentially to the surface of the membranes to prevent clogging of the filters [47,48]. Regenerated cellulose or polyolefin membranes can be used as filters. In practice, the raw nanoparticulate suspension is pumped into a CFF device. The filtrate, which contains the soluble undesired additives, is discarded and the suspension is recirculated several times in the CFF device (Fig. 8). The nanospheres are maintained in suspension by adding distilled water from the feed tank at the same rate as the filtration rate. This step is named diafiltration. The advantages of this purification technique are that it is fast, it does not alter the particle size, and upscaling the process is feasible by enlarging the filtering surface.

D. Gel Permeation

The use of gel permeation for the purification of polymeric colloidal carriers has been reported for poly(butyl cyanoacrylate) nanospheres [49,50], but this technique may also be suitable for polyester nanoparticles (Fig. 7d). With the use of Fractogel® HW55

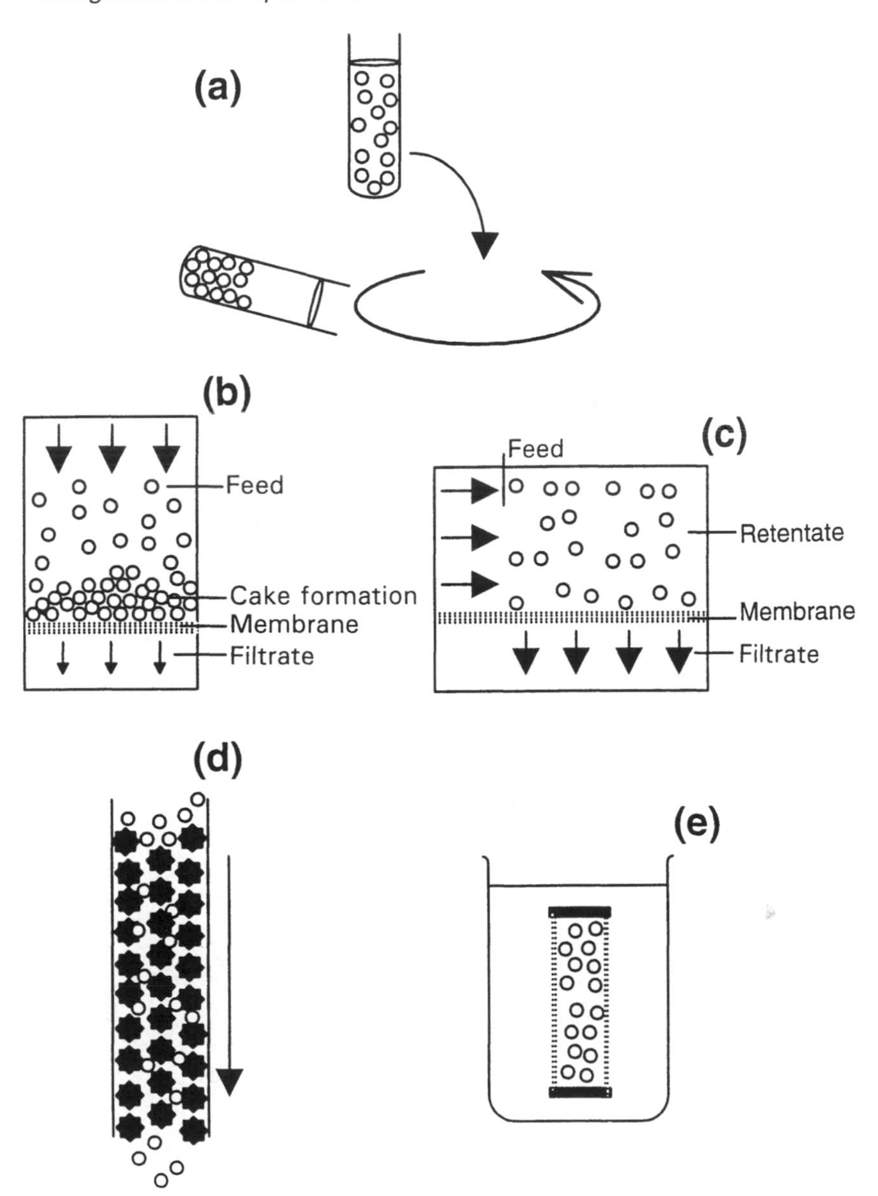

Fig. 7 Purification methods: (a) ultracentrifugation; (b) dead-end filtration; (c) cross-flow filtration; (d) gel permeation; and (e) dialysis.

(Merck, Darmstadt, Germany) or Sephacryl® S-200 (Farmacia, Freiburg, Germany), even high molecular weight compounds (e.g., poloxamers) molecules can be extracted from the raw suspensions. This technique has also been used by Beck et al. [51] with Sephadex® G50 (Farmacia) gel filtration media to separate free drug from particle-bound drug.

E. Dialysis

Dialysis may be a suitable purification method for laboratory-scale batches (Fig. 7e). It has been used by Rolland et al. [52] for eliminating residual polymerization initia-

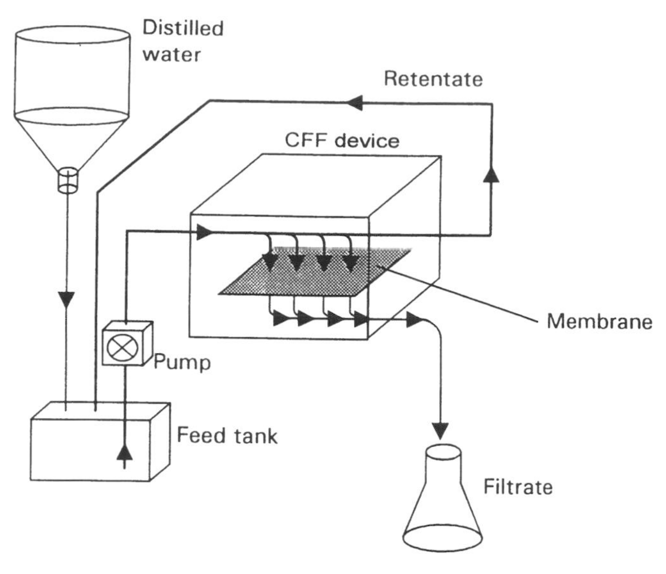

Fig. 8 Cross-flow filtration.

tors from poly(methyl methacrylate) nanospheres. The nanoparticulate suspension is dialyzed against a poloxamer solution through a cellophane membrane. This process is efficient but requires 24 h to purify 500 mg of particles. To quantitatively eliminate the undesired additives, the solution of the receptor compartment must be changed two or three times during the procedure.

V. FREEZE-DRYING

If nanoparticles are kept in an aqueous medium during storage, biodegradation of polymer, drug leakage, and/or drug degradation may occur [53]. To improve the physicochemical stability of the nanoparticles, a drying step is generally required. Heat-drying may be harmful to thermolabile active compounds. Moreover, due to the low glass transition temperature of poly(α-hydroxycarboxylic acids), during the particles at elevated temperatures induces fusion or aggregation. Therefore, freeze-drying represents the most applicable alternative.

Usually, for freeze-drying, the nanoparticulate suspensions are frozen at temperatures ranging from −40 to −60°C or in liquid nitrogen. Depending on the amount of water to be evaporated, they are freeze-dried for periods ranging from 24 to 90 h, at a pressure below 10 Pa.

Ease of redispersion of freeze-dried particles depends on the particle coating. Usually particles prepared in the presence of poloxamer, polysorbate, PVAL, or sodium dodecyl sulfate are readily redispersible in water. Residues of these stabilizers or surfactants display cryoprotective characteristics and therefore facilitate aqueous reconstitution.

If freeze-dried particles are not readily redispersible, addition of a cryoprotective agent may be necessary. Dextrose, lactose, trehalose, sucrose, and mannitol are the most common agents. One has to be aware that addition of such cryoprotective agents increases the tonicity of the final redispersed formulation. A careful calculation of the appropriate amount to be added is required.

Figure 9 shows the influence of a cryoprotective agent on the redispersibility of D,L-PLA nanospheres. They were prepared by the salting-out process and contained 2% chlorambucil. Addition of trehalose significantly favors the redipersion of the particles. For plain spheres, a 2 min hand-shaking is sufficient to ensure complete redipersion of the product. For PEG-coated nanospheres however, sonication is required to disperse particle aggregates. In this case, the addition of trehalose is essential to achieve redispersion [54].

For indomethacin-loaded nanocapsules of D,L-PLA, trehalose and mannitol were found unsuitable for stabilizing the preparation [55]. In this case, sucrose and glucose were recommended.

VI. STERILIZATION

Nanoparticles intended for injection must be sterile and apyrogenic, but, surprisingly, few researchers working on PLA and PLGA nanoparticles have addressed these issues. They must meet pharmacopoeial requirements relating to sterility and apyrogenicity. The nanosuspensions should be tested by the pharmacopoeial methods. Sterility testing using membrane filtration of the test specimens followed by incubation of the membrane in a culture medium is the method of choice. Pyrogens are tested by measuring the rise of temperature of rabbits following intravenous administration of the product, and bacterial endotoxins are assayed by Limulus Amebocyte Lysate (LAL), which is obtained from aqueous extracts of the circulating amebocytes of the horseshoe crab.

Aseptic processing in a clean room environment is costly, difficult to achieve, and inherently risky with respect to microbial contamination of the finished product. Generally, sterile filtration of a raw nanoparticulate suspension cannot be achieved, since

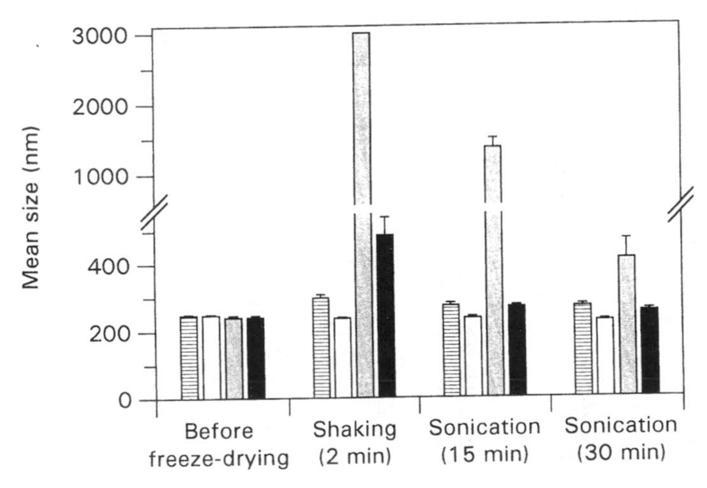

Fig. 9 Influence of redipersion method and cryoprotective agent on mean size of chlorambucil-loaded D,L-PLA nanospheres: □ plain nanospheres without trehalose; □ plain nanospheres with trehalose; □ PEG-coated nanospheres without trehalose; ■ PEG-coated nanospheres with trehalose (n = 3 ± SD). (Ref. 54.)

the particles have a similar size to that of the contaminants (0.25–1 μm). Up to now dry and moist heat have never been reported as methods for the sterilization of poly(α-hydroxycarboxylic acids). Recent trials have confirmed that the high temperature reached during the sterilization procedure (121 °C), causes irreversible fusion of nanoparticles if they are in a dry state [54]. However, if D,L-PLA particles that are suspended in water are sterilized at 121 °C for 15 min, their size remains almost unchanged [54]. Nevertheless, alteration of the active compound and of the release characteristics can be anticipated.

With ethylene oxide, toxicological problems may be encountered due to residual amounts of the agent that are retained in the product.

Gamma irradiation of poly(α-hydroxycarboxylic acids) may be the method of choice. However, it should be stressed, such a treatment can have a significant impact on the polymers. The weight-average molecular weights of the polymers (Mw) decreases with increasing irradiation dose. According to some authors working with microparticles, polydispersity indices (the weight-average molecular weight divided by the number-average molecular weight: Mw/Mn) are substantially unchanged after γ-sterilization [56], indicating a random chain cleavage mechanism, whereas Gildings et al. [57] observed an increase in polydispersity index, probably due to a terminal segment cleavage. This was termed the unzipping mechanism. Recently, the influence of γ-sterilization on the degradation of D,L-PLA nanoparticles was investigated by Leroux and Allémann [54] (Fig. 10). Particles were subjected to a dose of 25 kGy (2.5 Mrad) on dry ice to avoid degradation by heat. In this case, for both unloaded and CGP 57813 (a HIV protease inhibitor, Novartis, Switzerland) loaded nanoparticles, polymer degradation was observed. Polydispersity, however, remained unchanged, indicating a random chain cleavage. The size and morphology of the particles were not affected by the γ-radiation.

Independent of the degradation mechanism, one should keep in mind that a reduction in polymer molecular weight may significantly influence drug release patterns. In vitro, a neuroleptic compound (CGP 19486) is released faster from D,L-PLA nanospheres after γ-sterilization (Fig. 11) [30].

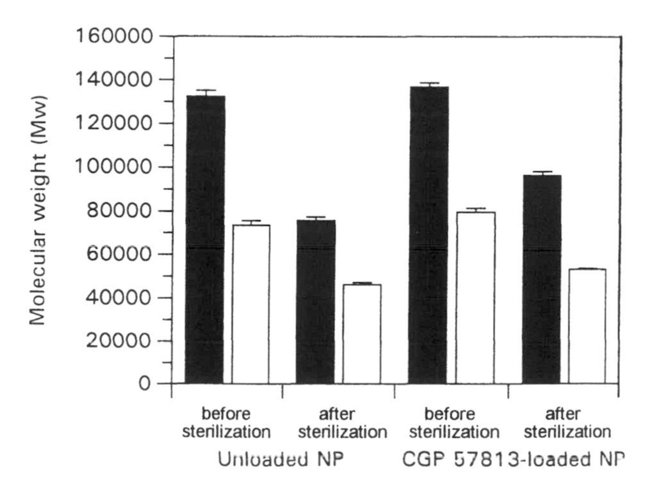

Fig. 10 Influence of γ-sterilization (25 kGy) on polymer molecular weight: ■ Mw; □ Mn (n = 5, ± SD) (NP: nanoparticles). (Ref. 54.)

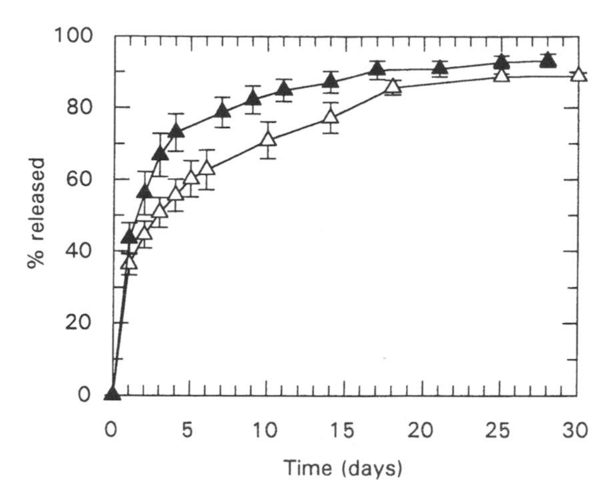

Fig. 11 In vitro release of CGP 19486 from D,L-PLA nanospheres before Δ and after ▲ sterilization by γ-radiation (25 kGy) (n = 4, ± SD). (From Ref. 29.)

Degradation of the polymer may not only affect drug release, but also the in vivo resorption of the drug carrier. Radiolytic degradation of the active compound can occur in certain cases with the formation of toxic or inactive by-products [56].

VII. CHARACTERIZATION

A. Yield, Drug Loading, Entrapment Efficiency

Before nanoparticles are produced, it is of prime importance to determine whether the preparation procedure chosen for incorporating a compound into the polymeric particles is efficient. This point is of great importance if one considers industrial applications with costly substances. To this end, three parameters are generally calculated: the weight yield, the drug loading, and the entrapment efficiency.

The overall yield of a procedure is simply expressed as

$$\text{Yield (\%)} = \frac{\text{initial amount of raw materials}}{\text{amount of nanoparticles}} \times 100 \qquad (1)$$

The initial amount of raw materials corresponds generally to the amount of active compound plus polymer. However, because of the adsorption of a certain amount of stabilizing agents or of surface active agents (e.g., PEG, PVAL, poloxamers, etc.) during the preparation of the particles, a correction factor should be introduced into Eq. (1):

$$\text{Yield (\%)} = \frac{\text{initial amount of raw materials}}{\text{amount of nanoparticles} \times (1 \quad \text{fraction of residual stabilizing agents})} \times 100 \qquad (2)$$

The drug loading, also called payload or drug content, is expressed as

$$\text{Drug loading} = \frac{\text{amount of drug in nanoparticles}}{\text{amount of nanoparticles}} \qquad (3)$$

Drug loading is very important with regard to release characteristics. Generally, increasing the drug loading leads to an acceleration of the drug release [30]. However, in particular cases, increasing the drug loading may slow down the release [12]. This can be explained by possible drug crystallization inside the nanospheres. In this case, the drug must dissolve prior to diffusing, which slows down the release.

Drug entrapment efficiency represents the proportion of the initial amount of drug, which has been incorporated into or adsorbed onto the particles. Sometimes as is defined as encapsulation efficiency, especially if nanocapsules are prepared. It is defined as

$$\text{Entrapment efficiency (\%)} = \frac{\text{percent drug loading}}{\text{percent of the initial content} \times (1 - \text{fraction of residual stabilizing agent})} \times 100 \quad (4)$$

Here again, a correction factor has been introduced to take into account the adsorbed additives.

Due to the small size of nanoparticles, the determination of drug loading is not always an easy task. Separation of free from bound drug has to be accomplished first. The available purification methods have previously been described in this chapter. Drug loading can be assessed after ultracentrifugation of a nanoparticulate suspension by dissolution of the sediment in an appropriate solvent and subsequent analysis. Additional analysis of the free drug in the supernatant may be used to confirm the results. This technique requires cautious separation of the sediment from the supernatant. If the particles are purified by cross-flow filtration and subsequently freeze-dried, the particles can be directly dissolved in an adequate solvent. Depending on the analyte, methods such as spectrophotometry, spectrofluorophotometry, high-performance liquid chromatography, or liquid scintillation counting can be used to determine the drug loading [58].

In short, if drug loading must be adjusted depending on the intended use of the particles (e.g., sustained release, burst release), high overall yield and entrapment efficiency are usually desired.

B. Size and Morphology

Particles size characteristics are of great importance for suspensions intended to be injected. To avoid embolism, particles of a large size cannot be present in the suspensions. In particular, these preparations must comply with the pharmacopoeial requirements indicated in the monograph: particulate matter in injections. The USP XXIII mentions that for emulsions, colloids, and liposomal preparations, a microscopic particle count test should be carried out. The preparation should not contain more than 12 particles/mL of more than 10 μm and not more than 2 particles/mL of more than 25 μm.

Modification of the size significantly influences the release rate of an incorporated substance. As presented in Table 9 an increase of the mean size from 303 to 671 nm significantly decreased the release rate. Contrary to poly(alkyl cyanoacrylate) nanoparticles, which generally allow sustained release over a maximum of 24 h, poly(α-hydroxycarboxylic acid) nanoparticles can display release profiles extending over several weeks. The size also strongly influences the organ distribution of intravenously

Table 9 In Vitro Release Characteristics of CGP 19486 from D,L-PLA Nanospheres as Function of Mean Particle Size

Mean size (nm)	Sw[a] (m^2/g)	$t_{25\%}$ (days)	$t_{50\%}$ (days)	$t_{75\%}$ (days)
303	13.2	< 1	2.7	13.9
671	6.0	3.2	18.1	> 30

[a]Specific surface area.
Source: Adapted from Ref. 30.

injected particles. Concerning the biodistribution of nanoparticles, Kanke et al. [59] demonstrated with beagle dogs after intravenous administration that particles with a size above 7 μm are filtered by the capillary bcd of thc lungs. Similar results have been obtained with rats [60] and rabbits [61]. Thus drug targeting to the lungs is possible with particles in the micrometer range, although this is rarely a goal. Smaller particles can also be found in the lung due to phagocytosis by lung macrophages, but the number of phagocytosed particles is relatively small. If the particles are not coated with a hydrophilic additives, the majority of them pass the lung capillary bed, but are extensively taken up by the Kupffer cells and by the macrophages of the spleen. As reviewed, by Davis et al. [62], there is also a clear relationship between particle size and spleen uptake; the smaller the size, the more efficient is the avoidance of this organ.

Therefore, a detailed knowledge of the size of the particles is required prior to conducting in vitro release kinetic and in vivo biodistribution studies. To this end, photon correlation spectroscopy, transmission electron microscopy, and scanning electron microscopy (SEM) are the most commonly used tools. Optical microscopy is not useful, since the detection limit is about 500 nm. These measuring techniques have been extensively reviewed by Müller [63] and Magenheim and Benita [58].

Photon correlation spectroscopy (PCS) is a laser light scattering method suitable for the measurement of particles ranging from 5 nm to 5 μm. The PCS device consists of a laser, a temperature controlled cell, and a photomultiplier. The light scattered from a colloidal dispersion is detected on a photomultiplier and transferred to a correlator for the calculation of a correlation function. This function is then processed to calculate the mean size of the particles. Powerful modern equipment allows the determination of the size distribution of mono- and multimodal populations of particles. The advantage of this technique is the ease of sample preparation. To be analyzed, the particles only need to be sufficiently diluted in a solvent, which is generally filtered water.

Scanning electron microscopy is widely used in the field of nanocarriers. It has high resolution and the sample preparation is relatively easy. However, to be analyzed, the samples must withstand a high vacuum. To visualize the particles, they have to be conductive and, therefore, coating of the surface of the sample with gold is required. The thickness of this coating is at least 20 nm. This must be taken into account in the size determination, especially for small nanoparticles (e.g., <200 nm). Removal of stabilizing agents added during the preparation of the particles is essential. Otherwise, depending on the amount of these additives, the particles are partially or completely hidden in a matrix of additives. Figure 12 shows a sample scanning electron micrograph.

Coating with gold and applying a high vacuum can modify the original aspect of the particles. New imaging tools, such as photon scanning tunneling microscopy and

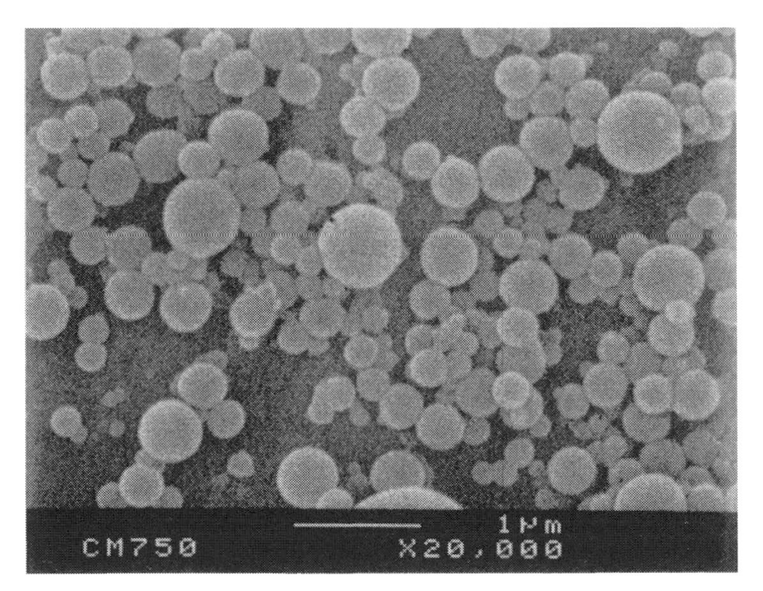

Fig. 12 Scanning electron micrograph of D,L-PLA nanoparticles loaded with CGP 57813. (Ref. 51.)

scanning force microscopy (also called atomic force microscopy), enable the visualization of nanoparticles at atmospheric pressure without gold coating [12,64]. Nevertheless, the resolution obtained with these new tools is still lower than that with SEM.

For size determination, transmission electron microscopy is not as widely used as PCS and SEM, but it is still a powerful method for determining the morphology of particles. With this technique, Fessi et al. [42] estimated the wall thickness of PLA nanocapsules. Krause et al. [18] described the highly porous structure of PLA nanospheres prepared by the emulsion-evaporation procedure.

VIII. IN VITRO RELEASE STUDIES

In vitro release studies should in principle be useful for quality control as well as for the prediction of in vivo kinetics. Unfortunately, due to the very small size of the particles, the release rate observed in vivo can differ greatly from the release obtained in a buffer solution. However, in vitro release studies remain very useful for quality control as well as for evaluation of the influence of process parameters on the release rate of active compounds. In vitro drug release from microdispersed systems has been extensively reviewed by Washington [65].

Depending on the type of polyester, drug release from nanoparticles can take place through several processes, of which the following appear to be the most important: (1) The drug may diffuse out of the carrier through the solid matrix; to allow complete release from the carriers, (the concentration of drug in the release medium should remain infinitely low, which condition is known as sink condition); (2) The solvent may penetrate the nanoparticles and dissolve the drug, which then diffuses out into the release medium. Depending on the physico-chemical characteristics of the particles, water can enter the particles through narrow pores or by hydration. Once the drug is dissolved, the drug diffuses out of the particles. Here again, since diffusion is driving the release, sink conditions are required to obtain meaningful information; (3) The carrier may be degraded by its surroundings; as long as this process is faster than diffusion, the release process is said to be erosion-controlled.

In vitro release kinetics of a drug entrapped in nanoparticles can be evaluated by several experimental methods.

A. Dialysis

In the dialysis approach, nanoparticles suspended in a small volume of release medium, are separated from a large bulk of sink medium by a dialysis membrane, which is permeable to the drug. The active substance is released from the nanoparticles and diffuses through the membrane. The released drug is assayed in the receptor phase. Washington [65] demonstrated that in most cases, unfortunately, this technique does not allow accurate determination of the real drug release rate. Indeed, since the donor compartment is not under sink conditions, the real release profile is completely obscured by partitioning between the sample and the donor compartment.

To solve this problem, Levy and Benita [66] proposed a technique based on reverse dialysis. The colloidal carrier is directly diluted in the release medium where perfect sink conditions are maintained. Prior to the addition of the nanoparticles, a number of dialysis bags are suspended in the release medium. At alloted times, released drug is assayed in the bags. The authors mentioned that this method is not sensitive enough to characterize rapid release rates, but assumed that this technique is suitable for releases lasting several days.

Nevertheless, the appearance rate of the drug in the sampling compartment depends on experimental factors, such as drug excipient interactions, drug membrane interactions, micelles, and osmosis. Therefore, dialysis methods are often not appropriate for determining the release characteristics of nanocarriers.

B. In Situ Method

For the in situ method the nanoparticles are dispersed in a buffer solution and the drug is directly assayed spectrophotometrically in a buffer solution in the suspension without separating the free from the particle-bound drug. To do this, some characteristics (e.g., spectroscopic) of the drug in solution must be different from the particle-bound drug. For example. Illum et al. [67] took advantage of the bathochromic shift, which occurred when the model compound, Rose Bengal, was adsorbed on poly(butyl cyanoacrylate) nanoparticles, to measure the release pattern. This substance undergoes a bathochromic shift when it is bound to the particles. This allows simultaneous determination, without separation, of free and bound drug at two different wavelengths. The disadvantages inherent to the dialysis method are therefore avoided. Unfortunately, most active compounds do not show significant differences in physico-chemical characteristics between the bound and the free state.

C. Sample and Separation Techniques

In the sample and separation techniques (Fig. 13), the carrier is diluted into a sink maintained at a constant temperature under agitation. Ideally, the sink should be a buffer solution with a pH of 7.4, but some active substances require other pH values or the addition of a surfactant to reach sink conditions. Usually, the working temperature is 37°C. Sampling is carried out at various time intervals. The continuous phase is then

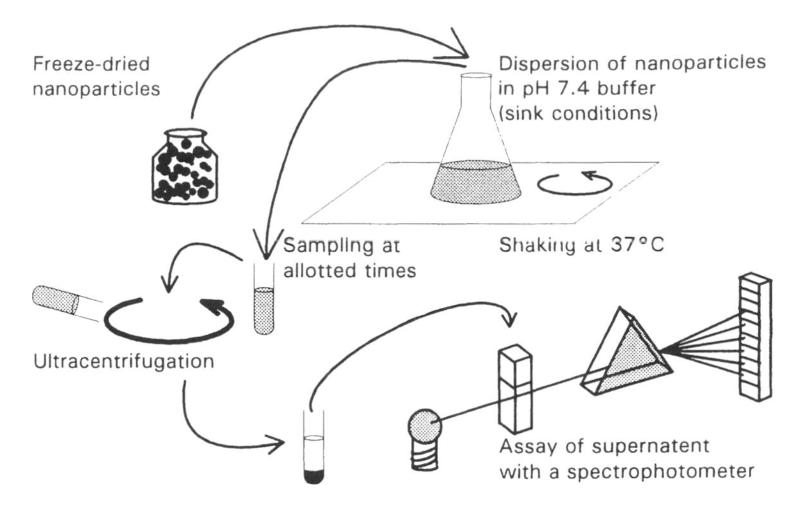

Fig. 13 Sample and separation technique.

separated from the dispersed phase by ultracentrifugation or centrifugal filtration [44], and the released drug is assayed [30,68]. This technique is satisfactory if the two phases can be separated sufficiently rapidly to avoid excessive drug release between sampling and assay. For example, if the release is expected to last several days or weeks, a 2 h ultracentrifugation of the samples is not considered to greatly influence the release profile. However, in the case of a release expected to last for 24 h or less, the time required for the separation should be taken into account in the interpretation of the release profiles.

For example Sánchez et al. [69] separated cyclosporin A loaded PLGA particles of variable sizes (285 nm to 30 μm) from free drug by a 1 h ultracentrifugation at 150,000 × *g*. Allémann et al. [30] separated CGP 19486 (a neuroleptic compound) loaded in D,L-PLA nanoparticles from the release compound by a 2 h ultracentrifugation at 55,000 × *g*.

D. Ultrafiltration at Low Pressure

This method is based on direct dilution of the nanoparticulate suspension in the release medium (in a sufficient amount to provide sink conditions) [10]. The amount of drug released from the carrier is determined by sampling the filtrate, which is collected at set time intervals. The small volume of sample required for spectroscopic analysis allows filtration at a slow rate and at low pressure (less than 0.5 bar), thus preserving the physical integrity of the colloidal dispersion. The ultrafiltration membrane should be characterized by low nonselective adsorption and a membrane cutoff point below the smallest particle size. For example, Magenheim et al. [10] used a membrane with a molecular cutoff point of 100,000 daltons.

IX. PHYSICAL AND CHEMICAL STABILITY

Nanoparticles of poly(α-hydroxycarboxylic acids) are biodegradable. After in vivo administration, complete degradation of the polymer matrix is desired in order to avoid accumulation in the body. However, during storage, the stability of the colloidal drug delivery sytems is a key factor. Modification of the integrity of the particles or of the molecular weight of the polymers significantly influences the drug release characteristics.

Polymer degradation can be followed by gel permeation chromatography, whereas particle integrity can be assessed by nephelometric studies.

A. Stability of Polymers Assessed by Gel Permeation Chromatography

In practice, after storage poly(α-hydroxycarboxylic acids) particles can be dissolved in an appropriate solvent, and the molecular weight can be determined by gel permeation chromatography (GPC). Methods for these determinations have been described elsewhere [70,71]. Usually, three size exclusion columns (Ultrastyragel®, Waters) with exclusion limits of 10^4, 10^3, and 10^2 are placed in series [54] and a differential refractometer is used as the detector. Operational conditions reported are chloroform as solvent, column and detector temperature of 30°C, injection volume of 20 μL, flow rate of 1 mL/min, and a solute concentration of 0.1% (w/v). Polystyrene standards are commonly used to calibrate the GPC system. The stability of PLA and PLGA nanoparticles under various conditions has been studied [7,71]. Table 10 shows that the molecular weight loss of the polymers is less important when nanoparticles are stored in a freeze-dried state. In suspension, degradation is faster in water than in a pH 7.4 buffer solution. The acidity of the degradation products (α-hydroxycarboxylic acids) catalyzes the degradation of the remaining polymers, whereas in a neutral buffer solution the degradation remains relatively low. If the particles are stored in water, the stability increases as the temperature decreases. The best stability is obtained when the suspension is frozen.

B. Stability of Nanoparticles Determined by Nephelometric Analysis

To analyze the physical stability of the particles, a nephelometric assay can be used [8]. The degradation of nanoparticles in aqueous suspension can be monitored with a spectrophotometer. The reduction in light transmission is caused by light scattering and not by molecular absorption. Absorption (or transmission) readings can be used to determine the turbidity of the suspension. A decrease in absorption (respectively, an increase in transmission) indicates a degradation of the particles.

Wallis and Müller [8] used this method to make comparative measurements of nanoparticle degradation rates of nanoparticles of poly(hydroxybutyric acid) (PHBA), PLA, and PLGA. To accelerate the degradation, experiments were conducted at pH

Table 10 Molecular Weight Loss (%) of PLA and PLGA in Nanoparticles Stored at Room Temperature under Different Conditions

	Molecular weight loss (%)		
Polymer	Water	pH 7.4 buffer	Freeze-dried
PLA	26.3	18.0	20.2
PLGA	67.0	44.4	35.4

Source: Adapted from Ref. 71.

10.5 and 13. The authors have shown that the physical stability of the particles ranked as follows: PLGA < PLA < PHBA.

X. INTERACTION OF BIODEGRADABLE NANOPARTICLES WITH BLOOD COMPONENTS AND CELL MEMBRANES

On exposure to blood, colloidal carriers become coated with plasma proteins. Some of these proteins are involved in the attachment of the particles to the membrane of mononuclear phagocytic cells and thus regulate the elimination of the carrier. The proteins responsible for this activity are known as opsonins. They are classified as specific and nonspecific opsonins depending on whether they interact with specific receptors on macrophages or just render the particle more adhesive to the cells [72]. The mechanisms involved in the opsonization of colloidal carriers are complex and still poorly understood. Recently, some basic in vitro investigations have been carried out, especially in the case of model polystyrene nanoparticles, in order to identify the plasma proteins involved in the elimination of the carriers [73–75]. The ultimate goal is to be able to predict the fate of nanoparticles relative to their adsorbed protein pattern. Generally, it has been demonstrated that nanoparticles coated with specific surfactants such as poloxamers are less prone to interact with plasma proteins. A simple approach to evaluate the protection offered by the coating agent is to measure the zeta potential of coated and noncoated nanoparticles in biological fluids [63]. As shown in Table 11, one can see that sterically stabilized polystyrene nanoparticles exhibit a lower surface charge when exposed to serum proteins. A low absolute zeta potential value indicates reduced interaction of the particle with serum proteins, but it does not necessarily imply that the uptake by the mononuclear phagocyte system will be significantly prevented. One has to take into account the surface physico-chemical characteristics of the carrier itself and the nature of the adsorbed plasma/serum components. The total amount of adsorbed proteins can also be directly quantified by spectrophotometric methods [76] or by densitometry on electrophoretic gels [74]. However, it is generally difficult to completely remove bound proteins from the surface of nanoparticles.

Studies carried out with polystyrene and more recently polyester nanoparticles have shown that a considerable amount of different proteins (e.g., IgG, complement components, fibrinogen, apolipoproteins) can adsorb onto particle surfaces [77]. The adsorbed protein patterns established by two-dimensional polyacrylamide gel electrophoresis have shown that the adsorption profile depends on several parameters such as

Table 11 Zeta Potential of Coated Polystyrene Nanoparticles in Serum

Poloxamer	Mw	EO[a] units	PO[b] units	Coating layer thickness (Å)	Zeta potential (mV)
—	—	—	—	—	-14.9
108	5,000	2 × 46	16	58	-14.4
188	8,350	2 × 75	30	76	-9.6
407	11,500	2 × 98	67	119	-4.6
338	14,000	2 × 128	54	154	-5.2

[a]Poly(ethylene oxide).
[b]Poly(propylene oxide).
Source: Adapted from Ref. 63.

the incubation medium (serum/plasma) [77] and the incubation time [78]. The function of the adsorbed protein is usually evaluated in vitro using cultured cells in order to determine which opsonins play a significant role in the elimination of the nanoparticles (Fig. 14). It appears that, in the case of polyester nanoparticles, heat labile opsonins, such as complement components and heat stable opsonins, such as IgG, may be strongly involved in the recognition mechanisms [79]. Coating polyester nanoparticles with poly(ethylene oxide) chains can reduce the activation of the complement system [80].

As shown in Fig. 14, several techniques can be used in vitro to evaluate and characterize the uptake of nanoparticles by the cells of the mononuclear phagocyte system (MPS). Nanoparticles can be physically or chemically labeled with various radioactive tracers (e.g., ^{14}C, ^{111}In-Oxine) [81,82]. After incubation with cells, the cell-associated radioactivity is measured [83]. The surface labeling of polyester nanoparticles is sometimes problematic. For instance, poly(lactic acid) nanoparticles can aggregate upon labeling with ^{131}I [81]. Among nonradioactive techniques, chemiluminescence and fluorimetric methods such as flow cytometry, are frequently used. Chemiluminescence

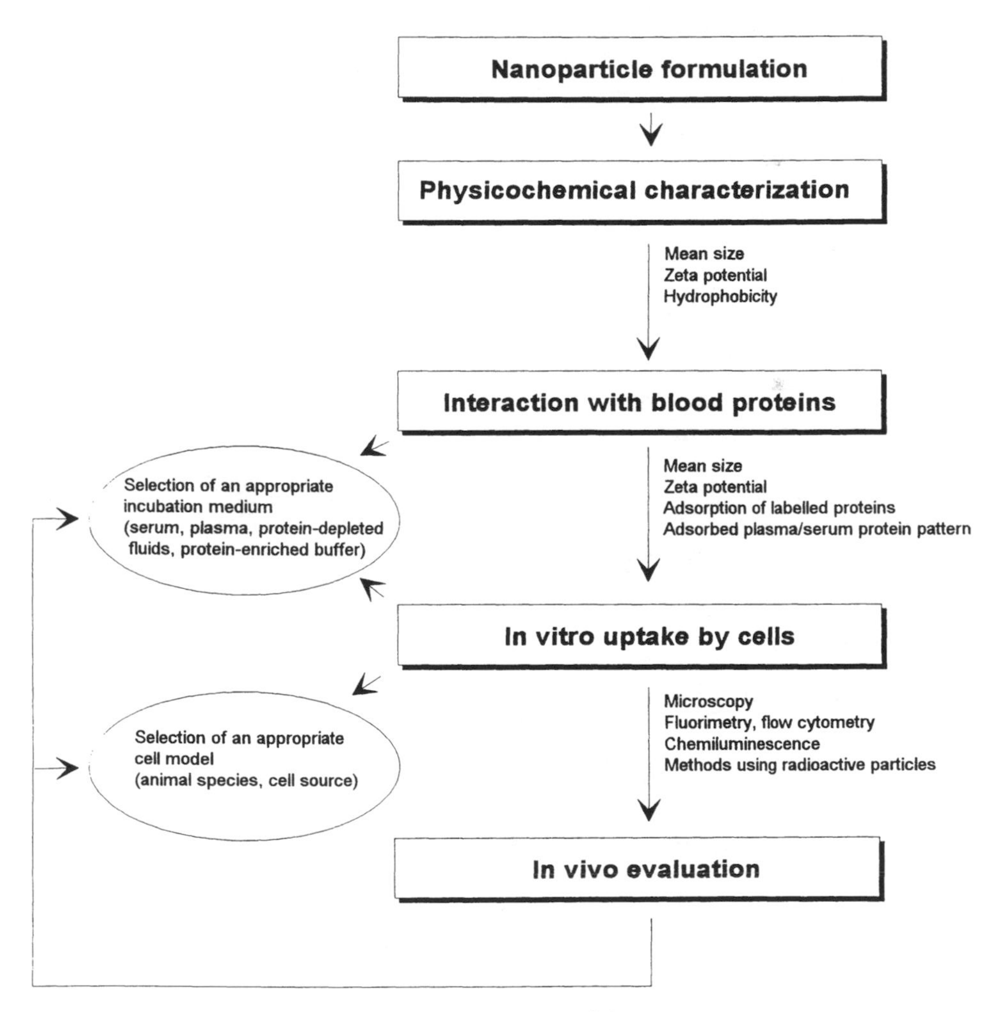

Fig. 14 In vitro evaluation of biodegradable nanoparticles.

is a technique that allows continuous measurement of the phagocytic uptake regardless of the nanoparticle type. It requires no labeling of the particles, and accordingly, the surface characteristics of the latter remain unchanged [84]. However, chemiluminescence gives no information on the adsorption process. Flow cytometry allows the simultaneous analysis of mixed cell populations by cell-by-cell fluorescence measurement [75]. The fluorescence emitted by the labeled nanoparticles is measured only in the gated cell population, which is distinguished from other cell populations by its intrinsic parameters, (e.g., size, cytoplasm content). The choice of an adequate fluorescent tracer is of prime importance. The fluorescent tracer must be hydrophobic or covalently attached to the polymeric matrix in order to avoid diffusion in the incubation medium, and has to retain its fluorescence properties when associated with the polymer. For instance, at relatively high concentrations fluorescein looses its fluorescence in part when entrapped in poly(lactic acid) nanoparticles, and the dye is rapidly released [54]. On the other hand, the fluorescent tracer, Nile red, remains strongly fluorescent under the same conditions. Furthermore, it is poorly soluble in aqueous media (high entrapment efficiency) and the fluorescence of the free dye is quenched in water [75,85]. Whatever the assay chosen, the incubation medium and cell model have to be carefully selected. Blood cells with phagocytic abilities such as monocytes or granulocytes can be easily obtained and purified, but they do not have exactly the same characteristics as extravascular macrophages such as Kupffer cells. Moreover, some coating agents can protect the particles from extensive phagocytosis in plasma but not in serum and buffer [79]. Therefore, depending on the in vivo results, the choice of the incubation conditions should be reevaluated continuously in order to define adequate in vitro models (Fig. 14).

In short, the existing methods help one understand the phagocytosis of uncoated and coated polymeric particles and represent promising in vitro tools for predicting the in vivo fate of the particles.

REFERENCES

1. S. J. Douglas, S. S. Davis, and L. Illum, Nanoparticles in drug delivery, *CRC Crit. Rev. Ther.*, 3:233–261 (1987).
2. E. Allémann, R. Gurny, and E. Doelker, Drug loaded nanoparticles—preparation methods and drug targeting issues, *Eur. J. Pharm. Biopharm.*, 39:173–191 (1993).
3. C. Schwarz, W. Mehnert, J. S. Lucks, and R. H. Müller, Solid lipid nanoparticles (SLN) for controlled drug delivery. I. Production, characterization and sterilization, *J. Controlled Release*, 30:83–96 (1994).
4. O. Vaizoglu and P. P. Speiser, The pharmacosome drug delivery approach, *Eur. J. Pharm. Biopharm.*, 38:1–6 (1992).
5. J. Kreuter, Evaulation of nanoparticles as drug-delivery systems, *Pharm. Acta. Helv.*, 58:196–208 (1983).
6. D. H. Lewis, Controlled release of bioactive agents from lactide/glycolide polymers. In: Biodegradable Polymers as Drug Delivery Systems. (M. Chasin, and R. Langer, eds.), Marcel Dekker, New York, 1990, pp 1–41.
7. A. Belbella, C. Vauthier, H. Fessi, J. P. Devissaguet, and F. Puisieux, In vitro degradation of nanospheres from poly(D,L-lactides) of different molecular weights and polydispersities, *Int. J. Pharm.*, 129:95–102.
8. K. H. Wallis, and R. H. Müller, Comparative measurements of nanoparticle degradation velocity using an accelerated hydrolysis test, *Pharm. Ind.*, 55:168–170 (1993).

9. R. A. Miller, J. M. Brady, and D. E. Cutright, Degradation rates of oral resorbable implants (polylactates and polyglycolates): Rate modification with changes in PLA/PGA copolymer ratios, *J. Biomed. Mater. Res.*, 11:711–719 (1977).
10. B. Magenheim, M. Y. Levy, and S. Benitz, A new in vitro technique for the evaluation of drug release profile from colloidal carriers—Ultrafiltration technique at low pressure, *Int. J. Pharm.*, 94:115–123 (1993).
11. S. Stolnik, S. E. Dunn, M. C. Garnett, M. C. Davies, A. G. A. Coombes, D. G. Taylor, M. P. Irving, S. C. Purkiss, T. F. Tadros, S. S. Davis, and L. Illum, Surface modification of poly(lactide-*co*-glycolide) nanospheres by biodegradable poly(lactide)-poly(ethylene glycol) copolymers, *Pharm. Res.*, 11:1800–1808 (1994).
12. R. Gref, Y. Minamitake, M. T. Peracchia, V. Trubetskoy, V. P. Torchilin, and R. Langer, Biodegradable long-circulating polymeric nanospheres, Sci., 263:1600–1603 (1994).
13. R. Gurny, G. S. Banker, and P. Buri, Développement d'un injectable de testostérone à action prolongeée à base de latex biodégradable, *Congr. hispano-français Biopharm. Pharmacocin.*, 173–179 (1977).
14. R. Gurny, N. A. Peppas, D. D. Harrington, and G. S. Banker, Development of biodegradable and injectable latices for controlled release of potent drugs, *Drug Dev. Ind. Pharm.*, 7:1–25 (1981).
15. J. W. Vanderhoff, M. S. El-Aasser, and J. Ugelstad, Polymer emulsification process. U.S. Patent 4,177,177 (1979).
16. H. Ibrahim, C. Bindschaedler, E. Doelker, P. Buri, and R. Gurny, Concept and development of ophtalmic pseudo-latexes triggered by pH, *Int. J. Pharm.*, 77:211–219 (1991).
17. R. Gurny, T. Boye, and H. Ibrahim, Ocular therapy with nanoparticulates systems for controlled drug delivery, *J. Controlled Release*, 2:353–361 (1985).
18. H. J. Krause, A. Schwartz, and P. Rohdewald, Polylactic acid nanoparticles, a colloidal drug delivery system for lipophilic drugs, *Int. J. Pharm.*, 27:145–155 (1985).
19. R. Bodmeier and H. Chen, Indomethacin polymeric nanosuspensions prepared by microfluidization, *J. Controlled Release*, 12:223–233 (1990).
20. G. F. Dawson and G. W. Halbert, Production of poly lactide-*co*-glycolide nanoparticulates and optimization of diameter, *J. Pharm. Pharmacol.*, 46 (supp 2):1037 (1994).
21. C. Bitz and E. Doelker, Influence of the preparation method on residual solvents in biodegradable microspheres, *Int. J. Pharm.*, 131:171–181.
22. C. Bindschaedler, R. Gurny, and E. Doelker, Process for preparing a powder of water-insoluble polymer which can be redispersed in a liquid phase, the resulting powder and utilization thereof, Eur Patent PCT/EP88/00281 (1993).
23. C. Bindschaedler, R. Gurny, and E. Doelker, Process for preparing a powder of water-insoluble polymer which can be redispersed in a liquid phase, the resulting powder and utilization thereof, U.S. Patent 4,968,350 (1990).
24. H. Ibrahim, C. Bindschaedler, E. Doelker, P. Buri, and R. Gurny, Aqueous nanodispersions prepared by a salting-out process, *Int. J. Pharm.*, 87:239–246 (1992).
25. H. Ibrahim, Concept et évaluation de systèmes polymériques dispersés (pseudo-latex) à usage ophtalmique, Ph.D. thesis, University of Geneva, Geneva, Switzerland.
26. E. Allémann, R. Gurny, and E. Doelker, Preparation of aqueous polymeric nanodispersions by a reversible salting-out process, influence of process parameters on particle size, *Int. J. Pharm.*, 87:247–253 (1992).
27. E. Allémann, E. Doelker, and R. Gurny, Drug loaded poly(lactic acid) nanoparticles produced by a reversible salting-out process: purification of an injectable dosage form, *Eur. J. Pharm. Biopharm.*, 39:13–18 (1993).
28. E. Allémann, N. Brasseur, O. Benrezzak, J. Rousseau, S. V. Kudrevich, R. W. Boyle, J. C. Leroux, R. Gurny, and J. E. van Lier, PEG-coated poly(lactic acid) nanoparticles for the delivery of hexadecafluoro zinc phthalocyanine to EMT-6 mouse mammary tumours, *J. Pharm. Pharmacol.*, 47:382–387 (1995).

29. E. Allémann, R. Gurny, E. Doelker, F. S. Skinner, and H. Schütz, Distribution, kinetics and elimination of radioactivity after intravenous and intramuscular injection of 14C-savoxepine loaded poly(D,L-lactic acid) nanospheres to rats, *J. Controlled Release*, 29:97–104.
30. E. Allémann, J. C. Leroux, R. Gurny, and E. Doelker, In vitro sustained release properties of drug loaded poly(D,L-lactic acid) nanoparticles produced by a salting-out procedure, *Pharm. Res.*, 10:1732–1737 (1993).
31. J. C. Leroux, E. Allémann, E. Doelker, and R. Gurny, New approach for the preparation of nanoparticles by an emulsification-diffusion method, *Eur. J. Pharm. Biopharm.*, 41:14–18 (1995).
32. S. Massen, E. Fattal, R. H. Müller, and P. Couvreur, Cell cultures for the assessment of toxicity and uptake of polymeric particulate drug carriers, *S. T. P. Pharma*, 3:11–22 (1993).
33. A. Martin, P. Bustamante, and A. H. C. Chun, Physical Pharmacy. Lea & Febiger, Malvern, PA, 1993, pp. 490–491.
34. H. Fessi, J. P. Devissaguet, F. Puisieux, and C. Thies, Procédé de préparation de systèmes colloïdaux dispersibles d'une substance, sous forme de nanoparticules., French Patent 2,608,988 (1986).
35. H. Fessi, J. P. Devissaguet, F. Puisieux, and C. Thies, Process for the preparation of dispersible colloidal systems of a substance in the form of nanoparticles, U.S. Patent 5,118,528 (1992).
36. A. de Labouret, O. Thioune, H. Fessi, J. P. Devissaguet, and F. Puisieux, Application of an original process for obtaining colloidal dispersions of some coating polymers. Preparation, characterization, industrial scale-up, *Drug Dev. Ind. Pharm.*, 21:229–241 (1995).
37. D. V. Bazile, C. Ropert, P. Huve, T. Verrecchia, M. Marlard, A. Frydman, M. Veillard, and G. Spenlehauer, Body distribution of fully biodegradable (14C)-poly(lactic acid) nanoparticles coated with albumin after parenteral administration to rats, *Biomaterials*, 13:1093–1102 (1992).
38. T. Verrecchia, G. Spenlehauer, D. Bazile, and M. Veillard, Poly(lactic acid)-poly(ethylene glycol) (PLAPEG): a new copolymer for injectable drug carriers, Proc. *Journées Galéniques Gatefossé*, 87:55–62 (1994).
39. T. Verrecchia, G. Spenlehauer, D. V. Bazile, A. Murry-Brelier, Y. Archimbaud, and M. Veillard, Non-stealth (poly(lactic acid / albumin)) and stealth (poly(lactic acid-polyethylene glycol)) nanoparticles as injectable drug carriers, *J. Controlled Release*, 36:49–61 (1995).
40. S. Stolnik, M. C. Davies, L. Illum, S. S. Davis, M. Boustta, and M. Vert, The preparation of sub-200 nm biodegradable colloidal particles from poly(beta-malic acid-*co*-benzyl malate) copolymers and their surface modification with poloxamer and poloxamine surfactants, *J. Controlled Release*, 30:57–67 (1994).
41. H. Fessi, F. Puisieux, and J. P. Devissaguet, Procédé de préparation de systèmes colloidaux dispersibles d'une substance, sous forme de nanocapsules., Eur. Patent 274 961 (1987).
42. H. Fessi, F. Puisieux, J. P. Devissaguet, N. Ammoury, and S. Benitz, Nanocapsule formation by interfacial polymer deposition following solvent displacement, *Int. J. Pharm.*, 55:R1–R4 (1989).
43. F. Fawaz, F. Bonini, M. Guyot, A. M. Lagueny, H. Fessi, and J. P. Devissaguet, Influence of poly(DL-lactide) nanocapsules on the biliary clearance and enterohepatic circulation of indomethacin in the rabbit, *Pharm. Res.*, 10:750–756 (1993).
44. N. S. Santos Megalhaes, H. Fessi, F. Puisieux, S. Benita, and M. Seiller, An in vitro release kinetic examination and comparative evaluation between submicron emulsion and polylactic acid nanocapsules of clofibride, *J. Microencapsulation*, 12:195–205 (1995).
45. S. S. Guterres, H. Fessi, G. Barratt, J. P. Devissaguet, and F. Puisieux, Poly(DL-lactide) nanocapsules containing diclofenac: I. Formulation and stability study, *Int. J. Pharm.* 113:57–63 (1995).

46. H. Pinto-Alphandary, O. Balland, and P. Couvreur, A new method to isolate polyalkylcyanoacrylate nanoparticle preparation, *J. Drug Targeting*, 3:167–169 (1995).
47. R. G. Gutman, *Membrane Filtration, the Technology of Pressure-Driven Crossflow Processes*, Adam Hilger, Bristol, U.K.
48. F. R. Gabler, Principles of tangential flow filtration: applications to biological processing. In: Filtration in the Pharmaceutical Industry. (T. H. Meltzer, ed.), Marcel Dekker, New York, 1987, pp. 453–490.
49. K. Langer, E. Seegmüller, A. Zimmer, and J. Kreuter, Characterization of polybutylcyanoacrylate nanoparticles: I. Quantfication of PBCA polymer and dextrans, *Int. J. Pharm.*, 110:21–27 (1994).
50. S. Pirker, J. Kruse, C. Noe, K. Langer, A. Zimmer, and J. Kreuter, Characterization of polybutylcyanoacrylate nanoparticles. Part II: Determination of polymer content by NMR-analysis, *Int. J. Pharm.*, 128:189–195 (1996).
51. P. Beck, D. Scherer, and J. Kreuter, Separation of drug-loaded nanoparticles from free drug by gel filtration, *J. Microencapsulation*, 7:491–496 (1990).
52. A. Rolland, D. Gibassier, P. Sado, and R. Le Verge, Purification et propriétés physicochimiques des suspensions de nanoparticules de polymère, *J. Pharm. Belg.*, 41:94–105 (1986).
53. M. Auvillain, G. Cavé, H. Fessi, and J. P. Devissaguet, Lyophilisation de vecteurs colloidaux submicroscopiques, *S. T. P Pharma*, 5:738–744 (1989).
54. J. C. Leroux and E. Allémann, unpublished data, 1995.
55. S. Chasteigner, H. Fessi, G. Cavé, J. P. Devissaguet, and F. Puisieux, Gastro-intestinal tolerance study of a freeze-dried oral dosage form of indomethacin-loaded nanocapsules, *S. T. P. Pharma Sciences*, 5:242–246 (1995).
56. C. Volland, M. Wolff, and T. Kissel, The influence of gamma-sterilization on captopril containing poly(D,L-lactide-co-glycolide) microspheres, *J. Controlled Release* 31:293–305 (1994).
57. D. K. Gildings and A. M. Reed, Biodegradable polymers for use in surgery—poly(glycolic)/poly(lactic acid) homo- and copolymers:1, Polymer, 20:1459–1464 (1979).
58. B. Magenheim and S. Benita, Nanoparticle characterization: A comprehensive physicochemical approach, *S. T. P. Pharma Sciences*, 1:221–241 (1991).
59. M. Kanke, G. H. Simmons, D. L. Weiss, B. A. Bivins, and P. P. DeLuca, Clearance of ^{141}Ce-labeled microspheres from blood and distribution in specific organs following intravenous and intraareterial administration in beagle dogs, *J. Pharm. Sci.*, 69(7):755–762 (1980).
60. R. M. Gesler, P. J. Garvin, B. Klamer, R. U. Robinson, C. R. Thompson, W. R. Gibson, F. C. Wheeler, and R. G. Carlson, The biologic effects of polystyrene latex particles administered intravenously to rats—a collaborative study, *Bull. Parent. Drug Assoc.*, 27:101–113 (1973).
61. L. Illum, S. S. Davis, C. G. Wilson, N. W. Thomas, M. Frier, and J. G. Hardy, Blood clearance and organ deposition of intravenously administered colloidal particles. The effects of particles size, nature and shape, *Int. J. Pharm.*, 12:135–146 (1982).
62. S. S. Davis, L. Illum, S. M. Moghimi, M. C. Davies, C. J. H. Porter, I. S. Muir, A. Brindley, N. M. Christy, M. E. Norman, P. Williams, and S. E. Dunn, Microspheres for targeting drugs to specific body sites, *J. Controlled Release*, 24:157–163 (1993).
63. R. H. Müller, *Colloidal Carriers for Controlled Drug Delivery and Targeting*, CRC Press, Boca Raton, FL, 1991.
64. M. Skiba, F. Puisieux, D. Duchêne, and D. Wouessidjewe, Direct imaging of modified betacyclodextrin nanospheres by photon scanning tunneling and scanning force microscopy, *Int. J. Pharm.*, 120:1–11 (1995).
65. C. Washington, Drug release from microdisperse systems: A critical review, *Int. J. Pharm.*, 58:1–12 (1990).

66. M. Y. Levy and S. Benitz, Drug release from submicronized o/w emulsion: A new in vitro kinetic evaluation model, *Int. J. Pharm.*, 66:29–37 (1990).
67. L. Illum, M. A. Khan, E. Mak, and S. S. Davis, Evaluation of carrier capacity and release characteristics for poly(butyl 2-cyanoacrylate) nanoparticles, *Int. J. Pharm.*, 30:17–28 (1986).
68. T. Niwa, H. Takeuchi, T. Hino, N. Kunou, and Y. Kawashima, In vitro drug release behavior of D,L-lactide/glycolide (PLGA) nanospheres with nafarelin acetate prepared by a novel spontaneous emulsification solvent diffusion method, *J. Pharm. Sci.*, 83:727–732 (1994).
69. A. Sánchez, J. L. Vila-Jato, and M. J. Alonso, Development of biodegradable microspheres and nanospheres for the controlled release of cyclosporin A., *Int. J. Pharm.*, 99:263–273 (1973).
70. M. D. Coffin and J. W. McGinity, Biodegradable pseudolatexes: The chemical stability of poly(D,L-lactide) and poly(epsilon-caprolactone) nanoparticlees in aqueous media, *Pharm. Res.*, 9:200–205 (1992).
71. D. Lemoine, C. François, V. Berlage, F. Kedzierewicz, P. Maincent, M. Hoffman, and V. Préat, Stability study of poly(D,L-lactide), poly(D,L-lactide-coglycolide) and poly(epsilon-caprolactone) nanoparticles, *Proc. 1st World Meeting APGI/APV*, 1995, 481–482.
72. H. M. Patel, Serum opsonins and liposomes: Their interaction and opsonophagocytosis, *Crit. Rev. Ther. Drug Carriers Syst.*, 9:39–90 (1992).
73. T. Blunk, D. F. Hochstrasser, J. C. Sanchez, B. W. Müller, and R. H. Müller, Colloidal carriers for intravenous drug targeting: plasma protein adsorption patterns on surface modified latex particles evaluated by two-dimensional polyacrylamide gel electrophoresis, *Electrophoresis*, 14:1382–1387 (1993).
74. M. E. Norman, P. Williams, and L. Illum, Influence of block copolymers on the adsorption of plasma proteins to microspheres, *Biomaterials*, 14:193–202 (1993).
75. J. C. Leroux, P. Gravel, L. Balant, B. Volet, B. M. Anner, E. Allémann, E. Doelker, and R. Gurny, Internalization of poly(D,L-lactic acid) nanoparticles by isolated human leukocytes and analysis of plasma proteins adsorbed onto the particlees, *J. Biomed. Mater. Res.*, 28:471–481 (1994).
76. J. S. Tan, D. E. Butterfield, C. L. Voycheck, K. D. Caldwell, and J. T. Li, Surface modification of nanoparticles by PEO/PPO block copolymers to minimize interactions with blood components and prolong blood circulation in rats, *Biomaterials*, 14:823–833 (1993).
77. J. C. Leroux, E. Allémann, F. De Jaeghere, E. Doelker, and R. Gurny, Biodegradable nanoparticles—From sustained release formulations to improved site specific drug delivery, *J. Controlled Release*, 39:339–350 (1996).
78. T. Blunk, M. Lück, D. F. Hochstrasser, A. Sanchez, B. W. Müller, and R. H. Müller, Kinetics of protein adsorption on model particles for drug targeting, *Eur. J. Pharm. Sci.*, 2:192 (1994).
79. J. C. Leroux, F. De Jaeghere, B. Anner, E. Doelker, and R. Gurny, An investigation of the role of plasma and serum opsonins on the internalization of biodegradable poly(D,L-lactic acid) nanoparticles by human monocytes, *Life Sci.*, 57:695–703 (1995).
80. D. Labarre, M. Vittaz, G. Spenlehauer, D. Bazile, and M. Veillard, Complement activation by nanoparticulate carriers, *Proc. Int. Symp. Control. Rel. Bioact. Mater.* 21:91–92 (1994).
81. R. H. Müller, F. Koosha, S. S. Davis, and L. Illum, In vitro and in vivo release of In-111 from PHB and PLA, *Proc. Int. Symp. Cont. Rel. Bioact. Mater.*, 15:378–379 (1988).
82. A. M. Le Ray, M. Vert, J. C. Gauthier, and J. P. Benoît, Fate of [14C]poly(DL-lactide-*co*-glycolide) nanoparticles after intravenous and oral administration to mice., *Int. J. Pharm.*, 106:201–211 (1994).
83. S. E. Dunn, A. Brindley, S. S. Davis, M. C. Davies, and L. Illum, Polystyrene-poly (ethylene glycol) (PS-PEG2000) particles as model systems for site specific drug delivery. 2.

The effect of PEG surface density on the in vitro cell interaction and in vivo biodistribution, *Pharm. Res.*, 11:1016–1022 (1994).

84. S. Rudt and R. H. Müller, In vitro phagocytosis assay of nano- and microparticles by chemiluminescence. I. Effect of analytical parameters, particle size and particle concentration, *J. Controlled Release*, 22:263–272 (1992).
85. P. Greenspan, E. P. Mayer, and S. D. Fowler, Nile red: A selective flourescent stain for intracellular lipid droplets, *J. Cell Biol.*, 100:965–973 (1985).

6

Aqueous Polymeric Dispersions

Vijay Kumar, Gilbert S. Banker, and Ganesh S. Deshpande*

University of Iowa, Iowa City, Iowa

I. INTRODUCTION

Aqueous polymeric dispersions are heterogeneous systems consisting of polymer particles as a dispersed phase and water as a dispersion medium. When the dispersed phase is a liquid polymer, the dispersion is called an emulsion. Although arbitrarily defined, dispersions containing particles ranging in size from 0.01 μm to 1.0 μm are referred to as colloids, whereas those with particle size greater than 1.0 μm are termed suspensions. The upper limit of 1.0 μm is set for colloids since it describes systems in which thermal convection and Brownian movement compensate for the sedimentation velocity of the particles and no sedimentation occurs on storage. Polymeric and other colloids containing very fine particles appear either milky or clear, and they show the Tyndall effect when light is shone through them.

Aqueous polymeric dispersions, depending on their method of preparation, are also classified as either latex or pseudolatex (false latex) systems. The former refers to dispersions that are developed from monomers using a catalyst and an initiator (e.g., Eudragits®, Rohm Pharma, Gmbh, Weiterstadt, Germany), whereas the latter represents dispersions that are produced by direct emulsification/dispersion of preformed polymers (e.g., Aquacoat®, FMC Corp., Philadelphia). Physically, latex and pseudolatex systems are indistinguishable. They both exhibit colloidal or nearly colloidal particle size, high solid concentrations, low viscosities (independent of molecular weight of the polymer), excellent film-forming properties. etc. Most latices and pseudolatices are lyophobic colloids, since they are prepared using hydrophobic monomers or polymers. Examples of lyophilic colloids include solutions of water-soluble polymers such as polyvinyl alcohol or methylcellulose, and suspensions of gel nanoparticles, such as cross-linked poly(acrylamide) latices [1].

**Current affiliation*: SmithKline Beecham Consumer Healthcare, Parsippany, New Jersey.

Because of their colloidal or near colloidal particle size range, high solid content, and nonhazardous nature (since no organic solvent is involved), aqueous polymer dispersions have found numerous applications in chemistry, engineering, material sciences, pharmaceuticals, and biomedical engineering. In pharmaceuticals, they have been used as a film-forming agents for coatings of solid dosages, as matrices and carriers for the entrapment of drugs, as adsorbents, including serving as carriers for ionic or molecular species. In this chapter, an overview of the methods of preparation and properties of various aqueous polymer dispersions and readily water-dispersible polymer powders useful in the field of pharmaceutics is presented.

II. METHODS OF PREPARATION

A. Latex Dispersions

The most common method used to prepare a latex is emulsion polymerization. The monomer must be water insoluble and it should be able to polymerize by free radical initiation. The procedure typically involves stirring (in some cases, both heating and stirring are required) the monomer in water containing an initiator, a surfactant, and other ingredients. The reaction can be performed in a batch process (all ingredients are added prior to heating and stirring the mixture), in a semicontinuous batch process (neat or pre-emulsified monomer is added continuously or incrementally to the reaction mixture), or in a continuous process (all ingredients are added continuously to the polymerization reaction vessel). The surfactant, which serves as an oil-in-water emulsifier, is usually used in excess of its critical micelle concentration. The polymerization, for the most part, occurs in the swollen micelles. As the polymerization progresses, micelles grow to become polymer particles. Vigorous agitation is used during the emulsion formation to produce particles having a size less than 1 μm, typically between 100 and 300 nm. Other factors that influence emulsion polymerization include the types and concentration of emulsifier, the rate of radical generation, the type and concentration of electrolyte, the temperature, the type and intensity of agitation, and parameters that effect nucleation of the polymer particles [2]. A schematic representation of latex preparation by emulsion polymerization is depicted in Fig. 1 [3].

The most common examples of pharmaceutically acceptable latex systems produced by emulsion polymerization are methacrylate ester copolymers (e.g., Eudragit NE 30D and Eudragit RS 30D), methacrylic acid copolymers (e.g., Eudragit L 30D), poly(vinyl acetate), vinyl acetate copolymers, polyacrylamide, and acrylamide copolymers.

B. Pseudolatex Dispersions

A pseudolatex is a mechanically produced colloidal dispersion of a polymer. Nearly all thermoplastic, water-insoluble polymers can be converted into a pseudolatex, providing the polymer is soluble in a water-immiscible volatile organic solvent. Since pseudolatices are prepared directly from an existing polymer, the dispersion is free of chemical contaminants, such as residual initiators, monomers, catalysts, and oligomers, commonly present in the synthetic latex systems. The various methods that have been used to prepare pseudolatices are shown in Fig. 2. A brief discussion of each method is presented below.

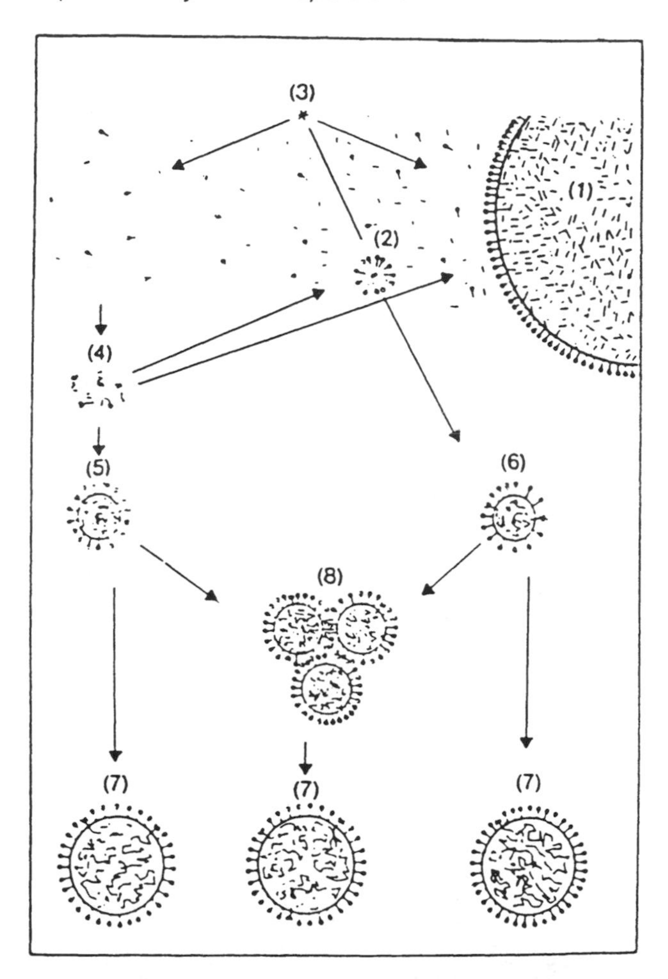

Fig. 1 Scheme of particle formation during emulsion polymerization (key: / monomeric molecule; (1) droplets of monomers; (2) micelles of emulsifier-containing monomer molecules; (3) water-soluble radical; (4) oligomer radicals; (5,6) newborn primary latex particles; (7) readily polymerized latex particles; and (8) stadium of secondary agglomeration). (From Ref. 3.)

1. Direct Emulsion

This method involves direct emulsification of the polymer using appropriate emulsifiers and emulsification methods. Typically, the polymer is first dissolved in a water-immiscible organic solvent and then emulsified in an aqueous phase containing a surfactant. For micro- and miniemulsions, a mixture of surfactant and an alcohol is often used as an emulsifying agent. The most common method used to prepare miniemulsions is to disperse a higher alcohol such as cetyl alcohol in the aqueous surfactant solution and then emulsify the polymer in this solution by stirring. For microemulsions, a lower alcohol such as ethanol, pentanol, hexanol, or heptanol is used. The procedure involves dissolving the lower alcohol in the polymer solution and then emulsifying this solution in the aqueous phase containing a suitable surfactant. Homogenization, followed by sol-

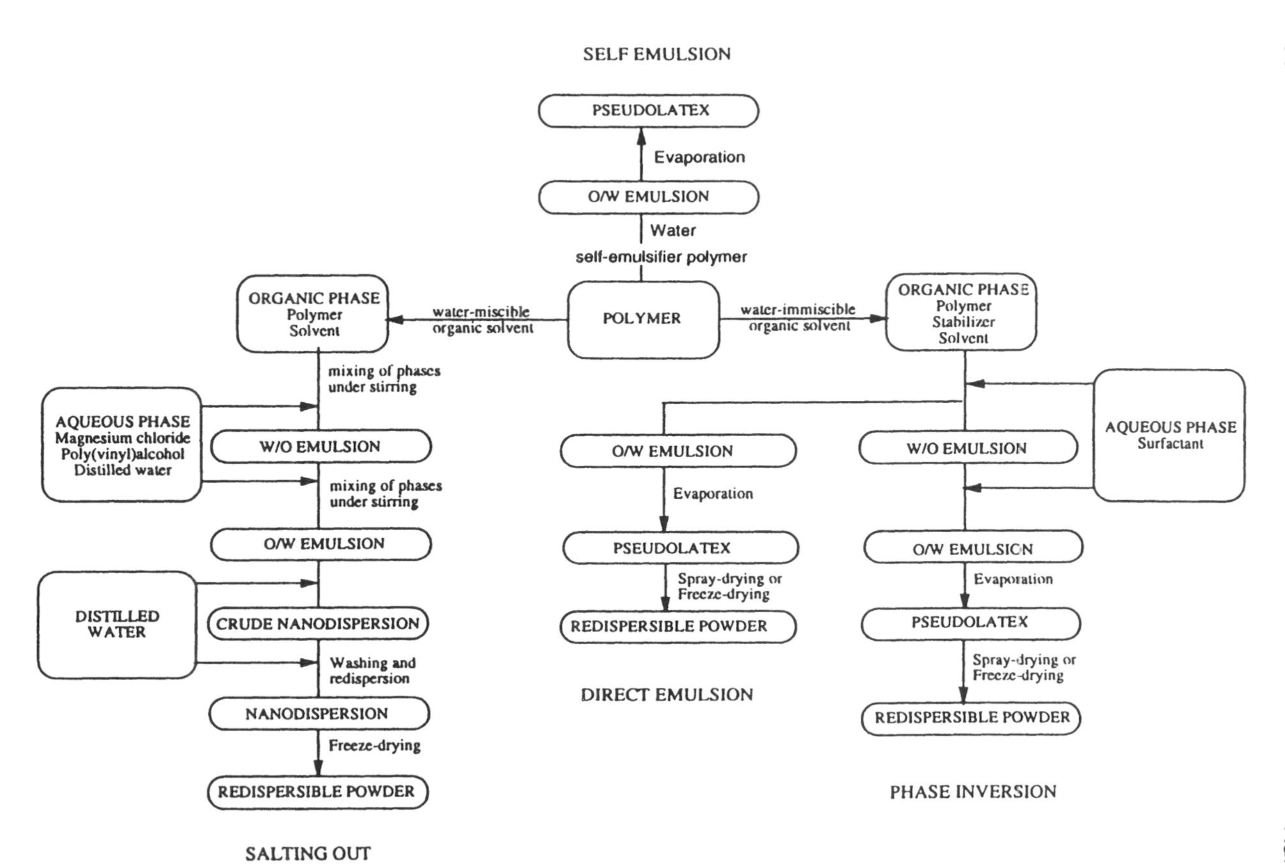

Fig. 2 Preparation of polymer aqueous dispersions prepared by salting out, direct emulsion, phase inversion, and self-emulsion processes.

vent removal under vacuum, produces the pseudolatex. As the solvent is removed in which the polymer was dissolved, by a vacuum treatment, small polymer "beads" are left behind. Since a trace of the solvent may also be left behind, it is important to select a water-immiscible solvent such as *n*-hexane, which is permitted in trace concentrations in foods. The recovered solvents are typically recycled for subsequent processing.

Various factors that influence the size of pseudolatex particles include the ratio of fatty alcohol to surfactant (if a mixed emulsifier is used), viscosity of the polymer solution, the types of solvent system (if used), an the emulsification methods. In general, pseudolatices prepared by direct emulsion produce particles ranging in size from 8 nm to 10 μm, usually in the submicron range. The most commonly and widely used example of a pseudolatex prepared by direct emulsion is Aquacoat®, an aqueous dispersion of ethylcellulose.

2. *Phase Inversion*

The phase inversion procedure involves adding an aqueous solution containing a long-chain fatty acid and an alkali, or a surfactant or a mixture of surfactants, to a solution of polymer in a water-immiscible organic solvent with vigorous agitation. The phase inversion, i.e., an initial water-in-oil emulsion, is converted to an oil-in-water emulsion when the aqueous phase that is added is capable of forming a continuous phase. Various emulsifier combinations can be used. Size reduction, followed by removal of the solvent, produces a pseudolatex. The particle sizes produced by this method may range between 10 and 100 Å. The best known example of the polymer dispersion prepared by this method is Surelease® (Colorcon, West Point, PA), an aqueous ammonical dispersion of ethylcellulose.

3. *Self-Emulsion*

Polymers containing acidic (e.g., carboxyl) or basic (e.g., amino) functional groups can be converted into a pseudolatex by the self-emulsion method [4]. Such polymers become self-emulsified upon dispersion in water. Parameters that are critical to the successful preparation of pseudolatices by this method are the amounts of these functional groups and their availability for self-emulsification. Typically, the method involves dissolving the polymer in a suitable organic solvent, and then dispersing this solution with vigorous agitation in water, optionally in the presence of a neutralizing agent. The removal of solvent by a conventional distillation method produces a pseudolatex. Because of the self-emulsifying properties of the polymers, no additional emulsifier is needed to produce an emulsion. Pseudolatices prepared by this method generally contain particles with sizes similar to those produced by emulsion polymerization. Examples of polymers used in the preparation of pseudolatices by this method include shellac, polyvinyl acetate phthalate, and methacrylic acid copolymers (Eudragit), and ethylene/α,-β-unsaturated carboxylic acid interpolymers [5].

4. *Salting Out*

In the salting out method [6,7], a viscous gel (viscosity 4–5 PaS), prepared by adding a water-soluble polymer (e.g., polyvinyl alcohol) to a highly concentrated aqueous solution of an electrolyte or nonelectrolyte, is added to a polymer solution in a water-miscible organic solvent (e.g., acetone and isopropanol) under vigorous mechanical stirring. Upon addition of a portion of the gel, a water-in-oil emulsion is formed. This occurs when the dissolved salt in the gel is capable of producing a salting out of the organic solvent. Further addition of the gel causes a change in phases (i.e., phase inversion), producing an oil-in-water emulsion. An adequate amount of water is then added to the emulsion to induce diffusion of acetone into the aqueous phase, causing the formation of nanospheres. The resulting polymer particles are then washed by centrifugation and then resuspended in water to produce a nanodispersion. Polyvinyl alcohol, used to prepare an aqueous viscous gel of electrolytes, serves as a stabilizing agent (or protective colloid) during the particle formation step and in the finished product. The mean diameter of particles produced may range between 172 and 1006 nm, depending on the type and molecular weight of the polyvinyl alcohol, the amount of polymer, the solvent, and the salting agent, and the type of homogenization method [6]. Since no heating is involved in the manufacture of nanodispersions, this method can be extended to biopolymers or heat sensitive substances. Other important features are (1) no surfactant is required; and (2) a single solvent miscible in all proportions in water is used. Examples of polymers converted to nanodispersions by this method include Eudragit E [7], poly(lactic acid) [7], cellulose acetate [6,8], cellulose acetate phthalate [6,8], and ethylcellulose [7].

C. Polymer Colloids/Suspensions

Stable colloids/suspensions of polymers can be prepared by direct homogenization of the polymer in water, optionally in the presence of a protective colloid or a suspending agent. A high-shear mixer or a colloid mill is often satisfactory for producing a homogeneous dispersion. This method is particularly useful for polymers that exhibit no or poor solubility in organic solvents and cannot be converted into a pseudolatex or water-dispersible powder form by direct emulsification or other methods listed above. Examples of suspending agents frequently used in the preparation of aqueous colloids include hydroxypropylcellulose, hydroxypropyl methylcellulose, acrylic acid polymer (Carbopol®, B. F. Goodrich, Cleveland), colloidal silica, etc.

D. Water-Dispersible Polymer Powders

Spray- and freeze-drying methods have been used to produce fine, free-flowing powders that when suspended in water readily form dispersions. Latices, pseudolatices, and mechanically produced polymer colloids/suspensions all can be spray- or freeze-dried to produce a water-dispersible powder. During spray-drying, the temperature at the surface of the spray droplets is usually kept below or near the minimum film-forming temperature (MFT) of the dispersion to prevent coalescence of particles, and, hence, the formation of films. The flow of drying air is also maintained such that the water is rapidly evaporated and there is no buildup of capillary forces to promote coalescence of particles. A barrier dispersant, such as acetylated monoglyceride (Myvacet® 9-40) and Polysorbate 60 (Tween® 60, ICI Americas, Wilmington, DE), is typically added prior to spray-drying to prevent the formation of films. The freeze-drying method, on the other hand, is particularly useful for polymers with low MFT values. Examples of water-dispersible powders prepared by spray-drying are Euduragit L 100-55 (methacrylic acid copolymer) and Aquateric® (cellulose acetate phthalate). The former is prepared from Eudragit L30D-55 latex and the latter from a proprietary pseudolatex system. Both

Eudragit L 100-55 and Aquateric® are used in enteric coatings. Polymers that have been converted into water-dispersible powders by freeze-drying methods include ethylcellulose; maleic anhydride copolymers, including poly(methylvinylether/maleic anhydride)-Gantrez® series (ISP Corp., New York) ethylene and styrene maleic anhydride; and a number of straight-chain and branched alkyl esters of maleic anhydride copolymers—acrylic/acrylate copolymers and acrylate polymers [9], and wax/lipids [10].

Compared to aqueous dispersions, freeze- and spray-dried powders offer several advantages. These include increased stability, reduction in storage space and shipping charges, and flexibility in adjusting the polymer content of the reconstituted dispersions.

III. DISPERSION ADDITIVES

Materials that are added to dispersions after they are formed are referred to as additives. The purpose of using additives is to alter the overall effectiveness of the dispersion. It is important that additives be nontoxic and compatible, i.e., they should not destabilize the dispersion and they must not alter the desired intrinsic properties of the polymer. The commonly used additives are plasticizers, flux regulators, and suspension stabilizers.

A. Plasticizers

Plasticizers typically are substantially nonvolatile, high-boiling substances that, when added in adequate amounts to a dispersion, alter certain physical and mechanical properties of the polymer films. For example, the addition of an effective amount of a plasticizer reduces the intermolecular forces along the polymer chains, and, consequently, lowers the glass transition temperature and modulus of elasticity. As a result, films exhibit more flexibility. The types and the amounts of a plasticizer that can be used depend on the nature and concentration of the polymer to be plasticized [11,12]. Typically, for maximum effectiveness, a plasticizer that is miscible with the polymer is used. With aqueous polymer dispersions, water-soluble plasticizers typically dissolve in the aqueous phase, whereas water-insoluble plasticizers are dispersed by mixing or emulsification in the aqueous phase of the dispersion. During plasticization, in the former case, the plasticizer may gradually diffuse into the polymer particles, whereas in the latter case it diffuses from the dispersed droplets through the water phase to the polymer [13]. Some of the most common and widely used plasticizers in film coatings are dialkyl phthalates, diacetin, triacetin, trialkylcitrates, diethyl tartarate, dialkyl sebacates, oleates including sorbitan monooleate, glycerin, propylene glycol, polyethylene glycols, ethylene glycol mono and diacetates, trimethyl phosphate, etc.

B. Flux Regulators

Flux regulators represent materials that aid in modifying drug release characteristics. They can be water-soluble (pH dependent or independent) or water-insoluble materials. Water-soluble regulators are used to increase the permeability of dispersion films/coatings, whereas water-insoluble regulators usually retard the release of drugs. Examples of water-soluble regulators frequently used are hydroxypropylcellulose, hydroxypropyl methylcellulose, polyethylene glycols, and a number of small molecules such as sucrose, urea, and mannitol. Water-insoluble materials, such as magnesium stearate, talc, and kaolin, also serve to reduce the tackiness of films during coatings. The amount of a flux regulator that is added typically depends on the drug release characteristics desired.

It should also be remembered that materials used as plasticizers often have multiple effects and will serve to influence drug release and water permeability in film-coated systems. If the plasticizer is water soluble, water dispersible, capable of hydration, and more polar than the polymer, one may see an increase in water permeation through the film and increase in drug release through the film as a result of plasticization. Insoluble plasticizers, those with no water reactivity, or those that are less polar than the polymer being plasticized, usually exhibit the opposite effect on water permeability and drug release rate.

Water-soluble flux regulators such as the three polymers noted above may also have multiple effects. In addition to altering permeability and diffusion characteristics of films, they may alter film mechanical properties, including film strength, elasticity, and other properties, in either a positive or negative manner depending on the polymer system being altered, ratios of materials being used, and the intended application and use of the coating. Because these film additives often have multiple effects, they are frequently termed film modifiers. This term connotes the fact that these additives typically have multiple effects, modifying not just one film property, but many. Investigators designing dosage forms using polymers and polymer film systems should be constantly alert to this fact.

C. Suspension Stabilizers

Surfactants and a number of water-soluble polymers have been used as suspension stabilizers. Surfactants regulate the surface energy of the dispersed polymer and, thereby, improve the physical stability of the dispersion during preparation and storage. Anionic, cationic, and nonionic surfactants are used for this purpose. Examples of surfactants commonly used include sodium lauryl sulfate, potassium laurate, Igepal®, Tweens, Spans®, and Brij® analogs (ICI Americas, Wilimington, DE). Water-soluble polymers improve stability by increasing the viscosity of the dispersion. Examples include hydroxypropylcellulose, hydroxypropyl methylcellulose, and a number of other viscosity-enhancing polymers. Films containing suspension stabilizers, in general, exhibit decreased elastic modulus and, hence, increased flexibility. In aqueous media, these agents dissolve or hydrate and form pores or a more open film structure, and, consequently, modify the release of drugs.

IV. CELLULOSE DISPERSIONS

Cellulose polymers have the longest history of use, the best regulatory status, and probably the widest use of any polymer class in foods and drugs. The various cellulose polymers, their chemical structures, and a summary of corresponding commercial aqueous dispersions/powders are listed in Table 1. The preparation, properties, and uses of each product are briefly summarized below.

A. Cellulose Acetate

Cellulose acetates (CAs) are partially or completely acetylated derivatives of cellulose. They are prepared by reacting cellulose with acetic anhydride/acetic acid in the pres-

Table 1 Cellulose Polymers and Their Aqueous Dispersions

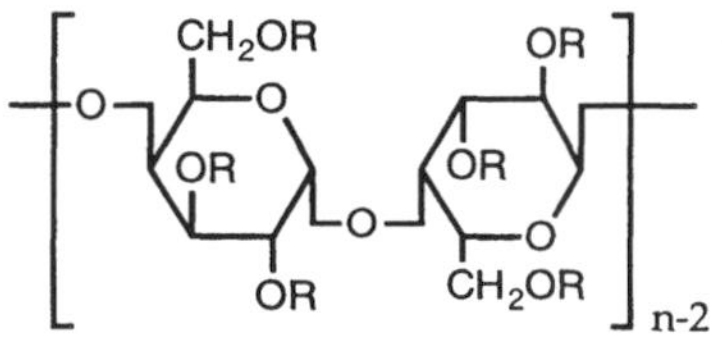

Polymer	R	Product (vendor)	Dispersion type	Additives	Typical applications
Ethyl cellulose (EC)	–H, $-C_2H_5$	Aquacoat-ECD (FMC)	Pseudolatex	Cetyl alcohol, sodium lauryl sulfate	Overcoat, taste masking, sustained release
		Surelease (Colorcon)	Pseudolatex	Dibutyl sebacate, oleic acid, ammonia, silica	
Cellulose acetate (CA)	–H, $-C(O)CH_3$	CA Latex (FMC)	—	—[b]	Taste masking, sustained release
Cellulose acetate butyrate (CAB)	–H, $-C(O)CH_2CH_2CH_3$	CAB latex[a]	—	—[b]	Taste masking, sustained release
Cellulose acetate phthalate (CAP)	–H, $-C(O)-C_6H_4-COOH$	Aquacoat-CPD[a] (FMC)	Pseudolatex	Poloxamer 124	Enteric
		Auateric (FMC)	Powder	Pluronic F-68, Tween 80, Myvacet 9-40	
Hydroxypropyl methylcellulose phthalate (HPMCP)	–H, $-CH_3$, $-CH_2CH(OH)CH_3$ $-C(O)-C_6H_4-COOH$ $-CH_2CH(OCO-C_6H_4-COOH)CH_3$	HPMCP latex[a]	—	—[b]	Enteric
Hydroxypropyl methylcellulose acetate succinate (HPMCAS)	–H, –Ch$_3$, $-C(O)CH_3$ $-C(O)CH_2CH_2COOH$ $-CH_2CH[OC(O)CH_3]CH_3$ $-CH_2CH(OH)CH_3$ $-CH_2CH(OCOCH_2CH_2COOH)CH_3$	AQOAT® (Shin-Etsu)	Powder/ suspension	—[b]	Enteric

[a]Current production status: CA latex, plant scale; CAB latex, lab scale; HPMCP, pilot scale.
[b]Not known.

ence of a strong acid catalyst (e.g., sulfuric acid) at elevated temperatures. The reaction is generally allowed to proceed to substitute all three hydroxyl groups. The fully substituted triester is then hydrolyzed to give the desired level of substitution. Currently, several different grades of cellulose acetates, varying in acetyl content from 29.0% to 44.8%, and molecular weights between 30,000 and 160,000, are commercially available. They exhibit a high cohesive energy density and are practically insoluble in water and nonpolar organic solvents [14]. Their solubility in polar organic solvents, however, depends on the level of acetyl substitution and distribution in the polymer chain. As the acetyl content in the product increases, the choice of solvent system to dissolve the polymer becomes limited. The moisture permeability of cellulose acetate, in general, increases with a decrease in the acetyl content.

Most of methods used for the preparation of CA pseudolatices are based on the direct emulsification approach. Because most solvents for CA are either slightly or completely miscible with water, binary solvent systems in which one solvent is water immiscible and the other water miscible are usually used to prepare CA emulsions. Seminoff and Zenter [15] dissolved CA in a 1:1 mixture of methylene chloride:methanol and then emulsified the solution in water using 0.1% sodium lauryl sulfate as an emulsifying agent. Ultrasonication followed by removal of the solvent produced a pseudolatex. A similar method was used by Bodmeier and Chen [16] to prepare a 10% CA pseudolatex with a particle size in the range between 250 and 300 μm. They dissolved the polymer in a 9:1 ratio of methylene chloride:methanol solvent and then emulsified the resulting solution in water using 0.5% sodium lauryl sulfate. Microfluidization, instead of ultrasonication, was used to reduce the particle size. Studies show that the stability of the homogenized emulsion is greatly dependent on the amount of solvent partitioned into the aqueous phase. In general, high concentrations of solvent in water cause either surfactant desorption or Ostwald ripening of the polymer and, consequently, result in the instability of the dispersion [14].

Because of the ready hydrolysis of the ester linkage in aqueous media, the stability of CA has been the subject of numerous studies [17–19]. Manley [20] reported that the breakdown of the polymer backbone may accompany CA hydrolysis. The latter is known to follow first-order kinetics. Recently, Bodmeier and Chen [16] investigated the physical and chemical stability of CA pseudolatices at different temperatures. They reported that the hydrolysis of a CA pseudolatex followed pseudo first-order kinetics. A plot between acetyl content remaining versus time as a function of temperature is shown in Fig. 3. The pseudo first-order hydrolysis rate constants ($\times 10{,}000$) determined at 4, 25, and 40°C were 3.8, 29.5, and 155 per day, respectively, corresponding to an activation energy of 16.1 kcal/mol. These results show that the stability of CA pseudolatices increases as the storage temperature decreases. At 4°C, there was no significant change in the acetyl content, suggesting that the pseudolatices can be stored at this temperature. However, at room temperature significant degradation over time occurs. Fig. 4 shows the pH profiles of pseudolatices during storage at different temperatures. The more rapid pH decrease with increasing temperature is consistent with the higher degradation rates observed at higher temperatures (Fig. 3). The relationship between pH of the dispersion and the acetyl content lost as a result of the hydrolysis of CA is depicted in Fig. 5. As is obvious from this figure the hydrolysis of CA is at a minimum at a pH of about 4. The lowering of pH at elevated temperatures, during storage, has also been reported to adversely affect the physical stability of CA pseudolatices. At temperatures above 4°C, CA pseudolatices converted to clear,

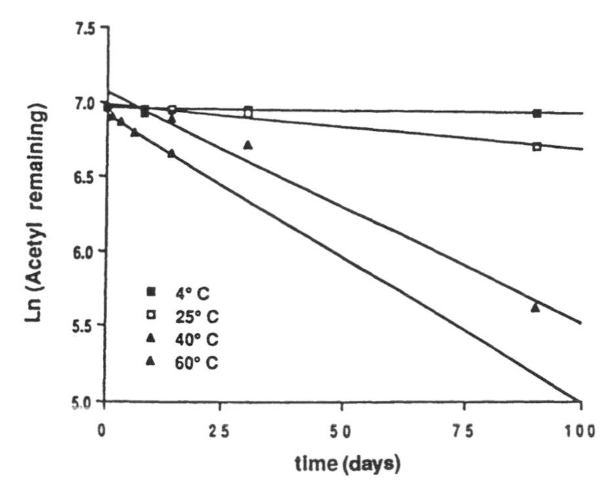

Fig. 3 Relationship between acetyl content lost versus time as a function of temperature. (From Ref. 16.)

viscous gels. Recently, FMC Corp. (Philadelphia) developed a reportedly more stable CA pseudolatex containing 25% (weight-by-weight) solids. It is recommended by the manufacture that this product be stored at 5°C upon receipt.

CA pseudolatices have been extensively investigated in the development of osmotic- and diffusion-controlled drug delivery systems [21–26]. Bindschaedler et al. [27] reported that plasticizer levels from 160% to 320%, by weight of CA, provide films that are comparable to those prepared from organic solutions. King and Wheatly [28] evaluated the effect of plasticizer levels on the glass transition temperature and found that the most effective plasticizers for CA pseudolatex are glyceryl diacetate, glyceryl triacetate, and triethyl citrate (Table 2). Films containing glyceryl diacetate and triethyl citrate exhibited the greatest change in the modulus of elasticity. The water permeability of the CA pseudolatex films is strongly dependent on the nature of the plasticizer

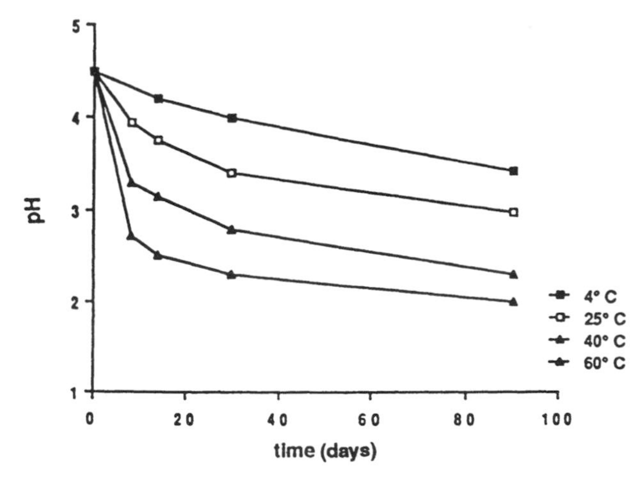

Fig. 4 pH profiles of CA pseudolatex over time at different temperatures. (From Ref. 16.)

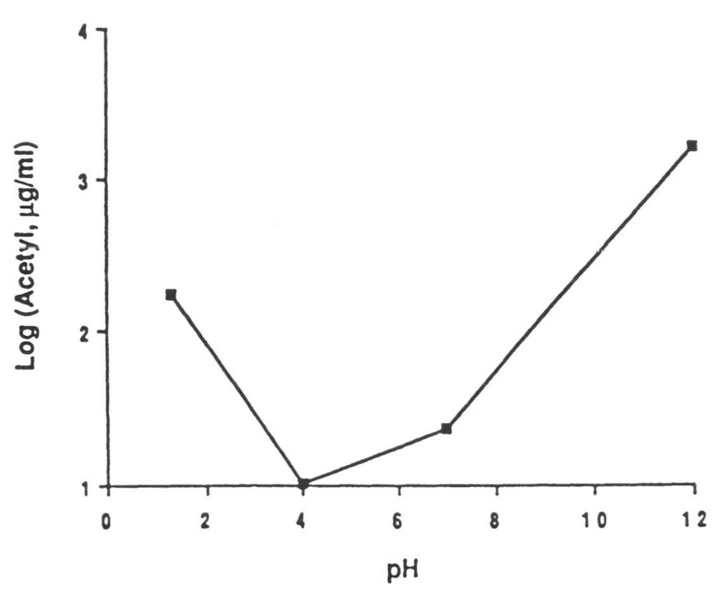

Fig. 5 Effect of pH on acetyl content lost. (From Ref. 16.)

and the processing conditions used. Generally, high boiling plasticizers produce more permeable membranes [26]. Kelbert and Bechard [23] evaluated the release of propranol HCl from tablets coated with a CA pseudolatex plasticized with triacetin and triethyl citrate, separately, with an without the presence of permeation enhancers. The results showed no release of KCl for the first 8 hr from tablets that were coated with 150% triacetin and 120% triethyl citrate even though the water had penetrated through them and the tablets swelled. Scanning electron photomicrographs revealed the films/coatings on the tablets to be dense, nonporous, homogeneous, and free of any defects at

Table 2 Influence of Plasticizer and Plasticizer Concentration on Glass Transition Temperature (T_g) of CA Latex Films

Plasticizer level (%)	T_g (°C)				
	GDA[a]	TEC[b]	GTA[c]	ATEC[d]	DBS[e]
0	164	164	164	164	164
20	114	—	—	—	—
40	99	107	109	131	—
60	65	—	—	—	—
80	31	62	86	109	157
120	26	37	79	103	—
160	< 20	< 20	71	89	152

[a]GDA = glycerin diacetate.
[b]TEC = triethyl citrate.
[c]GTA – glycerin triacetate.
[d]ATE = acetyl triethyl.
[e]DBS = dibutyl sebacate.
Source: Ref. 28.

tablet edges. In contrast, tablets coated with CA dispersions containing 40% sucrose provided a steady release of propranolol HCl over a 12 hr period time with a 2 hr lag time. The addition of 10% PEG 8000 to the pseudolatex, in addition to 40% sucrose, maintained the extended 12 hr release profile and reduced the lag time to 1 hr. It has been postulated that drug transport occurs mainly in the porous CA structure, and the mechanism responsible for it is a combination of molecular diffusion/osmotic pressure via water transport through the porous cellulose acetate membrane.

Recently, flux regulators such as urea and dibasic calcium phosphate, have been used to alter the release of KCl from CA pseudolatex-coated systems [24,25]. Urea is freely soluble in water, whereas calcium phosphate dibasic dissolves in an acidic medium. Coatings containing these agents when in contact with the dissolution media become microporous and consequently affect the release of KCl.

B. Cellulose Acetate Butyrate

Cellulose Acetate Butyrate (CAB) is available in a wide range of butyryl, acetyl, and hydroxyl levels from Eastman Chemical Products, Inc. (Kingsport, TN). Care must be taken in selecting a CAB since different products, depending on the butyryl, acetyl, and hydroxyl content, exhibit a wide range of properties. In general, the compatibility of CABs with plasticizers, and consequently their coating performance, improves with increasing butyryl content. CABs containing high hydroxyl levels are alcohol soluble.

A CAB pseudolatex has been prepared using a high-pressure direct emulsion method [16]. A polymer solution in a 9:1 methylene chloride:methanol mixture was preemulsified into an aqueous phase using sodium lauryl sulfate as an emulsifying agent. Microfluidization, followed by evaporation of the solvent at room temperature over a period of 48–72 hr produced the pseudolatex.

Compared to CA, CAB degrades slowly in water. The pseudo first-order rate constants and the activation energies for the acetate and butyrate groups in CAB pseudolatices at different temperatures are presented in Table 3. The higher activation energy (about 1.7 times) for the acetyl group in CAB compared to that in CA is due to the relatively greater hydrophobic character of CAB [16,29]. Bodmeier et al. [16] reported that CAB pseudolatices tend to flocculate during storage both at 4°C and at elevated temperatures.

Although no report of the use of CAB pseudolatices in the development of pharmaceutical products is currently known, CAB powder has been used in microencapsulation [30,31] and drug-loaded microsphere [32–34] preparations, by a precipitation method using a solvent/nonsolvent mixture or by evaporation of a volatile solvent from a nonaqueous or an aqueous emulsion. CAB is also useful in the preparation of contact lenses [35–37]. Sprockel et al. [38] reported that the water vapor permeability of

Table 3 Pseudo First-Order Rate Constants (×1000, 1/day) for Hydrolysis of CAB

	Temperature (°C)			Activation energy (kcal/mol)
	4	25	40	
Acetate	0.1	1.8	32.0	26.9
Butyrate	0.5	0.6	19.3	31.9

Source: Ref. 16.

solvent cast films of CAB, cellulose acetate propionate (CAPr), and CA follows the order CAB > CAPr > CA.

Recently, FMC Corp. developed a stable CAB latex dispersion containing 30% solids. It is currently supplied as a sample to interested pharmaceutical scientists.

C. Cellulose Acetate Phthalate

Cellulose acetate phthalate (CAP), also known as cellacephate, is the oldest phthalate-containing polymer used in enteric coatings [39]. It is prepared by reacting a partial acetate ester of cellulose with phthalic anhydride. The suitability of CAP for enteric coatings depends on the molecular weight of the polymer, degree of substitution of acidic functional groups, and p*Ka* values. To provide sufficient mechanical strength to coatings, use of a CAP with a molecular weight greater than 30,000 is recommended. The aqueous solubility of CAP increases with increasing phthalyl content. According to the U.S. Pharamcopeia/National Formulary (USP/NF) monograph, CAP must contain an acetyl content between 21.5% and 26.0% and a phthalyl content between 30.0% and 36.0%.

CAP has been converted into both pseudolatex and water-dispersible powder forms. The former is prepared using phase inversion [40–42], solvent change [43], and salting out [6] methods. The phase inversion method involves the drop-wise addition of an aqueous solution of a surfactant, with vigorous agitation, to a solution of CAP and a stabilizer in a 80:20 (v/v) mixture of ethyl acetate:isopropanol. Pluronic® F-68 (BASF Corp., Parsippany, NJ), sodium lauryl sulfate, and Myrj were used as effective emulsifiers. The stabilizers used included cetyl alcohol, polyethylene glycol 6000 (PEG 6000), or polyvinyl alcohol (Gelvatol® 20/30, Rhone-Poulenc Inc., Canbury, NJ). Also, *n*-Decane can be used as a stabilizer, but its use is not recommended because of toxicity concerns [40]. Davis et al. [41,42] found Pluronic F-68 and PEG 6000 to be the most effective emulsifier and stabilizer system for the preparation of a mixed CAP-ethylcellulose (equal parts) pseudolatex.

The salting out method, developed by Ibrahim et al. [6], involves adding a viscous gel containing poly(vinyl alcohol) and an aqueous concentrated salt solution (e.g., aluminum chloride, calcium chloride, magnesium chloride) to the CAP solution in acetone. The salt in the aqueous phase displaces the acetone from the polymer solution to produce a water-in-oil emulsion. As more gel is added to the organic solution, the water-in-oil emulsion changes to an oil-in-water emulsion. The latter is diluted with an adequate amount of water to ensure complete diffusion of the organic solvent into the aqueous phase, causing the formation of nanoparticles of CAP. Washing of the polymer particles by centrifugation (until no trace of chloride is detected) and subsequent redispersion of the washed particles in water produces a nanodispersion. The typical formula for the preparation of a CAP nanodispersion is presented in Table 4.

Recently, FMC Corp. developed a stable aqueous dispersion of CAP. It is currently marketed under the trade name Aquacoat-CPD. This product contains 29.0–32.0% solids, of which 19–27% is CAP. The other component present is a poly(oxypropylene)poly(oxyethylene) copolymer (Poloxamer or Pluronics). It is recommended that this product be stored at 5°C upon receipt. Recommended plasticizers for Aquacoat-CPD are diethyl phthalate, triethyl citrate, and triacetin. For most coating applications, a plasticizer amount equivalent to 20–24% of the latex solids is recommended.

A water-dispersible powder of CAP is commercially available under the trade name Aquateric (FMC Corp.). It is prepared from CAP pseudolatex by spray-drying.

Table 4 Typical Formula for Preparation of a CAP Nanodispersion

Emulsion step	
Organic phase	170.0 g
CAP	30.0 g
Acetone	140.0 g
Aqueous phase	250.0 g
Poly(vinyl alcohol)	9.2 g
Magnesium chloride. $6H_2O$	151.2 g
Distilled water	89.6 g
Particle formation step	
Addition of distilled water	150 g

Source: Ref. 6.

In general, the method involves preparation of the pseudolatex by direct emulsification of CAP using Tween 60 and Pluronic F-68 as emulsifiers. After subjecting the emulsion to high-shear processing (sufficient to produce polymer particles averaging less than about 0.5 μm), a barrier dispersant such as acetylated monoglycerides (Myvacet 9-40) is added, and the resulting dispersion is then spray-dried to produce a free-flowing powder [44]. As noted previously, the addition of a barrier dispersant is necessary during spray-drying to prevent coalescence of the particles and, consequently, the formation of agglomerates or a continuous film. The action of the barrier dispersant during spray-drying of the CAP pseudolatex is illustrated in Fig. 6. The spray-dried powder (Aquateric) is a loose cluster of latex particles containing 69.7% cellulose acetate phthalate, 20.0% Pluronic F-68, 10.0% Myvacet® 9-45, and 0.3% Tween 60. It is easily dispersed in water to form a pseudolatex of about the original particle size range

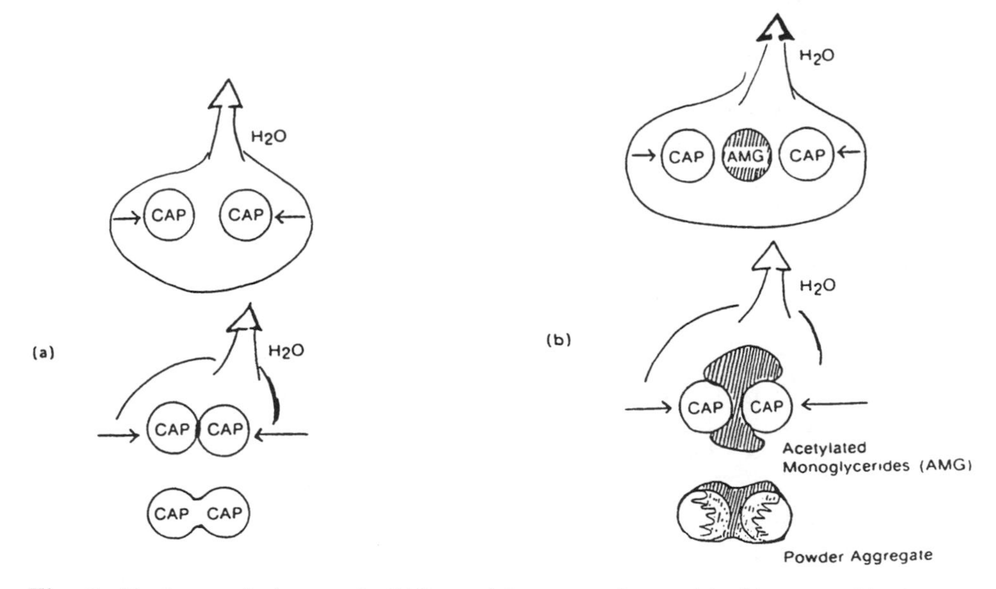

Fig. 6 Final spray-drying step in CAP pseudolatex manufacture (a) without and (b) with barrier dispersant. (From Ref. 54.)

(i.e., less than 1 μm). In preparing a dispersion, adequate amounts of Tween 80 (to aid dispersion) and a plasticizer are added to the water, and then the Aquateric is dispersed with gentle stirring for 60–90 min. Diethyl phthalate, triethyl citrate, propylene glycol, and triethyl citrate all have been reported to be compatible and effective plasticizers for CAP. Aquateric powder contains no more than 4% free phthalic acid and no more than 5% moisture.

CAP, like CA and CAB, is prone to hydrolysis. Davis et al. [41,42] investigated the stability of CAP pseudolatices and found that phthalate hydrolysis is the primary cause of instability of CAP pseudolatices. The hydrolysis rate increased with increasing temperature. The activation energy calculated from the Arrehenius equation was 23 kcal/mol. This value is somewhat higher compared to the activation energy of many esters. Davis et al. [41,42] attributed this to steric hindrance provided by the polymer backbone and to the presence of the polymer in a solid dispersed form in pseudolatex rather than being in solution.

The stability of CAP is also affected by the pH of the system. Davis et al. [41] studied the pH changes in unbuffered and pH-adjusted CAP pseudolatices over time at 3 and 25°C. No change in the pH of the unbuffered CAP pseudolatex at 3°C, or of the pH-adjusted CAP pseudolatices at 3 and 25°C were observed over a period of 400 days suggesting that products would be stable at these storage conditions. The marked decrease in the pH of the unbuffered pseudolatex at 25°C indicated an increased hydrolysis rate of the phthalate group. They concluded that the hydrolysis of phthalate can be minimized by adjusting the pH and storing the pseudolatex at refrigerator temperature (3°C). It must be noted, however, that the addition of hydrochloric acid or sodium hydroxide to adjust the pH may result in physical stability problems. For example, pH adjustment by sodium hydroxide cause an immediate separation of granular solid particles, whereas hydrochloric acid initially increased the viscosity and then the solid separates [41].

The nanodispersion prepared by the salting out method has also been freeze-dried to produce a redispersible CAP powder [6]. Due to the aqueous instability of CAP, reconstituted pseudolatex products are not recommended to be stored for prolonged periods of time.

D. Hydroxypropyl Methylcellulose Phthalate

Hydroxypropyl methylcellulose phthalate (HPMCP), also known as hypromellose phthalate, is one of the newest enteric coating materials. It is prepared by reacting hydroxypropyl methyl cellulose (HPMC) with phthalic anhydride. Although several grades of HPMCP are commercially available, only two types, 200731 and 220824, are currently listed in the USP/NF. The first two, the middle two, and the last two digits in this designation represent the average percentage content of the methoxy groups, the hydroxypropyl groups and the phthalyl groups, respectively. Products meeting these specifications are marketed under the trade names HP-50 and HP-55 by Shin-Etsu Chemical Co. (Chiyoda-ku, Tokyo, Japan) and HPMCP-50 and HPMCP-55 by Eastman Chemical Co. (Kingsport, TN). The numbers 50 and 55 in the names indicate the pH values (×10) at which the respective polymers dissolve. HP-50 and HP-55 are available in both granular (G grade) and fine powder (F grade) forms. The granular form is recommend for use with organic solvents, whereas the F grade is used for making aqueous dispersions.

Although no commercial HPMCP pseudolatex or dispersion is currently available, considerable effort has been made in recent years to prepare a stable pseudolatex. Davis et al. [41,42] used a phase inversion approach, wherein a plasticizer, water, and an insoluble excipient were dissolved in an 80:20 w/w mixture of ethyl acetate and isopropanol, and the resulting solution was then emulsified in water using Pluronic F-68 and Gelvatol 20/30 as the emulsifying/stabilizing agents.

Thioune et al. [45] prepared a series of HPMCP pseudolatices by mixing an aqueous solution containing one or more surfactants with a solution of the polymer in a water-miscible organic solvent, and then evaporating the organic solvent at 40–50°C under reduced pressure to the desired solid concentration. Acetone was used as a water-miscible organic solvent. Surfactants investigated included Tween 80, benzethonium chloride, and sodium lauryl sulfate. Of these, only sodium lauryl sulfate yielded stable HPMCP dispersions. Tween 80 produced large aggregates of HPMCP at the end of the evaporation step (when the polymer dispersion concentration reached 10–12%), whereas the use of benzethonium chloride resulted in a compact precipitate of the polymer during the concentration step. The increase in pseudolatex particle size occurred with increasing sodium lauryl sulfate concentration, however, it remains in the nanometers range (Table 5). The effect of the initial polymer concentration on dispersion particle size and polymer loss due to aggregation is illustrated in Figs. 7 and 8. As is evident up to a 1% initial polymer concentration, there was no loss of polymer due to aggregation. But beyond 1% a significant loss due to aggregation occurred. The physical properties of the optimized dispersion prepared using 1% polymer and 5% sodium lauryl sulfate are presented in Table 6. The smaller particle size, low viscosity, high polymer concentration values are all within the range of those usually encountered in a film-forming pseudolatex. Smaller particles prevent settling and require less driving force to coalesce to form films. The zeta potential value of –35.7 mV falls within the range of ±60 and ±30 mV, which, according to Nash [46], indicates the dispersion to be moderately stable. Dispersions with a zeta potential value between ±60 and ±100 mV exhibit excellent physical stability [46].

Recently, stable dispersions of HPMCP have also been prepared by simply dispersing HPMCP in water with the help of suspending agents and other additives [47]. Antifoam, Tween 80, and triethyl citrate were first dispersed/dissolved in water and then a suspension stabilizer (hydroxypropylcellulose or hydroxypropyl methylcellulose) was added. The addition of HPMCP to the resulting solution, under constant stirring, provides a stable dispersion.

Table 5 Effect of Sodium Lauryl Sulfate (SDS) Concentration on Particle Size of HPMCP Aqueous Dispersion

SDS concentration (%)	Particle size (nm)
2.5	108 ± 40
5.0	114 ± 38
7.5	133 ± 55
10.0	140 ± 60

Source: Ref. 45.

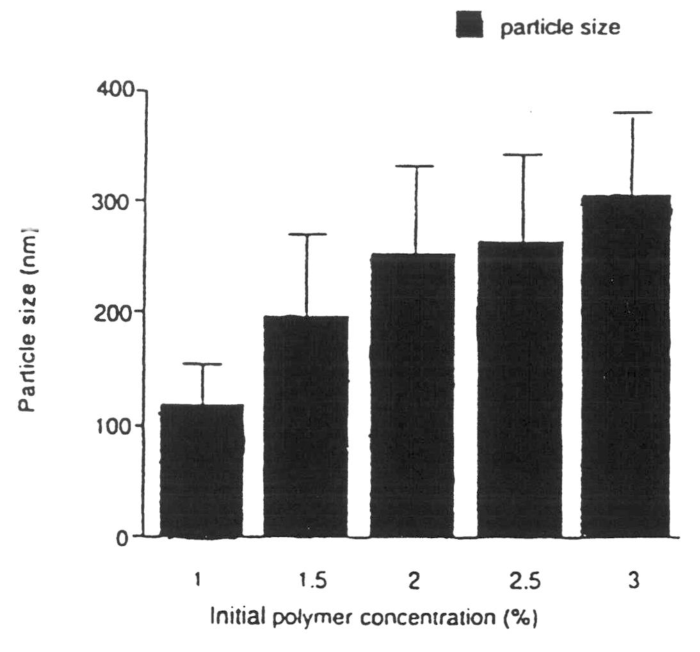

Fig. 7 Particle size of HPMCP dispersions as a function of initial polymer concentration. (From Ref. 45.)

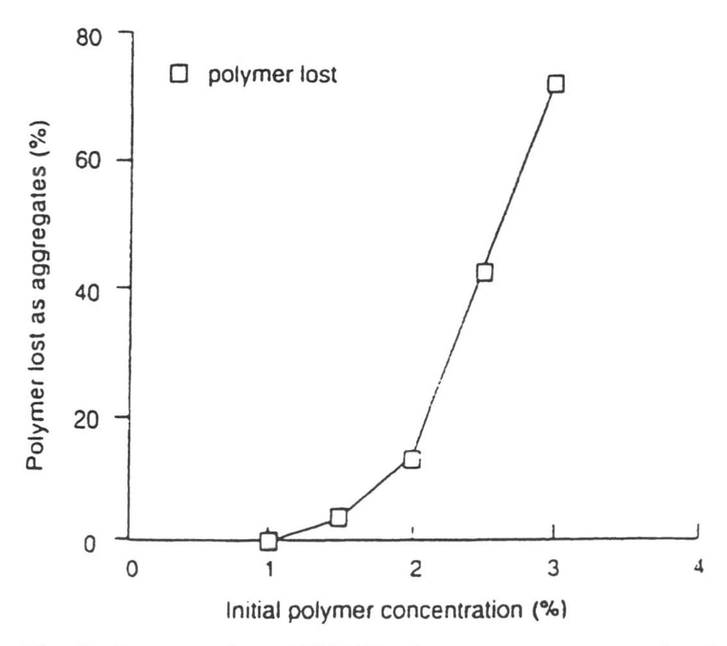

Fig. 8 Percentage loss of HPMCP polymer as aggregates as a function of initial polymer concentration. (From Ref. 45.)

Table 6 HPMCP Pseudolatex Characteristics

Appearance	very fluid, whitish, blue reflection
pH	2.34
Viscosity (mPa s)	2.926
Final polymer conc. (%)	18
Relative density (d_4^{20})	1.03
Particle size (nm)	114 ± 38
Zeta potential (mV)	-35.7

Source: Ref. 45.

HPMCP films prepared from aqueous dispersions are stronger (high tensile strength) but less elastic (high Young's modulus) compared to those prepared from organic solvent solutions [48]. An increase in the plasticizer level, however, increases the elasticity of the pseudolatex films. The effect of triethyl citrate on the Young's modulus of HPMCP pseudolatex films is shown in Fig. 9 [47]. This figure also shows that when water-soluble polymers are added to the aqueous dispersion, the films become even more elastic. Thioune et al. [45] compared the mechanical characteristics of films produced under the same conditions from a pseudolatex and from organic solvent solutions containing 20% and 30% polyethylene glycol 200 (PEG 200), and 30% polyethylene glycol 400 (PEG 400). They found that films from the organic solvent solution containing 20% PEG 200 were more elastic and less strong than those prepared from the corresponding pseudolatex. At a higher plasticizer level, similar results were obtained. With 30% PEG 400, no significant difference was observed between films prepared from an organic solvent or a pseudolatex. The ratio of tensile strength

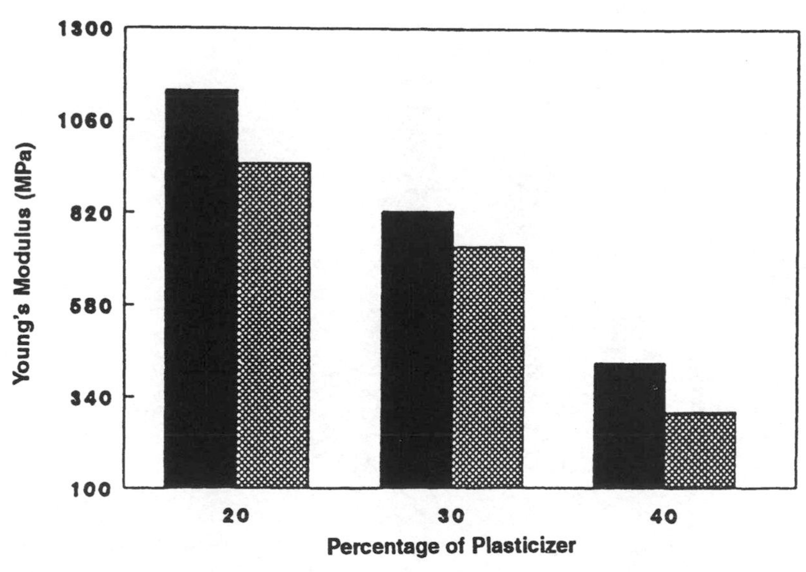

Fig. 9 The influence of hydroxypropylcellulose (HPC) on the elasticity of HPMCP 50 films (key: HPMCP: triethyl citrate (TEC) ■; HPMCP: TEC:HPC ▩. (From Ref. 47.)

to Young's modulus, which indicates film efficiency in table coating, however, was higher for films prepared from the pseudolatex, compared to those prepared from organic solvent solutions, when 20% PEG 200 and 30% PEG 400 were used (Fig. 10). With 30% PEG 200, films from the organic solutions exhibited a higher tensile strength to Young's modulus ratio. These results suggest that aqueous HPMCP dispersions could also be used in table coatings.

E. Hydroxypropyl Methylcellulose Acetate Succinate

Hydroxypropyl methylcellulose acetate succinate (HPMCAS) is a new enteric coating material developed by Shin-Etsu Chemical Co., and is marketed under the trade name Shin-Etsu (AQOAT® [49]. It is available in three grades: AS-L(F or G), AS-M(F or G), and AS-H(F or G), corresponding to an average succinyl to acetyl substitution ratio of 22.86%, 13.3%, and 0.5%, respectively. The LF, MF, and HF grades are fine powders and used to prepare aqueous dispersions, whereas LG, MG, and HG are granular and recommended for organic-based coating applications. The higher the SA ratio the lower the dissolution pH of the polymer. Table 7 lists the percentage compositions of various functional groups and the pH at which the polymer dissolves. Compared to CAP and HPMC, HPMCAS is more stable toward hydrolysis (Fig. 11) [50].

The coating dispersions of AS-LF, AS-MF, and AS-HF grades can be prepared by suspending the polymer in an aqueous solution containing a plasticizer (e.g., triethyl citrate) and an antifoaming agent (e.g., sorbitan sesquioleate SO 15, Nikkon Chemicals, Japan). It is important that the HPMCAS be added to water cooled to below 15°C. This prevents the formation of polymer aggregates and foaming. Typically, a suitable plasticizer is added first to cooled water. The antifoaming agent is added next.

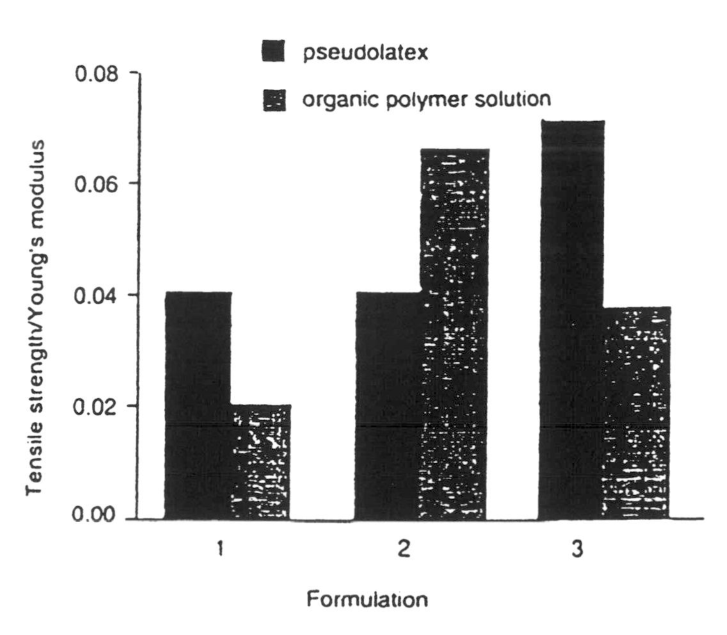

Fig. 10 Ratio of tensile strength to Young's modulus for films from HP-55 pseudolatex and HP-55 organic solution (key: formulation 1, 20% PEG 200; formulation 2, 30% PEG 200; and formulation 3, 30% PEG 400.) (From Ref. 45.)

Table 7 Compositions and Dissolution pH of HPMCAS

	HPMCAS grade		
	AS-LF (LG)	AS-MF (MG)	AS-HF (HF)
Methoxyl groups	20.0–24.0%	21.0–25.0%	22.0–26.0%
Hydroxypropoxyl groups	5.0–9.0%	5.0–9.0%	6.0–10.0%
Acetyl groups	5.0–9.0%	7.0–11.0%	10.0–14.0%
Succinoyl groups	14.0–18.0%	10.0–14.0%	4.0–8.0%
Dissolution pH	≥ 5.5	≥ 6.0	≥ 6.5

This is followed by the gradual addition of the HPMCAS with mixing. The amount of plasticizer that can be used depends on the type of HPMCAS and the amounts of other additives used. As a standard, 20%, 28%, and 35% of triethyl citrate is satisfactory for LF, MF, and HF grades, respectively. Talc or titanium dioxide is sometime added to the dispersion to prevent sticking of tablets to each other and aggregation of granules during coating. In some coating formulations, a high viscosity grade hydroxypropylcellulose is added to the dispersion to improve the stability of the coating dispersion [50]. The typical composition of HPMCAS coating dispersion is shown in Table 8 [50,51]. During the coating process, the suspension is kept below 20°C. Gentle stirring is typically used during coating to maintain a uniform dispersion of the HPMCAS.

Films prepared from aqueous dispersions and organic solvents showed no physical difference [50]. Their enteric performance was also similar, as illustrated in Fig. 12 [50].

F. Ethylcellulose

Ethylcellulose (EC) is widely used as a tablet binder, thickening agent, and coating material for tablets, microcapsules, and microparticles [52]. It is prepared by etherifi-

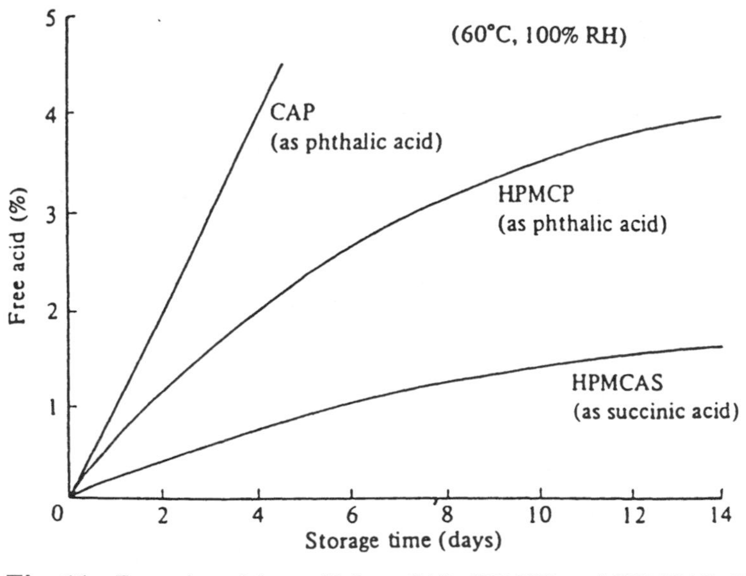

Fig. 11 Formation of free acid from CAP, HPMCP, and HPMCAS. (From Ref. 50.)

Table 8 Typical Composition of HPMCAS Coating Dispersions

	Coating dispersions	
	I	II
HPMCAS-MF	10%	10%
Triethyl citrate	2.8%	3%
Talc	3.0%	2%
Sorbitan sesquioleate	0.0025%	—
HPC-MF	—	0.05%
Water	84.2%	Up to 100%

Source: Column I, Ref. 49; Column II, Ref. 51.

cation of alkali cellulose with ethyl chloride. The degree of substitution (DS) of commercial EC products ranges from 2.2 to 2.6, corresponding to an ethoxy content of 44.5% to >49%. Each substitution grade is available in several viscosity grades. In general, the viscosity of EC increases as the length of the polymer chain increases. The most commonly and widely used EC grades in pharmaceutical products have a DS value between 2.44 and 2.58 (ethoxyl content 48.0% to 49.5%, respectively). These grades of EC have lower softening points, wider solubilities, and better water resistance, and are most commonly used for coating applications. EC is soluble in a wide variety of organic solvents. Although EC solutions of low viscosities can be prepared in low molecular weight aliphatic esters and ketones and aromatic solvent (e.g., toluene), bi-

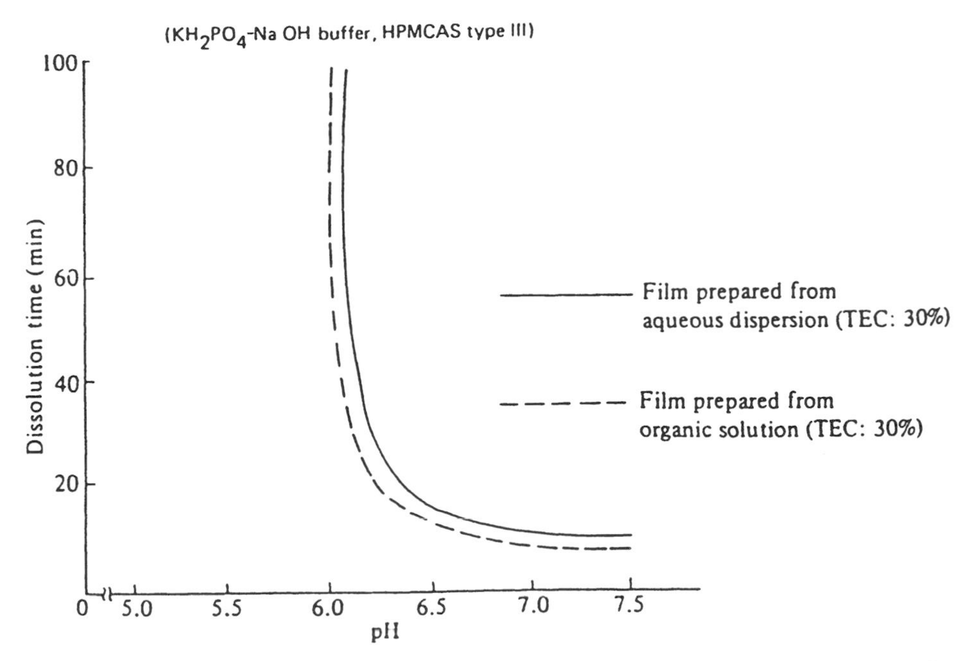

Fig. 12 pH dependency of solubility of films prepared from HPMCAS (key: TEC = Triethylcitrate.) (From Ref. 50.)

nary solvent systems in which one of the solvents is a lower molecular weight alcohol are frequently preferred for making solutions for use in coating.

EC is available as an aqueous dispersion under the trade names Aquacoat-ECD (formerly called Aquacoat) and Surelease. The former is now included in the NF 18 as ethylcellulose aqueous dispersion. This is the first aqueous polymeric dispersion developed by direct emulsification [53]. Aquacoat-ECD is sold as a 30% solids dispersion. The composition of the nonaqueous fraction is ethylcellulose 87.1%, cetyl alcohol 8.7%, and sodium lauryl sulfate 4.2%.

It is necessary to add a plasticizer to the Aquacoat® dispersion to obtain good films. The types of plasticizers that have been commonly used include dibutyl phthalate, diethyl phthalate, triethyl citrate, tributyl citrate, acetylated monglycerides, acetyl tributyl citrate, and triacetin. The optimum level for most of the plasticizers is 20–24% and for acetylated monglycerides about 30%. The effect of some selected plasticizers on the glass transition temperature of ethylcellulose pseudolatex is illustrated in Fig. 13 [28,54]. A decrease in the glass transition temperature with increasing plasticizer level is indicative of increased coalescence of the particles, and hence film formation. It must be noted that indiscriminate use of plasticizers could be contraindicated [55]. Studies showed that the water vapor transmission rates of films produced from a plasticized Aquacoat® pseudolatex were one-third the value of the mixed ethylcellulose-hydroxypropyl methylcellulose films produced from organic solvent solutions, and about one-half the value of an ethylcellulose polymer film from an organic solvent.

Guo et al. [56] investigated the effects of plasticizer, physical aging, and film-forming temperature on the mechanical and transport properties of films prepared from Aquacoat. They found that the water permeability first decreased and then increased with an increasing concentration of diethyl phthalate. The water vapor permeability was

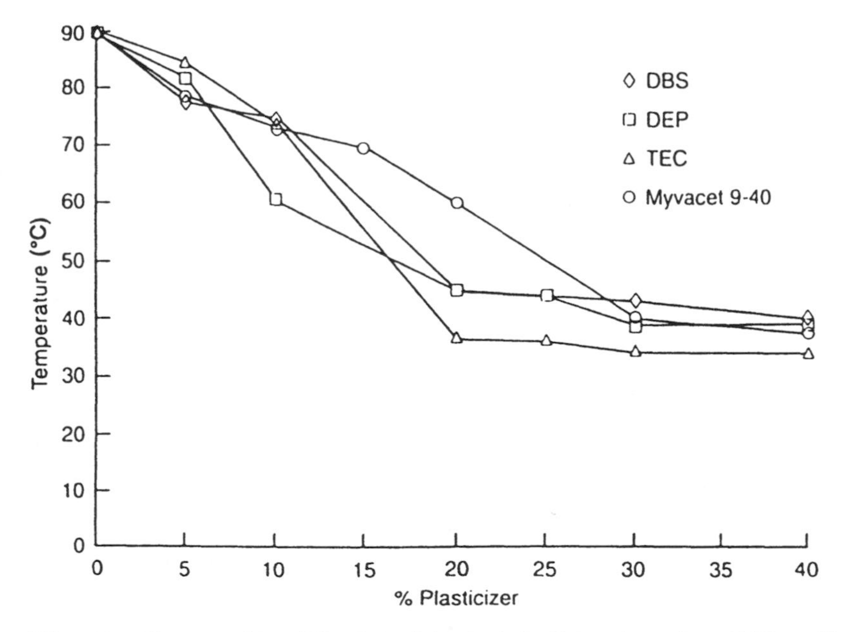

Fig. 13 Glass transition (T_g) of plasticized ethylcellulose latex (key: DBS = dibutyl sebacate; DEP = diethyl phthalate; and TEC = triethyl citrate). (From Ref. 54.)

higher for films prepared at 140°C than for those prepared at 100°C at all plasticizer levels. This was attributed to the presence of many voids or pinholes in films prepared at 140°C. No pinholes were seen in films prepared at or below 100°C.

Surelease is marketed as a 25% solid dispersion in an ammoniacal water solution (pH 9.5–11.5). It is prepared by a phase inversion in-situ emulsification technique. As supplied, Surelease contains ethylcellulose, oleic acid, esterified fatty acids, dibutyl sebacate, and fumed silica. The oleic acid, dibutyl sebacate, and esterified fatty acids all serve as plasticizers. They are incorporated during the manufacturing process. Compared to Aquacoat, which as noted above requires plasticization prior to use in coating, Surelease is ready to use. Like Aquacoat, it is used in tablet coating, granulation, granule coating, controlled and sustained release, taste masking, and barrier applications.

Recently, Nakagami et al. [57] reported the preparation of an aqueous coating suspension using micronized EC [58] and a nonionic dispersant. A comparative study of the dissolution temperature of EC in various plasticizers and the cloud point of the resulting polymer solutions indicated diethyl *d*-tartrate to be the most compatible plasticizer. With no plasticizer, the micronized EC suspension showed no film formation even at 140°C. In contrast, Aquacoat produced a transparent film at 100°C. Further, Aquacoat required much less plasticizer for film formation than the aqueous micronized EC suspension. These differences have been attributed to different film formation mechanisms for the two systems. In the case of Aquacoat, the driving force for film formation is the capillary force produced as water evaporates, whereas in the micronized EC aqueous suspension the film formation occurs through gelation of the micronized polymer caused by the plasticizer.

The use of Aquacoat-ECD and Surelease in coating has been covered in detail in a recent book [59]. Recently, Aquacoat has also been shown to be useful in microencapsulation of drugs [60] and in the preparation of microspheres [61]. The latter involves emulsification of Aquacoat into a heated oil phase to form a water-in-oil emulsion. The removal of water above the MFT then causes the polymer to coalesce and form microspheres with sizes varying between 5 and 250 μm.

G. Low Crystallinity Cellulose

Low crystallinity celluloses (LCCs) are materials that have a substantially reduced degree of crystallinity, typically ranging between 15% and 45%. They are prepared either by mechanical disintegration of the cellulosic source (dry or wet milling) or from a chemical reaction between cellulose and concentrated mineral acids. Recently, a new form of LCC that readily hydrates in water to form an aqueous dispersion has been prepared, using phosphoric acid under controlled sequenced temperature conditions [62,63]. The size of the primary particles produced by this method is in the submicron range. LCC dispersions containing about 3–7.5% solids are fluid and can be sprayed or cast to produce free films. Beyond 7.5% solid content, the dispersions are thixotropic lotions or creams. Dispersions containing a solid content of 10% or higher are creams or semisolids. The most effective plasticizers for LCC dispersions are polyhydric alcohols, such as glycerin and propylene glycol. A scanning electron micrograph of the free film prepared from an aqueous dispersion is depicted in Fig. 14. The film-forming property of LCC is due to the effect of both its low crystallinity and submicron particle size. The former allows more hydroxyl groups to be accessible and the latter

Fig. 14 SEM of free film prepared from an aqueous dispersion of LCC.

provides high surface energy for interaction with water molecules. As water evaporates during drying, fine particles of LCC are forced together, deformed, interacted, and coalesced into a film. LCC dispersions containing 3% or lower solids are physically unstable, unless stabilized with a secondary soluble polymer.

The aqueous dispersions can be converted into a fine powder (LCPC) or a bead form (LCBC) by using conventional dehydration and spray-drying techniques. The crystallinity of LCPC powder versus its hydrated form is compared in Fig. 15. It ap-

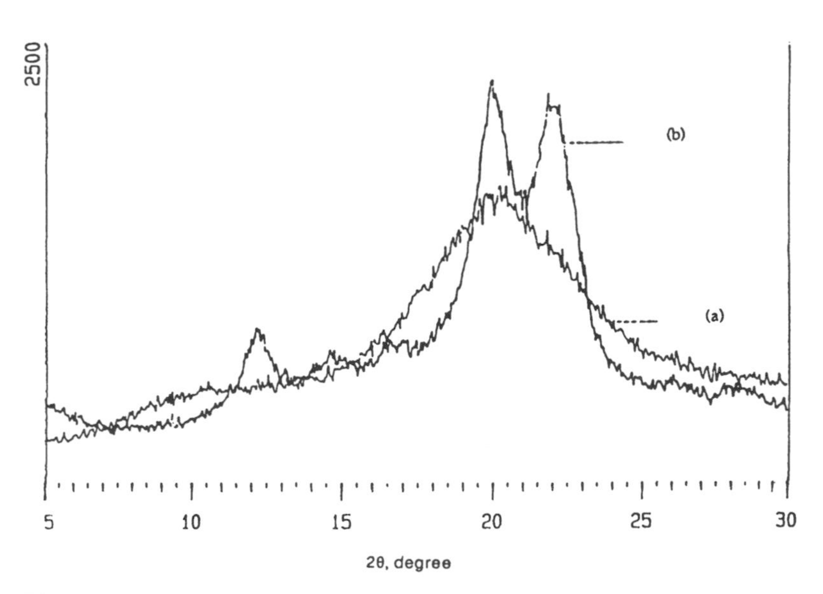

Fig. 15 Powder x-ray diffraction patterns of hydrated LCC (a) and LCPC (b).

pears that during drying, cellulose chains reorient themselves and establish an interchain hydrogen bonding network, causing an increase in the crystallinity of the powder. LCPC has been shown to be an effective binder in tableting, whereas LCBC serves as a superdisintegrant.

H. Oxidized Cellulose

Oxidized celluloses are water-insoluble materials prepared from a reaction between cellulose and an oxidant. Depending on the nature of the oxidant and the reaction conditions, oxidized cellulose containing carboxylic, aldehyde, and/or ketone groups can be prepared. Highly oxidized cellulose materials are bioresorbable. Oxidized celluloses containing 13% or higher carboxyl content are useful as a hemostatic agent and as a barrier to the formation of tissue adhesion.

Recently, oxidized celluloses that readily disperse in water to form colloids have been prepared using alkali and alkaline earth metal hypohalites and persulfates [64–66]. The white to off-white powder when suspended in water forms a stable colloidal or near colloidal thixotropic dispersion. The particle size distribution of an oxidized cellulose dispersion prepared using hypohalite is shown in Fig. 16 [66]. A wide variety of solid (crystalline or amorphous) and liquid bioactive compounds can be entrapped and loaded in such systems, thereby producing substantive controlled and/or sustained release formulations, having unique applications for the development of cosmetic, pharmaceutical, agricultural, and consumer products. The most effective plasticizer is glycerin. Appropriately plasticized aqueous colloidal systems of oxidized cellulose can be cast to develop transdermal patches or used to produce films or matrices. A scanning electron photomicrograph of a film produced from an oxidized cellulose dispersion plasticized with glycerin is shown in Fig. 17. Although the dispersions are microbiologically stable for several months, preservatives such as benzoates should be added to the dispersion to prevent and/or inhibit microbial growth.

V. SILICONE ELASTOMERS

Silicone elastomers, in general, are biocompatible and physiologically inert materials. They have been extensively used in many biomedical and controlled release applications [67–69]. Of particular interest to pharmaceutical applications is the cross-linked hydroxy end-blocked polydimethylsiloxane (PDMS) latex system. PDMS shows good thermal stability and possesses low surface energy. The solubility parameter and glass transition temperature are 7.4–7.6 $(cal/cm^3)^{1/2}$ [70–72] and −123°C [73], respectively. The cross-linked hydroxy end-blocked PDMS latex is prepared by first emulsifying the hydroxyl end-blocked polydiorganosiloxne, $HO(R_2SiO)_nH$, in water using an anionic surfactant, such as dodecylbenzene sulfonic acid, and then cross-linking the emulsified polymer with an alkoxyl silicone of the general formula $R'_aSi(OR^3)_{4-a}$, (where R'_a is an hydrocarbon radical having up to 16 carbon atoms, R^3 is an alkyl radical having from 1 to 6 carbon atoms, and a = 0 or 1), at 15–30°C for 5 hr at a pH of less than 5 [74,75]. Admixing sufficient base to raise the pH of the latex to greater than 7 then produces the cross-linked hydroxy end-blocked PDMS latex. Because no metallic catalyst is used in the process, the PDMS latex prepared in this matter is stable for a period of months without significant change in its properties. The removal of water yields an elastomer, which requires no further curing. It is suggested that the admixing of the alkoxy silicone

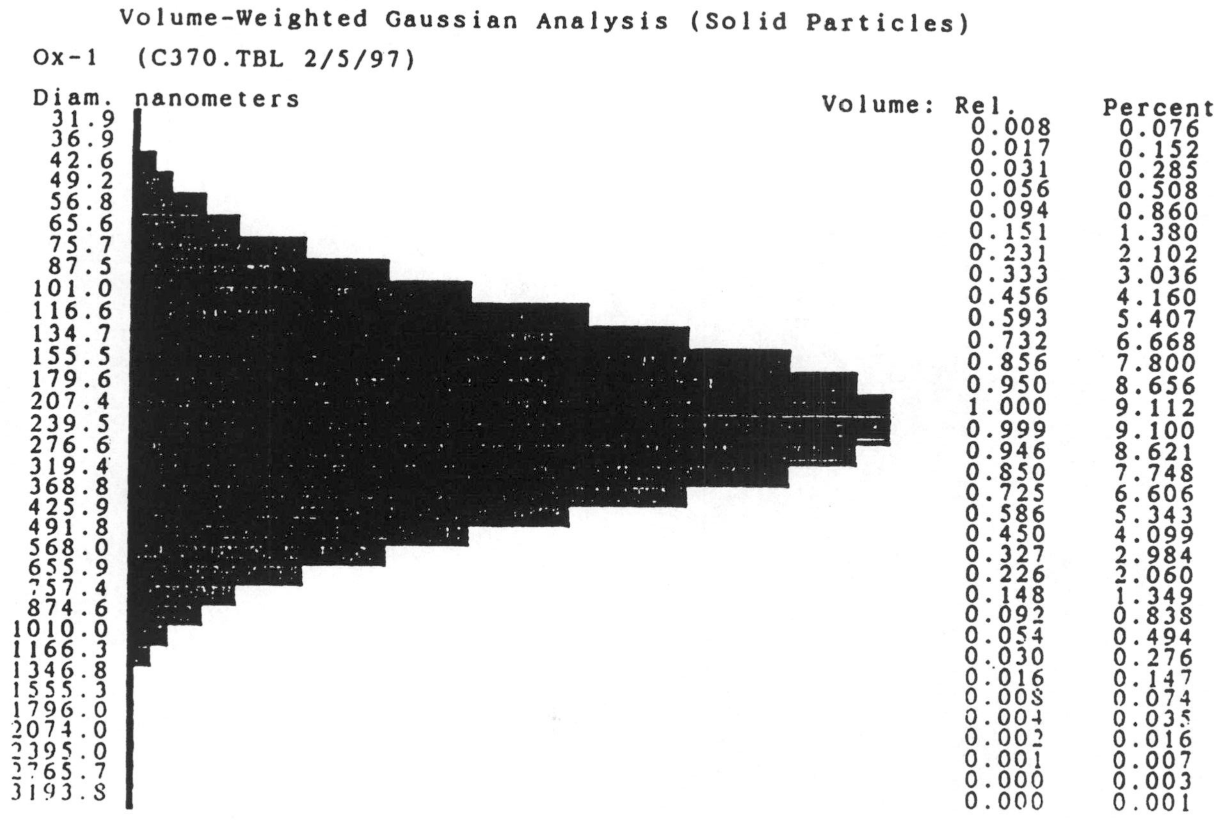

Fig. 16 Particle size distribution of an oxidized cellulose dispersion.

Fig. 17 SEM of free film prepared from an aqueous dispersion of oxidized cellulose.

compound into the homogenized polydiorganosiloxane, stabilized with a limited amount of surfactant, allows the alkoxy silicone compound to migrate from the water phase through the surfactant layer on the polydiorganosiloxane droplet surface into the surface of the droplets. Once inside the droplet surface, the alkoxy silicone compound can react with the hydroxyl end-blocked polydiorganosiloxane polymer to produce cross-links. The following reaction represents the major chemical reactions involved in the manufacture of the silicone elastomer latex [69]:

Since the cross-linked hydroxy end-blocked PDMS latex yields weak, continuous, cross-linked films (tensile strength and percentage elongation values range between 0.55 and 1.99 MPa and 149 and 325, respectively), colloidal silica sol (Nalcoag 1115, Nalco Chemical Co., Chicago) or silsesquioxane is typically added to the emulsion to produce mechanically strong films. The colloidal silica sols are commercially available as dispersions of colloidal silica in water. The colloidal silica content varies from 15% to 50%, by weight, with average particle sizes ranging from about 4 to 60 nm, and with pH values of about 3.2 and 8.5–10.5. Silsesquioxanes, such as methylsilsesquioxane having the unit formula $CH_3SiO_{3/2}$, are also available in the colloidal dispersion form. They are prepared by adding a silane to an aqueous solution of a surfactant with agitation under acidic or basic conditions [76]. Since both silica and silsesquioxane exhibit useful tensile strength, stretch under tension, and retract rapidly to recover their original dimensions, the silica- or silsesquioxane-reinforced cross-linked hydroxy end-blocked PDMS latices serve as excellent film-forming systems. The stress-strain profiles of dry free films prepared from the cross-linked hydroxyl end-blocked PDMS elastomer dispersion containing different amounts of silicone-to-silica ratios are shown in Fig. 18 [77]. The curves show that films with a low silicone-to-silica ratio are more brittle. As the silica content increases, the stress-strain curves become biphasic. The initial high modulus part of the curves corresponds to the response of the hydrophilic silica, whereas the lower modulus region is the response of the rubberlike silicone elastomer. As illustrated in Fig. 19, the hydrated films, compared to dry films (Fig. 18), exhibit con-

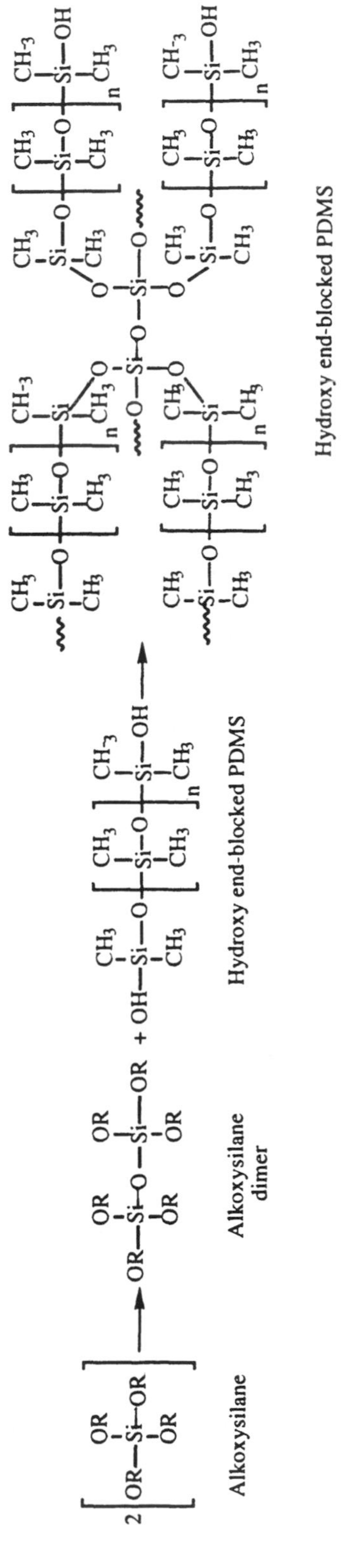

$$2\left[OR{-}Si(OR)_2{-}OR\right] \longrightarrow OR{-}Si(OR)_2{-}O{-}Si(OR)_2{-}OR \; + \; OH{-}Si(CH_3)_2{-}O{-}\left[Si(CH_3)_2{-}O\right]_n{-}Si(CH_3)_2{-}OH \longrightarrow$$

Alkoxysilane

Alkoxysilane dimer

Hydroxy end-blocked PDMS

Hydroxy end-blocked PDMS

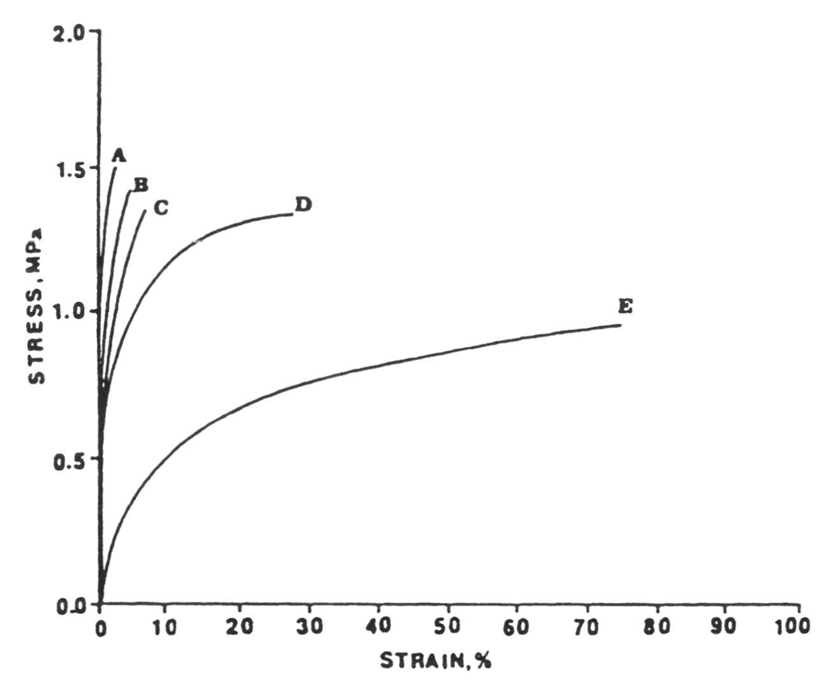

Fig. 18 Stress-strain profiles for silicone elastomer free films with five different silicone-to-silica ratios [key: (A) 2 to 1; (B) 3 to 1; (C) 4 to 1; (D) 6 to 1; and (E) 9:1]. (From Ref. 77.)

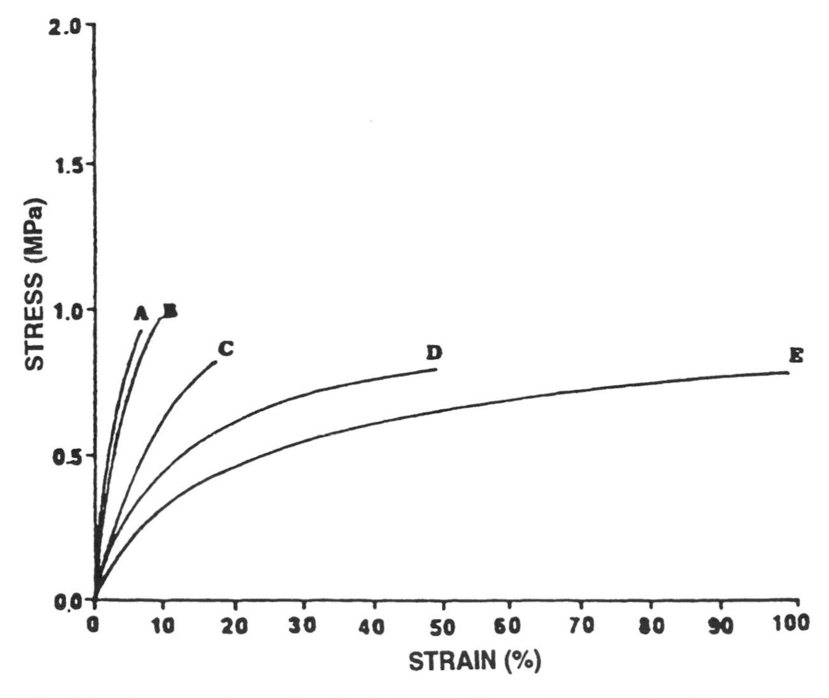

Fig. 19 Stress-strain profiles for hydrated silicone elastomer free films with five different silicone-to-silica ratios [key: (A) 2 to 1; (B) 3 to 1; (C) 4 to 1; (D) 6 to 1; and (E) 9:1]. (From Ref. 77.)

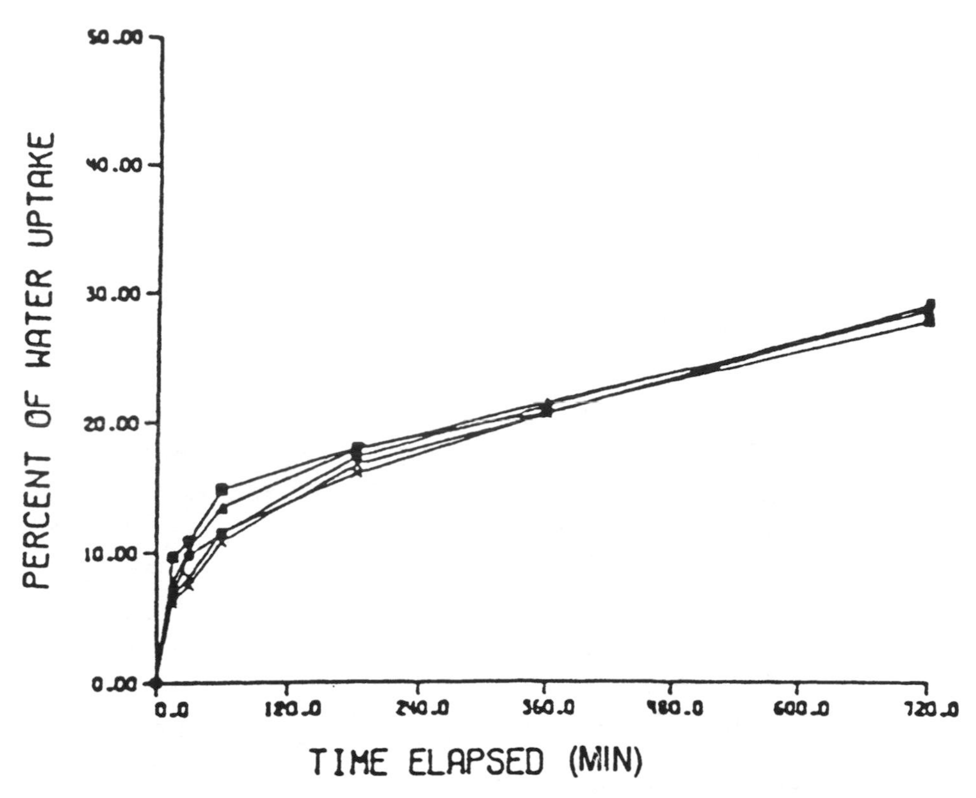

Fig. 20 Percent water uptake by silica-filled silicone elastomer free films with five different silicone-to-silica ratios [key: (■) 2 to 1; (▲) 3 to 1; (●) 4 to 1; (X) 6 to 1; and (x) 9:1]. (From Ref. 77.)

siderably reduced tensile strength and Young's modulus, but a marked increase in percent of elongation [77]. These changes in the stress-strain parameters have been attributed to the weakening of the silica-to-silicone linkages as well as the silica-to-silica linkages due to interaction between water and silica [77,78]. The hydration profiles of silica-filled elastomers containing five different ratios of silicone-to-silica ratios are shown in Fig. 20. As is evident, the content of silica incorporated in the films does not affect the overall hydration profiles, except that films that contain a high silicone-to-silica ratio exhibit a faster initial hydration [77]. A number of other additives, such as pigments or dyes, and heat-stability additives, such as iron oxide, have also been added to the aqueous latex to alter the properties of the elastomer films [79,80].

Because of the hydrophobic nature of polydiorganosiloxane, the cross-linked hydroxy end-blocked PDMS elastomer coatings are relatively impermeable to hydrophilic and ionic compounds [81]. The permeability to such compounds, however, can be increased by adding a water-soluble or pore-forming component to the dispersion [81,82]. Li and Peck [77,78] and Woodward et al. [75] used water dispersible/soluble or pore-forming agents (e.g., PEG 1450, 3350, or 8000) as additives to increase the permeability of the coatings to a variety of agents. Fig. 21 shows the effects of the molecular weight of polyethylene glycols and the solid content on the apparent viscosity of cross-linked hydroxy end-blocked PDMS-PEG dispersions. As is evident, with the same solid content, dispersions containing the higher molecular weight PEG are comparatively more viscous. As the solid content increases, the viscosity of the dispersion also

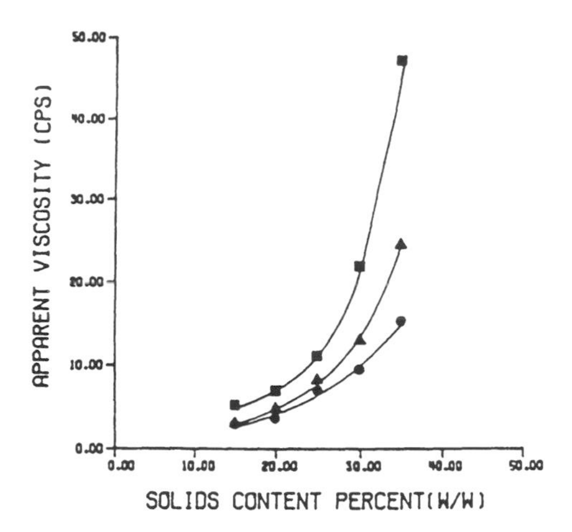

Fig. 21 The effect of total solids content on the apparent viscosity of dispersions with total solids consisting of silicone and silica in a ratio of 2 to 1 and 30% polyethylene glycol [key: (■) PEG 8000; (▲) PEG 3350; and (●) PEG 1450]. (From Ref. 77.)

increases. A more pronounced increase in viscosity occurred in the system that contained the highest molecular weight PEG. At a total of 30% PEG solid content, the viscosity of the dispersion was less than 25 cps. Other water-soluble substances such as low molecular weight PEGs, glycerin, propylene glycols, sucrose, and glucose, were also evaluated. These provided sticky and soft films, especially at higher concentration (>10%) and, therefore, were not regarded as useful.

The mean percent water uptake and weight loss of the free films prepared from the PEG-reinforced cross-linked hydroxy end-blocked PDMS free films, after soaking in water for 12 hr, are listed in Table 9 [69]. Fig. 22 shows the percent water uptake of PEG-reinforced cross-linked hydroxy end-blocked PDMS free films, containing three different levels of PEG 8000, as a function of time [77]. The data presented in Table 9 show that the water uptake and leaching of PEGs occur simultaneously. The percent of water uptake by the films containing 10% PEG was two to four times higher, depending on the molecular weight of the PEGs, than the weight loss as a result of the leaching of PEG. For PEG 8000 and 3350, the percent water uptake was significantly higher at all three loading levels than for films containing PEG 1450. Fig. 23 shows that the leaching of PEG 8000 at all three levels were nearly completed within the first hour. For the 10%, however, only 70% of the PEG loaded was leached [77]. The leaching of PEGs 1450 and 3350 followed the same pattern as PEG 8000. It has been sug-

Table 9 Mean Percent Weight Loss and Water Uptake for Various Silicone Elastomer-PEG Films Soaked in Water for 12 hr

PEG	Percent weight loss PEG loading level (%, w/w)			Percent weight uptake PEG loading level (%, w/w)		
(mol. wt.)	10	20	30	10	20	30
8000	8.95	21.45	31.85	32.10	28.27	37.63
3350	9.27	22.18	32.05	32.10	26.12	31.65
1450	10.43	21.95	32.23	25.83	20.33	26.77

Source: Ref. 69.

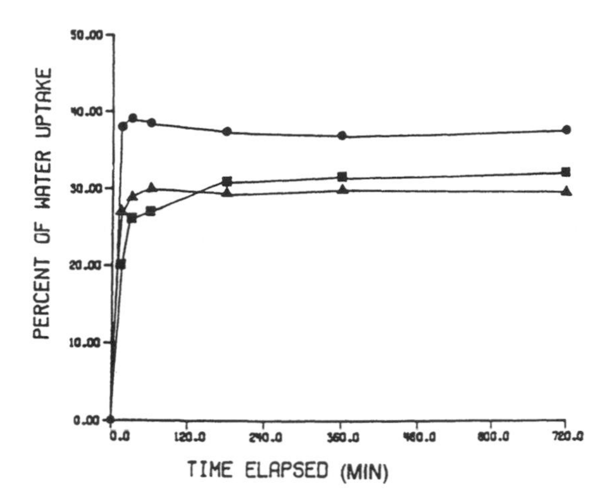

Fig. 22 The percent water uptake by hydrated polyethylene glycol-silicone elastomer mixed films containing silicone and silica in a ratio of 2 to 1 and polyethylene glycol 800 at three different loading levels [key: (■) 10%; (▲) 20%; and (●) 30%]. (From Ref. 77.)

gested that the PEG dissolves in the imbibed water and forms vesicles of concentrated solution. As a result, an osmotic pressure is generated within the hydrated films, causing more water to enter into the hydrated films.

In general, the cross-linked hydroxy end-blocked PDMS elastomer coatings exhibit elastic properties, and thus are suitable for applying on tablets with various shapes. Li and Peck [83] reported that capsule-shaped tablets, when coated with the cross-linked hydroxy end-blocked PDMS-PEG 8000 system, showed a faster drug release than round deep-cup shaped and oval-shaped tablets. They attributed this difference to the larger surface area associated with the capsule-shaped tablets.

Woodward et al. [74,75] reported that silicone elastomer coatings, in general, (1) provide sufficient tensile strength and elongation, allowing dosage forms to expand in the stomach without rupture; and (2) result in a substantial zero-order release of the active agent. Table 10 shows the rupture force and percent deformation for selected

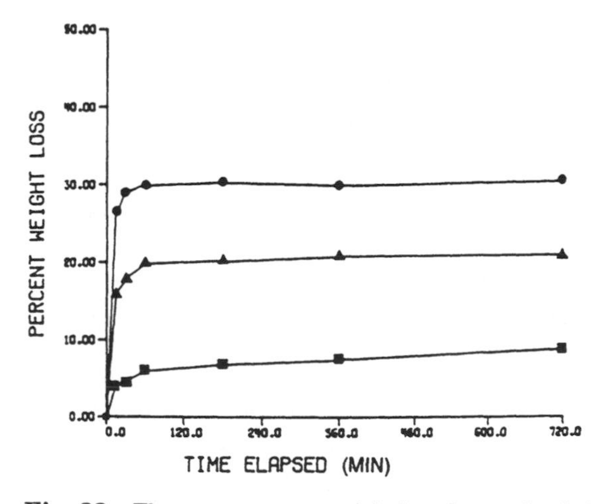

Fig. 23 The percent water weight loss from polyethylene glycol-silicone elastomer mixed films containing silicone and silica in a ration of 2 to 1 and polyethylene glycol 800 at three different loading levels [key: (■) 10%; (▲) 20%; and (●) 30%]. (From Ref. 77.)

Table 10 Rupture Force and Percent Deformation for Silicone Elastomer Coating Shells

Coating formulation				
PEG (mol. wt.)	PEG level	Silicone-to-silica ratio	Rupture force g[a]	Percent deformation[a]
8000	20	2	670.8 (41.1)	67.1 (3.4)
8000	30	2	394.2 (41.5)	46.1 (4.3)
8000	40	2	182.3 (20.4)	30.6 (1.2)
4450	30	2	415.9 (35.4)	42.7 (6.3)
1450	30	2	566.7 (64.4)	31.8 (4.3)
8000	30	3	409.2 (65.9)	49.7 (6.5)
8000	30	4	331.7 (21.1)	57.6 (3.6)

[a]Mean and standard deviation for six samples.
Source: Ref. 69.

silicone elastomer film coating formulations [69]. The films became weaker as the PEG content increased. At the same coating level, stronger and more rigid film coatings were produced by formulations consisting of the lower molecular weight PEGs. A higher silicone content (silicone-to-silica ratio) produces more elastic hydrated film coating as indicated by a greater percent of deformation. These data support the fact that silicone elastomer-PEG film coatings are generally strong and elastic in nature and are capable of maintaining physical integrity during in vivo drug release.

A cross-linked hydroxy end-blocked PDMS latex suspension, Q7-2837, marketed by Dow Corning, has a mean particle size of 250 nm, a pH of 7–8, and a total solids content of ~52% (w/w). The relatively low viscosity (58 cps) of the undiluted dispersion is an obvious advantage in tablet film coating because it allows a reduction in coating time, a decrease in labor and energy consumption, and a reduction in exposure time of drug to moisture and heat during the coating process [77].

VI. BIODEGRADABLE PSEUDOLATICES

Pseudolatices of biodegradable polymers, such as poly(*d,l*-lactide) (PLA), poly(glycolide) (PGA), poly(lactide-co-glycolide) (PLGA) and poly(ε-caprolactone) (PCL), have been prepared. These polymers are biocompatible. They degrade in vivo nonenzymatically to soluble byproducts that are either excreted or metabolized in the body. The degradation rate is greatly influenced by the composition of the polymers [84,85], and by pH and temperature of the degradation medium [86].

Three methods, the emulsion-evaporation approach [87], the salting out procedure [7,8], and the precipitation technique [88], have been used to prepare dispersions of biodegradable polymers. The emulsion-evaporation approach, used by Gurny et al. [87], involves an initial emulsification step, wherein an aqueous solution of Pluronic F-68, Tween 80, or sodium lauryl sulfate is added with constant stirring to a solution of PLA in a water-immiscible organic solvent or solvent mixture, followed by the evaporation of the organic solvent. The final dispersion contained 40% w/w polymer with an apparent viscosity value of 96 cps and an average particle size value of 0.45 ± 0.16 μm. Recently, Coffin [89] established a correlation between surfactant solubility in blends of organic solvents and the particle size of resultant PLA pseudolatices. Studies also show a time-dependent reduction in nanosphere size, to varying degrees, with different emulsification equipment/methods [89,90]. Since lactides degrade in water, the pseudolatex developed by Gurney and coworkers showed a loss of about 9–10% in molecular weight of PLA as a result of the latex manufacturing process. During storage for 120 days, the viscosity of the pseudolatex increased by about 10%. However, there was no significant change in the molecular weight of the PLA over the same storage period (Fig. 24). Recently, Coffin and McGinity [91] investigated the effects of surfactant system, temperature, and storage time on the stability of PLA and PCL pseudolatices. They reported that pseudolatices prepared using nonionic surfactants, polyoxyethylene 20 monolaurate and Pluronic F-68, exhibited greater physical and chemical stability than those containing the anionic surfactants potassium oleate or sodium dodecyl sulfate. This study also revealed that nonionic-surfactant-based PLA and PCL pseudolatices could be stored at 5°C for 4 months without any loss in the molecular weights of the polymer (Figs. 25 and 26). At 37°C, irrespective of the surfactant systems used, both PLA and PCL pseudolatices showed a decrease in the molecular weights during storage, suggesting hydrolysis of the polymer at this temperature. Since the hydrolysis of PLA and PCL results in the release of the corresponding free acid (for every hydrolysis reaction there is an additional carboxylic group produced), the periodic pH check of the pseudolatex may serve as a good measure of the stability of the system.

The salting out method, developed by Allémann et al. [7], uses a gel (prepared by mixing a saturated solution of a salt and a water-soluble polymer) and an acetone solution of PLA to prepare an emulsion. In the process, the acetone is salted out by the dissolved salt and the two-phase system is manufactured. Addition of water to the emulsion results in the preparation of nanodispersion.

Fessi et al. [92] reported a novel and simple procedure for the preparation of aqueous colloidal dispersions of PLA and PLGA copolymers by interfacial deposition following displacement of a semipolar solvent miscible with water from a lipophilic solution. Briefly, to a solution of PLA and phospholipid in acetone, a solution of benzyl benzoate or caprylic/capric triglyceride (Miglyol® 810) in the same solvent, containing an active, is added. The resulting mixed solution is then added to a solution of

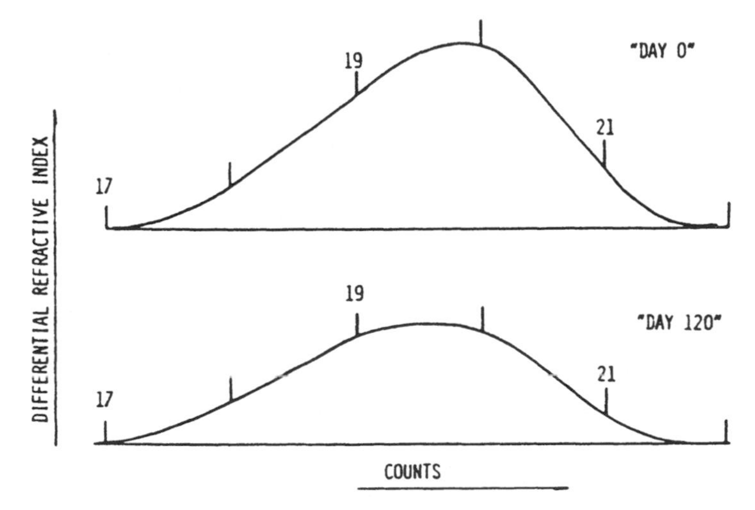

Fig. 24 GPC analysis of the neat polymer and the polymer latex after 120 days of storage at room temperature. (From Ref. 87.)

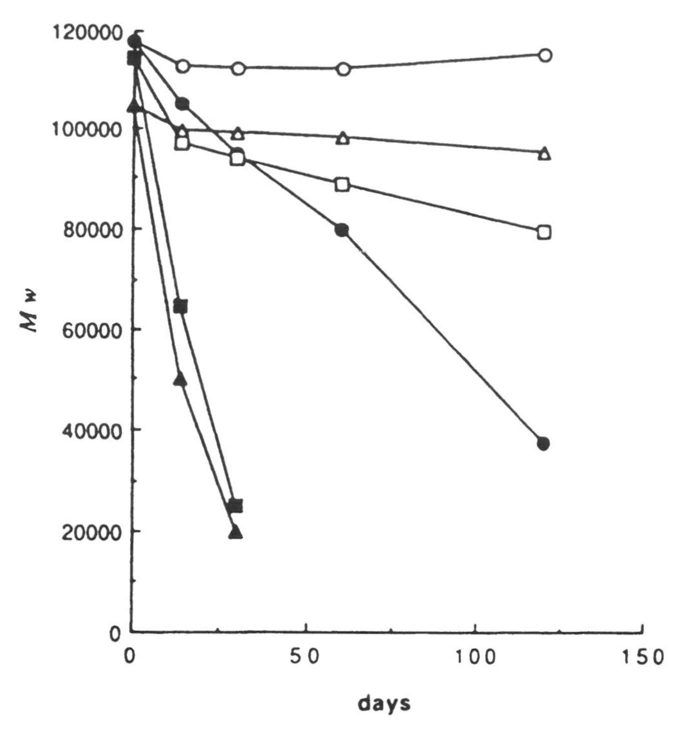

Fig. 25 Effect of pseudolatex surfactant system and temperature on degradation of PLA in unbuffered pseudolatices [key: (○) nonionic, 5°C; (●) nonionic, 37°C; (Δ) potassium oleate, 5°C; (▲) potassium oleate, 37°C; (□) sodium dodecyl sulfate, 5°C; (■) sodium dodecyl sulfate, 37° C]. (From Ref. 91.)

poloxamer in water under moderate agitation. The acetone is then removed from the aqueous phase concentrate to produce the colloidal dispersion. Guterres et al. [93] reported that PLA dispersion prepared using Miglyol 810 is stable, whereas those developed using benzyl benzoate showed considerable PLA degradation.

The effect of the concentrations of different nonionic surfactants on the glass transition temperature of films produced from PLA pseudolatices is shown in Fig. 27 [94,95]. As is evident from this Figure, Pluronic F-68 and Myrj® 52-S showed the most significant effect on the films. This is because polymers containing carboxylic groups are readily miscible with polyethers through hydrogen bonding. The ability of surfactants to optimally adsorb on the polymer surface enhances stabilization and coalescence processes. In general, PLA pseudolatices containing a high concentration of surfactants do not control drug release or maintain film integrity in the hydrated state.

Gurny et al. [87] investigated the in vitro and in vivo release of [4–^{14}C]-testosterone from the PLA pseudolatex and peanut oil. They reported that the oil-based formulation released the drug at a rate more than 10 times faster than that observed from the pseudolatex, where the drug was molecularly dispersed. The somewhat more rapid initial release of the drug was attributed to the drug located on the particle surface. The in vivo results, shown in Fig. 28, while not representing an ideal release profile, do show the potential of a biodegradable pseudolatex as a controlled release system. Recently, Frisbee and McGinity [90] investigated the release of chlorpheniramine maleate pellets coated with the PLA pseudolatex. As shown in Fig. 29, the release rates of

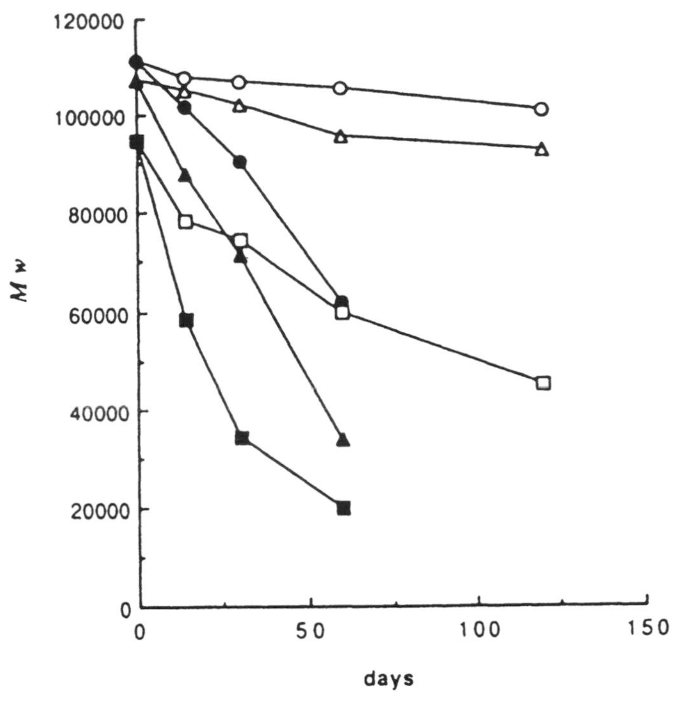

Fig. 26 Effect of pseudolatex surfactant system and temperature on degradation of PCL in unbuffered pseudolatices [key: (○) nonionic, 5°C; (●) nonionic, 37°C; (Δ) potassium oleate, 5°C; (▲) potassium oleate, 37°C; (□) sodium dodecyl sulfate, 5°C; (■) sodium dodecyl sulfate 37° C]. (From Ref. 91.)

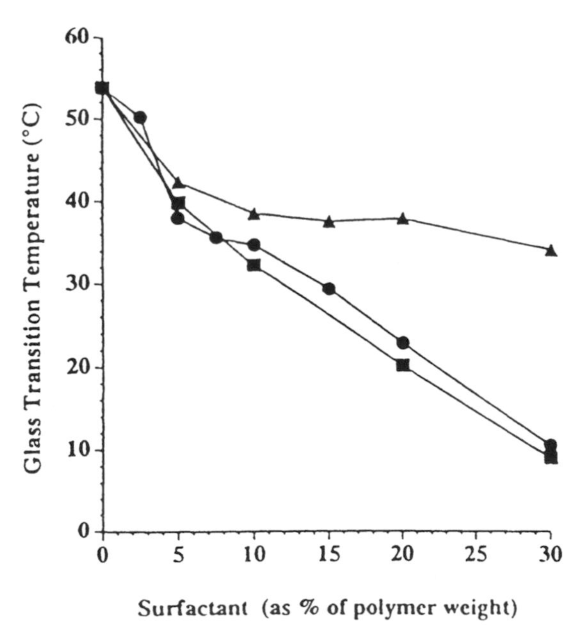

Fig. 27 Effect of three nonionic surfactants on glass transition temperature of films cast from poly(*d,l*-lactide) pseudolatices [key: (●) Pluronic F-68; (■) Myrj 52-S; (▲) Tween 60]. (From Ref. 95.)

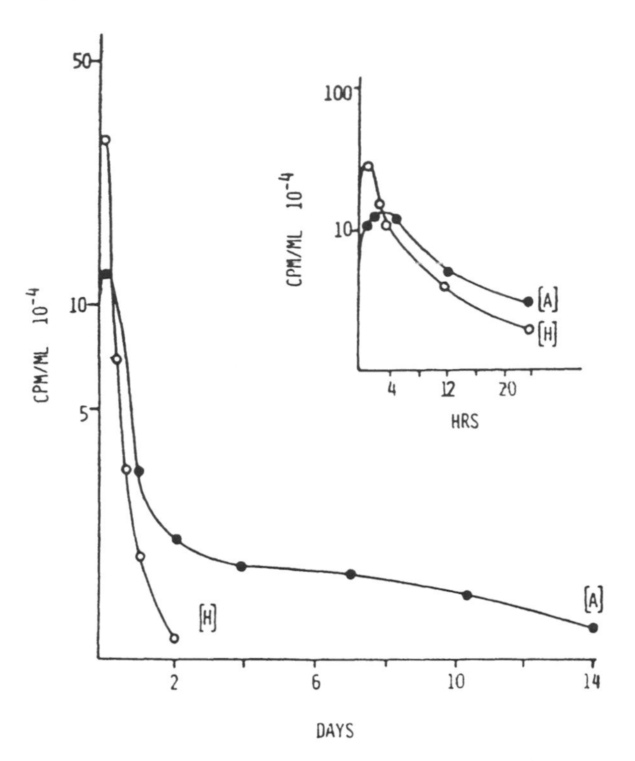

Fig. 28 In vivo release from latex (A) and solution (H) loaded with testosterone tested on six animals over 14 days. (From Ref. 87.)

chlorpheneramine maleate decreased with increasing PLA coating levels. A near zero-order release over 24 hr was obtained with the 10% PLA weight level. According to these workers there was no change in the release profiles from the coated pellets over 1 year of storage at room temperature. PLA pseudolatices have also been used as binders in tableting [96,97]. Studies showed that thermal treatment of tablets to temperature above the T_g of the PLA retarded the release of the drug [97]. This has been attributed to a more efficient distribution of the PLA throughout the tablet matrix and a stronger compact formation [97].

VII. POLYMETHACRYLATE DISPERSIONS

A number of chemically modified polymethacrylates (PMAs) are known. Many medical devices, including those that come in contact with blood, and various pharmaceutical products are made from PMA today. Polymethacrylates in general are well tolerated by the skin and the mucosa. Many that are used in pharmaceutical applications are known by the trade name Eudragit. Structurally, polymethacrylates consist of a continuous carbon chain as a backbone and methyl, ethyl, carboxylic, methyl ester, ethyl ester, or trimethylammoniumethylmethacrylate chloride as a side group. The parent polymer, poly(methyl methacrylate), is marketed under the trademark Plexiglas®. The structures, scientific names, composition, molecular weight, and characteristics of methacrylic and methacrylate ester copolymer latices commercially available from Rohm

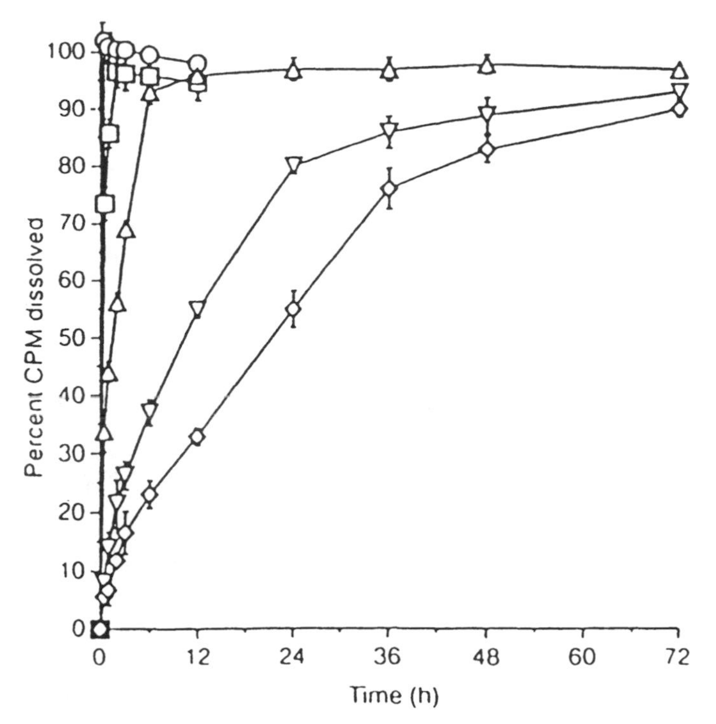

Fig. 29 Influence of coating level (% weight increase) on dissolution profiles of CPM beads with PLA pseudolatex dispersion [key: (○) 2%; (□) 4%; (Δ) 6%; (∇) 8%; (◇) 10%]. (From Ref. 95.)

Pharma (Gmbh Weiterstadt, Germany) for use in controlled and/or sustained release coatings of powders, crystals, granules, tablets, and dragées are presented in Table 11. Eudragit L 30 D is a 30% dispersion of an anionic copolymer based on methacrylic acid and methylmethacrylate. The code designation L represents the first letter of the German word *leichtlöslich* (freely soluble) and refers to this solubility characteristics in the intestine. Eudragit NE 30 D is a 30% aqueous dispersion of neutral poly-(ethylacrylatemethyl methacrylate) ester. The films produced by Eudragit NE 30 D are insoluble, but permeable and swellable in water and digestive juice. The latter provides a means to controlling active-substance release via diffusion processes. Eudragit RL 30 D and Eudragit RS 30 D are dispersions of copolymers of acrylic and methacrylic esters with a low content of quaternary ammonium groups. The code designations RS and RL refer to the low and high permeability of the polymer, respectively.

A stable hydroalcoholic colloidal dispersion of Eudragit RS100 (an acrylate-methacrylate copolymer containing a quaternary ammonium group) can also be prepared by dissolving the polymer in ethanol and then diluting with water and subsequently homogenizing the mixture at 1500 rpm for 15 min [98]. The addition of water to the alcoholic solution results in the formation of a nondispersible gel. The high-shear stirring then converts the gel into a stable, nonthixotropic colloidal dispersion. The lowest concentration of an electrolyte required to revert the dispersion (10% w/v) to a thixotropic gel was reported to be 0.2 M. Phenols, which are widely used as antiseptic agents, have been found to inhibit the gelation effect of the electrolyte. It has been suggested

Table 11 Methacrylic and Methacrylate Ester Copolymer Latices

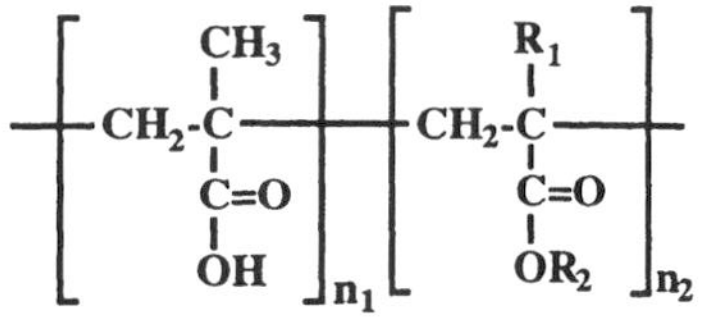

$$-\left[CH_2-C(CH_3)(C{=}O)(OH)\right]_{n_1}-\left[CH_2-C(R_1)(C{=}O)(OR_2)\right]_{n_2}-$$

Methacrylic acid copolymers

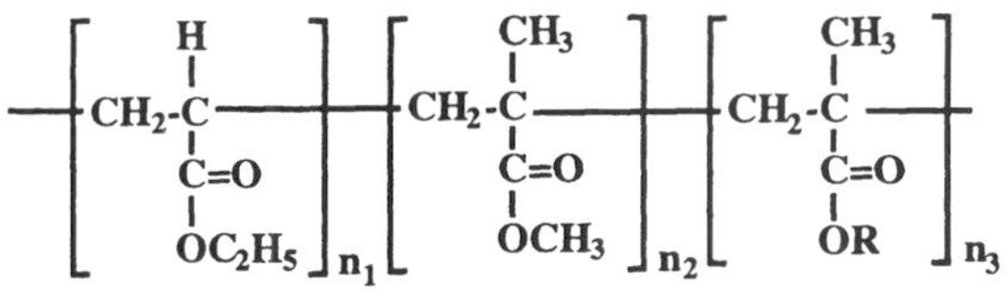

$$-\left[CH_2-C(H)(C{=}O)(OC_2H_5)\right]_{n_1}-\left[CH_2-C(CH_3)(C{=}O)(OCH_3)\right]_{n_2}-\left[CH_2-C(CH_3)(C{=}O)(OR)\right]_{n_3}-$$

Methacrylate ester copolymers

Scientific name, trade name, and corresponding monograph	$n_1:n_2:n_3$ (MW)	R_1, R_2, or R	Aqueous solubility and films permeability
Methacrylic acid copolymers			
Poly(ethylacrylate methylmethacrylate)	1:1	H, C_2H_5	Soluble > pH 5.5
Eudragit L 30 D	(250,000)		
Methacrylate ester copolymers:			
Poly(ethylacrylate.methylmethacrylate)	2:1:0	—	Insoluble, medium
Eudragit NE 30 D (Ph. Eur. polyacrylate dispersion 30%)	(800,000)		
Poly(ethylacrylate, methylmethacrylate)trimethyl-ammonioethylmethacrylate chloride	1:2:0.1	$CH_2CH_2N^+(CH_3)_3$[a]	Insoluble, high
Eudragit RS 30 D (USP/NF ammoni Methacrylate copolymer, type A)	(150,000)	Cl^-	
Eudragit RS 30 D (USP/NF ammoni methacrylate copolymer, type B)	1:2:0.2 (150,000)	$CH_2CH_2N^+(CH_3)_3$[a] Cl^-	Insoluble, low

[a]The molar ratio of this group to the neutral methacrylic ester in Eudragrit RS and RL are 1:20 and 1:40, respectively.

that the addition of phenol reduces the dielectric constant of the medium, causing an increase in the interparticle repulsive force, and hence inhibition of the gelation. Lehmann [99] reported that a stable acrylate aqueous dispersion can also be prepared by direct emulsification of the solid bulk polymer in hot water without any additives.

Nanocapsules, prepared using an interfacial polymerization process of alkylcyanoacrylate, have also been proposed as a new type of vesicular colloidal polymeric drug carrier [100,101]. Recently, Puglisi et al. [102] prepared a polyalkylcyanoacrylate latex as a colloidal drug carrier by polymerization of a cyanoacrylic monomer. The latter was added to an aqueous solution (adjusted to pH 2–5) containing a nonionic surfactant (Pluronic F68, Tween 80, or Triton® ×100). Polymerization of monomer occurred spontaneously at room temperature. The acid polymerization prevented the formation of agglomerates. After the polymerization, the milky suspension of nanoparticles was adjusted to a neutral pH with 0.2M NaOH solution. The mass average and number average molecular weights of the polyethyl-2-cyanoacrylate have been reported to be dependent on the nature and concentrations of the surfactant and the pH of the polymerization reaction medium.

Recently, pilocarpine-loaded pseudolatices of Eudragit RS-100, with and without polyvinylpyrrolidone, have been prepared by Vyas et al. [103] using the solvent removal method. It was reported that increasing the amount of the hydrophilic polymer in the pseudolatices increases the size of the particles and the viscosity of the pseudolatices, and retards the sedimentation on storage for a period of 12 weeks. There was no change in the pH of the pseudolatices during the storage period. The in vitro release study showed a marginal increase from 0.30 to 0.66 $\mu g/h^{1/2}$ per cm^2 in the release rate, with increasing concentration of PVP from 10% to 40%. This uniform, fast release was attributed to the uniform distribution of the drug at a molecular level. Compared to an aqueous solution of pilocarpine, pseudolatices provided increased bioavailability (1.5-2.15 folds) and I_{max} values, and induced a lower level of changes in the pupillary diameter in albino rabbits. The time for appearance of miosis was longer in rabbits treated with a pseudolatex versus a water solution of the drug. The half-lives of the appearance of miosis were 30 min and 7.3 min, respectively. The time corresponding to peak biological response was also significantly delayed. Neither corneal abrasions nor intraocular inflammatory signs were noted as a result of the instillation of pseudolatices.

VIII. SHELLAC

Shellac is a water-insoluble, natural polymer. It dissolves in most alcohols, and has been extensively investigated for enteric coatings. A pseudolatex of shellac has been recently developed by Chang et al. [4] using the self-emulsification approach. The method involves dissolving the shellac in a mixed water-miscible organic solvent (acetone:methanol, 2:1 v/v) and then dispersing it in water at 1500 rpm. The organic solvent and part of the water are then removed under mild agitation to produce a stable shellac pseudolatex. The pseudolatex particles produced were in a size range between 0.05 and 0.2 μm, with an average particle size of 0.15 μm. The smaller particle size and the negative charge on the surface of particles prevent sedimentation during storage. The dispersion, however, supports mold growth and proliferation and thus should be stored under preservation.

Unlike many aqueous polymeric dispersions, shellac pseudolatices do not require plasticization. This is probably because of its low glass transition temperature or self-plasticizing effect due to the presence of a mixture of lower molecular weight fatty acid esters in the interstices of the shellac structure [104].

Chang et al. [4] compared the release profiles of theophylline sustained-release tablets (prepared by layer coatings of the active drug and ethylcellulose on sucrose crystals), coated with a shellac pseudolatex and with shellac dissolved in organic solvent. Both pseudolatex and organic solvent coatings resisted dissolution in simulated gastric fluid. However, in the simulated intestinal juice, the organic-solvent-based coating provided a faster release than the original sustained release tablets, suggesting the partial destruction of the ethylcellulose seal coat during the organic-solvent-based shellac coating. The shellac-pseudolatex-coated theophylline tablets did not show any significant change in the dissolution profiles in the simulated intestinal fluid.

IX. POLY(VINYL ACETATE) EMULSION

A poly(vinyl acetate) emulsion was developed to overcome problems associated with cyanoacrylates (e.g., *n*-butyl 2-cyanoacrylate), the liquid embolization materials [105]. Cyanoacrylates have been reported to produce formaldehyde, a toxic product. Further, cyanoacrylates require that they be injected quickly and that the delivery catheter be withdrawn immediately after completion of the injection to prevent the formation of a rigid bioadhesive mass, which poses a risk of the injection needle being stuck within the vessel. Injection of a proper volume and concentration of cyanoacrylates is therefore very difficult.

The poly(vinyl acetate) emulsion is prepared by polymerizing vinyl acetate monomer in water in the presence of isobutyl amidine chloride at 60°C for 5 hr [105]. The latter serves as an initiator and gives a positive charge to the particle surface. The emulsion particles ranged in size between 0.3 and 0.7 μm.

X. POLYVINYL ACETATE PHTHALATE

Polyvinyl acetate phthalate (PVAP) is an enteric polymer produced by the esterification of a partially hydrolyzed polyvinyl acetate with phthalic anhydride. It is soluble in methanol, ethanol (95%), and mixed solvent systems such as methanol or ethanol:acetone (1:1), methanol:methylene chloride (1:1), etc. The solubility in water is pH dependent, readily dissolving at a pH of about 5. The disintegration/dissolution time in water increases as the phthalyl content in the polymer increases. Compared to CAP and HPMCP, PVAP is more resistant to hydrolysis [55], and hence is a superior enteric polymer from a stability standpoint. According to the NF specification, the phthalyl content in the polymer should be between 55% and 62%.

A PVAP pseudolatex has been prepared using the phase inversion method by Davis et al. [41]. The method involved preparing an oil-in-water emulsion by adding an aqueous solution of Pluronic F-68 to a PVAP solution in a 80:20 (v/v) mixture of methylene chloride:isopropanol, with vigorous agitation. Size reduction followed by removal of the organic solvent produced the pseudolatex.

A PVAP powder that readily disperses in an aqueous medium has been developed by Colorcon, and is currently marketed under the trade name Sureteric® (formerly called Coateric®) [55,106]. The product, as supplied, also contains a plasticizer (polyethylene glycol) and other ingredients to enhance the film-forming characteristics of the polymer. Because of the susceptibility of PVAP to hydrolysis, although minimal at ambient and elevated temperatures as shown in Fig. 30 [106], the aqueous dispersion of Sureteric is typically prepared freshly prior to use. The recommended procedure for preparing the dispersion involves adding the desired quantity of Sureteric powder to an appropriate quantity of water (the temperature of the water should not exceed 30°C) containing an antifoam emulsion (medical antifoam AF) at a level of 1% of Sureteric® solids, using an appropriate propeller-type mixer at a speed sufficient to prevent the formation of powder agglomerates. The dispersion is stirred for another 30–40 min and then is passed through a 60 mesh screen before spraying.

XI. GENERAL COMMENTS AND SUMMARY

Use of aqueous film coating has grown rapidly in the last several decades. Environmental and OSHA regulations as well as economic factors have shifted organic-solvent-based film coatings to water-based film coatings of pharmaceutical dosage forms. The water-based dispersions also offer various technical advantages over organic-solvent-based solutions. The advantages are in terms of film-forming and rheological properties. In spite of their high solid content, they have low viscosity, and hence can be easily sprayed. This also reduces the total processing time and cost of coating. The films prepared by aqueous dispersions exhibit, in general, comparable if not superior properties compared to those produced from organic solvent solutions.

Numerous polymers have been converted into aqueous polymeric dispersions, many as latex or pseudolatex colloidal or near-colloidal forms. Both of these classes

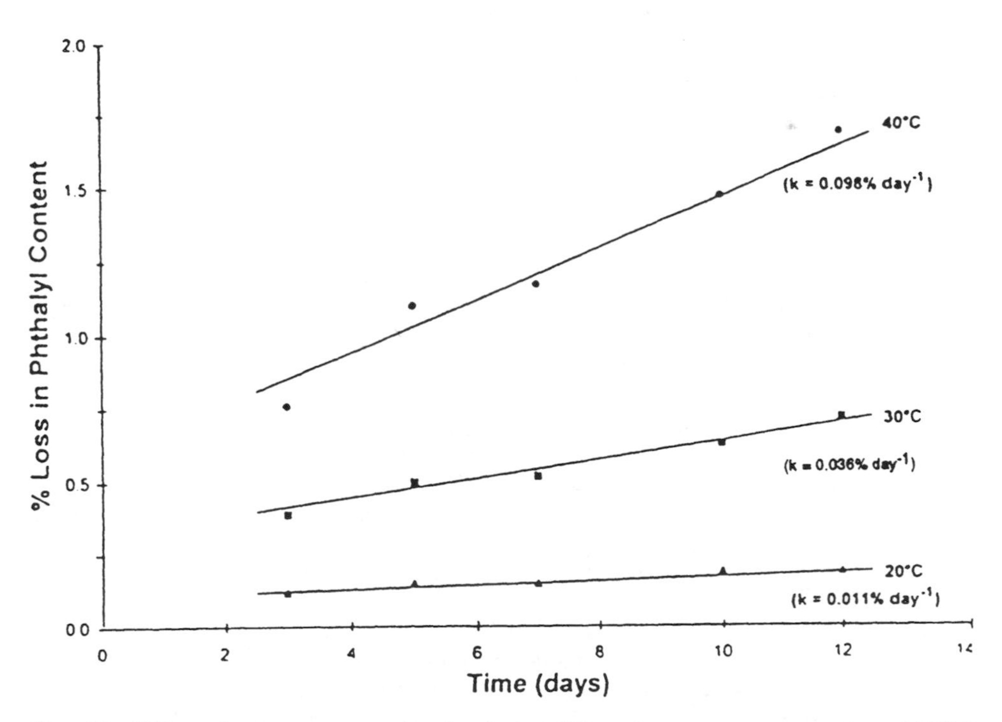

Fig. 30 Effect of temperature on the chemical stability of an aqueous suspension of PVAP (initial phthalyl content 57.6%). (From Ref. 106.)

of dispersions have very similar physicochemical and film-forming properties. Pseudolatices are mechanically produced colloidal dispersions of the polymer, whereas latices are prepared by emulsion polymerization. Another class of aqueous dispersion is the water-dispersible powder. These are prepared by spray- or freeze-drying of a latex or pseudolatex or by micronizing the polymer. Most of the aqueous dispersions require plasticization to produce a flexible film. Other additives are commonly used such as suspension or dispersion stabilizers and flux regulators. Flux regulators are used to alter water permeability and diffusion in the films or to alter drug diffusivity.

The aqueous dispersions described in this chapter are most commonly used in the pharmaceutical industry to coat tablets, nonpareil beads, and granules. The coatings are used for a variety of reasons, such as protecting coating (e.g., Aquacoat-ECD and Surelease), enteric coating (Aquateric-CPD), sustained-release coating (Aquacoat-ECD, Surelease, Eudragit), etc. These aqueous dispersions are also used as drug carriers, matrices, and vehicles for applications such as ophthalmic drug delivery systems.

Overall, because of the ease of handling, good processing characteristics, and good regulatory acceptance, aqueous polymeric dispersions have gained substantial acceptance in the pharmaceutical industry, and that acceptance and range of use in drug delivery in particular is expected to continue to grow in the years ahead.

REFERENCES

1. J. Vanderhoff, E. Bradford, H. Tarkowski, J. Shaffer, and R. Wiley, *Adv. Chem. Ser.*, 32:32 (1962).
2. J. W. Vanderhoff, Theory of colloids. In: *Pharmaceutical Dosage Forms: Disperse Systems* (H. A. Lieberman, M. M. Rieger, and G. S. Banker, eds.), Marcel Dekker, Inc., New York, 1996, p. 91.
3. N. Sutterlin, *Makromol. Chem. Suppl.*, 10/11:403 (1985).
4. R.-K. Chang, G. Iturriz, and C.-W. Luo, *Int. J. Pharm.*, 60:171 (1990).
5. M. E. Rowland and K. E. Springs, Dow Chemical Co., Midland, MI, U.S. Patent 5,387,635 (1995).
6. H. Ibrahim, C. Bindschaedler, E. Doelker, P. Buri, and R. Gurny, *Int. J. Pharm.*, 87:239 (1992).
7. E. Allemann, R. Gurny, and E. Doelker, *Int. J. Pharm.*, 87:247 (1992).
8. C. Bindschaedler, R. Gurny, and E. Doelker, U.S. Patent 4,968,350 (1990).
9. D. K. Mehra, C. I. Patel, and J. C. I. Bridges, FMC Corp., Philadelphia, PA, U.S. Patent 5,326,572 (1994).
10. S. C. Bagaria and N. G. Lordi, Research Corp, New York, NY, U.S. Patent 5,023,108 (1991).
11. O. Paeratakul, Ph.D. dissertation, University of Texas at Austin (1993).
12. R. Bodmeier and O. Paeratakul, *Int. J. Pharm.*, 103:47 (1994).
13. R. E. Dillon, E. B. Bradford, and R. D. Andrews, *Ind. Eng. Chem.*, 45:728 (1953).
14. C. Bindschaedler, R. Gurny, E. Doelker, and N. A. Peppas, *J. Coll. Interface. Sci.*, 108:75 (1985).
15. L. A. Seminoff and G. M. Zenter, Merck & Co. Inc., Rahway, NJ, U.S. Patent 5,126,146 (1992).
16. R. Bodmeier and H. Chen, *Drug Dev. Ind. Pharm.*, 19:521 (1993).
17. S. B. McCray and J. Glater, *ACS Symp. Ser.*, 281:141 (1985).
18. J. Glater and S. McCray, *Desalination*, 46:389 (1983).
19. K. D. Vos, F. D. Burris, and R. L. Riley, *J. Appl. Polym. Sci.*, 10:825 (1966).
20. R. S. J. Manley, *J. Polym. Sci., Polymer Physics Ed.*, 11:2303 (1973).
21. G. M. Zentner, G. S. Rork, and K. J. Himmelstein, *J. Control. Release*, 1:269 (1985).

22. A. G. Thombre, G. M. Zentner, and K. J. Himmelstein, *J. Membr. Sci.*, 40:279 (1989).
23. M. Kelbert and S. R. Bechard, *Drug Dev. Ind. Pharm.*, 18:519 (1992).
24. R. Bodmeier and O. Pearatakul, *Pharm. Res.*, 8:355 (1991).
25. L. E. Appel, J. H. Clair, and G. M. Zenter, *Pharm. Res.*, 9:1664 (1992).
26. C. Bindschaedler, R. Gurney, and E. Doelker, *J. Pharm. Sci.*, 76:455 (1987).
27. C. Bindschaedler, R. Gurny, and E. Doelker, *J. Pharm. Pharmacol.*, 39:335 (1986).
28. V. L. King and T. A. Wheatley, *ACS Symp. Ser.*, 520:80 (1993).
29. Cellulose esters-polymer characterization, *FMC Bull. CE-2*, PMC Corp., Philadelphia, 1987.
30. T. Kristmundsdottir and K. Invarsdottir, *J. Microencap.*, 11:633 (1994).
31. M. V. Cattaneo and T. M. Chang, *ASAIO Trans.*, 37:80 (1991).
32. S. B. Bhardwaj, A. J. Shukla, and C. C. Collins, *J. Microencap.*, 12:71 (1995).
33. P. Giunchedi, B. Conti, L. Maggi, and U. Conte, *J. Microencap.*, 11:381 (1994).
34. C. S. Chiao and J. C. Price, *J. Microencap.*, 11:153 (1994).
35. V. Compan, J. Garrido, J. A. Manzanares, J. Andres, J. S. Esteve, and M. L. Lopez, *Optometery and Vision Sci.*, 69:685 (1992).
36. P. D. Knick and J. W. Huff, *CLAO J.*, 17:177 (1991).
37. M. L. McDermott and J. W. Chandler, *Survey of Opthalmol.*, 33:381 (1989).
38. O. L. Sprockel, W. Prapaitrakul, and P. Shivanand, *J. Pharm. Pharmacol.*, 42:152 (1990).
39. G. D. Hiatt, Eastman Chemical Co., Rochester, NY, U.S. Patent 2,196,768 (1940).
40. A. M. Ortega, Ph.D. thesis, Purdue University West Lafayette, IN (1977).
41. M. B. Davis, G. E. Peck, and G. S. Banker, *Drug. Dev. Ind. Pharm.*, 12:1419 (1992).
42. M. B. Davis, Ph.D. thesis, Purdue University West Lafayette, IN (1983).
43. B. Pandyra, *Farm. Pol.*, 50:95 (1994).
44. E. J. McGinley and B. J. Tuason, FMC Corp., Philadelphia, PA, U.S. Patent 4,518,433 (1985).
45. O. Thioune, H. Fessi, D. Wouessidjewe, J. P. Devissaguet, and F. Puisieux, *STP Pharma. Sci.*, 5:367 (1995).
46. A. Nash, Pharmaceutical Suspensions. In: *Pharmaceutical Dosage Forms: Disperse Systems* (H. A. Lieberman, M. M. Rieger, and G. S. Banker, eds.), Marcel Dekker, New York, 1988, vol. 1.
47. N. A. Muhammad, W. Boisvert, M. R. Harris, and J. Weiss, *Drug Dev. Ind. Pharm.*, 18:1787 (1992).
48. H. C. Fessi, J. P. Devissaguct, F. Puisieux, and C. Thies, Centre National de la Recherche Scientifique, Paris, France, U.S. Patent, 5,118,528 (1992).
49. Hydroxypropyl methylcellulose acetate succinate Shin-etsu AQOAT, *Tech. Bull.*, Shin-Etsu Chemical Co., Chiyoda-ku, Tokyo, Japan (1996).
50. T. Nagai, F. Sekigawa, and N. Hoshi, Applications of HPMC and HPMCAS aqueous film coating of pharmaceutical dosage forms. In: *Aqueous Polymeric Coatings for Pharmaceutical Dosage Forms* (J. W. McGinity, ed.), Marcel Dekker, Inc., New York, 1997, p. 177.
51. G. A. Agyilirah and G. S. Banker, Polymers for enteric coating applications. In: *Polymers for Controlled Drug Delivery* (P. J. Tercha, ed.), CRC Press, Boca Raton, FL, 1991, p. 39.
52. G. S. Rekhi and S. S. Jambhekar, *Drug. Dev. Ind. Pharm.*, 21:61 (1995).
53. G. S. Banker and G. E. Peck, *Pharm. Tech.*, April 55 (1981).
54. T. A. Wheatly and C. R. Steuernagel, Latex emulsion for controlled drug delivery. In: *Aqueous Polymeric Coatings for Pharmaceutical Dosage Forms* (J. W. McGinity, ed.), Marcel Dekker, Inc, New York, 1997, p. 1.
55. S. C. Porter, *Pharm. Res.*, 4:67 (1980).
56. J. H. Guo, R. E. Robertson, and G. L. Amidone, *Pharm. Res.*, 10:405 (1993).
57. H. Nakagami, T. Keshikawa, M. Matsummura, and H. Tsukamoto, *Chem. Pharm. Bull.*, 39:1837 (1991).
58. Micronized EC N-10F, Shin-Etsu Chemical Co., Chiyoda-ku, Tokyo, Japan.

59. *Aqueous Polymeric Coatings for Pharmaceutical Dosage Forms*, (J. W. McGinity, ed.), Marcel Dekker, Inc., New York, 1997, p. 79.
60. R. Bodmeier and J. Wang, *Pharm. Res.*, 82:191 (1993).
61. C. M. Chang and R. Bodmeier, *Int. J. Pharm.*, 130:187 (1996).
62. S. Wei and G. S. Banker, Biocontrol Inc., Iowa City, IA, U.S. Patent 5,417,984 (1995).
63. S. Wei, V. Kumar, and G. S. Banker, *Int. J. Pharm.*, 142:175 (1996).
64. G. S. Banker and V. Kumar, Biocontrol Inc, Iowa City, IA, U.S. Patent 5,405,953 (1995).
65. G. S. Banker and V. Kumar, Biocontrol Inc, Iowa City, IA, U.S. Patent 5,414,079 (1995).
66. V. Kumar, Y. Yang, and G. S. Banker, College of Pharmacy, The University of Iowa, Iowa City, IA, Unpublished results.
67. E. L. Warrick, O. R., Pierce, K. E. Polamanteer, and J. C. Saam, *Rubber Chemistry and Technol.*, 52:437 (1979).
68. Y. W. Chien, *Novel Drug Delivery Systems: Fundamentals, Development, Concepts and Biomedical Assesments*, Marcel Dekker, New York, 1982.
69. L. C. Li and G. E.Peck, Silicon elastomer latex dispersions. In: *Aqueous Polymeric Coatings for Pharmaceutical Dosage Forms* (J. W. McGinity, ed.), Marcel Dekker, Inc., New York, 1997, p. 473.
70. A. S. Michaels, P. S. L. Wong, and R. P. M. Gale, *AIChE J.*, 21:1073 (1973).
71. R. Humcke-Bogner, J. C. Liu, and Y. Chien, *Int. J. Pharm.*, 42:199 (1988).
72. I. Yilgor and J. E. McGrath, Polysiloxane containing copolymers: A survey of recent developments. In: *Advances in Polymer Sciences: Polysiloxane Copolymers/Anionic Polymerization*, Springer-Verlag, New York, 1988.
73. W. Noll, *Chemistry and Technology of Silicones*, Academic Press, New York, 1986.
74. J. T. Woodward, M. C. Musolf, and J. P. Miller, Dow Corning, Midland, MI, U.S. Patent 5,358,900 (1987).
75. J. T. Woodward, M. C. Musolf, and P. J. Miller, Dow Corning, Midland, MI, U.S. Patent 5,310,572 (1994).
76. J. Cekada and D. R. Weyenberg, U.S. Patent 3,433,780 (1969).
77. L. C. Li and G. E. Peck, *Drug Dev. Ind. Pharm.*, 15:65 (1989).
78. L. C. Li and G. E. Peck, *Drug. Dev. Ind. Pharm.*, 15:499 (1989).
79. D. J. Huebner and J. C. Saam, Dow Corning, Midland, MI, U.S. Patent 4,584,341 (1986).
80. D. J. Huebner and J. C. Saam, Dow Corning, Midland, MI, U.S. Patent 4,568,718 (1986).
81. R. J. Kostelnik, *Polymeric Delivery Systems*, Gordon and Breach Publishers, New York, 1978.
82. G. Kallstrand and B. Ekman, *J. Pharm. Sci.*, 72:722 (1983).
83. L. C. Li and G. E. Peck, *Drug Dev. Ind. Pharm.*, 18:333 (1992).
84. M. C. Julienne, M. J. Alonso, J. L. G. Amoza, and J. P. Benoit, *Drug Dev. Ind. Pharm.*, 18:1063 (1992).
85. R. Bodmeier and J. W. McGinity, *Int. J. Pharm.*, 43:179 (1988).
86. K. Makin, H. Ohshima, and T. Konodo, *Chem. Pharm. Bull.*, 33:1195 (1985).
87. R. Gurny, N. A. Peppas, D. D. Harrington, and G. S. Banker, *Drug Dev. Ind. Pharm.*, 7:1 (1981).
88. H. Fessi, J.-P. Devissaguet, F. Puisieux, and C. Thies, French Patent 2,608,988 (1988).
89. M. D. Coffin, Ph.D. dissertation, University of Texas at Austin (1990).
90. S. E. Frisbee and J. W. McGinity, *Eur. J. Pharm. Biopharm.*, 40:355 (1994).
91. M. D. Coffin and J. W. McGinity, *Pharm. Res.*, 9:200 (1993).
92. H. Fessi, F. Puisieux, J. P. Devissaguet, N. Ammoury, and S. Benita, *Int. J. Pharm.*, 55:R1 (1989).
93. S. S. Gueterres, H. Fessi, G. Barratt, J.-P. Devissaguet, and F. Puisieux, *Int. J. Pharm.*, 113:57 (1995).
94. S. E. Frisbee, Ph.D. dissertation, University of Texas at Austin (1994).

95. S. E. Frisbee, M. D. Coffin, and J. W. McGinity, Properties of aqueous pseudolatex dispersions of biodegradable polymers. In: *Aqueous Polymeric Coatings for Pharmaceutical Dosage Forms* (J. W. McGinity, ed.), Marcel Dekker, Inc, New York, 1997, p. 441.
96. M. O. Omelczuk and J. W. McGinity, *Pharm. Res.*, 10:542 (1992).
97. M. O. Omelczuk and J. W. McGinity, *Pharm. Res.*, 9:26 (1993).
98. R. S. Okor, *Pharm. Res.*, 10:220 (1993).
99. K. Lehmann, *Acta Pharm. Technol.*, 32:146 (1986).
100. M. S. El-Samaligy, P. Rohdevald, and H. A. Mahmoud, *J. Pharm. Pharmacol.*, 38:216 (1986).
101. N. A.-K. Fallouh, L. Roblot-Treupel, H. Fessi, J. P. Devisaguet, and F. Puisieux, *Int. J. Pharm.*, 28:125 (1986).
102. G. Puglisi, G. Giammona, M. Fresta, B. Carlisi, N. Micalis, and A. Villari, *J. Microencapsul.*, 10:353 (1993).
103. S. P. Vyas, S. Ramachandraih, C. P. Jain, and S. K. Jain, *J. Microencapsul.*, 9:347 (1992).
104. H. S. Cockeram and S. A. Levine, *J. Soc. Cosmetic Chem.*, 12:316 (1961).
105. A. Sadato, W. Taki, Y. Ikada, I. Nakahara, K. Yamashita, K. Matsumoto, M. Tanaka, H. Kikuchi, Y. Doi, T. Noguchi, and T. Inada, *Neuroradiology*, 36:634 (1994).
106. K. L. Zak, Physicochemical Properties of Opadry, Sureteric®, and Surelease®. In: *Aqueous Polymeric Coatings for Pharmaceutical Dosage Forms* (J. W. McGinity, ed.), Marcel Dekker, Inc., New York, 1997, p. 373.

7

Polymeric Pharmaceutical Excipients

Joseph A. Ranucci

Roche Vitamins, Inc., Nutley, New Jersey

Irwin B. Silverstein

International Specialty Products, Wayne, New Jersey

I. INTRODUCTION

Polymers are a class of natural or synthetic chemicals that are characterized by having repetitive structural units. They are formed by a polymerization or depolymerization process, have a large molecular size (MW), and are composed of repetitive molecular groups. These compounds may be cross-linked and often contain substituted chemical groups along the polymer chain to modify their properties. Polymeric excipients are unique and have physical/chemical properties that differentiate them from other pharmaceutical excipients. This is the result of their large molecular size, which can hinder them from crossing biological membranes, resulting in a lack of systemic absorption.

The science of polymer chemistry began in the early part of the 20th century; methyl cellulose was first prepared in 1905 by Suida [1] and was followed by a patent issued for the preparation of sodium carboxymethylcellulose in 1918 [2]. Cellulose derived excipients are a family of polymers that have been used in the manufacture of pharmaceuticals since the 1960s as film formers for film-coating tablets and to a lesser extent as suspension and emulsion stabilizers. Presently the use of polymeric excipients in the manufacture of pharmaceuticals has expanded significantly and is attributed in large part to the availability of new and more versatile polymers and the benefits they offer the formulator. Polymer properties can be readily modified and improved, and are virtually unlimited due to their unique chemistry, which allows modification of their structure using the following variables:

- Monomer selection
- Number of repeating units (degree of polymerization)
- Use of different functional groups and degree of substitution
- Use of copolymers
- Degree of cross-linking

Beyond chemical synthesis, polymers can also be obtained by extraction from marine and terrestrial plants (e.g., alginates, and guar gum, respectively) or microbiologically (e.g., xanthan gum), or modified chemically from naturally obtained products (e.g., propylene glycol alginate, modified starches).

II. FUNCTION OF POLYMERS IN DISPERSE SYSTEMS

Polymeric excipients function to physically stabilize disperse systems, which can be described as one or more discontinuous phases homogeneously distributed in a continuous phase. The size of the dispersed phase (s) is used as a nonabsolute criterion in the classification of these systems. The sizes of dispersed phases are as follows:

- Suspensions, >1000 nm
- Emulsions, 50–1,000 nm
- Liposomes, 10–500 nm
- Nanoparticles, 10–100 nm
- Microemulsions, 10–75 nm

Disperse systems are metastable due to their nonhomogeneity and tend to equilibrate to a decreased energy state with time (e.g., destabilize via sedimentation, aggregation, or coalescence). Polymers can be used to stabilize these systems to increase their shelf life, consumer appeal, and effectiveness. They stabilize these systems (often in conjunction with polyelectrolytes to regulate zeta potential) by adsorbing at the interface of the dispersed phase, where they modify surface properties and reduce surface free energy. At the surface they prevent the close approach of dispersed phase particles, which can result in physical instability. Achieving particle separation of the dispersed phase is accomplished by electrical effects (zeta potential) and/or steric hindrance. A third stabilization mechanism uses the viscosity increasing effects of polymers in the continuous phase. The reader is referred to other chapters of this text series for a more complete and in depth discussion of this topic.

III. HISTORY

A. Regulatory History of Excipients

The process by which pharmaceutical excipients are deemed safe for use in formulations in the United States and other countries has been evolutionary, focusing mainly on determining that the new drug dosage product is safe and effective rather than evaluating the dosage form components singly. The new drug application (NDA) approval process evaluates the drug formulation, composed of active and excipient components, in animal and human clinical studies to determine its safety and efficacy. The FDA does not directly evaluate the safety and quality of the excipients in a formulation during clinical studies or the approval process. A prerequisite is that only safe and well-recognized excipients are included in a new formulation to be evaluated in humans in the NDA process. Acceptable excipients are used to avoid delays in the approval of the NDA due to the time required to review the safety of previously untried excipients (s); the NDA is not a suitable forum for evaluating the safety of excipients. The result is the absence of a de facto approval process for excipients and the inclusion of only previously used or well-recognized excipients in a NDA formulation. Regulatory Authorities can indirectly regulate excipients by using a provision of the Federal Food and Cosmetic Act in reviewing a NDA but have no mandate to evaluate excipients.

Excipients that may freely be used are termed "generally recognized as safe" (GRAS) and have a long proven history of safety in pharmaceuticals and foods. The use of non-GRAS excipients generally requires that they are officially sanctioned for use, have previously been approved in a similar formulation, or have delivery characteristics and potential reactivity and interactions that are well known and have been published in the literature [4] as well as a drug master file (DMF) that may be referenced by a regulatory authority [5] (see section on excipient DMFs later in this chapter). Determining the acceptability of excipients for international use further complicates the drug development and approval process due to a lack of consistency in different countries regarding excipient acceptability and uniform specifications for acceptance when an excipient is recognized internationally.

B. History of Excipient Use

Excipients used to manufacture the first commercial pharmaceutical formulations before 1938 were those that were safely used in apothecaries, pharmacies, or processed foods. Current governmental regulations require that all excipients used to formulate and manufacture a new dosage are listed. In 1983, the regulatory authorities attempted to clarify the use of acceptable excipients in new drug formulations by distributing a list of excipients previously approved for prescription products [6].

The excipient list is periodically updated and used by regulatory reviewers to assess the type of data needed for a particular excipient. The list provides the number of the last approved NDA containing the excipient and the dosage ranges. A published list of those excipients being used in NDAs is available to the public, and unofficial excipient lists have also been compiled in an effort to fill any information void [7].

Over a decade after the list of excipients that may officially be used was published, it has not been significantly enlarged, and though some new excipients have gained acceptance, the process still remains unchanged. The selection of excipients by a formulation scientist is still largely a static process in which excipient selection for a formulation is made from a list that is not increasing rapidly enough to meet and accommodate the new dosage form technologies presently being developed. In light of these constraints, excipient selection for a formulation must be judicious, balancing acceptability and safety with functionality and effectiveness in the dosage form. The process is further complicated and made more expensive when companies seek approval of their products in different countries, where the acceptability of these excipients may be viewed differently.

The recently formed International Pharmaceutical Excipients Council (IPEC), is a group of excipient manufacturers and product users formed in 1991. It has been actively seeking to initiate an official mechanism for new excipient approval that will allow sufficient expansion of the recognized excipient list to accommodate accelerating advances in the drug development process. They are also active in promoting international excipient harmonization by working with American, European, and Japanese government officials and pharmacopoeias. Their ultimate goal: harmonization of excipient specifications and other regulatory requirements in the international marketplace.

IV. SPECIAL REGULATORY CONSIDERATIONS FOR POLYMERS

Polymeric excipients are noted for their large molecular size, which often prevents their penetration of biological membranes and oral absorption. This proves to be an advantage that minimizes the toxicological effects of these compounds when orally administered.

The presence in polymers of low molecular weight impurities that may be absorbed does represent a potential source of toxicity. This aspect underscores the need to use only highly purified polymers for food and pharmaceutical applications. Since polymers are not chemically homogeneous, they often cannot be characterized as single chemical products. Rather, they represent a statistical distribution or combination of related structures (e.g., molecular weight distribution, side chain specificity, extent and location of cross-links, etc.).

In this context, an important property is the molecular weight distribution of a polymer, which, if skewed to the low molecular weight moieties, can result in oral absorption and possible adverse effects. Considering the difficulty in obtaining a consistent high-purity single-entity polymer, toxicological evaluations are performed on the most consistent, reproducible, and representative lot. To effectively ensure that the toxicological profile of future lots are the same as that tested earlier, all marketed batches must be representative and reproducible regarding the initially tested and approved material. They must also meet carefully selected specifications that effectively control product performance and safety.

The impurities present in all polymeric excipients must be characterized and minimized according to their toxicology and end-use application. In the event a trace impurity in a food excipient, such as a polymer, is determined to be carcinogenic, the FDA will not apply the Delaney Clause [8], but may apply a general safety clause of the act instead and request a *de minimis* standard.

V. POLYMER EXCIPIENT SAFETY TEST CONSIDERATIONS

Polymers first used in manufacturing pharmaceuticals had a high degree of safety and gained acceptability from initial animal toxicity and feeding studies. More recently, however, polymers, like any other excipient, require extensive acute and chronic toxicity studies in a number of different animal species as well as additional safety tests, such as teratogenicity, irritation, pharmacokinetics, mass balance, etc., before they may be approved for administration to humans.

Significant progress has recently been made by the International Pharmaceutical Excipients Council (IPEC) in the area of excipient safety by streamlining and unifying the toxicological criteria for testing newly developed excipients in determining their safety [9]. These toxicological guidelines now provide a standardized framework for evaluating new excipients and represent a more rational and scientific approach in the testing and safety requirements of these materials. This should result in improved test efficiencies and lower cost of the product. These protocols consider the route of administration of the excipient in the formulation as well as the anticipated exposure regarding the duration of treatment and the quantity of material ingested by the patient.

The IPEC *Guidelines* specify the need for background information prior to starting toxicological testing. This includes a literature review, definition of the chemical, a list of its physical properties, manufacturing process, product specifications (including impurities), projected exposure conditions, user population, and an assessment of pharmacological activity. This is followed by an assessment of an adequate set of animal toxicity data intended to demonstrate the safety of the excipient. The testing is designed to determine the effects in humans of acute and chronic exposure, mutagenicity, etc.

Polymeric excipients exhibiting satisfactory animal safety profiles are next evaluated for safety using single dosages in humans. This is followed by subchronic testing, embryo-fetal developmental effects, and in vitro and in vivo assays, after which the excipient is ready to be used in products intended for clinical trials.

The protocol specified in the IPEC *Guidelines* also requires additional tests depending on the route of administration as well as the nature of the patient dosing.

The low toxicological profile and high degree of safety exhibited by many polymeric pharmaceutical excipients (nonmetabolized) is due not only to their large molecular weight, which results in poor or no oral absorption, but also to the stringent purity requirements of these materials used for pharmaceutical applications. The toxicity exhibited by some polymers is often associated with low molecular weight, readily absorbed impurities, such as monomers, a catalyst, initiator residues, etc., that have not been effectively removed from the product during manufacture. It becomes obvious then that manufacturers must not only characterize the impurities in their product, but also determine their concentrations and toxicity in order to minimize their levels to safe and acceptable limits in the final product.

Finally, test substances that are used to perform toxicological studies must be representative in all respects of the product to be sold in commerce. This is particularly important for polymeric compounds whose composition is normally variable and therefore defined on a statistical distribution of molecular weight components and reactants.

VI. MANUFACTURING SPECIFICATIONS

An important requirement for polymers or any other excipient is that they are produced under well-controlled and validated conditions (current good manufacturing practice) to ensure that each manufactured lot is uniform, consistent within established parameters, and representative of all previously manufactured lots. This last point is essential in producing an excipient that functions safely and consistently in the products in which it is a component. Manufacturers' specifications represent minimal control criteria that are necessary to ensure consistent and reproducible product performance in a formulation. New specifications may be necessary for the same excipient when it is used for a different application (e.g., pyrogen-free for ophthalmic use, direct compression vs. encapsulation use, etc.).

VII. NEW EXCIPIENT AVAILABILITY

A look at the history of pharmaceutical polymeric excipients reveals that the introduction of safe, new polymers with unique properties has driven and broadened the technology of dosage delivery systems. The availability of new polymers provides an opportunity for designing new improved dosages to deliver drug substances to the body in a more predictive and efficient manner (e.g., controlled release oral, intramuscular, topical, ocular, and nasal delivery dosage forms).

The current paucity of acceptable *new* pharmaceutical excipients stems in large part from the extensive toxicity testing required prior to their being regarded as safe for use in a new formulation. An excipient manufacturer must first demonstrate excipient safety by performing extensive animal toxicological testing prior to making the excipient available to the pharmaceutical industry. After having been determined safe for use in drugs for human consumption, it must also prove economically attractive in its application. However, only after the polymer has been "accepted" in an approved NDA, is it "officially" sanctioned for that particular application. This limited "approval" does not signify, however, that it is "generally acceptable" for use in other formulations to be recognized as safe in humans. Under these conditions, most pharmaceutical companies are reluctant to include a totally new and previously untried excipient in a formulation for the following reasons:

- Long-term clinical testing may uncover undesirable effects that can jeopardize the product's approval.
- More "acceptable" excipients may be used if new excipients do not provide any novel advantages.
- The excipient manufacturer may determine that the excipient is not sufficiently profitable and discontinue marketing.
- The new excipient may exhibit excessive lot-to-lot variability that adversely impacts the product's performance.
- Additional time and cost are required in demonstrating excipient safety.
- There may be only one supplier.

VIII. GOOD MANUFACTURING PRACTICE

The issue of lot-to-lot variability of polymeric excipients leads to a discussion of the concept of current good manufacturing practice (GMP). Many polymeric excipients are produced by chemical companies that do not routinely practice current GMP principles. These companies are required to implement current GMP procedures in their facility for materials sold to pharmaceutical customers. The same considerations apply in the case of extracted and fermented polymers, which products' manufacturers may also supply to nonpharmaceutical customers, whose use of the product may be unregulated. Polymer manufacturers have recognized in the past that the regulatory authorities have not been stringent in applying GMP requirements to the manufacture of excipients. This ambiguity has changed, however, in recent years. Regulatory authorities have begun to broaden their interpretation of the application of the current GMP guidelines from finished pharmaceuticals to excipients as well as active ingredients.

In 1987 the FDA issued to their inspectors a *Guide to the Inspection of Bulk Pharmaceutical Manufacturers*. (Bulk pharmaceuticals include vitamins, drug substances, excipients, etc.) This document was updated in 1990 and has been used as guidance for the inspection of bulk pharmaceutical producers for evaluating current GMP compliance by the manufacturer. Polymer manufacturers have recognized, however, that these requirements are excessively costly and stringent for their purpose and are primarily directed at manufacturers' of bulk active pharmaceuticals made exclusively for the pharmaceutical industry. Polymeric as well as other excipients manufactured for the pharmaceutical industry often have significant material applications outside this industry, where GMP compliance is not required. More important, pharmaceutical polymer sales of many of these products are dwarfed by nonpharmaceutical applications. Therefore, excipient manufacturers are often reluctant to assign significant financial resources and commitment for pharmaceutical polymers because of costly and demanding product and production requirements.

The recently founded International Pharmaceutical Excipients Council recognized the need to more clearly define and elucidate the extent of GMP compliance requirements in the excipient industry and compiled a GMP guide in 1995 [10]. It is specifically for excipient manufacturers and more clearly defines the requirements for manufacturing excipients. The document, a collaborative effort between pharmaceutical users and excipient manufacturers, is explicit enough to ensure that producer compliance with the guide will meet user and regulatory authority requirements.

The IPEC guide is specific for excipient manufacturers and combines many of the requirements delineated in 21CFR 211 (*FDA Bulk Pharmaceutical Chemicals Inspection Guide*) as well as certain ISO 9000 (an international quality system for manufacture to a specification) requirements. The latter requirements are not specifically required for GMP conformance but were introduced to produce an up-to-date relevant document that was internationally acceptable. Since the chemical industry is familiar with, and in many instances compliant with, the ISO 9000 quality system standards for trade purposes, there is an added advantage for incorporating these principles into the IPEC guide.

The IPEC guide has several sections covering all aspects of excipient manufacturing, including general guidance, general auditing considerations, and discussion of excipient quality systems. In the last and most extensive section the requirements for GMP compliance are organized within the framework of the 19-part ISO 9002 standard requirements (as noted below, design control is not required for ISO 9002) which are as follows:

- Management responsibility indicates the requirements for management commitment to current GMP and ISO 9002.
- Quality System requires the development of a quality manual describing the quality system, the procedures for ensuring conformance to the quality system requirements, and the records to support compliance to the procedures.
- Contract review requires that the company ensures that customer contract requirements are met.
- Design control is not required by this standard, since it relates to new product development.
- Document and data control requires a system to ensure documents are properly approved and that only the most current are used.
- Purchasing ensures the proper acquisition of products and services.
- Control of customer supplied products is generally not applicable to excipient manufacturers, since it stipulates procedures to maintain customer-supplied ingredients to be properly used and kept confidential.
- Product identification and traceability mirrors the GMP requirement that all materials be properly identified and traceable both for raw materials and finished product.
- Process control also corresponds with a GMP objective, this one for process documentation and evidence of adherence.
- Inspection and testing reflects the FDA desire for documented testing of all materials.

- Control of inspection, measuring, and test equipment specifies preventive maintenance and calibration of equipment critical to the manufacture of the excipient.
- Inspection and test status requires that the status of all materials be unequivocally established.
- Control of nonconforming product is a quality unit responsibility for identification and disposition.
- Corrective and preventive action covers investigation of complaints, process failures, and service failures as well as their prevention.
- Handling, storage, packaging, preservation, and delivery requires procedures to ensure the product does not deteriorate or is otherwise adversely impacted after manufacture, and is delivered according to the customer contract requirements.
- Control of quality records demands that all documents stipulated by the quality system be properly maintained and available for review.
- Internal quality audits complements GMP requirements for self-assessment.
- Training covers not only GMP but all work-related skills such as safety, technical, etc.
- Servicing technical assistance generally does not apply to excipient manufacture unless covered by contractual requirements.
- Statistical techniques are required to be used for the optimization of process capability.

Thus, the GMP requirements have been complemented with ISO 9000 to define an internationally harmonized quality system for excipient manufacturers that will improve product uniformity and customer satisfaction by conformance with GMP and ISO 9002.

Adherence to the IPEC guidelines, while not a regulatory requirement, will ensure GMP compliance, minimize process variability, and increase product uniformity. It will also ensure that the facility is in compliance with the GMP standard regarding record keeping and procedural requirements. Finally, conforming to the guidelines will assure the regulatory authorities and the customer that manufacturing controls are in place, and that the product satisfactorily adheres to current GMP requirements.

IX. EXCIPIENT DRUG MASTER FILES (DMF)

While conformance of excipient manufacture to current good manufacturing practice regulations is mandatory, the establishment of a drug master file with the regulatory authorities is not. A manufacturer of a new excipient often files a DMF as a service to his customers and then authorizes the regulatory authorities to reference the DMF when submitting an IND, NDA, or ANDA [11]. This enables the FDA reviewer to reference the excipient DMF instead of having the new drug application filer provide the necessary information in their filing.

There are five types of drug master file, but only type IV, excipients, colorants, flavors, essences, and other additives are applicable to this discussion.

A DMF has several sections, namely a manufacturing process, release specifications, test methods, an environmental assessment, and product stability. The most extensive section is the one describing the manufacturing process. This section often contains proprietary manufacturing details from the raw material specifications to the finished product testing, including descriptions of the buildings, facilities, and equipment. Other information, which is often included, is the drug establishment registration number, if applicable, organization and personnel, toxicity studies, and product brochures.

One should remember that the regulatory authorities do not approve the excipient or the manufacturing process as detailed in the drug master file. An FDA reviewer will report any deficiencies in the information provided to the DMF holder and will also inform the new drug application filer of the need for additional information [12]. The DMF holder must submit the requested information to allow the review to continue. DMF holders are obligated to review the document annually and provide the regulatory authority with updates where appropriate.

X. CONCLUSION

Polymers are used extensively in the formulation and manufacture of dispersed pharmaceutical dosages, because they satisfy a number of unique needs (e.g., lowering surface tension, thickening, stabilization, etc.). Their wide-ranging physical/chemical properties, lack of reactivity, taste, and irritation make them preferred excipients for formulating dispersed dosage forms as well as advanced dosage delivery systems, such as topical, oral, and nasal controlled-release dosages.

The criteria for parenterally administered polymers are considerably different than those used for nonparenteral applications. Polymers for parenteral use must be pyrogen free, have sufficiently low molecular weight to allow elimination from the body, and, if of such a high molecular weight that will not allow excretion, they must be biodegradable. In the case of biodegradable polymers, the polymer itself must be nontoxic and nonirritating, as the biodegraded moieties must be also.

Polymers are regulated for use in pharmaceuticals just as any other excipient. The requirements for human pharmaceutical consumption are that the polymer must have prior approval for use in a pharmaceutical dosage form or in food. Absent this requirement, the IND applicant must justify the safety of the polymer in the formulation by using extensive animal toxicological studies. This process presents a dilemma for manufacturers who wish to use a new polymer that has no prior approval or previous inclusion in an approved NDA formulation. It presents the user of a new polymer with an uncertain undertaking regarding the NDA approval process. This results in companies not using new polymers without a proven track record unless they are absolutely necessary. In these rapidly moving times of fast identification of new drug entities and quick-to-market pharmaceuticals, a new avenue is needed for gaining rapid approval of safe new polymers. One proposal to resolve this situation would be to place the acceptance of new pharmaceutical excipients with the United States Pharmacopeia (USP), an independent unofficial group that presently oversees specifications and analytical methods for approximately 360 excipient monographs [13].

Enlarged membership of the existing Pharmaceutical Excipients Subcommittee would focus on the toxicological and safety aspects of new excipients. This two-tiered approach to excipient regulation has precedent in the Joint Expert Committee on Food

Table 1 Compendium Listings

U.S. Pharmacopeia/ NF	Food Chemical Codex	PhEUR	Japanese Pharm Pharmacopeia
Carbomer	Not listed	Not listed	Not listed
Carbomer 1342	Not listed	Not listed	Not listed
Carboxymethyl cellulose sodium	Sodium carboxy methylcellulose	Cellulosum natricum	Carboxymethyl carmellose sodium
Carrageenan	Carrageenan	Not listed	Not listed
Crospovidone	Polyvinylpoly pyrrolidone	Crospovidone	Not listed
Ethylcellulose	Ethyl cellulose	Ethylcellulosum	Not listed
Guar gum	Guar gum	Not listed	Not listed
Hydroxyethyl cellulose	Not listed	Hydroxyethyl cellulosum	Not listed
Hydroxypropyl cellulose	Hydroxypropyl cellulose	Hydroxypropyl cellulosum	Hydroxypropyl cellulose
Hydroxypropyl methylcellulose	Hydroxypropyl methylcellulose	Methylhydroxy propyl cellulosum	Hydroxypropyl methyl cellulose
Lactide-glycolide polymers[a]	Not listed	Not listed	Not listed
Methylcellulose	Methylcellulose	Methyl cellulosum	Methyl cellulose
Microcrystalline cellulose/CMC	Not listed Sodium	Not listed	Not listed
Pectin	Pectin	Not listed	Not listed
Poloxamer	Poloxamer	Not listed	Not listed
Polycarbophil	Not listed	Not listed	Not listed
Polyethylene glycols	Polyethylene glycols	Macrogolum	Macrogol
Polyethylene glycol monomethyl ether	Not listed	Not listed	Not listed
Poly(ethyleneoxide)	Not listed	Not listed	Not listed
Polyoxyl 35 castor oil	Not listed	Polyoxyl 35 castor oil	Not listed
Polyoxyl 40 hydrogenated castor oil	Not listed	Polyoxyl 40 hydrogenated castor oil	Not listed
Polyvinyl alcohol	Not listed	Not listed	Not listed
Povidone	Polyvinyl pyrrolidone	Povidone	Povidone
Propylene glycol alginate	Propylene glycol alginate	Not listed	Not listed
Sodium alginate	Sodium alginate	Not listed	Not listed
Xanthan gum	Xanthan gum	Not listed	Not listed

[a]Absorbable surgical suture.

Table 2 Ionic Charges

Compendial name	Chemical name	Charge	CAS no.
Carbomer 934	High MW cross-linked acrylic acid	Anionic	9007-16-3
Carbomer 934P	High MW cross-linked acrylic acid	Anionic	9003-01-4
Carbomer 940	High MW cross-linked acrylic acid	Anionic	76050-42-5
Carbomer 1342	High MW cross-linked acrylic acid	Anionic	Not available
Carboxymethyl cellulose sodium	Cellulose carboxymethylether sodium salt	Anionic	9004-32-4
Carrageenan	Sulfated polysaccharide	Anionic	9000-07-1
Crospovidone	Crosslinked 1-ethenyl-2-pyrrolidinone homopolymer	Nonionic	9003-39-8
Ethylcellulose	Cellulose, ethyl ether	Nonionic	9004-57-3
Guar gum	Galactomannan polysaccharide	Nonionic	9000-30-0
Hydroxyethyl cellulose	Cellulose, 2-hydroxyethyl ether	Nonionic	9004-62-0
Hydroxypropyl cellulose	Cellulose, 2-hydoxypropyl ether	Nonionic	9004-64-2
Hydroxypropyl methylcellulose	Cellulose, 2-hydroxypropyl methyl ether	Nonionic	9004-65-3
Lactide-glycolide polymers	Not applicable	Anionic	Not available
Methylcellulose	Cellulose, methyl ether	Nonionic	9004-67-5
Microcrystalline cellulose/CMC Sodium	Microcrystalline cellulose/carboxymethyl cellulose sodium	Anionic	9004-34-6 9004-32-4
Pectin	Partial methylesters of polygalacturonic acid	Anionic	9000-69-5
Poloxamer	Oxirane, methyl-, polymer with oxirane	Nonionic	9003-11-6

(*continued*)

Table 2 Continued

Compendial name	Chemical name	Charge	CAS no.
Polycarbophil	Polyacrylic acid cross-linked with divinyl glycol	Nonionic	9003-97-8
Polyethylene glycol	Polyethylene glycol	Nonionic	25322-68-3
Polyethylene glycol monomethyl ether	Methoxypolyethylene glycol	Nonionic	9004-74-4
Polyethylene glycol 660 hydroxystearate	Polyethylene glycol 660 12-hydroxystearate	Nonionic	Not available
Poly(ethylene oxide)	Polyethylene oxide	Nonionic	25322-68-3
Polyglycerol esters of fatty acids	Not applicable	Nonionic	See monograph
Polyoxyl 35 castor oil	Polyethoxylated castor oil	Nonionic	61791-12-6
Polyoxyl 40 hydrogenated castor oil	Polyethoxylated hydrogenated castor oil	Nonionic	61791-12-6
Polyvinyl alcohol	Ethenol, homopolymer	Nonionic	9002-89-5
Povidone	2-pyrrolidinone, 1-ethenyl-homopolymer	Nonionic	9003-39-8
Propylene glycol alginate	Propane-1,2-diol alginate	Nonionic	9005-37-2
Sodium alginate	Alginic acid, sodium salt	Anionic	9005-38-3
Xanthan gum	Polyanionicheteropoly saccharide	Anionic	11138-66-2

Table 3 CFR References

Compendial name	Chemical name	CFR reference	Status
Carbomer 934, 934P, 940, and 1342	High MW cross-linked polyacrylic acid	173.34	Indirect food additive
		175.105	
		175.300	
		175.320	
		176.170	
		176.180	
		176.200	
		177.1210	
		177.2260	
Carboxymethyl cellulose sodium	Cellulose carboxymethyl ether Na salt	173.310	Indirect food additive
		175.105	
		182.70	GRAS
		182.1745	
Carrageenan	Carrageenan	172.620	Direct food additive
		172.623	
		172.626	
		182.7255	GRAS
Crospovidone	cross-linked 1-ethenyl-2-pyrrol idinone homopolymer	173.50	Direct food additive
Ethylcellulose	Cellulose, ethyl ether	73.1	Direct food additive
		73.1001	
		172.868	
		182.90	
		573.420	
Guar gum	Galactomannan polysaccharide	184.1339	GRAS

(continued)

Table 3 Continued

Compendial name	Chemical name	CFR reference	Status
Hydroxyethyl cellulose	Cellulose, 2-hydroxyethyl ether	175.105 175.300 176.170 176.180 177.1210	Indirect food additive
Hydroxypropyl cellulose	Cellulose, 2-hydoxy-propyl ether	172.870	Direct food additive
Hydroxypropyl methylcellulose	Cellulose, 2-hydroxy propyl methyl ether	172.874 175.105	Direct food additive Indirect food additive
Lactideglycolide polymers	Not applicable	No reference	No listing
Methylcellulose	Cellulose, methyl ether	182.1480	GRAS
Microcrystalline cellulose/CMC sodium	Microcrystalline cellulose/carboxy methyl cellulose sodium	182.1745	GRAS
Pectin	Pectin	184.1588	GRAS
Poloxamer	Oxirane, methyl polymer with oxirane	No reference	No listing
Polycarbophil	High MW cross-linked poly-acrylic acid	See Carbomer	See Carbomer
Polyethylene glycols	Polyethylene oxide	172.770 172.820 175.300 175.380 175.390 176.170 176.180 177.1210 177.1350 178.3750	Direct food additive Indirect food additive

Polyethylene glycol monomethyl ether	Methoxy polyethylene glycol	No reference	No listing
Polyethylene glycol 660 hydroxystearate	Not applicable	176.170 176.180	Indirect food additive
Poly (ethylene oxide)	Poly(oxyethylene)	175.300 175.380 175.390 177.1210 177.1350	Indirect food additive
Polyglycerol esters of fatty acids	Not applicable	166.110 169.115 169.140 169.150 172.854 172.860	Direct food additive
Polyoxyl 35 castor oil	Polyethoxylated castor oil	No reference	No listing
Polyoxyl 40 hydrogenated castor oil	Polyethoxylated hydrogenated castor oil	No reference	No listing
Polyvinyl alcohol	Ethenol, homopolymer	73.1 175.105 175.300 175.320 176.170	Indirect food additive
Povidone	2-pyrrolidinone, 1-ethenyl homopolymer	73.1 73.1001 172.210 173.55 175.300 176.170	Direct food

(*continued*)

Table 3 Continued

Compendial name	Chemical name	CFR reference	Status
Propylene glycol alginate	Hydroxypropyl alginate	172.210	Direct food additive
		172.858	
		173.170	
		173.340	
Sodium alginate	Alginic acid, sodium salt	150.141	GRAS
		150.161	
		173.310	
		184.1133	
		184.1187	
		184.1610	
		184.1724	
Xanthan gum	Polysaccharide B-1459	172.695	Direct food additive
		176.170	

Additives of the FAO/WHO (JECFA), which subdivides the review of specifications and toxicology of food additives by using this approach.

XI. INTRODUCTION TO EXCIPIENT MONOGRAPHS

A comprehensive, easy-to-use overview of polymeric excipients for disperse systems is provided below preceded by a section of typical formulations. The monographs are presented in alphabetical order by the U.S. Pharmacopeia name wherever possible. They list some important facts for research scientists and regulatory managers. While the listings are not totally inclusive, the goal was to present currently recognized and used polymeric excipients. Where available, residual solvents and their concentrations have been listed to provide the reader with information that is timely and is presently being reviewed by some pharmacopeial bodies. Included in the chapter are some polyethoxylated excipients that, while not truly polymeric, are very important in the manufacture of many disperse systems. The reader is urged to verify the regulatory status of excipients in this chapter for their specific application and determine that only approved grades and manufacturers are used.

There are tables provided for quick reference listing for each excipient presented herein, their compendia listing (Table 1), ionic charges (Table 2), Code of Federal Regulation (CFR) references that add to major listings in the monographs for easy access (Table 3), trade names, and suppliers (Table 4). The tables are followed by sections for each excipient presented alphabetically according to the USP compendial name in the order listed in Table 1. The glossary at the end of the chapter defines some terms and acronyms.

XII. FORMULATIONS

A. Antacid Suspension (Provided by FMC Corp., Philadelphia, PA)

Component	Percent
Aluminum hydroxide compressed gel (Type F-500)	24.46
Magnesium hydroxide	12.94
Sorbitol 70% solution	5.00
Methyl paraben	0.10
Propyl paraben	0.01
Avicel RC-591	0.90
Xanthan gum	0.10
Deionized water	56.49

Directions

1. Heat the total volume of water, add and dissolve the preservatives with mixing.
2. Cool step 1 and add Avicel RC-591 with mixing until dispersed.
3. Slowly add the xanthan gum to step 2 and mix until dissolved.
4. To step 3 add sorbitol solution with mixing.
5. Slowly add the aluminum hydroxide to step 4 and mix until uniformly dispersed.
6. Add the magnesium hydroxide slowly to step 5 and mix until uniformly dispersed.
7. Bring to volume with deionized water and mix to uniformity.

B. Ampicillin Trihydrate Reconstitutable Suspension (Provided by FMC Corp., Philadelphia, PA)

Component	Percent
Ampicillin trihydrate	5.77
Sucrose	35.00
Avicel CL-611	2.25
Potassium sorbate	0.10
Sodium citrate	0.91
Citric acid	0.09
Xanthan gum	0.20
FD&C 40 red	0.01
Deionized water qs ad	100.0 mL

Directions

1. Pass the ampicillin trihydrate through a #50 U.S. standard sieve.
2. Dry blend the dye, preservative, and buffers, and mill to uniformity in a Wiley® mill.
3. Dry blend the Avicel CL-611 with ampicillin to a uniform mix.
4. Add the xanthan gum to step 3 and blend to uniformity.
5. Add 50% of the sucrose to step 4 and blend to uniformity.
6. Combine step 2 and step 5, blend, add the balance of the sucrose, and mix to uniformity.

Reconstitution: 75 mL of water added to 44.33 g of powder for reconstitution yields 100 mL of suspension.

Table 4 Trade Names and Suppliers

Excipient	Trade name	Manufacturer[a]
Carbomer 934, 934P, and 940	Carbopol 934, 934P, and 940	BFGoodrich Co.
Cabomer 1342	Pemulen	BFGoodrich Co.
Carboxymethylcellulose sodium	Natrosol, Blanose	Hercules, Inc.
Carrageenan	Genu Carrageenan	Hercules Inc.
	Stamere	Meer Corp.
Crospovidone	Polyplasdone	ISP Technologies Inc.
	Divergan, Kollidon CL	BASF Corp.
Ethylcellulose	Ethylcellulose	Hercules Inc.
	Ethocel	Dow Chemical Co.
Guar gum	Supercol	Aqualon Co.
	Guar Gum	Meer Corp.
	Merezan	Guinness Chemical
Hydroxyethyl cellulose	Natrosol	Aqualon Co.
	Cellosize	Union Carbide Corp.
	Bermocol	Berol Inc.
	Tylose	Hoechst Inc.

Table 4 Continued

Excipient	Trade name	Manufacturer[a]
Hydroxypropyl cellulose	Klucel	Aqualon Co.
Hydroxypropyl methylcellulose	Methocel E, F, K	Dow Chemical Co.
	Metolose, Pharmacoat	Shin-Etsu Chemical Co Ltd. (Biddle Sawyer)
	Tylose	Hoechst Inc.
	Benecel	Aqualon Corp.
	Culminal	Henkel Inc.
	Celocol HPM	British Celanese Co.
Lactide-glycolide polymers	Medisorb	Alkermes
Methylcellulose	Methocel A	Dow Chemical Co.
	Benecel	Aqualon Co.
	Metolose	Shin-Etsu Chemical Co. Ltd. (Biddle Sawyer)
	Culminal	Henkel Inc.
	Tylopur	Hoechst Inc.
	Celacol	British Celanese Co.
Microcrystalline cellulose/CMC sodium	Avicel RC	FMC Corp.
Pectin	Genu Pectin	Hercules Inc.
	Pectin	Meer Corp.
Poloxamer	Pluronic	BASF Corp.
	Lutrol	BASF PLC
Polycarbophil	Noveon	B.F. Goodrich Co.
Polyethylene glycol	Carbowax	Union Carbide Corp.
	Lutrol	BASF PLC
	Hodag	Calgene Inc.
Polyethylene glycol monomethyl ether	Carbowax Sentry	Union Carbide Corp.
Polyethylene glycol 660 hydroxystearate	Solutol HS 15	BASF Corp.
Poly (ethylene oxide)	Polyox WSR	Union Carbide Corp.
Polyglycerol esters of fatty acids	Caprol	Abitec Corp.
	Hodag	Calgene Inc.
Polyoxyl 35 castor oil	Cremophor EL	BASF Corp.
Polyoxyl 40 hydrogenated castor oil	Cremophor RH 40	BASF Corp.
Polyvinyl alcohol	Airvol	Air Products
	Elvanol	DuPont
Povidone	Plasdone	ISP Technologies Inc.
	Kollidon	BASF Corp.
Propylene glycol alginate	Kelcoloid	Kelco Corp.
	Pronova	Protan Ltd.
	Propylene glycol alginate	Meer Corp.
Sodium alginate	Kelcosol	Kelco Corp.
	Pronova	Protan Ltd.
	Sodium alginate	Meer Corp.

(*continued*)

Table 4 Continued

Excipient	Trade name	Manufacturer[a]
Xanthan gum	Keltrol	Kelco Corp.
	Rhodigel	Rhone-Poulenc Corp.
	Merezan gum	Meer Corp.

[a]Manufacturers' locations are as follows: Abitec Corp., Columbus, OH; Air Products and Chemicals Inc., Allentown, PA; Alkermes, Blue Ash, OH; Aqualon, Wilmington, DE; BASF Corp., Mount Olive, NJ and Germany; BASF PLC, England; Berol Inc., Sweden; BFGoodrich Co., Cleveland OH; British Celanese Co., England; Calgene Inc., Skokie, IL; Dow Chemical Corp., Midland, MI; DuPont, Wilmington, DE; FMC Corp., Philadelphia, PA; Guinness Chemical, Ireland; Henkel Inc., Germany; Hercules Inc., Wilmington, DE; Hoechst Inc., Germany; ISP Technologies Inc., Wayne, NJ; Kelco Corp., San Diego, CA; Meer Corp., North Bergen, NJ; Mendell Inc., Patterson, NY; Protan Ltd., England; Rhone-Poulenc Corp., Cranbury, NJ; Shin-Etsu Chemical Co., Japan, Biddle Sawyer, NY; and Union Carbide Corp., Danbury, CT.

C. Kaolin and Pectin Suspension (Provided by FMC Corp., Philadelphia, PA)

Component	Percent
Kaolin	19.00
Pectin	0.40
Glycerin	2.00
Sodium benzoate	0.05
Potassium sorbate	0.05
Avicel RC-591 (FMC Corp.)	0.90
Xanthan gum (Keltrol F)	0.10
Deionized water	77.50

Directions

1. To 94% of the water, slowly add the Avicel RC-591 with mixing until dispersed.
2. Add the xanthan gum slowly to step 1 with mixing until dissolved.
3. Weigh and blend the pectin with an equal quantity of kaolin.
4. Heat step 2 to 70–80°C; add the kaolin/pectin blend from step 3; mix until dispersed and remove from heat.
5. To step 4, while cooling, add the balance of the kaolin and mix until thoroughly dispersed.
6. Add glycerin to step 5 and mix to uniformity.
7. To the remaining 6% water, add and dissolve the sodium benzoate and potassium sorbate.
8. Add the preservative solution in step 7 to step 6 with mixing, bring to volume with water, and mix to uniformity.

D. Aqueous Vitamin a Palmitate solution (150,000 i.u./mL) (Provide by BASF Corp., Mount Olive, NJ)

Component	Quantity
Vitamin A palmitate 1.7 million i.u./g	8.8 g
Cremophor RH40	25.0 g
Water qs ad	100.0 mL

Directions

1. Mix the Vitamin A palmitate with Cremophor RH40 at 60–65°C.
2. Heat the water to 60–65°C and add *slowly* to step 1 with thorough mixing. Note: The first half of the water produces a viscous solution and further addition reduces the viscosity.

E. Vitamin A, D, and E Emulsion (Provided by BASF Corp., Mt. Olive, NJ)

Component	Quantity
Vitamin A propionate 2.5 million i.u./g	23.0 g
Vitamin D3 40 million i.u./g	0.2 g
Vitamin E acetate	5.5 g
Solutol® HS 15 (BASF Corp., polyethylene glycol 660 hydroxystearate)	15.0 g
Butyl hydroxytoluene (BHT)	1.5 g
Benzyl alcohol	1.0 g
Water for injection qs ad	100.0 mL

Directions

1. Add and disperse vitamins A, D, and E in Solutol® at 60°C.
2. Add and dissolve BHT in benzyl alcohol and add slowly to water.
3. Add step 2 slowly to step 1 at 60°C with mixing and bring to volume with water.
4. Mix to uniformity.

F. Ampicillin/Amoxicillin Trihydrate Reconstitutable Powder (Provided by BASF Corp., Mt. Olive, NJ)

Component	Quantity
Ampicillin trihydrate or amoxicillin trihydrate	5.0 g
Sodium citrate (fine powder)	5.0 g
Citric acid	2.1 g
Sodium gluconate	5.0 g
Sorbitol	40.0 g
Kollidon® CL-M (BASF Corp., Crospovidone NF)	6.0 g
Orange flavor powder	1.5 g
Lemon flavor powder	0.5 g
Saccharin sodium	0.4 g
Deionized water qs ad	100.0 mL

Reconstitution with deionized water yields 250 mg active ingredient per 5 cc.

G. Anti-Inflammatory Cream (Provided by BFGoodrich Corp., Cleveland, OH)

Component	% by Weight
Part A	
Carbopol® 1382 resin (BFGoodrich)	0.70
Purified water	25.00

Component	% by Weight
Part B	
Betamethasone dipropionate (microfine powder)	0.05
Mineral oil	15.00
Span® 80	0.50
Methyl paraben	0.10
Propyl paraben	0.01
Part C:	
Tris amino (Trihydroxyamino methane) to pH 6–7 and desired thickness	qs
Deionized water qs ad	100.00

Directions

1. Mix the Carbopol in the deionized water by submerging the impeller until it is very close to the bottom of the vessel. Angle the impeller to generate a vortex that is 1–1.5 impeller diameters. Slowly sift the resin into the vortex of the rapidly agitating liquid (about 800–1500 rpm). Increase the agitation as the viscosity of the dispersion increases to maintain a vortex.
2. Reduce the agitation to 400–600 rpm and reposition the mixer to a standard vertical position to avoid or minimize air entrapment. Continue the agitation for about 20 min, or until a lump-free dispersion is attained.
3. Separately combine the betamethasone dipropionate, Span 80, methyl paraben, and propyl paraben in the mineral oil and disperse well.
4. Add this mixture to the Carbopol dispersion under rapid agitation.
5. Reposition the mixer to a standard vertical position to avoid or minimize air entrapment.
6. Neutralize with the Tris amino and continue mixing at a speed of 1000 rpm for 15–20 min or until the polymer has swelled to produce a smooth product.

H. Liposome Moisturizer with Pemulen TR-1 & Carbopol Ultrez 10 (Provided by BFGoodrich Corp., Cleveland, OH)

Component	Percent
Part A	
Pemulen® TR-1 (BFGoodrich)	0.25
Carbopol® Ultrez 10 (BFGoodrich)	0.20
Octyl stearate	8.00
Mineral oil	10.00
Part B	
Deionized water	75.55
Glycerin	2.00
Part C	
Sodium hydroxide (18%)	0.50
Part D	
Parabens (Nipa, Wilmington, DE)	0.5
Part E	
Lecithin and evening primrose oil liposomes (Brooks Industry, So. Plainfield, NJ)	3.00

Directions

1. Combine the ingredients in part A using high shear mixing in a sufficiently large container to hold the final product.
2. Combine ingredients in part B and mix to uniformity.
3. Add ¼ of Part B to part A slowly with good mixing for 15 min to swell resins.
4. When the emulsion in step 3 is smooth and white, add part of the sodium hydroxide solution (to pH of ~7.0) part C and mix to uniformity.
5. Slowly add the balance of part B with moderate mixing.
6. Add the balance of the sodium hydroxide solution.
7. Add the parabens part D and mix to uniformity.
8. Add part E using slow agitation to avoid rupturing the liposomes and mix to uniformity.

XIII. COMPENDIAL LISTINGS

Carbomer

Structure:

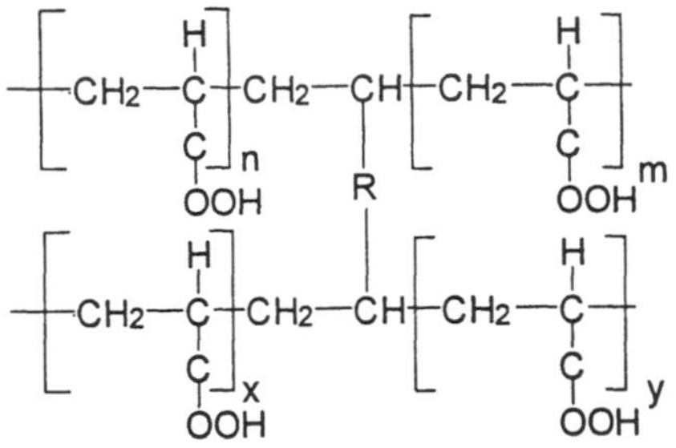

R=Diallyl (Pentaerythritol or sucrose)

Description: High molecular weight polyacrylic acid polymers cross-linked with polyalkenyl polyethers (iallylsucrose or iallylpentaerythritol).

CAS Number:		
	Carbopol 934	9007-16-3
	Carbopol 934P	9003-01-4
	Carbopol 940	76050-42-5

Generic nomenclature: Carbomer, carboxypolymethylene, carboxyvinyl polymer.

Appearance: Fluffy, white, hygroscopic powder having a mild acetic acid odor.

Solubility: The free acid is not truly soluble but swells in water when neutralized with sodium or potassium hydroxide to produce viscous, colloidal dispersions.

Pharmacopeial listing: NF, JSPI, JPE. (Umbrella monographs are in the process of being submitted covering all carbomer products in the United States, the United Kingdom, Germany, France, and Japan).

Applications: Topical 934 NF, JSPI (JPE), BP. Oral/topical 934P NF JSPI (JPE), BP. (934P is the pharmaceutical grade of 934.) Oral/topical 971P NF, BP. (Meets all standards for 941 but is manufactured using ethyl acetate. The USP is presently being petitioned to change the labeling notation for 941NF in the current monograph to accommodate this new oral grade by defining acceptable residual solvents and levels.) Topical 941 NF, JSPI (JPE), BP. Oral/topical 974P NF, BP. (Meets all NF standards for 934P but is manufactured using ethyl acetate.) Topical 940P NF, JSPI (JPE), BP.

Regulatory: Carbomer 934, 934P, 940, and 941 Drug Master File #153. Carbomer 971P, 974P Drug Master File #7170. Carbomer 934 and 974P Drug Master Files are currently being filed in the United Kingdom.

Health and safety: Residual Acrylic Acid Test (BFGoodrich Standard Test Procedure 005, Nov. 1994).

Residual Solvents

Carbomer	Concentration
Carbopol 934, 941	0.2% max benzene
Carbopol 934P	100 ppm max benzene
Carbopol 971P	0.5% max ethyl acetate
Carbopol 974P	0.7% max ethyl acetate

Acute oral toxicity: LD_{50} Carbopol 934 (rats, mice, guinea pigs, and dogs) range from 2.5–40g/kg [14].

Subacute/subchronic oral toxicity: Carbomer 934P administered to rats at dietary levels in daily doses of 0 or 5% for 21 days demonstrated no adverse effects [14]. Carbomer 934 administered to rats at dietary levels of 5 gm/kg for 49 days demonstrated no adverse effects [14]. Carbomer 974P (similar toxicity is expected with Carbopol 971P), administered to rats and dogs at dietary levels of 0–50,000 ppm for at least 13 weeks demonstrated no adverse effects [15].

Chronic oral toxicity: Carbomer 934 administered to rats at dietary levels of 0–5.0% w/w for 14 months demonstrated no adverse effects [15]. Carbomer 934 administered to dogs at 0.1, 0.5, and 1.0 g/kg for 6 months demonstrated no adverse effects [15].

Use: Emulsifier, suspending agent, suspension stabilizer, viscosity enhancer.

Carbomer 1342

Structure:

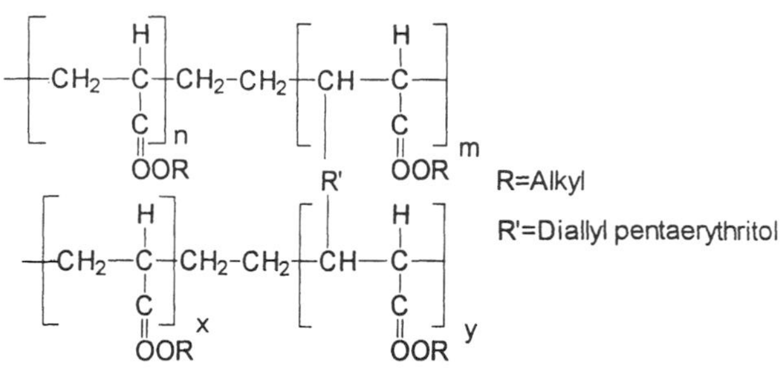

R=Alkyl

R'=Diallyl pentaerythritol

Description: High molecular weight polyacrylic acid polymers containing a C10-C30 alkyl acrylate cross polymer.

Generic nomenclature: Carboxyvinyl polymer, Carbomer X, carboxypolymethylene, C10-C30 alkyl acrylate co-monomer.

Appearance: Fluffy, white, hygroscopic powder having a mild acetic odor.

Solubility: Not truly soluble but swells in water to produce a colloidal dispersion.

Pharmacopeial listing: NF, JPE, JSCI.

Regulatory status: Drug Master File No. 7757.

Health and Safety:

Residual Solvent	
TR-1	<0.3% ethylacetate, cyclohexane
TR-2	<0.3% ethylacetate, cyclohexane

Acute oral toxicity: Carbomer 1342 at 5.0 g/kg (carbomer 1342 is chemically similar to Pemulen TR-1 and TR-2) exhibited no deaths or other abnormalities (rats) [15]. Residual acrylic acid test [16].

Use: Emulsifier, suspension stabilizer, thickener.

Carboxymethylcellulose Sodium

Structure:

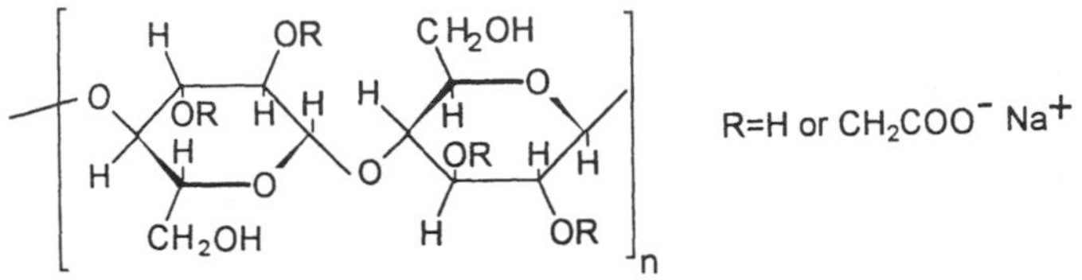

R=H or $CH_2COO^- Na^+$

Description: An anionic cellulose prepared by treating cellulose with caustic soda followed by reaction with sodium monochloroacetate under controlled conditions.

CAS Number: 9004-32-4.

Generic nomenclature: Sodium CMC, cellulose carboxymethyl ether, sodium salt, cellulose gum, carmellose sodium, sodium cellulose glycolate.

Appearance: White to off-white odorless hygroscopic powder.

Solubility: Soluble in hot and cold water, insoluble in most organic solvents.

Pharmacopeial listing: USP, EP, JP, FCC, Codex Alimentarius.

Regulatory status: GRAS as a miscellaneous and/or general purpose food additive in the U.S. Code of Federal Regulations Title 21 CFR 182.1745 (an "FDA proposal Feb. 23, 1979, would affirm" GRAS under 184.1745). EC product number is E 466 and the specification reference is FAS/IV/19.

Health and safety: Acceptable daily intake of "not specified" (no quantitative limit) [17].

Acute oral toxicity: CMC administered to rats, rabbits, and guinea pigs in single oral doses of 5 g/kg. exhibited no toxic effects [18]. Acute oral LD_{50} in rats and guinea pigs were 27 and 16 g/kg, respectively [19].

Radioactive oral absorption: In rats CMC (1.3 g/kg) demonstrated no absorption [19].

Subacute oral toxicity: CMC administered to dogs and guinea pigs at 1–10% and 1–2%, respectively, of dietary levels for 6 months demonstrated no adverse effects at 5% with retarded growth at 10% [19].

Chronic oral toxicity: CMC administered to rats at 0.2–2% of dietary levels for 2 years demonstrated no adverse effects [19].

Use: Suspending agent, stabilizer, thickener.

Carrageenan

Structure:

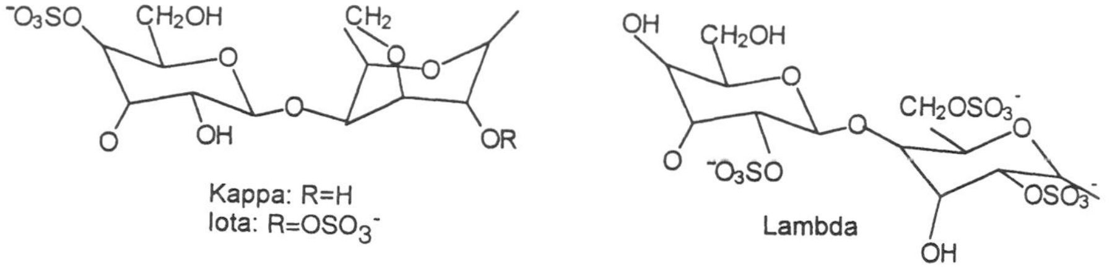

Kappa: R=H
Iota: R=OSO_3^-

Lambda

Description: An anionic linear water-soluble polysaccharide gum extracted from seaweed and composed of galactan sulfate esters, with galactose residues linked in alternating α ($1 \rightarrow 3$) and β ($1 \rightarrow 4$) positions.

CAS number: 9000-07-1.

Generic nomenclatures: Irish moss, sulfated polysaccharide.

Appearance: A white to brownish nearly odorless powder that may contain added sugars for standardization.

Solubility: Soluble in water-forming colloidal, viscous solutions.

Pharmacopeial listing: NF, FCC, Codex Alimentarius.

Regulatory status: As a direct food additive in the U.S. Code of Federal Regulations, Title 21, CFR Section 172.620 and 172.623 (as an emulsifier, stabilizer, or thickener) 172.626 and as a GRAS food substance in 21 CFR 182.7255. The international number is 407 (Codex Alimentarius, 2nd Ed., Vol. 1, Rome 1992).

Health and safety: Acceptable daily intake of "not specified" [20] (no quantitative limit has been established).

Use: Emulsifier, thickener, suspension stabilizer.

Crospovidone

Structure:

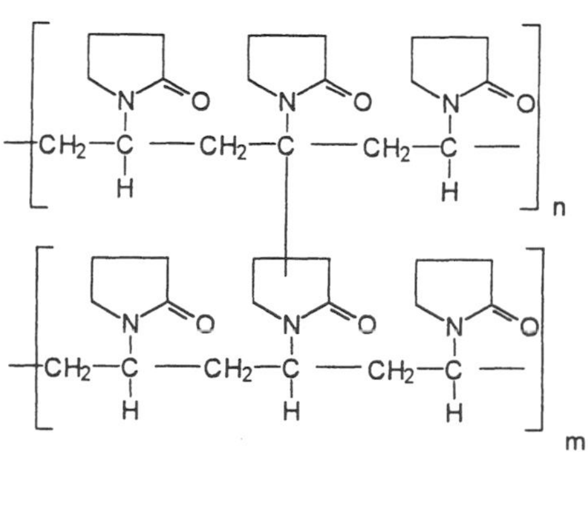

Description: An insoluble grade of polyvinylpyrrolidone prepared by a polymerization process that cross-links polyvinylpyrrolidone resulting in an insoluble powder with a highly porous "popcorn morphology."

CAS number: 9003-39-8.

Generic nomenclature: poly(polyvinylpyrrolidone), cross-linked polyvinylpyrrolidone, insoluble PVP.

Appearance: White- to- off-white hygroscopic, odorless free-flowing powder.

Solubility: Insoluble in practically all solvents due to its cross-linked structure.

Pharmaceopeial listing: NF, EP, JPE, FCC, Codex Alimentarius.

Health and safety: Acceptable daily intake of "not specified" (no quantitative limit) [21].

Acute oral toxicity: Fed at 5 g/kg to mice and rats no abnormal symptoms or deaths were observed [22]. After oral administration (rats) using ^{14}C-labeled Crospovidone no absorption, gastrointestinal accumulation, or biliary excretion was detected [22].

Subchronic oral toxicity: Administered at 300, 1200, or 4888 mg/kg for 26 weeks (dogs) with no compound related effects observed [22].

Regulatory: Approved for use in U.S. Code of Federal Regulations Title 21 CFR 173.50 as a clarifying agent in beer. The international Number is 1202 (Codex Alimentarius, 2nd Ed., Vol. 1, Rome 1992).

Use: Suspension stabilizer, clarifying agent.

Ethylcellulose

Structure:

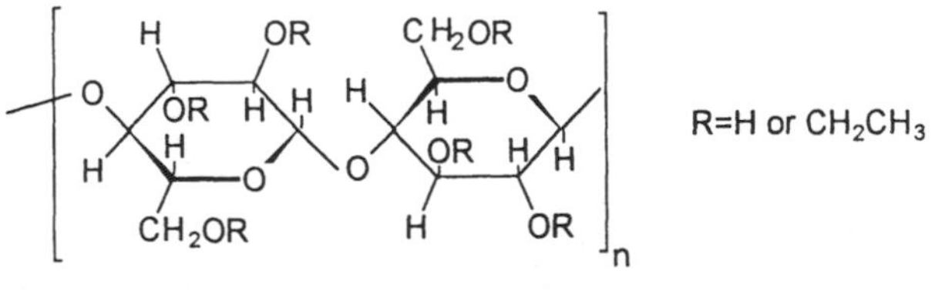

R=H or CH_2CH_3

Description: An organosoluble thermoplastic cellulose ethyl ether polymer prepared by the alkali hydrolysis of cellulose with caustic soda followed by reaction with chloroethane.

CAS number: 9004-57-3.

Generic nomenclature: EC, cellulose ethyl ether.

Appearance: White to off-white, odorless, tasteless powder.

Solubility: Insoluble in water and aliphatic hydrocarbons, soluble in alcohol and most organic solvents.

Pharmacopeial listing: NF, EP, FCC, Codex Alimentarius, JSPI.

Regulatory status: Approved as a pharmaceutical additive in dry vitamin preparations in the U.S. Code of Federal Regulations, Title 21 CFR, 172.868 as a binder, filler, or component as specified; 21CFR 73.1 in color mixtures for food; 21CFR 73.1001 in diluents in color additive mixtures for drug use exempt from certification; and 21CFR

573.420 as a binder or filler in dry vitamin preparations for animal feeds. GRAS for food packaging contact in the U.S. Code of Federal Regulations, Title 21 and CFR 182.90. The international number is 462 (Codex Alimentarius, 2nd Ed., Vol. 1, Rome 1992).

Health and safety: Acceptable daily intake of "not specified" (no quantitative limit) [23].

Subacute toxicity: Ethyl cellulose administered at 1.2% of dietary intake for 8 months (rats) resulted in no observed toxicity [24].

Use: Coacervation (U.S. patent 3,567,650), microencapsulation.

Guar Gum

Structure:

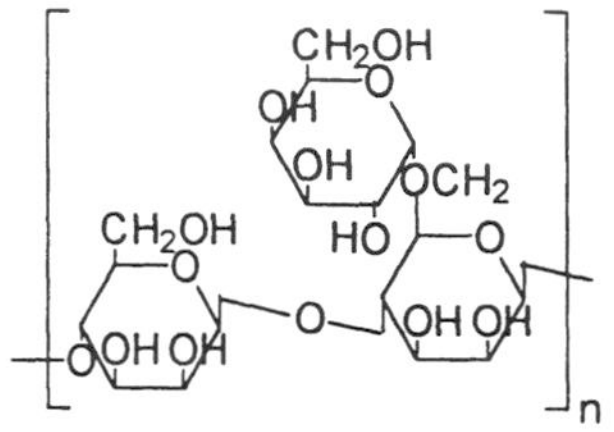

Description: A highly purified, nonionic, water-soluble, linear galactomannan polysaccharide gum (MW 1–2 million) substituted with galactose on every other mannose unit obtained by extraction of the seeds of a leguminous plant.

CAS number: 9000-30-0.

Generic nomenclature: Galactomannan polysaccharide.

Appearance: White to off-white odorless powder.

Solubility: Insoluble in organic solvents, soluble in water-forming high-viscosity colloidal dispersions.

Pharmacopeial listing: NF, FCC, Codex Alimentarius.

Regulatory status: GRAS in the U.S. Code of Federal Regulations Title 21 CFR Section 184.1339 in specified foods up to specified levels as an emulsifier, stabilizer, or thickener. Accepted as a food additive in Europe [25]. The international Code number is 412 (Codex Alimentarius, 2nd Ed., Vol. 1, Rome 1992).

Health and safety: Acceptable daily intake of "not specified" [26] (no quantitative limit).

Use: Thickener, suspension stabilizer.

Hydroxyethyl Cellulose

Structure:

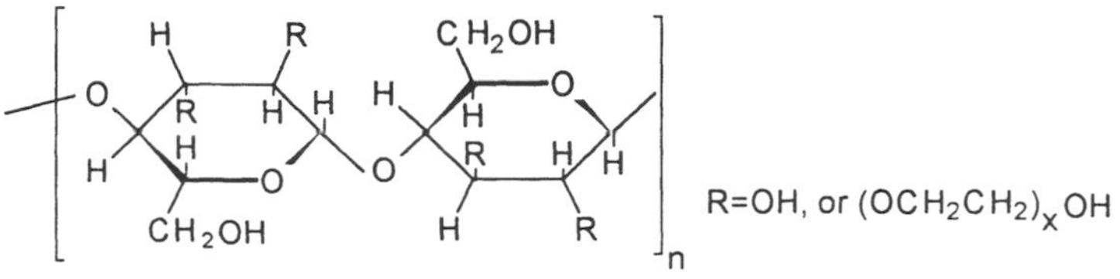

R=OH, or $(OCH_2CH_2)_xOH$

Description: A nonionic polymer prepared by treating cellulose with alkali to form alkali cellulose followed by etherification with ethylene oxide.

CAS number: 9004-62-0.

Generic nomenclature: Cellulose 2-hydroxyethyl ether.

Appearance: White to light tan, odorless, tasteless, hygroscopic powder.

Solubility: Soluble in cold and hot water; insoluble in most organic solvents.

Pharmacopeial listing: NF, EP.

Regulatory status: Approved as an indirect food additive in packaging materials under U.S. Code of Federal Regulations Title 21, CFR 175.300, resinous and polymeric coatings; CFR 176.170, component of paper and paperboard for aqueous and fatty foods; CFR 176.180, component of paper and paperboard for contact with dry foods; CFR 177.1210, component of sealing gaskets for food containers.

Health and safety:

Acute oral toxicity LD_{50}: Values are: 7.74 g/kg (250L), 0.774 g/kg (250H), and 23.07 g/kg (250 LR) (rats) [27].

Subacute oral toxicity: High- and low-viscosity administered at 0.2–5.0% of dietary levels for 3 months (rats) demonstrated no adverse effects [27,28].

Use: Suspending agent, suspension stabilizer, thickener.

Hydroxypropyl Cellulose

Structure:

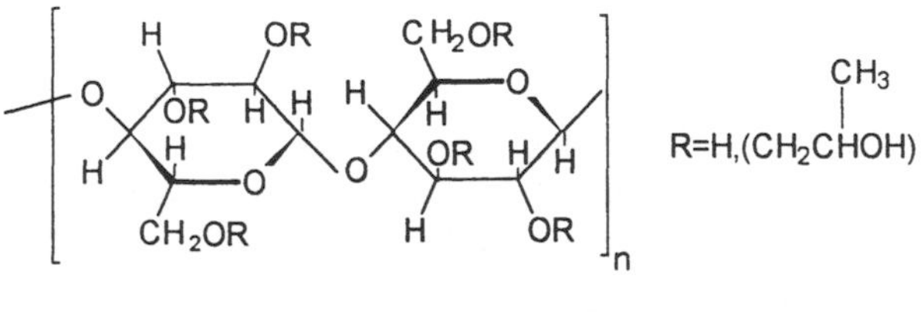

H OR CH₂OR
O OR H H H H O
H H O O OR H H
CH₂OR H OR
n

R=H,($CH_2CH(CH_3)OH$)

Description: A nonionic polymer prepared by treating cellulose with caustic soda followed by etherification with propylene oxide.

CAS number: 9004-64-2.

Generic nomenclature: Cellulose 2-hydroxypropyl ether, HPC.

Appearance: White to off-white, odorless, tasteless hygroscopic powder containing silicone dioxide.

Solubility: Soluble in water below 40°C, alcohol, and most polar organic solvents.

Pharmacopeial listing: NF, EP, JP, FCC, Codex Alimentarius.

Regulatory status: Approved as a direct food additive in the U.S. Code of Federal Regulations, Title 21 CFR 172.870 as an emulsifier, film former, thickener, protective colloid, stabilizer, and suspending agent. The international number is 463 (Codex Alimentarius, 2nd Ed., Vol. 1, Rome, 1992).

Health and safety: Acceptable daily intake of "not specified" (no quantitative limit) [29].

Acute toxicity: Low-, middle-, and high-viscosity solutions had an oral LD_{50} of >5.0 g/kg [30], 10.2 g/kg [31] >15.0 g/kg [32].

Subacute toxicity: HPC (substitution of 4.6) fed to rats at 0.2–5.0% of the diet for 90 days exhibited no effects compared to controls [31].

Gastrointestinal absorption: ^{14}C propylene oxide labeled HPC absorption studies found 99.1% of the administered dose excreted in the feces (rats) [31].

Use: Suspending agent, stabilizer, thickener.

Hydroxypropyl Methylcellulose

Structure:

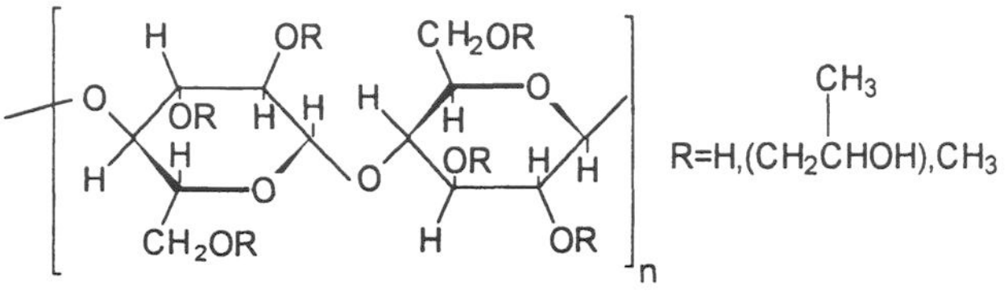

R=H,(CH_2CHOH),CH_3 (with CH_3 on the CHOH carbon)

Description: Nonionic cellulose ethers prepared by treating cellulose with caustic soda, followed by etherification with chloromethane and reaction with propylene oxide.

CAS number: 9004-65-3.

Generic nomenclature: HPMC, MHPC, cellulose 2-hydroxypropyl methyl ether.

Appearance: White to off-white, practically odorless, fibrous or granular powder.

Solubility: Soluble in cold water; insoluble in hot water; soluble in most alcohol/organic solvent blends.

Pharmacopeial listing: USP, EP, JP, FCC, Codex Alimentarius.

Regulatory status: Approved as a direct food additive in the U.S. Code of Federal Regulations, Title 21 CFR 172.870 as an emulsifier, film former, protective colloid stabilizer, suspending agent, or thickener. EC product code is E 464 in council directive 78/663.

Health and safety: Acceptable daily intake of "not specified" (no quantitative limit) [33]. Long- and short-term feeding and radioactive metabolic studies have been performed [34].

Use: Suspending agent, stabilizer, thickener.

Lactide-Glycolide Polymers

Structure:

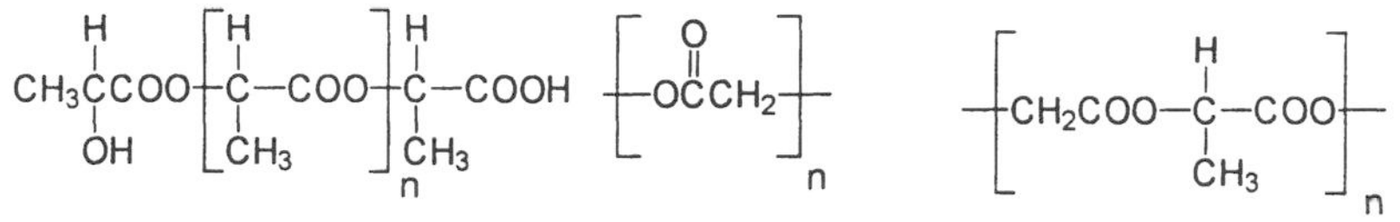

$CH_3C(H)(OH)COO-[C(H)(CH_3)-COO]_n-C(H)(CH_3)-COOH$

Polylactic Acid

$-[OC(=O)CH_2]_n-$

Polyglycolic Acid

$-[CH_2COO-C(H)(CH_3)-COO]_n-$

Polylactic-Glycolic Acid

Description: Homo- and copolymers of lactic and glycolic acids synthesized by ring-opening melt condensation of cyclic dimers (lactide and glycolide) at approximately 175°C [35,36].

Generic nomenclature:	MW
Poly(L)lactic acid, propanoic acid, 2-hydroxy-polymer	80–500M
Poly(DL)lactic acid, propanoic acid, 2-hydroxy-polymer	70–160M
Polyglycolic acid, hydroxyacetic acid polymer	Not determined
Poly(DL)lactic-glycolic acid, propanoic acid, 2-hydroxy copolymer with hydroxyacetic acid	50–100M

Appearance: Odorless, white to tan pellets

Solubility:	Ethyl acetate	Methylene chloride	Dimethyl formamide	Tetra-hydrofuran	Acetone
Poly(L)lactic acid	NS	S	NS	NS	NS
Poly(DL) lactic acid	S	S	S	S	S
Polyglycolic acid	NS	NS	NS	NS	NS
Poly(DL)lactic -glycolic acid	SS	S	S	SS	S

(NS=not soluble S=soluble SS=slightly soluble)

Pharmacopeial listing: USP (absorbable surgical sutures).

Regulatory: The FDA has approved specific applications using this family of polymers in medical products.

Health and safety: These polymers are hydrolytically degraded to glycolic or lactic acids in the body within weeks or months depending on polymer type and molecular weight.

Use: Absorbable controlled release polymer [37,38].

Methylcellulose

Structure:

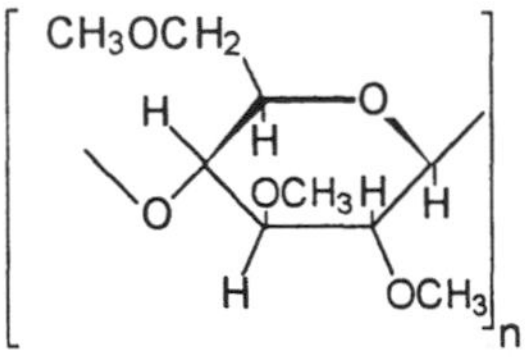

Description: Nonionic cellulose ethers prepared by treating cellulose with caustic soda followed by etherification with chloromethane.

CAS number: 9004-67-5.

Generic nomenclature: Methylcellulose, cellulose methyl ether.

Appearance: White to off-white, practically odorless, tasteless powder or granules.

Solubility: Insoluble in hot water; soluble in cold water and most polar alcohol-organic solvent blends.

Pharmacopeial listing: USP, EP, JP, FCC, Codex Alimentarius.

Regulatory status: Highly purified grades are GRAS under U.S. Code of Federal Regulations Title 21, CFR 182.1480. The EC product code is E461 in Council Directive 74/329.

Health and safety: Acceptable daily intake of "not specified" (no quantitative limit) [39]. Long- and short-term feeding and radioactive metabolite studies have been performed [40], and methylcellulose is generally regarded as being non toxic [41].

Use: Suspending agent, suspension stabilizer, thickener.

Microcrystalline Cellulose and Carboxymethylcellulose Sodium

Structure:

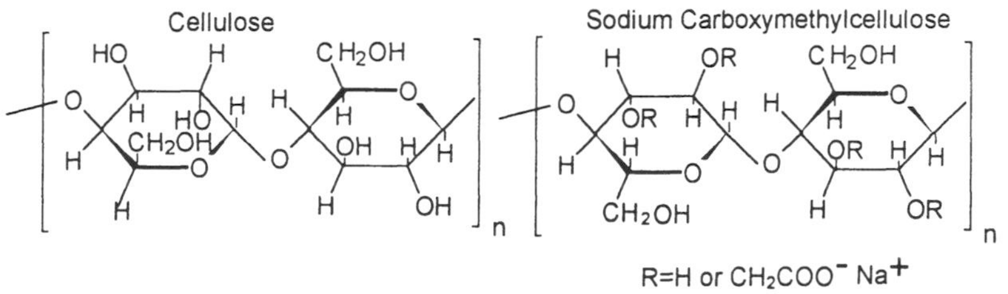

Cellulose

Sodium Carboxymethylcellulose

R=H or CH_2COO^- Na^+

Description: A colloidal form of microcrystalline cellulose intimately blended with sodium carboxymethylcellulose

CAS number: 9004-34-6 Microcrystalline Cellulose; 9004-32-4 Sodium Carboxymethylcellulose.

Generic nomenclature: Microcrystalline cellulose and sodium carboxymethylcellulose.

Appearance: White to off-white, odorless hygroscopic powders.

Solubility: Partially soluble in water and dilute alkali; insoluble in organic solvents and dilute acids.

Pharmacopeial listing: NF: Avicel RC 501, Avicel RC 581, Avicel RC 591, Avicel CL 611, and BP.

Regulatory: Microcrystalline cellulose and sodium carboxymethylcellulose combinations are GRAS according to the U.S. Code of Federal Regulations Title 21 CFR 182.1745 (CMC), and (MCC) (Communication FMC to FDA April 19, 1966 and USDA from the FDA March 31, 1986).

Health and safety:

Teratogenicity: Negative in the Ames Test [42].

Acute oral toxicity LD_{50}: >5 g/kg (rats) [43].

Subchronic toxicity—90 day diet: No significant health effects up to and including 50,000 ppm [44]. Teratology (maternal and fetal), no observable effect level >50,000 ppm (rats) [45].

Use: Suspending agent, emulsion stabilizer.

Pectin

Structure:

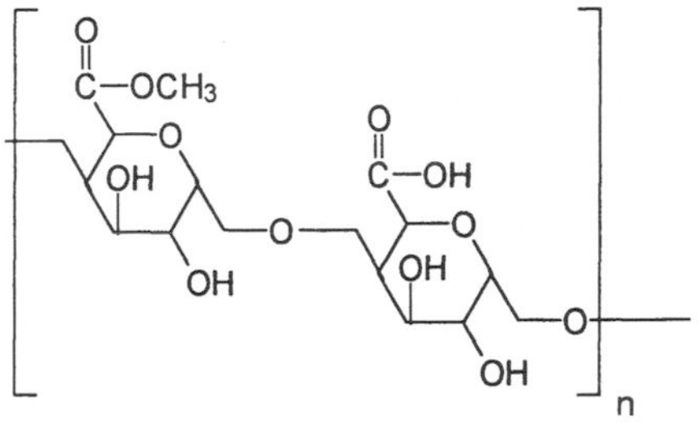

Description: An anionic linear polysaccharide extracted from the peel of citrus fruits comprised of partially esterified (methyl groups) α(1-4)-D-galacturonopyranosyl units. This leaves some remaining free acid groups that can be neutralized with metal hydroxides.

CAS number: 9000-69-5.

Generic nomenclature: Polygalacturonic acid methyl ester.

Appearance: Light cream to tan colored, odorless powders.

Solubility: Soluble in water producing viscous, colloidal solutions.

Pharmacopeial listing: USP, FCC, Codex Alimentarius.

Regulatory status: GRAS in the U.S. Code of Federal Regulations, Title 21 CFR Section 184.1588 (as an emulsifier, stabilizer, or thickener.) The international number is 440 (Codex Alimentarius, 2nd Ed., Vol. 1, Rome 1992).

Health and safety: Acceptable daily intake of "not specified" [46] (no quantitative limit).

Use: Thickener, suspension stabilizer.

Poloxamer

Structure:

$$HO(CH_2CH_2O)_x(CH_2\overset{CH_3}{\overset{|}{C}}HO)_y(CH_2CH_2O)_xH$$

Description: Block copolymers of ethylene and propylene oxide prepared by reacting propylene oxide with propylene glycol followed by ethoxylation to form the block copolymer.

CAS number: 9003-11-6.

Generic nomenclature: Polyoxyethylene-polyoxypropylene glycol copolymer, poloxalkol.

Appearance: White flakes, powders, or liquids.

Solubility: Soluble in water and 95% ethanol.

Pharmacopeial listing: NF: 124, 188, 237, 338, 407 (Equivalent to L44, F68, F87, F108, and F127, respectively); FCC: 331, 407, EP, JPE.

Regulatory status: Drug Master File #100, Food Master file #8.

Health and safety: 124 oral rat LD_{50} (practically nontoxic) 5 mL/kg [47], 188 oral human clinical excretion studies demonstrated no polymer appears in the urine [47] and is used in artificial blood substitute [48,49] Oral rat, dog, and mouse LD_{50} (practically nontoxic) is >15g/kg [47], 237 IV mouse LD_{50} is 3.75 g/kg [47], and the oral rat LD_{50} (slightly toxic, practically nontoxic) >5g/kg [47]. 338 IV mouse LD_{50} is 1.25 g/kg [47]. 407 oral rat LD_{50} is slightly toxic, practically nontoxic >10 g/kg [47].

Use: Emulsifier, dispersant, wetting agent, gel-forming agent.

Polycarbophil

Structure:

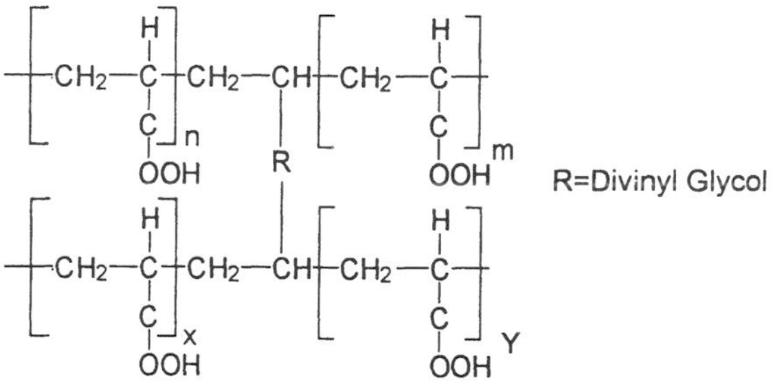

$$\left[-CH_2-\underset{COOH}{\overset{H}{C}}-\right]_n CH_2-\underset{R}{CH}-\left[-CH_2-\underset{COOH}{\overset{H}{C}}-\right]_m$$

R=Divinyl Glycol

$$\left[-CH_2-\underset{COOH}{\overset{H}{C}}-\right]_x CH_2-CH-\left[-CH_2-\underset{COOH}{\overset{H}{C}}-\right]_Y$$

Description: High molecular weight polyacrylic acid slightly cross-linked with divinyl glycol.

CAS number: 9003-01-4, free acid; 9003-97-8, calcium salt.

Generic nomenclature: Polycarbophil, calcium polycarbophil.

Appearance: White to off-white powder having a mild acetic odor.

Solubility: Both forms (free acid and calcium salt) are not truly soluble but swell in water to form colloidal dispersions.

Pharmacopeial listing: Free acid: USP, calcium salt: USP, EP, JP, FCC.

Regulatory status: Drug Master File (free acid) #7618 for oral/topical application and Drug Master File (calcium salt) #6542 for oral laxative and antidiarrheal. Found safe and effective as an over-the-counter bulk laxative and antidiarrheal by the U.S. FDA (Laxative Drug Products for Over-the-Counter Human Use) tentative final monograph 51FR 35136, Oct. 6, 1986.)

Health and safety: AA-1 (free acid) <0.45% ethylacetate (residual solvent); CA-1, CA-2 (calcium salt) water (residual solvent); residual acrylic acid typically 0.5% from MSDS [50].

Acute oral toxicity LD_{50}: >2.5 g/kg (rats) [51]. (Not directly tested but this data was obtained using similar polyacrylic acid resins.)

Chronic oral toxicity: Exhibited no significant effects when administered at 5% of dietary intake levels for 6 1/2 months (rats and dogs) [51].

Use: Bioadhesive, emulsion stabilizer, thickener.

Polyethylene Glycols

Structure:

$$HO\left(CH_2CH_2O\right)_n H$$

Description: A family of linear polymers prepared by the addition reaction of ethylene oxide and water.

CAS number: 25322-68-3.

Generic nomenclature: Poly(oxyethylene), oxyalkylene polymer, PEG.

Appearance: Clear, colorless, viscous liquid, waxy solid, powder, or flakes.

Solubility: Soluble in water and most polar organic solvents.

Pharmacopeial listing: NF (300-8000), EP (300, 400, 1000, 1500, 3000, 4000, 6000, 20,000, 35,000), JP (400, 1500, 4000, 6000, 20,000), FCC (300-9500). U.S., European, and Japanese monographs have not been harmonized and thus are not fully congruent. Some products will meet all three monographs but compliance should be determined by the seller's label claims.

Regulatory status: As a direct food additive in the U.S. Code of Federal Regulations Title 21, CFR 172.820 for coating, binding, plasticizing, or lubricating tablets used for food (only PEG 300-8000 meet the requirements of this regulation that they contain a maximum of 0.2% w/w ethylene glycol and diethylene glycol.) As an indirect food additive in the U.S. Code of Federal Regulations Title 21, CFR 175.105 components for adhesives used for food packaging (200-8000), 178.3910 component of surface lubricant used in drawing, stamping, or forming metallic articles for food contact applications (PEG 300-8000), and 178.3750 component of articles intended for use in contact with food. (PEG 300-8000). The acceptable daily intake is 10 mg/kg [52]. The Drug Master File exists for commercially available product.

Health and safety:	Acute oral toxicity (rats) [53]
PEG 300	27.5–38.9 g/kg
PEG 400	30.2–43.6 g/kg
PEG 1000	42.0 g/kg
PEG 1450 (1540)	51.2 g/kg
PEG 3350 (4000)	>50 g/kg
PEG 8000 (6000)	>50 g/kg

Note that previous designation for 540, 1450, 3350, and 8000 were 1500, 1540, 4000, and 6000, respectively.

Subacute (90 Day) oral toxicity: Administration of PEG (200, 300, 400, 600, 1000, 1500, 1540, 4000, and 6000) at 0–24% of dietary intake demonstrated no significant toxicity (rats), but the 6000 molecular weight was less toxic than the lower molecular weights [54].

Chronic (1 year) toxicity: Administration of PEG 400, 1540, and 4000 at 2% of dietary intake (dogs) demonstrated no effects [54].

Chronic (2 year) toxicity: Administration of PEG 400, 1540, and 4000 at 0–4% of dietary intake (rats) demonstrated no effects at 2 and 4% but minor toxicity at 8% [54].

Residual chemical impurities: Most PEGs contain water, typically less than 0.5% w/w, and residual unreacted ethylene oxide from 1–10 ppm, depending on the product grade and commercial requirements. They also contain 1,4 dioxane, acetaldehyde, and formaldehyde ranging from 1–10 ppm. The concentration of acetaldehyde, formaldehyde, other higher aldehydes, and organic acids can increase due to degradation during use of prolonged storage [55].

Use: Suspending agent, solubilizer, emulsifier, emulsion stabilizer.

Polyethylene Glycol Monomethyl Ether

Structure:

$$CH_3O\left(CH_2CH_2O\right)_n H$$

Description: Made by the addition reaction of monomethyl glycol ether [starter] with ethylene oxide and methanol, resulting in a family of ethylene oxide polymers having only one reactive hydroxyl group.

CAS number: 9004-74-4.

Generic nomenclature: Methoxypolyethylene glycol, oxyalkylene polymer monomethyl ether.

Appearance: Clear, colorless viscous liquid, waxy solid, powder, or flakes.

Solubility: Soluble in water and most polar organic solvents.

Pharmacopeial listing: NF (350-10,000).

Regulatory status: Drug Master File #9385.

Health and safety:

Peroral acute toxicity: Male rats: fed 16.0 g/kg of body weight—killed 0 of 5; females: fed 16.0 g/kg of body weight—killed 0 of 5 [56].

Percutaneous, rabbit (24-hr occluded): Males: 16.0 g/kg—killed 0 of 5; females: 16.0 g/kg—killed 0 of 5 [56].

Skin irritation, rabbit (4-hr occluded): Minor erythema on 2 of 6 rabbits, no edema on any of 6 from 0.5 mL (dosed as received). Irritation subsided within 24–48 hr [56].

Eye irritation, rabbit: No corneal injury or iritis in any of 4 rabbit eyes, minor conjunctival irritation in 4 from 0.1 mL. Irritation subsided within 24–72 hr [56].

Health and safety: MPEGs contain water, typically at less than 0.5 wt. %, and residual unreacted ethylene oxide at levels ranging from less than 1 to 10 ppm, depending on the product grade and commercial requirements. MPEGs may also contain glycol ethers, 1,4-dioxane, acetaldehyde, and formaldehyde at levels ranging up to 10 ppm. The concentrations of acetaldehyde, and formaldehyde, and other higher aldehydes and organic acids can increase in the product due to degradation during prolonged storage and use [57].

Use: Diverse medicament solubilizer in topical dosages.

Polyethylene Glycol 660 Hydroxystearate

Structure:

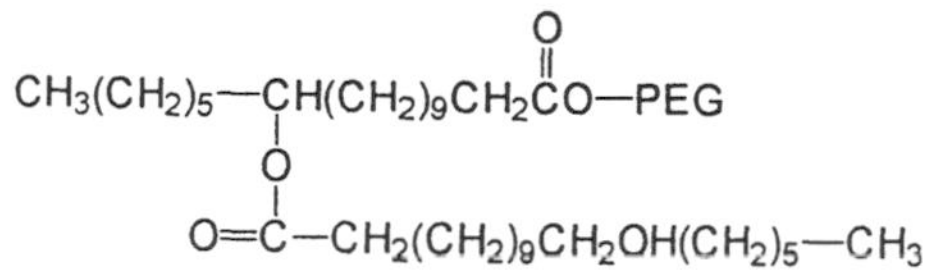

$$\begin{array}{l} CH_3(CH_2)_5-\underset{\displaystyle |}{C}H(CH_2)_9CH_2\overset{\displaystyle O}{\overset{\|}{C}}O-PEG \\ \qquad\qquad\quad\; O \\ \qquad\qquad\quad\; | \\ \qquad\quad O{=}C-CH_2(CH_2)_9CH_2OH(CH_2)_5-CH_3 \end{array}$$

Description: A nonionic surfactant prepared by reacting 12-hydroxystearic acid with ethylene oxide. The product includes di-ester, free polyethylene glycol, and its monoesters and 2–3% of the product has the 12-hydroxy group etherified with ethylene oxide.

Generic nomenclature: Polyethylene glycol 660 12-hydroxystearate.

Appearance: White paste at room temperature; liquid at 30°C.

Solubility: Soluble in water and most alcohols with decreasing solubility at higher temperatures.

Regulatory status: Cleared under U.S. Code of Federal Regulation Title 21 CFR 176.170 and 176.180 for food contact paper and paperboard. Drug Master File #9501.

Health and safety:

Acute toxicity (LD_{50}): mouse i.p. > 8.5 mg/kg; rabbit i.v. > 1.0 g/kg < 1.4 g/kg; dog i.v. > 3.1 g/kg; rat oral ~ 20 g/kg; rat i.v. > 1.0 g/kg < 1.47 g/kg [58].

Subacute toxicity i.v.: Administered at 25, 75, and 125 mg/kg for 4 weeks (dogs) followed by blood histamine determination after the first and 19th dose indicated no symptoms or histamine observed at 25 mg/kg, but histamine was detected in two of three dogs at 75 mg/kg after the first dose and all dogs after the 125 mg/kg dose.

Use: Emulsifier, solubilizer, wetting agent.

Poly(ethylene oxide)

Structure:

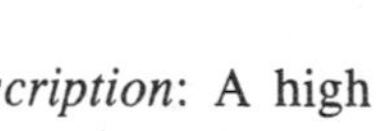

$$\left(OCH_2CH_2 \right)_n OH$$

Description: A high molecular weight nonionic water-soluble polymer prepared by polymerizing ethylene oxide (n = 2,000–135,000).

CAS number: 25322-68-3.

Generic nomenclature: Poly(oxyethylene).

Appearance: White to off-white powder having a slightly ammoniacal odor.

Solubility: Soluble in water, chlorinated hydrocarbons, and many organic solvents.

Pharmacopeial Listing: NF.

Regulatory status: As a direct additive to beer, according to U.S. Federal Regulation Title 21 CFR 172.770. For food packaging use: 175.300 (xxxiii) component of polymer coating for food use; 175.380 adjuvant for xylene-formaldehyde resin for food contact; 175.390 component for zinc silicone dioxide coatings for food contact; 176.170 component of paper-contacting aqueous and fatty foods; 176.180 component of paper-contacting food; 177.1210 sealing gasket/closure ingredient for food container contact; and 177.1350 adjuvant for EVA copolymer for food contact.

Health and safety: (Polyox WSR -301). The acute oral toxicity LD_{50} (rats) is > 2.0 g/kg [59]. Subchronic toxicity (rats): Fed at dietary levels of up to 4% for 96 days no adverse effects were observed; however, at 8% marginal toxicity effects were observed [59]. Chronic toxicity (rats, dogs): In two year feeding studies at dietary levels of up to 5% and 2% (rats and dogs, respectively) no adverse effects were observed [59]. Oral absorption (rats): ^{14}C radioactive labeled polymer administered orally at 65 mg/kg showed essentially no absorption (0.72% excreted in the urine, 90.5% in the feces) [59].

Use: Thickener, suspending agent.

Polyglycerol Esters of Fatty Acids

Structure:

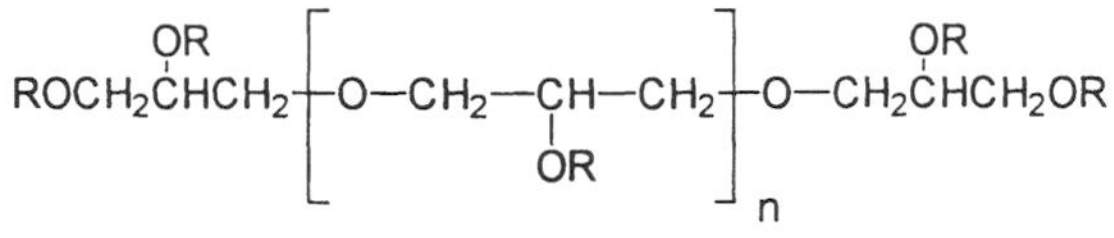

$$ROCH_2\underset{}{\overset{OR}{C}}HCH_2\left[O{-}CH_2{-}\underset{OR}{CH}{-}CH_2\right]_n O{-}CH_2\overset{OR}{C}HCH_2OR$$

R=H,Stearate,Oleate

Description: A series of mainly linear compounds ranging from diglycerol (3 hydroxy groups) to triacontaglycerol (32 hydroxyl groups) prepared by heating glycerin to over 320°C [60]. The polymerization that occurs produces a family of polyglycerols containing polyol groups that can be esterified with free organic fatty acids.

CAS number: Polytriglycerol oleate, 9007-48-1 polyglycerol-3-monooleate; polytriglycerol stearate, 37349-34-1 polyglycerol-3-stearate; polyhexaglycerol dioleate, 9007-48-1 polyglycerol-6-dioleate; polyhexaglycerol distearate, 61725-93-7 polyglycerol-6-distearate; polydecaglycerol tetraoleate, 34424-98-1 polyglycerol-10-tetraoleate; polydecaglycerol decaoleate, 11094-60-3 polyglycerol-10-decaoleate.

Appearance: Amber to dark brown viscous liquid, semisolid or waxy solid. HLB properties range from approximately 4–13 and are determined by the size of the glycerol moiety, chain length of the fatty acid, and number of free hydroxyl groups.

Solubility: Solubility ranges from water soluble to oil soluble depending on the polyglycerol ester.

Pharmacopeial listing: FCC.

Regulatory status: Approved as a direct food additive up to and including decaglycerol in the U.S. Code of Federal Regulations 172.854 and 172. 860.

Use: Emulsifier, stabilizer.

Polyoxyl 35 Castor Oil and Polyoxyl 40 Hydrogenated Caster Oil

Structure: Idealized structure; the structure shown is somewhat simplified.

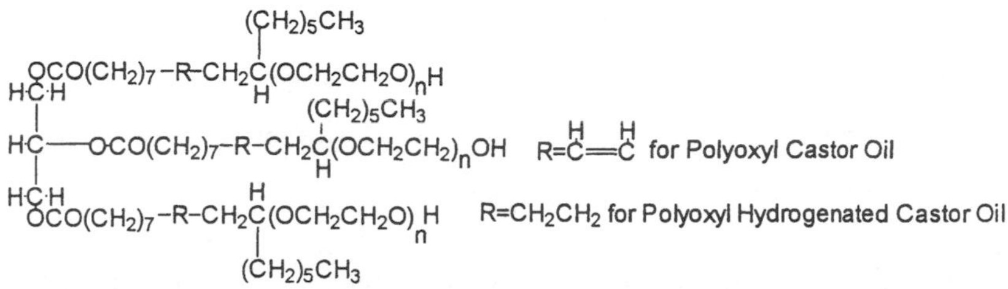

$(CH_2)_5CH_3$

HCH — $OCO(CH_2)_7\text{-}R\text{-}CH_2CH(OCH_2CH_2O)_nH$

$(CH_2)_5CH_3$

HC — $OCO(CH_2)_7\text{-}R\text{-}CH_2CH(OCH_2CH_2)_nOH$ $R{=}CH{=}CH$ for Polyoxyl Castor Oil

HCH — $OCO(CH_2)_7\text{-}R\text{-}CH_2CH(OCH_2CH_2O)_nH$ $R{=}CH_2CH_2$ for Polyoxyl Hydrogenated Castor Oil

$(CH_2)_5CH_3$

Description: A complex nonionic surfactant prepared by reacting 35 mols of ethylene oxide with castor oil (Cremophor EL) or 40 mols of ethylene oxide with hydrogenated castor oil (Cremophor RH40).

CAS number: 61791-12-6.

Generic nomenclature: Cremophor EL-(polyethyoxylated castor oil, polyoxyethylene glycerylricinoleate 35), Cremophor RH40-(polyethoxylated hydrogenated castor oil, macrogol glycerol hydroxystearate).

Appearance: Cremophor EL—clear, yellow-amber viscous liquid at 28°C. Cremophor RH40—almost tasteless, white-yellow thin paste at room temperature.

Solubility: Cremophor EL and RH 40 are soluble in water, alcohols, and most organic solvents with decreased solubility at elevated temperatures.

Pharmacopeial listing: NF, EP.

Regulatory status: Cremophor EL Drug Master File #2981, Cremophor RH40 Drug Master File #3420.

Health and safety:

Cremophor EL

Acute oral toxicity: LD_{50} > 6.4 mL/kg (practically nontoxic) (rats) [61].

Subacute oral toxicity: Repeated administration of 0–5.3 mL/kg (five times a week) for four weeks (dogs) did not result in any clinically detectable toxicity [61].

Chronic oral toxicity: Administered up to 6 months at up to 1% of dietary intake (rats and dogs) did not cause any clinically detectable toxicity [61].

Cremophor RH40

Acute oral toxicity: LD_{50}: >20 g/kg (rats) [62]; I.P. 12.5 g/kg (mice).

Chronic/subchronic oral toxicity: Administration for 26 weeks at 10,000-100,000 ppm of dietary intake (rats and dogs) demonstrated no adverse effects [62].

Use: Solubilizer [63–65], emulsifier, wetting agent.

Polyvinyl Alcohol

Structure:

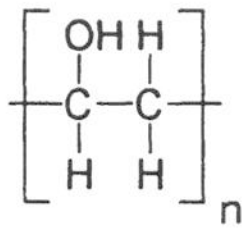

$$\left[-\overset{\text{OH}}{\underset{\text{H}}{\text{C}}} - \overset{\text{H}}{\underset{\text{H}}{\text{C}}} - \right]_n$$

Description: Polyvinyl alcohol, a linear polyhydroxy compound, manufactured by polymerizing polyvinyl acetate that is converted to polyvinyl alcohol by saponification using methanol in the presence of sodium hydroxide.

CAS number: Fully hydrolyzed, 9002-89-5; partially hydrolyzed, 25213-24-5.

Generic nomenclature: Ethenol homopolymer, vinyl alcohol polymer, PVA.

Appearance: White-cream-colored granular powder.

Solubility: Soluble in hot and cold water producing colloidal solutions. The solubility and water sensitivity are reduced with increased molecular weight and decreased hydrolysis.

Pharmacopeial listing: USP.

Regulatory status: Approved as an indirect food additive.

Use: Opthalmic lubricant, thickener.

Povidone

Structure:

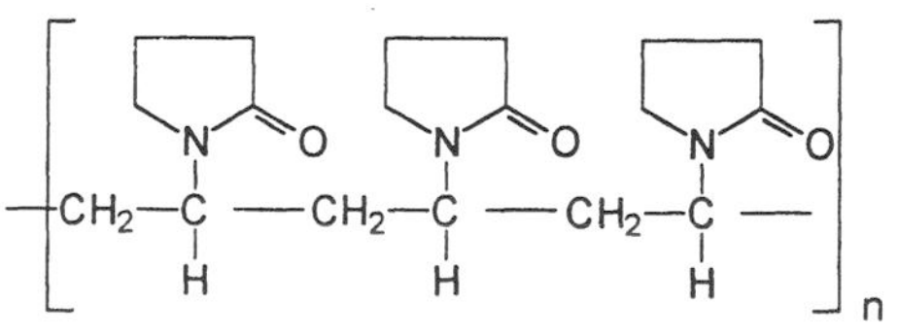

N O N O N O

$$\left[-CH_2-\underset{\text{H}}{\text{C}} -CH_2-\underset{\text{H}}{\text{C}} -CH_2-\underset{\text{H}}{\text{C}} - \right]_n$$

Description: A homopolymer of vinylpyrrolidone, manufactured using different initiator systems depending upon the molecular weight. The low molecular weight polymer, having a *K*-value of 30 or less (*K*-value represents the average molecular weight of soluble povidone grade and is calculated from the relative viscosity in water), is polymerized in water using a hydrogen peroxide initiator system or in isopropanol using an organic peroxide. Higher *K*-value polymer is made by aqueous homopolymerization using an organic azo or peroxide type initiator. When polymerization is carried out in isopropanol, the alcohol solvent is exchanged with water prior to drying.

CAS number: 9003-39-8.

Generic nomenclature: PVP, polyvinylpyrrolidone, 1-Vinyl-2-pyrrolidinone polymer.

Appearance: Free-flowing, white-yellowish, hygroscopic, tasteless powder or flakes having a slight amine odor.

Solubility: Freely soluble in water and most commonly used pharmaceutical solvents including alcohol and polyglycolated vehicles.

Pharmacopeial Listing: USP, EP, JP, FCC, Codex Alimentarius.

Regulatory: Povidone K-30 (MW 40,000) is approved in the U.S. Code of Federal Regulations Title 21 CFR, 173.55 (as a clarifying agent, stabilizer, dispersant, or tableting adjuvant as specified). The international code number is 1201 (Codex Alimentarius, 2nd Ed., Vol. 1, Rome 1992).

Health and safety: Reported impurities in Povidone are residual monomer, *N*-vinyl-2-pyrrolidinone, acetaldehyde, and hydrazine. Limit for residual *N*-vinyl-2-pyrrolidinone is less than 10 ppm, whereas acetaldehyde content is below 500 ppm and hydrazine below 1 ppm. Peroxide functionalities are also known to be present, with typical concentrations well below 400 ppm. No USP organic volatile impurities should be present. The use of organic initiators leads to trace quantities of organic materials. The acceptable daily intake is 0–50 mg/kg [66]. Numerous publications have demonstrated the safety of Povidone[67].

Acute oral toxicity: Doses of 300–2700 mg/kg were administered with no significant adverse effects (rabbits) [68].

Subchronic oral toxicity: PVP K-90 fed at 2.5 or 5.0% of the diet for 28 days (dogs) demonstrated no toxic, pathological, or histological abnormalities [69].

Chronic oral toxicity: No toxic effects were observed in a two-year study at 0, 5.0, or 10% PVP K-30 [70] and in a 138-week study at 1, 2.5, and 5% PVP K-90 (rats) [71].

Grades: K-12, C-15 (ISP), K-17 (BASF), and C-30 (ISP) (all of these grades are pyrogen tested), K-25, K-29/32 (ISP), K-30 (BASF), K-90, and K-120. Their molecular weights are 2500 (K-12), 10,000 (C-15 and K-17), 30,000 (K-25), 50,000 (K-29/32 and K-30), 1×10^6 (K-90), and 3×10^6 (K-120).

Use: Binder, complexing aid, suspension stabilizer, thickener.

Propylene Glycol Alginate

Structure:

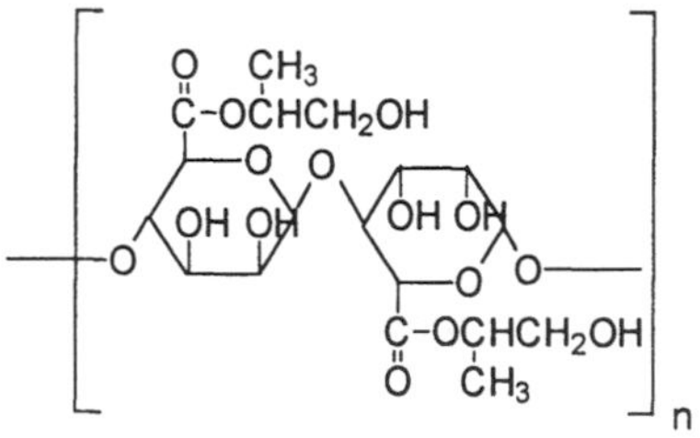

Description: An organic derivative of alginic acid manufactured as the propylene glycol ester of alginic acid.

CAS number: 9005-37-2.

Generic nomenclature: Propylene glycol alginate, alginic acid propylene glycol ester, propane-1,2-diol alginate, hydroxypropyl alginate.

Appearance: White to yellowish colored practically odorless powder.

Solubility: Soluble in water and dilute acids forming viscous colloidal solutions.

Pharmacopeial listing: NF, FCC, Codex Alimentarius.

Regulatory status: Approved as a food additive emulsifier, stabilizer, thickener, or defoamer in the U.S. Code of Federal Regulations Title 21 CFR 172.858. The acceptable daily intake is 70 mg/kg [72]. On the European Community List as an approved emulsifier/stabilizer. The EC product code number is E-405 in Annex IV to Directive 95/2/EC.

Health and safety: The toxicological properties of propylene glycol alginate have been extensively investigated and summarized [73–76].

Use: Thickener, stabilizer.

Sodium Alginate

Structure:

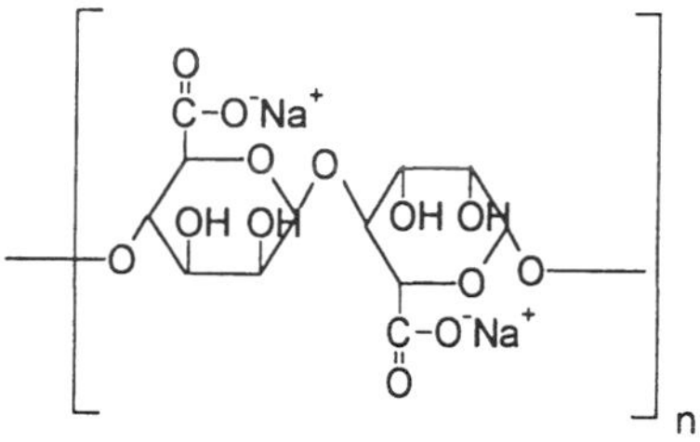

O
$C{-}O^{-}Na^{+}$
O O
OH OH OH OH
O O O
$C{-}O^{-}Na^{+}$
O
n

Description: A linear chain polymer extracted from seaweed composed of mannuronic and guluronic acids linked in the 1 to 4 positions and neutralized to form the sodium salt.

CAS number: 9005-38-3.

Generic nomenclature: Algin, sodium polymannuronate, alginic acid sodium salt.

Appearance: Ivory-colored practically odorless powder.

Solubility: Soluble in water forming viscous colloidal solutions.

Pharmacopeial listing: NF, FCC (ammonium alginate, calcium alginate, potassium alginate, sodium alginate), Codex Alimentarius.

Regulatory status: The ammonium, potassium, sodium, and calcium salts of alginic acid are GRAS in the U.S. Code of Federal Regulations Title 21, CFR 184.1133, 184.1610, 184.1724 (as a firming agent, flavor adjuvant, stabilizer, and thickener up to specified levels, and 184.1187, respectively). Accepted in Europe as a food additive. There is accepted daily intake of "not specified" (no quantitative limit) for alginic acid and its edible salts [77]. The European Economic Community lists alginic acid and its edible salts as an approved emulsifier/ stabilizer in Annex I to Directive 95/2/EC. Product code number is E401, E402, E403, and E404 for sodium, potassium, ammonium, and calcium alginates, respectively.

Health and safety: The toxicological properties of alginates have been extensively investigated and summarized [78–81].

Use: Thickener, suspension stabilizer.

Xanthan Gum

Structure:

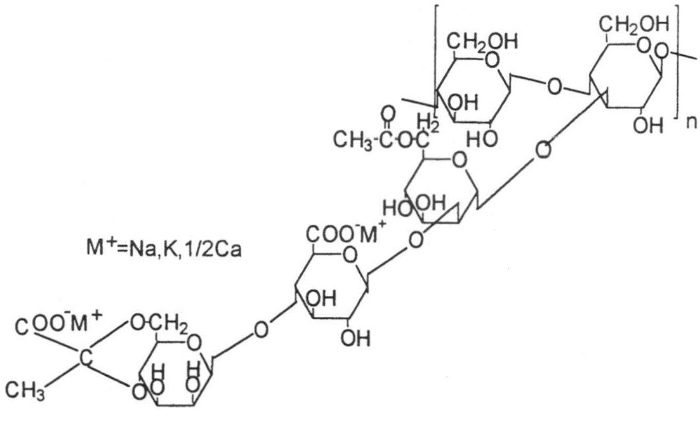

Description: A polyanionic heteropolysaccharide polymer produced by the aerobic fermentation of glucose having a cellulose backbone (β1,4 - *D*-glucose linked) consisting of five sugar residues—two glucose, two mannose, and one glucuronic acid [82,83].

CAS number: 11138-66-2.

Generic nomenclature: Xanthan gum, corn sugar gum.

Appearance: Cream- to white colored odorless powder.

Solubility: Soluble in hot and cold water.

Pharmacopeial listing: NF, FCC, Codex Alimentarius.

Regulatory status: A direct food additive in the U. S. Code of Federal Regulations Title 21 CFR 172.695 for use as a stabilizer, emulsifier, thickener, suspending agent, or foam enhancer. Acceptable daily intake of "not specified" (no quantitative limit) [84]. The EC product code number is E 415 in Annex I to directive 95/2.

Health and safety: Chronic toxicity studies in rodents and other animals indicated no adverse effects from the high dose ingestion of xanthan gum [85,86].

Use: Thickener, suspension stabilizer, suspending agent.

XIV. GLOSSARY

CAS: Chemical Abstract Service
DMF: Drug Master File
EP: European Pharmacopeia
FAO/WHO: Food & Agriculture Organization/World Health Organization
FCC: Food Chemicals Codex

FDA: Food & Drug Administration
GRAS: Generally recognized as safe
IPEC: International Pharmaceutical Excipients Council
JECFA: Joint Expert Committee on Food Additives
JP: Japanese Pharmacopeia
JPE: Handbook of Japanese Pharmaceutical Excipients
JSPI: Japanese Standard of Pharmaceutical Ingredients
LD_{50}: A lethal dose for 50% of a specified organism
MSDS: Material Safety Data Sheet
NDA: New Drug Application
NF: National Formulary
PhEUR: Pharmacopeia Europa (European Pharmacopeia)
USDA: United States Department of Agriculture
USP: United States Pharmacopeia
USP/NF: National Formulary to the USP

ACKNOWLEDGMENT

We wish to acknowledge the assistance provided by Dr. Khurshid Iqbal.

REFERENCES

1. W. Suida, *Monatsch*, 26:413 (1905).
2. German Patent 332,203 (Jan. 10, 1918) to Deutsche Celluloid Fabrik Eilenberg.
3. 21 C.F.R. Sec. 170.35 (1990). Affirmation of generally recognized as safe (GRAS) status.
4. D. Monkhouse, *Drug development and industrial Pharmacy*, 15(13): 2115–2116 (1989).
5. 21 C.F.R. Sec. 314.420 (1990) Drug master files. See also *Guidelines for Drug Master Files*, U.S. Dept. of Health and Human Services, Public Health Service, Food and Drug Administration, Sept. 1989.
6. J. L. Brown, Incomplete labeling of pharmaceuticals: A list of "inactive" ingredients, *New England J. of Medicine*, 439, Aug. 18, 1983. (FDA list is reprinted in the article.)
7. *Handbook of Pharmaceutical Excipients*, 2nd Ed., American Pharmaceutical Association, Washington, DC, and Pharmaceutical Society of Great Britain, London, England, 1994.
8. Scott vs. FDA, 728 F2d. 322 (6th Cir. 1984).
9. *New Excipient Evaluation Guidelines*, International Pharmaceutical Excipients Council, 1995. Wayne, NJ.
10. *Good Manufacturing Practices Guide for Bulk Pharmaceutical Excipients*, International Pharmaceutical Excipients Council, 1995. Wayne, NJ.
11. *Draft Guideline for Drug Master File*, Center for Drugs and Biologics, Food and Drug Administration, Washington, DC, Oct. 1987.
12. *Draft Guideline for Drug Master File*, Center for Drugs and Biologics, Food and Drug Administration, Washington, DC, Oct. 1987, p. 16.
13. Pinco, Robert G., Hurdling international barriers to existing and new excipients, *World Pharmaceutical Standard Review*, p. 14–19, Vol 1(5) 1991.
14. Final assessment report of the safety of carbomer 934, 934P, 940, 941, and 962, Prepared by expert board of Cosmetic Review board article available from administrative cosmetic ingredient review suite f10, 1110 Vermont Ave. N.W., Washington, DC 1005. *J. Am. Coll. Toxic.*, 1(2):109–141 (1982).
15. *Worldwide Monograph for Carbopol and Pemulen Resins TDS 229*, PHS Section 4, BFGoodrich Cleveland, OH, June 1994.

16. *Standard Test Procedure SA-005*, BFGoodrich, Cleveland, OH, Nov. 1994.
17. FAO/WHO evaluation of certain food additives and contaminants, *35th Rep. of the FAO/WHO Expert Committee on Food Additives, Tech. Rep. Ser. Wld. Hlth. Org. No. 789*, Geneva, 1990.
18. V. K. Rowe, H. C. Spencer, E. M. Adams, and D. D. Irish, "Response of laboratory animals to cellulose, glycolic acid and its sodium and aluminum salts, *Food Res.* 9:175–182 (1944).
19. *Aqualon Toxicological Data Bulletin T123C*, Aqualon Corp., Wilmington, DE, Dec. 1990.
20. Toxicological evaluation of certain food additives and food contaminants, 28th Rep. of *FAO/WHO Expert Committee on Food Additives, WHO Food Additives Ser. No. 19 WHO Tech. Rep. Ser. No. 710*, International Programme on Chemical Safety (IPCS), Geneva, 1984.
21. *FAO/WHO Expert Committee on Food Additives Rep. 27*, International Programme on Chemical Additives and Food Contaminants, The Toxicological Evaluation of Certain Food Additives and Contaminants No. 696 Geneva, 1983.
22. *Summary of toxicity information*, International Specialty Products Corp., Wayne, NJ, Nov. 16, 1994.
23. FAO/WHO evaluation of certain food additives and contaminants, *35th Rep. of the FAO/WHO Expert Committee on Food Additives, Tech. Rep. Ser. Wld. Hlth. Org. No. 789,* Geneva, 1990.
24. W. Deichmann and S. Witherup, Observations on the ingestion of methyl and ethyl cellulose by rats, *J. Lab. Clin. Med.*, 28(14) :1725–1727 (1943).
25. *Guide to the Safe Use of Food Additives*, 2nd Ser., Codex Alimentarius Commission, FAO/WHO Standards Programme, Rome, 1979.
26. T. S. Chen, *Guar Gum, Handbook of Pharmaceutical Excipients*, 2nd Ed., (A. Wade, P. J. Weller eds.), American Pharmaceutical Association, Washington, DC, 1994, pp. 215–216.
27. *Aqualon Toxicological Bulletin T 101F* Aqualon Corp., Wilmington, DE, April, 1987.
28. Meerman Santos, E. Final report on the safety assessment of hydroxyethylcellulose, hydroxypropylcellulose, methylcellulose, hydroxypropylmethylcellulose, and cellulose gum, *J. Am. Coll. Toxicol.,* 5(3):1–60 (1986).
29. [29]FAO/WHO evaluation of certain food additives and contaminants, *35th Rep. of the FAO/WHO Expert Committee on Food Additives, Tech. Rep. Ser. Wld. Hlth. Org. No. 789*, Geneva, 1990.
30. H. Kitagawa, T. Tokunaga, S. Ebihara, H. Kawana, and T. Satoh, Acute toxicity of HPC in mice and rats, *Oyo Yakuri*, 4(6):1013–1015 (1970).
31. *Klucel Toxicological Data Bulletin T122D*, Aqualon Corp., Wilmington, DE, June, 1990.
32. H. Kitagawa, H. Yano, H. Saito, Y. Fukuda, Acute, subacute and chronic toxicities of low substituted HPC in rats, *Oyo Yakuri*, 12(1):41–66 (1976).
33. FAO/WHO evaluation of certain food additives and contaminants, *35th Rep. of the FAO/WHO Expert Committee on Food Additives, Tech. Rep. Ser. Wld. Hlth. Org. No. 789*, Geneva, 1990.
34. S. B. McCollister, R. J. Kociba, and D. D. McCollister, Dietary feeding studies of MC and HPMC on rats and dogs, *Food Cosmet. Toxicol.*, 17(6):943–953 (1973).
35. R. K. Kulkarni, E. G. Moore, A. F. Hegyelli, and F. Leonard, Biodegradable polylactic acid polymers, *J. Biomed. Mater. Res.,* 5:169 (1971).
36. V. W. Dittrich and R. C. Schulz, Kinetics and mechanism of the ring-opening polymerization of L-lactide, *Angew. Makromol. Chem.*, 15:109 (1971).
37. C. G. Pitt, M. M. Gratzel, G. L. Kimmel, J. Surles, and A. Schindler. Aliphatic polyesters, 2, The degradation of poly(D,L-lactide), poly(*e*-caprolactone) and their copolymers in vivo, *Biomaterials*, 2:215 (1981).
38. T. R. Tice, D. H. Lewis, R. L. Dunn, W. E. Myers, R. A. Casper, and D. R. Cowsar, Biodegradation of microcapsules and biomedical devices prepared with resorbable polyesters, *Proc. Int. Symp. Control Rel. Bioact. Mater.*, 9:21 (1982).

39. FAO/WHO evaluation of certain food additives and contaminants, *35th Rep. of the FAO/WHO Expert Committee on Food Additives, Tech. Rep. Ser. Wld. Hlth. Org. No. 789*, Geneva, 1990.
40. S. B. McCollister, R. J. Kociba, and D. D. McCollister, Dietary feeding studies of MC and HPMC in rats and dogs, *Food Cosmet. Toxicol.*, 11(6):943–953 (1973).
41. Meerman Santos, E. Final report on the safety assessment of hydroxyethylcellulose, hydroxypropylcellulose, methylcellulose, hydroxypropyl methylcellulose, and cellulose gum, *J. Am. Coll. Toxicol.*, 5(3):1–60 (1986).
42. FMC Corp. Study Number 191-1188, Philadelphia, PA.
43. FMC Corp. Study Numbers 182-605, 182-610, 182–615, Philadelphia, PA.
44. FMC Corp. Study Number 192-1711, Philadelphia, PA.
45. FMC Corp. Study Number 192-1712, Philadelphia, PA.
46. Copenhagen Pectin product brochure 03.94.1500. Division of Hercules Inc., Lille, Skensved, Denmark.
47. BASF Material Hazard Report, Mt. Olive, NJ, March 9, 1994.
48. R. P. Geyer, Bloodless rats through the use of artificial blood substitutes, *Fedn. Proc.*, 34:1499–1505 (1975).
49. K. C. Lowe and C. Washington, Emulsified perfluorochemicals as respiratory gas carriers: Recovery of perfluorodecalin emulsion droplets from rat tissue, *J. Pharm. Pharmacol.*, 45:938–941 (1993).
50. *Standard Test Procedure SA-005*, BFGoodrich, Cleveland, OH, Nov. 1994.
51. *Noveon AA MSDS*, BFGoodrich Corp., Cleveland, OH, Oct. 1991.
52. *23rd Rep. of the FAO/WHO Expert Committee on Food Additives, Tech. Rep. Ser. WHO. No. 648*, FAO/WHO, 1980.
53. H. F. Smyth Jr. C. P. Carpenter, and C. S. Weil, The toxicology of polyethylene glycols, *J. Am. Pharm. Assoc.*, 39(6):349–354 (1950).
54. H. F. Smyth Jr., C. P. Carpenter, and C. S. Weil, The chronic oral toxicity of polyethylene glycols, *J. Am. Pharm. Assoc.*, 44(1):27–30 (1955).
55. *PEG 8000 MSDS*, Union Carbide Corp., Danbury, CT, Nov. 2, 1993.
56. S. M. Christopher, Carbowax methoxy PEG 5000: Acute toxicity and primary irritancy using rat (peroral toxicity) and rabbit (cutaneous and ocular), *Lab Project No. 93U1348*, Union Carbide Corp., Danbury, CT, 1993.
57. Material Safety Data Sheet, Union Carbide Corp., Danbury, CT, July 20, 1994.
58. Solutol HS-15, *Technical Information Bulletin MEF 151e*, April 1992 BASF Corp., Mount Olive, NJ.
59. Polyox resins BB-TL 4004, *Toxicology Summary*, Union Carbide Corp., Bound Brook, NJ.
60. B. R. Harris, U.S. Patent 2,258,892, Oct 14, 1941; U.S. Patent 3,637,774, V. K. Babayan, et al., Jan. 19, 1967.
61. *Internal Toxicology and Environmental Data Summary Communication*, BASF Corp., Mount Olive, NJ, Dec. 16, 1986.
62. *Cremophor RH-40 Tech. Bull.* D-205, June 1991, BASF Corp., Mt. Olive, NJ.
63. K. Woodburn and D. Kessel, The alteration of plasma lipoproteins by cremophor EL, *J. Photochem Photobiol.*, 22:197–201 (1994).
64. S. V. Balasubramanian, J. L. Alderfer and R. M. Straubinger, Solvent and concentration dependent molecular interactions of taxol, *J. Pharm. Sci.*, 83(10):1470–1477 (1994).
65. E. Sykes, K. Woodburn, D. Decker, and D. Kessel, Effects of Cremophor EL on distribution of taxol to serum lipoproteins, *British J. Cancer*, 70(3):401–404 (1994).
66. Evaluation of Certain Food Additives and Contaminants. WHO Technical Report Series, No. *751 FAO/WHO Rep.* 30th report, 30–31, Geneva, 1987.
67. B. V. Robinson, F. M. Sullivan, J. F. Borzelleca, S. L. Schwartz PVP, *A Critical Review of the Kinetics and Toxicology of Polyvinylpyrrolidone*, Lewis Publishers, 1990, Chelsea, MI.

68. A. Neumann, A. Leuschener, W. Schwertfeger, and W. Dontenwill, Study on the acute oral toxicity of PVP in rabbits, unpublished report to BASF, 1979.
69. P. Kirsch, F. Dati, K. O. Freisberg, H. Birnstiel, D. Mirea, and H. Zeller, Report on the study of the effects of Kollidon 90 when applied orally to rats over a 28 day period, BASF Gewerbehygiene und Toxikologie, 1972. Ludwigshafen, Germany.
70. M. V. Shelanski, Two year chronic oral toxicity study with PVP K-30 in rats, unpublished report, Industrial Biological Research and Testing Laboratories (USA) for GAF Corp., 1957.
71. Chronic oral toxicity of Kollidon K-90 USP XIX VERS NR 77-244 in Sprague-Dawley rats: Repeated dosage over 129/138 weeks, unpublished report, BASF Corp., 1980, submitted to JECFA, 1983.
72. FAO/WHO Expert Committee on Food Additives, *41st Rep. of JECFA, WHO Tech. Rep. Ser. No. 837*, Geneva, 1993.
73. C. F. Morgan, The effects of algin products on the rat, unpublished report, Georgetown University Medical School, Washington, DC, 1959.
74. J. W. Johnston, B. J. Lobdell, and G. Woodard, Safety evaluation of Kelcoloid and related algin products by repeated oral administration to laboratory animals: A review, Woodard Research Corp., unpublished report, 1964.
75. GRAS (generally recognized as safe food ingredients)—alginates, *National Technical Information Service 221-226.*, N.T.I.S., 1972.
76. W. H. McNeely and P. Kovacs, The physiological effects of alginates and xanthan gum. In: *Physiological Effects of Food Carbohydrates*, (A. Jeanes and J. Hodges, eds.), ACS Symp Ser. 15, American Chemical Society, Washington, DC, 1975, 269–281.
77. FAO/WHO Expert Committee Report on Food Additives, *39th Rep. of JECFA, WHO Tech. Rep. Ser. No. 828*, 1992.
78. J. W. Johnston, B. J. Lobdell, and C. Woodard, Safety evaluation of Kelcoloid and related algin products by repeated oral administration to laboratory animals: A review, Woodard Research Corp., unpublished report, 1964.
79. GRAS (generally recognized as safe food ingredients)—alginates, *National Technical Information Service 221-226.*, N.T.I.S., 1972.
80. C. E. Morgan, The effects of algin products on the rat, unpublished report, Georgetown University Medical School, Washington, DC, 1959.
81. W. H. McNeely and P. Kovacs, The physiological effects of alginates and xanthan gum. In: *Physiological Effects of Food Carbohydrates*, (A. Jeanes and J. Hodges, eds.), ACS Symp. Ser. 15, American Chemical Society, Washington, DC, 1975, 269–281.
82. P. E. Jansson, L. Kenne, and B. Lindberg, Structure of the extracellular polysaccharide from Xanthomonas campestris, *Carbohydrate Research*, 45:275–282 (1975).
83. L. D. Melton, L. Mindt, D. A. Rees, and G. R. Sanderson, Covalent structure of the extracellular polysaccharide from Xanthomonas campestris: evidence from partial hydrolysis studies, *Carbohydrate Research* 46(2):245–257 (1976).
84. Evaluation of Certain Food Additives and Contaminants FAO/WHO Expert Committee on Food Additives, *30th Rep. of JECFA, WHO Tech. Rep. Series No. 751*, Geneva, 1987.
85. Woodard G., Woodard M.W., McNeely W. H., Kovacs P. and Cronin M. T. I., Xanthan gum: safety evaluation by two-year feeding studies and a three-generation reproduction study in rats. *Toxicol. and Appl. Pharm.*, 24:30–306 (1973).
86. W. H. McNeely and P. Kovacs, The physiological effects of alginates and xanthan gum. In: *Physiological Effects of Food Carbohydrates* (A. Jeanes and J. Hodges, eds.), ACS Symp. Ser. 15, American Chemical Society, Washington, DC, 1975, 269–281.

8

A Practical Guide to Equipment Selection and Operating Techniques

Roy R. Scott

ARDE Barinco, Inc., Norwood, New Jersey

S. Esmail Tabibi

National Cancer Institute, National Institutes of Health, Bethesda, Maryland

I. INTRODUCTION

The scale-up of a liquid-dispersed pharmaceutical product from the laboratory beaker to the pilot plant and then to the production facility is a major undertaking, where early intelligent decisions can yield significant economic benefits and also prevent major headaches and delays. There are many types and variations of mixing, dispersion, emulsification, and size-reduction equipment that can be used to prepare these dispersed systems. By selecting a given type of equipment, certain operating techniques are made possible as a result of the individual capabilities of that equipment. Since it is essential that the product manufactured on a large scale be exactly the same as that made in the laboratory and in the pilot plant, it is important to select laboratory equipment that can be effectively scaled up without production limitations. In this chapter we assume that a product has been successfully produced on a small scale in a laboratory. Given this promising start, the next decision—how to make the product and with which type of equipment—will bear heavily on the chances of the product being produced at the lowest cost, with the best reproducibility, and with the least chance of production delays.

Dispersed systems fall into two categories. First, there are the emulsions, which are defined as physically semistable mixtures of two or more immiscible liquids with a combination of surface active agents, stabilizers, and emulsifiers that promote physical stability and the function of the product. Second, there are suspensions, which are physically semistable mixtures of finely suspended solids in a liquid with other functional ingredients, including stabilizers and dispersing agents. In addition, pharmaceutical products contain an active agent, which may be solid or liquid, which is dispersed into the emulsion or dispersion base. Since this active agent provides the desired therapeutic effect, it is very important that the active agent be uniformly dispersed throughout the product. This means that each individual sample removed from the bulk of the preparation vessel must have the same concentration of the active agent if the therapeutic effect is to be predictable.

Finally, in order for a pharmaceutical product to be approved in the United States, the entire process must be carefully controlled and recorded for subsequent evaluation. This is the procedure of validation of a production process.

II. DISCUSSION OF PARAMETERS

It is clearly impossible to devise a logical and organized method of planning a procedure to mix, disperse, or emulsify a multicomponent system without realizing that certain properties of these components determine how well a given procedure will work. There are many different factors that influence the process, but the most important parameter is the viscosity. The viscosity of the original components, the final viscosity, and the viscosity at any point in the process must be known. The next parameter required is the density of the individual components and also the average density. If a suspension is to be made, the particle size, the particle size distribution, the particle shape, and also the strength of the individual particles must be known. If an emulsion is to be made, the surface tension of each component must be known, along with the required information of chemical activity of the liquid phases, surfactants, and stabilizers. Another parameter to be considered is the order of combination, particularly with emulsions [1]. These parameters must be known first, if a production method with a high probability of success is to be tried.

A. Relationship of Mixing Equipment to Viscosity

Mixing equipment uses a mechanical device that moves through a liquid at a given velocity. This creates velocity gradients or shear rates that will produce—depending on the viscosity—a given amount of flow. In addition, the shear rate combined with the viscosity causes shear stresses that will act on the liquid and any solid particles that are in the liquid.

If a product development technologist knew the strength of the solid particles that needed to be broken to a given size, it would be a simple matter to specify a machine that, at a given shear rate and at a given viscosity, produces the required amount of force to break the particle. Unfortunately, in actual practice, this is not possible for a number of reasons. First, there are very few liquid mixtures that have only a single viscosity. Liquids in which the ratio of the shear rate to the shear stress is constant are known as Newtonian, giving credit to the person who first put these concepts into practice. Most liquids used in pharmaceutical dispersed products are non-Newtonian, meaning that the shear rate to shear stress ratio varies depending on the shear rate. Fortunately, a large number of the dispersed products (emulsions and suspensions) exhibit viscosity phenomena known as pseudoplasticity, Bingham plasticity, or thixotropy. A comprehensive discussion of the flow properties of emulsions is provided by Walstra [2]. Many dispersed products exhibit a combination of these and, for this reason, the description "shear thinning" is used.

But, what will be the shear rate developed by the mixer? More accurately, what different shear rates exist in the mixing vessel during the mixing process? This is the second problem, leading to the specification of the type of mixing equipment based entirely on a straightforward interpretation of physics. It is not this simple. As mentioned above, shear rate is defined as the velocity gradient or the rate of change of velocity. A mixing impeller causes the liquid to flow away from itself at a given velocity. In a simple example, velocity decreases as the distance from the impeller increases. Mixing equipment designed for dispersion or emulsification is more complicated. The impeller or impellers are fashioned in a way to ensure that the product flows evenly to all areas of the vessel. As a consequence, many different shear rates exist at different localities in the mixing vessel.

Dispersing and emulsifying equipment is most often described as "high shear" mixing equipment. This refers to the maximum shear rate, which usually occurs very close to the mixing impeller. Where the shear rate is high enough, the viscosity may be low enough with a shear-thinning mixture to provide very good flow. However, if the shear rate is low in another location, the viscosity may be high enough to halt flow altogether. This is referred to as the dreaded "dead spot" in an improperly specified mixer into which components of the mixture can flow and remain. It is clearly a situation to be avoided.

B. Microscale Mixing Versus Macroscale Mixing

An understanding of microscale and macroscale mixing is required for a logical discussion of the specialized capabilities of equipment used for disperse delivery systems. Macroscale mixing refers to the requirements of adequate flow in all areas in the mixing vessel. There must be a sufficient flow of the components from side to side and top to bottom of the vessel to prevent any stratification that could cause unacceptable concentrations of components at different locations within the process vessel (lack of dose uniformity). This process is also known as blending. Suffice it to say that there is no chance of producing an acceptable product if macroscale mixing is inadequate. But, even if the macroscale mixing is vigorous, with product components constantly pumped from top to bottom and throughout the mixing vessel, this is no guarantee that an acceptable product will result. This mixing will be successful only if, on a microscale, the individual components are separated or dispersed to the correct particle size and particle size distribution established by product development. For example, the solid particles in an oral product must be small enough to preclude grittiness, or the droplet size distribution of an oil-in-water emulsion might need to be sufficiently small to attain the required degree of physical stability. When the mixer is capable of producing the correct qualities on a microscopic scale then the microscale mixing is successful [3].

C. Mixer Power Equation

As discussed previously, all mixers must be able to pump the contents of the mixing vessel and also shear the contents in such a way as to produce a desired microscopic result. Pumping can be defined as the work of lifting a given mass of liquid through a vertical distance. The amount of work required to lift a given volume of liquid also depends on the density of the liquid. Most pharmaceutical dispersed delivery systems have densities similar to that of water. However, there are some high solid-content suspensions that do have a specific gravity higher than water. Shear stress—force exerted on an area—is also produced by the mixer. The amount of shear stress depends on the viscosity of the product and also the ability of the mixer to produce a shear rate.

Both the pumping work and the shearing work are produced mechanically by the mixer impellers or within the pipelines and passages of the high-shear homogenizers. In the impeller mixers, the power required to do this work is transmitted down the mixer drive shaft from another device, usually an electric motor.

Estimating the power requirement of a mixer, even when the properties of the components to be mixed are available, is difficult. If the data are available from the laboratory mixing experiments, the extrapolation to pilot plant or scale-up conditions is possible only if one maintains the geometric, dynamic, and kinetic similarity. The geometric similarity means the constant ratio of any pair of the dimensions between the small and the large mixer. The kinetic similarity relates to the geometric similarity of the flow path of the liquid. On the other hand, dynamic similarity is the constant ratio of forces acting on the liquid at all points in both small and large mixers. Thus, any analysis is complicated. Since the only parameters one can readily challenge are the impeller shape, size, and speed of rotation, the use of recommendations resulting from the analysis may be difficult [4–6].

The forces involved in mixing are the inertial forces due to the impeller, the viscous force due to resistance to flow, and the gravitational force related to the weight of the liquid [4,6]. For a given geometrical arrangement, the power consumption can be expanded by the following reasonable relationship:

$$P=f\,(g,D,\eta,\rho,N) \tag{1}$$

where P = power; D = impeller diameter, N = angular velocity (speed of rotation); g = acceleration due to gravity; η = liquid viscosity; and ρ = liquid density. By simple dimensional analysis one obtains:

$$\frac{P}{\rho N^3 D^5} = C\left(\frac{\rho N D^2}{\eta}\right)^{-a} \cdot \left(\frac{N^2 D}{g}\right)^{-b} \tag{2}$$

where C = a proportionality constant; and exponents a and b are determined experimentally. The work done by the mixer, the power number, is the group on the left side of the equation. Those on the other side, are the Reynolds number—representing the ratio of inertial force to viscous force—and the Froude number—dealing with the gravitational effects. The Reynolds number in its normal form consists of a velocity element proportional to the "peripheral linear velocity" of the impeller tip, i.e., ND. Therefore, the diameter of the impeller and the impeller speed have a large influence on power P due to their exponential effects as expressed in Eq. (2).

Examination of Eq. (2) reveals that at lower impeller rotation the flow is laminar and at that the exponent a of the Reynolds number is very close to 1 in this range. Since there is no vortex formation under laminar flow conditions, the exponent b of the Froude number is zero. However, upon reaching some turbulent flow conditions (the Reynolds number in the range of 10,000–100,000) both of the exponents a and b in Eq. (2) approach zero, rendering the power number equal to the experimentally determined constant. The following example will demonstrate the point. Consider a disperse delivery system of a 100 centipoise viscosity and a density of 1.02 g/cm^3, which is agitated in an unbaffled mixing tank by a centrally mounted 10 cm propeller at 1500 rpm. What type of flow exists in this process and what are the values of exponents a and b in this system?

*Reynolds number $= \rho ND^2/\eta = 1.02 \times 1500 \times (10)^2/1 \times 60 = 2550$

Froude Number $= N^2D/g = (1500/60)^2 \times 10/981 = 6.4$

The calculated values of the Reynolds and Froude numbers indicate that the flow is laminar and the greater contribution is provided by the Reynolds number. Therefore, the value of the exponent *a* is close to 1 and that of the exponent *b* is close to zero. To provide faster mixing, the formulator decides to use a 20-cm propeller at 3500 rpm. Similar calculations reveal that the values of Reynolds and Froude numbers are 23,800, and 70, respectively. A turbulent flow condition results from the large value of the Reynolds number and exponent *a* is obviously much closer to zero.

The work done by the mixer divided between pumping and shearing can be easily assessed by rearranging the power number [6]

$$P = \underset{\text{(shear)}}{C_1N^2 D^2} \; \underset{\text{(pumping)}}{C_2 N D^3} \tag{3}$$

The diameter of the mixing impeller has an important role in the power consumption with an overriding influence on pumping. Any change in the speed of the impeller affects the linearity of the pumping rate. The design of the different types of mixing equipment for the production of dispersed delivery systems must consider these relationships. The impeller speed and its diameter should be set in order to provide the required amount of pumping for macroscale mixing, and they provide at the same time the shear that is required to produce the desired microscale dispersion, emulsification, or size reduction.

This change in design requires a more powerful motor to drive the propeller shaft. Using similar conditions, the increase in the power requirement may be calculated based on Eq. (3)

$$\frac{P_n}{P_o} = \frac{20^3\left[\dfrac{3500}{60}\right]^5}{10^3\left[\dfrac{1500}{60}\right]^5} = 553$$

The result indicates that the increased mixing speed based on this change in design cannot be justified.

The diameter of the mixing impeller and the rotation speed combine to develop the maximum speed or tip speed of a mixer. The tip speed (peripheral linear velocity) of the impeller is a function of the diameter *D* of the rotating impeller and the speed of rotation *N*

$$\text{Tip speed} = ND\pi \tag{4}$$

On the other hand, if different types of mixing impellers are compared, the amount of pumping capability and the amounts and level of shear are not entirely dependent on these measurements. Indeed, different types of impellers create different amounts of shear or pumping even though they may have the same diameter and speed of rotation. These are reflected in the constant factors of the above power equation. In other words, this correlating equation hardly solves all the problems. However, it does enable mixing data, for a mixer geometry on hand, to be presented coherently.

A great deal of research has been completed on open impeller mixers of the radial type, known as the Rushton turbine, and also on axial flow mixers, from simple pitched-blade mixers to the computer-designed variable pitch impellers. One conclusion of these analyses is that an axial flow, pitched-blade impeller is able to produce four times the flow of a Rushton turbine for the same amount of horsepower [7]. Therefore, one cannot compare mixers, especially high-shear and high-speed mixer dispenser/homogenizers, on the basis of horsepower draw alone. A mixer with an axial flow impeller should be able to provide the same pumping level with 5 horsepower as that supplied by a radial flow mixer with 20 horsepower.

Examination of Eq. (2) reveals that at low values of the Reynolds number (turbulence free mixing) the power number is linear and independent of the Froude number. This suggests the potential difficulties of increasing the diameter of the impeller. In other words, a small increase in diameter causes a large increase in power consumption [4].

Pumping capability is the most important factor for to combining and blending components. The shear rate and, specifically, the maximum shear rate, determines the microscale dispersing and emulsifying capabilities of the machine. A wide selection of different mixing blades, rotor/stator gap configurations, and hybrid equipment, combining high-pressure pumps with special cavitating orifices, is available. There are even machines that use the forces of ultrasonic vibrating piezoelectronic crystals to enhance shear rates in liquid mixtures.

III. METHODS

Scheme 1 represents a simplified flow sheet for dispersed delivery systems where different mixing equipment at various stages of processing—depending on the components' parameters—may be involved [1]. However, details of the parameters' effect on the equipment selection to process disperse delivery systems are considered in this chapter.

A. Blending of Miscible Liquids

No dispersed delivery system—suspension or emulsion—can be made unless all of the product's components can be combined and evenly distributed. When the components are all liquid and miscible, the process is known as *blending*. This type of mixing is less demanding on equipment than high-shear dispersion, but this is not to say that no significant mistakes can be made on this, the "easy" part of many dispersed delivery systems procedures. A knowledge of blending is often required to complete the first step of preparing an emulsion.

The measure of success of a blending operation is whether the different components have been combined and whether the resultant mixture is homogeneous to a required degree. Since the liquids are miscible, it does not take a high level of shear to produce a homogeneous mixture. For this reason, blending applications are mostly related to pumping. That is, if the mixer can successfully pump the product's components around the mixing vessel, blending will result.

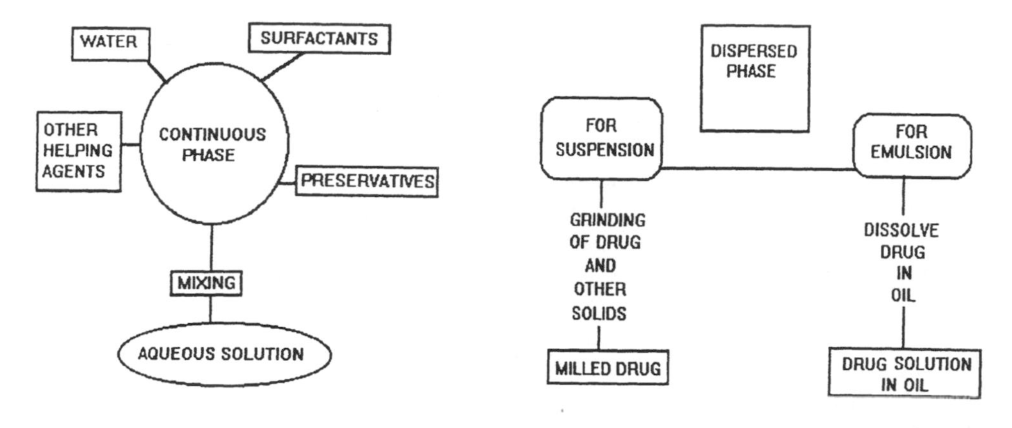

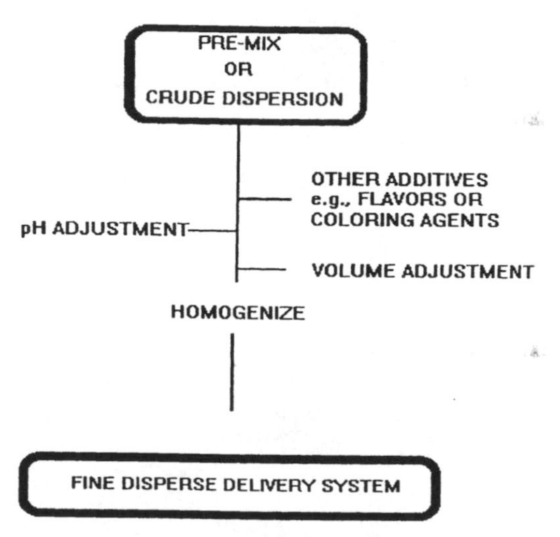

Scheme 1 Processing of Disperse Delivery System

The most often-used equipment for blending are pitched-blade turbines or marine-type impellers (Fig. 1). These designs combine good pumping capability without causing a great deal of turbulence or shear.

Whenever possible, components should be added together while mixing. This reduces power requirements and mixing time. For this to be possible, the impeller of the mixer must be covered by the first component before the second component can be blended in. For this reason, the impeller of the mixer is usually located near the bottom of the vessel.

Propeller mixers (Fig. 2) are used for most small batches. These can be top mounted or placed in a side-entry position. By orienting the shaft at an angle of 5–15° off the vertical, a marine propeller can be installed in a vessel without baffles with good

Fig. 1 Marine-type propeller. (From Ref. 8.)

blending results. Depending on the depth of the vessel, a second propeller may be required. Side-entry mixers normally provide good pumping action with a low initial cost. However, they are not currently in great demand in the pharmaceutical industry since they require an in-product seal mechanism that can be avoided with the top-entering designs.

For blending applications in larger batches over 500 gal. marine propeller mixers are no longer the optimum choice. Turbine agitators that operate in the 20–120 rpm range are best for a good combination of low-shear mixing and lowest initial cost. These types of mixers should be installed on the centerline of tanks. To prevent vortexing and to optimize mixing, the vessel should be equipped with four vertical baffles that are 1/12 the diameter of the vessel. For miscible-liquid blending and also solids-suspension

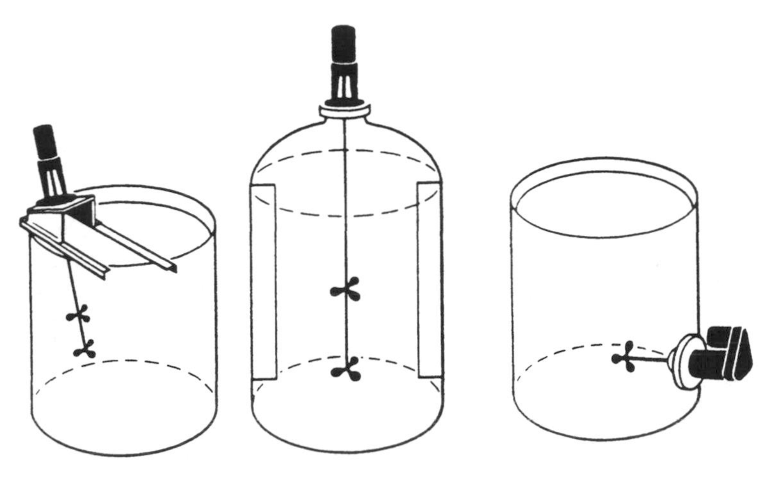

Fig. 2 Propeller mixer arrangements. (From Ref. 8.)

applications, axial flow impellers or the more modern variable pitch impellers are chosen for their ability to supply high pumping rates at low power requirements.

When the viscosity of the process exceeds 50,000 centipoise, turbine and propeller mixers are rarely able to provide the pumping action necessary for good blending. It is at this point that a number of other mixing problems may arise, as for example, an excessively long residence time of the product near the heat transfer surfaces. If the product remains close to the heating jacket of the vessel, product degradation could occur. Consequently, for the higher viscosity type of products, even slower and wider blade devices, such as anchor agitators, are best employed. These operate as large paddling mixers that actually push the product around the blending tank.

Every mixing process must at least pass the macroscale requirements of blending. In emulsification applications, separate batches of each component may only require the action of a propeller with high-shear agitation being unnecessary. Higher viscosities require the positive action of paddle or anchor agitators.

B. Suspension of Solids

In all dispersion applications and almost all emulsification applications, there is a requirement to suspend solids in liquid. If the solid is more dense than the liquid—as is usually the case—then the mixer must provide flow velocities high enough to overcome the settling velocity (based on the principles of Stokes' law) of the solid particles. If the solids tend to float on the surface, the mixer must be able to provide a flow pattern that causes the particles to enter the fluid. If the mixer is able to incorporate all of the required solids and hold them so that they are evenly distributed in the mixing vessel while the mixer is operating, then a successful suspension results.

If the settling velocity is low (less than 1 ft/min), the flow velocities exerted in a typical blending operation are usually sufficient to prevent stratification. Since the majority of suspension applications are for mixtures that will at some point become semistable (with settling velocities approaching zero), the premixing equipment required for such a suspension type of product is normally the same as that required for liquid blending.

If the percent of solids is above 50%, the suspension does not behave in a Newtonian way. Here Stokes' law does not closely apply, since the particles are hindering each other in their tendency to settle. In essence, Stokes' law is a worst case. If the mixing machinery can provide the flow necessary to overcome settling, then a satisfactory suspension can be attained.

Not all suspension problems are caused by settling. Some powdered solids actually resist breaking through the surface of the liquid when they are added to the batch. A variety of techniques exist to solve this problem. The most common approach uses a propeller mixer or offset turbine mixer in order to form a vortex in the liquid to provide a flow that could drive the solids from the top surface of the liquid downward toward the impeller of the mixer. Another method uses a submerged impeller in an axial-flow rotor/stator mixer, which pumps downward, thereby drawing the solids into the mixing head. Processes operating under vacuum can draw solids in from the bottom upward, allowing immediate dispersion of the solids if the solids are fed directly into the vicinity of the mixer. Finally, there are devices that are specifically designed to "wet out" solids continuously so they can subsequently be stirred into a batch (Fig. 3).

C. Dispersion of Solids

After all the solids have been incorporated and suspended, some dispersed products may be considered completed. However, if the final product must remain well dispersed for a desired time interval or if a smaller particle size is necessary, then additional pro-

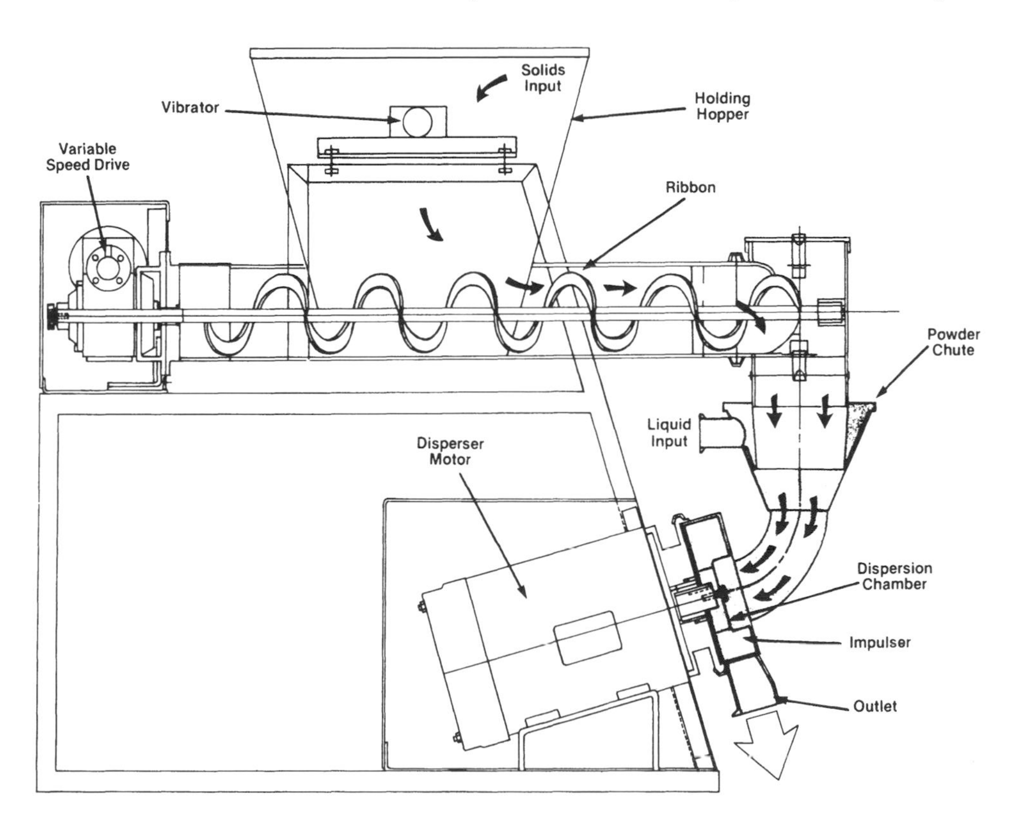

Fig. 3 Dilumelt, a device to wet out solids. (From Ref. 9.)

cessing must be done to reach the satisfactory end product. The simplest plan is to continue processing or mixing until the optimal particle size is achieved. This is possible only if the mixers used to incorporate the solids are also capable of exerting the shear rates required to subject the particles to the level of stress necessary to fracture discrete particles, or, at least, separate an agglomeration of particles.

After all of the components necessary to form a stable solids dispersion have been combined and are evenly blended, additional physical action may be necessary to reach the desired level of dispersion (see Scheme 1). This is done using a special device, commonly called a disperser. After all of the components are premixed in a vessel, which by itself is not capable of finishing the operation, then the product at this point is subjected to a last step of high-shear mixing, e.g., passing through a colloid mill or piston homogenizer. The disperser equipment is not always required to form a satisfactory product. Often, the machine used to combine and blend the components is also used to provide the necessary shear that will grind or separate the particles to the required size distribution.

The simplest method of producing dispersions is to load all of the components into a mixing vessel and "stir until done." If the equipment is capable of blending all of the components and then provides the required level of microscale mixing, this is the most direct approach. However, this is not necessarily the optimal method. If no single item or equipment can be found with all of the required capabilities, then the direct processing approach cannot be used. On the other hand, even if the equipment is available, there are often faster and more economical approaches.

For instance, a large-batch processing vessel having a high-speed rotor/stator mixer combined with a slow-moving, counter-rotating, scraped-surface agitator may be capable of combining, blending, and dispersing a product successfully. The initial capital cost for this type of equipment, however, may be too high for consideration. In such a case, using a series of less expensive premixing and blending vessels and a colloid mill may prove capable of producing an acceptable product at reduced cost. On the other hand, a multiple equipment arrangement may have unacceptable operating costs due to increased labor and floor space requirements. In addition, multiple steps in a process can increase the risk of microbial contamination. Isolation in one sterilizable, controllable, and monitorable piece of equipment may have an overriding influence on the selection of equipment, particularly if sterile processing is necessary.

Most decisions on equipment selection and methodology are not made based on the choice of new equipment. Usually a product is formulated by using equipment existing in the laboratory or the pilot plant. Ideally, the results of such processing can be scaled up since similar production-type equipment that has previously proven its capability of following from laboratory to pilot lab is available. This underscores the importance of using laboratory and pilot-scale machinery that can be readily applied to production equipment.

Some of the methods of dispersion of solids are described in Table 1.

If the solid particles are small and only loosely agglomerated, all that is required to effect dispersion is blending equipment. Sometimes this low-speed type of equipment is not able to incorporate one of the solid components—usually the stabilizer—and auxiliary equipment is required in order to aid in the initial incorporation of the ingredient. When the solids are difficult to disperse, high shear is required. This can be installed in a batch vessel or provided as a downstream second step in the process. Finally, these in-line devices can be used to provide all of the mixing and dispersion in a con-

Table 1 Dispersion Methodology

Process type	Method	Required equipment
Batch	Premix/blend/disperse with easy-to-disperse powders	Low-shear blending batch vessels, e.g., propeller or turbine or anchor mixers
	Premix/blend/disperse with easy-to-disperse powders that do not incorporate easily and tend to form large lumps	Option 1: above equipment with auxiliary equipment to provide initial solids wet out Option 2: replace low-shear equipment with high-speed equipment capable of drawing in powder and separating lumps Option 3: install mixer in vacuum vessel and draw solids into bottom of the mixer
	Premix/blend/disperse with difficult-to-disperse solids	Batch mixing and dispersion equipment capable of handling blending and incorporation while also dispersing, e.g., rotor/stator mixers or rotor/stator mixers combined with anchor agitators
Multiple steps	Premix/blend/disperse with difficult-to-disperse powders	Premix and blend as above and then pass mixture through an in-line high-shear device such as an in-line rotor/stator, colloid mill, three-roll mill, piston homogenizer, agitated bead mill, or hammer mill
Continuous	Premix/blend/disperse	Use continuous feeding devices to dose solids and liquids into an in-line rotor/stator, colloid mill, three-roll mill, piston homogenizer, agitated bead mill, or hammer mill

tinuous system if accurate feeding equipment is provided to feed and meter the various components of the product.

D. Emulsification of Immiscible Liquid Systems

An emulsion is a semistable mixture of two immiscible liquids combined with surfactants to change their compatibility with each other to enhance the temporary stability of emulsion. In some delivery systems, stabilizers are also added to increase viscosity. The equipment and methods needed to make the emulsion must provide for the addition of all the components in such a way that the desired droplets of the internal phase are dispersed evenly and possess the required droplet size and size distribution for stability. The process requires that all of the components be thoroughly blended and that the dispersed (internal) phase be sheared in the right way to yield the desired stable emulsion.

Formulations start in the laboratory, using emulsification equipment that can be applicable to scaled-up production-size batches. After setting up the laboratory with the emulsifying equipment that is deemed most suitable, one can begin formulation and processing studies. Given that there are three variables—formulation, equipment type, and method of addition—there are many things that can go wrong [1]. Fortunately, there is usually more than one formula and procedure that can provide an acceptable product.

The temperature of the emulsification process is often of paramount importance. If some of the components of one phase are solid at ambient temperature, they will need to be melted. This is a common requirement for oil phases that contain waxy or petrolatum-type components. Likewise, the temperature during combination and dispersion is often a key to success. Finally, the cooling step to bring the product back to ambient temperature must also be controlled. Typically, a trial will begin with some previous methodology to build on. But if no previous experience is available, the following heat transfer sequence is commonly used:

1. Melt all solid components of oil phase and bring all components of both phases to 70°C.
2. Maintain 70°C throughout dispersion step.
3. Cool to 40°C slowly and package.

Heating is normally supplied by condensing steam in the jacket (Fig. 4) of the premixing kettles, and cooling is supplied by circulating cool water through the same jacket. If the viscosity of the emulsion is over 10,000 centipoise, it is worthwhile to consider scraped-surface agitation in order to remove liquid that may become too hot or cold by residing near the heat transfer surface too long.

The phases must be prepared separately and then combined, and the internal phase must be dispersed. This can be done, for instance, by dispersing the oil into the water to make an oil-in-water emulsion. One recommendation that should be remembered is to carefully meter one phase into the other. If the two phases are haphazardly thrown together, there is no control during the processing as to what type of emulsion—O/W or W/O— is being formed in a given area of the vessel. It is beneficial to disperse one

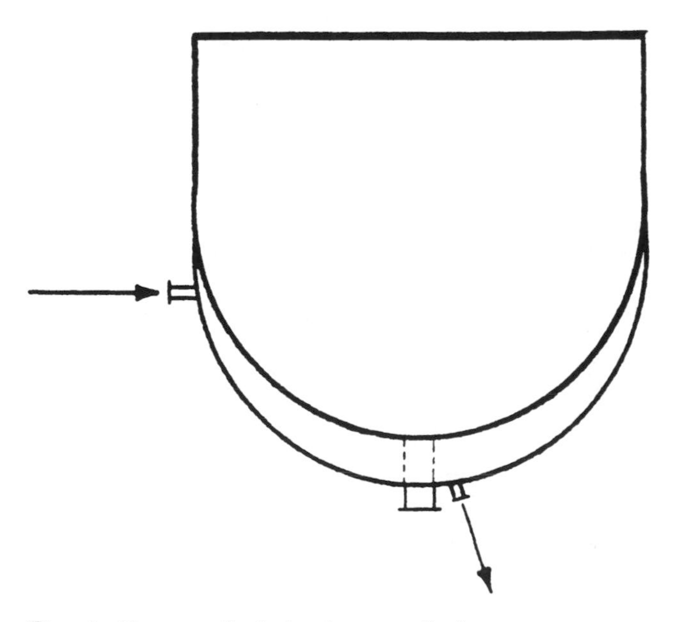

Fig. 4 Heat-transfer jacket (cross-section).

phase into the other by feeding it into the vicinity of the mixing/dispersing element. In this way, the phase being added is quickly dispersed into the continuous phase.

Although it is widely accepted that the higher the shear rate produced by the mixer the smaller the droplets and, hence, the more stable the emulsion, there is a major problem that must be avoided if good results are to be obtained with high-speed mixing equipment. Every effort should be made to avoid incorporating air into the mix. Air forms a third phase that could ruin emulsion stability in a number of ways. Air usually reduces the viscosity. The addition steps should be organized such that the impeller of the mixture is always submerged deeply enough to avoid surface turbulence or splashing. The arrangement of the mixer angle and/or baffles should avoid vortexing. Another alternative is to perform all of the emulsion-making steps in a vacuum-processing vessel. An additional method is to premix the components at low speeds and shear rates and then subsequently execute the high-shear portion of the process with in-line equipment in the absence of air. In short, aeration should be avoided.

Sometimes the direct approach is not the most effective one. When one phase is first added to another, the small amount of liquid being added forms the internal phase. If more of this liquid is added there comes a point where the continuous phase loses its ability to hold all of the internal phase and the emulsion inverts to the opposite type, e.g., from O/W to W/O. Since it has been found that this practice (phase inversion) can yield small droplet sizes, this method is widely used in batch processing. To execute this maneuver, one needs to begin mixing with only a small amount of liquid in a batch that will later increase to usually more than four times the starting volume. Therefore, the mixer has to extend well to the bottom of the vessel. One way to avoid this small volume of starting liquid is by using an in-line mixer in a recirculation loop attached to the main mixing vessel as illustrated in Fig. 5.

The initial phase is recirculated through the in-line high mixer and the phase to be inverted is then carefully metered directly into the recirculation line. This avoids

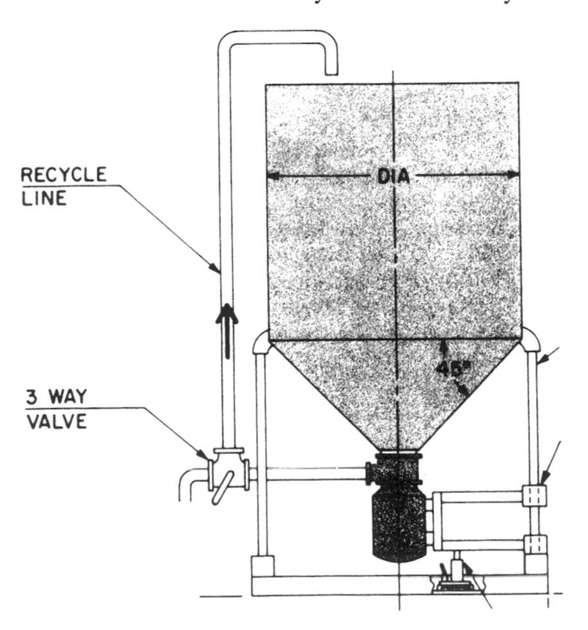

Fig. 5 In-line mixer in recirculation loop to kettle.

the problem of too little volume in the mixing vessel; no air is incorporated by a splashing mixer operating too near the surface, and the addition of one phase into another is easily accomplished.

An emulsion-forming process can reach several different levels of viscosity during manufacture. Often the range is too wide to specify one simple mixing device to handle all requirements. A number of different options are available to address this problem. If scraped-surface and/or counter-rotating agitation is required due to viscosities near the 50,000–100,000 centipoise level, there are designs that combine this type of mixer with a high-shear rotor/stator mixer. This enables a single vessel to work throughout the entire viscosity range. On the other hand, the process might be carried out by conducting the high shear in one kettle and then, for example, performing the cooling step in a second kettle. Usually, the high-shear mixer—whatever type—is used only during the important dispersion portion of the process. The slow cooling step uses only the low-shear mixer.

A summary of a number of different methods of preparing emulsions is shown in Table 2.

Just as with the dispersion of solids, the making of emulsions requires the blending of many components and then the microscale shearing of one phase into another.

Table 2 Emulsion Methodology

Production type	Method	Required equipment
Batch	Disperse oil or water directly into opposite continuous phase (low viscosities) after preparing individual phases	Batch mixing equipment necessary to give required shear rate ranges from propeller mixers or turbines to rotor/stator or other high-speed mixers
	As above (high viscosities)	As above except that anchor agitators with scraper agitators are probably required
	Batch inversion method	Option 1: use dispersion vessel with dispersing head located near bottom of batch to allow dispersion of major phase into minor phase Option 2: use recirculation through in-line device, e.g., continuous rotor stator
Multistep	Prepare phases separately and then premix into a lot that supplied as a feed to high-shear in-line device	Batch mixing equipment required to prepare and blend phases together and then use any of the following options: colloid mill, high-pressure piston type homogenizer, ultrasonic homogelnizer, or in-line rotor/stator mixer
	Prepare phases separately and then meter them continuously into the dispersion area.	Very similar to above except that the phases are not preblended; accurate liquid metering pumps are required to ensure correct concentrations of each phase
Continuous	Continuously feed components of both phases into a dispersing mixer	As above except that accurate metering devices are required to dose components (using the in-line devices in the multi-operation, but the phases are still initially prepared by the batch method)

The stability of the emulsion or lack of it depends on the chemistry of the components, the composition, the levels of shear exerted by the dispersing equipment, and the method, manner, and order of addition of the components [1].

E. Preparation of Liposomes

Liposomes are closed spherical structures delimited by one or more amphiphile bilayer(s). The bilayer arrays are more stable than micelles, because stronger and more specific cohesive molecular interactions such as hydrophobic effects, van der Waals attraction, and hydrogen bonding exist in the bilayer. Bilayer membranes form spontaneously due to the combinations of these forces and the unfavorable interactions of amphiphiles, particularly phospholipids, with water [10,11]. However, formation of the vesicles with desired size, size distribution, and structure, to entrap a sufficient quantity of materials without leakage after formation, is the paramount task. Various methods of manufacturing liposomes (lipid vesicles) involve the following three basic stages. First comes drying down the liquid soluble components of the liposomes from an organic solution onto a solid support. Second is the mechanical dispersion of the dried-down lipids after addition of the aqueous medium that contains all the water-soluble ingredients of the liposomes. Third is the purification of the manufactured liposomes by either centrifugation or gel permeation chromatography. Due to the specialized nature of this disperse delivery system, examining various techniques used in the manufacturing of the liposomes is beyond the scope of this chapter. However, a description of some new manufacturing techniques is described later. Martin has given an excellent review [12].

F. Scale-Up

After a successful batch of product has been produced using applicable laboratory or pilot-plant equipment, after a stable formula has been derived, and after the methods and order of addition of the components have been decided, the next step is to increase the scale of the process so that the desired product can be made more economically. This is known as scale-up. Inherent in any desire to scale-up must be an underlying requirement to produce a quantity of product in an acceptable amount of time. Only after a production rate is known can decisions on the size of the required equipment be made.

If the marketing department requires 1 million L of an ointment per year, the processing equipment must be found to achieve this goal in a desired time processing period. The pilot laboratory found that they could make 25 L of the ointment in 1.5 hr with a half-hour required to clean up; this would be a production rate of 100L/day on a one-shift basis. Given a 200-day year, only 20,000 L could be produced. One way to scale-up the process would be to install an additional 49 units of exactly the same size equipment and hire the additional staff to run it! Clearly, this does not seem reasonable. However, this is one way to be certain that the product made in the pilot plant is exactly the same as the product made for sale.

Another way to scale-up for the above example would be to make the product in a batch that is 50 times as large or 1250 L. This would work only if the 1250 L could be made in 1.5 hr with a half-hour of cleaning time. This is unlikely since larger batches almost always require longer production times. If the 1250 L batch took twice as long as the lab-size batch to produce and clean up, then something else must change. Two ideas would be to install two of the 1250 L units or operate the one on a two-shift basis.

The speed of the impeller has a direct relationship on the shear rates and shear stresses exerted on the product by the mixer. Examination of the second part at the right-hand side of Eq. (3)—the Froude number relates to the power required to pump the liquid—helps one understand the proportional relationships of scale-up described in Table 3.

The actual tip speed (peripheral linear velocity) of the impeller is a function of the diameter D of the rotating impeller and also the speed of rotation N. These factors combine in the power equation to give the amount of work per time or power that is required to run the mixer in the product. If the volume of the vessel is divided by the power being used, a ratio of power to unit volume P/V can be found.

As is described in the second part at the right-hand side of Eq. (3), the power required for pumping rate Q is proportional to the speed times the impeller diameter to the third power.

$$Q \propto N\,D^3 \tag{5}$$

Different impeller designs provide different amounts of flow or pumping rate, and this is related to the amount of time required for complete blending. If the pumping rate is divided by the total volume in the vessel, the pumping per unit volume factor Q/V can be calculated. Another useful factor combines the impeller diameter and speed with viscosity and density and is known as the Reynolds number, i.e., the first part of the right-hand side of Eq. (3). A final "factor" used in scale-up is the principle of geometric similarity shown in Fig. 6. Using geometric similarity, if the tank diameter is tripled, the diameter of the mixing impeller is also multiplied by 3. All of the above factors are used in deciding how to increase the size of a batch mixer. As was described earlier, attainment of dynamic and kinetic similarities in the scale-up of disperse delivery systems are difficult. Consequently, many of the factors are used when scaling up continuous equipment.

There is an often-held misconception that each unit of mixing machinery has a single scale-up factor that is easily multiplied in order to specify a larger mixer to do a larger volume batch. For instance, if a batch of 1000 L is to be made, and it is known

Table 3 Properties of Scale-up

Property	Pilot scale 5 gal. (0.019 m^3)	Plant scale 625 gal. (2.37 m^3)			
Power	1.0	125.0	3125.0	25.0	0.2
P/V	1.0	1.0	25.0	0.2	0.0016
Speed (N)	1.0	0.34	1.0	0.2	0.04
Turbin diameter (D)	1.0	5.0	5.0	5.0	5.0
Pumping (Q)	1.0	42.5	125.0	25.0	5.0
Q/V	1.0	0.34	125.0	0.2	0.04
Tip speed (ND)	1.0	1.7	5.0	1.0	0.2
Reynolds number	1.0	8.5	25.0	5.0	1.0

Source: Ref. 11.

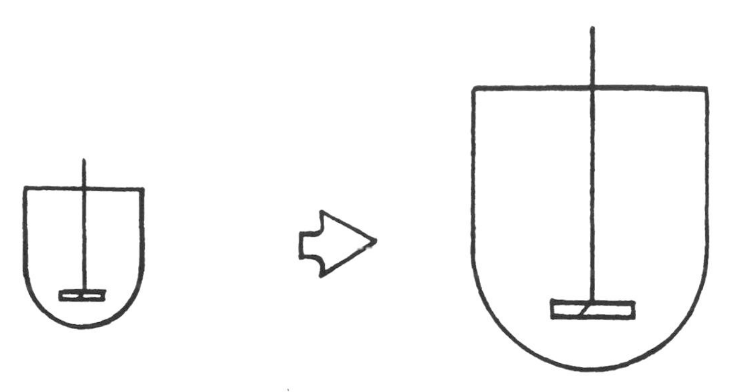

Fig. 6 Scale-up using geometric similarity.

that it required a mixer with 10 KW of power to make 100L, does it follow that a 100 KW motor will be required? Scale-up of process-mixing machinery is based on the comparison of not one single scale-up factor but, rather, a combination of factors. What clouds the issue is the fact that for a given process, a different scale-up factor may be dominant in assuring success. In addition, designing a machine with respect to the factors of power consumption; power per nit volume ratio; rotation speed; impeller diameter; conditions of geometric, dynamic, and kinetic similarity; tip speed; impeller flow Q; flow per unit volume Q/V; maximum shear rate; or average shear rate may yield a successful scale-up but also specify a system that is not economically feasible. Table 3 describes the proportional relationships of the above parameters for an open impeller turbine mixer with all pilot-plant values held to unity [13].

As shown in row one, the amount of power supplied must increase by the same amount as the increase in volume in order to maintain the same power to unit volume ratio. In this case, it is 125 times as much from a pilot-scale test of 5 gal. to a production test of 625 gal.

If pumping per unit volume Q/V is used as a direct scale-up factor, the power must be increased by an inordinate amount when speed is held constant. Pumping per unit volume is the factor that must be held constant if mixing time is to be held constant. Hence, most scaled-up batches are not completed in the same incremental mixing time. Insistence on the same mixing time for a pilot batch and a larger production-size batch is rarely feasible.

If the peripheral linear velocity (tip speed) is held constant, many would believe that perfect results are ensured. The tip speed does determine the maximum shear rate, but it does not have the same effect on the pumping per unit volume. Thus, as alluded to in the previous paragraph, mixing time must be increased.

An additional factor often used in scale-up is the concept of conditions of geometric, kinetic, and dynamic similarity. This method increases the diameter of the impeller in the same ratio as the size of the mixing vessel increases; this method works only in some cases.

In any scale-up, the key is to find the factor or factors that control the outcome of the process. This depends on the specific process. For dispersions, the controlling factor is often the tip speed, which determines the maximum shear rates. These rates in turn determine the ability of the mixer to exert the shear stresses in the liquid necessary to separate particles or make small droplets.

Several problems arise during the selection of a mixing machine, especially one of the specialized machines used for dispersed delivery systems. First, the machine chosen may be economically unfeasible. It may have too high a power requirement or too high a speed requirement. It may be a machine design that can only be custom-made at a high price. Manufacturers of equipment typically have standard designs capable of processing a volume of a given type of product. Optimization of a scale-up around only one parameter is not usually feasible. Some compromises normally are necessary.

Given that economic and commercial realities may obstruct an ideal scale-up, the most important thing to remember is that there are many different relationships at work. Oversimplification may lead to erroneous decisions.

Selection of the correct type of laboratory or pilot equipment is extremely important if usable results are to be obtained. Never use laboratory-scale or pilot-scale equipment that cannot be built and operated in larger sizes. The largest size that is available is a function of the manufacturer's capabilities and the market demand for equipment of that size. Fortunately, there are many suppliers of dispersion and emulsification equipment that can satisfy almost any scale-up requirements [14].

G. Vacuum Processing

As discussed in the section on emulsions, the presence of dispersed air is almost always a detrimental factor to emulsion stability. All mixing steps should be conducted so as to incorporate the least amount of air. Unfortunately, it is often the aim of the mixing equipment to exert shear at highest levels in order to achieve the desired dispersed particle size or emulsion droplet size. With open-top vessels, a compromise between shear rate and avoiding aeration or foaming is often difficult to attain. Frequently, the initial goal of preparing an emulsion or dispersion is the incorporation of components that tend to float on the surface of the liquid. Use of a high-speed, more powerful mixer should solve this problem but may cause downstream problems of instability or low-viscosity readings due to incorporated air.

One way of combating aeration problems is to perform the entire process in a vacuum. If the batch vessel is sealed, all of the mixing can take place while a vacuum pump withdraws the air originally in the vessel and the air introduced to the vessel when components were added to it.

As shown in Fig. 7, vacuum processing allows the option of drawing in components such as powders into the batch from the bottom outlet. This technique is very useful when the powder or the liquid phase can be drawn directly into the eye of the impeller for immediate dispersion.

Foam may form even during processing under vacuum. If a product tends to foam during atmospheric, open mixing, some interesting effects can be observed during vacuum mixing. First, if there is any air in the components as they are added, this air will cause a layer of foam to grow as the vacuum pump is pulling the initial vacuum on the process vessel. *Remember that, when pulling a vacuum on a batch of product that contains air, so much foam may be propagated that the foam may reach the top of the vessel and be drawn out by the vacuum pump.* Care must be taken to avoid this situation by varying the amount of vacuum and allowing the air in the foam bubbles to escape. Gentle mixing is best for this initial deaeration step. By decreasing the vacuum (increasing the absolute pressure), the pressure outside of the foam bubble is made

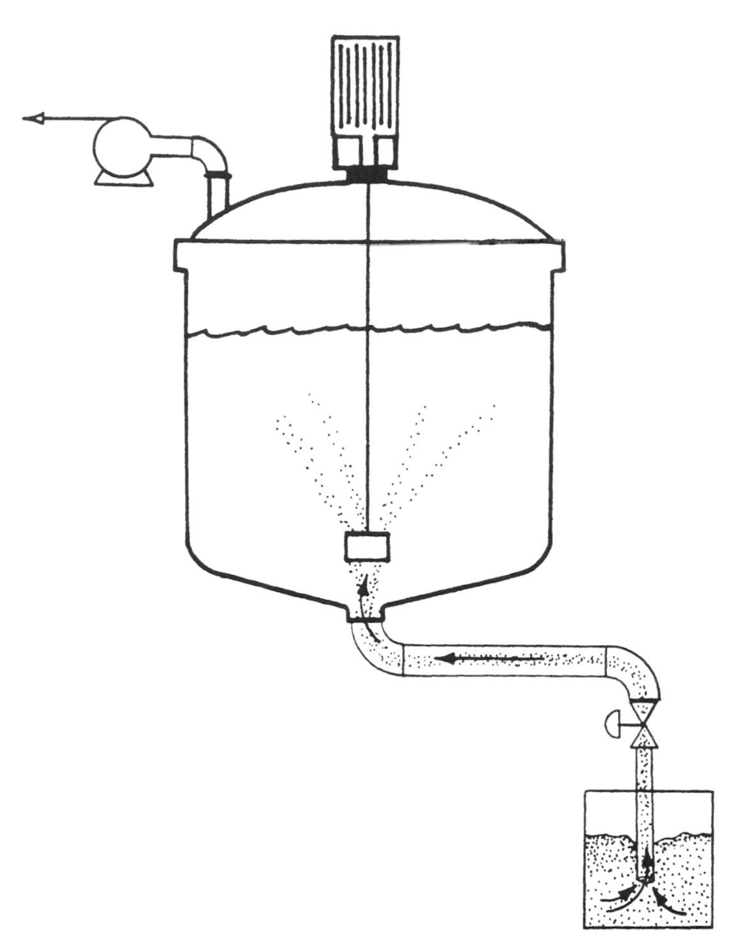

Fig. 7 Drawing solids directly into the bottom outlet under vacuum.

greater than that inside the bubble, and the result is a crushed bubble. By increasing and then decreasing the vacuum, the bubbles that contain the air can be removed, i.e., the batch can be deaerated. There are also devices available that perform deaeration on a continuous basis. For small batches this route may be a lower capital investment than mixing under vacuum in the dispersion equipment. Finally, as shown in Fig. 8, there are devices that can continuously incorporate solids into liquids while minimizing or totally eliminating air incorporation.

IV. MIXING/DISPERSION EQUIPMENT AND TECHNIQUES

A. Mixers

There is no reason to purchase and use complicated and expensive equipment to formulate and manufacture emulsions and dispersions, unless it is found through trials that the more capable equipment is required or offers an economically justified advantage. There are several questions that should be asked. Can this product be made using only simple open-impeller mixers? Is a propeller mixer able to incorporate all of the solids

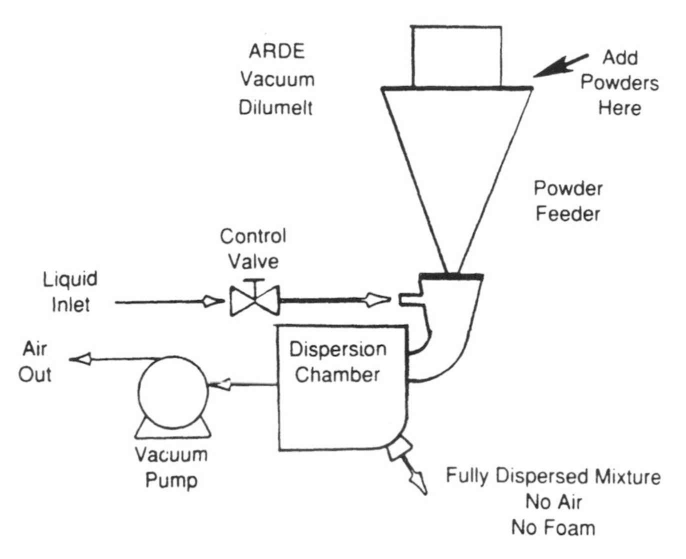

Fig. 8 Vacuum Dilumelt, a device for wetting out solids without incorporation of air. (From Ref. 15.)

and liquids? Can a turbine mixer blend all of the components? Finally, will open-impeller mixers offer the level of shear rate that is required to give the microscale mixing that is necessary for the process? In other words, will the impeller provide the separation of solid particle agglomerates and the shearing of liquid droplets to the microscopic level that is deemed necessary?

Powdered raw materials can be manufactured that require only very low shear rates for even dispersion into liquids. Such materials usually are called "high quality" and most often command a premium price. The advantage of using these materials, along with their convenience, must be compared to the alternative of using lower cost materials that must be processed in higher capability equipment.

Likewise, with emulsions, choices in the desired formula's (including surfactant mix and stabilizer type and amount) droplet size and its root of administration can have an overriding influence on equipment requirements. By changing surfactants, the difference in interfacial tension between the two immiscible liquids can be reduced, and this facilitates the formation of an emulsion. The stabilizer increases the viscosity, depending on the amount added. Viscosity, of course, increases the stability, since the unhindered settling rate in Stokes' law decreases with increased viscosity. By adding stabilizer, an acceptable, stable product may be obtainable using rudimentary equipment. These tactics must be measured against the capital and operating costs of higher speed, higher shear, and expanded-capability equipment.

1. Propeller Mixers

The most often used mixing implement is the ubiquitous marine propeller mixer. Often called a "prop" mixer for short, these machines use a rounded, pitched, three- (or four-) blade design that produces mostly axial flow. Propeller mixers provide good flow and, hence, blending capabilities in small batches of low to medium viscosities.

Propeller mixers (Fig. 9) are made in quantities by manufacturers, and hence they are offered at an attractive price. The impellers come in a range of diameters from a

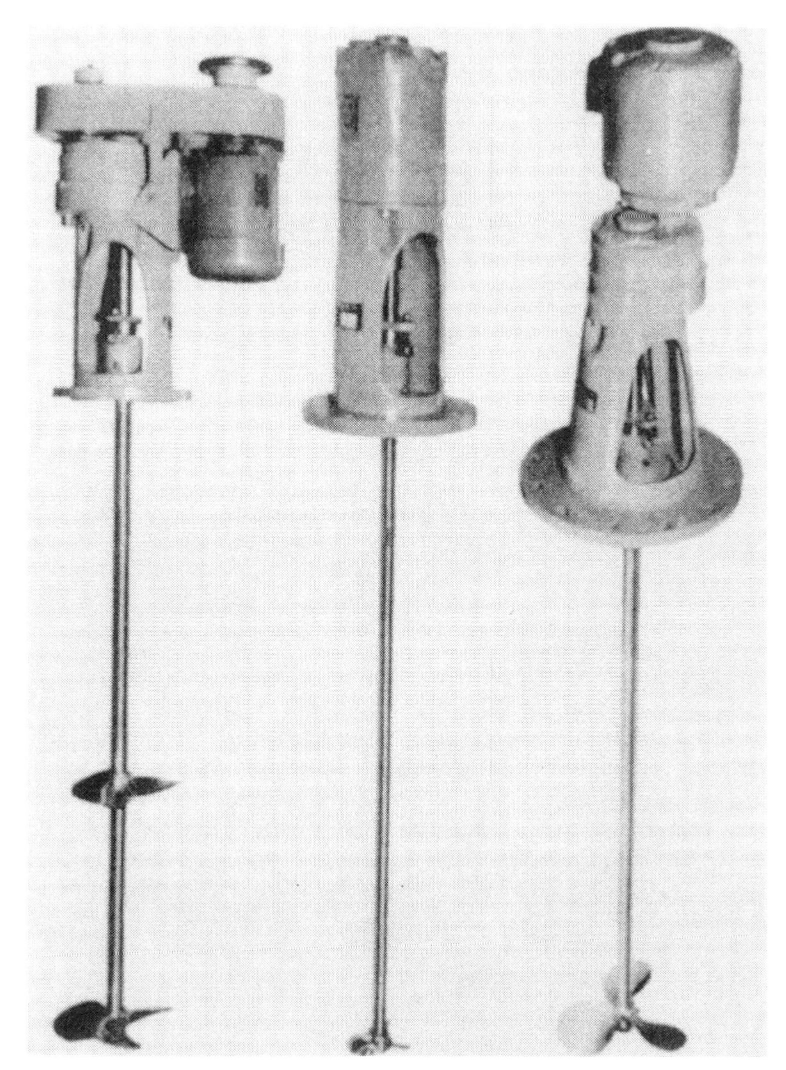

Fig. 9 Marine propeller mixer impeller. (From Ref. 16.)

few inches up to around 3 ft depending on the manufacturer. The impellers are of cast construction, which reduces the cost but also obviates customizing of the impeller without incurring added cost.

Propeller mixers can be installed on a vertical centerline, on an angle off the vertical, or through the sidewall of the process vessel. Centerline installation requires the use of baffles in the vessel. Side-entering designs offer the advantage of very effective pumping for low horsepower and low initial cost in large vessels over 1000 gal. However, side-entering designs must have a seal that has to contend with side loads of the cantilevered shaft. This has been a problem in some critical sanitary applications. However, if the batches are very large, a side-entering prop mixer may be a logical alternative to top-centerline-installed axial- and radial-flow turbines.

The most often used installation configuration is the top-entering angled method. This method produces a good circulation pattern in the process mixing tank without the installation of baffles. Angle-mounted mixers are usually clamp-mounted onto a support plate or channel affixed to the sidewall of the vessel. These are the "portable" mixer designs that are widely used. They offer ease of installation and low cost (Fig. 10).

Propeller mixers, typically, are operated at around 300–400 rpm or at high speeds around 1750 rpm, depending on the manufacturer's standards. Usually, lower speed units with larger impeller diameters are the best choices for blending applications. However, a high-speed propeller, which may not be optimized with respect to cost and horsepower, may provide the higher shear rates required for a level of emulsifying or of dispersing of solid agglomerates.

For drawing in solids from the top surface of the liquid, a second impeller is often specified. This works well in some instances. The propeller should not be so near the surface as to cause surface turbulence and aeration.

Fig. 10 Clamp-mounted portable mixer. (From Ref. 17.)

Styles of propeller marketing by different manufacturers vary. Hence, it is not possible to show a sizing chart for all prop mixers. These are available from the respective manufacturers, but care should be taken since most information on propeller sizing is for lower viscosities.

Since different impellers pump and shear differently, it is important to make sure that the size, speed, and power for the mixer be checked by the manufacturer of the mixer. However, a calculation cannot be made unless all of the information regarding viscosity, density, vessel configuration, and description of the desired process is obtained. This is true for all types of mixer specifications.

Propeller mixers are used mostly for liquid-liquid blending applications and some easily producible suspensions. They are often an important part of the process, since subsequent mixing steps may never be able to even out inadequately blended liquids or lumps of unmixed solids.

2. *Turbine Mixers*

The most versatile of all mixers in the entire span of mixing equipment are the axial and radial flow turbines. These machines can be designed flexibly with respect to impeller diameter, rotation speed, blade angle, number of blades, and blade shape. More research and development has been done on this group of mixers than any other. Unfortunately, not much of the research has been conducted on the subject of pharmaceutical dispersed products, but much of the knowledge is applicable to solving the specific problems of dispersing and emulsifying. Turbine mixers can be made to handle huge batches, even up to 500,000 gal., which are clearly out of the realm of pharmaceutical dispersed products. They are used optimally for large batches where propeller mixers are not the best choice. If it is known that the production scale batches will be over 1000 gal., then turbine mixers should be considered for use in the formulation laboratory and pilot plant.

Basic research on axial-flow turbines versus radial-flow turbines established that the axial-flow designs were able to produce higher pumping flow rates than radial-flow turbine mixers of similar horsepower requirements. The converse finding of the research was that radial-flow turbines provide a greater amount of shear to the product, i.e., their power is going toward shear instead of pumping. The maximum shear rate is based on the tip speed. Although axial-flow impellers are the best design for blending, they still do exert the same level of maximum shear rate of radial-flow impellers with the same diameter and rotation speed (rpm). Radial-flow impellers cause more turbulence behind the blades, which is also the site where the greatest amount of shear is occurring [18].

Axial flow impellers have been redesigned in the last decade to minimize turbulence and, hence, shear. This means that more of their power can be devoted to pumping. For blending applications, these variable-pitch airfoil impellers are very effective (Fig. 11).

Turbine agitators are best used in the preparation of large batches of over 500–1000 gal. For pure blending applications, axial-flow impellers are suggested. If a higher amount of shear than that provided by axial-flow impellers is desired, radial-flow or Rushton-type impellers (Fig. 12) are the best choice. Since there are many combinations of the variables of blade type, impeller diameter, rotation speed, blade width, and

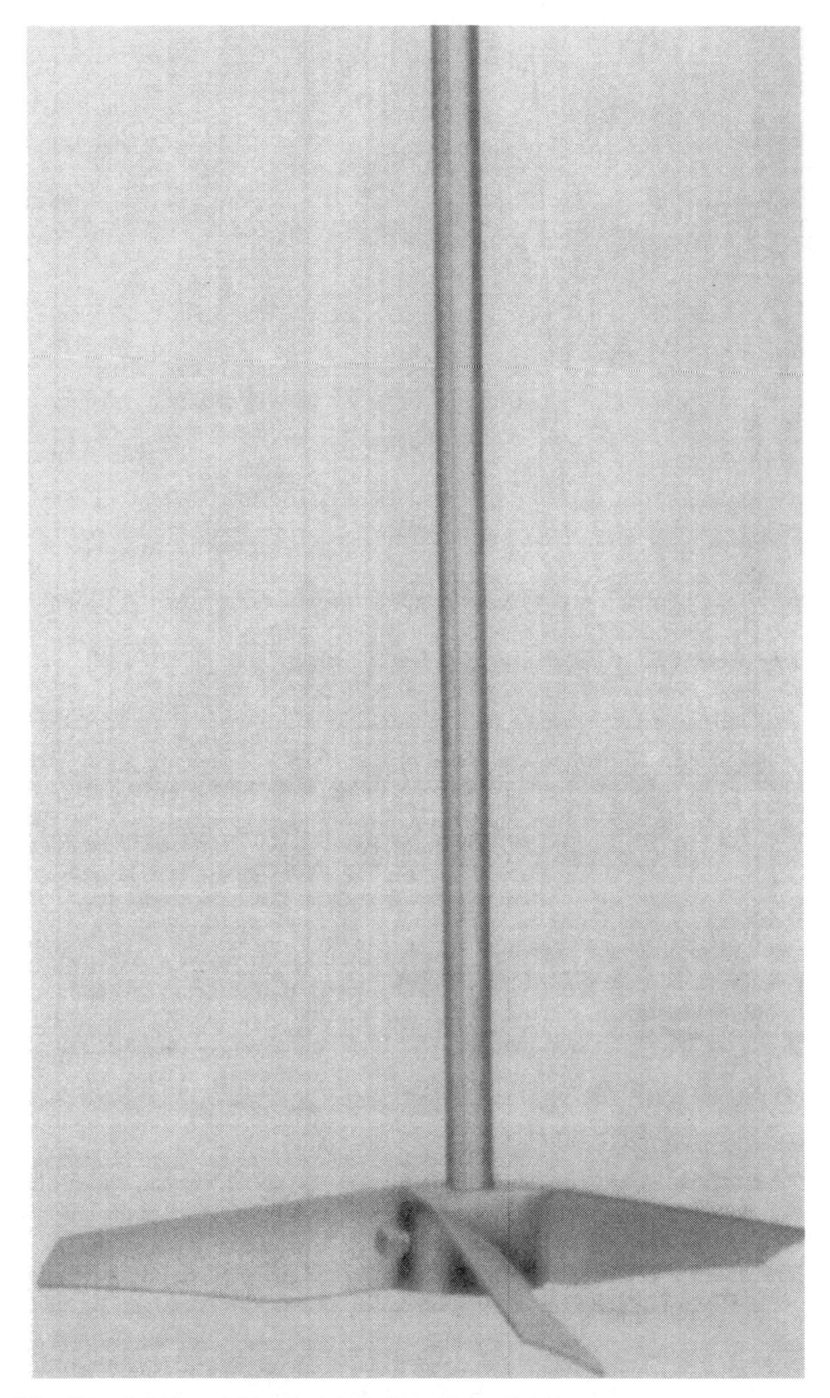

Fig. 11 A 310 variable-pitch impeller. (From Ref. 19.)

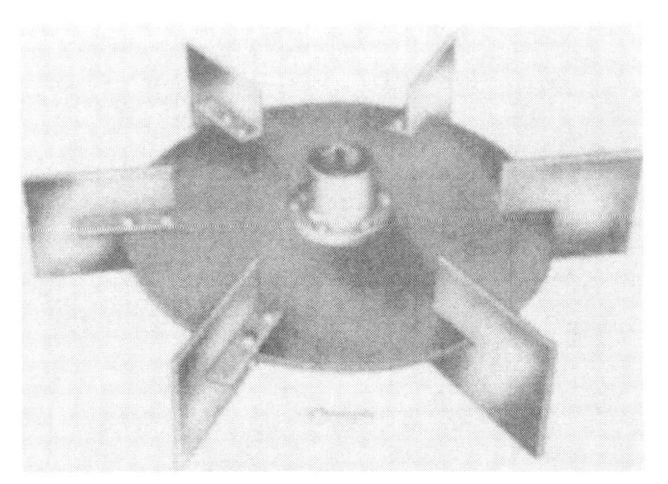

Fig. 12 Radial (Rushton) type impeller.

blade angle, it is best to work closely with the manufacturers of the mixer to specify an optimum design for the process.

The preceding discussion of axial- and radial-flow turbines has been a very cursory survey of what can be a very involved and detailed study. As mentioned above, a large amount of research on these types of mixers is available [13,14]. A detailed discussion of this subject would be beyond the scope of this work. If a blending or suspension problem occurs in large production batches, consultation of the references on mixing included at the end of this chapter or, even better, consulting the experts at the major manufacturers of this type of mixer, would be the best place to start.

3. Anchor Mixers

An often overlooked mixing device, which is low speed and considered low capability, is the anchor agitator, so named for its anchorlike shape, as illustrated in Fig. 13. However, this slowly moving agitator makes it possible for many dispersion and emulsification processes to be accomplished without overshear, aeration, and heat transfer problems.

The anchor agitator is a slow (up to 50 rpm) device whose sole function is to rotate the contents of a batch in a radial direction without providing any significant shear. These are high-torque devices that must be designed sturdily to withstand the forces of the high viscosities. Anchor agitators are typically designed to be able to withstand a maximum viscosity beyond which they might actually bend or break. That is, the anchor itself is built of materials strong enough to withstand the drag of the viscous liquid as it passes by the mixer. In addition, the motor has to supply the very high torque requirement that arises when the anchor is stirring viscous materials. When designing the mixer it is important not to understate the viscosity. This is especially important if there is a point in the process where the anchor must be stopped. If this happens, in the case of shear thinning materials, the agitator has to start up from rest in a viscosity much higher than that normally occurring during the process. Products exhibiting pseudoplastic or Bingham plastic behavior are very difficult to move when at rest.

Fig. 13 Anchor agitator.

The great flexibility of the anchor agitator arises from the fact that it can be combined with auxiliary agitators and mixers. Propeller, turbine, and high-speed mixers, including rotor/stator mixers, can be located directly beside an anchor agitator to provide multi-shear-level capabilities.

Variable speed is a valuable option with anchor agitators. They can be used to give vigorous rotation or a very gentle stirring for holding without causing aeration.

When anchor agitators are built with the anchor very close to the walls of the vessel they are known as swept-surface agitators. When flexible or movable blades are attached to the anchor for the purpose of actually scraping the sidewalls, they are known as scraped-surface agitators.

4. Scraped-Surface Agitators

By affixing a flat blade made of a polymeric material, such as nylon or Teflon, to an anchor agitator in such a way that the blade actually contacts the inner walls of the vessel, a number of distinct advantages can be obtained. High-viscosity materials do not readily flow away from the surface of the mixing vessel unless they are acted on by an impeller in very close proximity to the wall. Even with swept-surface agitators, the viscous product may be allowed to reside near the wall for too long and become too hot or too cold. This can cause the problems of "burn on," overheating, or "freeze on" whenever heat transfer is used. To avoid these problems, scraped-surface agitation is almost always *absolutely necessary* in emulsification vessels (Fig. 14).

In the initial emulsifying stage of the process, the temperature is typically maintained somewhere between 40 and 80°C. Heat is most efficiently supplied by condensing pressurized steam in an annular space that has as one surface the inner wall of the mixing vessel. Even though there is a temperature drop through the vessel sidewall, the temperature at the sidewall is higher than the desired temperature in the batch. If the product is allowed to remain at the sidewall too long, the prolonged heat on the product may

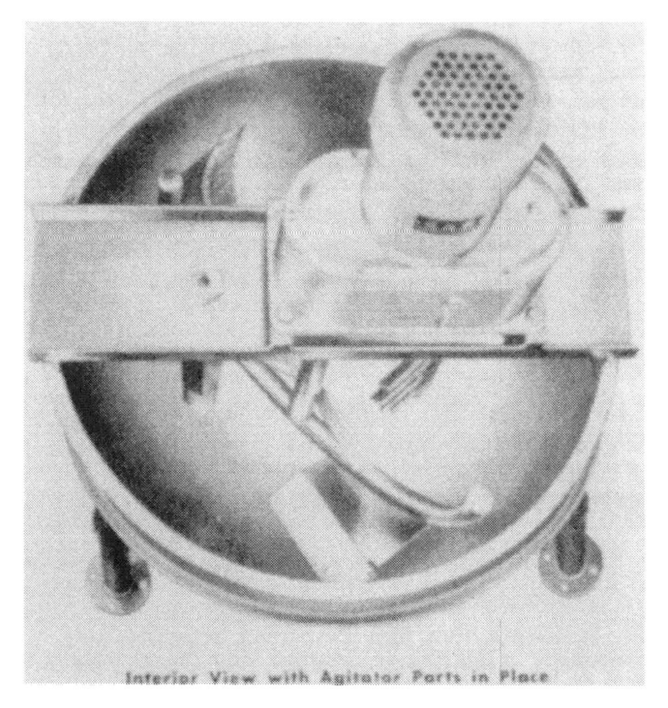

Fig. 14 Scraped-surface anchor agitator with auxiliary crossbar agitator. (From Ref. 20.)

have many deleterious effects on it. First, the emulsion may have components that cannot stand the wall temperature, which may be as high as 110–125°C. This is even more important if the dosage has active ingredients that decompose at these temperatures. Second, if the temperature is hot enough, the product may actually stick or burn on the sidewall.

Cooling of product through sidewall heat transfer can cause almost as many problems as heating. During cooling, the viscosity of a product almost always increases. A viscous product that is not physically removed from the sidewall builds up and forms an insulating layer than resists efficient heat transfer. Again, once this condition occurs, it is very difficult to reverse it.

There is a variety of different designs of scraper blades. Some are arranged in rows. Some are offset on either side of the anchor, allowing some overlap as an anchor makes a complete revolution. Some actually are designed to allow the anchor to revolve in opposite directions, which can prevent the buildup of product on the following edge of the anchor. Some designs use a spring to force the blade against the wall. Most modern designs use the force of the liquid flowing into the blade to bring it close to the wall.

Scraped-surface agitators are definitely required in emulsification equipment where heat transfers are necessary. These anchor agitators with scraping blades can be just as simple anchors or part of complex multishaft mixers.

5. Counterrotation

Anchor-type agitators have a decided weakness when handling high-viscosity products of more than about 75,000–100,000 centipoise. They tend to rotate only the product, without producing any appreciable velocity differences, so that almost no mixing occurs. By installing a stationary baffle, some mixing capability is added. This works well on materials with viscosities from 5000–25,000 centipoise. For the best results, with products between 100,000 and 250,000 centipoise, a counterrotating set of crossbars provides excellent blending. Such equipment is illustrated in Fig. 15.

Some designs use the same motor, turning a pinion gear between two opposing bevel gears to provide rotation in opposite directions. Others provide a greater degree of flexibility by driving the two shafts on separate motors. In either case, there will be a hollow shaft driving the anchor agitator and an additional shaft located inside the hollow shaft to drive the inner crossbars [21].

B. High-Speed Dispersers

A simple yet powerful device used extensively in industries other than pharmaceutical manufacturing for dispersion of solid particles in liquids is the high-speed disperser. Sometimes called a saw-blade disperser for the shape of the mixing impeller, this machine consists of a variable-speed shaft connected to an impeller with a serrated edge. The mixer is designed to rotate at a high speed in order to produce shear and pumping (Fig. 16).

This type of equipment is designed specifically to disperse powders, usually pigments, into liquids. Much has been written that high-speed dispersers are capable only of dispersing "easy" pigments [23]. This is true if the particles are hard agglomerates or individual hard particles with some strength. Furthermore, the high-speed disperser design is ineffective if the viscosity is low. The only shear stress that is delivered to particles is due to the hydraulic shear that is a product of the shear rate and the viscos-

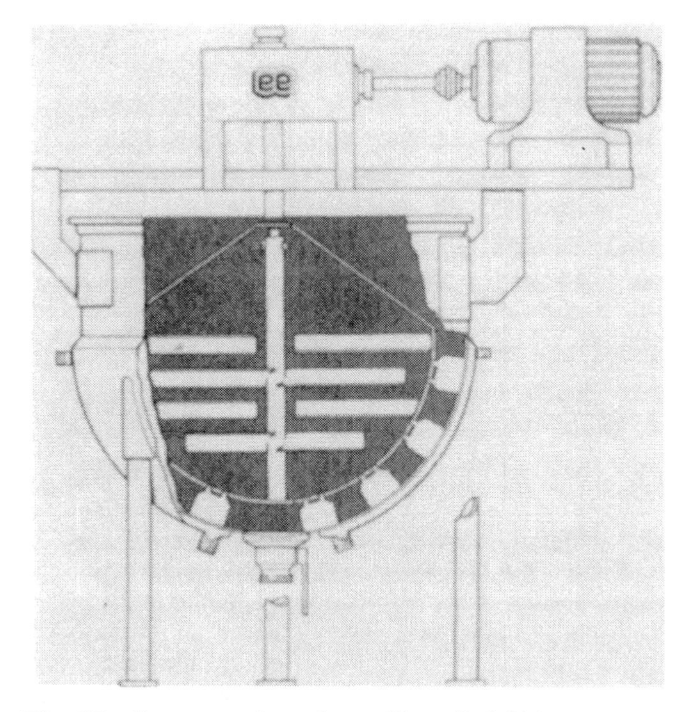

Fig. 15 Counterrotating agitator. (From Ref. 21.)

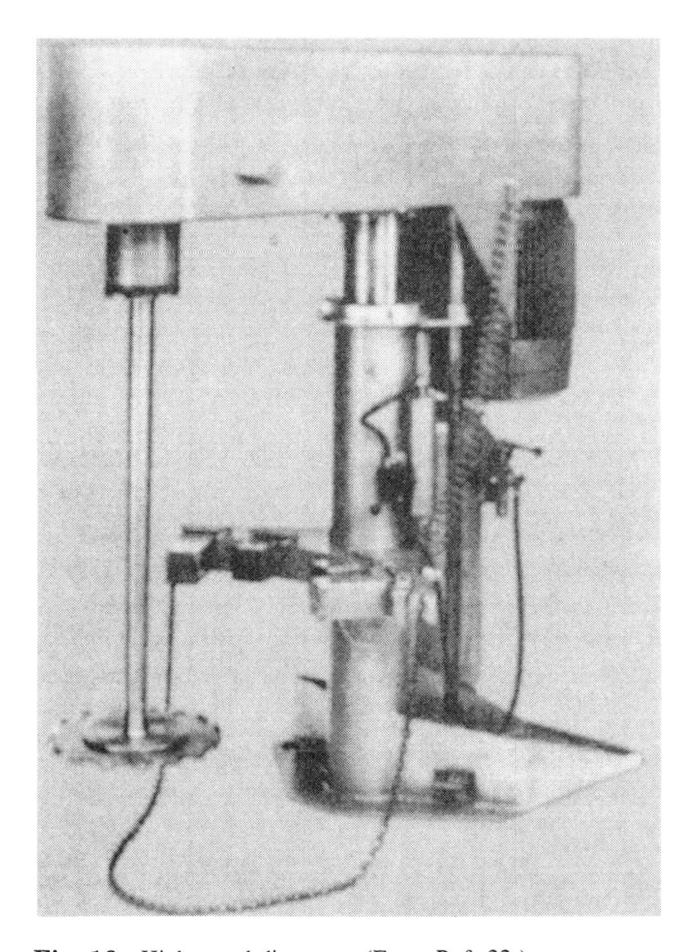

Fig. 16 High-speed disperser. (From Ref. 22.)

ity. Hence, the high-speed disperser does its best job of deagglomerating particles when the viscosity is between 10,000 and 20,000 centipoise. If the shear rate is calculated in a fashion similar to the method used for a rotor/stator, it is found to be very low, since the "gap" between the disperser blade (rotor) and the vessel bottom (stator) is usually around 30 cm.

For a disperser with a 30 cm blade running at 2000 rpm located 30 cm off the bottom of a vessel

$$dv/dx = (\pi)\,(30)(2000)/(30)(60) = 104 \text{ sec} \tag{6}$$

where dv = velocity difference between the moving impeller and the stationary object (bottom of the vessel); and dx = distance between moving impeller and stationary object.

Clearly, the maximum shear rates are higher than this in the vicinity of the blade tip, but there has not been much research into the velocity gradients set up by high-speed dispersers. Much of this lack of research is no doubt because the bulk of the commercial applications for the disperser deal with viscous liquids that are completely opaque, making the measurement of the various velocities difficult. However, there has been a large amount of trial-and-error testing, and several rules of thumb for optimum operation have arisen from this Edisonian study.

First, the tip speed of the high-speed disperser should be set around 4000 ft/min (2000 cm/sec) [23]. Second, the diameter of the impeller should be around 1/3 of the diameter of the dispersion vessel. And third, the impeller should be located one impeller diameter off the bottom of the vessel. These are rules of thumb that have proved to work well. It is recommended that the tip speed of both laboratory and full-size production units be kept constant. This suggests that the tip speed does have an overriding effect on the dispersing abilities of the machine.

High-speed dispersers set up a flow pattern in the vessel that is very conducive to the addition and incorporation of powders. This flow pattern is effective up to the point for high-viscosity liquids, where additional mixing is suggested. Air incorporation also has been a problem with this type of flow pattern. Thus, the high-speed disperser is best used for suspensions and not emulsions.

C. Rotor/Stator Mixers

One of the most important machines used for dispersing and emulsifying is the rotor/stator mixer. This mixing machine uses an impeller that is installed at a close tolerance to a stationary housing, which baffles and restricts the flow caused by the rotation of the impeller in the liquid. Particles caught between the rotor and stator are crushed and separated by either the mechanical action of the impeller hammering the particle into the stator or the hydraulic shear caused by the very high flow gradients resulting from the liquid having to flow into and around the stationary parts of the mixing assembly often known as the *mixing head*.

Since the gap between the rotor and stator is usually around 1/2–2 mm (500–2000 μm), particles as small as 50 μm, or liquid droplets as small as 1 μm, are not likely to be directly stressed by the rotor and stator. However, the rotor/stator mixer can produce fine dispersions and emulsions. The reason lies in the fact that extremely high shearing rates are generated inside the rotor/stator mixing head.

By using the definition of viscosity and assuming that the shear rate inside the mixing head corresponds to the speed of the impeller in centimeters per second divided by the rotor/stator gap, a method for approximating the maximum shear rate in a rotor/stator mixer can be derived. For example

rotor/stator gap = 1 mm = 0.1 cm
rotor speed = 3600 rpm
rotor diameter = 10 cm

Assuming the flow rate at the stator to be zero and the shear velocity at the impeller to equal the tip speed

$$\begin{aligned} \text{Shear rate} &= dv/dx = v(\text{velocity})/x(\text{gap}) \\ &= (\pi)\ (10)[(3600)/(60\ \text{seconds/minute}]/0.1 \\ &= 1884\ \text{centimeters/second}/0.1\ \text{centimeters} \\ &= 18{,}840/\text{second} \end{aligned} \tag{7}$$

All rotor/stator design variations are made to provide a combination of flow and high-shear rates. Depending on the design, different amounts and levels of each are generated by the mixer.

1. Radial Flow with Stator

The rotor/stator mixer, which uses a radial-flow impeller to pump outward from the stator to the sides of the tank, is optimized for shear versus flow. Excellent results have been obtained for emulsions and suspensions, since the shear rates obtained are high enough to effect suspensions below 150 μm and emulsions in the 1–5 μm range. As always, the specifics of the solid particles and the liquid viscosities and surface tensions can vary these results. In addition, this type of mixer is capable of positive mechanical grinding, since the gap between the rotor and stator ranges form 0.015–0.060 in., i.e., 380–1500 μm.

The radial-flow impeller expels the liquids and solid particles out through the stator and sets up the circulation pattern that is illustrated in Fig. 17. As is shown in Fig. 17, the liquid is pumped into the bottom and horizontally out of the sides of the stator at a high velocity. This exiting flow impinges on the wall of the mixing vessel. From there the flow deflects upward and downward. This can have the effect of swirling the liquid radially, if the viscosity is low enough. Care should be taken during scale-up to ensure that an adequate flow is present at the top surface of the liquid. This is especially important if stabilizers, which tend to float, are to be added. Often an auxiliary

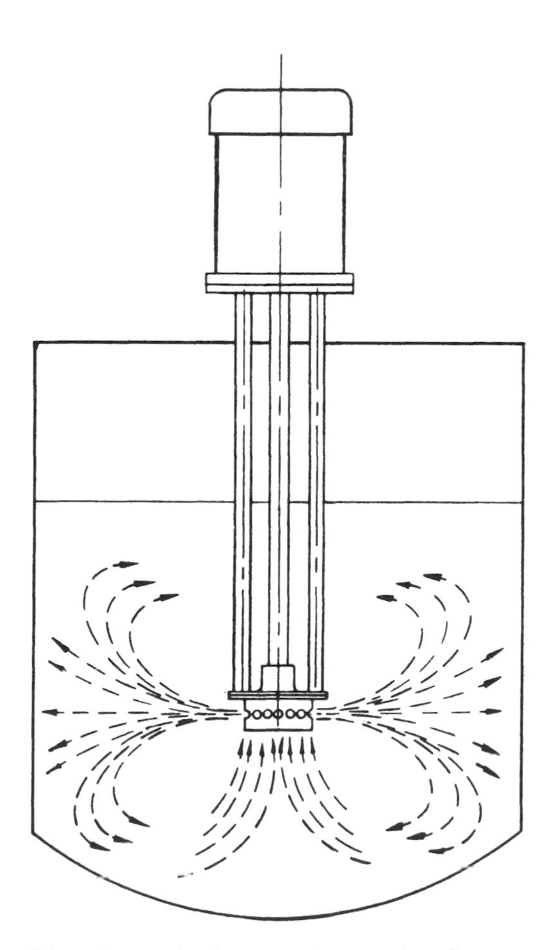

Fig. 17 Radial-flow rotor/stator mixer flow pattern. (From Ref. 24.)

marine-type impeller is added to provide some axial-flow pumping along with the radial-flow high-shear mixing. This impeller may be located at a level that can cause splashing if only a portion of the batch is mixed, e.g., during the addition of one emulsion phase to another.

A wide variety of mixing-head configurations are available. Some have vertical teeth, a stator with round holes, a stator with vertical slots, or a finer screen stator, which type is often specified for emulsions, as shown in Fig. 18.

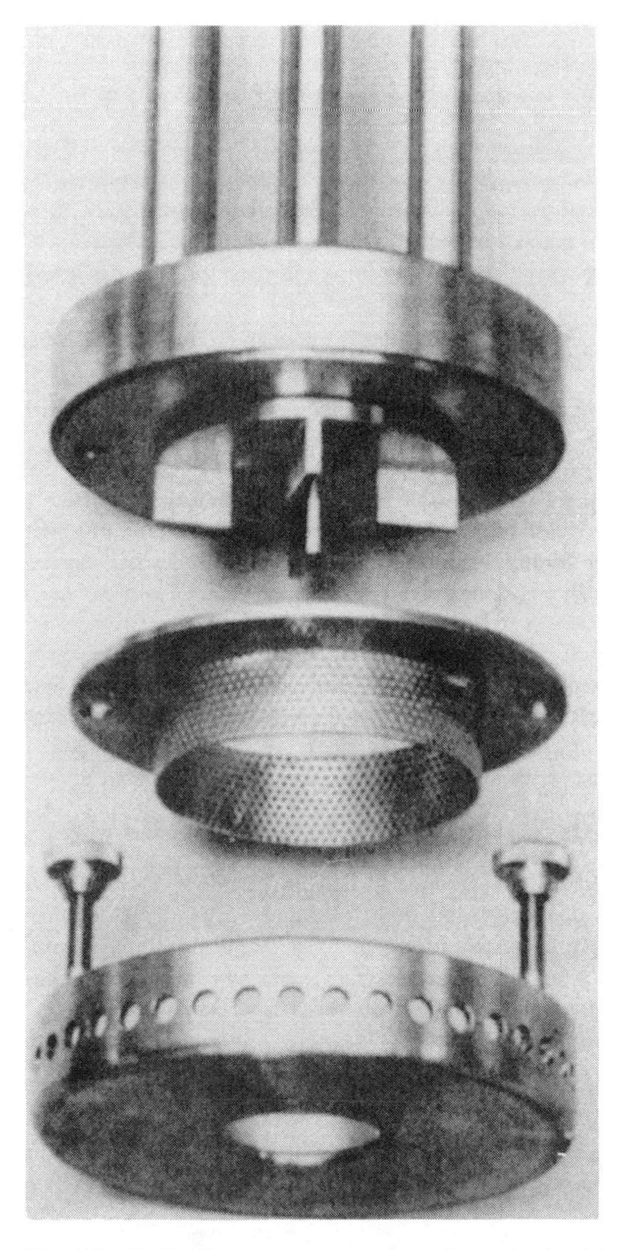

Fig. 18 Radial-flow rotor/stator mixer. (From Ref. 25.)

Since the particle or droplet sizes of dispersed delivery systems are always several orders of magnitude smaller than the stator gaps of whatever geometry, the effectiveness of one type over another has not yet been documented. The formulator or the scale-up engineer may find it beneficial to test various configurations.

Radial-flow rotor/stator mixers can be fed directly through a bottom valve in order to disperse liquid phases. The major limitation is the inability to handle viscosities over 5000–10,000 centipoise depending on design. Hence, these types of rotor/stator mixers are frequently used with auxiliary mixers, such as scraped-surface agitators, which are described later.

2. Rotating Stator

An interesting hybrid of the rotor/stator mixer group is the rotating stator type. In this design the impeller and the "stator" both rotate on the drive shaft. For this reason there are no supporting rods necessary, and no steady bearing is required or possible.

There are several different configurations offered by the different manufacturers. One type, shown in Fig. 19, uses a conical impeller with vanes connecting the cone to the shaft. This cone will pump from the smaller open end of the cone through the larger opening.

If two of these cones are installed on the same shaft, they can be directed to pump fluids toward each other for a combined effect that is described as hydraulic shear. This is correct terminology since the two streams meet each other causing twice the shear rate that would occur if only one stream at a given velocity were provided.

If the two cones are connected with an enclosure with holes or slots, the flow of liquid is also acted on mechanically. However, since the particles are typically so much smaller than the slot openings, it is not likely that this is where the maximum shear rates occur. On the other hand, the high-velocity flow is impinging on the surfaces of the rotating stator, and some high shear should occur at these sites.

In another configuration, the stator is installed on a slightly impeded free-wheeling degree of freedom with the shaft. The radial-flow impeller rotates at high speeds of 1800–3600 rpm while the stator only rotates at 10% of this speed. Fig. 20 illustrates a rotating stator mixing head. This design provides for mechanical shear rates that should be close to those seen with stators that "rotate" at 0% of the speed of the impeller, i.e., true stators. The stators are designed with outside vanes, which are forced by the liquid to rotate the stator at a slower rate. These vanes are also credited as providing additional pumping for blending.

The chief advantage of rotating stator designs is that they achieve the levels of shear rate and pumping capability of most radial-flow rotor/stator mixers without requiring the support structure for a true stator. Hence, they are simpler and marginally easier to clean. They do not have a bearing in the product, which means less maintenance.

The main limitation or drawback of this type of design is mechanical instability due to the high speed of operation and the lack of the steady bearing. For this reason, long-shafted machines are difficult and expensive or impossible to build. This places a limitation on the size of batches that can be processed. Each manufacturer should be consulted as to how large a batch of a given viscosity is feasible.

The lack of steady bearings and supporting rods or quill pipe does give some advantage from a cleanliness standpoint. However, revolving stator types do have two

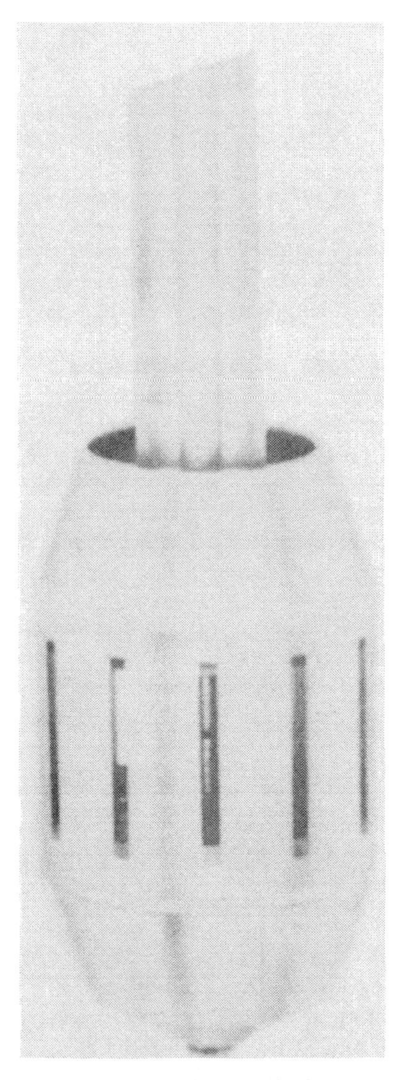

Fig. 19 Turbon cone-type impeller. (From Ref. 26.)

replaceable gaskets, which separate the revolving stator from the drive shaft. Dual cones with connecting enclosure designs are usually able to pump out any residual material as long as that material has not dried. In that case, cleaning may actually be more difficult than cleaning of other rotor/stator mixers since there is no way to open the head. Soaking the mixing head in hot soapy water while running is the best method of cleaning here.

The relative simplicity of rotating rotor/stator mixers provides several attractive points. If their capabilities can be shown through testing in the laboratory and pilot plant to meet the requirements of the process, they have their place in making dispersed products.

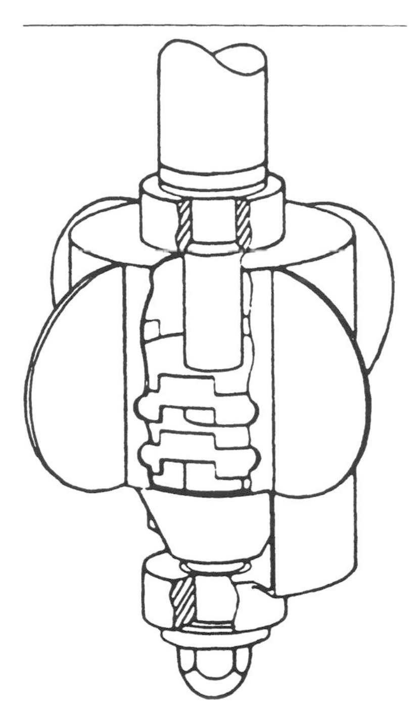

Fig. 20 Rotating stator mixer. (From Ref. 27.)

3. Axial-Flow Rotor/Stator Mixers

By installing an axial-flow impeller in a stationary housing, a combination of the high shear rate ability of rotor/stator mixers with the efficient pumping ability of axial-flow impellers can be obtained. The added pumping capability makes the handling of higher viscosities or larger batches possible with a given size (power) mixer. The high shear rates are generated in the gap between the high-speed rotor and the stator.

Axial-flow impellers are capable of producing four times the flow for a given horsepower as Rushton-type turbines. Hence, they are almost exclusively used in processes dominated by flow [7]. On the other hand, axial-flow impellers are not known for exerting high amounts of shear. Although the maximum shear rates are similar for axial- and radial-flow mixers (since this is determined by diameter and rotation speed), radial-flow impellers produce more turbulence and shear on their own than axial-flow impellers. This is why the original inventors of what is often called the Eppenbach mixer designed the rotor/stator mixing head in such a way that the amount of shear along with the maximum shear rate would be increased in order to make the machine capable of producing emulsions and suspensions.

As with radial-flow rotor/stator mixers, there are several manufacturers of axial-flow rotor/stator mixers. A cross-section of a typical mixing head is shown in Fig. 21.

The stator is a short, cylindrical enclosure sometimes called a "can" or a shroud. The axial-flow impeller or rotor is driven by a shaft, which extends into the shroud, and is usually steadied by a bushing-type bearing. The mixer acts as a submerged pump. The product is drawn in and expelled at a high rate into the other areas of the process vessel. Since the flow rate can be very high with low-viscosity materials, a baffle plate

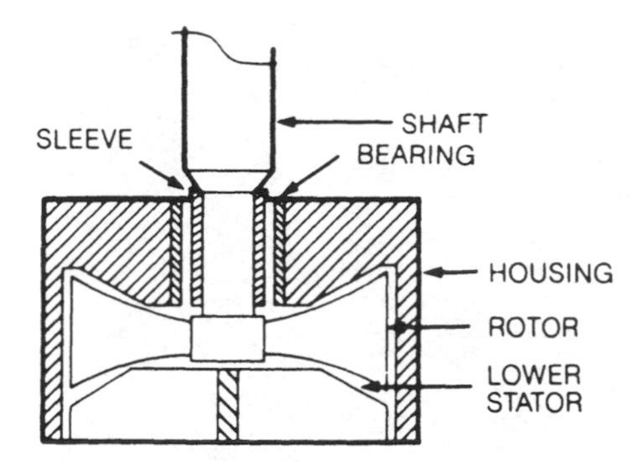

Fig. 21 Cross-section of axial-flow rotor/stator mixing head. (From Ref. 28.)

is necessary to deflect the flow so the product does not leave the vessel. As shown in Fig. 22, all axial-flow rotor/stator mixers can pump in an upward direction.

This baffle plate can then be used to optimize certain performance traits of the mixer. First, the plate can be set directly at the surface at rest. This will give the most active mixing on the surface. Fig. 23 shows the placement of the baffle plate above the surface. Second, if a lesser level of surface turbulence is desired (especially im-

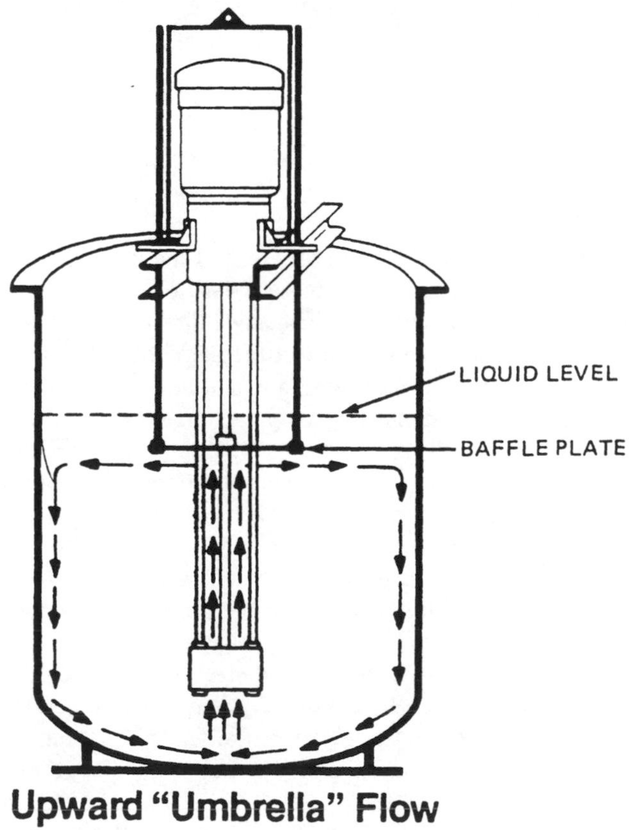

Fig. 22 Up-mode mixing flow diagram. (From Ref. 28.)

Fig. 23 Baffle plate above surface. (From Ref. 29.)

portant to avoid air incorporation or foaming), the baffle plate can be lowered to prevent splashing. This allows the mixer to be emulsifying at highest speed and, hence, highest shearing rates while avoiding aeration.

All mixers or mixing systems must provide flow to all areas of the process vessel if they are to be deemed successful. In the case of these axial-flow rotor/stator mixers, the flow emanates from the mixing head and flows in a single direction. In order for the flow to reach every area of the vessel, it must deflect off the baffle plate and then the sidewall. If the mixer cannot produce enough flow to reach the sidewall, then a dead spot exists. The amount of flow required and the amount of flow produced by a given size mixer depends on the viscosity and the design of the specific mixer. The manufacturer should know the pumping capabilities of their mixers at different viscosities in order to select equipment for different size mixing vessels. Table 4 shows the ability of a typical axial-flow rotor/stator mixer.

The batch size that can be handled on a macroscale basis can be determined from Table 4 for the axial-flow rotor/stator mixer if the diameter of the process vessel and the diameter of the rotor are known. This is a trial-and-error problem. By choosing a batch size, vessel diameters can be obtained by use of standard-size vessels. If a feasible mixer can be installed in a standard-size vessel, the total system capital cost can probably be lowered. The rotor diameters that are available for trial-and-error solution are usually set by the manufacturer. That is, various sizes are available but not an infinite variety.

As an example, take a 1000 gal. process tank with a 72 in. diameter. If a 6.5 in. diameter rotor unit is used, a viscosity of up to about 9000 centipoise can be pumped

Table 4 Axial-Flow Rotor/Stator Mixer Sizing Chart

Ratio of allowable tank diameter to rotor diameter	Maximum viscosity (centipoise)
1:4	30,000–70,000
1:6	20,000–30,000
1:8	10,000–20,000
1:10	5,000–10,000
1:15	2,000–5,000
1:20	<2,000

Source: Ref. 25.

throughout the vessel for good blending. If the same mixer were used in a 60 in. diameter vessel, a ratio of 1:9.2 would give a feasible viscosity operating level of 15,500 centipoise. Comparison with other high-shear mixers with less efficient pumping capabilities is the ability to reverse and pump in the downward direction. This option provides a flow that resembles a vortex, which can be used to draw in powders that tend to float on the top surface of the liquid. An illustration of downward "vortex" flow is presented in Fig. 24. This is especially valuable when drawing in stabilizer powders, which often tend to form lumps that are difficult to disperse completely after they are formed.

The added pumping efficiency of an axial-flow rotor/stator enables it to handle higher viscosities and larger batches with less horsepower and capital cost. These sav-

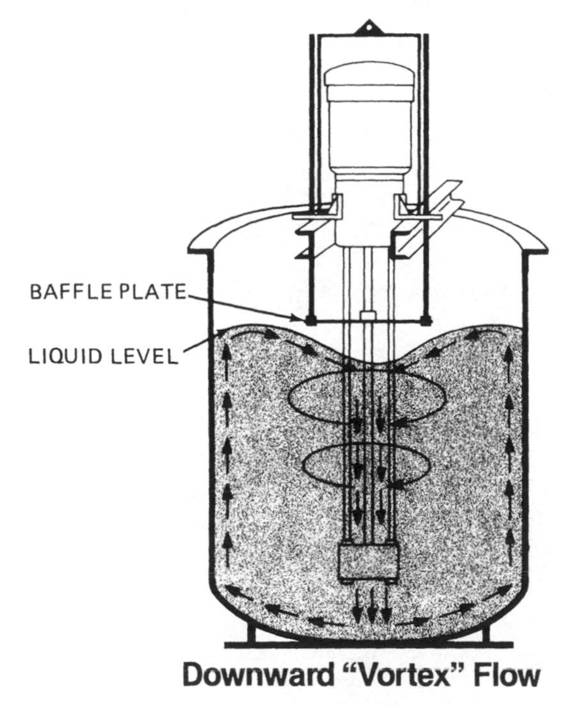

Fig. 24 Down-mode flow pattern (From Ref. 28.)

ings are more pronounced as the size of the batches increases. Hence, the capability is worthwhile to consider when selecting laboratory and pilot-scale equipment.

D. Combination Mixers

The viscosities during processing of a typical emulsion and many suspensions often change from water-thin to thick creams and pastes. Designing one simple agitator mixer to handle both high and low viscosities at the different high- and low-shear rates required is usually impossible. Some of the mixers and dispersers described earlier can be installed in the same process vessel to give multiviscosity and multi-shear-rate capability in one controllable batch. Completing all of the mixing steps in one vessel gives the advantage of better control of temperature and pressure, eliminates line losses caused by pumping from tank to tank, saves floor space, and simplifies the process, which can save on labor costs.

1. Anchor Plus Rotor/Stator

One of the most economical methods of obtaining multiple viscosity and shear-rate capability is to combine a simple anchor agitator with a rotor/stator mixer. The anchor supplies the slow moving of the viscous product from the heat transfer surfaces while the rotor/stator provides the high shear rates for dispersion and emulsification.

The advantages of this design are simplicity and low capital costs when compared to alternate multishaft designs. The high-shear mixers can be obtained from any of the rotor/stator companies, but certain designs may work better than others in specific situations. An axial-flow rotor/stator is beneficial if there is a requirement to pull down powders from the top. The flow rate produced by the rotor/stator mixer is not required to turn over the entire vessel since the anchor assists—especially near the bottom of the vessel. However, the flow rate of the high-shear mixer determines how much time is required to recirculate the batch through the highest shear zones, and this has a major bearing on mixing time.

One weakness of this configuration of mixers is that it is physically difficult to design the anchor such that it allows the placement of the high-shear mixer close to the bottom of the vessel. In addition, as shown in Fig. 25, a brace is usually required to strengthen the anchor, and this brace compounds the placement problem. This can be a serious limitation if emulsions are to be made, or if a low-volume phase must first be prepared in the vessel or, especially, if the inversion technique is to be used.

2. Anchor Plus Disperser

Almost everything that can be said about combining a rotor/stator mixer with an anchor agitator can be stated about combining a high-speed disperser with an anchor. The anchor gives the formulator the ability to work with high-viscosity liquids. This combination is less costly than more complicated designs using coaxial shafting.

As described in the section on high-speed dispersers, the combination design is used for handling suspensions that are either too high in viscosity for high-speel dispersers alone or when some other requirement, for example, slow cooling after dispersion, is required. The wide diameter of the disperser blade also contributes to the problem of locating the blade close to the bottom of the batch. As long as air incorporation is not a problem, the combination anchor with high-speed disperser works well on medium- to high-viscosity suspensions.

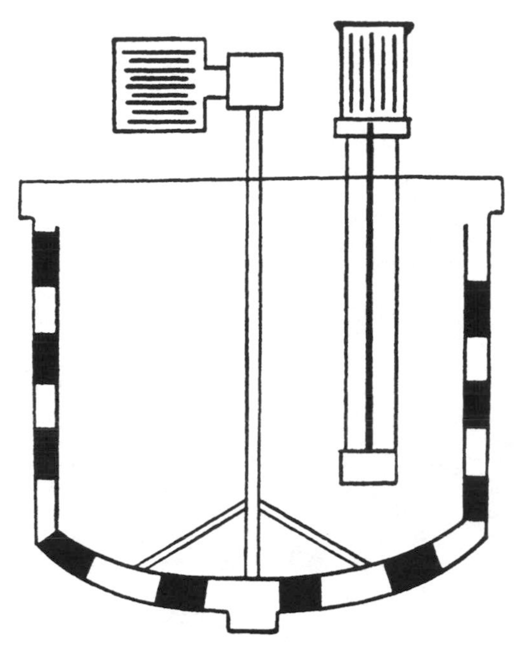

Fig. 25 Anchor mixer with rotor/stator mixer.

Counterrotating with Coaxial Rotor/Stator

One of the most complicated and powerful dispersion- and emulsion-making designs is the counterrotating scraped-surface agitator combined with a centerline-installed high-shear rotor/stator mixer (Fig. 26). This design, sometimes known as a triple-action mixer, combines the high-viscosity mixing capabilities of counterrotating axial-flow crossbars with a standard anchor-type scraped-surface agitator and also with a high-speed rotor/stator mixer, which is capable of generating fine dispersions and fine droplets of the internal phase for stable emulsions. These machines are built to pharmaceutical standards by a number of companies both in the United States and Europe. They can be outfitted for atmospheric or vacuum/pressure specifications. They are always jacketed for heating and cooling.

Triple-action mixers are multiviscosity mixers. They are typically justified only for processes that require all of these differing abilities in one process vessel. If a process starts at a low viscosity and then finishes at a high viscosity, or if there is a peak viscosity somewhere in the middle of the process which is higher than 50,000 centipoise, triple-action mixers may be economically justified. This design offers a major advantage over anchor-type agitators with the high-shear mixer located at the side of the centerline anchor scraped-surface agitator. First, the counterrotating mixer pumps high-viscosity products efficiently in both axial and radial directions. This is essential for good heat transfer, which is necessary during the cooling stage of most creams and ointments.

Even more important is the fact that, by locating the high-shear mixer on the centerline, the high-shear mixer head can be located on the bottom of the mixing vessel. A given vessel can then be operated at a much lower minimum fill level with the high-shear mixer running. Since the important dispersing mixing head is located in the

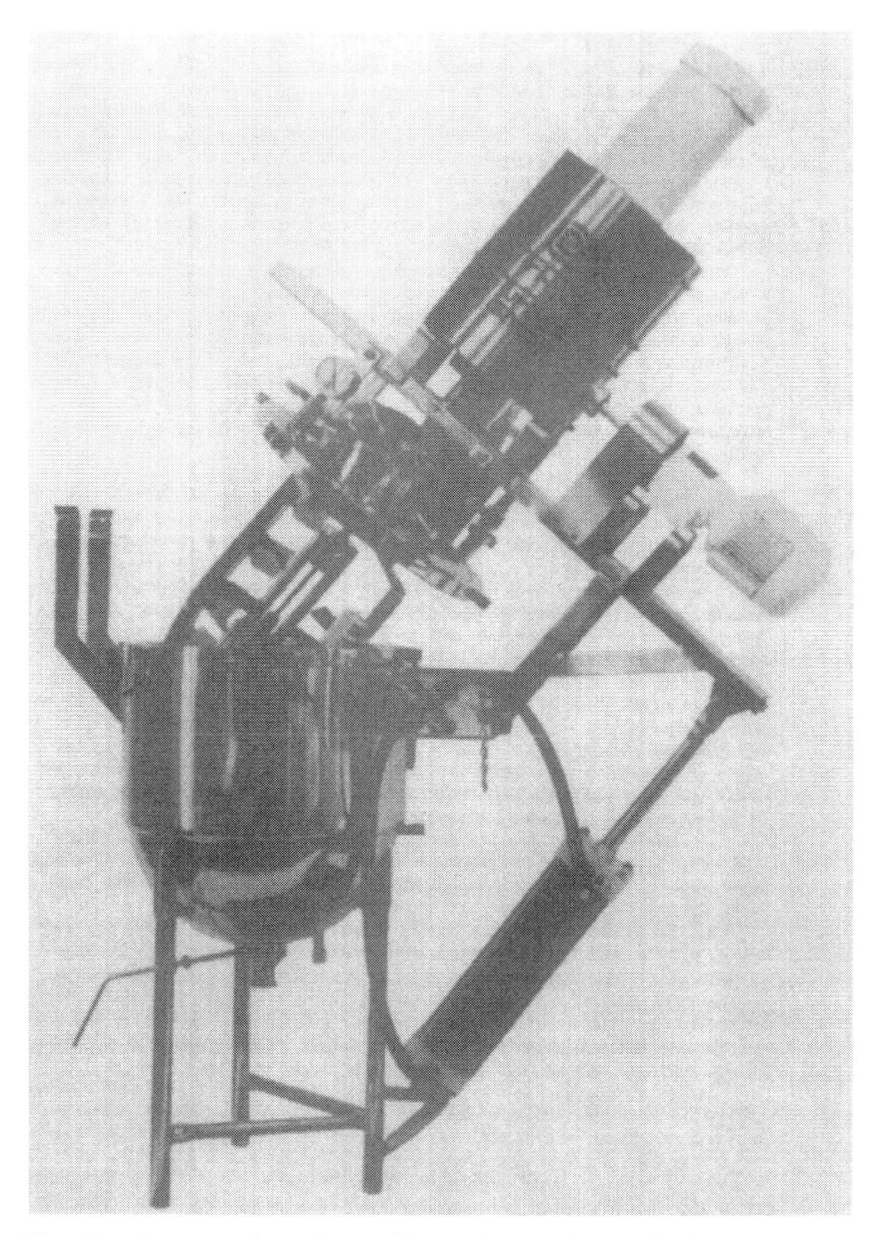

Fig. 26 Counterrotating agitator with rotor/stator. (From Ref. 30.)

bottom half of the process vessel, processing less than a full-size batch is possible. Equally important is the capability of beginning to disperse or emulsify with only a small portion of the final batch. If one desires to disperse the major phase of an emulsion into the minor phase, the mixer must be at the bottom of the kettle. The presence of the high-shear mixer at the bottom of the vessel makes it possible to draw liquids directly into the mixing head when they are pumped into the bottom outlet. The ability to draw one phase of an emulsion directly into another is always valuable in terms of control of the process, and this prevents any separation from occurring at the beginning of the process. The technique of inverting an emulsion by overloading the starting phase in the vessel can often yield superior results, but cannot be used if the mixer is not submerged in the original phase from the start.

Direct and immediate dispersion of powders can be used in triple action-mixers built for vacuum operation. Powders can be drawn directly into the eye of the high-shear mixer by opening the bottom valve. Powders can be drawn by vacuum through a pipe or flexible hose. This technique provides process-mixing time savings, since there is essentially no time required for shearing out powder agglomerations or lumps. If the process uses large quantities of powder, such as toothpaste, or if the powders have a propensity to lump, this method of addition can be a worthwhile feature.

The chief disadvantage of triple-action mixers is their complicated design. There are three separate rotational actions down one axis. This is done by using a solid shaft for the high-shear mixer and two hollow or "quill" shafts to power the two directions of the counterrotating mixer. A separate motor always powers the high-shear shaft. The anchor and counterrotating crossbars are sometimes driven by one motor turning a pinion gear between two facing bevel gears. On other designs, two separate motors drive the anchor and crossbars for a total of three different motors on one mixing machine. However, this three-motor design does give the ability of independent control of each motor speed.

Each shaft must be sealed using mechanical seals if vacuum or pressure is to be held during the process. These seals do work well, if they are maintained correctly. This is an important checkpoint for specifications of a machine that operates under vacuum or pressure. Mechanical seals are wear parts and do require routine maintenance. Triple-action mixers have, by necessity, three sets of mechanical seals.

As described with counterrotating equipment, no large solid object can ever be allowed to get between the anchor and crossbars. The damage done on a triple-action mixer by a paddle, drum, or unmelted block of wax can result in prolonged and expensive downtime.

A final drawback to triple-action mixers is that their complexity and extra mixing capability actually contribute to the difficulty of cleaning. Simply put, there are more crossbars, scrapers, and other components where high-viscosity products tend to accumulate and do not drain out of the machine. Most manufacturers are still affixing the crossbars and scraper blades with bolts and pins to facilitate easy repair. In short, these machines are some of the most complicated to clean.

Triple-action mixers offer the greatest capability for handling a wide range of viscosities, from water-thin to as high as 1 million centipoise on some designs. Along with the ability to handle high viscosity, these machines can disperse and emulsify very efficiently and quickly. This is especially true if a triple-action mixer is equipped with vacuum capabilities. The complicated design agitator must be run with care to avoid cleaning problems and unexpected breakdowns. Operated optimally, triple-action mixers can provide the shortest mixing time and a degree of process control that readily justifies their high capital cost.

E. In-Line Mixers

There is an entire class of mixers known as in-line or continuous, which are used to produce disperse delivery systems outside of batch processing vessels. These are the colloid mills, piston homogenizers, rotor/stator mixers, Microfluidizer™ (a registered trademark of Microfluidics International Corp.) technologies, ultrasonic mixers, and hybrid devices. Each uses a unique processing technique to shear a mixture or combine the flows of materials in order to form an emulsion or suspensions. Most of the time these devices are not used in a truly continuous process. Rather, after the components of a dispersed delivery system are combined and blended in a batch vessel, the components in the mixture are passed through the device, and the shearing and mixing that take place inside the device affect particle size reduction, dispersion, and emulsification.

1. Rotor/Stator Mixer Disperser Emulsifiers

"All mixers pump and all pumps mix." This is reflected in the earlier-shown power equation, Eq. (3). A type of in-line device that is very similar to a rotor/stator batch mixer is the rotor/stator continuous mixer disperser emulsifier. Indeed, most of the designs of this type of in-line high-shear device are essentially identical to the batch equipment designs of a given manufacturer. Since rotor/stator batch mixers are acting as submerged pumps, a design can be made that places the rotor/stator in a pump housing and allows for product to be pumped through itself (Fig. 27). During the time the product is inside the rotor/stator mixing pump, the droplets and particles are subjected to a wide variety of high shear rates. All pumps of any kind impart some level of shear to the product that passes through the pump. Rotor/stator mixing pumps are designed with fine tolerance rotor/stator gaps that promote the high shear rates and high amounts of shear per pass through.

Shear rates in a rotor/stator in-line mixer are equal to those in rotor/stator batch mixers. The maximum shear rates occur in the gap between the high-speed rotating

Fig. 27 Rotor/stator in-line mixer disperser emulsifier. (From Ref. 31.)

impeller and the stationary housing. Many ingenious designs are available that use various configurations with the purpose of increasing the probability that the solid particles or liquid droplets travel through the rotor/stator zones where maximum shearing occurs. Almost all designs have some form of teeth or blades, which are meshed into an accompanying stationary housing as illustrated in Fig. 28.

Of particular importance is the fact that, when a mixture containing solid particles or agglomerations or emulsion droplets is pumped through a fixed gap rotor/stator mixer, not all of the droplets or particles pass through the highest shearing zones. Some particles passing through the machine may escape the rotor/stator gaps and be exposed only to some of the lower shear zones. The more open the design of the rotor/stator, the

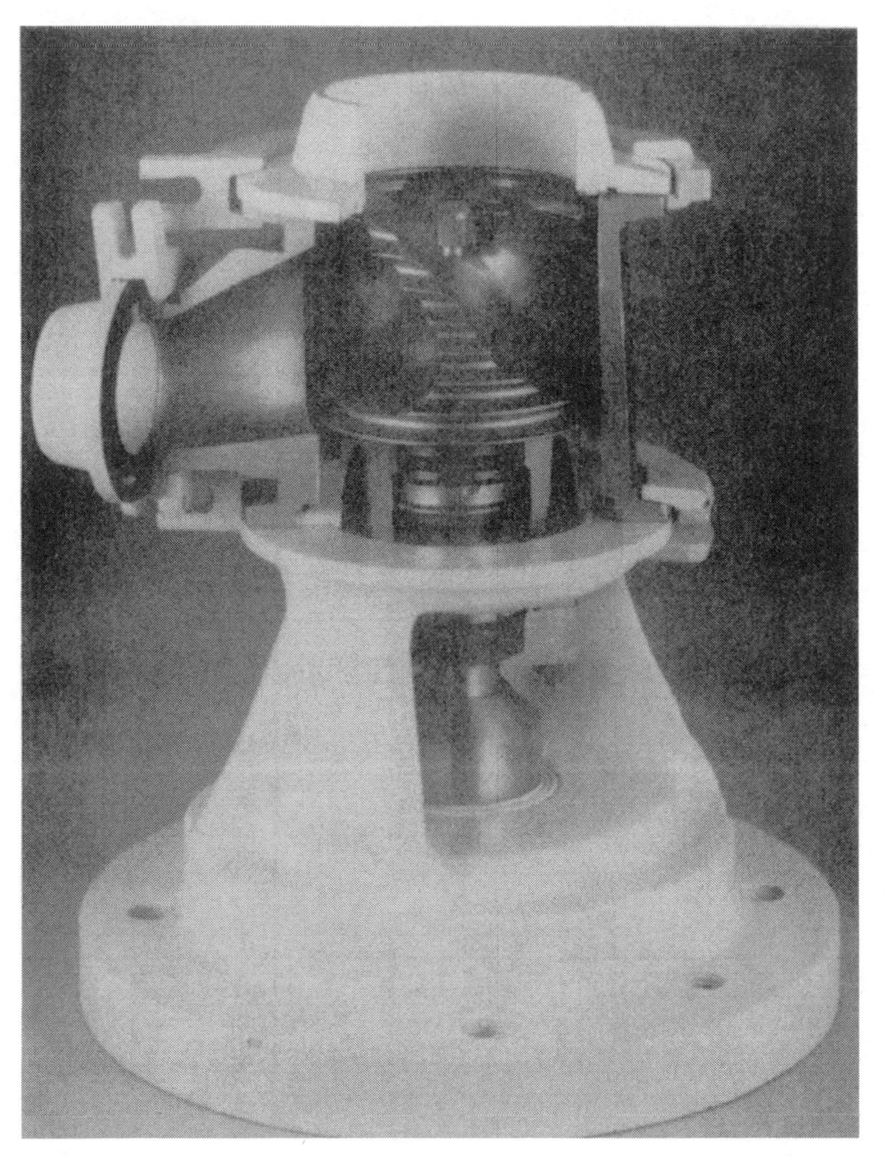

Fig. 28 Internal parts of rotor/stator mixer showing meshed teeth. (From Ref. 31.)

greater the probability that the particles will miss the high shear zones on a single pass through the device. For this reason the discharge of many rotor/stator devices is restricted with a series of holes or bars, or with a grid, which performs two separate tasks. First, a discharge grid absolutely limits the maximum particle size that can exit the mixer. If the holes on the discharge grid are 2 mm, it is virtually impossible for particles larger than that diameter to pass through the machine. Rotor/stator in-line mixers are made with restricting grids as small as 0.5 mm, and a wide range of larger sizes up to 50 mm or even unrestricted outlets. A sample of such grids is presented in Fig. 29.

But even the smallest usable discharge grid does not necessarily provide absolute control of the required size distributions in the range of fine dispersion. This is the second reason for the restriction on the discharge. By putting a more limiting discharge on the outlet, the flow rate through the device for a given set of pumping circumstances (viscosity, density, system suction, and discharge head) is reduced and the residence time in the machine for a given particle or droplet is increased. The longer residence time increases the probability that a particle has to travel through the highest shearing zones and, thereby, be reduced to the smallest size that the machine is capable of producing.

Fig. 29 Discharge grid opening on in-line high-shear device. (From Ref. 31.)

There is another method of increasing the residence time in a given rotor/stator mixing pump, and that is to change the pumping circumstances by increasing the discharge head or by restricting the outlet side of the pump by use of a valve. These rotor/stator pumps all work on a centrifugal pump basis. They are not positive displacement pumps, and their ability to produce flow decreases as viscosity and the required discharge head is increased. A discharge head can be increased merely by increasing the vertical distance that the pump is required to move the product, but the simplest method is to install a valve downstream and close it partially until the desired flow rate is obtained.

Restricting the outlet of a rotor/stator mixing pump does increase the probability that a particle will have to pass through the maximum shear zones. There is a simpler and more widely used method to increase that probability: recirculation or multiple recycle passes through the device. By installing the mixer pump in a recycle loop to the batch blending and mixing vessel, the contents of the batch can be subjected to the high shear rates in the rotor/stator mixer repeatedly. This processing method is easy to control since the machine can be allowed to recirculate until the desired amount of high shear has been imparted. By examining samples over a period of mixing time, the time required to produce the desired particle size can be determined.

The maximum particle size is controlled by the restricting discharge grid in the rotor/stator mixer. The minimum size and the size distribution is controlled by the maximum shear rates in the rotor/stator gap. Repeating the number of trips that a particle has to take through the device increases its probability of exposure to the maximum shear in the equipment.

A problem encountered often with combination batch mixers is the inability to start high-shear mixing when the batch processing vessel is only partially full. This is always the case for emulsions prepared by inversion. The phase in which the dispersion begins is only a minor portion of the final batch. With a rotor/stator mixer located in a recirculation loop to the dispersing vessel, the problem of too little starting volume is eliminated. In addition, the phase to be dispersed can be metered directly into the mixing head of the in-line device to prevent any separation. This method has been found to be very efficient from the standpoint of power consumption and mixing time.

Whenever an emulsion is prepared by the inversion method, it is important not to add too much of the dispersed phase at once since the emulsion could invert instantaneously in the in-line device. The addition of one phase into another is actually easier to control by recirculation than by batch mixing, since there is complete assurance that the two streams will be completely blended in the rotor/stator in-line mixing device (Fig. 30).

As shown in Fig. 30, this method of starting with a small amount in the mixing vessel can also be used to prepare small particle size distributions of solids in liquids. If only a portion of the total final desired liquid is combined with all of the desired solids to be dispersed, the viscosity of this base mixture will usually be much higher than that of the final mixture. If the in-line device is used to recirculate and mill this base, the shear stresses exerted on the particles will be higher, since the viscosity is usually higher at higher solids concentrations than at low solids concentrations. This procedure can also be used to mill or grind solids with batch mixers, but the same problem seen with the invert emulsion technique, i.e., getting the mixer to reach down into the bottom of the vessels is encountered.

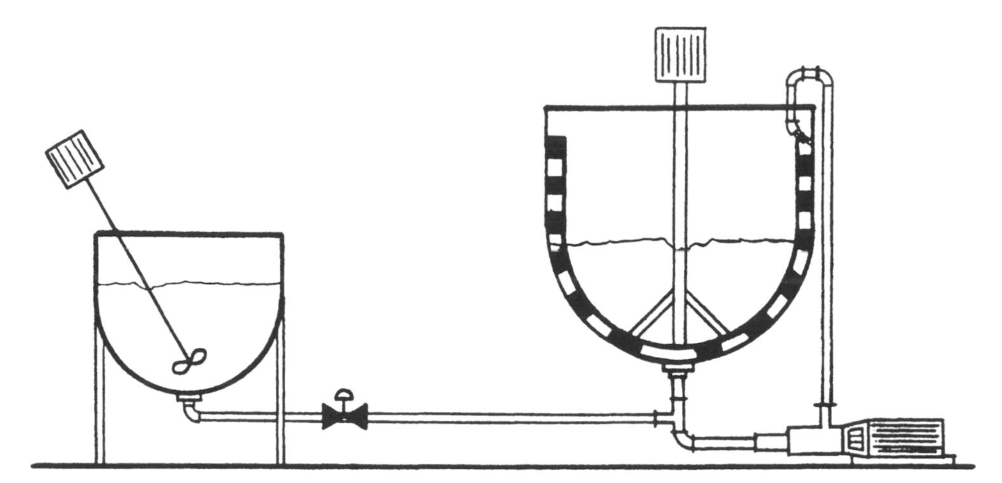

Fig. 30 Invert emulsion method using in-line high-shear device.

Installation of a recirculating high-shear device into an existing batch mixer, which is not providing adequate levels of shear, is almost always one of the least capital intensive options. If the unit can be used to invert an emulsion or process a small (high viscosity) portion of a solids dispersion, then the installation is even more useful. Finally, high-shear rotor/stator mixing pumps offer the option of use with multiple mixing vessels.

2. *Colloid Mills*

The colloid mill is a very widely used machine both to disperse solids into liquids and to emulsify immiscible liquid-liquid systems. In this device, all of the premixed components must pass through a fine gap located between a rotating truncated cone that fits closely into an opening. A mechanical control is used to adjust the gap between the rotating cone or rotor and the housing or stator (Fig. 31).

The rotor/stator gap is usually set between 0.030 and 0.001 in. The product is drawn into the rotor/stator gap and flung by centrifugal force to the open area below the rotor/stator.

Machines with rotors made of stainless steel or similar alloys are known as colloid mills. If the rotor and stator are made of carborundum, then the machine is usually called a stone mill, which brings up the similarity between this device and the grinding mills of the old water-powered grist mill. The stone or colloid mill is clearly a wet grinding device.

The shear rates generated in colloid mills can be significantly higher than those in fixed-gap rotor/stator machines, since the gaps may be set so much finer. However, with a colloid mill set at a fine gap, such as 0.002 in., there is hardly any annular space for the product to be pumped through. Hence, the flow rates are much lower than those through fixed rotor/stator machines of similar horsepower and price. Colloid mills can often generate sufficient amounts of heat to necessitate internal cooling of the contact parts of the machine, as shown in Fig. 32.

Production-size colloid mills typically operate at speeds of 3600 rpm, while laboratory and pilot units use higher speeds with smaller diameter rotors. Production-size

Fig. 31 Internals of colloid mill. (From Ref. 29.)

colloid mills, typically equipped with rotor diameters of 10–30 cm, provide flow rates in the area of 4000–6000 L/hr, depending upon the viscosity.

The key operating requirements of colloid mills are to feed the mill with a well-blended premix and to set the gap at the correct and reproducible setting. There is often some difficulty with setting the gap at exactly the required distance, since the calibration of the gap can only be done at the manufacturer. This is less of a problem if the mill is well made and the product is not abrasive. If abrasive wear attacks the rotor or stator, the gap may become larger than the setting on the machine indicates.

Colloid mills are generally used as "polishing" machines for emulsions or suspensions. That is, after the product has been totally and uniformly blended, the batch is passed through the colloid mill one or two times to further reduce the droplet or particle size. Whether or not multiple recycling passes are required depends on product requirements. Generally speaking, the colloid mill produces emulsions and suspensions with particle-size distributions smaller than the particle sizes obtainable using fixed gap rotor/stator mixers. They do represent an extra step in the process, and their use is suggested only when it is found that this added ability to disperse is necessary to produce a fine enough particle- or droplet-size product to enhance a product's stability.

3. *Piston Homogenizers*

The most powerful device for producing emulsions and suspensions is the piston homogenizer or high-pressure homogenizer. This device uses a high-power positive displacement piston-type pump to produce pressures of 3000–10,000 psig and then force

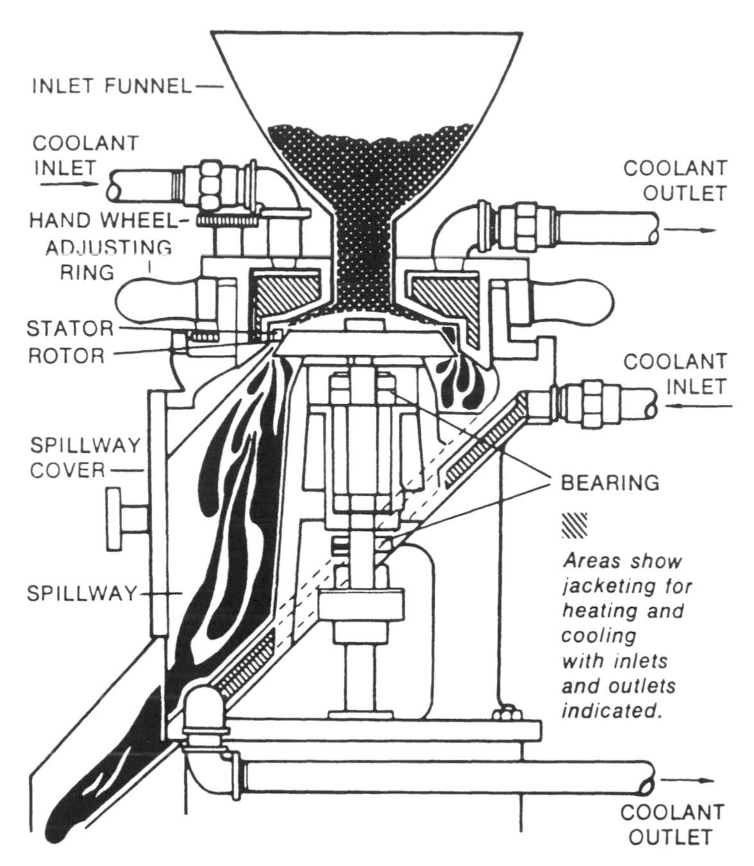

Fig. 32 Cross-section of colloid mill. (From Ref. 29.)

the premixed product through a specially designed restricting valve where extremely high shear forces are exerted. This machine, originally invented in 1899 in France by August Gaulin for the purposes of homogenizing the cream and skim phases of milk, has seen repeated application in the preparation of a wide variety of emulsions and dispersions [32].

The energy required to cause the high shear rates in the homogenizing valve shown in Fig. 33 is derived from a piston pump, which can have three to five stages. Typical units provide continuous capacities of 2500 L/hr at 15 horsepower to 50,000 L/hr at

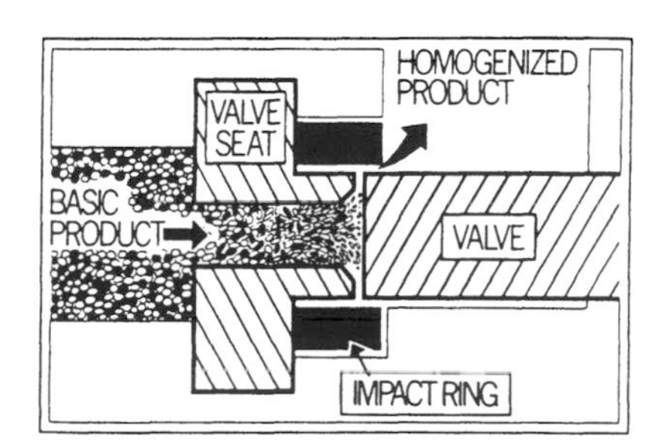

Fig. 33 Piston homogenizer homogenizing valve. (From Ref. 33.)

150 horsepower. As a consequence, these devices have high power requirements but are also able to process large quantities of finished product. The chief reason for using this type of equipment for emulsions or dispersions is not efficiency or throughput but, rather, the ability of the homogenizing valve to produce particle- or droplet-size distributions that are finer than those achievable by the use of alternative equipment. As is seen in Fig. 33, the product is forced through a precision-machined spring-loaded valve, which is set to a fine gap. The pressure of the liquid causes the valve to open slightly against the spring. The mechanism of reduction of particles or droplets is not clear. However, one may consider turbulence and high shear as the major parameters in size reduction [1,7]. Various types of valves, such as conventional plug, knife-edge, grooved, and knife-edge valve and seat, are available form the manufacturer [1]. Different types of valves have different natural harmonic frequencies, which are used for different applications recommended by the manufacturers. For example, the "knife-edge valve and seat" is recommended for cell disruption [1]. On some configurations, there can be more than one valve for a stepped-down release of pressure. The velocity of the product through the valve can approach speeds of 17,000 m/min [33].

Piston homogenizers usually are set up in two-step processes of premixing and subsequent homogenization. However, the design lends itself well to true continuous processing, since the intensity of mixing and shearing is all that is needed to mix and emulsify or disperse.

There are several limitations of high-pressure piston-type homogenizers. First, the high-pressure piston pumps are not designed to handle product feeds above 200 centipoise. This does not, however, mean that high viscosities cannot be prepared with piston homogenizers, but that the initial feed of the premix cannot be above this limit. Second, despite the fact that the homogenizing valve parts are made of very hard materials, such as stellite, appreciable wear can arise on these precision parts when abrasive solids are present in the formula. This can result in high maintenance costs and downtime. Third, the valve setting may not conform to specifications due to regular wear and spring fatigue. This may cause lack of product homogeneity and batch-to-batch variability [1]. Finally, there are two reasons to try to prevent the inclusion of air into the premix. First, the air causes the machine to run with vibration, and second, the air most certainly is very well dispersed into the product, which is almost always undesirable.

The piston homogenizer, or high-pressure homogenizer, is capable of producing dispersions down to the low (1–50) μm range and emulsions to the 0.10 μm area. High-powered models can turn out large quantities of product in a short time. For very fine particle size requirements and large production runs, their high power and high initial investment costs are justified.

4. *Ultrasonic Vibrating Homogenizers*

A unique and effective device used for dispersing and emulsifying is the ultrasonic vibrating homogenizer (Fig. 34). Having some similarity to the high-pressure homogenizer, the ultrasonic homogenizer uses a positive displacement pump to force the premixed liquid through an elliptical opening at a speed of 100 m/sec or more. This high-speed flow impinges on to the edge of a blade-shaped obstacle, called a vibrating knife. In some designs, the blade is caused to vibrate at the ultrasonic frequency by the action of the fluid. In other machines, this vibration is caused by an electrically powered piezoelectric crystal [1,34].

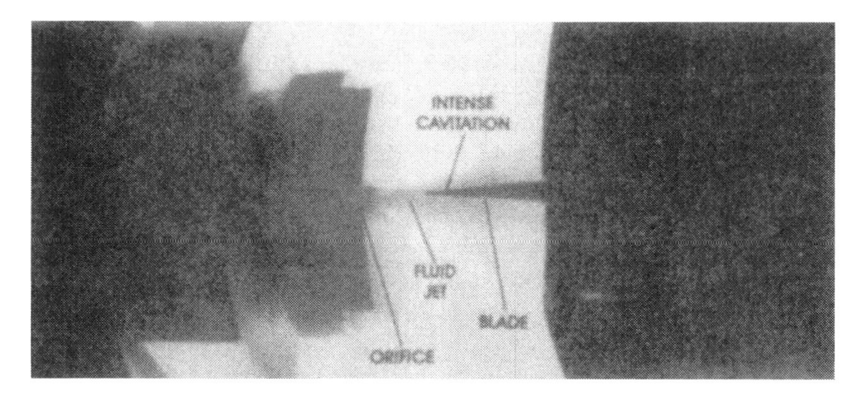

Fig. 34 Strobe photograph and ultrasonic homogenization. (From Ref. 35.)

These units have seen application in both a two-step premix-homogenization procedure and true multicomponent continuous processing. Capacities and pressures of systems range from laboratory units, producing 4–10 L/min at 1200–1700 psig requiring 2–5 horsepower, to full-scale production units with capacities of up to 450 L/min operating around 350 psig and requiring 60 horsepower.

Figure 35 presents the dual-feed system, in which the two separately prepared components are metered volumetrically into the area of the vibrating knife, where the intense cavitational effects occur.

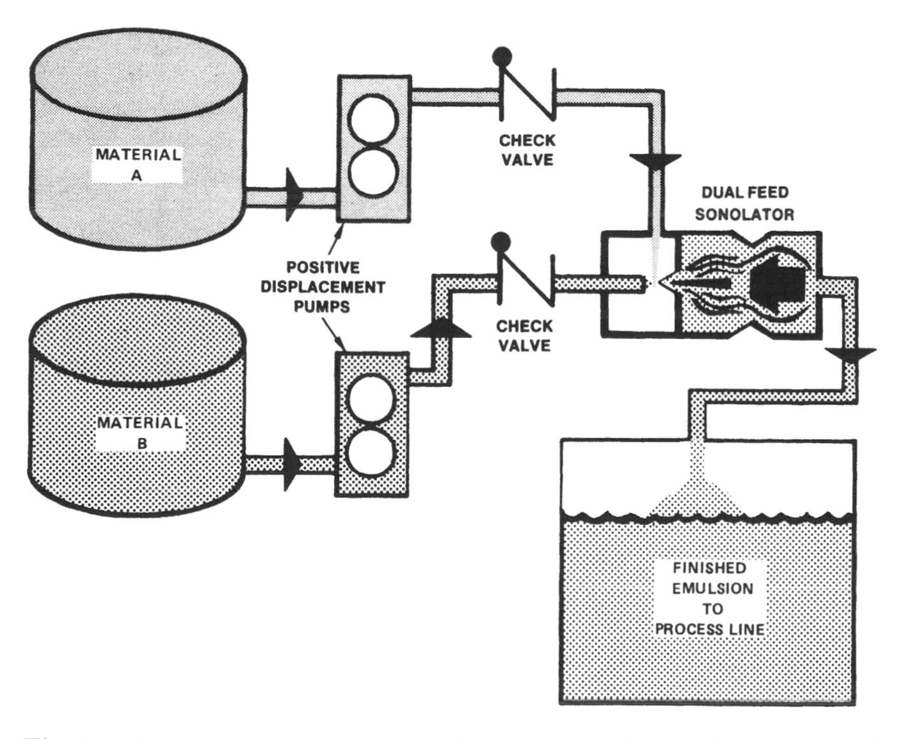

Fig. 35 Continuous dual-feed system configuration using ultrasonic homogenizers. (From Ref. 35.)

Ultrasonic homogenizing systems are able to produce particle-size and droplet-size distributions that approach those of piston homogenizers with a lower power requirement. In order to work, they must be fed a well-blended premix or a metered feed of the liquid components. The vibrating element is an extra maintenance item, especially in heavy or abrasive service. Overall, they offer an attractive option when fixed-gap rotor/stator devices do not produce the required size distributions.

5. *Homogenizer/Extruder*

Another high-pressure homogenizer/extruder with an adjustable valve having production capacities from 8 mL/hr to 12,000 LL/hr is available. A positive displacement pump produces pressures up to 30,000 psig. The manufacturer claims that no O-ring is used in the product pass and pump seal, and this homogenizer/extruder was approved by the U.S. Food and Drug Administration for pharmaceutical use [36]. At this writing, information concerning the internal structure is not available. The apparatus is capable of producing fine emulsions and liposomal dispersions. Figure 36 shows a laboratory unit.

6. *Microfluidizer Technologies*

A more recent invention to find wide use in specialized forms of dispersed system dosage forms is the microfluidizer. This device uses a high-pressure positive-displacement pump operating at a pressure of 500–20,000 psig, which accelerates the process flow to up to 500 m/min through the interaction chamber. The interaction chamber consists of small channels known as microchannels. The microchannel diameters can be as narrow as 50 μm and cause the flow of product to occur as very thin sheets. The configuration of these microchannels within the interaction chamber resembles Y-shaped flow streams in which the process stream divides into these microchannels, creating two separate *microstreams*. The sum of cross-sectional areas of these two microstreams is less than the cross-sectional area of the pipe before division to two separate streams. This narrowing of the flow pass creates an (axisymmetric) elongational flow to generate high

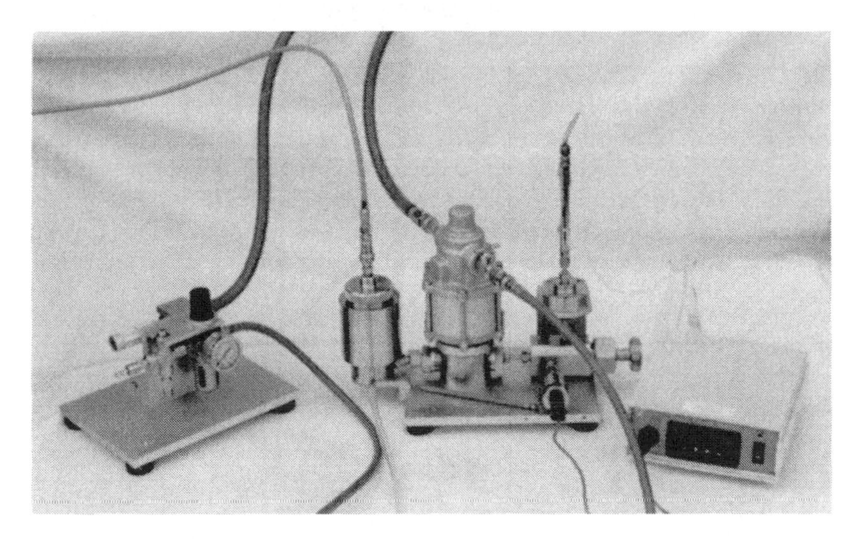

Fig. 36 Emulsiflex-C5, a high-pressure homogenizer. (From Ref. 36.)

flow velocity and a high shear stress [37]. Based on the "continuity equation" for flow in this process stream, the total mass rate of flow, which is a function of velocity times cross-sectional area, remains constant. Decrease in cross-sectional area, therefore, requires an increase in velocity to keep the mass flow rate constant. The microstreams are then brought together in an impingement area through which all of the product must flow. The bends in the flow pass cause shear-rate differences between the outer and inner layers of the bend to enhance the size reduction. At the impingement area, the collision of the two high-speed flow streams in a very tight spot creates various droplet size reductions and different mixing mechanisms such as cavitation, implosion, shears, and turbulence. An illustration of the interaction chamber and its internal configuration of microchannels and impingement areas is presented in Fig. 37 [1,37].

The microfluidizer technologies have been used either continuously or with recycling, as is shown in Fig. 38. There are no moving or required adjustment parts in microfluidizer technology, and the configuration of interaction chamber is the same for both laboratory and scale-up equipment.

Extremely fine emulsions in the submicron region have been obtained. The shear rates propagated in the interaction chamber are caused by the huge pressure drop that can be as high as 20,000 psig. Unit capacities range from a laboratory unit producing 50–800 mL/min to production units capable of 200 L/min [37,38]. The laboratory models of this machine offer the added convenience of being powered by air-operated pumps, while production models use the typical piston-type pumps used on other similar high-pressure homogenizers.

Very fine emulsions are obtainable using this new device, including the preparation of the difficult-to-form unilamellar liposomes. These structures offer a unique therapeutic value for the encapsulated drug within an encapsulating layers of phospholipids. These structures can easily be made when the aqueous phase and the polar phospholipid phase are combined in the microfluidizer device [39].

The microfluidizer technology satisfies the requirements for producing the finest emulsions, known as microemulsions. The equipment operates at high pressures to effect shear with very large pressure drops. The premix must be sufficiently small, i.e., devoid of any particles that can plug the small channels in the interaction chamber. The plugged chamber can be easily opened by reversing the flow direction, but the pump must be depressurized before reversing the interaction chamber. Sometimes, the feed to the microfluidizer must be preconditioned with another high-shear device, such as a rotor/stator mixer to prevent plugging.

The microfluidizer technology offers various configurations of this system for different applications. For example, a high-temperature/high-pressure microfluidizer (M-110ET) is designed to process hard-to-handle or very viscous products without using organic solvents. This model generates process stream temperatures to 315°C, and uses

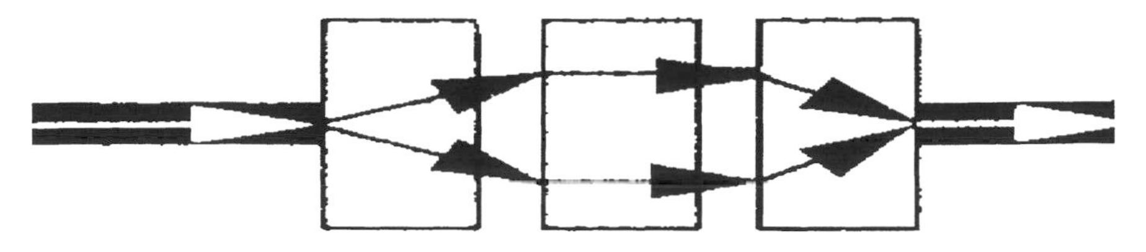

Fig. 37 Flow path through the "interaction chamber" of the Microfluidizer. (From Ref. 1.)

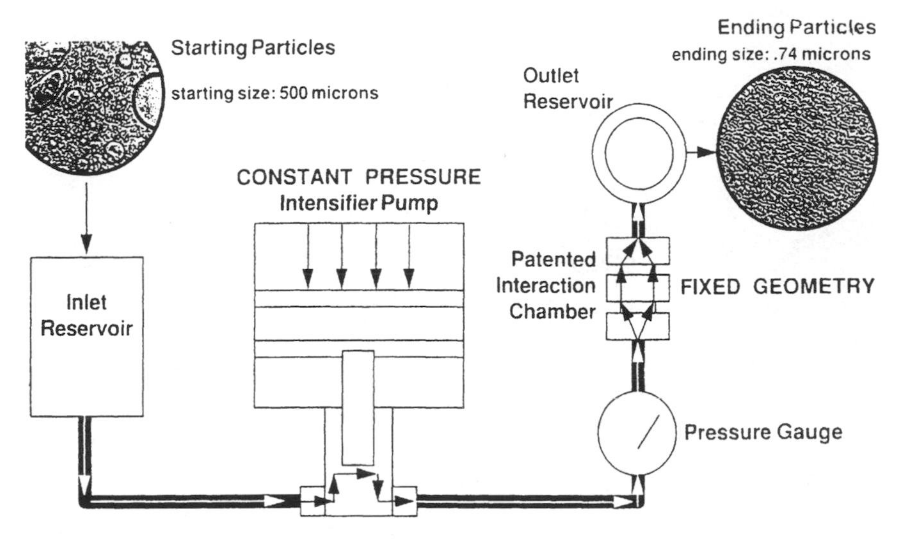

Fig. 38 Flow path through the Microfluidizer. (Adapted from Ref. 38.)

two in-line heat exchangers—one to maintain the product at the required temperature during the processing and the second one to cool the product. Therefore, consulting the manufacturer for a particular application is highly recommended [1,37,38]. Recently, a regular laboratory model microfluidizer was used to produce a wax-containing pharmaceutical emulsion without premixing [40].

7. *Low-Pressure Cyclone Emulsifiers*

Another unique apparatus used for the formation of emulsions and suspensions is the cyclone emulsifier. This device uses a positive-displacement pump to feed a special chamber with a tangential entry port. The product is forced to circulate in concentric layers toward the center and the ends of the chamber, where it is expelled. The shear arises from the difference in the velocity of the fluid (in layers of increasing velocity) as the fluid travels in a spiral toward the center. This principle is illustrated in Fig. 39.

These devices require substantially less pressure than high-pressure homogenizers, ultrasonic homogenizers, and microfluidizer technology. They operate in the 200 psig range and have operating capacities of 7.5–225 L/min. They are capable of producing emulsions in the 2–10 μm range, which is very similar to that obtained from fixed-gap rotor/stator machines and some settings of colloid mills. There are no moving or required adjustment parts in the cyclone-type emulsifier. The recommended viscosity limit is nearly the same as that for piston homogenizers, 1–2000 centipoise, but some acceptable results have been obtained at viscosities as high as 15,000 centipoise. In certain installation, the cyclone homogenizer is recommended for premixing of fluid streams before they are processed in high-shear devices, such as piston homogenizers.

Another cyclonelike device has been patented [42,43] in which the interaction chamber consists of a fixed-dimension cylinder having one closed end and the other end forming the outlet port, which has adjustable orifices. Depending on the size, two to eight small cylindrical entry ports are located tangential to the body of the interaction

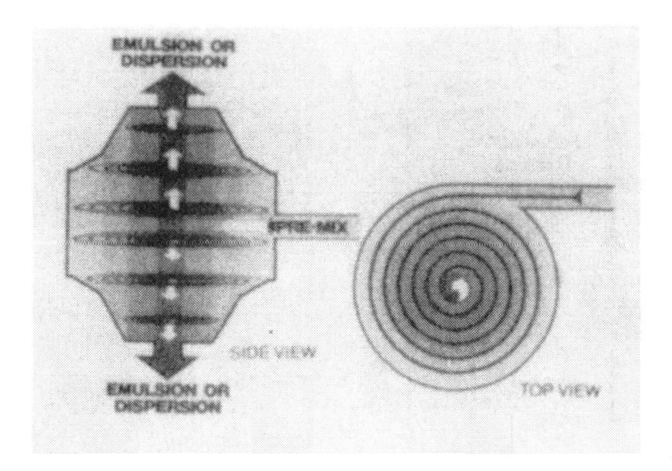

Fig. 39 Cyclone-type homogenizer mixing chamber. (From Ref. 41.)

chamber. The symmetry axes of these entry ports are perpendicular to the symmetry axis of the interaction chamber. This design is presented in Fig. 40, with only four entry ports. This machine is called Novamix® (a registered name for MicroVesicular Systems). It was originally designed to process and produce nonphospholipid lamellar microstructures or lipid vesicles.

The lipid vesicles are composed of two immiscible aqueous and lipid phases. The lipid phase consists, generally, of solid polyoxyethylene-derived amphiphiles that form micelles in aqueous media. Under the proper mixing conditions, i.e., a combination of shear, heat, and turbulence, followed by appropriate cooling, the micelles of these types of lipids fuse to form lipid vesicles. The two phases are metered carefully and heated in separate reservoirs and finally pumped to the interaction chamber for processing. The interaction chamber and pump heads are confined in an insulated compartment that is maintained at the required temperature for the production of the lipid vesicles. The outlet is attached to a chilling device that cools the product at the required rate [43].

The flow pattern is similar to that of a cyclone, i.e., the flow of liquid is in a vertically positioned rotating cylinder along its vertical axis. The streamlines are concentric circles with their radii decreasing toward the center of the cylinder. The decrease is a function of cylinder radius, flow rate of fluid (speed of rotation), and other parameters like viscosity, density, and surface tension of the formulation. In curved type of flow with changing radii, there exists a pressure gradient, i.e.

$$dP/dr = \rho V^2/r \tag{8}$$

where P = pressure; r = vessel (interaction chamber) radius; V = tangential linear velocity; and ρ= the liquid density. Since the change in pressure is positive for a positive radius change, the pressure at successive points increases from the concave to the convex side of the streamline [39]. The exact change in pressure depends on the variation in tangential linear velocity, which is proportional to the speed of the rotation and the radius.

The flow pattern in the interaction chamber is neither a free vortex, due to the presence of an initial momentum from the pumps, nor a forced vortex, for the stream-

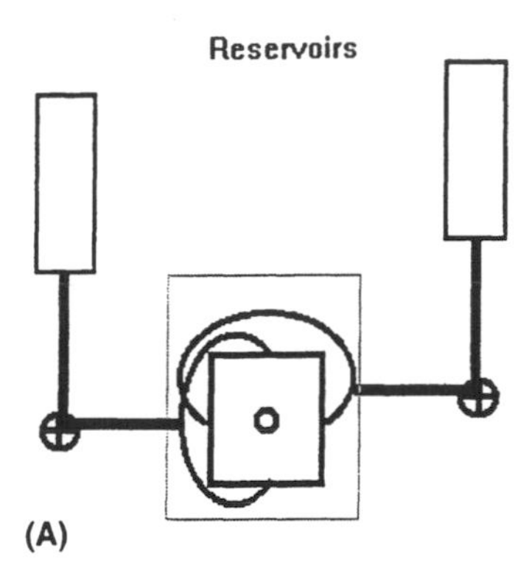

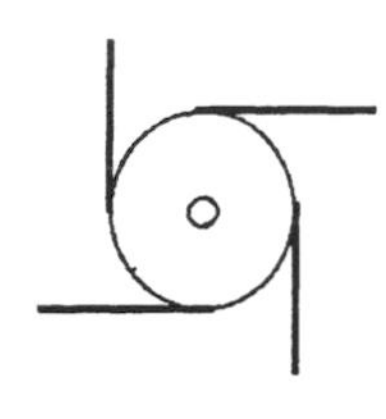

(B)

Fig. 40 Novamix® (A) cross-sectional view and (B) of the interaction chamber. (From Ref. 43.)

lines are concentric circles around a central axis. However, the combination of the two causes the required shear and turbulence without cavitation in the process stream. Figure 41 illustrates a cross-section of the flow pattern within the interaction chamber [43]. The opposing entry ports are used for the same phase.

As far as is known, this processing equipment has not been used in the production of other dispersed delivery systems. However, the design of a heated interaction chamber makes it useful as processing equipment for high-viscosity products like creams, lotions, and ointments. This processing technique has low shear with very high through-

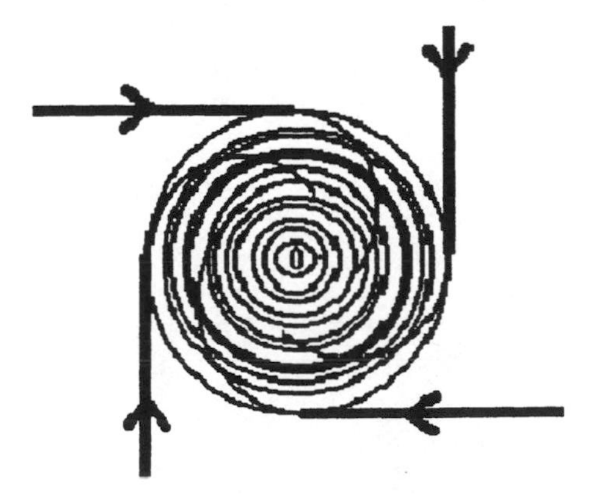

Fig. 41 Cross-section of the flow path through the "interaction chamber" of the Novamix®. (From Ref. 43.)

put capacity and does not require premixing; it is fairly inexpensive and suitable for continuous operation. Major drawbacks to this equipment are its lack of availability, the need for special heating and cooling control systems, no available laboratory model, and the need for many trial-and-error runs in order to scale-up to production.

8. *Static Mixers*

A true low-shear and low-energy requirement device for emulsifying immiscible liquid mixtures is the static mixer. Sometimes called a pipeline mixer, this device is actually a series of specially designed baffles in a cylindrical pipe as shown in Fig. 42. These simple devices are used extensively for the preparation of unstable emulsions for liquid-liquid extraction purposes. Droplet sizes, obtainable using static mixers, have been studied extensively and vary with viscosity, interfacial tension, pressure drop, and static mixer design [45]. Size distributions obtainable range from 1000-100 μm. Hence, although there are very few emulsions stable in this region, the static mixer has seen application as an in-line premixer in continuous processes or in recirculation loops to batch-processing equipment.

F. Nonmechanical Disperse Processing

Recently a new processing technique became available for the production of stable and uniform liposomes. It uses the physico-chemical properties of the supercritical liquids rather than the mechanical forces of the pumps. One such a process technology is presented in this section.

1. *Critical Fluids Liposome Process*

Near-critical or supercritical fluid solvents with or without polar cosolvents (SuperFluids™) (Aphios, Corp., Woburn, MA) for the formation of uniform and stable liposomes having high encapsulation efficiencies has been used [46–48]. Supercritical or near-critical fluids as shown by the pressure-temperature diagram in Fig. 43, are gases such as carbon dioxide and propane that have been processed under ambient conditions. When compressed at conditions above their critical temperature and pressure, these substances become fluids with liquidlike density and the ability to dissolve other materials, and gaslike properties of low viscosity and high diffusivity. The gaseous characteristics increase mass transfer rates, thereby significantly reducing processing time. Small added amounts of miscible polar cosolvents, such as alcohol, can be used to adjust polarity and to maximize the selectivity and capacity of the solvent.

Fig. 42 Static mixer. (From Ref. 44.)

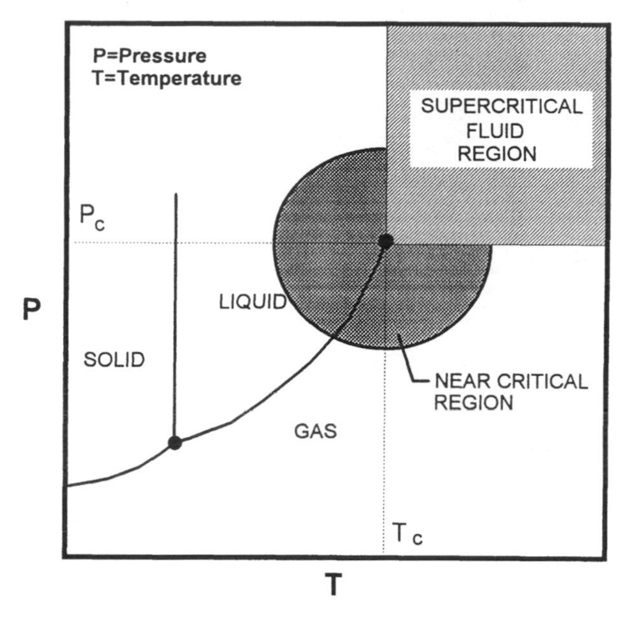

Fig. 43 Supercritical fluid diagram.

Using this process for liposome formation, supercritical fluids at appropriate conditions of pressure and temperature are used to solubilize phospholipids, cholesterol, and other liposomal raw materials in an apparatus as shown in Fig. 44 [49,50]. A circulating pump operating in a high-pressure loop ensures good mixing between the supercritical fluids and the liposomal raw materials. After a specified residence time, the resulting mixture is decompressed using a back-pressure valve attached to a dip tube with its nozzle in a decompression chamber (vessel B) that contains an aqueous preparation (solution or fine dispersion) of the therapeutic compound targeted for liposomal encapsulation.

As illustrated in Fig. 45, bubbles form at the tip of the injection nozzle and, as a result of the depressurization of the supercritical fluids, the solubilized phospholipids deposit on the phase boundary of the aqueous bubbles. As these bubbles detach from the nozzle into the aqueous solution, they rupture, causing the bilayer of the phospholipids to detach and encapsulate solute molecules. They also spontaneously seal themselves to form uniform liposomes. This injection technique is well suited for the liposomal encapsulation of recombinant proteins used for enhanced drug delivery, DNA in gene therapy, and other hydrophilic therapeutic agents.

In another procedure, the phospholipids and the target compound are solubilized simultaneously in a supercritical fluid "cocktail," which is continuously dispersed into an aqueous solvent. The process stream is decompressed, and the unstable phospholipid bilayer fragments collide and rapidly seal to form liposomes with the entrapped compound. The controlling parameters are the pressure and the rate of decompression. The decompression procedure is readily scaled up to larger production volumes. It is a "one-step" process, and the composition of the supercritical fluid stream can be designed to achieve feed streams containing concentrated phospholipid and target compound. The results yield high product-entrapment efficiencies and concentrated product-recovery streams.

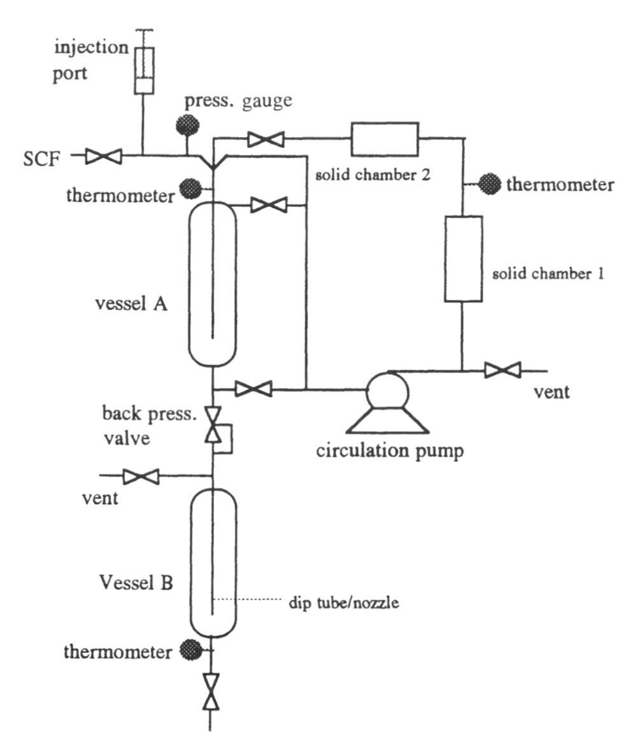

Fig. 44 Experimental supercritical fluid apparatus.

In still another procedure, supercritical fluids can be uniquely used to encapsulate very hydrophobic molecules such as potent anticancer drugs, for example, paclitaxel, camptothecin, (a very effective topoisomerase-I inhibitor), and other anti-infective (anti-HIV) agents [49]. In this case, the hydrophobic drug and the phospholipids are directly solubilized in the supercritical fluids prior to their injection into a phosphate buffered saline (PBS) or some other biocompatible solvents. After decompression through a nozzle, the supercritical fluids evaporate leaving an aqueous dispersion of liposomes containing hydrophobic molecules entrapped within their lipid bilayer.

G. Fine Suspensions and Size Reduction Equipment

The hydraulic shearing stresses set up in liquids by open impellers, rotor/ stator mixers, and even high-pressure homogenizing devices are often not high enough to break up hard and strong discrete solid particles or tightly agglomerated particles. Equipment with a more positive mechanical shearing action is required. A wide variety of different equipment exists for this general application, but relatively few designs are used widely in the pharmaceutical industry. These include three-roll mills, and to a lesser extent, ball mills or jar mills, and continuous stirred media mills.

1. Three-Roll Mills

An established machine capable of dispersing small tightly bound agglomerates and hard discrete particles is the three-roll mill. By causing the premixed suspension to travel

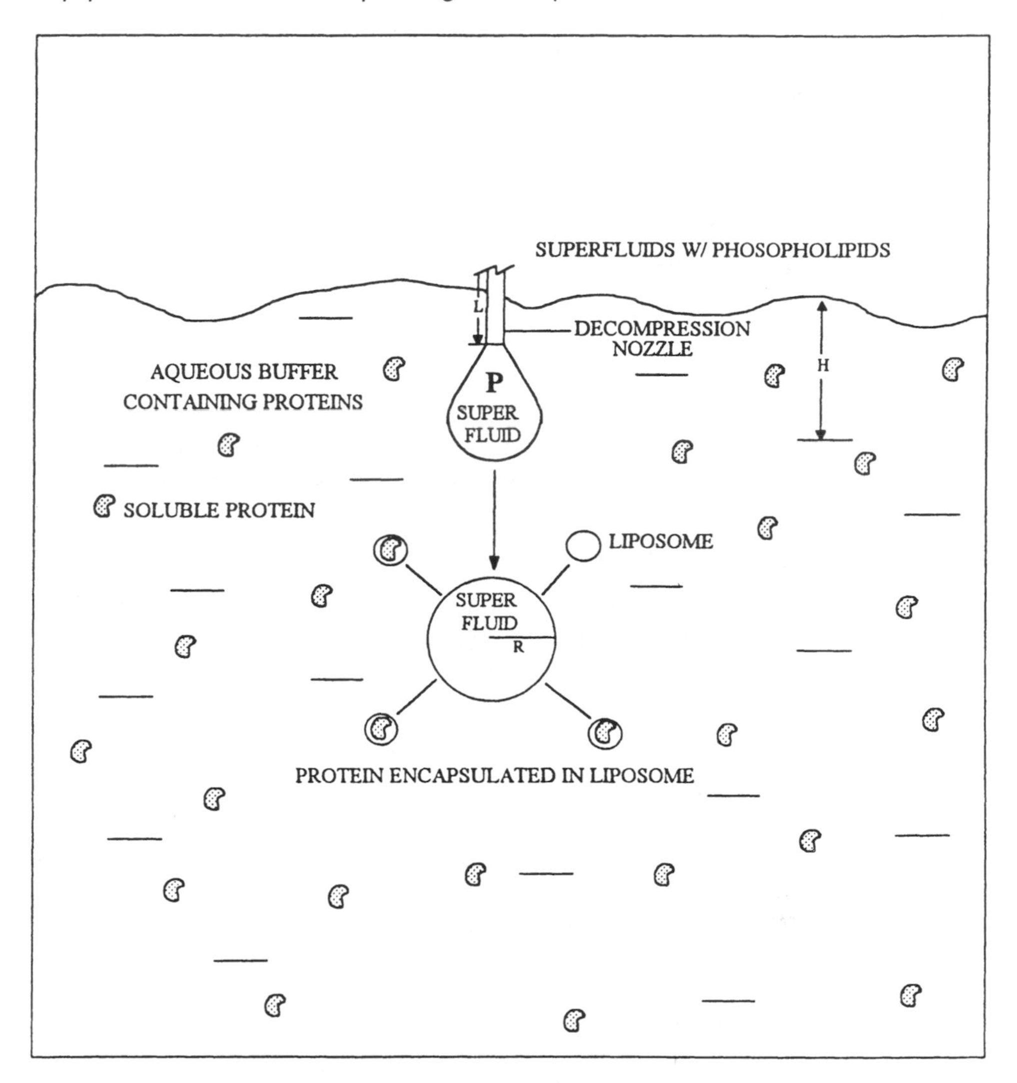

Fig. 45 Formation of liposomes with supercritical fluid.

between two rotating rolls that are located about 10–50 μm apart, the particles not only are subjected to very high shear rates but also mechanical crushing and smearing when the particle is larger than the clearance between the rotating rolls. Of course, if this device is to be adjustable in this range, it must be exceedingly sturdy and precision made.

The three rolls have different names to reflect their uses. The first is the feed roll, since it is the point at which the premixed product is fed to the machine. As shown in Fig. 46, the feed material resides in the V-shaped space between the feed roll and the next roll, known as the center roll. The product flows between the two rolls, down to the bottom of both rolls. The product that adheres to the feed roll comes back to the feed area and the product that sticks to the center roll comes to the even finer gap set between the center roll and the apron roll. A transfer blade is used to wipe the material off the apron blade, causing the material to flow down the apron as finished product.

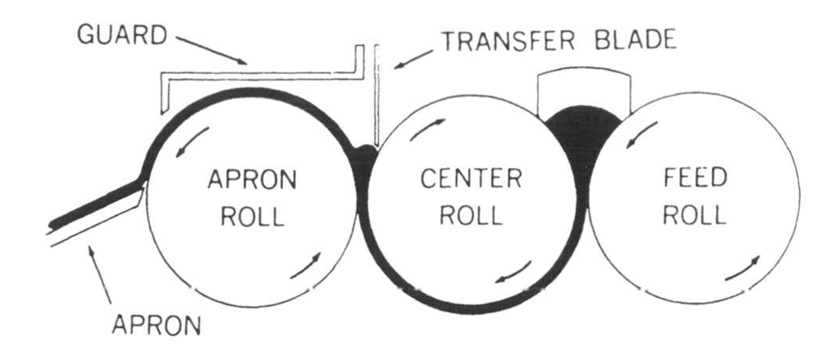

Fig. 46 Three-role mill. (From Ref. 51.)

The shear rates achievable in a three-roll mill are a function of the roll radius, the difference in the rpm of the rolls in contact, and the clearance between the rolls, known as the nip clearance [52]

$$\text{Shear rate} = (105\ R)\ (\delta\ \text{rpm})/z \tag{9}$$

where R = roll radius (inches); δ rpm = difference in roll speed; and z = nip clearance (mils). The estimated shear rates for a typical apron roll nip are in the range of 10,000–80,000/sec. For example, what shear rate would exist between the apron roll and the center roll in a three-roll mill with 10 in. rolls as described in Table 5?

$$\text{Shear rate} = (105 \times 10)\ (300 - 150)/4 = 39{,}375\ \text{sec}$$

This shear rate would occur if the nip clearance were set to 4 mils. Typical conditions and production rates for three-roll mills are shown in Table 5.

Three-roll mills are not typically used for high-capacity processes where large quantities of liquids are required. Three-roll mills are considered slow, but can perform tasks that other machines cannot finish in unlimited time. If the desired level of dispersion can be obtained using simpler equipment, there is almost always a time savings.

Three-roll mills require setting and tuning that is beyond the capability of untrained operators. However, if experienced personnel are available, the fine dispersing and grinding capabilities of the three-roll mill make the preparation of some products possible where almost no other method will succeed.

Table 5 Three-Roll Mill Capacities

Roll diameter (in.)	5	10	14
Roll length (in.)	12	22	30
Speed—apron roll (rpm)	158	300	300
Speed—center roll (rpm)	81	150	142
Speed—feed roll (rpm)	31.5	59	56.5
Capacity (lb/hr)[a]	50	350	700

[a]Based on a heavy paste mix, approximately 14 lb/gal. using an average film dispersion thickness of 0.0005 in.
Source: Ref. 51.

2. Ball Mills

For true size reduction of fine solid particles or for deagglomeration of very tightly bound agglomerates, the machine with the most successful record is the ball mill. A small version of the ball mill is known as the jar mill. As shown in Fig. 47, both machines include a cylindrical drum into which a charge of heavy spherical balls—usually metal or ceramic—is loaded along with the components of the dispersion. The cylinder with all of the components is rotated causing the balls to tumble and collide with each other. Between the balls the tiny particles are trapped and shattered by the compressive force caused by the collision of the balls.

The operation of these machines is relatively simple, which is one of the reasons for their widespread use in many industries. The procedure is simply to load all of the components and rotate the batch until the desired particle size is reached. The vessel is then dumped and the product strained from the ball charge. For several reasons, ball mills are not widely used in the pharmaceutical industry. First, ball milling is typically a time-consuming process. Milling times are often measured in days. Of course, this is very dependent on the size of the batch and on the difficulty of grinding the product, which is a function of strength, toughness, ductility, and other factors. For certain difficult-to-grind raw materials, though, the ball mill is still the machine of choice. In the case of fine solids grinding, there is almost always a second option, i.e., use of a finer grade of raw material. If it is available and does not ruin the economics of the process, then this finer grade of material, albeit more expensive, is probably the best alternative. However, if the suspension being made is very specialized, a source for this type of finely ground raw material may not exist, and ball milling is the best choice.

3. Agitated Bead Mills

A modernized improvement of the ball mill is the agitated bead mill. Just like the ball mill, the bead mill uses a charge of inert small balls around 2–8 mm in diameter. If the beads are made of ceramic, the machine is known as a media mill (Fig. 48). If the beads are made of steel balls, then the machine is called a shot mill. The original machine used special large grains of sand, about 3 mm in diameter, and is known as a sand mill. The design consists of a cylinder, which can be either horizontal or vertical, that has a high-speed agitator, which is capable of fluidizing the charge of beads, causing them to collide at very high speeds. The dispersion arises from the action of the beads hitting each other and also hitting the stirring impeller. The premix is pumped through the housing. There is a high probability that each particle must be repeatedly subjected to the high stresses that result when the beads collide with each other or the high-speed impeller.

The agitated bead mill is the workhorse of the pigment dispersion industry due to its fine-solids grinding and dispersing capabilities. It is not often used in the pharmaceutical industry, except when particle size requirements fall below 10 μm.

VI. ADDITIONAL TOPICS

A. Materials of Construction

The most commonly used material for the construction of mixing and dispersion equipment is stainless steel. There are many grades, but most of the time type 304 is all that

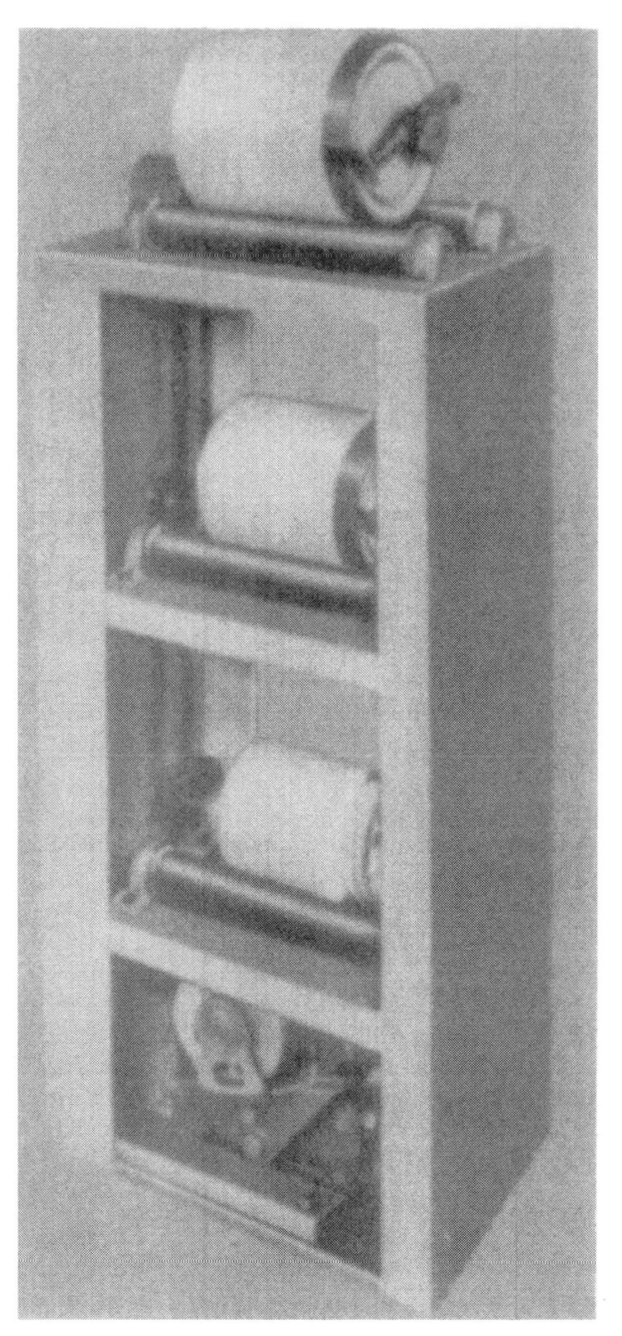

Fig. 47 Jar mill. (From Ref. 53.)

is needed. Occasionally, components of some formulas may require type 316 or even the more corrosion resistant type 316 L (for low carbon content) stainless steel. In some unusual instances, a product may require that the process vessel be even more resistant to corrosion, and exotic alloys or even glass-lined or coated equipment is necessary. As the degree of corrosion resistance increases, the availability of some of the

Fig. 48 Media mill. (From Ref. 54.)

designs described earlier is reduced. The cost of building some of the designs on a custom basis out of expensive materials can be prohibitive.

The surface finish of a material can have a major bearing on whether the material is able to withstand the corrosive conditions of a process. The most often seen finish on stainless steel is known as a #4 finish, which corresponds to a 125–180 grit finish, depending on the manufacturer. There is a mechanically produced series of scratches that gives a pleasing matte appearance. If this type of finish is electropolished, then the peaks between the scratches are smoothed out and a much easier to clean and also more corrosion resistant finish is the result. Electropolishing is an electrochemical process whereby some of the metal material is actually removed to form a very smooth finish. The manufacturer of the Microfluidizer has recently announced that the internal element of the interaction chamber made of diamond creates a highly corrosion resistant and durable interaction chamber [55].

B. Pilot Laboratory

A well-designed pilot laboratory is essential to avoid problems of scale-up and also to provide a means for answering questions that often arise after the process has been operating smoothly for months or even years. Ideally, the pilot lab will have the same type of equipment but in a smaller size, capable of making batches in the 20–100 L range. In many cases, doing experimental or troubleshooting work in this range can be very expensive. However, many of the problems that might arise during scale-up or troubleshooting can never be solved if only small batches measured in milliliters are studied.

The key is to be able to study the process in a large enough batch size to be representative of production yet not so large to make controlled testing cost prohibitive. A pilot-plant unit cannot provide all of the same mixer parameters as production. Since the vessel will be smaller, the pilot unit cannot possibly have the same diameter impeller. Hence, it is necessary to run the impellers of the pilot units at a higher rpm than the production units run in order to approximate the tip speeds or maximum shearing rates of the larger, wider diameter machines used in the plant. Generally, this is the way most pilot-plant equipment is designed, with smaller impellers operating at higher speeds. For example, Table 6 describes the impeller diameter, pumping rate, and maximum tip speed. The pilot or "laboratory" unit has a maximum impeller tip speed that falls in the middle of the range of the production equipment.

The pilot plant should be used to scale-up a new product from the laboratory beaker to a procedure that can be adapted to full-size production. This is possible only if the pilot equipment closely resembles the production facility. Fortunately, pilot-plant-scale designs are usually available from the manufacturers of the full-size equipment.

Table 6 Axial-Flow Rotor/Stator Mixer Capacities and Specifications

Model	HP	Rotor diameter (in.)	rpm	Tip speed (ft/min)	Flow rate of water (gal./min)
CJ-4C	3/4	1.88	10,000	4,920	88
C-1	5	3.875	3,600	3,655	235
C-2	7.5	5.00	3,600	4,710	464
C-3	1.5	5.75	3,600	5,400	905
C-5	25	6.5	3,600	6,120	1,730

Source: Ref. 25.

C. Troubleshooting

After a product has been successfully developed in the pilot laboratory, scaled up for production, and manufactured for a period of time, one could assume that the product can be duplicated as long as nothing in the procedure, equipment, or raw materials changes. Indeed, this is true, except that the task of making sure that nothing ever changes in the procedure, raw materials, or equipment is almost insurmountable over the long term. Operating personnel change, new supplies of raw materials are chosen, and equipment slows down and wears out. Troubleshooting is the job of asking the right questions and then investigating to find out what exactly has changed.

The place to start is with the product. What is different or unsatisfactory? Has the emulsion or dispersion lost its stability? Is the particle size distribution too large? Is the viscosity too low or too high? Or, even worse, has the dosage form been shown to be losing its desired effects? Key to productive troubleshooting is a logical plan of investigation that objectively eliminates suspects as they are found not to be the problem.

Which of the following—procedure, equipment, or raw materials—is the most likely area that has changed? As a start, the production staff, technical service, procurement, and quality assurance should be requested to answer the question as to whether the problem is a one-time occurrence or whether it has been happening often—since a certain date, or even intermittently from the beginning. At this point, there is no substitute for good records. Is it possible to ascertain when a problem first occurred? Probably the best place to start an investigation is in the production area itself, but not with the supervisors. Instead, the very best information usually comes from the personnel actually running the machinery.

If any degree of reproducibility is to be expected, the operating personnel must have a definitive plan for the addition of all components, the length of mixing time, and the speeds and flow rates that are required. This plan should be in writing, so that individual operators do not alter the mixing parameters that were determined during scale-up. A widespread trend in manufacturing is the use of programmable logic controllers and, more recently, small computers to actually control the addition of components and the running of mixing machines. However, even in the most modern plants, there are process steps that are left up to individuals.

A simple example drawn from the experience of the author is considered below.

A manufacturer of a topical lotion finds that the stability of the product is not what it used to be. The product is separating after one month. In addition, quality assurance has found the viscosity is not as high as it should be. Since the raw materials are coming from the same supplier and are always checked before processing, they are ruled out. One theory is that the mixing equipment is not performing as it did when it was new. Another theory is that the plant operators are mixing the final product far too long, and that this is ruining the stabilizer. A test run is set up with representatives of the departments of quality assurance, production, and technical service on hand.

The product is an oil-in-water emulsion with a polymeric stabilizer dispersed into the water phase. The mixing vessel has an anchor, scraped-surface agitator along with an axial-flow rotor/stator homogenizer. The water and surfactants are added to the tank and heated to 80°C. Next, 25 kg of the stabilizer is added by feeding the powder into the vortex created by the axial-flow rotor/stator mixer. When the stabilizer is fully dispersed, the oil phase, which has been prepared separately, is pumped into the bottom of the kettle so as to feed it directly into the head of the rotor/stator mixer. After all of the oil phase is dispersed, high-shear mixing is continued at 3550 actual rpm for 10 min. The high-shear mixer is then slowed to 1750 rpm and the product is cooled to 30°C in 1 hr using ambient groundwater in the cooling jacket.

The anchor/scraped-surface agitator is run at 25 rpm throughout. The product is then strained through a 200 μm filter and packaged in jars. Quality control checks the viscosity from the test run and finds it to be right on target. An accelerated stability study is conducted using a centrifuge, and the product is found to be very stable. After two to four weeks, samples are studied and found to be stable. The answer to the questions as to what is going wrong has not yet been found.

Perhaps, the operators are not switching to the lower speed on the homogenizer mixer for the cooling step. This could cause attack on the long-chain molecules that make up the stabilizer. A decision is made to install a programmable controller on the mixer. Essentially, all that the operator will do is turn on the mixer, add the stabilizer, and discharge the batch to the holding tank. The programmable controller regulates the temperature automatically and even informs the operator when to feed the stabilizer. Technical service justifies the cost of the controller and its installation against the lost product costs. Management approves the expenditure.

Despite these efforts, the same problem happens again! The vice president in charge of production demands a complete report. The report shows that most of the bad batches are produced late, on the second shift. But second shift personnel seems to be performing the procedure correctly. Another full-scale test batch is supervised by all parties on the second shift. The results are perfect. The problem still has not been identified. The unacceptable batches cannot be disposed of easily and at low cost, since they include certain active organic compounds. All of the bad batches, along with the filters used, are being held in drums until a method of disposal—certain to be costly—has been found. Technical service is also in charge of this project. It is likely that the water can be evaporated, and that the remaining viscous liquid is poured into the same drums as those holding the used filter cartridges. One of the operators entrusted with this job notices that a few of the drums contain cartridges that are totally covered with white sticky lumps, whereas most of the discarded cartridges look almost clean.

Could this deposit have anything to do with the stability problems? A sample of the white sticky material is sent to the laboratory and found to be the polymeric stabilizer. But why was the stabilizer caught in the filter only once in a while? There was only one job that the operators had in the new automated process: the addition of the stabilizer. The operators had found that the mixer would mix in all of the powder whether or not they sifted in the powder slowly. Unfortunately, when 25 Kg of this stabilizer was added at once, many sticky gum balls formed, and these agglomerations required more than 10 min of high-shear mixing to be smoothed out.

Since the operating procedures are carried out by human beings, there are more chances of error, negligence, and negative creativity causing problems than with the equipment or the raw materials. That is not to say that problems cannot occur as a consequence of the performance of the equipment or the varying quality of the ingredients.

Some procedural questions that should be asked when problems arise are:

1. Are all of the operators fully trained and aware of the consequences of changing procedures?

2. Is there a record-keeping system into which the production workers make entries showing that each process step is completed?
3. Have all of the workers actually handling the controls been interviewed about their methods?
4. Is there a way to observe performance without making it known that there is an official inquiry, in order to observe "usual" behavior?
5. To what extent have the variables of procedure been taken out of the hands of the operators through automation or other means?

There is no substitute for well-trained and motivated production personnel that care and diligently follow established procedures.

Other questions concern whether all of the equipment is operating at the same level of performance at which it was originally operating.

1. Are the speeds of the mixers and agitators the same as they were previously? Are the motors still operating at the same level? Has the electricity, e.g., voltage in the plant remained the same? Are there worn parts in the drive slipping?
2. Have the impeller parts worn, possibly enlarging rotor/stator gaps?
3. Is the heat transfer system operating the same as before? Are the heat transfer media, e.g., steam or water, at the same temperature and pressure? Has the heat transfer jacket fouled, causing decreased heat transfer per time and also lower surface temperatures?
4. Has the surface finish become scratched and, hence, difficult to clean?
5. Are pressure seals leaking? Is the vacuum level equal to previous runs? Is the system able to hold pressure during the steam sterilization step?

Another set of questions concern raw materials, whether there has been any change in the supply of ingredients.

1. Has the supplier of raw materials changed?
2. Has the grade of raw material changed?
3. Has the particle size distribution of a solid raw material changed?
4. Is the size distribution of all solid materials checked before use?
5. Is it possible that an off-specification batch of raw material was supplied?

In the previous example, full-scale batches were used in the troubleshooting procedure. This was feasible since acceptable product was made and not wasted. Often this is not the option and the preparation of a full-size batch is very expensive. In this case a well-equipped pilot laboratory is valuable. A pilot-size batch can be used to try different variables. Ideally, the pilot lab is used in advance whenever raw materials or procedures are changed to see whether there is any change in the product.

As a second example, one might consider an antacid product in which a raw material was changed to a grade of material exhibiting a particle size distribution greater than that previously used. Purchasing had found a source of magnesium hydroxide that would cost 20% less. Before committing to the purchase of large quantities of this raw material, a series of tests were run in the pilot laboratory. The standard procedure was to blend the magnesium hydroxide, aluminum hydroxide, stabilizers, and additives in a vessel using an axial-flow turbine and then pass the mixture through a colloid mill with a 0.0050 in. gap setting. The resulting suspension was not intended to be stable. Rather, the mixture needed to be dispersed and deagglomerated down to a particle size of 99% less than 25 μm.

Before trying an expensive 1000 gal. production-size batch, a 25 gal. batch was successfully prepared in the pilot laboratory. But, the first full-scale production batch exhibited a particle size of approximately 15% above 25 μm. Something was clearly different between the pilot- and full-scale results. This was made even more evident after the second full-scale test batch came out exactly like the first.

A 3 in. diameter 10,000 rpm mill was used in the pilot laboratory and a 10 in. 3600 rpm in the plant. The respective peripheral speeds were 7853 ft/min for the pilot colloid mill, and 9425 ft/min for the production machine. Since the production machine had a higher maximum tip speed, an assumption was made that it also had a higher maximum shear rate, and that any change in grinding ability from one machine to the other would probably mean that the production machine would grind even a little finer. This had also been seen in earlier controlled tests between the two machines.

At this point it would have been a "safe" decision to go back to using the original, more expensive raw material. The oversize particles in the test product were large enough to be discerned on the surface of the tongue, so it was definitely not going to be acceptable to users of the product accustomed to the smooth-mouth feel of the original product. The question of why the pilot machinery and procedure produced good product while the production plant could not was disturbing, since many products had been successfully scaled up in the past.

An answer was provided by a careful inspection of the machinery in operation and after disassembly for cleaning. The gap setting on a colloid mill is one of the prime determinants of the shear rate and, hence, grinding ability. This setting is made with an adjusting ring that must be set carefully. During the second test, the setting was checked to make sure it was the same as previously set. The problem was not a matter of incorrect setting by the operator. However, the setting did not absolutely guarantee that the actual gap was the required 0.0050 in. This would be true only if the surfaces of the rotor/stator were in their original shape. Inspection of the parts did not show any highly visible wear but the parts did "seem a lot shinier than when they were originally supplied," according to one operator.

Polishing is the first sign of wear. The manufacturer of the mill was asked to inspect the parts, and their opinion regarding the wear was much different. They were able to measure the parts of the rotor and stator very accurately and found that 2.5 thousandths of an inch had been removed from both the rotor and stator. This amount of wear is hardly visible to an untrained eye. However, this amount of wear caused the gap setting to be actually double than that necessary to obtain a successful scale-up. The parts were replaced with new ones faced with the hard material Stellite for increased resistance to wear, and the next batch made with the production-size equipment was successful.

One question remained though. Why had the wear problem not shown up before the new raw material was tried? The reason was that the tip speed of the production mill was high enough to disperse the higher grade raw material even at twice the gap setting. The higher grade raw material was easier to disperse. The prescribed rotor/stator gap settings and the procedure had been obtained from a successful test in the pilot plant and specified for the production process. The fact that a wider gap would also work was never tried.

Troubleshooting is a process of isolating the variable(s) that may have been changed in a process in order to determine a method to successfully produce acceptable product given the present realities. There is a definite requirement for good records in order to know whether anything has changed in the raw materials, procedures, or equipment. After consulting the records of previous processes, the best place to start an investigation is with the actual operators of the equipment. If several variables need to be studied, it is usually more cost-efficient to perform tests on a pilot scale. If this fails, full-scale trial runs are required. The key to any troubleshooting investigation is an attitude on the part of the investigator that does not exclude any potential variables—raw material, procedures, or equipment—from serious consideration.

REFERENCES

1. S. E. Tabibi, "Production of disperse drug delivery systems." In *Specialized Drug Delivery Systems: Manufacturing and Production Technology* (P. Tyle, ed.), Marcel Dekker, Inc., New York, 1990, pp. 317-332.
2. Walstra, P., "Formation of emulsions." In: *Encyclopedia of Emulsion Technology*, Vol. 1, (P. Becher, ed.), Marcel Dekker, New York, 1983, pp. 57–127.
3. J. Y. Oldshue, *Fluid Mixing Technology*, McGraw-Hill, New York, 1983, p. 222.
4. E. Shotton and K. Ridgway, *Physical Pharmaceutics*, Clarendon Press, Oxford, U.K., 1974, pp. 265–270.
5. R. C. Binder, *Fluid Mechanics*, 4th Ed. Prentice-Hall, Englewood Cliffs, NJ, 1962, pp. 93–103.
6. A. Szatmary, *Powder Liquid Mixing with High Speed-Shrouded Impeller*, American Institute of Chemical Engineers, March 1982, pp. 8–9.
7. J. Y. Oldshue, *Fluid Mixing Technology*, McGraw-Hill, New York, 1983, p. 187.
8. *Mixmor Mixers for Every Industry*, Mix Mor, Inc., Los Angeles, CA.
9. *The ARDE Dilumelt-Mix Difficult Solids into Liquids*, ARDE Barinco, Inc., Norwood, NJ.
10. R. R. C. New, Preparation of liposomes. In: *Liposome, a Practical Approach* (R. R. C. New, ed.), IRL Press, Oxford, U.K., 1992, pp. 33–103.
11. S. E. Tabibi and C. T. Rhodes, Disperse systems. In: *Modern Pharmaceutics*, 3rd ed., (G. S. Banker and C. T. Rhodes, eds.), Marcel Dekker, New York, 1995, pp. 299–332.
12. F. J. Martin. Pharmaceutical manufacturing of liposomes. In: *Specialized Drug Delivery Systems: Manufacturing and Production Technology* (P. Tyler, ed.), Marcel Dekker, Inc., New York, 1990, pp. 267–316.
13. J. Y. Oldshue, *Fluid Mixing Technology*, McGraw-Hill, New York, 1983, p. 197.
14. R. R. Scott, A practical guide to equipment selection and operating techniques. In: *Proc. Tech. Program*, Interfex USA, 1992, pp. 153–169.
15. *The ARDE Vacuum Dilumelt*, ARDE Barinco, Inc., Norwood, NJ.
16. *Chemineer Top Entering Agitators*, Chemineer, Inc., Dayton, OH.
17. *Industrial Mixing Equipment, Bull. No. 174*, Spring/Summer 1996, Indco, New Albany, IN.
18. Nagata, S., *Mixing Principles and Applications*, Wiley, New York, 1975, pp. 138–149.
19. *Lightnin Laserfoil Mixers*, Mixing Equipment Company, Inc., Rochester, NY.
20. *Model T A (Twin Shaft) Sanitary Agitator Kettles*, Groen Division of Dover Corp., Elk Grove Village, IL.
21. *Agitators by Lee*, Lee Industries, Philipsburg, PA.
22. *Mixers/Dispersers*, Premier Mill Corp., Reading, PA.
23. T. C. Patton, *Paint Flow and Pigment Dispersion*, Wiley, 2nd Ed., New York, 1975, pp. 468–469.
24. A. Szatmary, *Powder Liquid Mixing with High Speed-Shrouded Impeller*, American Institute of Chemical Engineers, March 1982, p. 2.
25. *Kalish Turbine Homogenizers, Turbine 92/1*, H. G. Kalish Co., Ltd., Somerville, NJ, 1992.
26. *Scott Turbon Mixers*, Scott Turbon Mixer Co., Canoga Park, CA.
27. *The Lee Multi-Mix*, Lee Industries, Philipsburg, PA.
28. *The ARDE Barinco Reversible Homogenizer*, ARDE Barinco, Inc., Norwood, NJ.

29. *Premier Colloid Mills*, Premier Mill Corp., Reading, PA.
30. *Tri Mix Turbo Shear*, Lee Industries, Philipsburg, PA.
31. *The ARDE Dicon In-Line Dispersing Grinder*, ARDE Barinco, Inc., Norwood, NJ.
32. L. H. Rees, Gaulin Homogenizer, *Chem. Engrg.*, 302(June):51 (1974).
33. *APV Gaulin-Innovators in Homogenization Technology*, APV Gaulin, Inc., Everett, MA.
34. M. Loncin, *Food Engineering Principles and Selected Applications*, Academic Press, New York, 1979, p. 256.
35. *Ultrasonic Mixing*, Sonic Corp., Stratford, CT.
36. *Emulsiflex™-C5, Publication C5961* Avestine World Wide, Ottawa, ON, Canada, 1996.
37. S. E. Tabibi, Microfluidizer processing: From research to production. In: *Controlled Release Society Workshop Handbook*, Reno, NV, 1990.
38. *Innovation Through Microfluidizer Technology, Publication # GB100-1*, Microfluidics International Corp., Newton, MA, 1996.
39. E. Mayhew, R. Lazo, W. J. Vail, J. King, and A. M. Green, Characterization of liposomes prepared using a microemulsifier, *Biochemica Acta*, 775:169–174 (1984).
40. S. E. Tabibi, N. P. Pathak, and R. W. Mendes, Controlled release aqueous emulsion, Nigerian patent no.,: RP12421, Nigeria, 1996 (Also see U.S. patent application no.: US 08/263.277 of 21 June 1994 and US 08/491,626 of 19 June 1995).
41. Company literature, APV Gaulin, Inc., Everett, MA.
42. D. H. F. Wallach and C. Yiournas, Method and apparatus for producing vesicles, U. S. Patent No. 4,895,452, (1989).
43. S. E. Tabibi, D. F. H. Wallach, and C. Yiournas, Theoretical consideration of lipid vesicle formation by Novamix. In: *INTERPHEX USA, Proc. Tech. Program*, New York, 1991, pp. 61-69.
44. *Chemineer Static Mixers*, Chemineer, Inc., Dayton, OH; also see *Bull. 712*, Chemineer, Inc., 1995.
45. R. J. Barbini and J. C. Arbo, Droplet studies in two motionless mixers. In: *Company Rep.*, Charles J. Ross & Sons, Inc., Hauppauge, NY.
46. T. P. Castor, An improved liposome manufacturing process, *NSF Phase I SBIR Rep. on Grant No. ISTI-8961217*, National Science Foundation, 1990.
47. T. P. Castor, Supercritical fluid liposome formulations, *Presented Paper*, Am. Inst. Chem. Engrs. Annu. Meeting, 1994.
48. T. P. Castor, Method and Apparatus for Making Liposomes, U.S. Patent, May 1996.
49. L. Chu and T. P. Castor, Solubility of phospholipids in supercritical fluids, *Presented Paper*, 1994.
50. T. P. Castor and L. Chu, Methods and apparatus for making liposomes containing hydrophobic drugs, U. S. patent pending, Nov. 18, 1994.
51. *Day Three-Roll Dispersion Mills*, Day Mixing Div., Littleford Brothers, Cincinnati, OH.
52. T. C. Patton, *Paint Flow and Pigment Dispersion*, 2nd Ed., Wiley, New York, 1975, p. 402.
53. *Grinding Mills*, U.S. Stoneware Corp., Mahwah, NJ.
54. *Premier Media Mills*, Premier Mill Corp., Reading, PA.
55. *Annual Rep.*, Microfluidics International Corp., Newton, MA, 1995.

9

Scale-Up of Disperse Systems: Theoretical and Practical Aspects

Lawrence H. Block

Duquesne University, Pittsburgh, Pennsylvania

I. INTRODUCTION

Scale-up of a manufacturing process, referred to as process scale-up or process translation, involves the transformation of a small-scale process occurring in the laboratory or in a pilot plant to a large-scale process occurring in a production plant. Tatterson [1] notes that the ultimate purpose of process translation is to make a product in large, commercially useful quantities, thereby ensuring mass production, cash flow, wealth, and profit. Dale [2] emphasizes that the proper design and development of the scale-up process reduces the time to market and allows for more rapid commercialization of a product. Unfortunately, process scale-up from the bench or pilot plant to commercial production is not a simple extrapolation, particularly for disperse systems. Few guidelines for the formulator have been available, up until now, in the pharmaceutical literature [3]. As yet, there is no scale-up algorithm that permits us to rigorously predict the behavior of a large-scale process based on the behavior of a small-scale process [4]. This is due to the complexity of the manufacturing process, which involves more than one type of unit operation (e.g., weighing, mixing, transferring). The successful linkage of one unit operation to another defines the functionality of the overall manufacturing process. Each unit operation per se may be scalable, in accordance with a specific ratio, but the composite manufacturing process may not be, as the effective scale-up ratios may be different from one unit operation to another. Unexpected problems in scale-up are often a reflection of the dichotomy between unit operation scale-up and process scale-up. Furthermore, commercial production introduces problems that are not a major issue on a small scale: e.g., storage and materials handling may become problematic only when large quantities are involved; heat generated in the course of pilot-plant or production-scale processing may overwhelm the system's capacity for dissipation to an extent not anticipated based on prior laboratory-scale experience.

By and large, scale-up tends to be viewed by formulators as little more than a ratio problem to be addressed in some empirical fashion. The operational scale-up ratio may be defined as follows:

$$\text{Scale-up ratio} = \frac{\text{large-scale production rate}}{\text{small-scale production rate}} \tag{1}$$

Disperse system scale-up ratios may vary from 10 to 100 for laboratory to pilot-plant process translation and 10 to 200 for scaling from pilot-plant to commercial production. Actual production rates may vary considerably from expected production rates, since overall process efficiency is dependent on a wide range of factors. The processing of disperse systems, whether liquid-liquid or liquid-solid, is still relatively empirical due to the substantial interfacial effects that predominate and control the relevant unit operations. Furthermore, unit operations may function in a rate-limiting manner as the scale of operation increases from the laboratory bench to the pilot plant to commercial production. Thus, although conventional wisdom suggests the necessity of scale-up studies, the appropriate approach is not necessarily initiated with miniaturized commercial processing systems [5].

The concept of scale-up has taken on a substantive regulatory aspect in more recent years with the issuance of Guidance 22-90 by the Food and Drug Administration's (FDA's) Office of Generic Drugs in September 1990 and the establishment of the Scale-Up and Post Approval Changes (SUPAC) Task Force by the FDA's Center for Drug Evaluation and Research. In May 1993, the American Association of Pharmaceutical Scientists, the Food and Drug Administration, and the United States Pharmacopeia cosponsored a workshop on the scale-up of liquid and semisolid disperse systems [6]. The primary finished product attribute to control during the scale-up of a disperse system, whether manufactured in identical, similar, or different equipment, is the degree of sameness of the finished product relative to previous lots. The consensus of the workshop committee was that four criteria be used to evaluate sameness: (1) adherence to raw material controls and specifications; (2) adherence to in-process controls; (3) adherence to finished product specifications; and (4) bioequivalence to previous lots.

The aim of this chapter is to provide the formulator with an appreciation, on the one hand, of the complexity of the scale-up problem associated with disperse systems, and an awareness, on the other hand, that scale-up problems can be resolved, to a great extent, by drawing on the vast literature and experience of chemical engineering. In 1964, H. W. Fowler [7] initiated a series of progress reports in pharmaceutical engineering that appeared over time in the periodical *Manufacturing Chemist*. Fowler's ouevre was distinguished by his focus on fundamentals, i.e., on material properties and on operation and process mechanisms. His intention was "to look at the literature of chemical engineering and to discuss developments which are relevant to pharmacy." It is the present author's intention (in part, through this chapter on scale-up of disperse systems) to validate the interdisciplinary process that Fowler began more than 30 years ago.

II. DISPERSE SYSTEMS: UNIT OPERATIONS AND PROCESS VARIABLES FROM A SCALE-UP PERSPECTIVE

The complexity of disperse systems and their manufacturing is daunting! At the outset, disperse systems, with few exceptions (e.g., microemulsions), are thermodynamically unstable. Furthermore, disperse systems are multiphase systems in which interfacial phenomena play a profound, often critical, role in process outcomes. The processing of dispersions almost always involves unit operations in which the area of contact between the phases and the heterogeneity of the interphases are crucial. In the course of scale-up, suspensions may be more unforgiving than emulsions due to the properties of the solid state, the most complex state of matter [8], which confound process translation.* Finally, the flow conditions and viscosities encountered during the course of processing of a disperse system can vary by several orders of magnitude, or more, depending on the scale of scrutiny, i.e., whether one examines flow on a macroscopic (bulk) scale or microscopic (molecular) scale, and the effective rate of shear (see Fig. 1). Thus, mixing, particle size reduction, material transfer (e.g., pumping), and heat transfer are unit operations that need to be examined in some detail.

A. Mixing

The unit operation of mixing is fundamental to the preparation of disperse systems. As an art, mixing predates recorded history. The science of mixing, however, has only been explored since the 1940s [10]. Dickey and Hemrajani [11] have referred to the erroneous perception that mixing is a mature technology. Unquestionably, mechanistic and quantitative descriptions of the mixing process are as yet incomplete [11–13]. Computer simulations of mixing often introduce artifacts, i.e., effects that are not mirrored in the real mixing state. The enormity of the computations results in what Ottino [12] has described as "caricatures of flows." Nonetheless, enough fundamental and empirical data are available to allow some reasonable predictions to be made. Given the centrality of mixing to other unit operations such as heat transfer and mass transfer,† the failure to use available information and insights is unacceptable.

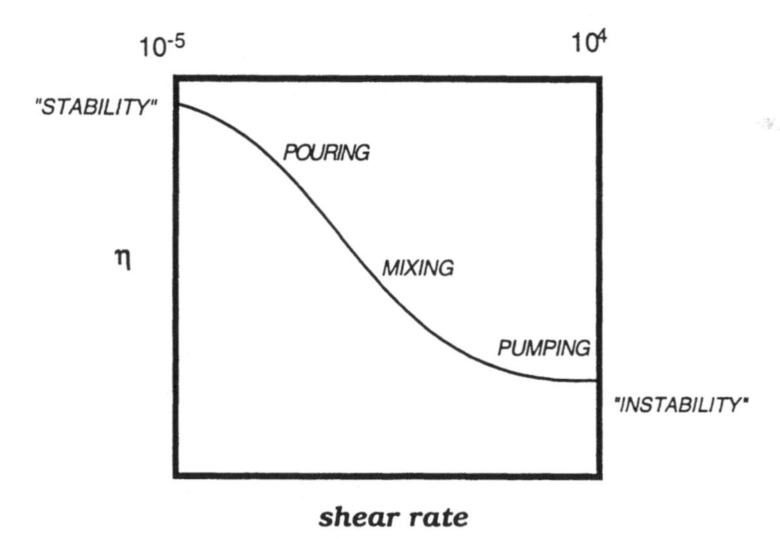

Fig. 1 The "universal" relationship between viscosity and shear rate for various unit operations. (Adapted from Ref. 9.)

*Since the molecules comprising solid particles cannot readily move to new positions, particle shapes are relatively unaffected by the interfacial tension (unlike liquid droplets), coalescence is uncommon (except under compression), and surface or bulk heterogeneities tend to persist as thermal diffusion is impaired [8].

† Mixing differs from other unit operations in that its indirect effects, e.g., on heat transfer, may be the basis for its inclusion in a process.

The terms mixing and agitation are often used, incorrectly, as synonyms [14]. However, the terms should not be confused [14]: mixing, or blending, refers to the random distribution of two or more initially separate phases into and through one another, whereas agitation refers only to the induced motion of a material in some sort of container. Agitation does not necessarily result in an intermingling of two or more separate components of a system to form a more or less uniform product. Some authors reserve the term blending for the intermingling of miscible phases and use mixing for materials that may or may not be miscible.

The diversity of dynamic mixing devices is unsettling: their dynamic, or moving, component's blades may be impellers in the form of propellers, turbines, paddles, helical ribbons, Z-blades, or screws. In addition, one can vary the number of impellers, the number of blades per impeller, the pitch of the impeller blades, and the location of the impeller, and thereby affect mixer performance to an appreciable extent. Furthermore, while dispersators or rotor/stator configurations may be used rather than impellers to effect mixing, mixing may also be accomplished by jet mixing or static mixing devices. The bewildering array of mixing equipment choices alone would appear to make the likelihood of effective scale-up an impossibility. However, as diverse as mixing equipment may be, evaluations of the rate and extent of mixing and of flow regimes* make it possible to find a common basis for comparison.

In low-viscosity systems, miscible liquid blending is achieved through the transport of unmixed material, via flow currents (i.e., bulk or convective flow), to a mixing zone (i.e., a region of high shear or intensive mixing). In other words, mass transport during mixing depends on streamline or laminar flow, involving well-defined paths, and turbulent flow, involving innumerable, variously sized eddies or swirling motions. Most of the highly turbulent mixing takes place in the region of the impeller, fluid motion elsewhere serving primarily to bring fresh fluid into this region. Thus, the characterization of mixing processes is often based on the flow regimes encountered in mixing equipment. Reynolds' classic research on flow in pipes demonstrated that flow changes from laminar to irregular, or turbulent, once a critical value of a dimensionless ratio of variables has been exceeded [15,16]. This ratio, universally referred to as the Reynolds number N_{Re} is defined by

$$N_{Re} = \frac{Lv\rho}{\eta} \tag{2}$$

where ρ = density; v = velocity, L = characteristic length; and η = Newtonian viscosity. N_{Re} represents the ratio of the inertia forces to the viscous forces in a flow. High values of N_{Re} correspond to flow dominated by motion, whereas low values of N_{Re} correspond to flow dominated by viscosity. Thus, the transition from laminar to turbulent flow is governed by the density and viscosity of the fluid, its average velocity, and the dimensions of the region in which flow occurs (e.g., the diameter of the pipe or conduit, the diameter of a settling particle). For a straight circular pipe, laminar flow occurs when $N_{Re} < 2100$; turbulent flow is evident when $N_{Re} > 4000$. For $2100 \leq N_{Re} \leq 4000$, flow is in transition from a laminar to a turbulent region. Other factors,

*The term flow regime is used to characterize the hydraulic conditions (i.e., volume, velocity, and direction of flow) within a vessel.

such as surface roughness, shape, and cross-sectional area of the affected region, have a substantial effect on the critical value of N_{Re}. Thus, for particle sedimentation, the critical value of N_{Re} is 1; for some mechanical mixing processes, N_{Re} is 10–20 [17]. The erratic, relatively unpredictable nature of turbulent eddy flow is further influenced, in part, by the size distribution of the eddies, which are dependent on the size of the apparatus and the amount of energy introduced into the system [15]. These factors are indirectly addressed by N_{Re}.

In turbulent flow, eddies move rapidly with an appreciable component of their velocity in the direction perpendicular to a reference point, e.g., a surface, past which the fluid is flowing [18]. Because of the rapid eddy motion, mass transfer in the turbulent region is much more rapid than that resulting from molecular diffusion in the laminar region, with the result that the concentration gradients existing in the turbulent region will be smaller than those in the laminar region [18]. Thus mixing is much more efficient under turbulent flow conditions. Nonetheless, the technologist should bear in mind potentially compromising aspects of turbulent flow, e.g., increased vortex formation [19] and a concomitant incorporation of air, and increased shear and a corresponding shift in the particle size distribution of the disperse phase.

Although continuous-flow mixing operations are used to a limited extent in the pharmaceutical industry, the processing of disperse systems most often involves batch processing in some kind of tank or vessel. Thus, in the general treatment of mixing that follows, the focus is on batch operations* in which mixing is accomplished primarily by the use of dynamic mechanical mixers with impellers, although jet mixing [22,23] and static mixing devices [24]—long used in the chemical process industries—are gaining advocates in the pharmaceutical and cosmetic industries.

Mixers share a common functionality with pumps. The power imparted by the mixer, via the impeller, to the system is akin to a pumping effect and is characterized in terms of the shear and flow produced

$$P \propto Q\rho H \tag{3}$$

or

$$H \propto \frac{P}{Q\rho}$$

where P = power imparted by the impeller; Q = flow rate (or pumping capacity) of material through the mixing device; ρ = density of the material; and H = velocity head, or shear, Thus, for a given P, there is an inverse relationship between shear and volume throughput. Power consumption approximations, in kWm^{-3}, are shown in Fig. 2 as a function of apparent viscosity for mixing processes in various systems.

The power input in mechanical agitation is calculated using the power number N_P,

$$N_P = \frac{Pg_c}{\rho N^3 D^5} \tag{4}$$

---

*The reader interested in continuous-flow mixing operations is directed to references that deal specifically with that aspect of mixing such as the monographs by Oldshue [20] and Tatterson [21].

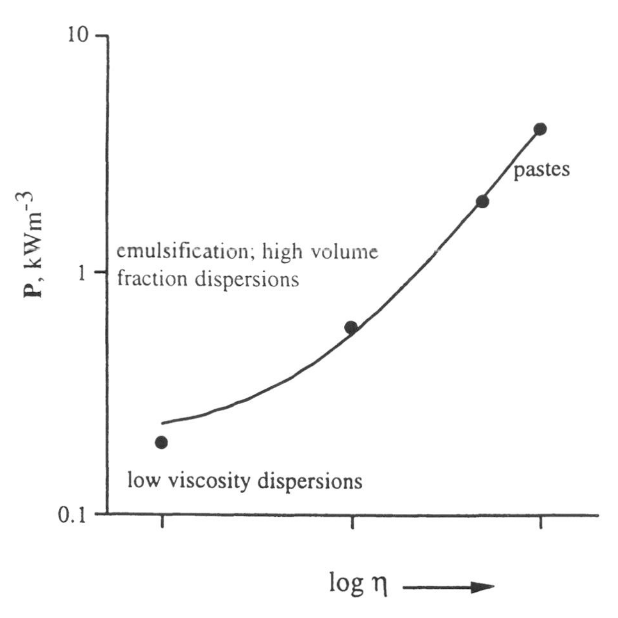

Fig. 2 Typical power consumption approximations for mixing processes in various systems. (Adapted from Ref. 25.)

where g_c = force conversion factor ($g_c = \frac{kg \cdot m \cdot s^{-2}}{\text{Newton}} = \frac{g \cdot cm \cdot s^{-2}}{\text{dyne}}$); N = impeller rotational speed (s^{-1}); and D = diameter of the impeller. For a given impeller/mixing tank configuration, one can define a specific relationship between the Reynolds number (Eq. 2*) and the power number (Eq. 4) in which three zones (corresponding to the laminar, transitional, and turbulent regimes) are generally discernible. Tatterson [26] notes that for mechanical agitation in laminar flow, most laminar power correlations reduce to $N_p N_{Re} = B$, where B is a complex function of the geometry of the system,† and that this is equivalent to $P \propto \eta N^2 D^3$; "if power correlations do not reduce to this form for laminar mixing, then they are wrong and should not be used." Turbulent correlations are much simpler: for systems using baffles,‡ $N_P = B$; this is equivalent to $P \propto \rho N^3 D^5$. Based on this function, slight changes in D can result in substantial changes in power.

Valuable insights into the mixing operation can be gained from a consideration of system behavior as a function of the Reynolds number N_{Re} [27]. This is shown schematically in Fig. 3, in which various dimensionless parameters (dimensionless velocity, v/ND; pumping number, Q/ND^3; power number, $N_P = (Pg_c)/(\rho N^3 D^5)$; and dimensionless mixing time, $t_m N$) are represented as a log-log function of N_{Re}. Although density, viscosity, mixing vessel diameter, and impeller rotational speed are often viewed by formulators as independent variables, their interdependency, when incorporated in

*Here, the Reynolds number for mixing is defined in SI-derived values as $N_{Re} = \frac{1.667 \times 10^{-5} ND^2 \rho}{\eta}$, where D, impeller diameter, is in mm; η is in Pa·s; N is impeller speed, in rpm.; and ρ is density.

†An average value of B is 300, but B can vary between 20 and 4000 [26].

‡Baffles are obstructions placed in mixing tanks to redirect flow and minimize vortex formation.

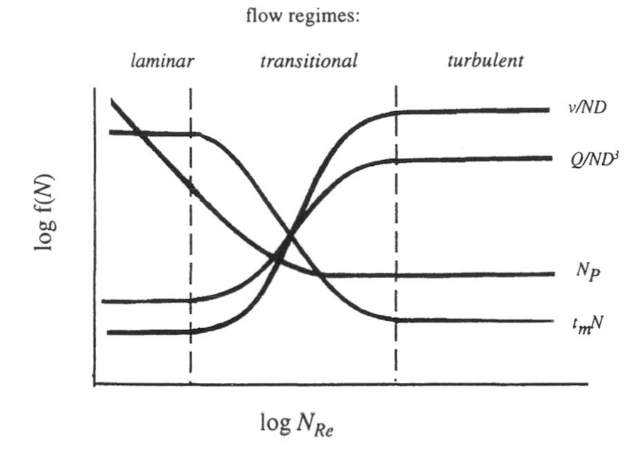

Fig. 3 Various dimensionless parameters (dimensionless velocity, $v^* = v/ND$; pumping number, $N_Q = Q/ND^3$; power number, $N_P = (Pg_c)/(\rho N^3 D^5)$; and dimensionless mixing time, $t^* = tmN$) as a function of the Reynolds number for the analysis of turbine-agitator systems. (Adapted from Ref. 27.)

the dimensionless Reynolds number, is quite evident. Thus, the schematic relationships embodied in Fig. 3 are not surprising.*

Mixing time is the time required to produce a mixture of predetermined quality; the rate of mixing is the rate at which mixing proceeds toward the final state. For a given formulation and equipment configuration, mixing time t_m depends on material properties and operation variables. For geometrically similar systems, if the geometrical dimensions of the system are transformed to ratios, mixing time can be expressed in terms of a dimensionless number, i.e., the dimensionless mixing time θ_m or $t_m N$

$$t_m N = \theta_m = f(N_{\mathrm{Re}}, N_{\mathrm{Fr}}) \Rightarrow f(N_{\mathrm{Re}}) \tag{5}$$

The Froude number, $N_{\mathrm{Fr}} = v/\sqrt{Lg}$, is similar to N_{Re}; it is a measure of the inertial stress to the gravitational force per unit area acting on a fluid. Its inclusion in Eq. (5) is justified when density differences are encountered; in the absence of substantive differences in density, e.g. for emulsions more so than for suspensions, the Froude term can be neglected. Dimensionless mixing time is independent of the Reynolds number for both laminar and turbulent flow regimes, as indicated by the plateaus in Fig. 3. Nonetheless, as there are conflicting data in the literature regarding the sensitivity of θ_m to the rheological properties of the formulation and to equipment geometry, Eq. (5) must be regarded as an oversimplification of the mixing operation. Considerable care must be exercised in applying the general relationship to specific situations.

Empirical correlations for turbulent mechanical mixing have been reported in terms of the following dimensionless mixing time relationship [29]:

$$\theta_m = t_m N = K\left(\frac{T}{D}\right)^a \tag{6}$$

*The interrelationships are embodied in variations of the Navier-Stokes equation, which describes mass and momentum balances in fluid systems [28].

where K and a = constants; T = tank diameter; N = impeller rotational speed; and D = impeller diameter. Under laminar flow conditions, Eq. (6) reduces to

$$\theta_m = H_0 \tag{7}$$

where H_0 is referred to as the mixing number or homogenization number. In the transitional flow regime

$$H_0 = C(N_{\mathrm{Re}})^a \tag{8}$$

where C and a = constants, with a varying between 0 and -1.

Flow patterns in agitated vessels may be characterized as radial, axial, or tangential relative to the impeller, but are more correctly defined by the direction and magnitude of the velocity vectors throughout the system, particularly in a transitional flow regime: while the dimensionless velocity v^* or v/ND is essentially constant in the laminar and turbulent flow zones, it is highly dependent on N_{Re} in the transitional flow zone (Fig. 3). Initiation of tangential or circular flow patterns, with minimal radial or axial movement, is associated with vortex formation, minimal mixing, and, in some multiphase systems, particulate separation and classification. Vortices can be minimized or eliminated altogether by redirecting flow in the system through the use of baffles* or by positioning the impeller so that its entry into the mixing tank is off-center. For a given formulation, large tanks are more apt to exhibit vortex formation than small tanks. Thus, full-scale production tanks are more likely to require baffles even though smaller (laboratory or pilot-plant scale) tanks are unbaffled.

Mixing processes involved in the manufacture of disperse systems, whether suspensions or emulsions, are far more problematic than those used in the blending of low-viscosity miscible liquids due to the multiphasic character of the systems and deviations from Newtonian flow behavior. It is not uncommon for both laminar and turbulent flow to occur simultaneously in different regions of the system (see the section on Macroscopic and microscopic scales of scrutiny, later in this chapter]. In some regions, the flow regime may be in transition, i.e., neither laminar nor turbulent but somewhere in between. The implications of these flow regime variations for scale-up are considerable. Nonetheless, it should be noted that the mixing process is only completed when Brownian motion occurs to a sufficient extent that uniformity is achieved on a molecular scale.

1. *Viscous and Non-Newtonian Materials*

Mixing in high-viscosity materials ($\eta > \sim 10^4$cps) is relatively slow and inefficient. In general, laminar flow is apt to occur rather than turbulent flow, and, as a result, the inertial forces imparted to a system during the mixing process tend to dissipate quickly. Eddy formation and diffusion are virtually absent. Thus, efficient mixing necessitates substantial convective flow, which is usually achieved by high-velocity gradients in the mixing zone. Fluid elements in the mixing zone, subjected to both shear and elongation, undergo deformation and stretching that ultimately result in the size reduction of the fluid elements and an increase in their overall interfacial area. The

*The usefulness of baffles in mixing operations involving disperse systems is offset by increased clean-up problems (due to particulate entrapment by the baffles or congealing of product adjacent to the baffles). Furthermore, "overbaffling"—excessive use of baffles—reduces mass flow and localizes mixing, which may be counterproductive.

repetitive cutting and folding of fluid elements also result in decreasing inhomogeneity and increased mixing. The role of molecular diffusion in reducing inhomogeneities in high-viscosity systems is relatively unimportant until these fluid elements have become small and their interfacial areas have become relatively large [30]. In highly viscous systems, rotary motion is more than compensated for by viscous shear, so baffles are generally less necessary [31].

Mixing equipment for highly viscous materials often involves specialized impellers and configurations. Propeller-type impellers are not generally effective in viscous systems. Instead, turbines, paddles, anchors, helical ribbons, and screws are resorted to, successively, as system viscosity increases. Multiple impellers or specialized impellers (e.g., sigma-blades, Z-blades) are often necessary along with the maintenance of narrow clearances, or gaps, between impeller blades and between impeller blades and tank (mixing chamber) walls in order to attain optimal mixing efficiency (30,31]. However, narrow clearances pose their own problems. Studies of the power input to anchor impellers used to agitate Newtonian and shear-thinning fluids showed that the clearance between the impeller blades and the vessel wall was the most important geometrical factor—N_P at constant N_{Re} was proportional to the fourth power of the clearance divided by tank diameter [32]. Furthermore, although mixing is promoted by these specialized impellers in the vicinity of the walls of the mixing vessel, stagnation is often encountered in regions adjacent to the impeller shaft. Finally, complications (wall effects) may arise from the formation of a thin, particulate-free, fluid layer adjacent to the wall of the tank or vessel that has a lower viscosity than the bulk material and allows slippage (i.e., nonzero velocity) to occur, unless the mixing tank is further modified to provide for wall-scraping.

Rheologically, the flow of many non-Newtonian materials can be characterized by a time-independent power law function (sometimes referred to as the Ostwald-de Waele equation)

$$\tau = K\dot{\gamma}^a$$

or

$$\log \tau = K' + a(\log \dot{\gamma}) \tag{9}$$

where τ = shear stress; $\dot{\gamma}$ = rate of shear; K' = logarithmically transformed proportionality constant K with dimensions dependent upon a, the so-called flow behavior index. For pseudoplastic or shear thinning materials, $a < 1$; for dilatant or shear thickening materials, $a > 1$; for Newtonian fluids, $a = 1$. For a power law fluid, the average apparent viscosity η_{avg}, can be related to the average shear rate by the following equation:

$$\eta_{avg} = K'\left(\frac{dv}{dy}\right)_{avg}^{n'-1} \tag{10}$$

Based on this relationship, a Reynolds number can be derived and estimated for non-Newtonian fluids from

$$\left[N_{Re} = \frac{Lv\rho}{\eta}\right] \Rightarrow \left[N_{Re,non-N} = \frac{ND_i^2\rho}{K'(dv/dy)_{avg}^{n'-1}}\right] \tag{11}$$

Dispersions that behave, rheologically, as Bingham plastics, require a minimum shear stress (the yield value) in order for flow to occur. Shear-stress variations in a system can result in local differences wherein the yield stress point is not exceeded. As a result, flow may be impeded or absent in some regions compared to others, resulting in channeling or cavity formation and a loss of mixing efficiency. Only if the yield value is exceeded throughout the system will flow and mixing be relatively unimpeded. Helical ribbon and screw impellers would be preferable for the mixing of Bingham fluids, in contrast to conventional propeller or turbine impellers, given their more even distribution of power input [33]. From a practical vantage point, monitoring power input to mixing units could facilitate process control and help to identify problematic behavior. Etchells et al. [34] analyzed the performance of industrial mixer configurations for Bingham plastics. Their studies indicate that the logical scale-up path from laboratory to pilot plant to production, for geometrically similar equipment, involves the maintenance of constant impeller tip speed, which is proportional to $N{\cdot}D$, the product of rotational speed of the impeller N and the diameter of the impeller D.

Oldshue [31] provides a detailed procedure for selecting mixing times and optimizing mixer and impeller configurations for viscous and shear thinning materials that can be adapted for other rheologically challenging systems.

Gate and anchor impellers, long used advantageously for the mixing of viscous and non-Newtonian fluids, induce complex flow patterns in mixing tanks: both primary and secondary flows may be evident. Primary flow or circulation results from the direct rotational movement of the impeller blade in the fluid; secondary flow is normal to the horizontal planes about the impeller axis (i.e., parallel to the impeller axis) and is responsible for the interchange of material between different levels of the tank [35]. In this context, rotating viscoelastic systems, with their normal forces, establish stable secondary flow patterns more readily than Newtonian systems. In fact, the presence of normal stresses in viscoelastic fluids subjected to high rates of shear ($\sim 10^4$ s^{-1}) may be substantially greater than shearing stresses, as demonstrated by Metzner et al. [36]. These observations, among others, moved Fredrickson [37] to note that "neglect of normal stress effects is likely to lead to large errors in theoretical calculations for flow in complex geometries." However, the effect of these secondary flows on the efficiency of mixing, particularly in viscoelastic systems, is equivocal. On the one hand, vertical velocity near the impeller blade in a Newtonian system might be 2–5% of the horizontal velocity, whereas, in a non-Newtonian system, vertical velocity can be 20–40% of the horizontal. Thus, the overall circulation can improve considerably. On the other hand, the relatively small, stable toroidal vortices that tend to form in viscoelastic systems may result in substantially incomplete mixing. Smith [35] advocates the asymmetric placement of small deflector blades on a standard anchor arm as a means of achieving a dramatic improvement in mixing efficiency of viscoelastic fluids without resorting to expensive alternatives such as pitched-blade anchors or helical ribbons.

Sidewall clearance, i.e., the gap between the vessel wall and the rotating impeller, was shown by Cheng et al. [38] to be a significant factor in the mixing performance of helical ribbon mixers not only for viscous and viscoelastic fluids but also for Newtonian systems. Bottom clearance, i.e., the space between the base of the impeller and the bottom of the tank, however, had a negligible, relatively insignificant effect on power consumption and on the effective shear rate in inelastic fluids. Thus, mixing efficiency in nonviscoelastic fluids would not be affected by variations in bottom clearance. For viscoelastic fluids, on the other hand, bottom clearance effects were negligible only at lower rotational speeds (≤ 60 rpm); substantial power consumption increases were evident at higher rotational speeds.

The scale-up implications of mixing-related issues such as impeller design and placement, mixing tank characteristics, new equipment design, and the mixing of particulate solids are beyond the scope of this chapter. However, extensive monographs are available in the chemical engineering literature (many of which have been cited herein*) and will prove to be invaluable to the formulator and technologist.

B. Particle Size Reduction

Disperse systems often necessitate particle size reduction, whether it is an integral part of product processing, as in the process of liquid-liquid emulsification, or an additional requirement insofar as solid particle suspensions are concerned. It should be noted that solid particles suspended in liquids often tend to agglomerate. Although milling of such suspensions tends to disrupt such agglomerates and produce a more homogeneous suspension, it generally does not affect the size of the unit particles constituting the agglomerates. In the case of emulsions, the dispersion of one liquid as droplets in another can be expressed in terms of the dimensionless Weber number N_{We}

$$N_{\mathrm{We}} = \frac{\rho v^2 d_0}{\sigma} \tag{12}$$

where ρ = density of a droplet; v = relative velocity of the moving droplet; d_0 = diameter of the droplet; and σ = interfacial tension. The Weber number represents the ratio of the driving force causing partial disruption to the resistance due to interfacial tension [10]. Increased Weber numbers are associated with a greater tendency for droplet deformation (and consequent splitting into still smaller droplets) to occur at higher shear, i.e., with more intense mixing. This can be represented by

$$N_{\mathrm{We}} = \frac{D_i^3 N^2 \rho_{\mathrm{cont.}}}{\sigma} \tag{13}$$

where D_i = diameter of the impeller; N = rotational speed of the impeller; and $\rho_{\mathrm{cont.}}$ = density of the continuous phase. For a given system, droplet size reduction begins above a specific critical Weber number [39]; above the critical N_{We}, average droplet size varies with $N^{-1.2}D_i^{-0.8}$, or, as an approximation, with the reciprocal of the impeller tip speed. In addition, a better dispersion is achieved, for the same power input, with a smaller impeller rotating at high speed [40].

As the particle size of the disperse phase decreases, there is a corresponding increase in the number of particles and a concomitant increase in interparticulate and interfacial interactions. Thus, in general, the viscosity of a dispersion is greater than that of the dispersion medium. This is often characterized in accordance with the classical Einstein equation for the viscosity of a dispersion

$$\eta = \eta_o(1 + 2.5\phi) \tag{14}$$

*The reader is directed to previously referenced monographs by Oldshue and by Tatterson as well as to standard textbooks in chemical engineering, including the multivolume series authored by Coulson et al. and the one volume work by McCabe et al.

where η = viscosity of the dispersion; η_o = viscosity of the continuous phase; and ϕ = volume fraction of the particular phase. The rheological behavior of concentrated dispersions may be demonstrably non-Newtonian (pseudoplastic, plastic, or viscoelastic) and its dependence on ϕ more marked due to disperse phase deformation and/or interparticulate interaction.

Maa and Hsu [41] recently investigated the influence of operation parameters for rotor/stator homogenization on emulsion droplet size and temporal stability in order to optimize operating conditions for small- and large-scale rotor/stator homogenization. Rotor/stator homogenization effects emulsion formation under much more intense turbulence and shear than that encountered in an agitated vessel or a static mixer. Rapid circulation, high-shear forces, and a narrow rotor/stator gap (< 0.5 mm) contribute to the intensity of dispersal and commingling of the immiscible phases, since turbulent eddies are essential for the breakup of the dispersed phase into droplets. Maa and Hsu's estimates of the circulation rates in small- and large-scale rotor/stator systems—based on the total area of the rotor/stator openings, the radial velocity at the openings (resulting from the pressure difference within the vortex that forms in the rotor/stator unit), and the centrifugal force caused by the radial deflection of fluid by the rotor—appear to be predictive for the scale-up of rotor/stator homogenization [41].

Particle size reduction of solids, or comminution, is considerably different from that of the breakup of one liquid by dispersal as small droplets in another. Comminution is generally achieved by one of four mechanisms: (1) compression; (2) impact; (3) attrition; or (4) cutting or shear. Equipment for particle size reduction or milling includes crushers (which operate by compression, e.g., crushing rolls), grinders (which operate principally by impact and attrition, although some compression may be involved, e.g., hammer mills, ball mills), ultrafine grinders (which operate principally by attrition, e.g., fluid-energy mills), and knife cutters. Accordingly, a thorough understanding of milling operations requires an understanding of fracture mechanics, agglomerative forces (dry and wet) involved in the adhesion and cohesion of particulates, and flow of particles and bulk powders. These topics are dealt with at length in the monographs by Fayed and Otten [42] and Carstensen [43].

As Austin [44] notes, the formulation of a general theory of the unit operation of size reduction is virtually impossible given the multiplicity of mill types and mechanisms for particulate reduction. The predictability of any comminution process is further impaired given the variations among solids in surface characteristics and reactivity, molecular interactions, crystallinity, etc. Nonetheless, some commonalities can be discerned. First, the particle size reduction rate is dependent on particle strength and particle size. Second, the residence time of particles in the mill is a critical determinant of mill efficiency. Thus, whether a given mill operates in a single pass or a multiple pass (retention) mode can be a limiting factor insofar as characterization of the efficacy of comminution is concerned. Third, the energy required to achieve a given degree of comminution is an inverse function of initial particle size. This is due to (1) the increasing inefficiency of stress or shear application to each particle of an array of particles as particle size decreases; and (2) the decreasing incidence of particle flaws that permit fracture at low stress [44].

If monosized particles are subjected to one pass through a milling device, the particle size distribution of the resultant fragments can be represented in a cumulative form. Subsequent passes of the comminuted material through the milling device often result in a superimposable frequency distribution when the particle sizes are normalized, e.g., in terms of the weight fraction less than size y resulting from the milling of particles of larger size x. The mean residence time τ of material processed by a mill is given by

$$\tau = \frac{M}{F} \tag{15}$$

where M = mass of powder in the mill; and F = mass flow rate through the mill. Process outcomes for retention mills can be described in terms of residence time distributions defined by the weight fraction of the initial charge at time $t = 0$, which leaves between $(t + dt)$. If the milling operation is scalable, the particle size distributions produced by a large and a small mill of the same type would be comparable and would differ only in the time scale of operation, i.e., the operation can be characterized as a $f(t/\tau)$. The prospect for scalability may be further enhanced when the weight fraction remaining in an upper range is a log-linear (first-order) function of total elapsed milling time [44].*

One estimate of the efficacy of a crushing or grinding operations is the crushing efficiency E_c, described as the ratio of the surface energy created by crushing or grinding to the energy absorbed by the solid [45]

$$E_c = \frac{\sigma_s(A_{wp} - A_{wf})}{W_n} \tag{16}$$

where σ_s = specific surface or surface per unit area; A_{wp} and A_{wf} = areas per unit mass of product particulates and feed particulates, i.e., after and before milling, respectively; and W_n = energy absorbed by the solid, per unit mass. The energy absorbed by the solid, per unit mass, is less than the energy W supplied to the mill, per unit mass, i.e., $W_n < W$. While a substantial part of the total energy input W is needed to overcome friction in the machine, the rest is available for crushing or grinding. However, of the total energy stored within a solid, only a small fraction is converted into surface energy at the time of fracture. As most of the energy is converted into heat, crushing efficiency values tend to be low, i.e., $0.0006 \leq E_c \leq 0.01$, principally due to the inexactness of estimates of σ_s [45].

A number of quasi-theoretical relationships have been proposed to characterize the grinding process. Rittinger's "law" (1867) is

$$\frac{P}{\dot{m}} = K_R\left(\frac{1}{\overline{D}_p} - \frac{1}{\overline{D}_f}\right) \tag{17}$$

which states that the work required in crushing a solid is proportional to the new surface created, and Kick's "law" (1885)

$$\frac{P}{\dot{m}} = K_K \ln\frac{\overline{D}_f}{\overline{D}_p} \tag{18}$$

*Total elapsed milling time encompasses the time during which solids are subjected to a milling operation, whether the particulates undergo single or multiple passes through the mill.

which states that the work required to crush or grind a given mass of material is constant for the same particle size reduction ratio. In Eqs. (17) and (18), $\overline{D}_p$ and $\overline{D}_f$ represent the final and initial average particle sizes,* P is the power (in kilowatts); $\dot{m}$ is the rate at which solids are fed to the mill (in tons/hr), and K_R and K_K are constants for the Rittinger equation and the Kick equation, respectively.

Bond's "law" of particle size reduction provides an ostensibly more reasonable estimate of the power required for crushing or grinding of a solid [46]:

$$\frac{P}{\dot{m}} = \frac{K_B}{\sqrt{D_p}} \tag{19}$$

where K_B = a constant that is mill-dependent and solids-dependent; and D_p = particle size (in millimeters) produced by the mill. This empirical equation is based on Bond's hypothesis that the work required to reduce very large particulate solids to a smaller size is proportional to the square root of the surface to volume ratio of the resultant particulate product. Bond's work index W_i is an estimate of the gross energy required, in kilowatt hours per ton of feed, to reduce very large particles (80% of which pass a mesh size of D_f mm) to such a size that 80% pass through a mesh of size D_p mm

$$W_i = \frac{K_B}{\sqrt{D_p}} \tag{20}$$

Combining Bond's work index (Eq. 20) with Bond's law (Eq. 19) yields

$$\frac{P}{\dot{m}} = W_i \cdot \sqrt{D_p}\left(\frac{1}{\sqrt{D_p}} - \frac{1}{\sqrt{D_f}}\right) \tag{21}$$

which allows one to estimate energy requirements for a milling operation in which solids are reduced from size D_f to D_p. (Note that W_i for wet grinding is generally smaller than that for dry grinding; $W_{i,\text{wet}}$ is equivalent to $(W_{i,\text{dry}})^{3/4}$ [45].)

Although the relationships in Eqs. (17–21) are of some limited use in scaling up milling operations, their predictiveness is limited by the inherent complexity of particle size reduction operations. Virtually all retentive or multiple-pass milling operations become increasingly less efficient as milling proceeds, since the specific comminution rate is smaller for small particles than for large particles. Additional complications in milling arise as fines build up in the powder bed [44]: (1) the fracture rate of all particle sizes decreases, the result, apparently, of a cushioning effect by the fines that minimizes stress and fracture; and (2) fracture kinetics become nonlinear. Other factors, such as coating of equipment surfaces by fines, also affect the efficiency of the milling operation.

Nonetheless, mathematical analyses of milling operations, particularly for ball mills, roller mills, and fluid energy mills, have been moderately successful. There continues to be a pronounced need for a more complete understanding of micromeritic

*In this section, particle size refers to the nominal particle size, i.e., the particle size based on sieving studies or on the diameter of a sphere of equivalent volume.

characteristics, the intrinsic nature of the milling operation itself, the influence of fines on the milling operations, and phenomena including flaw structure of solids, particle fracture, particulate flow, and interactions at both macroscopic and microscopic scales.

C. Material Transfer

The transfer of material from mixing tanks or holding tanks to processing equipment or to a filling line, whether by pumping or by gravity feed, is potentially problematic. Instability (chemical or physical) or further processing (e.g., mixing, changes in the particle size distribution) may occur during the transfer of material (by pouring or pumping) from one container or vessel to another due to changes in the rate of transfer or in shear rate or shear stress. While scale-up–related changes in the velocity profiles of time-independent Newtonian and non-Newtonian fluids due to changes in flow rate or in equipment dimensions or geometry can be accounted for (see "Suppository Development and Production," in *Pharmaceutical Dosage Forms: Disperse Systems*, 2nd Ed., vol. 2, pp. 464–465), time-dependency must first be recognized in order to be accomodated.

Changes in mass transfer time as a consequence of scale-up are often overlooked. As Carstensen and Mehta [47] note, mixing of formulation components in the laboratory may be achieved almost instantaneously with rapid pouring and stirring. They cite the example of pouring 20 mL of liquid A, while stirring, into 80 mL of liquid B. On a production scale, however, mixing is unlikely to be as rapid. A scaled-up batch of 2000 L would require the admixture of 400 L of A and 1600 L of B. If A were pumped into B at the rate of 40 L min^{-1}, then the transfer process would take at least 10 min, and additional time would also be required for the blending of the two liquids. If, for example, liquids A and B were of different pH (or ionic strength, or polarity, etc.), the time required to transfer all of A into B and to mix A and B intimately would allow some intermediate pH (or ionic strength, or polarity, etc.) to develop and persist, long enough for some adverse effect to occur such as precipitation, adsorption, or change in viscosity. Thus, transfer times on a production scale need to be determined so that the temporal impact of scale-up can be accounted for in laboratory or pilot-plant studies.

D. Heat Transfer

On a laboratory scale, heat transfer occurs relatively rapidly as the volume to surface area ratio is relatively small; cooling or heating may or may not involve jacketed vessels. However, on a pilot-plant or production scale, the volume to surface area ratio is relatively large. Consequently, heating or cooling of formulation components or product takes a finite time during which system temperature T °C may vary considerably. Temperature-induced instability may be a substantial problem if a formulation is maintained at suboptimal temperatures for a prolonged period of time. Thus, jacketed vessels or immersion heaters or cooling units with rapid circulation times are an absolute necessity. Carstensen and Mehta [47] give an example of a jacketed kettle with a heated surface of A cm^2, with inlet steam or hot water in the jacket maintained at a temperature of T_0 °C. The heat transfer rate (dQ/dt) in this system is proportional to the heated surface area of the kettle and the temperature gradient, $T_0 - T$, (i.e., the difference between the temperature of the kettle contents T and the temperature of the jacket T_0) at time t

$$\frac{dQ}{dt} = C_p\left(\frac{dT}{dt}\right) = kA(T_0 - T) \tag{22}$$

where C_p = heat capacity of the jacketed vessel and its contents; and k = heat transfer coefficient. If the initial temperature of the vessel is T_1°C, Eq. (22) becomes

$$T_0 - T = (T_0 - T_1)e^{-at} \tag{23}$$

where $a = kA/C_p$. The time t required to reach a specific temperature T_2 can be calculated from Eq. (23) if a is known or estimated from time-temperature curves for similar products processed under the same conditions. Scale-up studies should consider the effect of longer processing times at suboptimal temperatures on the physico-chemical or chemical stability of the formulation components and the product. A further concern for disperse system scale-up is the increased opportunity in a multiphase system for nonuniformity in material transport (e.g., flow rates and velocity profiles) stemming from nonuniform temperatures within processing equipment.

III. HOW TO ACHIEVE SCALE-UP

Full-scale tests using production equipment, involving no scale-up studies whatsoever, are sometimes resorted to when single-phase systems are involved and processing is considered to be predictable and directly scalable. By and large, these are unrealistic assumptions when dispersions are involved. Furthermore, the expense associated with full-scale testing is substantial: commercial-scale equipment is relatively inflexible and costly to operate. Errors in full-scale processing involve large amounts of material. Insofar as disperse systems are concerned then, full-scale tests are not an option.

On the other hand, scale-up studies involving relatively low scale-up ratios and few changes in process variables are not necessarily a reasonable alternative to full-scale testing. For that matter, experimental designs employing minor, incremental changes in processing equipment and conditions are unacceptable as well. These alternative test modes are inherently unacceptable as they consume time, an irreplaceable resource [2] that must be used to its maximum advantage. Appropriate process development, by reducing costs and accelerating lead times, plays an important role in product development performance. In *The Development Factory: Unlocking the Potential of Process Innovation*, author Gary Pisano [48] argues that while pharmaceuticals compete largely on the basis of product innovation, there is a hidden leverage in process development and manufacturing competence that provides more degrees of freedom in developing products to more adroit organizations than to their less adept competitors. Although Pisano focuses on drug synthesis and biotechnology process scale-up, his conclusions translate effectively to the manufacturing processes for drug dosage forms and delivery systems. In effect, scale-up issues need to be addressed jointly by pharmaceutical engineers and formulators as soon as a dosage form or delivery system appears to be commercially viable. Scale-up studies should not be relegated to the final stages of product development, whether initiated at the behest of FDA (to meet regulatory requirements) or marketing and sales divisions (to meet marketing directives or sales quotas). The worst scenario would entail the delay of scale-up studies until after commercial distribution (to accommodate unexpected market demands).

Based on the classification scheme developed by Kline et al. [49], acceptable approaches to scale-up for disperse systems include the following:

- Modular scale-up (limited scale-up)
- Known scale-up correlations (limited scale-up)
- Fundamental approach (high scale-up ratios)

Modular scale-up involves the scale-up of individual components or unit operations of a manufacturing process. The interactions among these individual operations constitute the potential scale-up problem, i.e., the inability to achieve sameness when the process is conducted on a different scale. When the physical or physico-chemical properties of dispersion components are known, the scalability of some unit operations may be predictable.

Known scale-up correlations thus may allow scale-up even when laboratory or pilot-plant experience is minimal. The fundamental approach to process scaling involves mathematical modeling of the manufacturing process and experimental validation of the model at different scale-up ratios. In a recent paper on fluid dynamics in bubble column reactors, Lübbert et al. [50] noted

> Until very recently fluid dynamical models of multiphase reactors were considered intractable. This situation is rapidly changing with the development of high performance computers. Today's workstations allow new approaches to . . . modeling.

Insofar as the scale-up of disperse systems is concerned, virtually no guidelines or models for scale-up have generally been available. Not surprisingly, then, the trial-and-error method is the one most often used by formulators. As a result, serendipity and practical experience play large roles in the successful pursuit of the scalable process.

A. Principles of Similarity

Irrespective of the approach taken to scale-up, the scaling of unit operations and manufacturing processes requires a thorough appreciation of the principles of similarity. "Process similarity is achieved between two processes when they accomplish the same process objectives by the same mechanisms and produce the same product to the required specifications." Johnstone and Thring [51] stress the importance of four types of similarity in effective process translation: (1) geometric similarity; (2) mechanical (static, kinematic, and dynamic) similarity; 3) thermal similarity; and (4) chemical similarity. Each of these similarities presupposes the attainment of the other similarities. In actuality, approximations of similarity are often necessary due to departures from ideality (e.g., differences in surface roughness, variations in temperature gradients, changes in mechanism). When such departures from ideality are not negligible, a correction of some kind has to be applied when scaling up or down: these scale effects must be determined before scaling of a unit operation or a manufacturing process can be pursued. It should be recognized that scale-up of multiphase systems, based on similarity, is often unsuccessful since only one variable can be controlled at a time, i.e., at each scale-up level. Nonetheless, valuable mechanistic insights into unit operations can be achieved through similarity analyses.

1. Geometric Similarity

Point-to-point geometric similarity of two bodies (e.g., two mixing tanks) requires three-dimensional correspondence. Every point in the first body is defined by specific x, y, and z coordinate values. The corresponding point in the second body is defined by specific x', y', and z' coordinate values. The correspondence is defined by the following equation:

$$\frac{x'}{x} = \frac{y'}{y} = \frac{z'}{z} = L \tag{24}$$

where the linear scale ratio L is constant. In contrasting the volume of a laboratory scale mixing tank V_1 with that of a geometrically similar production scale unit V_2, the ratio of volumes (V_1/V_2) is dimensionless. However, the contrast between the two mixing tanks needs to be considered on a linear scale: e.g., a 1000-fold difference in volume corresponds to a 10-fold difference on a linear scale, in mixing tank diameter, impeller diameter, etc.

If the scale ratio is not the same along each axis, the relationship among the two bodies is of a distorted geometric similarity and the axial relationships are given by

$$\frac{x'}{x} = X, \frac{y'}{y} = Y, \frac{z'}{z} = Z \tag{25}$$

Thus, equipment specifications can be described in terms of the scale ratio L or, in the case of a distorted body, two or more scale ratios (X, Y, Z). Scale ratios facilitate the comparison and evaluation of different sizes of functionally comparable equipment in process scale-up.

2. *Mechanical Similarity*

The application of force to a stationary or moving system can be described in static, kinematic, or dynamic terms that define the mechanical similarity of processing equipment and the solids or liquids within their confines. Static similarity relates the deformation under constant stress of one body or structure to that of another; it exists when geometric similarity is maintained even as elastic or plastic deformation of stressed structural components occurs [51]. In contrast, kinematic similarity encompasses the additional dimension of time, and dynamic similarity involves the forces (e.g., pressure, gravitational, centrifugal) that accelerate or retard moving masses in dynamic systems. The inclusion of time as another dimension necessitates the consideration of corresponding times t' and t, for which the time scale ratio $\mathbf{t}$, defined as $\mathbf{t} = t'/t$, is a constant.

Corresponding particles in disperse systems are geometrically similar particles that are centered on corresponding points at corresponding times. If two geometrically similar fluid systems are kinetically similar, their corresponding particles will trace out geometrically similar paths in corresponding intervals of time. Thus, their flow patterns are geometrically similar and heat- or mass-transfer rates in the two systems are related to one another [51]. Pharmaceutical engineers may prefer to characterize disperse systems' corresponding velocities, which are the velocities of corresponding particles at corresponding times

$$\frac{v'}{v} = \mathbf{v} = \frac{L}{\mathbf{t}} \tag{26}$$

Kinematic and geometric similarity in fluids ensures geometrically similar streamline boundary films and eddy systems. If forces of the same kind act on corresponding particles at corresponding times, they are termed corresponding forces, and conditions for dynamic similarity are met. While the scale-up of power consumption by a unit operation or manufacturing process is a direct consequence of dynamic similarity, mass and heat transfer—direct functions of kinematic similarity—are only indirect functions of dynamic similarity.

3. *Thermal Similarity*

Heat flow, whether by radiation, conduction, convection, or the bulk transfer of matter, introduces temperature as another variable. Thus, for systems in motion, thermal similarity requires kinematic similarity. Thermal similarity is described by

$$\frac{H'_r}{H_r} = \frac{H'_c}{H_c} = \frac{H'_v}{H_v} = \frac{H'_f}{H_f} = \mathbf{H} \tag{27}$$

where H_r, H_c, H_v, and H_f = heat fluxes or quantities of heat transferred per second by radiation, convection, conduction, and bulk transport, respectively; and $\mathbf{H}$, the thermal ratio, is a constant.

4. *Chemical Similarity*

This similarity state is concerned with the variation in chemical composition from point to point as a function of time. Chemical similarity, i.e., the existence of comparable concentration gradients, is dependent on both thermal and kinematic similarity.

5. *Interrelationships Among Surface Area and Volume upon Scale-Up*

Similarity states aside, the dispersion technologist must be aware of whether a given process is volume dependent or area dependent. As the scale of processing increases, volume effects become increasingly more important while area effects become increasingly less important. This is exemplified by the dependence of mixing tank volumes and surface areas on scale-up ratios (based on mixing tank diameters) in Table 1 (adapted from Tatterson [52]). The surface area to volume ratio is much greater on the small scale than on the large scale; surface area effects are thus much more important on a small scale than on a large one. Conversely, the volume to surface area ratio is much greater on the large scale than on the small scale: volumetric effects are thus much more important on a large scale than on a small scale. Thus, volume-dependent processes are more difficult to scale up than surface-area-dependent processes. For example, exothermic processes may generate more heat than can be tolerated by a formulation,

Table 1 Area- and Volume-Dependence on Scale-Up Ratios[a]

Scale	Tank diam. (m)	Area (m^2)	Volume (m^3)	Area/volume	Volume/area
1	0.1	0.0393	0.000785	50	0.02
10	1	3.93	0.785	5	0.2
20	2	15.7	6.28	2.5	0.4
50	5	98.2	98.2	1	1

[a]Assumptions: tank is a right circular cylinder; batch height = tank diameter; area calculations are the sum of the area of the convex surface and the area of the bottom of the cylinder.

leading to undesirable phase changes or product degradation unless cooling coils, or other means of intensifying heat transfer, are added. A further example is provided by a scale-up problem involving a 10-fold increase in tank volume, from 400 to 4000 L, and an increase in surface area from 2 to 10 m^2. The surface area to volume ratios are 1/200 and 1/400, respectively. In spite of the 10-fold increase in tank volume, the increase in surface area is only fivefold, necessitating the provision of additional heating or cooling capacity to allow for an additional 10 m^2 of surface for heat exchange.

As Tatterson [52] notes, "there is much more volume on scaleup than is typically recognized. This is one feature of scaleup that causes more difficulty than anything else." For disperse systems, a further mechanistic implication of the changing volume and surface area ratios is that particle size reduction (or droplet breakage) is more likely to be the dominant process on a small scale, whereas aggregation (or coalescence) is more likely to be the dominant process on a large scale [52].

6. *Interrelationships Among System Properties upon Scale-Up*

When a process is dominated by a mixing operation, another gambit for the effective scale-up of geometrically similar systems involves the interrelationships that have been established for impeller-based systems. Tatterson [53] describes a number of elementary scale-up procedures for agitated tank systems that depend on operational similarity. Thus, when scaling up from level 1 to level 2

$$\frac{(P/V)_1}{(P/V)_2} = \begin{cases} \left(\dfrac{N_1}{N_2}\right)^3 \left(\dfrac{D_1}{D_2}\right)^2 & \text{for turbulent flow} \\ \left(\dfrac{N_1}{N_2}\right)^2 & \text{for laminar flow} \end{cases} \tag{28}$$

power per unit volume is dependent principally on the ratio N_1/N_2, since impeller diameters are constrained by geometric similarity.

A change in size on scale-up is not the sole determinant of the scalability of a unit operation or process. Scalability depends on the unit operation mechanism(s) or system properties involved. Some mechanisms or system properties relevant to dispersions are listed in Table 2 [54]. In a number of instances, size has little or no influence on processing or on system behavior. Thus, scale-up does not affect chemical kinetics or thermodynamics, although the thermal effects of a reaction could perturb a system, e.g., by affecting convection [54]. Heat or mass transfer within or between phases is indirectly affected by changes in size, whereas convection is directly affected. Thus, since transport of energy, mass, and momentum are often crucial to the manufacture of disperse systems, scale-up can have a substantial effect on the resultant product.

B. Dimensions, Dimensional Analysis, and the Principles of Similarity

Just as process translation or scaling up is facilitated by defining similarity in terms of dimensionless ratios of measurements, forces, or velocities, the technique of dimensional analysis per se permits the definition of appropriate composite dimensionless numbers whose numeric values are process-specific. Dimensionless quantities can be pure numbers, ratios, or multiplicative combinations of variables with no net units.

Table 2 Influence of Size on System Behavior or Important Unit Operation Mechanisms

System behavior or unit operation mechanisms	Important Variables[a]	Influence of Size
Chemical kinetics	C, P, T	None
Thermodynamic properties	C, P, T	None
Heat transfer	Local velocities; C, P, T	Important
Mass transfer within a phase	N_{Re}, C, T	Important
Mass transfer between phases	Relative phase velocities; C, P, T	Important
Forced convection	Flow rates, geometry	Important
Free convection	Geometry; C, P, T	Crucial

[a] $C, P,$ and T are concentration, pressure, and temperature, respectively.
Source: Adapted from Ref. 54.

Dimensional analysis is concerned with the nature of the relationship among the various quantities involved in a physical problem. An approach intermediate between formal mathematics and empiricism, it offers the pharmaceutical engineer an opportunity to generalize from experience and apply knowledge to a new situation [55,56]. This is a particularly important avenue as many engineering problems—scale-up among them—cannot be solved completely by theoretical or mathematical means. Dimensional analysis is based on the fact that if a theoretical equation exists among the variables affecting a physical process, that equation must be dimensionally homogeneous. Thus, many factors can be grouped, in an equation, into a smaller number of dimensionless groups of variables [56].

Dimensional analysis is an algebraic treatment of the variables affecting a process; it does not result in a numerical equation. Rather, it allows experimental data to be fitted to an empirical process equation that results in scale-up being achieved more readily. The experimental data determine the exponents and coefficients of the empirical equation. The requirements of dimensional analysis are that (1) only one relationship exists among a certain number of physical quantities; and (2) no pertinent quantities have been excluded nor extraneous quantities included.

Fundamental (primary) quantities, which cannot be expressed in simpler terms, include mass M, length L, and time T. Physical quantities may be expressed in terms of the fundamental quantities: e.g., density is ML^{-3}; velocity is LT^{-1}. In some instances, mass units are covertly expressed in terms of force F in order to simplify dimensional expressions or render them more identifiable. The MLT and FLT systems of dimensions are related by the equations

$$\left.\begin{aligned} F &= Ma = \frac{ML}{T^2} \\ M &= \frac{FT^2}{L} \end{aligned}\right\} \tag{29}$$

According to Bisio [57], scale-up can be achieved by maintaining the dimensionless groups characterizing the phenomena of interest constant from small scale to large scale. However, for complex phenomena this may not be possible. Alternatively, dimensionless numbers can be weighted so that the untoward influence of unwieldy variables can be minimized. On the other hand, this camouflaging of variables could lead to an inadequate characterization of a process and a false interpretation of laboratory or pilot-plant data.

Pertinent examples of the value of dimensional analysis have been reported recently in a series of papers by Maa and Hsu [24,41,58]. In their first report, they successfully established the scale-up requirements for microspheres produced by an emulsification process in continuously stirred tank reactors (CSTRs) [58]. Their initial assumption was that the diameter of the microspheres d_{ms} is a function of phase quantities, physical properties of the dispersion and dispersed phases, and processing equipment parameters:

$$d_{ms} = f(D\omega, D/T, H, B, n_{imp}, g_c, g, c, \eta_o, \eta_a, \rho_o, \rho_a, v_o, v_a, \sigma) \tag{30}$$

Gravitational acceleration g is included to relate mass to inertial force. The conversion factor g_c was included to convert one unit system to another. The subscripts o and a refer to the organic and aqueous phases, respectively. The remaining notation is as follows: D = impeller diameter (cm); ω = rotational speed (angular velocity) of the impeller(s) (s^{-1}); T = tank diameter (cm); H = height of filled volume in the tank (cm); B = total baffle area (cm^2); n = number of baffles; n_{imp} = number of impellers; v_o and v_a = phase volumes (mL); c = polymer concentration (g/mL); η_o and η_a = phase viscosities (g cm^{-1} s^{-1}); ρ_o and ρ_a = phase densities (g mL^{-1}); and σ = interfacial tension between organic and aqueous phases (dyne cm^{-1}).

The initial emulsification studies used a 1 L "reactor" vessel with baffles originally designed for fermentation processes. Subsequent studies were successively scaled up from 1 L to 3, 10, and 100 L. Variations due to differences in reactor configuration were minimized by utilizing geometrically similar reactors with approximately the same D/T ratio (i.e., 0.36–0.40). Maa and Hsu contended that separate experiments on the effect of the baffle area B on the resultant microsphere diameter did not significantly affect d_{ms}. However, the number and location of the impellers had a significant impact on d_{ms}. As a result, to simplify the system, Maa and Hsu always used double impellers (n_{imp} = 2) with the lower one placed close to the bottom of the tank and the other located in the center of the total emulsion volume. Finally, Maa and Hsu determined that the volumes of the organic and aqueous phases, in the range they were concerned with, played only a minor role in affecting d_{ms}. Thus, by the omission of D/T, B, and v_o and v_a, Eq. (30) was simplified considerably to yield

$$d_{ms} = f(D\omega, g_c, g, c, \eta_o, \eta_a, \rho_o, \rho_a, \sigma) \tag{31}$$

Equation (31) contains 10 variables and four fundamental dimensions (L, M, T, and F). Maa and Hsu were able subsequently to define microsphere size d_{ms} in terms of the processing parameters and physical properties of the phases

$$\frac{g(\rho_o - \rho_a)d_{ms}^2}{\sigma} = \Pi_2^{-0.280}\Pi_3^{-0.108}\Pi_4^{0.056}(0.0255\Pi_5^e + 0.0071) \tag{32}$$

where Π_i = dimensionless multiplicative groups of variables. [The transformation of Eq. (31) into Eq. (32) is described by Maa and Hsu [58] in an appendix to their paper.] Subsequently, linear regression analysis of the microsphere size parameter, $[g(\rho_o - \rho_a)d_{ms}^2]/\sigma$, as a function of the right-hand side of Eq. (32), i.e., $[\Pi_2^{-0.280}\Pi_3^{-0.108}\Pi_4^{0.056}(0.0255\Pi_5^e + 0.0071)]$, resulted in $r \approx 0.973$ for 1, 3, 10, and 100 L reactors at two different polymer concentrations. These composite data are depicted graphically in Fig. 4.

Subsequently, Maa and Hsu [24] applied dimensional analysis to the scale-up of a liquid-liquid emulsification process for microsphere production using one or another of three different static mixers that varied in diameter, number of mixing elements, and mixing element length. Mixing element design differences among the static mixers were accommodated by the following equation:

$$d_{ms} = 0.483 d^{1.202} V^{-0.556} \sigma^{0.556} \eta_a^{-0.560} \eta_o^{0.004} n^h c^{0.663} \tag{33}$$

where d_{ms} = diameter of the microspheres (μm) produced by the emulsification process; d = diameter of the static mixer (cm); V = flow rate of the continuous phase (mL s^{-1}); σ = interfacial tension between the organic and aqueous phases (dyne/cm); η_a and η_o = viscosities (g cm^{-1} s^{-1}) of the aqueous and organic phases, respectively; n = number of mixing elements; h = an exponent the magnitude of which is a function of static mixer design; and c = polymer concentration (g mL^{-1}) in the organic phase. The relative efficiency of the three static mixers was readily determined in terms of emulsification efficiency ε, defined as equivalent to $1/d_{ms}$; better mixing results in smaller microspheres. In this way, Maa and Hsu were able to compare and contrast CSTRs with static mixers.

C. Mathematical Modeling and Computer Simulation

Basic and applied research methodologies in science and engineering are undergoing major transformations. Mathematical models of real-world phenomena are more elabo-

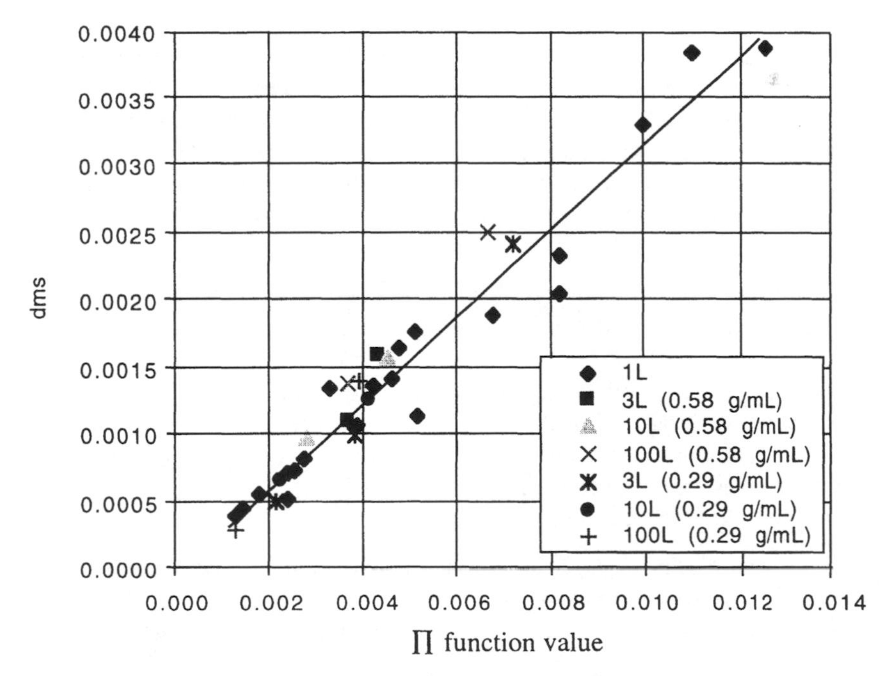

Fig. 4 Microsphere diameter parameter d_{ms} as a function of processing parameters and physical properties of the phases [∏ functions of the right-hand side of Eq. (32)]. (After Ref. 58.)

rate than in the past, with forms governed by sets of partial differential equations that represent continuum approximations to microscopic models [59]. Appropriate mathematical relationships would reflect the fundamental laws of physics regarding the conservation of mass, momentum, and energy. Euzen et al. [60] list such balance equations for mass, momentum, and energy (e.g., heat), for a single-phase Newtonian system (with constant density ρ, viscosity η, and molar heat capacity at constant pressure C_p) in which a process takes place in an element of volume ΔV (defined as the product of *dx, dy*, and *dz*):

$$\left.\begin{array}{c}
\frac{\partial C_i}{\partial t} = -\left\{v_x \frac{\partial C_i}{\partial x} + v_y \frac{\partial C_i}{\partial y} + v_z \frac{\partial C_i}{\partial z}\right\} + \left\{D_{ix} \frac{\partial^2 C_i}{\partial x^2} + D_{iy} \frac{\partial^2 C_i}{\partial y^2} tD_{iz} \frac{\partial^2 C_i}{\partial z^2}\right\} + R_i \\
\text{Mass balance} \\
\rho\left\{\frac{\partial v_x}{\partial t} + v_x \frac{\partial v_x}{\partial x} v_y \frac{\partial v_x}{\partial y} + v_z \frac{\partial v_x}{\partial z}\right\} = -\frac{\partial P}{\partial x} + \eta\left\{\frac{\partial^2 v_x}{\partial x^2} + \frac{\partial^2 v_x}{\partial y^2} + \frac{\partial^2 v_x}{\partial z^2}\right\} + \rho g_x \\
\text{Momentum balance (e.g., in x − direction)} \\
\rho C_p\left\{\frac{\partial T}{\partial t} + v_x \frac{\partial T}{\partial xt} v_y \frac{\partial T}{\partial y} + v_z \frac{\partial T}{\partial z}\right\} = \left\{k_x \frac{\partial^2 T}{\partial x^2} + k_y \frac{\partial^2 T}{\partial y^2} + k_z \frac{\partial^2 T}{\partial z^2}\right\} + S_R \\
\text{Energy balance}
\end{array}\right\} \qquad (34)$$

where P = pressure; T = temperature; t = time; v = fluid flow velocity; k = thermal conductivity; and R_i, g_x, and S_R = kinetic, gravitational, and energetic parameters, respectively. Equation (34) is presented as an example of the complex relationships that are becoming increasingly more amenable to resolution by computers rather than for its express use in a scale-up problem. Rice [61] estimated that solutions to real-world problems in the past required equipment mock-ups or prototypes, trial-and-error experimentation, and noncomputational analog methods, approaches which are time-consuming, expensive, and suboptimal. He gives the example of the location of cooling pipes for an automobile engine block. A computed solution, he claims, should not be difficult as it involves a well-understood physical phenomenon (heat flow) and the solving of

> A Poisson problem for a complicated three-dimensional object. Methods and machines were available in 1940 that could, in principle, solve such a problem. Yet as a practical matter . . . , I estimate that this computation for just one engine block would have cost the entire wealth of the United States in 1940. When I first encountered the problem in 1963, there had been enormous progress in both computing hardware and algorithms since 1940. Nevertheless, the computation was still not economically feasible. Today the computer time to solve it costs a few tens of dollars. [61]

Scale-up via the exact solution of Eq. (34) and its non-Newtonian counterparts is not yet feasible, let alone routine. However, increasingly more effective algorithms for computation coupled with the advent of computer systems capable of teraflops* of power

---

*The prefix "tera" refers to 2^{40} ($\approx 10^{12}$); the term "flops" refers to floating point operations.

and terabytes of memory will eventually allow simulation techniques and virtual reality systems to replace scale models and pilot plants.

D. Scale-Up Paradigm

A paradigm for scale-up that is applicable to all scale-up problems, not just those involving disperse systems, is presented in Fig. 5. At the outset, as part of process analysis, the formulator must know and understand the physical and physico-chemical principles that are relevant to the unit operation(s) that are to be scaled up. If the relevant operation variables and parameters are inadequately known or characterized (path I),

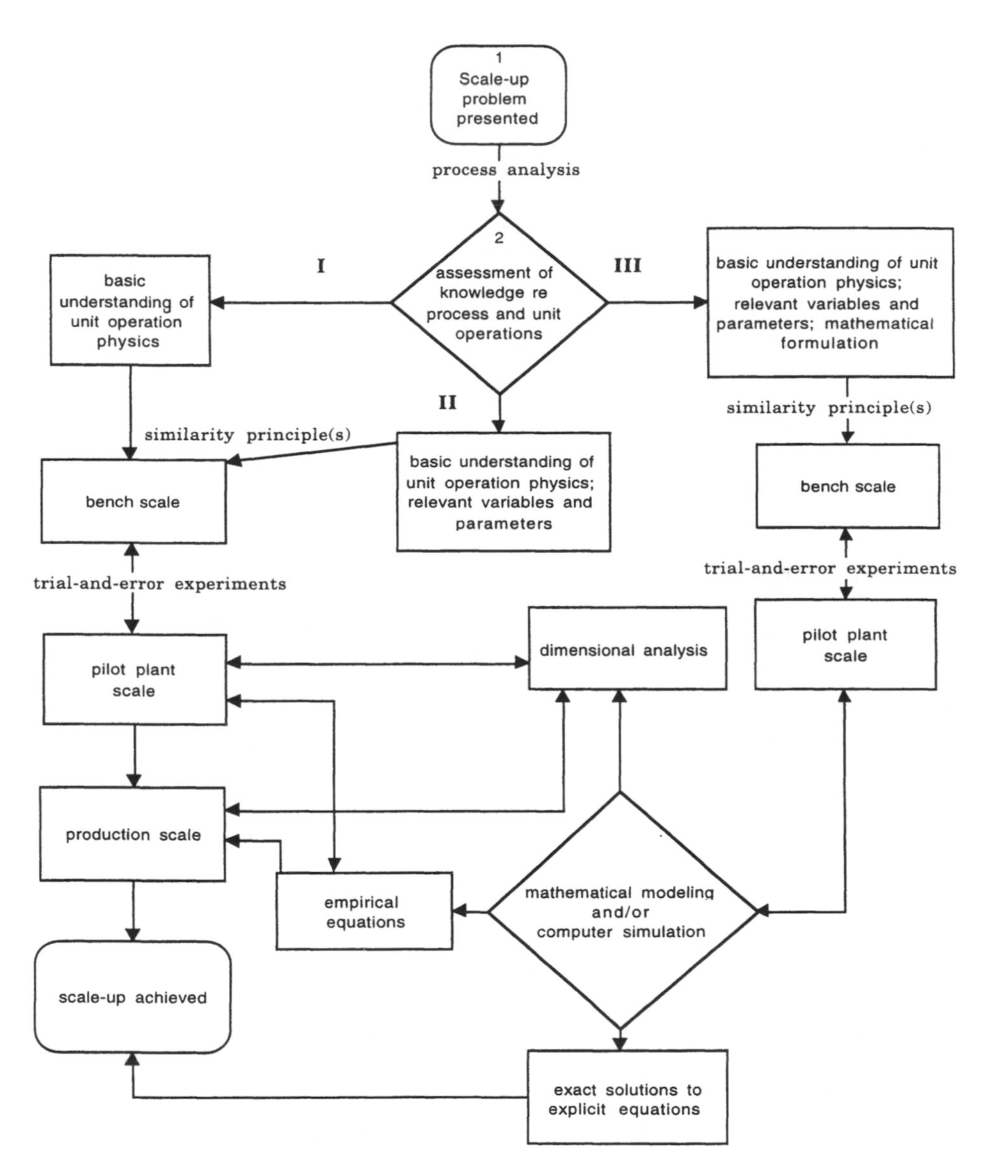

Fig. 5 Paradigm for scale-up.

scale-up is accomplished through empirical experiments conducted in a trial-and-error manner at all scales (bench, pilot, and production), with little more than similarity principles to guide the technologist. Even if the relevant operation variables and parameters are known or characterized (path II), and similarity principles used, scale-up is still accomplished through empirical experiments conducted in a trial-and-error manner at all scales (bench, pilot, and production), although the path length is likely to be shorter. Path III, which involves mathematical modeling and/or computer simulation, corresponds to an engineering approach to unit operations, and, arguably, has the shortest path length of all. "Hands-on" experimentation early on is still required in order to eliminate redundant or inconsequential variables or functions from equations. Relatively simple models and mathematical relationships can be developed—particularly through the application of dimensional analysis—so that reasonable approximations of reality can be generated, although, as noted above, exact solutions to explicit fundamental equations, which fully describe pharmaceutical systems, are not yet attainable. The progression of pharmaceutical technology toward a more definitive characterization of unit operations and manufacturing processes remains an impractical goal at the present time.

No matter what path is followed, small-scale experimentation is necessary in the beginning. As Baekeland [62] pointed out, "Commit your blunders on a small scale and make your profits on a large scale." However, the vital nature of effective process translation (scale-up) in the pharmaceutical industry and its economic repercussions necessitate a commitment to this interdisciplinary field by pharmaceutical scientists and technologists so that trial-and-error methodologies do not remain the modus operandi.

IV. PROCESS TRANSLATION/SCALE-UP PROBLEMS

A. Macroscopic and Microscopic Scales of Scrutiny

One problematic aspect of scale-up via dimensional analysis involves the scale of scrutiny. Molecular diffusion and microscopic viscosity influence transport on a so-called microscopic scale. These parameters, among others, help to characterize the molecular environment, i.e., the environment in the vicinity of a diffusing molecule or of a droplet or particle of a dispersion. On a very much larger (macroscopic) scale, these parameters may not appear to have a noticeable effect, yet they cannot be ignored: were there no mass or momentum transport at the microscopic (molecular) scale, larger scale processes would not function properly [50].

Heat transfer and mixing processes are typically characterized on a microscopic scale, whereas a system's flow patterns, or regimes, have an effect at both the microscopic and macroscopic level. Flow regimes may be laminar, turbulent, or transitional. Laminar flow regimes are encountered when one layer of fluid moves over another. Turbulent flow involves the rapid tumbling and retumbling of relatively large portions of fluid, or eddies, characterized by random swirling motions superimposed on simpler flow patterns. Turbulence, encountered to some degree in virtually all fluid systems, tends to be isotropic on a small scale but anisotropic on a large scale. Transitional flow regimes are often considered to be a combination of laminar and turbulent flow. Imagine a group of particles in a dispersion at the origin (0,0,0) of a three-dimensional coordinate system at zero time. If these particles dissipate as a function of time, the extent to which they spread out is indicative of the flow pattern in the system. Their gradual dispersal and movement away from each other can be described in terms of the width of the particle "cloud." The change in the standard deviation σ of the particles' coordinates as a function of time $\sigma(t)$ is indicative of the mixing efficiency of the system [50]

$$\sigma(t) = at^n \tag{35}$$

When transport occurs by molecular diffusion (Brownian motion), $\sigma(t) \propto t^{1/2}$ [63]; for convective transport (bulk flow), $\sigma(t) \propto t$ [64]; and, for turbulent flow, $\sigma(t) \propto t^{3/2}$ [65]. Effective scale-up, then, mandates an awareness of the relative importance of various process parameters at different scales of scrutiny. This is absolutely crucial if computer modeling or process simulation is to be successful in process translation.

B. Complexity of Overall Manufacturing Process

In far too many instances, process translation and scale-up issues involving disperse systems are addressed in a trial-and-error manner. This is due, in no small measure, to the complexity of the manufacturing process for disperse systems. Examination of a typical processing cycle for an O/W emulsion (Fig. 6) reveals the multiplicity of inter-

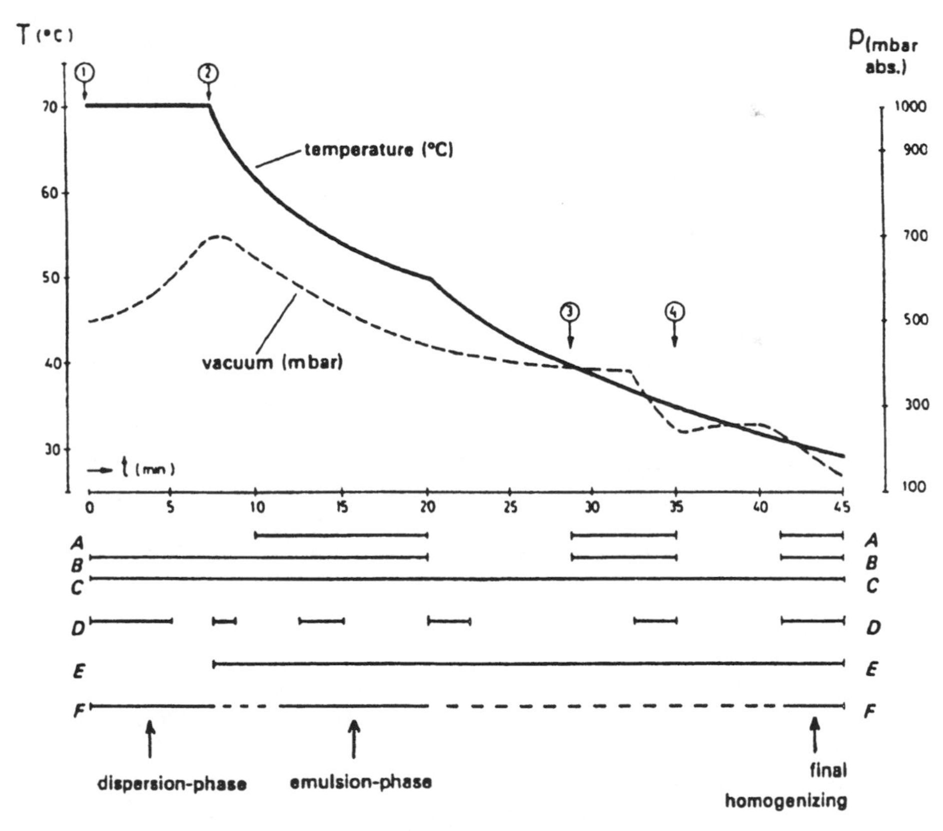

Fig. 6 Processing cycle for O/W topical emulsion manufacture: (A) running time of colloid mill; (B) running time of dissolver; (C) running time of scraper-stirrer; (D) running time of vacuum pump; (E) running time of cooling; (F) cycle of the process; (1) start addition of water phase (with thermostable components); (2) end addition of water-phase components, begin addition of oil-phase components; (3) addition of thermolabile component; (4) addition of volatile components (e.g., perfume). (From Ref. 62.).

related operations that occur episodically, rather than continuously, during manufacturing, e.g., agitation at two or three different levels of intensity, deaeration, heat exchange (cooling), particle size reduction, and emulsification. The temporal characterization of these interrelated operations presents a challenge to the pharmaceutical technologist. One can appreciate that major problems arise when unit operations or process fundamentals are ignored or improperly interpreted. Among some of the more grievous and egregious scale-up errors are

- Scaling based on wrong unit operation mechanism(s)
- Use of different types of equipment at different levels of scale-up
- Insufficient knowledge of process; lack of important process information
- Unrealistic expectations (e.g., heat dissipation)
- Changes in product or process (e.g., altered formulation, phase changes, changes in order of addition, etc.) during scale-up
- Incompletely characterized equipment, e.g., multishaft mixer/ homogenizers

As to the last point, the increasing acceptance in the industry of multifunctional "one pot" processing equipment, e.g., multishaft mixers/homogenizers (Fig. 7), is paradoxical. On the one hand, it is a reflection, in part, of efforts to meet GMP standards by minimizing product loss during the course of manufacturing and processing by eliminating the need for material transfer until processing is complete. On the other hand, the quantitative characterization of the velocity profiles and thermal gradients in such inherently complex multifunctional equipment would appear to be far more difficult than with more conventional "monofunctional" equipment. Nonetheless, computer simulations of unit operations and manufacturing processes are increasingly more realizable through engineering and mathematical modeling software, although the actual limitations of process simulations are often ignored or masked by the terminology used.*† Clearly, there is a need for further research into the scalability of the whole gamut of pharmaceutical manufacturing equipment.

The collaborative development (by the pharmaceutical industry and the Food and Drug Administration) of guidelines for equipment selection and use for scale-up is an outgrowth of regulatory concerns regarding SUPAC. Efforts must be continued to define unit operations and to classify manufacturing equipment on the basis of operating principles and design characteristics. This will encourage pharmaceutical manufacturers to use more scalable equipment throughout, from the laboratory to the pilot plant to the production facility, thereby facilitating process translation and scale-up.

V. CONCLUSIONS

Unfortunately, monographs on the scale-up of pharmaceutical dispersions are noticeably absent from the literature. Most of the papers on scale-up that have been published in the pharmaceutical literature to date, with the few exceptions noted herein, have dealt with the scale-up of solid dosage forms or single-phase liquids. Although process trans-

*One should bear in mind that process simulations are significantly affected by overestimates of (1) the accuracy and precision of physical property data or process parameters; and (2) the range of applicability of the models used.

†A "rigorous" model, to some, is one that can be used over a range of conditions, although not necessarily with a high degree of accuracy. To many others, however, rigorous means "correct" [67].

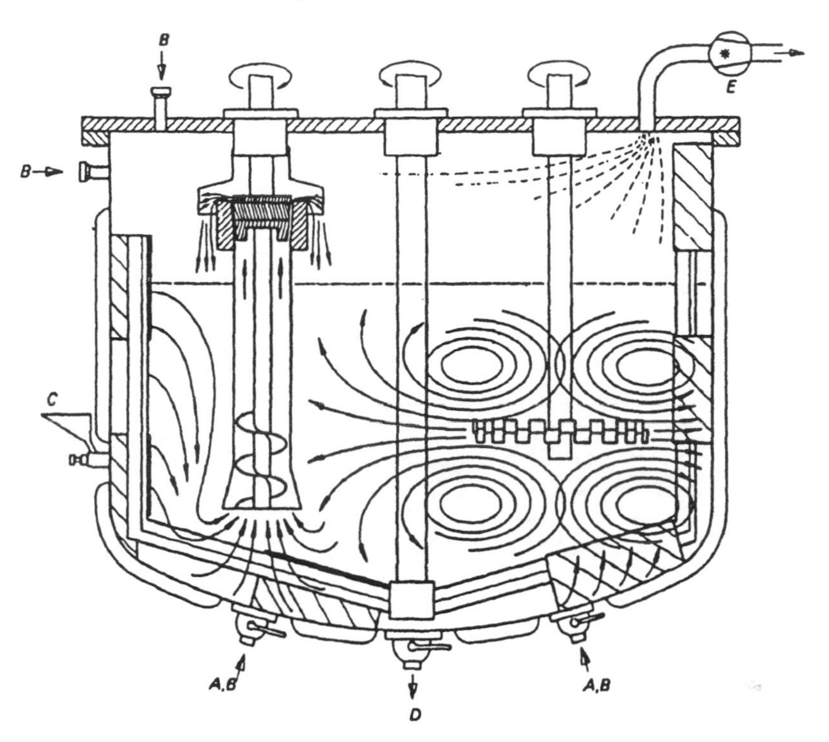

Fig. 7 Multishaft-multifunctional processing unit: (A) inlet powder; (B) inlet liquid; (C) inlet active agents; (D) outlet end product; (E) vacuum pump. (From Ref. 66).

lation and scale-up are a fundamental part of pharmaceutical manufacturing, formulators and technologists have often been uncertain as to how to proceed once a commercially viable formulation is successfully prepared in the laboratory. Only in recent years, as a result of a regulatory impetus, has scale-up begun to be addressed in a more formal, scientific manner. Certainly neither the regulatory nor the financial implications of rapid process translation and scale-up can be ignored any longer. Clearly, the perspective of pharmaceutical engineers [68] will, of necessity, be interwoven with that of pharmaceutical formulators and technologists.

REFERENCES

1. G. B. Tatterson, *Scaleup and Design of Industrial Mixing Processes*, McGraw-Hill, New York, 1994.
2. W. J. Dale, The scale-up process: Optimize the use of your pilot plant, Abstract 108c, Session 108 on Experimental Strategies for Pilot Plants, 1996 Spring Meeting, Am. Inst. Chem. Engrs., New York.
3. S. Harder and G. Van Buskirk, Pilot plant scale-up techniques, In: *The Theory and Practice of Industrial Pharmacy*, 3rd Ed., (L. Lachman, H. A. Lieberman, and J. L. Kanig, eds.), Lea & Febiger, Philadelphia, 1986, pp. 681–710.
4. G. Astarita, Scaleup: Overview, closing remarks, and cautions, In: *Scaleup of Chemical Processes: Conversion from Laboratory Scale Tests to Successful Commercial Size Design* (A. Bisio and R. L. Kabel, eds.), Wiley, New York, 1985, p. 678.
5. A. Bisio, Introduction to Scaleup, In: *Scaleup of Chemical Processes: Conversion from Labo-*

ratory Scale Tests to Successful Commercial Size Design (A. Bisio and R. L. Kabel, eds.), Wiley, New York, 1985, p. 11.
6. AAPS/FDA Workshop Committee, Scale-up of liquid and semisolid disperse systems, *Pharm. Technol.*, 19 (6):52, 54, 56, 58, 60 (1995).
7. H. W. Fowler, Pharmaceutical engineering, *Mfg. Chem.*, 35(4):63–64 (1964).
8. R. D. Nelson Jr., *Dispersing Powders in Liquids*, Elsevier, Amsterdam, The Netherlands, 1988, pp. 4–5.
9. P. Whittingstall, The importance of rheology, *Am. Lab.*, 25(3):41–42 (1993).
10. J. A. Skidmore, Some aspects of the mixing of liquids, *Am. Perf. Cosmetics* 84(1):31–35 (1969).
11. D. S. Dickey and R. R. Hemrajani, Recipes for fluid mixing, *Chem. Eng.* 99(3):82–89 (1992).
12. J. M. Ottino, The mixing of fluids, *Sci. Am.*, Jan. 260(1):56–57, 60–67 (1989).
13. U. Frisch, *Turbulence: The legacy of A. N. Kolmogorov,* Cambridge University Press, Cambridge, U.K. 1995.
14. W. L. McCabe, J. C. Smith, and P. Harriott, *Unit Operations of Chemical Engineering*, 5th ed., Mc-Graw Hill, Inc., New York, 1993, p. 235.
15. D. Bershader, Fluid physics, In: *Encyclopedia of Physics*, 2nd Ed., (R. G. Lerner and G. L. Trigg, eds.), VCH Publ., New York, 1991, pp. 402–410.
16. R. J. Stokes and D. F. Evans, *Fundamentals of Interfacial Engineering*, Wiley-VCH, New York, 1997, pp. 88–89.
17. Z. Sterbacek and P. Tausk, *Mixing in the Chemical Industry*, Pergamon Press, Oxford, UK, 1965, p. 8.
18. R. E. Treybal, *Mass-Transfer Operations*, 3rd Ed., McGraw-Hill, New York, 1980, p. 45.
19. H. J. Lugt, Vortices and vorticity in fluid dynamics, *Am. Sci.*, 73:162–167 (1985).
20. J. Y. Oldshue, *Fluid Mixing Technology*, McGraw-Hill, New York, 1983, pp. 338–358.
21. G. B. Tatterson, *Scaleup and Design of Industrial Mixing Processes*, McGraw-Hill, New York, 1994, pp. 125–131.
22. G. B. Tatterson, *Scaleup and Design of Industrial Mixing Processes*, McGraw-Hill, New York, 1994, pp. 15–18.
23. H. Gladki, Power dissipation, thrust force and average shear stress in the mixing tank with a free jet agitator, In: *Industrial Mixing Fundamentals with Applications* (G. B. Tatterson, R. V. Calabrese, and W. R. Penny, eds.), Am. Inst. Chem. Engrs., New York, 1995, pp. 146–149.
24. Y. F. Maa and C. Hsu, Liquid-liquid emulsification by static mixers for use in microencapsulation, *J. Microencaps.*, 13:419–433 (1996).
25. J. M. Coulson, J. F. Richardson, J. R. Backhurst, and J. H. Harker, *Chemical Engineering: Fluid Flow, Heat Transfer and Mass Transfer*, 4th Ed., Vol. 1, Pergamon Press, Oxford, UK, 1990, p. 242.
26. G. B. Tatterson, *Scaleup and Design of Industrial Mixing Processes*, McGraw-Hill, New York, 1994, pp. 67–74.
27. D. S. Dickey and J. G. Fenic, Dimensional analysis for fluid agitation systems, *Chem. Eng.*, 83(1):139–145 (1976).
28. R. B. Bird, W. E. Stewart, and E. N. Lightfoot, *Transport Phenomena*, Wiley, New York, 1960, pp. 71–122.
29. G. B. Tatterson, *Scaleup and Design of Industrial Mixing Processes*, McGraw-Hill, New York, 1994, pp. 117–125.
30. J. M. Coulson, J. R. Richardson, J. R. Backhurst, and J. H. Harker, *Chemical Engineering: Fluid Flow, Heat Transfer and Mass Transfer*, 4th Ed., Vol. 1, Pergamon Press, Oxford, UK, 1990, pp. 227–230.
31. J. Y. Oldshue, *Fluid Mixing Technology*, McGraw-Hill, New York, 1983, pp. 72–93, 295–337.

32. J. L. Beckner and J. M. Smith, *Trans. Instn. Chem. Engrs.*, 44:T224 (1966); through H. W. Fowler, Progress report in pharmaceutical engineering, *Mfg. Chemist Aerosol News*, 37(11):60 (1966).
33. G. B Tatterson, *Scaleup and Design of Industrial Mixing Processes*, McGraw-Hill, New York, 1994, pp. 97–101.
34. A. W. Etchells, W. N. Ford, and D. G. R. Short, Mixing of Bingham plastics on an industrial scale, In: *Fluid Mixing III*, Symposium Series 108, Instn. of Chem. Engrs., Hemisphere Publ. Corp., Rugby, UK, 1988 pp. 271–285.
35. J. M. Smith, The mixing of Newtonian and non-Newtonian fluids, *J. Soc. Cosmet. Chem.*, 21:541–552 (1970).
36. A. B. Metzner, W. T. Houghton, R. A. Sailor, and J. L. White, A method for the measurement of normal stresses in simple shearing flow, *Trans. Soc. Rheol.* 5:133–147 (1961).
37. A. G. Fredrickson, Viscoelastic phenomena in thick liquids: Phenomenological analysis, In: *Modern Chemical Engineering*, Physical Operations, Vol. 1, (A. Acrivos, ed.), Reinhold Publ., New York, 1963, pp. 197–265.
38. J. Cheng, P. J. Carreau, and R. P. Chhabra, On the effect of wall and bottom clearance on mixing of viscoelastic fluids, In *Industrial Mixing Fundamentals with Applications* (G. B. Tatterson, R. V. Calabrese, and W. R. Penny, eds.), Am. Inst. Chem. Engrs., New York, 1995, pp. 115–122.
39. J. Hinze, *Am. Inst. Chem. Eng. J.*, 1:289–295 (1995); through Z. Sterbacek and P. Tausk, *Mixing in the Chemical Industry*, Pergamon Press, Oxford, UK, 1965, pp. 47–48.
40. W. L. McCabe, J. C. Smith, and P. Harriott, *Unit Operations of Chemical Engineering*, 5th Ed., McGraw Hill, Inc., New York, 1993, p. 276.
41. Y.-F. Maa and C. Hsu, Liquid-liquid emulsification by rotor/stator homogenization, *J. Contr. Rel.*, 38:219–228 (1996).
42. *Handbook of Powder Science and Technology*, (M. E. Fayed and L. Otten, eds.), Van Nostrand Reinhold, New York, 1984.
43. J. Carstensen, *Solid Pharmaceutics: Mechanical Properties and Rate Phenomena*, Academic Press, New York, 1980.
44. L. G. Austin, Size reduction of solids: crushing and grinding equipment, In: *Handbook of Powder Science and Technology*, (M. E. Fayed and L. Otten, eds.), Van Nostrand Reinhold, New York, 1984, pp. 562–606.
45. W. L. McCabe, J. C. Smith, and P. Harriott, *Unit Operations of Chemical Engineering*, 5th Ed., McGraw Hill, New York, 1993, pp. 960–993.
46. F. C. Bond, Crushing and grinding calculations, *Brit. Chem. Eng.*, 6:378 (1965).
47. J. T. Carstensen and A. Mehta, Scale-up factors in the manufacture of solution dosage forms, *Pharm. Technol.*, 6(11):64, 66, 68, 71, 72, 77 (1982).
48. G. P. Pisano, *The Development Factory: Unlocking the Potential of Process Innovation*, Harvard Business School Press, Boston, 1997.
49. P. E. Kline, A. J. Vogel, A. E. Young, D. I. Towsend, M. P. Moyer, and F. G. Aerstin, Guidelines for process scale-up, *Chem. Eng. Progr.*, 70(10):67–70 (1974).
50. A. Lübbert, T. Paaschen, and A. Lapin, Fluid dynamics in bubble column bioreactors: Experiments and numerical simulations, *Biotechnol. Bioeng.*, 52:248–258 (1996).
51. R. E. Johnstone and M. W. Thring, *Pilot Plants, Models, and Scale-up Methods in Chemical Engineering*, McGraw-Hill, New York, 1957, pp. 12–26.
52. G. B. Tatterson, *Scaleup and Design of Industrial Mixing Processes*, McGraw-Hill, New York, 1994, pp. 112–113.
53. G. B. Tatterson, *Scaleup and Design of Industrial Mixing Processes*, McGraw-Hill, New York, 1994, pp. 243–262.
54. J. P. Euzen, P. Trambouze, and J. P. Wauquier, *Scale-Up Methodology for Chemical Processes*, Editions Technip, Paris, 1993, p. 15.

55. E. S. Taylor, *Dimensional Analysis for Engineers*, Clarendon Press, Oxford, UK, 1974, p. 1.
56. W. L. McCabe, J. C. Smith, and P. Harriott, *Unit Operations of Chemical Engineering*, 5th Ed., McGraw Hill, New York, 1993, pp. 16–18.
57. A. Bisio, Introduction to scaleup, In: *Scaleup of Chemical Processes: Conversion from Laboratory Scale Tests to Successful Commercial Size Design* (A. Bisio and R. L. Kabel, eds.), Wiley, New York, 1985, pp. 15–16.
58. Y.-F. Maa and C. Hsu, Microencapsulation reactor scale-up by dimensional analysis, *J. Microencaps.,* 13(1):53–66 (1996).
59. G. E. Karniadakis, Simulation science, Internet publication, 1997.
60. J. P. Euzen, P. Trambouze, and J. P. Wauquier, *Scale-Up Methodology for Chemical Processes*, Editions Technip, Paris, 1993, pp. 13–15.
61. J. R. Rice, Computational science and the future of computing research, *IEEE Comp. Sci. Eng.*, 2(4):35–41 (1995).
62. L. H. Baekeland, Practical life as a complement to university education—medal address, *J. Ind. Eng. Chem.*, 8:184–190 (1916).
63. A. Einstein, *Investigations on the Theory of Brownian Movement*, (R. Fürth, ed.; A. D. Cowper, trans.) Dover, New York, 1956.
64. L. F. Richardson, Atmospheric diffusion shown on a distance neighbour graph, *Proc. Roy. Soc. London*, A110:709–737 (1926).
65. O. Levenspiel and T. J. Fitzgerald, A warning on the misuse of the axial dispersion model, *Chem. Eng. Sci.*, 38:489–491 (1983).
66. *Modern Process Techniques for the Pharmaceutical and Cosmetic Industry*, Fryma, Inc., Edison, NJ.
67. W. B. Whiting, Effects of uncertainties in thermodynamic data and models on process calculations, *J. Chem. Eng. Data*, 41:935–941 (1996).
68. M. J. Groves, Some engineering aspects of process scale-up, AAPS/FDA/USP Workshop on the Scale-up of Liquid and Semisolid Disperse Systems, Arlington, VA, May, 1993.

10

Scale-Up of Dispersed Parenteral Dosage Forms

Matthew Cherian

Pharmacia and Upjohn, Nerviano, Italy

Joel B. Portnoff

Fujisawa USA, Melrose Park, Illinois

I. INTRODUCTION

Disperse dosage forms, which include emulsions, microemulsions, liposomes, lipid complexes, suspensions, and colloidal dispersions are increasingly used in the pharmaceutical industry. Apart from the fact that these systems make possible the delivery of some molecules that would be difficult or impossible to administer in soluble-solution dosage forms, disperse dosage products offer certain potential therapeutic advantages, e.g., decreased toxicity, longer half-life, tissue or organ targeting, control of biodistribution, and at times enhanced efficacy. Beyond the formulation and processing problems associated with solutions or freeze-dried dosage forms, disperse dosage forms offer additional and sometimes enormous challenges during scale-up from small laboratory batches of a few milliliters to hundreds of liters necessary for full-scale batches. The reasons for these challenges are at least threefold:

- The techniques used for making small-scale batches are not practical for larger batches. For example, when preparing a batch of 100 mL, solvents can be readily removed by using a rotovap/water bath and vacuum system, or simple evaporation by heating and/or sparging with inert gasses. To scale up to say 500 L (a typical production batch size), this would entail a rotovap of about 5000 times the size used for a lab batch.
- Disperse systems are nonhomogeneous on a microscopic scale. When one considers that the formation of the dispersed phase was governed by such microscopic factors as local pressure, temperature, concentration of active and inactive ingredients, concentration of solvents, turbulence, contact with materials, and any number of physical chemical parameters, one may easily see why scale-up of disperse systems can be a formidable challenge.

- There are a number of factors controlled by rates. Dispersed systems have structures to a lesser or greater degree. Emulsions have a structure, because an interfacial film of surfactants is oriented in a predefined way. Liposomes and lipid complexes are even more structured because they are composed of bilayers. The formation of these structures is governed to varying degrees by rates (what happens in unit time), e.g., rate of addition of dispersed phase into continuous phase, rate of shearing in a homogenizer, rate of removal of solvents, rates of pumping, and rates of mixing and subdividing.

In attempting to develop disperse dosage forms, it is advisable to use techniques at the small scale that can potentially be used on a larger scale. Small-scale processing that uses forces or processes/procedures that simply do not exist with production equipment may be impossible to satisfactorily scale up. Early development itself must be creative, but always with an eye on how scale-up of process and equipment parameters will evolve. At the same time, it is important to understand the dynamics of formation of the systems on a microscopic scale. If one has sound knowledge of what takes place in a laboratory scale (especially at microscopic levels), then the possibilities of success of scaling up the formulation(s) are vastly enhanced.

In this chapter some general issues pertaining to scale-up, such as mixing and homogenization, filtration, freeze-drying, removal of solvents, aseptic processing, and sterilization are addressed.

Table 1 is a list of emulsions, suspensions, and liposomes/lipid complexes that are currently approved for marketing in the United States. They represent a broad range of products, ranging from anesthetics to oncolytics, and cover a broad array of routes of administration—from intralesional to subcutaneous to intravenous.

A. Typical Materials Used

A wide range of materials is used for the dispersed phase, in addition to the drug, and other materials are used as emulsifying or stabilizing agents.

Used in the dispersed phase are

Soybean oil
Sunflower oil
Medium chain triglycerides (MCTs)
Polyethyleneglycol
Propyleneglycol

Emulsifying agents used include

Soy phosphatidylcholine (Soy PC)
Egg phosphatidylcholine (EPC)
Dimyristoylphosphatidylcholine (DMPC)
Dimyristoylphosphatidylglycerol (DMPG)
Dipalmitoylphosphatidylcholine (DPPC)
Dipalmitoylphosphatidylglycerol (DPPG)
Distearoylphosphatidylcholine (DSPC)
Distcoroylphosphatidylglycerol (DSPG)
Tween-80
Cremophor-EL

Table 1 Emulsions, Suspensions, and Liposomes/Lipid Complexes

Trade name	Active ingredient	Route of administration	Processing	Nature of formulation
Hydeltra-TBA	Prednisolone	Intralesional, soft tissue	Aseptic	Suspension
Decadron-LA	Dexamethasone	Intramuscular, intralesional, intra-articular, soft tissue	Aseptic	Suspension
Solganal	Aurothioglucose	Intramuscular	Aseptic	Suspension
Exosurf	Dipalmitoylphosphatidylcholine (colfosceril palmitate)	Intratracheal	Aseptic	Suspension
Survanta	Dipalmitoylphospatidylcholine (colfosceril palmitate)	Intratracheal	Aseptic	Suspension
Diprivan	Propofol	Intravenous	Aseptic	Emulsion
Doxil	Doxorubicin	Intravenous	Aseptic	Liposome
Amphocil	Amphotericin	Intravenous	Aseptic	Liposome
Abelcet	Amphotericin	Intravenous	Aseptic	Lipid complex
Humulin	Insulin	Subcutaneous, intravenous	Aseptic	Suspension
NPH Isletin	Insulin	Subcutaneous, intravenous	Aseptic	Suspension
Lente Insulin	Insulin	Subcutaneous, intravenous	Aseptic	Suspension
Lupron Depot	Leuprolide	Intramuscular	Aseptic	Suspension
Depo-Medrol	Methylprednisolone	Intramuscular, soft tissue	Aseptic	Suspension
Depo-Provera	Medroxy-progesterone	Intramuscular	Aseptic	Suspension
Intralipid	Soybean oil	Intravenous	Heat sterilized	Emulsion
Daunoxome	Daunorubicin	Intravenous	Aseptic	Liposome

Typical solvents used include

Methylenechloride
Chloroform
Dimethylsulfoxide
Dimethylformamide
Dimethylacetamide
Freons

B. Some Generic Formulations

While the proportions of active and inactive ingredients may vary widely, some general guidelines are provided for the novice.

Formulation 1: (Freeze-dried)—powder		
Active ingredient	=	1 mg
Dimyristoylphosphatidylcholine (DMPC)	=	7 mg
Dimyristoylphosphatidylglycerol (DPPG)	=	3 mg
Mannitol	=	25 mg
Formulation 2: (Liquid)—suspension		
Active ingredient	=	1 mg
Egg PC	=	10 mg
Formulation 3: (Liquid)—emulsion		
Active ingredient	=	1 mg
Soybean oil	=	10 mg
Egg PC	=	10 mg
Formulation 4: (Liquid)		
Active ingredient	=	1 mg
Polysorbate 80	=	2 mg
Polyethyleneglycol 4000	=	3 mg
Sodium chloride	=	8.5 mg
Benzyl alcohol	=	9 mg
Sodium hydroxide or hydrochloric acid	For pH adjustment	

II. MIXING AND HOMOGENIZATION

A. Mixing

Mixing is an essential unit operation in the preparation of dispersions, liposomes, and emulsions. Mixing may be accomplished by rotating or reciprocating action of devices within the tank, by sparging gases, or by a shaking motion of the entire vessel. For large-scale manufacture of product, use of mechanically driven impellers is by far the most common means of agitation. Mixing by sparging gas is rarely used, partly because the amount of power that can be imparted to the system is very limited, and also many processes involve at least one ingredient that is surface active, and this may lead to excessive foaming.

Mixing tanks are usually vertical cylinders with a smooth internal surface. The pharmaceutical industry typically avoids the use of baffled vessels, especially for injectable products, because of the difficulty in adequately cleaning such vessels. The tanks are often fitted with a dished bottom. Flat-bottomed tanks give regions of stagnant fluid and more difficult to clean 90° angles, and are not usually recommended [1]. The effect of liquid depth on effectiveness of mixing has been extensively studied. For best results, the depth of liquid should be at least equal to the tank diameter.

A variety of stirring devices is used by the pharmaceutical industry. These include propellers, turbines, paddles, and variations thereof (see Fig. 1). The usual practice is to design the drive for top entry. Side-entry propellers, though used by the chemical industry, do not find much use in the pharmaceutical industry, especially for making injectables. Propellers that produce axial flow are used for low-viscosity liquids and are operated in a manner that produces downward flow. Turbines and paddles produce radial flow if baffles are used, or rotational flow in the absence of baffles. For unbaffled vessels, eccentrically mounted impellers are often used. The asymmetric flow patterns thus generated often result in the desirable radial flow with both horizontal and vertical movement necessary for good mixing.

When liquids are mixed by an axially located impeller in an unbaffled vessel, a vortex develops at the center of the vessel. This is true except at very low impeller

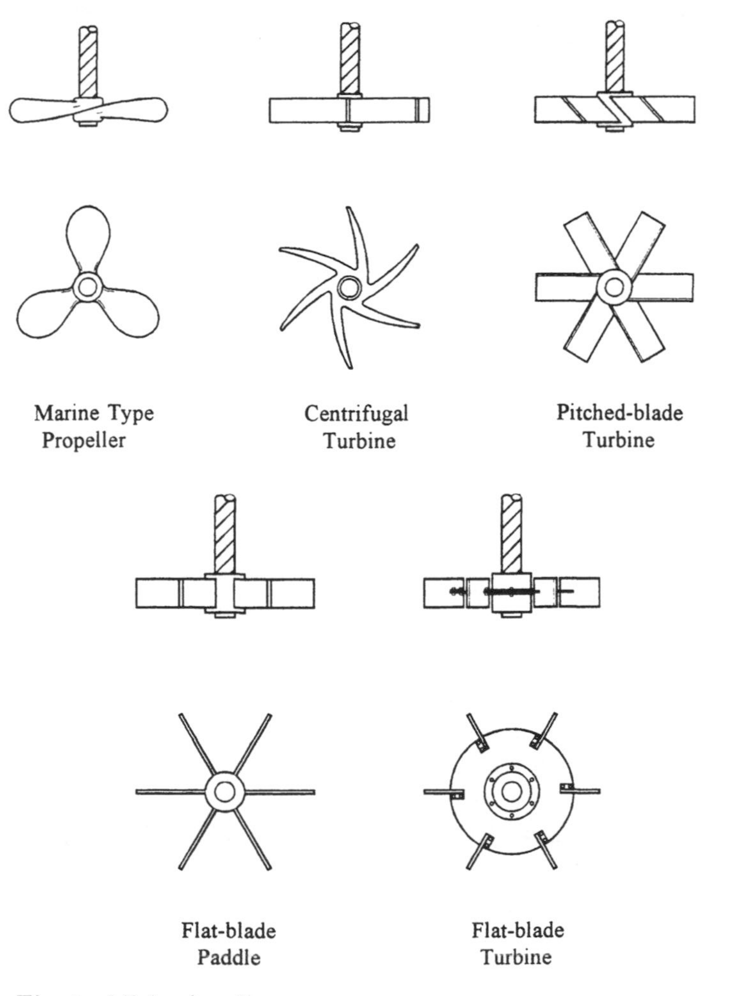

Fig. 1 Mixing impellers.

speeds or at very high liquid viscosities (>20,000 cp), neither of which is normally encountered in practice in the pharmaceutical industry [2]. When a vortex is formed, gas is drawn into the impeller and is dispersed into the liquid—an undesirable situation as it may lead to foaming, especially if surfactants are present, and also because the full power of the impeller is not imparted to the liquid. Apart from the above two problems, vortexing makes it difficult to scale up as it is impossible to achieve geometric and kinematic similitudes. (Geometric similarity is achieved when corresponding dimensions of the small- and large-scale vessels are in the same ratio. For example, line drawings of such vessels are superimposable on each other. For fluid systems that are geometrically similar, there is said to be kinematic similarity when flow patterns in one vessel are similar to flow patterns in the other.)

Vortices may be avoided by off-center location of the impeller. This is the preferred method in the industry. Alternatively, baffles may be used, but difficulty of cleaning baffled vessels is an impediment in their use for preparing sterile product. A stator ring that fits around the impeller can be used. This arrangement gives a high degree of turbulence and may be desirable, especially if emulsification is required. A vortex can be avoided by operating a closed, airtight vessel at full capacity. This method is ideal for achieving geometric and kinematic similarities.

High-velocity mixing results in splashing of the liquids onto the sides of the tank above liquid level and on the cover. This vexatious problem may result in poor reproducibility of the batches. This annoyance can be ameliorated by intermittent spraying of the continuous phase through nozzles located inside the tank.

Power for mixing is an important consideration when designing mixing systems. Power correlations are plentiful for single liquids [3]. These correlations provide a functional relationship between power number and impeller Reynolds number. Power number is a dimensionless number as defined below

$$P_o = \frac{Pg_c}{\rho N^3 d_i^5} \tag{1}$$

and an impeller Reynolds number is defined by

$$N_{\mathrm{Re}} = \frac{di^2 N\rho}{\mu} \tag{2}$$

where di = impeller diameter; N = rpm for impeller; ρ = density of the liquid(s); μ = viscosity of the liquids; g_c = gravitational constant; P = power; and Po = power number.

The power correlations of single liquids may be used for two or more immiscible liquids by computing density and viscosity of such systems using the following formula:

$$\rho_{\mathrm{m}} = \rho_1\phi_1 + \rho_2\phi_2 + \ldots\ldots \tag{3}$$

where ρ_m = density of the mixture; ρ_i = densities of each liquid, i = 1, 2, . . . ; and ϕ = volume fraction of each liquid.

Also

$$\mu_m = \frac{\mu_c}{\varphi_c}\left(1 + \frac{1.5\mu_D\varphi_D}{\mu_D + \mu_C}\right) \tag{4}$$

where μ_M = viscosity of the mixture; μ_C = viscosity of the continuous phase; φ_C = volume fraction of the continuous phase; and φ_D = volume fraction of the dispersed phase.

1. *Example 1*

An emulsified product has 20 mL of oil dispersed in 100 mL of emulsion. The viscosity of the oil is 2.1 centipoises, and the viscosity of the continuous phase (water) is 1 centipoise. What is the effective viscosity of the mixture?

$$\mu_D = 2.1 \text{ centipoise}$$
$$\mu_C = 1.0 \text{ centipoise}$$
$$\phi_C = \frac{(100-20)}{100} = 0.8$$
$$\phi_D = \frac{20}{100} = 0.2$$
$$\mu_M = \frac{1.0}{0.8}\left[1 + \frac{1.5 \times 2.1 \times 0.2}{1.0 + 2.1}\right]$$
$$= 1.50 \text{ centipoise}$$

2. *Example 2*

What is the power required to mix two liquids whose properties are given below?

Density of continuous phase	=	62.5 lb/ft^3
Density of dispersed phase	=	52.1 lb/ft^3
Impeller diameter	=	4 in.
rpm of the impeller	=	200

The viscosity and proportions of continuous and dispersed phase are as in the previous example.

$$\text{Density of mixture} = (62.5 \times 0.8 + 52.1 \times 0.2))\frac{\text{lb}}{\text{ft}^3}$$
$$60.4\frac{\text{lb}}{\text{ft}^3}$$
$$\text{Impeller Reynolds number} = \frac{di^2 N\rho}{\mu}$$

(Note: Reynolds number should be calculated using a consistent set of units. Impeller diameter in ft, density in lb/ft^3, and viscosity in lb/ft · sec. Rpm has dimension min^{-1}.)

$$N_{Re} = \frac{(4/12)^2 \times (200 \times 60) \times 60.4}{0.0151 \times 241.69} = 22{,}066$$

$$\left(1 \text{ centipoise} = 0.01 \text{ poise} = 0.01\frac{\text{g}}{\text{cm sec}}\right)$$

So 2.42 is the conversion factor for poise to lb/ft · hr.

Referring to power correlation for the type of impeller, at the above Reynolds number the power number is 5.7

$$\mathrm{Po} = \frac{\mathrm{P}g_c}{\rho \mathrm{N}^3 di^5}$$

$$5.7 = \frac{\mathrm{p}g_c}{60.4 \times (200 \times 60)^3 \times (4/12)^5}$$

$$\mathrm{P} = \frac{5.7 \times 60.4 \times (200 \times 60)^3 \times (4/12)^5}{4.17 \times 10^8}$$

$$= 5841 \text{ ft} \cdot \text{lb/hr.}$$

B. Homogenization

Homogenization of dispersed products is often necessary because of the limits placed on the size of the largest particle that can be administered intravenously. For large-volume parenterals this is often an average no greater than 500 nm. The current regulatory requirements on particle size specify only an average, and not an upper limit on the particle size distribution. A distribution can have an average of say 500 nm, yet contain particles that are several microns in size. Larger particles, particularly in large volumes of fluids for administration, can cause emboli and other complications when administered to an already compromised patient. For small-volume injections, administered intravenously, regulatory agencies are more likely inclined to allow a larger average particle size. Currently there are products approved in various countries that have average particle sizes of 1–2 microns and an upper limit of 5 microns.

What happens when a parenteral dispersion is homogenized [4]? Prior to homogenization, the particle size distribution more often than not has no well-defined character. However, after a single pass one can observe a "leptokurtic" or left-leaning distribution with a long tail. The left-leaning nature of the distribution stems from the fact that in any size-reduction step, there are always more smaller particles than larger ones. If the product is passed through the homogenizer a second time, the distribution on the whole moves to the left, giving a smaller average size and a shortened tail (see Fig. 2). Further passes reduce the average size by a small amount, but the tail can be seen to be shortened. Multiple passes ultimately lead to a near symmetric distribution, which many instrumental particle analyzers fit to a Gaussian curve [5].

A variety of dispersion equipment is available today that fits many different applications. These include colloid mills, high-pressure reciprocating type of homogenizers, rotor-stator high-shear mixers and submerged jet dispersers (Fig. 3). Ultrasonicators have been used in the laboratory but are not used for the manufacture of injectables.

The rotor-stator mixer and the colloid mill have a similar design, consisting of a central rotor and a stationary ring around the rotor. The gap between the rotor and stator is usually of the order of 1 mm. The width of the gap is an important determinant of these high-shear-type equipments. The dispersing efficiency is closely related to the shear

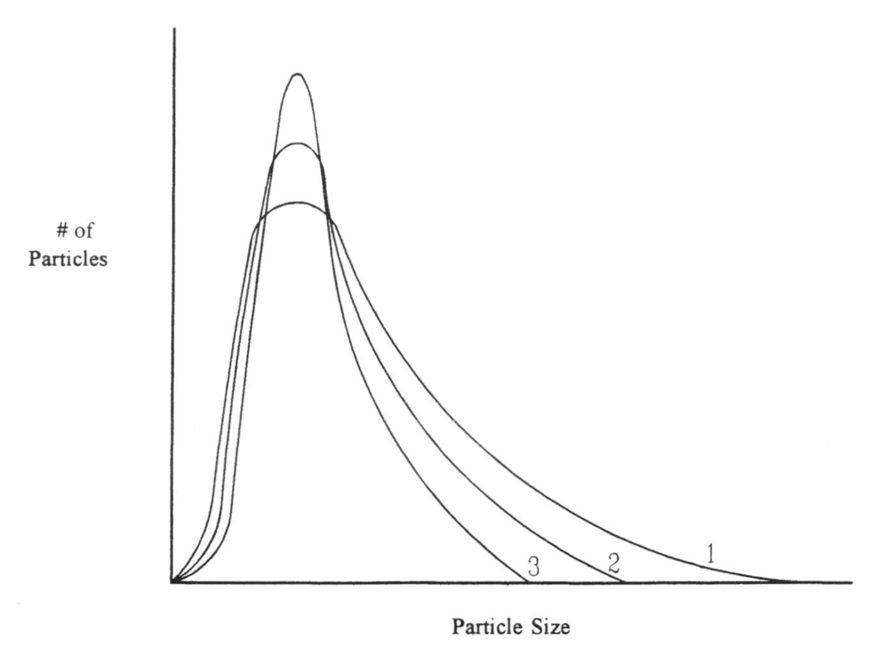

Fig. 2 Particle size as a function of number of homogenization passes.

gradient, which, in turn, is proportional to the peripheral velocity of the rotor and inversely proportional to the gap width between the rotor and stator (Fig. 4). A particle caught in the gradient is subjected to shear forces that in turn break it up into smaller particles. Apart from the tangential shear forces, the fluid is also subject to strong centrifugal forces, which tend to "pump" the fluid away from the dispersing element. This pumping action causes a suction effect, thereby keeping the dispersing element supplied with fresh fluid. This type of mixer disperser is extremely efficient and is capable of generating particles of the order of 100 nm or less. The main disadvantage of this type of homogenizer is the generation of heat. The circumferential speed of these rotors can be as high as 80–100 ft/sec. Such high speeds lead to viscous dissipation of heat, which typically increases in proportion to the square of the rpm. Another disadvantage of this type of high shear mixer is the requirement for gaskets and O-rings to seal off the rotor shaft from the dispersing element. These gaskets and O-rings tend to shed particu-

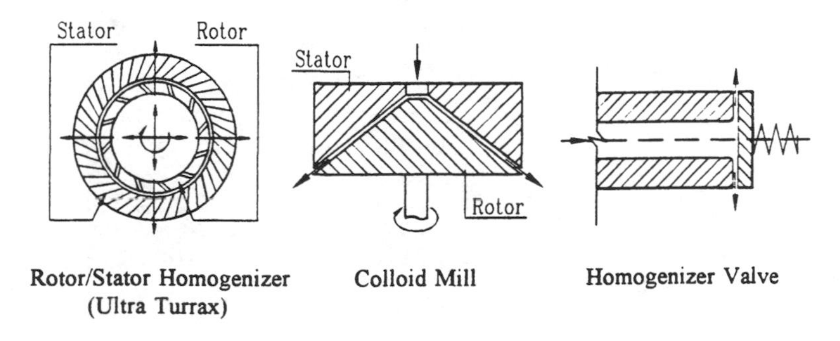

Fig. 3 Some types of homogenizers.

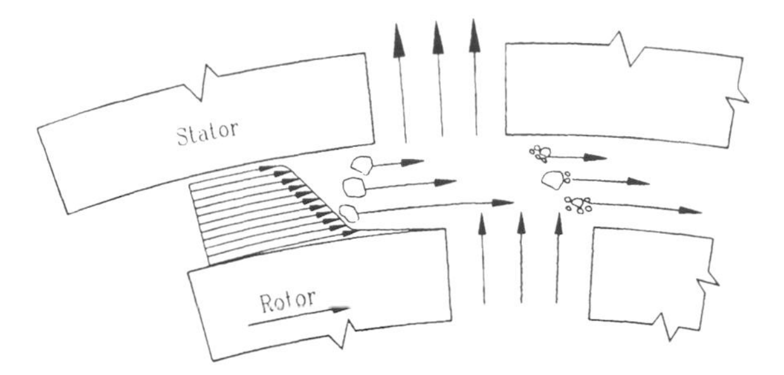

Fig. 4 Principle of operation of rotor/stator homogenizer.

late matter and can have compatibility problems with the product. [These sealing aids are now being made of polytetrafluoroethylene (PTFE), which improves the situation in most cases.] For manufacture of sterile products, the dispersing elements need to be sterilized, preferably by using steam. Experiments have shown that these elements are steam sterilizable in 15–30 min at 121°C. These dispersers are available as in-line models or vertically mounted "probe"-style units (see Fig. 5).

A colloid mill has several similarities with the rotor-stator type mixer. It has a stationary "stator" through the center of which the product is pumped at high pressure. The rotor moves close to the stator. The gap is often adjustable and is of the order of 0.1 mm. The adjustable feature of the gap often makes reproducibility of batches a challenge, as the screw type of adjusting mechanism for the rotor often is not reproducible to a high degree of accuracy.

In a reciprocating design homogenizer, a piston moves in a cylinder and pushes the product through a nozzle whose opening is adjustable by means of a spring loaded

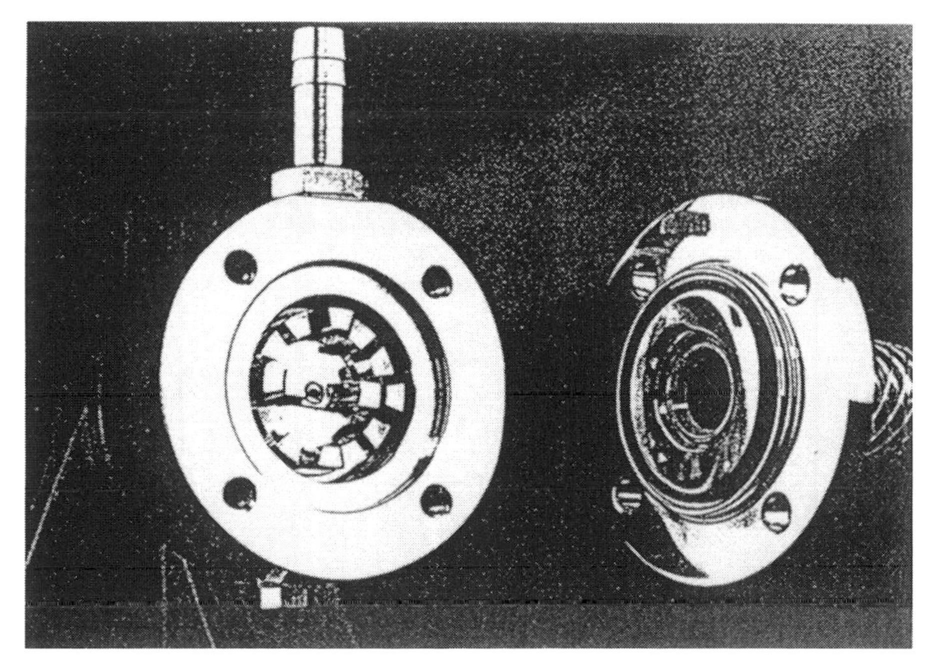

Fig. 5 Dispersing element of an in-line unit. (Photo courtesy of Ika-Werke, Germany.)

screw type of mechanism (see Fig. 3). This type of disperser has a high throughput and is widely used in the industry for making large-volume fat emulsions.

The submerged-jet-type disperser is a new design that has recently been introduced. This type of equipment makes use of "free turbulence," created when two jets of fluid impinge on each other. The product is pumped at pressures ranging from 3000 to 25,000 psi through a narrow channel, which is then bifurcated into two channels of smaller size. The product streams passing through these channels are then directed at each other in an interaction chamber, where they impinge upon each other (see Fig. 6). The resulting free turbulence (as opposed to the wall turbulence created when moving parts create turbulence) results in size reduction. Considerable heat is generated and in general this calls for intercooling between passes. While these types of dispersers can be steam-sterilized, the extremely small size of their channels can provide a challenge when trying to validate the sterility by seeding with spores. Sterilization using hydrogen peroxide may be a highly efficient and pharmaceutically acceptable method for this type of equipment. This type of equipment is available from lab-scale to full-scale size (see Fig. 7).

Yet another ingenious device is the hydroshear processor. In this device a premixed formulation enters the processing chamber tangentially, at pressures up to 600 psi, generated by centrifugal variable speed pumps. As the fluid layers move toward the center, the diameter decreases while the height increases, and the fluid velocity increases proportionately. The velocity difference between adjacent concentric layers result in steep velocity gradients and consequently intense shear. When the product reaches the axis, it exits the chamber through special outlet nozzles (see Fig. 8).

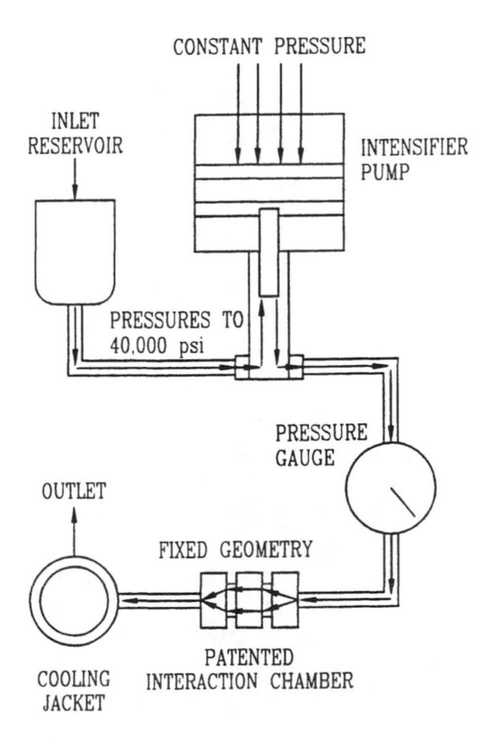

Fig. 6 Principle of a submerged jet homogenizer.

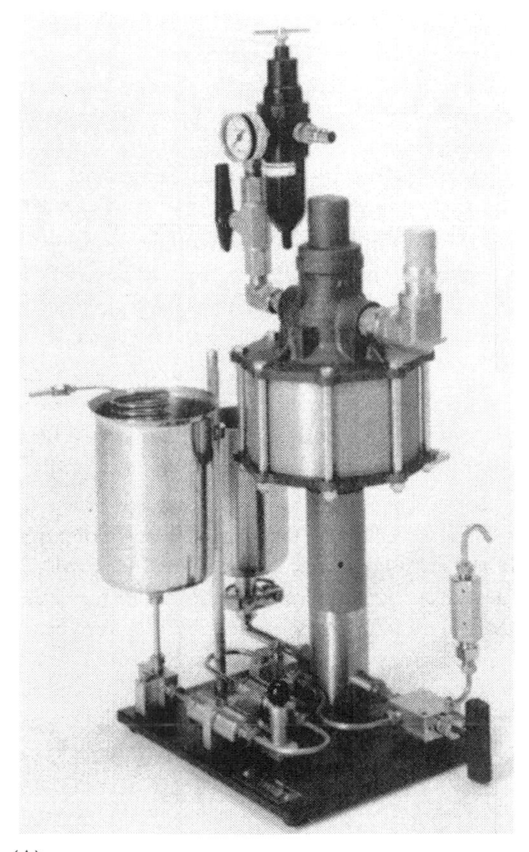

(A)

Fig. 7 Models of submerged jet homogenizers: (A) lab-scale model; (B) full-scale model. (Photo courtesy of Microfluidics Corporation, Boston, MA.)

III. SCALE-UP OF FILTRATION SYSTEMS

Filtration is used in the manufacture of disperse systems to sterilize, to remove solvents and extraneous viable and nonviable particulate matter, to concentrate dilute mother liquor, to obtain desired particle size distributions, to remove endotoxins, and to generate dispersions [6].

Sterilization of aseptically manufactured products is achieved by sterile dead-end filtration using 0.2 μm filters. If the final product has a particle size less than 0.2 μm, then it can, and should, be sterile-filtered. One should remember that, although the average particle size is less than 0.2 μm, the particle size distribution may include particles larger than 0.2 μm, which will result in fouled filters. However, if the final product has a significant proportion of its particle size greater than 0.2 μm, then, obviously, filtration using 0.2 μm filters is not practical. In such cases, the ingredients can be aseptically compounded by sterile filtration of their solutions in suitable vehicles in previously cleaned and sterilized containers under class 100 conditions. Aseptic processing is done in class 100 rooms. (Air in class 100 rooms has less than 100 particles of size ≥ 0.5 microns/ft^3).

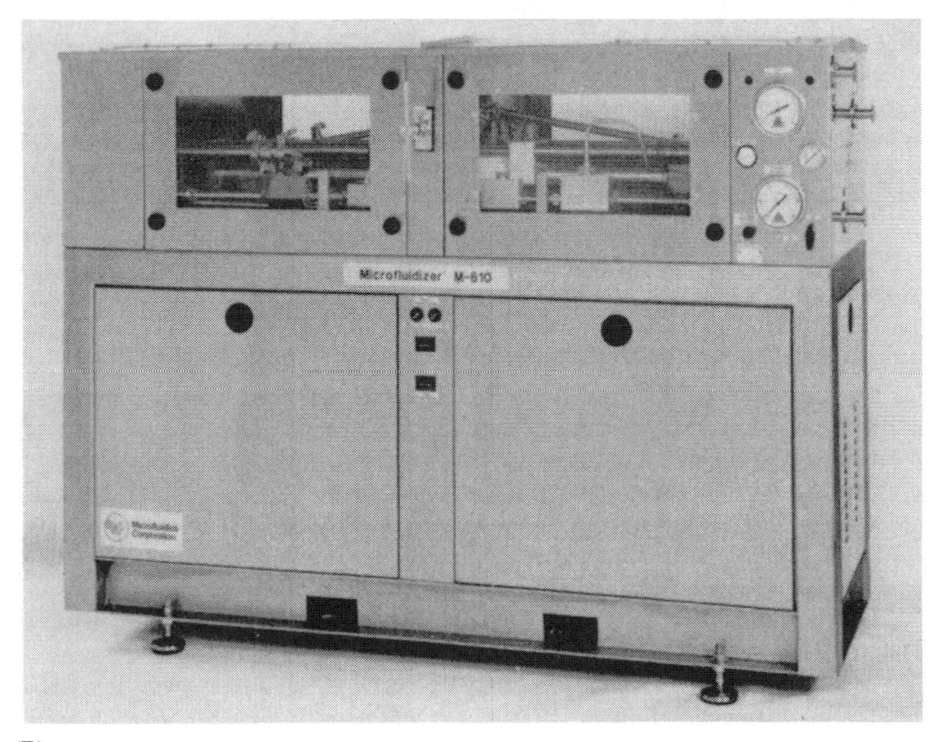

(B)

Fig. 7 Continued.

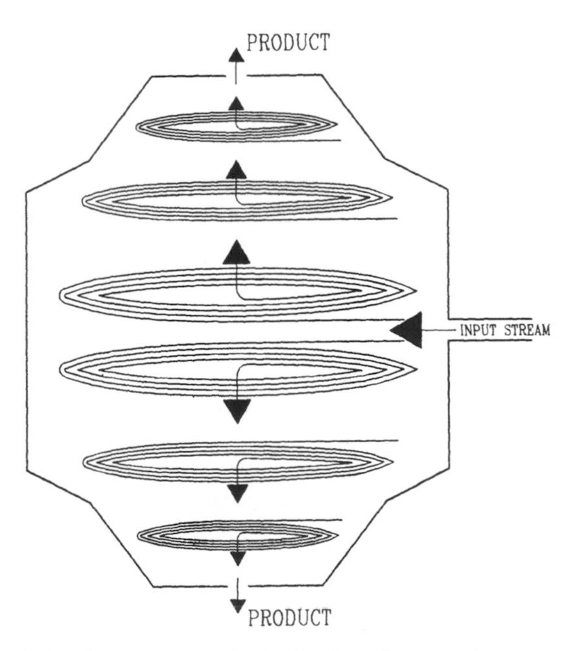

Fig. 8 Principle of a hydroshear homogenizer.

Scaling up of sterile filtration is often dictated by the configuration of filters available from manufacturers. These days, all leading manufacturers provide filters that range in surface area over factors of 100 or more. While scaling up sterile filtration it generally is desirable to maintain the same bulk flow velocity through the filters. Often this can be achieved only by assembling a filtration battery with multiple filters in parallel. Such batteries can be tricky to operate, as fluids tend to take paths of least resistance, which can result in some of the parallel filters not getting any product. Judicious use of valves to throttle the flow is necessary. In today's regulatory climate, media runs are required to validate any aseptic process. (In media runs, a growth medium like tryptic soy broth is used to simulate the process.) During such media runs, a filtration battery with the maximum number of parallel filters that will ever be used simulate the worst-case scenario.

As has been industry practice, all filters need to undergo compatibility testing on the different active and inactive ingredients in the formulation. Sometimes filters of different surface area, though they may be made of the same materials of construction, may have different bonding materials. This means compatibilities may have to be determined with different filters even from the same manufacturer. Needless to say, all compatibility testing must be done at the highest temperature and pressure conditions that might be encountered. Testing individual solvents and excipients may be necessary as testing all ingredients and solvents in one single fluid stream sometimes makes it difficult to detect leachables from filters or to observe sorption to filters. As a result of the filter industry's increasing use of thermal bonding, without using any bonding material, the compatibility problem is primarily confined to the filter medium and material used for the filter shell.

Tangential flow (also called cross-flow) filtration systems have been in use in the pharmaceutical industry for quite some time. Tangential flow solves one of the impediments in dead-end filtration, namely concentration polarization. Concentration polarization is a result of accumulation of dispersed particles at the filter surface, which consequently results in decreased or no flux. Tangential flow rectifies this situation by directing this flow parallel to the filter medium. Tangential flow filters should be operated in a turbulent regime. Laminar flow tends to be less effective in preventing concentration polarization.

In tangential flow filtration, a stream of fluid with suspended particles and dissolved molecules flows parallel to a surface. The surface can be tubular, as is the case of hollow-fiber membranes, flat sheets used in flat-plate devices (e.g., Mini-Tan), or spirally wound cartridges, which consist of multiple layers of a spirally wound membrane with a feed screen, and a permeate screen wound around a central collection tube. Irrespective of the type of tangential flow system, flow is parallel to the surface of the membrane.

As can be seen in Fig. 9, except in entrance regions, flow is two-dimensional—with one component parallel to the surface due to the bulk flow of the feed, and a radial or normal component due to the flow of the permeate. Conceptually, this radial component of the velocity results in the suspended particles being carried to the surface and deposited there. This leads to a certain degree of concentration polarization at the surface. According to basic principles of fluid flow parallel to a solid surface, the layer immediately adjacent to the membrane is stationary (no-slip boundary condition). The layer next to it has a finite velocity, which increases as you move away from

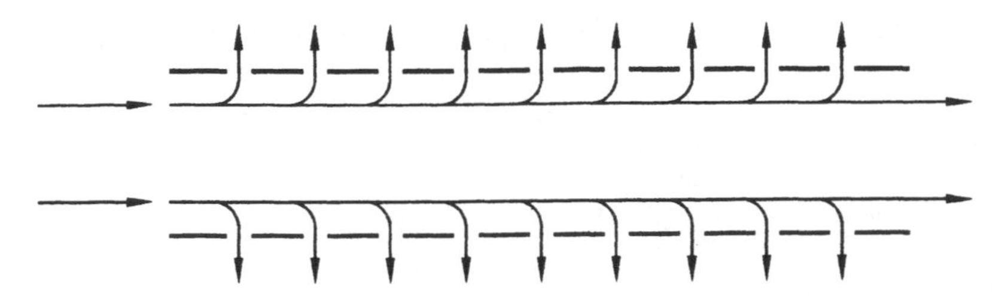

Fig. 9 Two-dimensional flow pattern in hollow fibers.

the surface until you reach regions where bulk flow is encountered. The thickness of this layer, called the boundary layer, is a function of the bulk flow velocity and is inversely proportional to velocity.

Polarization adversely affects the operation of the tangential flow filters. A measure of the degree of polarization is the polarization modulus C_w/C_b, where C_w and C_b are the concentrations of the suspended, membrane-impermeable particles at the wall and bulk, respectively. The polarization modulus is related to the transmembrane solvent flux by the relation $C_w/C_b = \exp(J/K)$, where C_w = concentrations of the membrane impermeable solute at the wall; C_b = concentration of the membrane impermeable solute in the bulk stream; J = transmembrane solvent flux; and K = local mass transfer coefficient.

The mass transfer coefficient can be approximately by the relation $K = D/\delta$, where D = diffusivity of the solution; and δ = boundary layer (polarization layer) thickness.

As can be seen from the above relationships, an increase in flux J will result in an increase in C_w, due to an increase in transmembrane pressure that consequently results in a decrease in permeate flow. Thus, there is room to optimize the transmembrane pressure and the permeate flow rate.

When using tangential flow filtration of a feed that consists of particles greater than 100 nm, the above models based on diffusion and convection are not applicable. Particles of that size have very low diffusivities, and consequently a diffuse boundary layer with the resultant concentration polarization cannot exist. For particles much smaller than 100 nm, (it is not unusual to find particles of size 10 nm in today's pharmaceutical products), concentration polarization can still be a problem. Attempts should be made to optimize the filter operation balancing permeate flow rate and transmembrane pressure drop.

Tangential flow filters are available in tubular, hollow-fiber, pleated, spirally wound, or annular and flat channel designs. Of these, the first three tend to give the desired turbulent flow when operated in an appropriate manner. Spirally wound or annular designs are typically operated in the laminar regime and are desirable when turbulent flow can be damaging to the dispersion, e.g., small peptides.

In the plate and frame design, membrane sheets are mounted on frames with openings for inlet, permeate, and retentate. This design is modular and can be stacked to provide surface areas ranging from 0.5 to 50 ft^2 or more. The addition of modules increases filtration rate but not the flux. Plate and frames are available with fixed and variable flow channels. These devices have the disadvantage that they can cause problems with leakage. On the other hand, they can be operated at very low flow rates because of the ease with which they can be depolarized. Thus, they are extremely flexible at all stages of scale-up.

Hollow-fiber tangential flow filters are small diameter tubes with diameters of 50–100 μm. Filtrate flows radially outward while retentate flows out at the other end of the tube [7]. A bundle of tubes, typically 100 or more, is held together in a tubular casing to provide adequate surface area. The operating pressures of hollow-fiber cartridges are in the range of 10–30 psi. The hollow fibers typically do not have the mechanical strength beyond this and should not be operated above these pressures.

Pleated cartridge membranes pack considerable surface area in a given volume, but so far have only been available in the microporous range and not in ultrafiltration or microfiltration ranges.

Temperature and pH have a strong influence on the performance of tangential flow filters. Apart from compatibility problems associated with a very low or very high pH, operating close to a material's isoelectric point can lead to agglomeration and fouling of membranes. The pH level of the product can alter the electric charge on the membrane surface, which in turn can lead to binding and a consequent drop-off in filtration rates. Temperature can have a strong bearing on the filtration rate. Elevated temperatures result in lower viscosity, a thinner boundary layer and higher flow rates. Tangential flow filtration systems should be operated at the highest temperature compatible with the membrane and product.

The pore size of ultrafiltration membranes are characterized by their molecular weight cutoff (MWCO). Molecular weight cutoff refers to the molecular weight of a globular solute at which 90% of the solute is rejected by the membrane. Apart from molecular weight, rejection is affected by molecular shape, electrical characteristic, etc. Typically, commercially available filters have a MWCO in the range of 1 to 1000 kilodaltons.

Tangential flow filtration has definite advantages over dead-end filtration due to flow not being a strong function of the size of retained particles, and the ability to backflush the surface to depolarize it.

Tangential flow filters must be qualified for a given separation before a scale-up is attempted. Compatibilities with all solvents and active and inactive ingredients must be established. One way to do this is to pump the feed through the filters of interest, collect permeate and retentate, and run UV-VIS assays using the feed as a blank. But being a disperse system, the dispersion must be separated into clear liquids before UV-VIS assays can be performed. Alternatively, the feed can be held in the filter for prolonged periods of time (12–24 hr or more) and then examined for the presence of leachants. After these experiments, the filters should be cut apart and examined under a microscope for physical deformation, change of color, and pore size distribution.

Integrity of tangential flow filters can be as important as the integrity of sterilizing filters. This is especially so if the tangential flow filter happens to be the last filter in a process stream, and if either the permeate or the retentate streams are collected in a nonsterile area. This is because of the possibility of ingress of bacteria from such areas into the product. Integrity of the filter before and after use must be tested to ensure asepsis of the process.

Several filter manufacturers ship their filters filled with a preservative. The filters should be cleaned using large volumes of Water for Injection, USP, making sure that there are no areas that are not wetted. A sample of the permeate or retentate should be tested using a UV-VIS spectrophotometer or by HPLC to make sure that all the preservative has been washed out. After cleaning, the filters should be wrapped in paper and steam-sterilized.

All filters used in the process should be validated. As a minimum, these procedures should have physico-chemical compatibility and be biologically efficient.

To ensure physico-chemical compatibility, pump the process stream through the filter at the highest temperature and pressure that is expected to be encountered in the process. Periodically draw samples of the fluid and perform an HPLC or UV-VIS assay to make sure that no leachables are entering the stream. At the end of the process, remove a part of the filter medium, clean and dry it to constant weight and determine its weight loss compared to an equal area of membrane that has not ben subjected to testing. There should not be more than 1% weight loss. Examine several samples of the filtration medium under a light or electron microscope. Observe any dimensional changes in the medium when compared with an unused membrane. There should be no dimensional change.

To achieve biological efficiency, challenge the filters using *Pseudomonas diminuta* at a level of at least 10^6. Sometimes the process stream can be lethal to the bacteria, in which case the placebo or another medium with similar fluid properties may be used. A fluid stream incubated with bacteria is circulated through the filters at conditions that simulate the process conditions. A minimum 6 log reduction in bacterial load is expected.

In addition, selection and design of filter holders and housings is critical in designing scaled-up systems. Filter holders and housings may be made of stainless steel 316 or polypropylene. The material of construction obviously is dependent on the compatibility of the product with the material. Preferably, stainless steel should not be used for oxygen-sensitive products. All fittings and connections to the system must be sanitary, i.e., no threads should come in contact with the product. The filter holder should be equipped with an air vent, be mounted at the highest point of the holder, and have a liquid drain at the lowest point. Steam-in-place sterilization of the filter results in condensation of steam in the filter, which can result in dilution of the product. Thus, it should be possible to drain the condensate from the filter. Sealing of the holders is done using O-rings. These rings may be made of rubber, polymers, or Teflon. The housing should be torqued to the specification set by the manufacturer. Too high a torque results in compression setting of the O-ring or cold flow in the case of Teflon. This results in loss of sealing capability. Extreme care should be taken in installing O-rings, as cuts or abrasion of the ring can lead to loss of integrity and consequent loss of product. O-ring grooves should be carefully examined prior to installation to make sure that the groove is not deformed or cut. The O-rings should be installed wet because in a dry condition they tend to become pinched. They may be lubricated with water or ethanol or another suitable liquid that is compatible with the product.

IV. FREEZE-DRYING

Parenteral disperse systems are often freeze-dried for stability reasons. If the active ingredients or the excipients undergo hydrolysis during the expected shelf life of the product, then freeze-drying offers an avenue to enhance the long-term stability of the product.

As the name suggests, in the freeze-drying process, the product is first frozen. During this stage, the drug material and excipients are confined to the space between ice crystals. The drug may be in crystalline or amorphous form, depending on how the product was frozen or on the chemical nature of the compound. The size distribution of the ice crystals is an important determinant in the structure, porosity, and reconstitution characteristics of the final product. Water is removed while the product is frozen by applying a pressure that is less than the vapor pressure of water at that temperature. Under these conditions, water sublimes and freezes on the condenser. This stage of drying is known as primary drying. The heat of sublimation that is required is supplied by heat put into the shelves. Care must be taken to see that this heat input does not exceed the sublimation requirement. If it does, melt-back of the product occurs, resulting in an unacceptable product. Once primary drying is complete, the shelf temperature is raised to remove bound water. The degree to which bound water is removed depends on stability requirements for the product. In some cases, bound water is required to enhance product stability.

Scale-up of laboratory batches requires careful experimental work starting with the smallest batches. The eutectic temperature of the formulation is useful when developing a freeze-drying cycle. Typically, primary drying is conducted below the eutectic temperature in order to avoid collapse of the cake. Excipients such as mannitol and lactose, which are used as bulking agents or cryoprotectants, affect the eutectic temperature of the product. If a choice is possible, select excipients that result in higher eutectic temperatures, as this results in shorter lyophilization drying cycles. When alcohols like tertiary butanol are used as vehicles, they tend to creep up the side of vials during primary drying (an interfacial effect). This is detrimental in that product ends up dried on the sides of the vial, or, in extreme cases, on the lip of the opening. Threshold temperatures above which this occurs should be determined, and primary drying should be done below this temperature.

Difficulty in translating a drying cycle from the laboratory scale to full scale is due to several reasons. Sometimes the two freeze-dryers may have entirely different designs, e.g., one with an internal condenser and the other an external one. Often, operating parameters of the two pieces of equipment do not match. In order to avoid this, the following parameters should be considered.

- Shelf temperature mapping. Shelf temperature should be mapped over a range of -40° to + 50°C, using calibrated thermocouples. Such mapping shows the range of temperatures that occur over the shelves at various times during the drying cycle. This temperature range may vary depending on the capability of the equipment and the needs. Mapping should be done at several temperatures over the range. An acceptable specification should be no more than $\pm 1°C$ from the mean temperature of the shelves. The process is repeated for every shelf over the range of temperatures. Shelf-to-shelf variation (mean temperature) should be no more than $\pm 2°C$ from the mean for all shelves.
- Rate of freezing. Rate of freezing is a function of the load on the shelves. For comparison purposes, similar loads (same-size vials filled to the same level with a placebo or water) should be used. The rate of freezing, as defined by the slope of the time vs. temperature plot, should be comparable for the lab/pilot scale lyophilizer and the full-scale unit.
- Rate of heating. This operation is identical to the above. The slope of the temperature vs. time curve between pilot-scale and full-scale units should be comparable.
- Pressure range and vacuum sweep. This is a range of pressure over which the cycle is run. The highest pressure in the range has to be below the sublimation

pressure of the vehicle. The capability of maintaining the required pressure range throughout the cycle should be available on both small-scale and large-scale units.

Other parameters that may be tested include leak rate (not more than 3 μm/min), evacuation rate, and condenser capacity.

Reproducibility and portability of lyocycles depend to a great extent on the degree of control that can be exercised. In general, shelves with no load or partial load tend to exert a significant influence on the cycle. Therefore, the placebo for the formulation should be used to fill the chamber during the development of small-scale batches and scale-up.

Today's newer freeze-dryers come with trayless loading systems, which use automatic devices to load vials from filling lines onto shelves, without use of trays. These are useful for at least three reasons: (1) There is better contact between vials and shelves; (2) the robotic loading system allows the entire load to be loaded into the freeze-dryer in 10–30 min—thus, all vials have essentially the same thermal history, as opposed to the results of loading one tray at a time with the potential time differential of up to 8 hr or more between loading the first vial and the last one; (3) microbial contamination is minimized as human intervention during filling and loading is minimal. It is needless to say that if trayless loading systems are intended to be used, then cycle development should be done without using trays, by placing filled vials directly on shelves.

Primary packaging components have a strong bearing on freeze-drying cycles. Molded glass vials with thick glass and an arched bottom tend to offer more thermal resistance than thin-walled tubing vials with flat bottoms. Complaints about more breakage with tubing vials than with molded vials have abated with the introduction of better quality tubing. Stoppers are available with a single vent (the "igloo" type) or two or more vents. Multiple-vent stoppers are preferable, because they seat better and offer less mass transfer resistance.

V. REMOVAL OF SOLVENTS

Manufacture of parenteral disperse systems often involves removal of organic solvents. Chlorinated hydrocarbons like methylene chloride and chloroform, and other solvents such as dimethylsulfoxide and dimethylformamide are commonly used for making liposomes and lipid complexes. Tertiary butanol is often used as a vehicle for freeze-drying due to its high freezing point (+25°C). Although Freons can be used too, their cost, plus impending withdrawal from worldwide markets due to environmental contamination, precludes their use. These solvents, of necessity, have to be removed to very low levels. Regulatory requirements on the levels and solvents allowed vary with different countries. For example, the European regulatory agencies do not allow chloroform to be used in the manufacture of injectable products, whereas in the United States it can be used as long as products do not contain more than 50 ppm in the finished dosage form. Following are the USP requirements on some solvents [8] used in parenteral products: benzene, 100 ppm; chloroform, 50 ppm; methylene chloride, 500 ppm; trichloroethylene, 100 ppm.

Although removal of the major amounts of solvents in products is relatively easy, removal to ppm levels can be a challenge. For small-scale batches, application of a vacuum to a system, e.g., rotovap, is adequate. However, for larger batches, this method does not provide the required results. This stems from the fact that the solvent molecules need to diffuse through a pool or matrix of material before they come to the

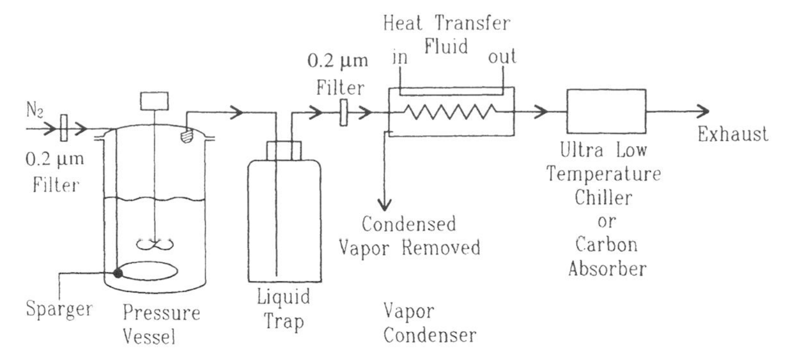

Fig. 10 Solvent removal by sparging.

surface, where the vacuum provides effective removal, and most of the mass transfer resistance is in the liquid phase [9]. Therefore, increasing vapor/liquid interfacial area increases the rate of solvent removal proportionately. This is accomplished by sparging the liquid with a gas, preferably nitrogen (see Fig. 10). The inert gas argon should be avoided, as it forms clathrates in an aqueous medium. The exiting gas stream is then passed through a heat exchanger, where the solvent can be condensed and removed. The speed of removal of solvents by this method is a function of temperature, viscosity of the liquid, and the interfacial area created. Creation of the interfacial area is accomplished by using a sparger. These are made of sintered metals or glass frits. Compatibility of the formulation with a sintered metal sparger should be established before its use. Higher temperature does favor speed of removal because of the decreased viscosity of the liquid phase, and the consequently increased diffusivity of the solvent molecules [10]. However, parenteral dispersed systems, like liposomes and lipid complexes, are highly structured formulations, and their structure can be adversely modified or destroyed by elevated temperatures. Sparging is useful as a tool for removing solvents only if they are volatile. Sparging can be done aseptically by filtering the gas using 0.2 μm filter. The exit gas/vapor stream should have a 0.2 μm filter, too, in order to prevent ingress of microbial contamination by that route.

For less volatile solvents, (e.g., dimethyl sulfoxide), tangential flow filtration, (also called diafiltration), is a preferred choice (see Fig. 11). The liquid formulation con-

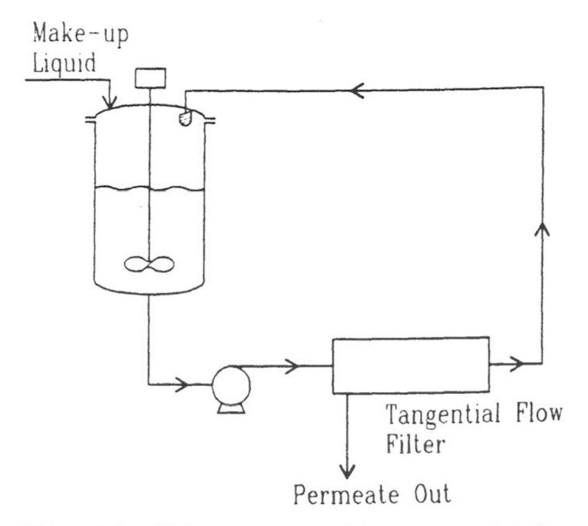

Fig. 11 Solvent removal by tangential flow filtration.

taining the undesirable solvent is passed through one or more parallel tangential flow filters. A low MWCO filter is selected in order to prevent loss of product through this permeate. The permeate laden with the solvent is removed and an equal volume of water, or other suitable vehicle, is added to the system. The process is continued until solvents reach the desired low levels. Tangential flow filtration can be done aseptically. Compatibility of solvents with the filter needs to be tested before a filter is selected. Pumps used can potentially destroy the product by the high turbulence that is created. Hence "gentle" pumps, like peristaltic or grear-lobe pumps, should be used.

Aseptic centrifuges can be useful in removing solvents. The formulations can be spun down and the supernatant removed, and then resuspended in water or other vehicles. The process if repeated until the desired low level of solvent is achieved. Aseptic centrifuges are expensive and difficult to validate.

VI. ASEPTIC PROCESSING AND STERILIZATION METHODS

Aseptic processing is widely used in the manufacture of parenteral emulsions, suspensions, and liposomes. Aseptic processing can be defined as a method by which the drug formulation and the primary packaging components are sterilized separately and assembled in a final filling/stoppering operation [11].

Parenteral emulsions, suspensions, and liposomes are often heat sensitive and cannot tolerate terminal sterilization by heat. (One notable exception is the intravenous fat emulsion, which, being a large-volume parenteral, is required by U.S. law to be heat-sterilized; however, special autoclaving techniques are used for this product.) Therefore, aseptic processing is preferred in the manufacture of these types of parenteral products.

Aseptic processing, by its very nature, entails precleaning; wrapping and sterilizing all processing vessels, pumps, filters, tubes, hoses, and other components; and steam-sterilizing them, usually in a steam autoclave [12]. These components of the process are then assembled in a class 100 clean room, using aseptic technique. In many aseptic processes, there is no final filtration, as the product is not amenable to a final filtration with a 0.2 μm filter. Therefore the compounding is done aseptically in a class 100 room by sterile-filtering the ingredients in appropriate vehicles into a compounding vessel. Absence of final filtration also means that extreme care should be taken so that particulate matter, both viable and nonviable, is not introduced into the process during the unwrapping of presterilized parts and their assembly, and also during processing. A cleanliness test, performed by breaking up the emulsion/dispersion, should be conducted to make sure that no nonviable particulate matter from wrappings or from the process system is carried into the product. This is done by filtering the broken-down emulsion or dispersion onto a membrane. Often the breakdown can be accomplished by the addition of a suitable cosolvent, like ethanol, after prefiltration.

Encapsulated parenteral products pose special problems regarding sterility. During the encapsulation process, there is a potential for bacteria to be encapsulated in the product, which may not be detected by the usual USP sterility test. This is a weak link in the aseptic processing of encapsulated products. The USP sterility test should be suitably modified to ensure that encapsulated bacteria do not go undetected. Needless to say, such modified sterility testing needs to be validated. Incoming process streams may have to be seeded with bacteria and demonstrate that they are effectively removed as a result of the processing.

Aseptic compounding poses special problems with regard to the material balance of active and inactive ingredients of the formulation. The ingredients are dissolved in appropriate vehicles and then sterile-filtered into a class 100 room. Line losses and losses in the filters may cause the active and inactive ingredients initially weighted not to end up in the compounding vessel. The result is that the yield of product based on actual quantities weighted is less than the tolerance limit set as the yield for the usual injectable products. In such cases, master batch records should be suitably modified so that the quantity of ingredients that actually enters the compounding vessel as measured by weighing the vessel on a scale forms the basis for calculations of yields.

During early stages of scale-up, dispersed products are best compounded in a clear glass vessel. Due to the heterogeneous nature of these products, the operator should visually observe the process in the vessel. For products that call for prolonged compounding, light exposure of the product in a clear glass vessel can be a problem. Low-UV lights (these lights emit low levels of ultraviolet radiation) must be used in the processing area. Once the process has been well understood, type 316 stainless steel vessels can be used, provided the product is not metal sensitive.

Media runs in which a bacterial growth medium like tryptic soy broth is used to compound, process, and fill the product should be conducted to validate the setup and assembly of the process in the clean room as well as during the actual process and ensuing filling operations. During media runs, the actual setup and operation should be replicated in all respects, including compounding, freeze-drying, processing time, the number of personnel involved in the preparation, and the batch size. However, any heating step involved in the operation should be omitted for obvious reasons. A minimum of three media runs should be conducted, with at least one repeat run every year. In general, only personnel who have successfully taken part in the above-mentioned media runs should be involved in the aseptic processing of the actual product.

Parenteral emulsions, suspensions, and liposomes sometimes tend to be heavier than the suspending aqueous medium, and, hence, they tend to settle on standing. This can be a problem during the filling operations because content uniformity may suffer. Therefore, the product must be kept suspended by stirring using a suitable means. Gentle stirring by sparging nitrogen or air would be helpful, but entrainment of air should be avoided at this stage.

Motors and electric/electronic controls sometimes have to be located in the sterile core, where processing is done. Only totally sealed motors and controls can be used in such areas. These items, not being autoclavable, are wiped with alcohol, phenol, glutaraldehyde, hydrogen peroxide, or other suitable sterilants before they are introduced into the clean room. The current technology of hydrogen peroxide vapors is ideal for this purpose.

Sterilization using hydrogen peroxide is a useful technique, especially for complex machinery. Steam sterilization of such machinery, like homogenizers and pumps, is often difficult because of the convoluted pathway that the steam must take. Long and tortuous paths result in poor thermal delivery at the downstream side. In such situations, sterilization using hydrogen peroxide may be desirable. Hydrogen peroxide has proven to have excellent sporicidal activity against a wide range of pathogens including bacteria, fungi, and viruses. Sporicidal activity of hydrogen peroxide is a strong function of temperature and concentration of hydrogen peroxide. For sterilization, a concentration of 15% or more may be required at room temperature, whereas considerably lower concentrations are adequate at temperatures of 35°C. This mode of sterilization is validated by using either spore strips inserted in the tortuous pathway or by having a spore suspension dried onto the surfaces. Recovery of such organisms may need to be proved in order to assure validation of the sterilization. One advantage of using hydrogen peroxide is that it degrades into water. However, residual amounts of peroxide can be extremely damaging to drugs and excipients. Therefore, the machinery needs to be tested for residual hydrogen peroxide. There are highly sensitive assays available in the literature that can detect ppb levels of hydrogen peroxide. Some metals and certain plastics can be degraded by the action of hydrogen peroxide. Therefore, the compatibility of these materials with hydrogen peroxide need to be established prior to its use.

Protracted aseptic processing has the potential for sterility failure. This is especially true early in the development of products, as the process is not yet well understood and typically it takes longer to do the first batches. During process optimization, steps should be taken to reduce the duration of the processing steps so that total processing time can be reduced. Processes that take significantly more than 24–48 hr pose special problems with respect to sterility assurance.

All filters used in processing must be suitably presterilized and integrity-tested before use in the process. Whenever possible, redundant filters should be used, as failure of postfiltration integrity can be disastrous for structured products like emulsions, suspensions, and liposomes. (There is no possibility of reprocessing these types of products once they have failed sterility or any other product parameters.)

Endotoxin levels in these products are a combination of input from incoming process streams, processing vessels, hoses, etc. Incoming process streams should be monitored for bioburden, and all efforts should be taken to keep the bioburden low. When multiple streams enter a process, monitoring endotoxin levels may be helpful for monitoring the incoming streams during process development. Depyrogenation (endotoxin removal) of processing vessels and related items is a more complex task. Pyrogens are difficult to remove from large process vessels by dry heat due to the size limitations of dry heat ovens. In such cases, depyrogenation by dilution, (washing with Water for Injection, USP), is the obvious choice.

All in all, the integrity of a complex aseptic operation is the sum of its parts, and each step should be carefully thought out and designed to achieve the desired end results.

VII. TERMINAL STERILIZATION BY STEAM

Scientists who work in the field have accepted the fact that emulsions, liposomes, and suspensions typically cannot be terminally sterilized by autoclaving. This because all disperse systems are stabilized by surfactants or dispersing agents, and application of heat tends to dislocate or remove these from where they were originally located. Application of heat also tends to favor film formation and drainage, which plays a major part in coagulation of emulsions and results in their breakage.

The first major challenge to this state of affairs came when Kabi, a Swedish pharmaceutical company, developed intravenous lipid fat emulsions for the U.S. market. As U.S. laws do not allow large-volume parenterals to be aseptically manufactured, a method had to be devised for heat-sterilizing these emulsions. As was expected, the emulsions started breaking apart on application of heat. Further investigation showed that if the emulsion was kept agitated during heat application, there was no problem of emulsion breaking. In fact, this idea was refined and special autoclaves were designed with a mechanism for keeping the emulsion containers moving. The type of movement, rotatory, double-planetary, or other variants thereof, have been patented by various manufacturers of intravenous fat emulsions. During autoclaving, considerable extraction of materials from stoppers can occur. These leachants, apart from being harmful, can lead to stability problems. Of interest, none of the small-volume emulsion products on the market today are heat-sterilized.

Liposomes are being developed by several pharmaceutical companies. Liposomes, because of their bilayer structure, were believed to be especially fragile, and therefore, could not undergo the stress associated with heat sterilization. However, experiments have shown that these assumptions are not necessarily true [13]. Certain liposomes have been shown to be terminally heat-sterilized. This might be because the hydrophobic tails of the phospholipids reassemble due to hydrophobic forces. In general these forces are enhanced by an increase in temperature.

VIII. VALIDATION OF SCALE-UP

Validation is an integral part of scale-up. What validation purports to do is provide documented assurance that the process does what it is supposed to do. FDA regulations and the need for constant quality make it imperative for pharmaceutical companies to validate all scale-up efforts. Process validation helps set up acceptable limits for process variables and is necessary when setting up in-process controls. An interesting outcome of scale-up validation is optimization of the process. This results in maximizing efficiency while maintaining quality standards.

Before actual scale-up runs are carried out, all equipment, instruments, tanks, etc., must be tested to ensure that they meet manufacturer's specifications, including extremes of operating conditions, such as pressure, input voltage, or any parameters influencing equipment or process performance. Materials of construction of tanks, piping, and other parts that come into contact with the product should be validated by doing compatibility studies (on test coupons of the metals) prior to fabrication of the items. Cleanability of all equipment in product contact should be carefully evaluated. This is especially true if clean-in-place (CIP) is used, because this method of cleaning may not be adequate to clean hard-to-reach areas. If CIP is desired, as it would be for dedicated process setups, care should be taken during design to avoid such areas. Degree of cleanliness should be validated by taking rinse samples of the final rinse using Water for Injection, USP, and doing product-specific assays for active and inactive ingredients. Often, this is done using HPLC. The undesirable residues should be no more than three log units below the no-effect dose for the active ingredient. Sometimes the levels of quantitation on an HPLC, using UV detection, do not permit this, in which case other modes of detection, such as fluorescence, may be used. A placebo batch, comparable in size to the minimum batch size, should be made after cleaning, and the end product tested for the presence of the drug. All instruments used should be calibrated. For manufacture of sterile products, the ability to sterilize the equipment as it is set up needs to be validated. The steam-in-place method is highly desirable, since it prevents having to make many aseptic connections. For certain processes, the equipment has to be dried prior to processing the product. This is especially difficult to validate as the test for dryness is often too subjective.

The incoming raw materials should be validated with respect to different vendors, for conformance to specifications with regard to maximum impurities/related substances, moisture content, crystal form, particle size, water content, and cleanliness. This is accomplished by making scale-up batches using raw materials at the extremes of the specifications and conducting stability studies of these batches.

Freeze-drying cycles call for validation. The parameters to investigate are the range of shelf temperatures for a given set point, chamber pressure ranges, condenser temperature, the load on shelves, and the temperature of shelves at loading. All of the above parameters can be individually controlled except condenser temperature. Batches at extremes of these parameters may have to be made, and a study of the stability of these products must be done.

Scale-up of lyophilization poses special problems, especially for dispersed products. The lyophilization process is aseptic and the parts of the process that can potentially lead to contamination problems must be identified. To do this, media fills should be conducted in stages. About 3000 vials are first filled with the drug product and stoppered. This part of the filling step should identify any problems associated with this step in the operation. Next, another 3000 vials are filled and loaded on to the lyophilizer and stoppered. This part of the operation is useful in validating the loading part of the operation. Next, another 3000 vials are filled with growth media (e.g., tryptic soy broth) and loaded and left in the chamber with doors closed for several hours. During this time, nitrogen (or other gases) are bled into the chamber to simulate "nitrogen sweep" during the actual freeze-drying process. When the capacity of the freeze dryer is less than 3000 or when the batch size is less than 3000, as it is for early clinical batches, only fill as many vials as fill the chamber. Though some manufacturers mimic the operation of the freeze-dryer by freezing the product to be lyophilized, this is not recommended, as freezing reduces the level of some microbial contaminants and thereby make it more difficult to identify contamination in the processing. Another area of still greater potential for contamination is in the transport of vials from the filling line into the chamber. All manufacturers have downward laminar flow of air in the area between the filling line and the freeze-dryer. There is always a potential for the air to pick up contaminants from the floor and deposit them in the vials that are only partially closed (stopper in lyophilization position). This area should be investigated with smoke.

After sterilization of the freeze-dryer, the chamber should be dried with a ring pump. Water remaining on the shelves will freeze during the freezing stage of the lyophilization cycle and can lead to poor contact of the shelves with the vials. This, in turn, can lead to uneven drying of some vials. In the worst-case scenario, when the clearance between the vials and the shelf above is small, formation of ice and the consequent expansion can lead to partial or incomplete stoppering of the vials. This results in some vials having liquid product at the end of a cycle.

Scale-up and changes of the lyophilization cycle have posed special problems. When a product is scaled up without modifying lyophilization cycles the rates of freezing tend to be different. For example, a lower rate of freezing tends to give larger crystals and leads to faster rates of sublimation. On the other hand, slower rates of freezing tend to produce a frozen mass that is nonuniform. Also, the rate of freezing has an effect on the types of polymorphs of active and inactive ingredients. While scaling up it may be desirable to hold the rate of freezing constant.

The leak rate of air into the freeze-dryer should be monitored and controlled. Rates of <1 μm/min are desirable. Leakage into a freeze-dryer may originate from outside the chamber through gaskets, seals, thermocouple harnesses, from refrigerant through the refrigeration system, from heat transfer fluid from shelves and condenser, and from oil from the vacuum pump and hydraulic ram used for stoppering. All but the oil from the vacuum pump can be prevented from contamination by proper mechanical maintenance and monitoring. Ingress of oil from the vacuum pump can be avoided by operating the freeze-dryer above the vapor pressure for the oil. The leak rate should be checked prior to each run.

Components of dispersed system products often tend to settle during the loading process, resulting in a nonuniform cake. Some manufacturers tend to overcome this situation by loading on to the frozen shelves. This practice is objectionable because humidity from the filling room can condense onto frozen shelves causing all the aforementioned problems that are associated with the formation of ice on shelves. Also, such condensing vapor can potentially result in microbial contamination.

IX. HYPOTHETICAL PRODUCTION DEVELOPMENT CASE STUDY

Problem: To formulate a drug as an injectable, dispersed drug product. The following properties of the drug are known:

Solubility in water	—	less than 100 ppm and
	—	not stable at a pH greater than 7
Solubility in soybean oil	—	25 mg/mL
Solubility in sunflower oil	—	23 mg/mL
Solubility in PEG400	—	12 mg/mL
Sensitivity to oxygen	—	not oxygen sensitive
Sensitivity to visible light	—	light sensitive
Sensitivity to heat	—	not sensitive

Therapeutic considerations require that 200 mg of the drug be delivered to pediatric patients and 500 mg of the drug be delivered to adults weighing 75 kg.

Solution: The active moiety shows only limited water solubility. Although not stable in water at pH greater than 7, there is the possibility of buffering the drug solution in water at a pH below 7. However, doses of 200 and 500 mg have to be delivered to children and adults, respectively. The drug solubility of 100 ppm in water makes it impractical to deliver the drug as an aqueous solution.

The next logical step is to attempt to make a salt of the drug, which usually increases a drug's water solubility. Previous information on the stability of the drug at an alkaline pH indicates that a sodium salt of the drug is not practical, and an attempt to make an acid-salt of the drug has not been successful.

Looking at the solubility of the drug in soybean oil and PEG400 shows that there are possibilities in this area. At a solubility of 25 mg/mL in soybean oil, an emulsion-type product is capable of delivering 200–500 mg of drug in a total volume of less than 100 mL. The total volume must be kept at less than 100 mL, as otherwise the product will be categorized as a large-volume parenteral requiring terminal heat sterilization. On the other hand, neither a simple oil solution or a PEG solution can be injected. Thus, making a disperse system dosage form appears attractive. Of the two vehicles, soybean oil is preferable from a safety point of view, although PEG in small volumes in drug products has been used and approved in the United States.

It appears feasible to develop a prototype formulation of the drug as an emulsion in soybean oil. (There is an excellent safety record for soybean oil, which has been widely used in intravenous fat emulsions). One of the first tasks is to select an emulsifying agent. Among the several that area available, the first choices are egg phosphatidylcholine and dimyristoylphosphatidylcholine (DMPC). Egg PC is available in purified form and is far less expensive than DMPC. From a stability perspective, egg PC is less desirable because of the presence of multiple double bonds on it, which tend to make it susceptible to oxidation. Yet when the total cost of a dosage form is considered, the incremental cost due to the use of the synthetic phospholipid DMPC is relatively small. Therefore, the decision is to use DMPC as the emulsifying agent.

One of the first tasks is to determine experimentally the optimal oil/phospholipid ratio. The drug is first dissolved in the oil. Then DMPC is added to this solution in varying proportions before the system is emulsified using a laboratory-scale emulsifier. Samples are autoclaved in a laboratory autoclave at 121°C for 20 min. Samples are then placed on short-term accelerated stability for three months at 2–8, 25, and 37°C and tested for particle size distribution, apparent pH, potency of the drug, and related substances. Based on the test results the optimum oil/phospholipid ratio is determined.

The next attempt is to make medium-scale, 10–100 L batches, using the same oil/phospholipid/drug ratio. Care is taken to maintain all process parameters as closely as possible to those of the laboratory batches. The object is to obtain the same particle size distribution while maintaining potency, pH, and related substances constant. Several batches are made and placed on accelerated stability. Parameters to be tested include potency of the drug, pH, particle size distribution, chemically related substances, sterility, and presence of endotoxins. Tests are carried out on at least three temperature storage conditions (say 2–8, 25, and 37°C) for a minimum of three months. Test results are used to project shelf life of the product by using Arrhenius kinetics. A minimum of 18 months of shelf life is required for a product to be marketable.

If the above scale-up is successful, then it is time to attempt manufacture of full-scale batches. At this time, careful thought should be given to process validation (some organizations call it product characterization) in order that a reproducible, robust process may result. All process parameters that are expected to have a bearing on the product are listed. These include temperature, pressure, rpm of the mixer, operating pressure of the homogenizer, operating temperature in the homogenizer, concentrations of active and inactive ingredients, temperature distribution in the autoclave, ramp-up time (the time it takes for the temperature to reach set point), ramp-down time, etc. From the above variables, those that are expected to have the most effect on the product are selected. During full-scale runs, the process is operated at a combination of either the maxima or minima of the variables such that product batches at extremes of the most sensitive variables are obtained.

For a terminally heat-sterilized product, the sterilization process is validated with a view to deliver sufficient microbial lethality to the product while at the same time avoiding thermal drug degradation. *Bacillus stearothermophilus*, a heat-sensitive organism, is added to the product. The product, being an emulsion, needs a certain proportion of the bacillus in an encapsulated form in the emulsion (i.e., the bacillus must be in the emulsion droplets) in order to be sure that all possible pathogens are killed during heat sterilization. (*Bacillus stearothermophilus* is more resistant to heat when it is out of contact with water.) Experiments are conducted to determine D and Z values of the bacillus in the formulation (The *D*-value represents the time required for a 1 log reduction in the microbial population at the conditions of the experiment. The *Z*-value is the number of degrees of temperature change necessary to change the *D* value by a factor of 10. Based on this data, a heat-sterilization cycle is developed that is appropriate for the product.

A minimum of three stability batches are manufactured under conditions that are representative of full-scale manufacturing. These batches are made at the extremes of the more sensitive parameters, such as the highest temperature and pressure. These batches are placed on long-term stability at a minimum of three temperature and time conditions. The products are stored both upright and inverted—the inverted position for simulating extended contact with stoppers. The batches are monitored for drug potency, its chemically related substances, potency of the emulsifier and its related substances, particle size distribution, apparent pH, sterility, and presence of endotoxins. Usually the studies continue beyond the expected shelf life of the product.

REFERENCES

1. E. S. Bissell, H. C. Hesse, H. J. Everett, and J. H. Rushton, *Chem. & Mat. Eng.*, 53: 118 (1946).
2. J. H. Rushton, and J. Y. Oldshue, *Chem. Eng. Progr.*, 49: 161, 267 (1953).
3. J. H. Rushton, E. W. Costich, and H. J. Everett, *Chem. Eng. Progr.*, 46: 395, 467 (1950).
4. P. Becher, *Emulsions: Theory and Practice*, ACS Monograph Series, 1957.
5. P. Becker, *Encyclopedia of emulsion technology*, Marcel Dekker, New York, NY, 1988.
6. M. Cheryan, *Ultrafiltration*, Technomic Publishing Co., Lancaster, PA, 1986.
7. T. Meltzer, *Filtration in the Pharmaceutical Industry*, Marcel Dekker, New York, NY, 1987.
8. *U.S. Pharmacopeia XXIII*, U.S. Pharmacopeial convention, Rockville, MD, 1995.
9. R. E. Treybal, *Mass Transfer Operations*, McGraw Hill Publishing Co., New York, NY, 1980.
10. T. K. Sherwood, R. L. Pigford, and C. R. Wilke, *Mass Transfer*, McGraw Hill Publishing Co., New York, NY, 1977.
11. M. J. Groves, and R. Murty, *Aseptic Pharmaceutical Manufacturing*, Interpharm Press, Buffalo Grove, IL, 1995.
12. F. J. Carleton, and J. P. Agalloco, *Validation of Aseptic Pharmaceutical Processes*, Marcel Dekker, New York, NY, 1986.
13. M. Cherian, Heat treating liposomes, Patent, PCT Pub W090-03808.

11

Quality Assurance

Samir A. Hanna

Bristol-Myers Squibb Company, New Brunswick, New Jersey

I. INTRODUCTION

Total quality control is a concept that strives to produce a perfect product by a series of measures requiring organized effort by an entire company to prevent or eliminate errors at every stage in production. Though the responsibility for assuring product quality belongs principally to quality assurance, many other departments and disciplines within a company must also be involved. A team effort is required for an effective quality assurance program. The quality of a drug product must be built in during product and process design, as well as during the production and packaging of the product. The physical plant design, space, ventilation, cleanliness, sanitation, etc., during routine production of a product all play an integral role in the assurance of producing a quality product. The product and process design begins with, and includes, research and development, preformulation, and physical, chemical, therapeutic, and toxicologic considerations. Product quality, then, considers materials, in-process and product control (including specifications and tests for the actives and excipient of the product), specific inspection procedures for the product, containers, packaging, and labeling to ensure that the container closure system provides functional protection of the product against such factors as moisture, oxygen, light, volatility, and drug/package interaction. Provision is required for a cross-referencing system to allow for any batch of a product to be traced from its raw materials to its final destination in the event of the need to investigate unexpected difficulties.

The assurance of product quality depends on more than just proper sampling, adequate testing of various components, and finished dosage form. The prime responsibility of maintaining product quality during production should rest primarily with the manufacturing department. Removal of this responsibility from manufacturing can result in imperfect composition such as a missing ingredient, sub- or superpotent addition of ingredients, or mix-up of ingredients; mistakes in packaging or filling, such as product contamination, mislabeled or deficient packaging; and lack of conformance to product registration. A quality assurance team must establish control or checkpoints to monitor the quality of the product as it is processed and upon completion of manufacture. These begin with raw material and component testing and include in-process packaging and labeling, finished product testing, and batch auditing and stability monitoring.

A quality assurance system is diagrammed in Table 1. This system can vary in details but not in principle from company to company and depends on the nature and size of the manufacturing facility and on the types of disperse system dosage form produced.

The increasing complexity of pharmaceutical disperse systems' manufacturing, resulting from unique and different kinds of drugs and dosage forms, has introduced various ethical, legal, and economic responsibilities on those concerned with their manufacture.

II. GOOD MANUFACTURING PRACTICES REQUIREMENTS

Most governments promulgate regulations governing the manufacture, processing, packaging, and distribution of finished pharmaceuticals. International standards have been

Table 1 Quality Assurance System

	Purchase of manufacture raw material to specifications ↓	
Analytical lab\half arrows	Storage of raw material in quarantine area ↓	
	Transfer of raw material to released area ↓	
Calculate and weigh		
	raw material quantities needed for manufacturing ↓	
	Check working formula procedures & batch file ↓	
Analytical lab ⇌	In-process ↓	Checking of equipment ⇌ and ingredients
Analytical lab ⇌	Complete processing ↓	Working formula ⇌ procedure audit
	Packaging ↓	Checking of equipment ⇌ finished product, labels, containers
Analytical lab ⇌	Finished product in quarantine area ↓	⇌ Batch audit
Analytical lab for stability ← monitoring	Storage in released finished product area	→ Marketing

published by the World Health Organization, but each country prefers to promulgate regulations that fit its own needs. Examples of such regulations in the United States are found in Parts 210 and 211, Title 21, of the Code of Federal Regulations, *Current Good Manufacturing Practices in Manufacturing, Processing, Packaging, or Holding Drugs* (CGMPs). These regulations were originally published in the *Federal Register* of June 20, 1963, by the Food and Drug Administration (FDA); and over the past 12 years these regulations have been revised several times. On February 13, 1976, the FDA published In the *Federal Register* proposals for revising drug CGMPs to update them in light of current technology and to adopt more specific requirements to better assure the quality of finished products. The final regulation was published in the *Federal Register* on March 28, 1979. The current CGMPs are enforced by the FDA, and it is on the basis of these regulations that the FDA has insisted on the proper manufacturing of drugs. The regulations extend into the area of finished pharmaceuticals, buildings, equipment, personnel, components, master production and control records, batch production, production and control procedures, product containers and their components, laboratory controls, distribution records, stability, expiration dating, and complaint files.

In January 1989, the Commission of the European Communities (EEC) issued the final regulation, the *EEC Guide to Good Manufacturing Practice for Medical Products*. This regulation incorporates a chapter and is similar to the FDA current GMPs, covering quality management, personnel, premises and equipment, documentation, production, quality control, contract manufacture and analysis, complaints and product recall, and self-inspection. Japan issued its "Provisions of the Pharmaceutical Affairs Law Concerning GMP" in January 1990. Japan's International Drug GMP's contains seven sections, covering provisions purpose, regulations for manufacturing control and quality control drugs, quality control, regulations for buildings and facilities for drug manufacturing plant, manufacturing control and manufacturing hygiene control for sterile preparation, etc., and drug manufacture's self-inspection manual as to conformity with GMP requirements.

Regulations for good manufacturing practices during disperse system manufacturing are aimed at assuring that only drug products that have met the established specifications and are packaged and labeled under proper controls are distributed.

III. COMPENDIAL REQUIREMENT

Disperse systems in pharmaceutical dosage forms are, primarily, emulsions and suspensions. Since there is such a wide diversity in the formulation of finished dosage forms that might compromise disperse systems, it is impossible to adequately describe the testing of each of these products. The current editions of the USP XXIII and NF XVIII contain individual monographs that detail specifications and testing methods for each of the included products. Specifications published in the official *Compendia* are designed to assure a pharmaceutically elegant and therapeutically effective dosage form. It is a compendial requirement that any sampling within a specific batch of product would reveal the product to be in compliance with respect to the individual monograph specifications. An emulsion is a two-phase system in which one liquid is dispersed in the form of small droplets throughout another liquid. It is defined as oil-in-water (O/W) if the continuous phase is aqueous and water-in-oil (W/O) if, conversely, the continuous phase is oleaginous material. Emulsions are usually stabilized by emulsifying agents of diverse types. Emulsifying agents commonly used are sodium lauryl sulfate, polysorbate 80, benzalkonium chloride, gum as acacia, carboxymethylcellulose sodium, carboxypolymethylene, and sorbitan monostearate. The consistency of emulsions varies widely ranging from liquids to semisolids. All emulsions require an antimicrobial agent with fungistatic and bacteriostatic properties. Effectiveness of the preservative system should always be tested in the final product. Preservatives commonly used are methyl-, ethyl-, propyl-, and butyl-parabens, benzoic acid, quaternary ammonium compounds, and alcohol. Emulsion properties depend on the physical and chemical characteristics and phase-volume ratio of the oil and water phases, the concentration of the emulsifying agents, the order of addition of ingredients, the temperature of emulsification, the type of mechanical emulsifier used, and the method and rate of cooling. Emulsion systems are described in the pharmacopeia for the following pharmaceutical forms: parenteral nutrition products, artificial blood substitutes, vehicles for vaccines, oral emulsions, topical creams, and ointments.

A suspension is a preparation of finely divided, undissolved drug(s) dispersed in a liquid phase. Powders for suspension are preparations of finely powdered drug(s) intended for suspension in liquid vehicles. Many of the antibiotics are administered in the form of suspensions either by oral or parenteral route. Antibiotics for oral suspension are finely divided antibiotic powders mixed with suspending and dispersing agents. They are generally intended to be reconstituted with the prescribed volume of purified water and mixed before dispensing. Antibiotic suspensions for parenteral use are sterile preparations. Their preparations may be for immediate use or may require mixing before use with a prescribed amount of water for injection or other suitable sterile vehicle. Suspensions should never be injected intravenously or intrathecally. Suspending agents commonly used include those used in emulsifying agents in addition to many natural and synthetic gums (e.g., pectin, alginates and xanthan gums, or synthetic cellulose derivatives).

IV. RAW MATERIALS SPECIFICATIONS

Good raw material specifications must be written in precise terminology and be complete, providing specific details of test methods, test instruments, and the manner of sampling, and materials must also be properly identified. Table 2 lists general tests, limits, and other physical or chemical data for raw materials related to identity, purity, strength, and manner of quality assurance.

In the development of raw material specifications, the analytic research and development chemists should strive for the following:

- To ascertain which chemical, physical, and biological characteristics are critical for assuring reproducibility from lot to lot of raw materials to be used in evaluating each lot of raw material produced or purchased
- To establish the test methods and acceptable tolerance for the attributes to be evaluated
- To establish the supplier's ability to supply raw materials of consistent quality.

The current FDA good manufacturing practices covering raw material handling procedures are found in the Code of Federal Regulations, Title 21, Section 211.42. It states simply that "components" be received, sampled, tested, and stored in a reasonable way, that rejected material be disposed of, that samples of tested components be retained, and that appropriate records of these steps be maintained. In practice, the

Table 2 Raw Material Quality Assurance Monograph

A. (Raw material name)
   1. Structural formula, molecular weight
   2. Chemical name(s)
   3. Item number
   4. Date of issue
   5. Date of superseded, if any, or new
   6. Signature of writer
   7. Signature of approval
B. Samples
   1. Safety requirement
   2. Sample plan and procedure
   3. Sample size and sample container to be used
   4. Reservation sample required
C. Retest program
   1. Retesting schedule
   2. Reanalysis to be performed to assure identity, strength, quality, and purity
D. Specifications (where applicable)
   1. Description
   2. Solubility
   3. Identity
      a. Specific chemical tests; organic nitrogenous basis; acid moiety or inorganic salt tests; sulfate, chloride, phosphate, sodium, and potassium; spot organic and inorganic chemical tests
      b. Infrared absorption
      c. Ultraviolet absorption
      d. Melting range
      e. Congealing point
      f. Boiling point or range
      g. Thin-layer, paper, liquid, or gas chromatography
   4. Purity and quality
      a. General completeness of solutions, pH, specific rotation, nonvolatile residue, ash, acid-insoluble ash, residue on ignition, loss on drying, water content, heavy metals, arsenic, lead, mercury, selenium, sulfate, chloride, carbonates, acid value, iodine value, saponification value
      b. Special quality tests, particle size, crystallinity characteristics, and polymorphic forms
      c. Special purity tests, ferric in ferrous salts, peroxides and aldehydes in ether, and related degradation products
   5. Assay, calculated either on anhydrous or hydrous basis
   6. Microbial limits, especially for raw materials from natural sources
E. Test procedures
   1. Compendial, USP, or NF references
   2. Noncompendial, detailed analytical procedure, weights; dilutions; extractions; normality; reagents; instrumentation used and procedure, if any; calculations
F. Approved suppliers
   1. List of prime suppliers and other approved alternative suppliers, if any

manufacturer physically inspects and assigns lot numbers for all raw materials received and quarantines them until they are approved for use. Each raw material is sampled according to standard sampling procedures. Samples of raw materials are to be collected in containers, using a disinfected sampling "thief" or scoop, observing aseptic technique for microbiological analysis or clean container, and clean technique for the analytical laboratory. The number of containers to sample in a given lot can be determined by using MIL-STD-105D, as shown in Table 3.

Samples are to be labeled as to lot number, receiving number, supplier, container size and type, name of raw material, and date of receipt. Samples are then submitted to quality assurance analytical and microbiological laboratories for testing according to written prescribed procedures. If acceptable, the raw material is moved to the release storage area with a properly applied sticker to indicate the item number, name of ma-

Table 3 Number of Containers of Raw Materials to Be Sampled per Lot

	Containers	Number of samples	
Inactive raw materials	1	All	
	2–8	2	
	9–15	3	
	16–90	5	
	91–150	8	
	151–280	13	
	281–500	20	
	501–1200	32	
	1201–3200	50	
Active raw materials	1–5	All	
	6–10	6	
	11–18	7	
	19–28	8	
	29–100	9	
	>101	10	
	Cases, rolls, or boxes (no. per lot)	Cases, rolls, or boxes (no. of samples)[a]	Units sample number
Packaging components	1–8	2	—
	9–15	3	—
	16–90	5	—
	91–150	8	—
	151–280	13	—
	281–500	20	—
	501–1,200	32	—
	1,201–3,200	50	—
	3,201–10,000	80	—
	10,001–35,000	125	315
	35,001–150,000	200	500
	105,001–500,000	315	1250
	>500,001	500	1250

[a]Across pallet.

terial, lot number, date of release, reassay date, and signature of the quality assurance inspector. At time of use, after prolonged storage, the raw material is retested as necessary according to an established schedule to assure that it still conforms to specifications. The quality assurance department should preserve samples of the active and inactive raw materials that consist of at least twice the necessary quality to perform all tests required to determine whether the material meets the established specifications. These samples should be retained for at least seven years. Approved components shall be rotated in such a manner that the oldest stock is used first. Any raw material not meeting specifications must be isolated from the acceptable materials and have a reject stick applied and be returned to the supplier, or disposed of promptly. To verify the supplier's conformance to specifications, further supporting assurance by means of on-site periodic inspections are pertinent to the total quality of raw materials. Inspecting the supplier's premises ensures that cross-contamination does not take place due to improperly cleaned equipment or poor housekeeping practices, since contaminants may go undetected in sample testing, since specifications are not designed to control the presence of unrelated materials.

In general, raw materials may be classified into two broad groups: those that are active or therapeutic, and those that are inactive or inert.

A. Active Ingredients

The current editions of the USP XXIII and NF XVIII contain monographs on most therapeutically active materials used in pharmaceutical dosage forms. Since there is such a wide variance in the nature of the active ingredients used in dosage forms using disperse systems in manufacturing, it is impossible to summarize briefly the testing of those raw materials. Compendial monographs state allowable purity, yet manufacturing may require still purer material(s). One of the most important decisions to be made in raw material control is the degree of purity that will be maintained for each material. It is not uncommon to find an appreciable variation in the degree of purity between samples of the same raw material purchased from different commercial sources. The selection, then, must be one that results in the highest purity practical for each raw material and is consistent with safety and efficacy of the final dosage form. A typical raw material currently existing in a compendium has a purity requirement generally of not less than 97%. Its specifications normally consist of a description; a solubility, identification; information on melting range, loss on drying, residue on ignition, special metal testing, specific impurities that are pertinent to the method of synthesis of each individual raw material; and an assay. The methods of assay are usually chemical in nature. However, it should be indicated that these compendial tests are intended as the minimum required from the legal point of view. For certain products, it may be necessary to obtain an active ingredient with special specification far tighter than those of the comparable compendial standard. Raw materials cannot be adequately evaluated and controlled without special instrumentation, such as spectrophotometry; infrared spectrophotometry, potentiometric titrimetry; column, gas, paper, tin-layer and high-pressure liquid chromatography; polarography; x-ray diffraction; x-ray fluorescence; spectrophotofluorimetry; calorimetry; and radioactive tracer techniques. No less demanding are the tests required for microbiological assay, pharmacological assay, and safety testing. For certain products, even when high-purified and well-characterized raw materials are involved, specifications should include additional critical features such as crystalline versus amorphous forms. Any of these characteristics could have an effect on the safety or effectiveness of the final dosage form. It is a GMP requirement that all raw materials, active or inactive, be assigned a reassay date, meaningful or indicative, that would assure purity and potency. Tests are performed at reassay times to confirm suitability of each raw material.

B. Inactive Ingredients

1. Emulsifying and Suspending Agents

There is a wide variety of emulsifying and suspending agents available to the pharmaceutical scientist. Emulsifiers or suspending agents can be anionic (sodium lauryl sulfate), nonionic (polysorbate 80), or cationic (benzalkonium chloride). The origin can be natural (xanthan gum), semisynthetic (carboxymethylcellulose sodium), or synthetic (polyvinyl alcohol). The ideal emulsifying or suspending agent should be of low toxicity, nonirritating, tasteless, chemically stable, microbiologically clean, and compatible with all other excipient and drugs. Emulsifying and suspending agents are used to prevent coalescence or sedimentation; therefore, testing for this specific function is important. A wide variety of polymers, including polyoxyethylene derivatives of sorbitan stearate, oleate, laurate or palmitate, and lanolin and beeswax play a significant role in formulating disperse system products. A suitable test or combination of tests for such parameters as viscosity, saponification value, iodine value, hydroxyl value, acid value, water absorption, gel formation, specific gravity, and swelling power will ascertain such properties. Other physical and chemical tests of emulsifying and suspending agents are tests for foreign matter, heavy metals, residue on ignition, water content, and polymers' physical characteristics such as odor, color, and taste. For natural compounds, total microbial count and test for *Salmonella* species must be performed.

2. Antioxidants

Emulsions commonly contain antioxidants to prevent rancidity due to oxidative degradation. Such reactions are mediated either by free radicals or by molecular oxygen, and often involve the addition of oxygen or the removal of hydrogen.

Antioxidants are added to emulsions alone or in combination, and function either, preferably, by being oxidized and thereby gradually consumed, or by blocking the oxidative chain reaction. The most commonly used antioxidants are ascorbic acid, butylated hydroxyanisole (BHA), butylated hydroxytoluene (BHT), cysteine, sodium bisulfite, sodium metabisulfite, sodium formaldehyde sulfoxylate, and tocopherols. A typical analysis of an antioxidant should include identity tests, tests of solubility, heavy metals, arsenic, residue on ignition, specific impurity, and an assay.

3. Preservatives

Emulsions usually require a preservative system, since the aqueous phase is susceptible to the growth of microorganisms. A suitable preservative system should be included in suspension preparations intended for multiple use to protect them from bacterial and mold contamination.

Consideration must be given to the stability and effectiveness of the preservative system in combination with active, inactive, and container-closure systems of the disperse system. A combination of, or a single preservative in suitable concentration, depending on the disperse system, can be used. Preservatives used are methyl-, propyl-, ethyl-, and butylparabens, benzalkonium chloride, benzyl alcohol, phenol, phenyl-mercuric nitrate, thimerosol, and chlorbutanol.

Preservatives usually are tested for identity, water insolubles, heavy metals, arsenic, residue on ignition, melting range, water content, and special impurities and assayed.

4. *Coloring Agents*

Approved certified water-soluble food, drug, and cosmetics dyes, or mixtures thereof, or their corresponding likes, may be used to color an oral disperse system. Color serves mainly as a means of identification. The FDA determines and approves colorants for use in foods and drugs with recommendation of limits, if any. Table 4 lists selected colors and FDA restrictions on their use. The FDA also certifies and releases colors batch by batch for human use. A typical analysis of a color may include tests for identity, total volatile matter, heavy metals, water-insoluble matter, synthesis impurities, arsenic, lead, and total color. An FD&C color-lake analysis may include additional tests for chloride, sulfate, and inorganic matter.

5. *Flavoring Agents*

If a flavor is desired, flavors or volatile oils may be added, directly or as an alcoholic solution to the bulk, or dry flavors may be used by blending with other formula constituents. If dry flavors are used, a tight limit for water content assures the quality of the dry flavors. For the same reason, dry flavors should be stored in tightly closed containers away from excessive heat and retested for water content very six months.

Flavors are usually tested for refractive index, specific gravity, solubility, and alcohol content, if any. A gas liquid chromatogram can be used as a "fingerprint" for each specific flavor and helps in assuring the supplier's continuous compliance to specifications. A knowledge of any synthetic FD&C dyes in the flavor formula is important for the formulator to keep up with FDA colorants regulations.

6. *Water*

Water is an ingredient in the vast majority of disperse system products. Water is used for initial washing and rinsing, and as a vehicle. An adequate supply of water must be assured, one that will meet all criteria of quality for different needs in production, from the feed water to the final step. A good water-system design must consider equipment stability, material selection, operational controls, component compatibility, construction practices, cleaning procedures, sanitary methods, sampling procedures, preventive maintenance, and compliance with control specifications. The quality aspects of a water system are affected by the quality of the raw or potable water, any prior processing, and the distribution system. That microorganisms can exist in water means that the production of water poses special problems of preparation, storage, and distribution.

The microbial and chemical quality of water is of great importance in disperse system products. Most raw or potable water used in pharmaceutical processes contains a wide variety of contaminating electrolytes, organic substances, gross particle matter, and dissolved gases such as carbon dioxide and microorganisms. Bacteria indigenous to fresh raw water include *Pseudomonas* sp., *Alcaligenes* sp., *Flavobacter* sp., *Chromobacter* sp., and *Serratia* sp. Bacteria that are introduced by soil erosion, rain, and decaying plant matter include *Bacillus subtilis*, *B. magaterium*, Klebsiella aerogenes,

Table 4 Coloring Agents

	Colorant	Restriction on use
FD&C	Blue #1	Permanent listing for use in foods, drugs, and cosmetics
	Blue #2	Permanent listing for use in foods, drugs, and cosmetics
	Green #2	Provisional listing for use in foods, drugs, and cosmetics
	Red #3	Provisional listing for use in foods, drugs, and cosmetics
	Rcd #40	Provisional listing for use in foods, drugs, and cosmetics
	Yellow #5	Provisional listing for use in foods rugs, and cosmetics
	Yellow #6	Provisional listing for use in foods, drugs, and cosmetics
D&C	Blue #6	Provisional listing for use in drugs and cosmetics with requirement for label declaration
	Green #5	Provisional listing for use in drugs and cosmetics
	Green #6	Provisional listing for use in drugs and cosmetics
	Orange #5	Provisional listing for use in drugs and cosmetics
	Orange #10	Provisional listing for use in drugs and cosmetics
	Orange #17	Provisional listing for use in drugs and cosmetics
	Orange #17	Permanent listing for use in drugs and cosmetics
	Red #6	Provisional listing for use in drugs and cosmetics
	Red #7	Provisional listing for use in drugs and cosmetics
	Red #21	Provisional listing for use in drugs and cosmetics
	Red #22	Provisional listing for use in drugs and cosmetics
	Red #27	Provisional listing for use in drugs and cosmetics
D&C	Red #28	Provisional listing for use in drugs and cosmetics
	Red #30	Provisional listing for use in drugs and cosmetics
	Red #8	Provisional listing for use in drugs and cosmetics with restriction on NMT 0.75 mg to be ingested on a daily basis
	Red #12	Provisional listing for use in drugs and cosmetics with restriction on NMT 0.75 mg to be ingested on a daily basis
	Red #19	Provisional listing for use in drugs and cosmetics with restriction on NMT 0.75 mg to be ingested on a daily basis
	Red #33	Provisional listing for use in drugs and cosmetics with restriction on NMT 0.75 mg to be ingested on a daily basis
	Red #36	Provisional listing for use in drugs and cosmetics with restriction on NMT 0.75 mg to be ingested on a daily basis
	Yellow #10	Provisional listing for use in drugs and cosmetics
	Annatto	Permanent listing for use in foods, drugs, and cosmetics
Lakes of the above		
	Carotene	Permanent listing for use in foods, drugs, and cosmetics
	Caramel	Permanent listing for use in foods, drugs, and cosmetics and provisional listing for use in cosmetics
	Carmine	Permanent listing for use in foods, drugs, and cosmetics
	Titanium dioxide	Permanent listing for use in drugs and cosmetics

Source: *Federal Register*, Food and Drug Administration, Washington, D.C.

and Enterobacter cloacae. Bacteria that are introduced by sewage contamination include *Proteus* sp., *Escherichia coli* and other enterobacteria, *Steptococcus faecalis*, and *Clostridium* sp. Stored-water bacteria contamination include mainly gram-negative bacteria and other microorganisms, such as *Micrococcus* sp., *Cytophaga* sp., yeast, fungi, and *Actinomycetes*. The reliance on a sampling program as a means of monitoring the quality of the water is practical only if the sample is truly representative of the water quality. Sampling points, frequency of sampling, and type of testing should be considered from the standpoint of the water system. Type and size of water treatment and pretreatment equipment and its operational characteristics have a direct effect on the chemical and microbial quality of water. Bacteria may gain access to a water distribution system at any outlet, such as a tap or sampling point, especially those fitted with a hose, if they are not regularly disconnected and disinfected. Microbial infection and chemical contamination may build up in any unused section of pipeline "dead legs," booster pumps, and water meters. A standard water-sampling procedure for total microbial count is shown in Table 5. Similar procedures without strict microbiological cleanliness are followed for water chemical testing. A suggested water-sampling program is shown in Table 6. Purified water produced from an approved drinking water supply is usually used for formulation of oral or topical disperse systems.

a. Drinking Water. The quality of water from the main supply varies with the source, the type of treatment it is subjected to, and the local authority. Essentially, it should be free from known pathogens and from fecal contamination, such as *E. coli*, but it may contain other microorganisms and must meet certain chemical purity specification. When the supply is derived from surface waters, microorganisms are usually greater in number and faster growing those from deepwater sources, such as a well or spring. Due to the variabilities of source, temperature, season, organic level, and complexity of distribution systems, the bactericidal effect of the initial chlorine addition can be decreased and, if used, leads to less chemical contamination. Drinking water frequently contains significant levels of microorganisms and a variety of chemical impurities. Chemical and microbiological testing of drinking water usually includes pH, free chlorine, chloride, sulfate, ammonia, calcium and magnesium, carbon dioxide, heavy metals, oxidizable substances, total solids, and bateriological purity for total microbial count and *E. coli*. United States Public Health Services regulations describe the testing procedures and limits for each locality.

Table 5 Water-Sampling Procedure

1. Prepare sufficient number of 120 mL Pyrex sample containers according to the sampling program and cap loosely.
2. For sample containers to be used for drinking-water samples, add 0.1 mL of a 10% sodium thiosulfate solution to deactivate any residual chlorine.
3. Autoclave all sample containers at 121°C for 15 min.
4. Open sampling points such as tap fixtures, and allow water to run for not less than 500–1000 mL.
5. Hold the sample container by the base and remove the screw cap, taking precautions not to touch the lip edge of the sample container.
6. Collect not less than 100 mL of the water sample and immediately secure the cap to the sample container.
7. Label the sample container, source of water, sample point location and number, type of water, and time and date sampled.
8. Transfer the sample within 1 hr of sampling to the microbiology laboratory.
9. Refrigerate the sample in the microbiology laboratory until testing with 24 hr of sampling.
10. Follow USP total aerobic microbial count test.

Table 6 Water-Sampling Program

Location sample point	Test	Frequency
Raw water (potable water)	Microbial	Daily
	Chlorine residual	Daily
	Conductivity	Continuous
	Chemical, USP	Weekly
	pH	Daily
Carbon filter	Microbial	Daily
	Chlorine residual	Weekly
Deionized water equipment	Conductivity	Continuous
	Total solids, USP	Daily
	pH	Daily
	Microbial	Daily
	Pyrogen	Weekly
	Chemical, USP	Weekly
	Resin analysis	Six months
Reverse osmosis equipment	Microbial	Daily
	pH	Continuous
	Chlorine residual	Continuous
	Pyrogen	Daily
	Conductivity	Continuous
	Chemical, USP	Daily
	Feedwater hardness	Daily
Distillation equipment	Microbial	Daily
	pH	Daily
	Pyrogen	Daily
	Conductivity	Continuous inlet and outlet
Location sample point	Test	Frequency
	Chemical, USP	Daily
	Particulates	Weekly
Storage	Microbial	Daily
	pH	Daily
	Pyrogen	Daily
	Chemical, USP	Daily
Distribution use points	Microbial	Weekly
	Pyrogen	Weekly
	Conductivity	Weekly
	Chemical, USP	Weekly
	Particulates	Monthly
	pH	Weekly
Clean steam generator	Chemical, USP	Weekly

b. Purified Water. Purified water is usually produced by passing the water through anion- and cation-exchange resin beds or by reverse osmosis. It should be prepared from drinking water, complying with the limits and requirements of the United States Public Health Services.

Ion-exchange treatment removes dissolved ionic impurities. Deionization does nothing to improve the microbiological quality of the water. Ion-exchange beds that are not frequently regenerated with strong acid and alkali contribute significantly to bacteriological contamination, leading often to pyrogenic problems. Ion-exchange equipment should ensure frequent regeneration independent of the chemical quality-regeneration requirements, which usually take longer, thus encouraging the growth of a bacteria and possible pyrogenic problems. Intermittent and low-flow conditions can be minimized by installing a recirculation cycle on the ion-exchange system. The flow rate of this system should approach the rated service flow of the ion-exchange system.

Reverse osmosis treatment removes a large portion of the dissolved minerals, particulates, bacteria, viruses, and pyrogens. However, procedures must be written carefully to ensure that the reverse osmosis system is properly monitored, maintained, and sanitized on a regular basis, as it has been shown that bacterial contamination can occur.

Chemical and microbiological testing of purified water includes determination of pH, chloride, sulfate, ammonia, calcium, carbon dioxide gas, heavy metals, oxidizable substances, total solids, and bacteriological purity for total microbial count and *E. coli*. Testing procedures and limits are shown in the USP.

c. Water for Injection. Water for injection, used for antibiotic sterile suspensions, is intended to conform not only to a high degree of chemical purity, but also to be free from pyrogenic substances. Water for injection is prepared by distillation or reverse osmosis. Distillation is the most widely used and acceptable method of producing sterile pyrogen-free water. As the water leaves the still, it is free of microorganisms, but contamination may occur as a result of a fault in the cooling system, heat exchanger design, vent filter installation, storage vessel, or the distribution system. The bacterial contaminants of distilled water are usually gram-negative bacteria. The heating and storing of water for injection at 80°C prevents bacterial growth and the production of pyrogenic substances that accompany such growth. Certain drug components cannot be formulated at this temperature, and water has to be cooled before use, which may lead to microbial growth. In these cases, it is better to plan the production schedule so that storage for more than a few hours at room temperature is avoided, especially if the products cannot be terminally sterilized. Chemical and microbiological testing of water for injection include pH, chloride, sulfates, ammonia, calcium, carbon dioxide, heavy metals, oxidizable substances, total solids, and pyrogen. Testing procedures and limits are shown in the USP.

C. Packaging Components

The microbial flora of packaging components is affected by its composition, transportation exposure, and storage conditions.

Glass containers, particularly those transported in cardboard boxes, often contain mold spores of *Penicillin* sp. and *Aspergillis* sp., and bacteria, such as *Bacillus* sp. and *Micrococcus* sp. Other packaging and closure system components, such as aluminum, Teflon, metal foils, and other polymeric materials, all of which usually have a smooth impervious surface free from crevices or interstices, are usually free from microbial contamination.

Disperse system products are packaged and sealed in a variety of containers with different closure systems that comprise a wide range of chemical compounds. These include glass, various polymeric materials, aluminum, and assorted closure systems.

Most disperse system products today are in liquid form, although a number of products must be packaged as powders until administered, at which time they are reconstituted into the proper liquid form.

Disperse systems containers intended to provide protection from light must meet the requirements for the USP light-transmission test (Table 7). The light-resistant amber color of containers results from an interaction between iron and sulfur for greenish amber or iron and titanium for brownish amber.

1. *Glass Containers*

Glass containers are still in use for disperse system because of their chemical resistivity. Glass (particularly types III and NP), is commonly used because it resists the corroding action of water, acids, bases, and salts to varying degrees. Dry materials do not react chemically with glass. However, glass can be chemically active under certain conditions; for example, the formation of flakes in neutral saline solutions.

Table 8 lists the three glass types commonly used for disperse systems. An incoming shipment of glass containers should be checked by the quality assurance department to assure that it meets the appropriate preestablished tests, as shown in Table 9.

2. *Polymeric Containers*

During the past decade, more and more polymeric containers have been used. Advantages claimed for the use of synthetic plastic containers include reduction of particulate material, elimination or reduction of the possibility of airborne contamination, reduction in breakage, economy in space during transportation and storage, simplified disposal, and a reduction in weight and noise that simplifies handling. However, polymeric containers are not necessarily totally inert and can present a number of problems to the development pharmacist. Problems that have occurred or may occur include permeation, leaching, sorption, chemical reaction, and instability of polymeric material used. Loss of drug potency and antimicrobial activity due to sorption have been reported. Polymeric materials generally used are chemically related to polyolefin, vinyl resins, or polystyrene. Table 10 lists some of the characteristics of these plastics. Regardless of end use or fabrication method of the polymer, additives must be compounded or dry-blended into the base resin. These additives can be classified as stabilizers, plasticizers, lubricants, colorants, filters, impact modifiers, and processing aids. Not all polymers contain all these types of additives. Polyvinyl chlorides, polypropylenes, and polyethylenes possess good thermal stability and other desirable processing and packaging properties that make them the most commonly used polymers.

Table 7 Glass and Plastic Light-Transmission Limits for Containers

	Maximum percent of light transmission[a]	
Size	Flame sealed	Closure sealed
1	50	25
2	45	20
5	40	15
10	35	13
20	30	12
50 and more	15	10

[a]At any wavelength between 290 and 450 nm (USP).

Table 8 Identification of Glass Types

Type	Description	Major chemical composition		USP test	Limit size (mL)	0.02 N acid (mL)
		Component	Average %			
I	Borosilicate glass, highly resistant	SiO_2	80	Powdered glass	All	1.0
		Al_2O_3	5	—		
		Na2O	—	7		
		K_2O	—	0.5		
		B_2O_3	—	12		
	CaO	—	1			
III	Soda-lime glass, somewhat average, chemical resistance	SiO_2	75	Powdered glass	All	8.5
		Al_2O_3	2	—		
		Na_2O	15	—		
		K_2O	0.5	—		
		B_2O_3	3	—		
		CaO	12	—		
NP	Soda-lime glass	SiO3	73	Powdered glass	All	15.0
		Al_2O_3	11	—		
		Na_2O	17	—		
		K_2O	0.5	—		
		GO	5	—		
		MgO	4	—		

Table 9 Glass Bottle, 100 mL

Item number: ________	Date of issue: ________	Superseded: ________
Written by: ________	Approved by: ________	
Sampling plan: (see Table 1)	Preservation sample (none)	Retest program (2 years)
Description:	100 mL flint clear glass bottle	
Physical measurements	cm	
Height	60–62	
Width, outer diameter	21.0–23.0	
Lip, height	4–5	
Neck, inside diameter	10.5–11.5	
Neck, outer diameter	15–16	
Seal test		
Visual properties		
Spikes or "bird-swings"		
Learners out of plumb		
Height inconsistency		
Free of blemishes or scratches		
Chemical resistance:	USP glass power test	

USP physico-chemical tests are designed to measure some of the physical and chemical properties of plastic containers using an extract in purified water. These tests include determination of nonvolatile residue, residue on ignition, heavy metals, and buffering capacity. Other physical and chemical test techniques used to identify and characterize polymers include infrared by attenuated total reflectance, ultraviolet spectrophotometry, nuclear magnetic resonance, differential scanning calorimetry, and thermogravimetric methods. For molecular weight or molecular-weight distribution, melt viscosity, gel permeation, or exclusion chromatography is recommended.

3. *Aluminum Containers*

Aluminum containers normally are made of high purity grade aluminum 3003 alloy which has 1.0–1.5% manganese content. Impact-extruded collapsible aluminum tubes are used extensively for packaging a wide range of pharmaceutical and cosmetic disperse system formulas, e.g., creams, pastes, ointments, and semiliquids. They are convenient for patient use and, as the contents are expelled by squeezing the tube, there is no tendency for the walls to recover their original shape when the pressure is released. Consequently, the risk of air entering the tube and reacting with the product is minimized.

As aluminum tubes are not inert and subject to attack, especially from acidic products, internal coating may be necessary to prevent possible chemical reactions. A wide range of internal coating systems, which are sprayed into the tube immediately after forming it, are used, including vinyl, epoxy and phenolic resins, and *n*-microcrystalline wax.

Aluminum has a number of excellent properties that account for its use as a packaging container. Aluminum containers reflect 90–95% of infrared radiation; resist rust corrosion; are impermeable to moisture vapor; possess gas, oil, and heat resistance; and are lightweight by comparison with other containers.

Table 10 Characteristics of Some Polymers Used in Disperse System Containers

Polymer	Clarity	Permeability				Effect of laboratory reagents			
		O_2	N_2	CO_2	H_2O	Weak acid	Strong acid	Weak alkali	Strong alkali
Polyethylene	Opaque	High	Low	High	Low	Resistant	OAA[a]	Resistant	Resistant
Polypropylene	Translucent	High	Low	High	Low	Resistant	OAA	—	Resistant
Polyvinyl chloride	Clear	High	—	High	High	—	—	—	—
Polystyrene	Clear	High	Low	High	High	—	OAA	—	—

[a]Oxidizing acid attack.

V. FINISHED PRODUCT

A. Compounding

A working formula card and procedure should be prepared for each batch size. To attempt expansion or reduction of a batch size of a disperse system product by manual calculations at the time of production cannot be considered good practice. Quality assurance must review and check the working formula card and procedures for each production batch before, during, and after production operation for

- Signature and dating when issued by a responsible production person
- Proper identification by name and dosage form, item number, lot number, effective date of document, reference to a superseded version (if any), amount, lot, code number, and release date of each of raw material used
- Calculations of both active and inactive materials, especially if there were any corrections for 100% potencies for actives used
- Reassay dates of components used
- Starting and finishing times of each operation
- Equipment to be used, record of its cleanliness, and specifications of its setup
- Initialing of each step by two of the operators involved
- Proper labeling of release components and equipment indicating product name, strength, size, lot number, and item number

Only released and properly labeled raw materials are allowed in the manufacturing area. Quality assurance should check and verify that the temperature, humidity, microbial monitoring, airborne particulates, and pressure differential in the manufacturing area are within the specified limits. Tables 11–15 show protocols for quality assurance testing for air velocity measurement, high efficiency particulate air (HEPA) filter leak testing, temperature and humidity control tests, pressure differential measurement, and particle count tests, respectively.

Particle counts are needed to verify air cleanliness, and probes should be temporarily or permanently located at the critical working levels, like filling needles, and readings should be taken during normal working activity. Normal accepted maximum levels are 100 and 100,000 particles of 0.5 μm or larger per cubic foot of air for fill-

Table 11 Air Velocity Measurement

Written by _______ Approved by _______ Superseded _______ Date of issue _______

Procedure

1. Adjust fans, dampers, registers, and any other control devices to deliver the air quantities specified for each and every component of the air system.
2. Using a calibrated anemometer, measure the ft/min velocity during the steady portion of the cycle for
   Supply air in ft^3/min (ft/min × area ft^2)
   Static pressure
   Fresh air cfm
   Return velocity
   Discharge air at every supply outlet
3. A limit of ± 10% of specification is usually accepted.

Table 12 HEPA Filter Leak Test

Written by ______ Approved by ______ Superseded ______ Date of issue ______

Procedure

1. Use polydisperse dioctylphthalate (DOP) aerosol generated by blowing air through liquid DOP at room temperature into the airstream ahead of the HEPA filter to produce a uniform concentration.
2. Measure the upstream concentration immediately upstream of the HEPA filter.
3. Scan the filter by holding the light-scattering photometer probe not more than 1 in. from the filter face and passing the probe in slightly overlapping strokes across the filter face so that the entire face of the filter is sampled. Then, make separate passes around the entire periphery of the filter and along the band between filter media and frame and other joints in the installation.
4. Using a light-scattering photometer with a threshold sensitivity of not less than 10^{-3} μg/L for 0.3 μm diameter DOP particles and capable of measuring concentrations in the range of 80–120 μg/L, and an air ample flow rate of 1 ft^3 ± 10% per minute, the approximate light-scattering mean droplet size distribution of the aerosol is
   99% smaller than 3 μm
   95% smaller than 1.5 μm
   92% smaller than 1 μm
   50% smaller than 0.72 μm
   25% smaller than 0.45 μm
   11% smaller than 0.36 μm
5. A linear readout light-scattering photometer reading more than 0.01% is considered unacceptable.
6. A HEPA filter should have an efficiency of 99.97% to particules 0.3 μm and larger. Most of the known visible contaminants are above those limits.

Table 13 Temperature and Humidity Measurement

Written by ______ Approved by ______ Superseded ______ Date of issue ______

Procedure

1. The air conditioning system shall be in continuous operation for a period of at least 24 hr.
2. During the test, all the lights and equipment in the test area should be turned on and operational personnel should be present.
3. Read simultaneously the temperature and humidity every 15 min for a period of 2 hr using dry and wet bulb thermometers and sling psychrometer.
4. Take different readings at different locations in the area, especially the air inlet and outlets next to each module.
5. Set thermostats and humidistats to the upper limit (75°F and 55–60% RH), the lower limit (65°F and 40–50% RH), and the operational temperatures (68°F and 50% RH). Initial temperatures and humidities and the time required for the area to reach the control level at the various locations in the area are recorded.
6. The upper and lower levels achieved at every testing location give the operational performance data of the air system. If the deviation from the control points exceeds the tolerances for the system, proper corrective action should be taken. The out-of-tolerance variances could be indicative of inadequate air distribution or improper airflow pressure differentials.

Table 14 Pressure Differential Measurement

Written by ______ Approved by ______ Superseded ______ Date of issue ______

Procedure
1. Verify the airflow pattern using a visible smoke generator.
2. Measure the differential pressure among the various areas in parenteral manufacturing with the use of a calibrated manometer.
3. First obtain base data for each area with minimum activity, the normal data with normal activity, including number of people, equipment in operation, and so on.
4. Take readings from various points at the openings of each area.
5. A positive pressure differential should always be kept between clean and nonclean areas, especially those that are adjacent to and connected by wall openings for conveyors.

ing and all other parenteral manufacturing areas, respectively. Monitoring of nonviable airborne particles in aseptic areas is required even through some sterile products are subjected to terminal sterilization. Certain air cleanliness levels are required by good manufacturing practices to assure product quality and absence of particulate matter. The specified air cleanliness level should be validated at the working level during actual production and filling operations for sterile liquids and without actual product for sterile solids. Air sample acquisition frequency and sampling time depends on the type of manufacturing operation being monitored and the data required to assure the air cleanliness level specified. Routine monitoring of nonviable airborne particles at frequencies of not less than once per shift is usually performed. The light-scattering optical measurement technique is most commonly used for single or continuous monitoring of nonviable airborne particles. A protocol to monitor nonviable airborne particles in aseptic areas is shown in Table 16.

Common methods for checking the microbial quality of the environment include the exposure of nutrient agar medium for a given period of time, or drawing a measured quantity of air into a sampler with a vacuum pump and impinging it on a nutri-

Table 15 Particle Count Measurement

Written by ______ Approved by ______ Superseded ______ Date of issue ______

Procedure
1. Calibrate the light-scattering particle counter and its recorder every six months.
2. Measure particle counts at various points in the parenteral production area as close as possible to the filling area.
3. Take the particle count without operational personnel first to obtain a threshold level, then a second reading with personnel and other equipment in operation to obtain the operational level.
4. Collect not less than 1 ft^3/min air volume at each sampling location
5. At any location, if the particle count is 10 particles of 0.5 μm or larger per cubic foot of air, then repeat the readings at least five times and record the average.
6. Record the particle count data at the hour of readings, day, location sampled, and average number of readings used at each point.
7. Maximum of 100 or 100,000 particles of 0.5 μm diameter and larger per cubic foot of air at filling points and all other parenteral manufacturing areas, respectively, are acceptable.

Table 16 Nonviable Airborne Particle Monitoring in Aseptic Areas

Written by ______	Approved by ______	Superseded ______	Date of issue ______

Purpose:	To demonstrate that at the filling location within the aseptic area, a count of less than 100 particles per cubic foot of air, 0.5 μm in diameter or larger is maintained.
Equipment:	1. Obtain the baseline data in static condition, with personnel absent and equipment at rest, using the particle analyzer to count particles greater than or equal to 0.5 μm in diameter at heights of 30–40 in. from filling needles taking a minimum of five counts at each point.
	2. Repeat the above mentioned testing at the dynamic working condition with personnel present and the equipment in operation.
Acceptance criteria	The nonviable airborne particles at the point of filling in the aseptic area should not exceed 100 particles 0.5 μm in diameter and larger per cubic foot of air.

ent agar medium. Samples should be taken as close to the working area as possible. The spore-forming bacteria *Bacillus* sp. and *Clostridium* sp., the nonsporing bacteria *Staphylococcus* sp. and *Streptococcus* sp., the molds *Aspergillus* sp. and *Mucor* sp., and the yeast *Rhodoturula* sp. are commonly found in untreated air environments. A microbial air count of less than one microorganism per 1000 cu ft of air should be maintained at the parenteral production filling areas. An assessment of the indigenous bioburden in the aseptic areas should include seasonal variations to determine common as well as unusual organisms present in the areas. Monitoring of the environment is accomplished by various techniques including Rodac plates, settling plates, air samplers impinging on broad spectrum media, and swabs. An example of a microbial-viable microorganisms form is shown in Table 17. Table 18 shows the expected limit of microorganisms in accordance with air cleanliness classification. Other environmental control assessment required in aseptic areas includes air flow. Temperature monitoring protocol is shown in Table 19.

Quality assurance should verify and document the use of proper equipment, the proper addition of ingredients, proper mixing time, proper drying time, and proper filter type and size. At certain points, samples are to be taken for the analytic and microbiological laboratories for potency assay and any other testing necessary to ensure batch uniformity. In-process released bulk materials waiting for filling should be labeled with product name, item number, lot number, size, strength, gross, and tare weight and net weight or volume of contents.

B. Filling and Packaging

Good manufacturing practices require that in-process quality assurance testing be adequately planned throughout all stages of manufacturing. The number of samples taken for testing and the type of testing are obviously dependent on the size of the batch and the type of product. If deviation from specified limits occurs, the necessary corrective action is taken and recorded, and a resample is taken and tested to determine whether the quality attribute of the product is now within limits. In some instances, as in the case of volume checking, if the deviation is excessive, all material produced prior to the corrective action must be isolated, accounted for, and rejected.

Table 17 Microbial Monitoring in Aseptic Areas

Name	Frequency	Microbial fallout plates (no. CFU/plate)	Rodac (no. CFU/plate)			Air Sampler (Biotest RCS) (no. CFU/ft^3)[b]
			Floor	Walls	Equipment	
Filling room 1	Once/shift	1	2	3	2	1
Filling room 2	Once/shift	1	2	3	2	1
Washing room	Twice/week	5	5	10	5	5
Bulk manuf. room	Once/shift	5	5	5	3	5

[a]Colony forming unit.
[b]Calculated.

Table 18 Expected Number of Microorganisms According to Air Cleanliness Classification

Class	Microorganisms per cubic foot
100	0.1
10,000	0.5
100,000	2.5

A variable group of tests, including checking for volume for solutions and weight variation for powders, are used widely for in-process production control. Certain excess volumes that are sufficient to permit administration of the labeled volumes of liquids are recommended.

The use of control charts is increasingly becoming an essential part of any quality assurance operation. Control charts may be classified as portraying attributes or variables. Variable charts are based on the normal distribution; attribute charts are based on binomial distribution. Variable charts are applied when actual numerical measurements of quality attributes are available; attribute charts refer to some other attributes of quality that are present or absent in which each sample inspected is tested to determine whether it conforms to the requirements. Variable charts, or the X and R (mean and range) charts are undoubtedly the most generally used charts in the quality assurance. Routinely, in-process results are plotted on a control chart so that a complete picture of any possible fluctuation during the entire filling operation can be readily detected. The control limits or process capability can be determined by sampling, measuring, and recording results in subgroups that cover the filling operation. The range within each group, i.e., the absolute number difference between the lowest and highest individual product reading and the average reading, is calculated for the total number of groups. The average reading plots can detect movements toward limits that allow making necessary corrections before limit values are exceeded. Although the subgroup's sample range plots allow the monitoring of the sample range trend, an increase in sample range values or generally high variability indicates possible control problems.

Table 19 Temperature Monitoring in Aseptic Areas

Written by ______ Approved by ______ Superseded ______ Date of issue ______	
Purpose:	To demonstrate the ability of the air conditioning system to control temperature at 72 ± 20°F.
Equipment:	Calibrated dry-bulk thermometer or thermocouples and recorder.
Procedure:	1. Monitoring should be at static and dynamic working conditions with all lights on in the aseptic areas. 2. Measure and record temperature at 30 min intervals during dynamic and static working conditions.
Acceptance criteria:	The air conditioning system should be able to maintain the temperature at 72 ± 20°F at static and dynamic working conditions with the specified occupancy and light generation levels.

If the quality control laboratory analysis confirms that the product complies with specifications and that the quality assurance audit of manufacturing operations was satisfactory, the bulk product is released to the packaging department, and production control is notified. Production control issues a packaging form that carries the name of the product, item number, lot number, number of labels, number of inserts, packaging materials to be used, operations to be performed, and the quantity to be packaged. A copy of this form is sent to the supervisor of label control who, in turn, counts out the required number of labels and inserts. Since labels and inserts may be spoiled during the packaging operation, a definite number in excess of that actually required is usually issued. However, all labels and inserts must be accounted for at the end of the operation, and unused labels and inserts must be accounted for before their destruction. If the lot number and expiration date of the product are not going to be printed directly on the filling line, the labels are run through a printing machine that imprints the lot number and expiration date. The labels are recounted and placed in a separate container with proper identification for future transfer to the packaging department. The packaging department then requests, according to the packaging form, the product to be packaged and all packaging components, such as labels, inserts, bottles, caps, seals, cartons, and shipping cases. Quality assurance inspects and verifies all packaging components and equipment to be used for packaging operations to ensure that they have the proper identification, that the line has been thoroughly cleaned, and that all materials from the previous packaging operation have been completely removed. Packaging operations should be performed with adequate physical segregation from product to product. Products of similar size should not be scheduled on the neighboring packaging lines at the same time. Quality assurance should periodically inspect the packaging line and check filled and labeled containers for compliance with written specification, e.g., absence of foreign drugs and labels, adequacy of the containers and closure system, and accuracy of labeling. Some packaging operations, especially those using high-speed equipment, are fitted with automated testing equipment to check each container for fill and label placement. Alternatively, an operator may visually inspect all packages fed into the final cartons. Proper reconciliation and disposition of the unused and wasted labels should occur at the end of the packaging operation.

Quality assurance should examine specific sample numbers per specific time from each phase of packaging and labeling operation for the following defects:

- Label. Incorrect identification band; incorrect or missing code bars; incorrect color; whether it is loose or torn, soiled or defaced; and registration
- Printing. Lot number/expiration date, whether it is wrong or missing, illegible or mislocated; lot number on shipper missing; printing missing or skips; incorrect shipping label; and smeared printing
- Insert. Whether it is missing, incorrect, torn, dirty, or poorly folded
- Intermediate shipper. Labeling requirements and physical appearance requirements
- Outer shipper. Whether it is soiled internally or externally, incorrect top, and incorrect count

A proper action level and acceptable quality level (AQL) categories should be assigned to each defect according to its effect on the quality of the product, ranging from fatal, critical, and major to minor defects.

Quality assurance should select finished preservation samples at random from each lot. The samples should consist of at least twice the quantity necessary to perform all tests required to determine whether the product meets its established specification. These samples should be retained for at least 2 years after the expiration date and stored in their original package under conditions consistent with product labeling.

Quality assurance should also select a finished sample and send it to the analytical control laboratory for final testing, which is usually an identification test.

C. Sampling Procedure

Sample procedures of finished products can be based either on attribute inspection that grades the product as defective or nondefective, or inspection by variables for percentage defective. The focal point of any sampling plan is the acceptable quality level. The second important step is to decide on the inspection level of the sampling plan, which determines the relationship between the lot size and the sample size (N/n). The principal purpose of the sampling plan is to assure that products manufactured are of quality at least as good as the designed AQL. This means that, as long as the product fraction defective (r) is less than the AQL designated for a specific production procedure, then a large percentage of lots produced will be accepted. Sampling procedures for inspection by variables for percentage defective may be used if a quality characteristic can be continuously measured and is known to be normally distributed, such as mean of the sample or the mean and standard deviation of the sample. The assumption of a specific distributional form is a special feature of variable sampling. A separate plan must be employed for each quality characteristic that is being inspected or a common sampling plan is used, but the allowable number of defects varies for each quality characteristic; that is, no critical defects are allowed (c), but some minor defects are allowed. Also, the fraction defective yielded by a given process mean and standard deviation should be calculated to assure a normal distribution of sample statistics.

For practical purposes, MIL-STD-414 for inspection by variables for percentage defective and MIL-STD-105D for inspection by attributes for defective or nondefective products are often used to design a sample plan.

In manufacturing, sampling procedures for inspection by attributes are generally used, for the following reasons:

- Variables sampling, as compared with attributes sampling, requires more mathematical understanding and clerical calculation.
- Switching procedures from different inspection levels in variables sampling is morecumbersome.
- For large lot sizes, which is the ususal case in manufacturing, producer's risk is larger in variables sampling than in an attributes sampling plan.
- The smaller sample size required by variables sampling sometimes costs more, depending on the type of quantitative test performed, than a large sample sized required by a comparable attributes plan because of the precise measurements required by the variable plan.
- Variables data can be converted to attributes data, but the reverse is not possible.

There are three types of attributes sampling plans: single sampling, double sampling, and multiple sampling.

1. Single-Sampling Plan

A single-sampling plan specifies the sample size that should be taken from each lot and the number of defective units that cannot be exceeded in this sample. For example, a sample of 100 (n) is taken from a lot; if 2 (c) or fewer defective units are found, the lot is accepted. The discriminatory power of a sampling plan is explained by its operating characteristics (OC) curve. This curve serves to show how the probability of accepting a lot varies with the quality of the sample inspected. The operating characteristic curves for a single sampling plan that gives the probability of accepting a lot from a randomly operating process turning out products of average quality at various defective levels for samples of different sizes and different acceptance numbers are given in Fig. 1.

From this figure, the OC curve of the above-mentioned example of a single-sampling plan, $n = 100$ and $c = 2$, shows that if the injectable quality (percentage defective) is 5, the probability of lot acceptance is 12; if it is 1, the probability of acceptance is 92. Again, Fig. 1 shows that OC curves vary with the number of n as c in this example is kept proportional with the size of a sample (n). The three OC curves for sampling plan, $n = 100$, $c = 2$, $n = 100$, $c = 1$; and $n = 100$, $c = 0$ illustrate that a plan varies with the acceptance number alone. The smaller the c, the tighter is the plan; as c is increased, the plan becomes more lax and the OC curve is raised. The schematic instructions for a single-sampling plan are shown in Fig. 2.

2. Double-Sampling Plan

For a double-sampling plan the first sample is smaller than a comparable single-sample plan. The second sample size is generally twice the size of the first. Consequently, if the lot is accepted or rejected on the first sample, there may be a considerable savings

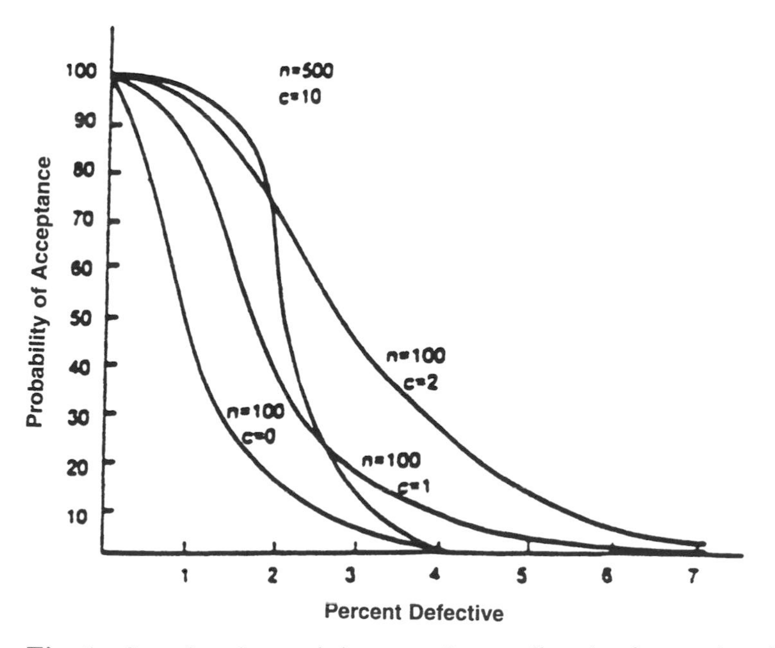

Fig. 1 Operating characteristic curves for sampling plan for samples of different sizes and different acceptance numbers.

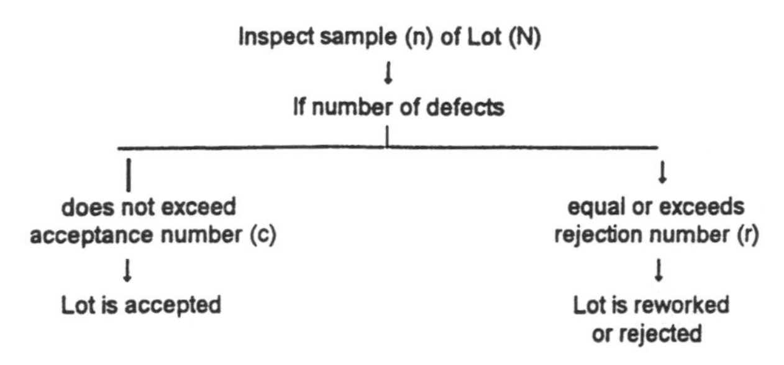

Fig. 2 Schematic instructions for MIL-STD-105D single-sample plan.

in total inspection cost. If the results of the first sample fall within the acceptance and rejection values, a second sample is taken. The results of the two samples are combined and compared with the final acceptance or rejection values. For example, a first sample of 50 ($n_1 = 50$) is taken from a lot; if 2 ($c_1 = 2$) or fewer defective units are found, the lot is accepted; if 7 or more defective units are found, the lot is rejected. If the number of defective units is 3, but not more than 6 ($c_2 = 6$), a second sample of 100 ($n_2 = 100$) is taken; if in the combined sample ($n_1 + n_2 = 150$), the number of defective units is 6 or less, the lot is accepted; if 7 or more defective units are found, the lot is rejected. Operating characteristic curves of double-sampling plans showing the probability of acceptance or rejection on the first sample and combined first and second samples for the above-mentioned example are shown in Fig. 3. Curve II in this figure gives the principal operating characteristic curve for the plan, since it gives the probability of final acceptance. The difference between curve II and curve I gives the probability of rejection on the second sample. To illustrate, for the previously mentioned example, for a lot with a fraction defective of 5, the probability of acceptance

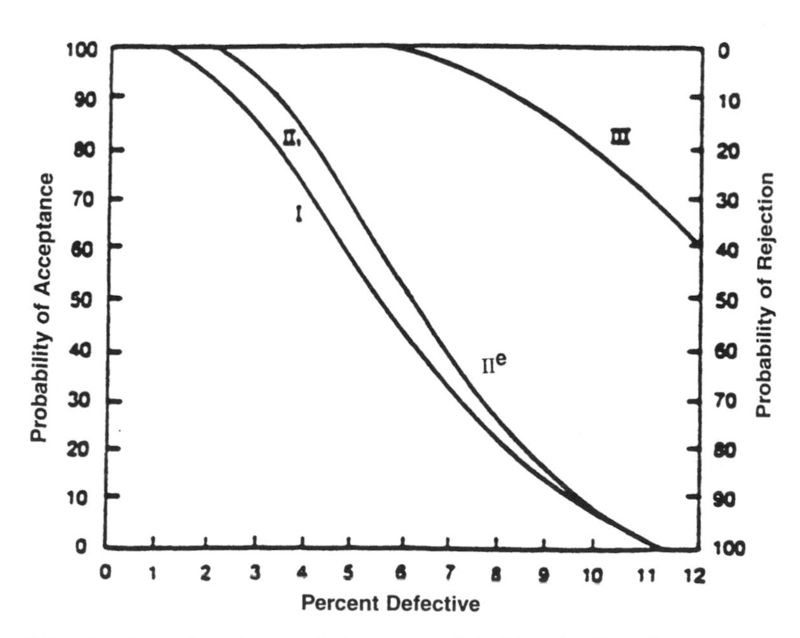

Fig. 3 Operating characteristic curves of double-sampling plan.

on the second sample is 59 and the probability of final acceptance is 63.5. The schematic instructions for double-sampling plan are shown in Fig. 4.

3. Multiple-Sampling Plan

This plan allows for more than two samples when necessary for a final decision. For standardized sampling, the plan is tied to a maximum of seven equal samples. For nonstandardized sampling, the sample size may vary between inspection checks, depending on the proximity of the sample results to the acceptance or rejection values. For example, in a multiple standardized sampling plan, if from a given lot the cumulative sample sizes, acceptances, and rejection numbers of 20, 40, 60, 80, 100, 120, and 140; 0, 1, 3, 5, 8, 9, and 10; 4, 5, 6, 8, 10, 11, and 12 are assigned respectively, the lot is rejected if the number of defective units at any sampling stage equals or exceed the rejection number. If not, the multiple sampling procedure continues until at least the seventh sample is taken when a decision to accept or reject the lot is to be made. The schematic instructions for a multiple-sampling plan are shown in Fig. 5.

Military standard sampling procedures for inspection by attributes (MIL-STD-105D) was issued by the U.S. Government in 1963. The focal point of MIL-STD-105D is the acceptable quality level. In applying MIL-STD-105D, it is necessary also to decide on the inspection level. This determines the relationship between the lot size and the sample size. For a specified AQL and inspection level and a given lot size, MIL-STD-105D gives a reduced, a normal, or a tightened plan. The switch from the normal plan to the tightened plan is made if two of five consecutive lots have been rejected on original inspection. Switching back from tightened to normal plan is made if five consecutive lots have been accepted on original inspection. Switching from normal to reduced sampling plan is made if 10 consecutive lots have been accepted on original normal inspection and the total number of defectives is less than a value set forth in a special table. Fig. 6 shows the operation characteristic curves for both normal and tightened plans for a single sampling pan with an AQL of 1% and sampling

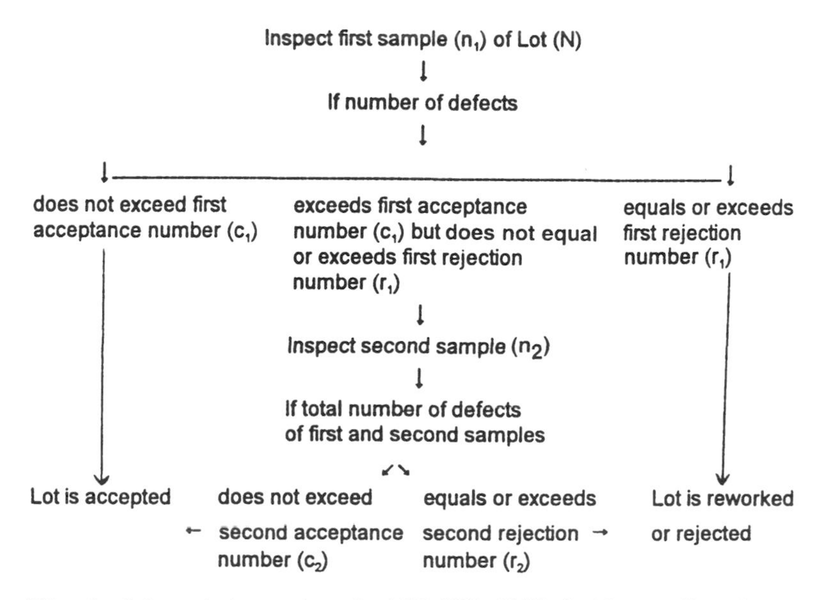

Fig. 4 Schematic instructions for MIL-STD-105D double-sampling plan.

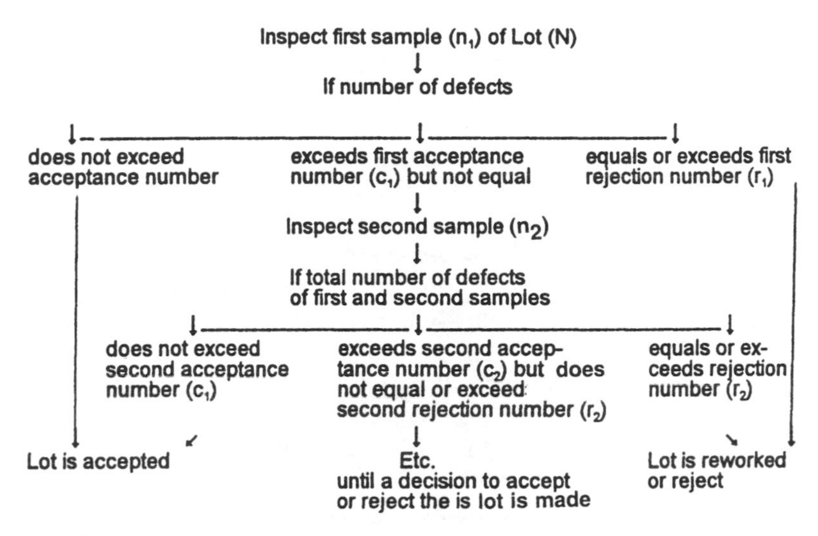

Fig. 5 Schematic instructions for MIL-STD-105D multiple-sampling plan.

size of 50 ($n = 50$). If the lot has a fraction defective of 1, the probability of acceptance on the normal inspection is 92.5; if the tightened inspection is used, this probability will decrease to 82.

The construction of a sampling plan normally requires that four quality standards be specified: acceptable quality level, unacceptable quality level (UQL), producer's risk (α), which is the probability of rejecting good quality, and consumer's risk (β), which

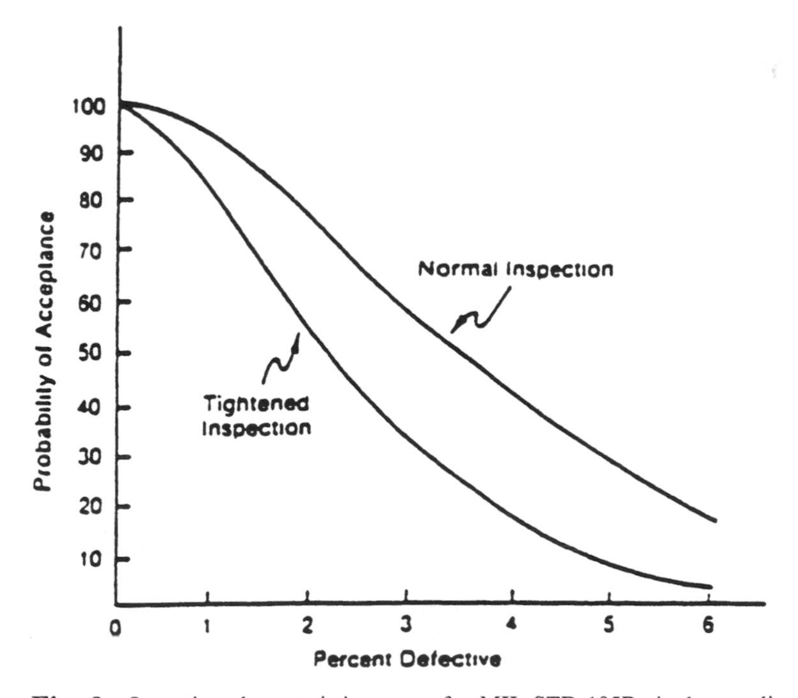

Fig. 6 Operating characteristic curves for MIL-STD-105D single-sampling plan.

is the probability of accepting poor quality. Figure 7 defines these parameters for projections for a sampling plan. The usual approach is the determination of desirable AQL, UQL, α, and a subsequent computation of sample size and acceptable values by applying the tables of MIL-STD-105D. For low-dosage or highly toxic products, as in the case of steroid suspension injections, it is desirable that the AQL and UQL be kept close together and α and β be large, as in the case of antacid suspensions. For example, a lot of 50,000 bottles may not contain more than 1% defective bottles, and the single normal inspection level of MIL-STD-105D is used. Entering Table 20 of MIL-STD-105D, find letter N under column II for the general inspection levels for lot size of 35,000–150,000. Entering Table 21, find the sample size of 500 and at AQL 1; for acceptance the value is 10 and for rejection it is 11.

In practice, this means that a 500-bottle sample is taken from the lot at random and tested; the lot is accepted if 10 or fewer are defective, and rejected if 11 or more are defective. If tightened inspection is used for the sample example, it would call for a sample size code of P for the general inspection level III. From Table 22, a sample size of 800 would now have to be used, instead of 500, and at an AQL of 1, the values are 12 for acceptance and 13 for rejection. On the other hand, if reduced inspection is to be used for the same example, it would call for a sample size code of L for the general inspection level III. From Table 23, a sample size of only 80 is required and at an AQL 1, with a value of 2 for acceptance and 5 for rejection.

To examine the kind of sampling job one accomplishes with these three sampling plans, one has to examine the OC curves for code letters N, P, and L at an AQL level of 1.0% defective in the same MIL-STD-105D. These curves indicate that, at sampling size code letter N, one would accept lots containing 3.0% defectives 12% of the time.

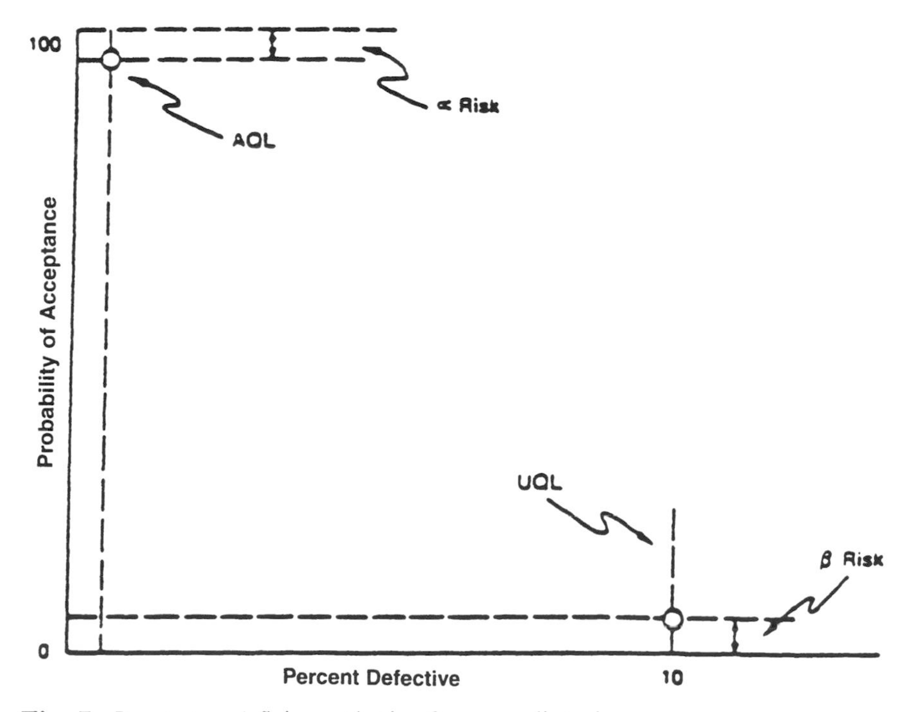

Fig. 7 Parameters defining projection for a sampling plan.

Table 20 Sample Size Code Letters

Lot or batch size	Special inspection levels				General inspection levels		
	S-1	S-2	S-3	S-4	I	II	III
2–8	A	A	A	A	A	A	B
9–15	A	A	A	A	A	B	C
16–25	A	A	B	B	B	C	D
26–50	A	B	B	C	C	D	E
51–90	B	B	C	C	C	E	F
91–150	B	B	C	D	D	F	G
51–280	B	C	D	E	E	G	H
281–500	B	C	D	E	F	H	J
501–1,200	C	C	E	F	G	J	K
1,201–3,200	C	D	E	G	H	K	L
3,201–10,000	C	D	F	G	J	L	M
10,001–35,000	C	D	F	H	K	M	N
35,001–150,000	D	E	G	J	L	N	P
150,001–500,000	D	E	G	J	M	P	Q
500,001 and over	D	E	H	K	N	Q	R

Source: From Military Standard, Sampling Procedures, and Tables for Inspection by Attributes, MIL-STD-105D, U.S. Department of Defense, Washington, D.C., 1963.

At a sample size code letter P, however, one would accept lots with 3% defectives only 2% of the time. In reduced inspection, with a sample size code letter L, one would accept lots with 3% defectives 46% of the time.

MIL-STD-105D gives four additional special inspection levels—S-1, S-2, S-3, and S-4—which may be used when relatively small sample sizes are necessary, such as might be the case with costly destructive testing.

In summary, the steps necessary for the use of MIL-STD-105D are as follows:

1. Choose the AQL.
2. Choose the inspection level.
3. Determine lot size.
4. Find sample size code letter from the table.
5. Choose the type of sample plan and find its table.
6. Use the tightened or reduced inspection table for the same type of plan whenever it is required.

A continuous sampling plan with in-process testing clearly can yield more valuable information on the homogeneity of the production procedure and increase the opportunity to detect and correct any production difficulties. Such testing is facilitated by the fact that the entire lot is accessible and the sample may be obtained entirely at random. Actually, the same procedures described before for sampling plans may be applied to continuous sampling as well.

D. Specification and Testing

Final testing of disperse system products is made in the analytical and microbiological quality control laboratories. These tests are designed to determine compliance with

Table 21 Master Table for Normal Inspection

Sample size code letter	Sample size	Acceptable quality levels (normal inspection, single sampling)																									
		0.010	0.015	0.025	0.040	0.065	0.10	0.15	0.25	0.40	0.65	1.0	1.5	2.5	4.0	6.5	10	15	25	40	65	100	150	250	400	650	1000
		A R	A R	A R	A R	A R	A R	A R	A R	A R	A R	A R	A R	A R	A R	A R	A R	A R	A R	A R	A R	A R	A R	A R	A R	A R	A R
A	2	↓	↓	↓	↓	↓	↓	↓	↓	↓	↓	↓	↓	↓	↓	0 1	↓	↓	1 2	2 3	3 4	5 6	7 8	10 11	14 15	21 22	30 31
B	3	↓	↓	↓	↓	↓	↓	↓	↓	↓	↓	↓	↓	↓	0 1	↑	↓	1 2	2 3	3 4	5 6	7 8	10 11	14 15	21 22	30 31	44 45
C	5	↓	↓	↓	↓	↓	↓	↓	↓	↓	↓	↓	↓	0 1	↑	↓	1 2	3 3	3 4	5 6	7 8	10 11	13 15	21 22	30 31	44 45	↑
D	8	↓	↓	↓	↓	↓	↓	↓	↓	↓	↓	↓	0 1	↑	↓	1 2	2 3	3 4	5 6	7 8	10 11	↑	↑	↑	↑	↑	↑
E	13	↓	↓	↓	↓	↓	↓	↓	↓	↓	↓	0 1	↑	↓	1 2	2 3	3 4	5 6	7 8	10 11	14 15	↑	↑	↑	↑	↑	↑
F	20	↓	↓	↓	↓	↓	↓	↓	↓	↓	0 1	↑	↓	1 2	2 3	3 4	5 6	7 8	10 11	14 15	21 22	↑	↑	↑	↑	↑	↑
G	32	↓	↓	↓	↓	↓	↓	↓	↓	0 1	↑	↓	1 2	2 3	3 4	5 6	7 8	10 11	14 15	21 22	↑	↑	↑	↑	↑	↑	↑
H	50	↓	↓	↓	↓	↓	↓	↓	0 1	↑	↓	1 2	2 3	3 4	5 6	7 8	10 11	14 15	21 22	↑	↑	↑	↑	↑	↑	↑	↑
J	80	↓	↓	↓	↓	↓	↓	0 1	↑	↓	1 2	2 3	3 4	5 6	7 8	10 11	14 15	21 22	↑	↑	↑	↑	↑	↑	↑	↑	↑
K	125	↓	↓	↓	↓	↓	0 1	↑	↓	1 2	2 3	3 4	5 6	7 8	10 11	14 15	21 22	↑	↑	↑	↑	↑	↑	↑	↑	↑	↑
L	200	↓	↓	↓	↓	0 1	↑	↓	1 2	2 3	3 4	5 6	7 8	10 11	14 15	21 22	↑	↑	↑	↑	↑	↑	↑	↑	↑	↑	↑
M	315	↓	↓	↓	0 1	↑	↓	1 2	2 3	3 4	5 6	7 8	10 11	14 15	21 22	↑	↑	↑	↑	↑	↑	↑	↑	↑	↑	↑	↑
N	500	↓	↓	0 1	↑	↓	1 2	2 3	3 4	5 6	7 8	10 11	14 15	21 22	↑	↑	↑	↑	↑	↑	↑	↑	↑	↑	↑	↑	↑
P	800	↓	0 1	↑	↓	1 2	2 3	3 4	5 6	7 8	10 11	14 15	21 22	↑	↑	↑	↑	↑	↑	↑	↑	↑	↑	↑	↑	↑	↑
Q	1250	0 1	↑	↓	1 2	2 3	3 4	5 6	7 8	10 11	14 15	21 22	↑	↑	↑	↑	↑	↑	↑	↑	↑	↑	↑	↑	↑	↑	↑
R	2000	↑	↑	1 2	2 3	3 4	5 6	7 8	10 11	14 15	21 22	↑	↑	↑	↑	↑	↑	↑	↑	↑	↑	↑	↑	↑	↑	↑	↑

A = Acceptance number.
R = Rejection number.
↓ = Use first sampling plan below arrow. If sample size equals or exceeds lot or batch size, do 100% inspection.
↑ = Use first sampling plan above arrow.
Source: Mil. Std. 105D.

Table 22 Master Table for Tightened Inspection

Sample size code letter	Sample size	Acceptable quality levels (tightened inspection, single sampling)																									
		0.010	0.015	0.025	0.040	0.065	0.10	0.15	0.25	0.40	0.65	1.0	1.5	2.5	4.0	6.5	10	15	25	40	65	100	150	250	400	650	1000
		A R	A R	A R	A R	A R	A R	A R	A R	A R	A R	A R	A R	A R	A R	A R	A R	A R	A R	A R	A R	A R	A R	A R	A R	A R	A R
A	2	↓	↓	↓	↓	↓	↓	↓	↓	↓	↓	↓	↓	↓	↓	↓	↓	↓	↓	1 2	2 3	3 4	5 6	8 9	12 13	18 19	27 28
B	3	↓	↓	↓	↓	↓	↓	↓	↓	↓	↓	↓	↓	↓	↓	0 1	↓	↓	1 2	2 3	3 4	5 6	8 9	12 13	18 19	27 28	41 42
C	5	↓	↓	↓	↓	↓	↓	↓	↓	↓	↓	↓	↓	↓	0 1	↓	↓	1 2	2 3	3 4	5 6	8 9	12 13	18 19	27 28	41 42	↑
D	8	↓	↓	↓	↓	↓	↓	↓	↓	↓	↓	↓	↓	0 1	↓	↓	1 2	2 3	3 4	5 6	8 9	12 13	18 19	27 28	41 42	↑	↑
E	13	↓	↓	↓	↓	↓	↓	↓	↓	↓	↓	↓	0 1	↓	↓	1 2	2 3	3 4	5 6	8 9	12 13	18 19	27 28	41 42	↑	↑	↑
F	20	↓	↓	↓	↓	↓	↓	↓	↓	↓	↓	0 1	↓	↓	1 2	2 3	3 4	5 6	8 9	12 13	18 19		↑	↑	↑	↑	↑
G	32	↓	↓	↓	↓	↓	↓	↓	↓	↓	0 1	↓	↓	1 2	2 3	3 4	5 6	8 9	12 13	18 19	↑	↑	↑	↑	↑	↑	↑
H	50	↓	↓	↓	↓	↓	↓	↓	↓	0 1	↓	↓	1 2	2 3	3 4	5 6	8 9	12 13	18 19	↑	↑	↑	↑	↑	↑	↑	↑
J	80	↓	↓	↓	↓	↓	↓	↓	0 1	↓	↓	1 2	2 3	3 4	5 6	8 9	12 13	18 19	↑	↑	↑	↑	↑	↑	↑	↑	↑
K	125	↓	↓	↓	↓	↓	↓	0 1	↓	↓	1 2	2 3	3 4	5 6	8 9	12 13	18 19	↑	↑	↑	↑	↑	↑	↑	↑	↑	↑
L	200	↓	↓	↓	↓	↓	0 1	↓	↓	1 2	2 3	3 4	5 6	8 9	12 13	18 19	↑	↑	↑	↑	↑	↑	↑	↑	↑	↑	↑
M	315	↓	↓	↓	↓	0 1	↓	↓	1 2	2 3	3 4	5 6	8 9	12 13	18 19	↑	↑	↑	↑	↑	↑	↑	↑	↑	↑	↑	↑
N	500	↓	↓	↓	0 1	↓	↓	1 2	2 3	3 4	5 6	8 9	12 13	18 19	↑	↑	↑	↑	↑	↑	↑	↑	↑	↑	↑	↑	↑
P	800	↓	↓	0 1	↓	↓	1 2	2 3	3 4	5 6	8 9	12 13	18 19	↑	↑	↑	↑	↑	↑	↑	↑	↑	↑	↑	↑	↑	↑
Q	1250	↓	0 1	↓	↓	1 2	3 3	3 4	5 6	8 9	12 13	18 19	↑	↑	↑	↑	↑	↑	↑	↑	↑	↑	↑	↑	↑	↑	↑
R	2000	0 1	↑	↓	1 2	2 3	3 4	5 6	8 9	12 13	18 19	↑	↑	↑	↑	↑	↑	↑	↑	↑	↑	↑	↑	↑	↑	↑	↑
S	3150		↑	1 2								↑	↑	↑	↑	↑	↑	↑	↑	↑	↑	↑	↑	↑	↑	↑	↑

A = Acceptance number.
R = Rejection number.
↓ = Use first sampling plan below arrow. If sample size equals or exceeds lot or batch size, do 100% inspection.
↑ = Use first sampling plan above arrow.
Source: Mil. Std. 105D.

Table 23 Master Table for Reduced Inspection

Sample size code letter	Sample size	Acceptable quality levels (reduced inspection,* single sampling)																									
		0.010	0.015	0.025	0.040	0.065	0.10	0.15	0.25	0.40	0.65	1.0	1.5	2.5	4.0	6.5	10	15	25	40	65	100	150	250	400	650	1000
		A R	A R	A R	A R	A R	A R	A R	A R	A R	A R	A R	A R	A R	A R	A R	A R	A R	A R	A R	A R	A R	A R	A R	A R	A R	A R
A	2	↓	↓	↓	↓	↓	↓	↓	↓	↓	↓	↓	↓	↓	↓	0 1	↓	↓	1 2	2 3	3 4	5 6	7 8	10 11	14 15	21 22	30 31
B	2	↓	↓	↓	↓	↓	↓	↓	↓	↓	↓	↓	↓	↓	0 1	↑	↓	0 2	1 3	2 4	3 5	5 6	7 8	10 11	14 15	21 22	30 31
C	2	↓	↓	↓	↓	↓	↓	↓	↓	↓	↓	↓	↓	0 1	↑	↑	0 2	1 3	1 4	2 5	3 6	5 8	7 10	10 13	14 17	21 24	↑
D	3	↓	↓	↓	↓	↓	↓	↓	↓	↓	↓	↓	0 1	↑	↓	0 2	1 3	1 4	2 5	3 6	5 8	7 10	10 13	14 17	21 24	↑	↑
E	5	↓	↓	↓	↓	↓	↓	↓	↓	↓	↓	0 1	↑	↓	0 2	1 3	1 4	2 5	3 6	5 8	7 10	10 13	14 17	21 24	↑	↑	↑
F	8	↓	↓	↓	↓	↓	↓	↓	↓	↓	0 1	↑	↑	0 2	1 3	1 4	2 5	3 6	5 8	7 10	10 13	↑	↑	↑	↑	↑	↑
G	13	↓	↓	↓	↓	↓	↓	↓	↓	0 1	↑	↓	0 2	1 3	1 4	2 5	3 6	5 8	7 10	10 13	↑	↑	↑	↑	↑	↑	↑
H	20	↓	↓	↓	↓	↓	↓	↓	0 1	↑	↓	0 2	1 3	1 4	2 5	3 6	5 8	7 10	10 13	↑	↑	↑	↑	↑	↑	↑	↑
J	32	↓	↓	↓	↓	↓	↓	0 1	↑	↓	0 2	1 3	1 4	2 5	3 6	5 8	7 10	10 13	↑	↑	↑	↑	↑	↑	↑	↑	↑
K	50	↓	↓	↓	↓	↓	0 1	↑	↓	0 2	1 3	1 4	2 5	3 6	5 8	7 10	10 13	↑	↑	↑	↑	↑	↑	↑	↑	↑	↑
L	80	↓	↓	↓	↓	0 1	↑	↓	0 2	1 3	1 4	2 5	3 6	5 8	7 10	10 13	↑	↑	↑	↑	↑	↑	↑	↑	↑	↑	↑
M	125	↓	↓	↓	0 1	↑	↓	0 2	1 3	1 4	2 5	3 6	5 8	7 10	10 13	↑	↑	↑	↑	↑	↑	↑	↑	↑	↑	↑	↑
N	200	↓	↓	0 1	↑	↓	0 2	1 3	1 4	2 5	3 6	5 8	7 10	10 13	↑	↑	↑	↑	↑	↑	↑	↑	↑	↑	↑	↑	↑
P	315	↓	0 1	↑	↓	0 2	1 3	1 4	2 5	3 6	5 8	7 10	10 13	↑	↑	↑	↑	↑	↑	↑	↑	↑	↑	↑	↑	↑	↑
Q	500	0 1	↑	↓	0 2	1 3	1 4	2 5	3 6	5 8	7 10	10 13	↑	↑	↑	↑	↑	↑	↑	↑	↑	↑	↑	↑	↑	↑	↑
R	800	↑	↑	0.2	1.3	1 4	2 5	3 6	5 8	7 10	10 13	↑	↑	↑	↑	↑	↑	↑	↑	↑	↑	↑	↑	↑	↑	↑	↑

*If the acceptance number has been exceeded, but the rejection number has not been reached, accept the lot, but reinstate normal inspection.

A = Acceptance number.

R = Rejection number.

↓ = Use first sampling plan below arrow.

↑ = Use first sampling plan above arrow.

specifications. Thus, the testing of the finished product for compliance with a predetermined standard prior to release of the product and subsequent distribution is a critical factor in quality assurance. The purpose of establishing these specifications and standards is to assure that each unit contains the amount of drug claimed on the label, that all the drug in each unit is available for complete absorption, that the drug is stable in the formulation in its specific final container for its expected shelf life, and that it contains no toxic foreign substances.

Normally the design of test parameters, procedures, and specifications is done during product development. It is a good manufacturing practice to base such parameters on experience developed during processing of several pilot and production batches. Furthermore, the results of these studies should be subjected to statistical analysis in order to appraise correctly the precision and accuracy of each procedure for each characteristic. In the long run, with additional production experience, it is possible that specifications be modified for perfection and upgrading of product quality. The various disciplines of quality control testing can be clearly understood from the quality control attributes outlined in the quality measurements that follow.

1. Appearance and Feel

A good disperse system formulation should have both physical and chemical stability and cosmetic acceptability. Disperse systems can vary in appearance and feel, due to viscosity, glass, smoothness, and texture. The appearance and feel of a disperse system is dependent on particle size of the dispersed phase, on the phase-volume ratio of the oil and water phases, on the surfactants used, on the order of addition of ingredients, on the temperature and emulsification, and on the method and rate of cooling.

Disperse systems may range in appearance from clear to milky white lotions or creams. Topical creams and lotions may have a smooth, flat, or pearly appearance.

Appearance and touch properties include tackiness, wetness, grittiness, oiliness, spreading qualities, and drying time.

Visual assessment of the instability of disperse system products have been tried, and a qualitative system for this purpose is given in Table 24. To visually rate the observed instability phenomena numerically, a maximum rating of 9 is given for no signs of instability, while a rating of 0 is given to completely separated disperse system products.

Table 24 Visual Stability Rating for Disperse System Products

9	No visual separation, completely homogeneous
8	No visual separation, virtually homogeneous
7	Very indistinct separation, no clear layer bottom or top
6	Indistinct separation, no clear layer at bottom or top
5	Distinct separation, no clear layer at bottom or top
4	Homogeneous top or bottom layer, clear layer at top or bottom
3	Distinct separation, clear layer at top or bottom with no coalescence
2	Distinct separation with slight coalescence
1	Distinct separation with strong coalescence
0	Complete separation and complete coalescence

2. Measurement of Surface and Interfacial Tension

Measurement of surface and interfacial tension is achieved by the determination of surface or interfacial free energy. Thus, surface tension expressed as dyn/cm and surface free energy expressed as ergs/cm^2 are dimensionally equivalent and results are reported as γ. There are several indirect methods for measuring the surface free energy from which the surface tension is calculated according to a mathematical formula. Surface free-energy measurement changes with time, temperature, and other factors.

Methods used for surface and interfacial tension determination can be classified according to the speed of attainment of surface tension equilibrium. In the case of pure liquids, equilibrium is reached quickly and static methods are used. On the other hand, if equilibrium is attained slowly the drop weight, ring, or bubble pressure method are used as dynamic methods.

a. Capillary Rise Method. Although the capillary rise method is slow for surface tension measurement, its accuracy is high. Best results are obtained when capillary tubes of small diameters are used. For very narrow capillary tubes, surface tension of a liquid γ is measured using Eq. (1)

$$\gamma = \tfrac{1}{2}\, r\, h\, d\, g \tag{1}$$

where r = radius of the capillary tube, or one-half of the diameter; h = distance from liquid level to bottom of meniscus; d = density of the liquid; and g = acceleration of gravity.

Interfacial tension γ_1 is measured using Eq. (2)

$$\gamma_1 = \tfrac{1}{2}\, r\, h\, g\, (d_1 - d_2) \tag{2}$$

where d_1 and d_2 = densities of the two liquid phases.

Capillary rise apparatus (Fig. 8) consists of a calibrated, uniform base radius capillary tube mounted through a holder into a wide-diameter outside tube. A magni-

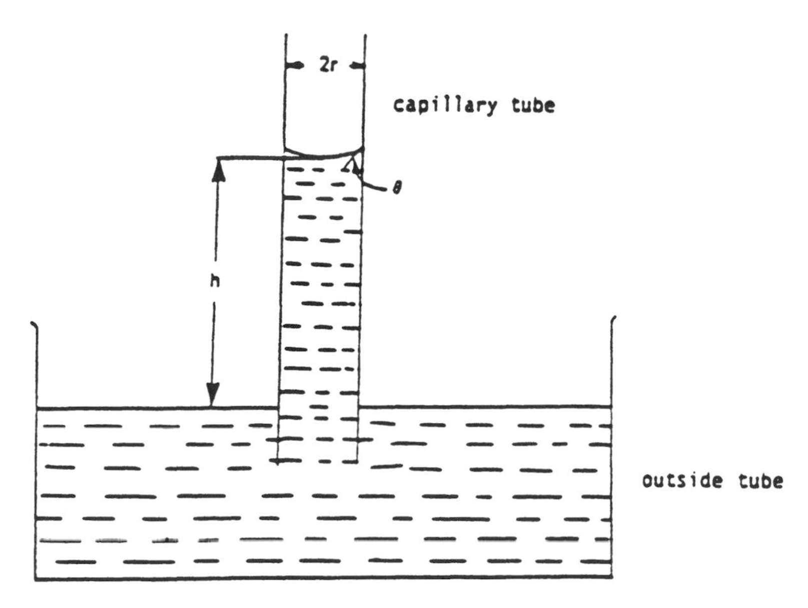

Fig. 8 Capillary rise method apparatus.

fying lens or cathetometer is normally used in measuring the capillary rise. Capillary tubes must be steamed or cleaned thoroughly before each use. A modification of the method by measuring the pressure needed to force the meniscus in the capillary tube back to the liquid level of the outside tube is also available commercially. This modified method is used to measure the surface and interfacial tension in aerosol propellant disperse systems.

b. Drop Weight Method. The drop weight method is the preferred general method for surface and interfacial tension determination. It provides the best accuracy and speed over other procedures. Using a capillary tube, the surface tension in the drop weight method can be calculated using Eq. (3)

$$\gamma = \frac{mg}{r}F \tag{3}$$

where mg = weight of liquid drop; r = capillary radius; and F = correction factor.

Interfacial tension γ_1 is measured using Eq. (4)

$$\gamma_1 = \frac{V(d_1 - d_2)g}{r}F \tag{4}$$

where V = volume of the drop; and d_1 and d_2 = densities of the two liquid phases (Fig. 9).

The drop weight method apparatus consists of a capillary tube A attached to bulk acting as reservoir for the liquid whose surface tension is to be measured. The liquid drops into a tared weighing bottle B place in bottle C. The whole apparatus is immersed in a constant-temperature water bath. A slight vacuum is applied until a droplet from capillary tube A reaches full size and drops off by itself. This is repeated a minimum of 25 times, and the average weight and volume of a single drop is calculated.

c. Ring Method. The ring method determines the surface tension by measuring the force required to detach a wire from a liquid.

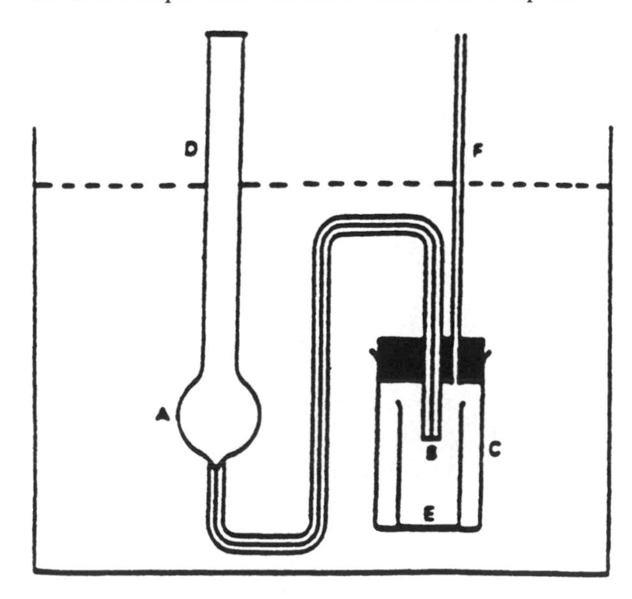

Fig. 9 Drop weight method apparatus.

The surface tension in the ring method can be calculated using Eq. (5)

$$\gamma = \frac{Mg}{4\pi R} F \tag{5}$$

where M = mass of liquid; R = average radius of the wire ring; F = correction factor.

The value of the correction factor can be obtained from standard tables or, for each particulate ring, they can be replaced by a value from a calibration curve.

The best-known commercial ring method apparatus is the du Nouy tensiometer, which consists of a tension-wire balance to measure the pull required to detach a ring from the liquid to be measured and is placed in shallow, wide dish. The dish is raised slowly until the ring contacts the liquid surface and tension is then applied until it is equal to the pull needed to detach the ring where the lever arm springs up. The tension is read directly on a calibrated scale. Control of the liquid's temperature, avoidance of vibration, and complete wetting of the ring are necessary for reproducible and accurate results. A modification of this procedure, using a flat plate rather than a ring, is referred to as the Wilhelmy plate method.

d. Bubble Pressure Method. In the bubble pressure test method for measuring surface tension, a small air bubble is blown at the bottom of a capillary tube into the liquid. The maximum pressure occurs when the air bubble is a hemisphere just before it bursts. A modification of this procedure, where the surface tension is measured from the shape of a liquid drop resting on a solid surface, is referred to as the sessile drop method.

3. Measurement of Viscosity

Viscosity is defined as the fluid's resistance to change in form due to internal friction.

The flow characteristics of low molecular weight pure substance and solutions obey Newtonian flow in that the velocity of flow is directly proportional to the shearing stress. Most disperse systems do not show Newtonian flow behavior but rather follow any of the other flow behaviors for fluids, namely pseudoplastic, plastic, or dilatant behavior (Fig. 10).

Disperse systems are formulated with a desired viscosity. As a result, the viscosity value should not change appreciably with age. A wide variety of techniques and many devices have been developed to measure viscosity properties of disperse systems. These procedures can be classified as either absolute or relative. The absolute either directly or indirectly measures specific components of shear stress and shear rate to define an appropriate rheological function. Methods used for absolute viscosity measurements are flow through a tube, rotational methods, or surface viscosity methods. Methods used for relative viscosity measurement are those using orifice viscometers, falling balls, or plungers. Such instruments, although they do not measure stress or shear rate, offer valuable quality control tests for relative comparison between different materials.

a. Rotational Methods. These methods measure the shearing forces between two rotational cylinders. The two commercial apparatuses commonly used are the Couette and the Brookfield viscometers.

The Couette viscometer consists of an outer vessel that contains the sample to be tested, jacketed for temperature control, and geared to a driving motor. An inner cylinder, or forked paddle, is submerged in the liquid sample whose viscosity is to be

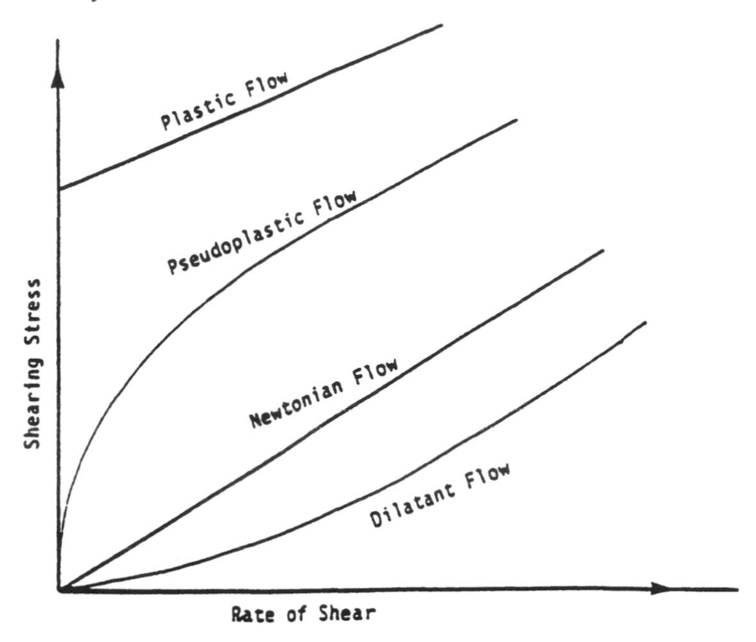

Fig. 10 Types of flow behavior.

measured. The cylinder, or the forked paddle, is caused to rotate using certain torque and specific weights. The cup is rotated at a standard speed and the torque required on the cylinder or forked paddle is measured by means of a calibrated wire. The data are reported as the weight required to obtain a standard rate of rotation.

The Brookfield viscometer consists of a low-speed synchronous motor that drives a specially shaped spindle through the liquid whose viscosity is to be measured. The torque exerted by the liquid viscosity on the spindle is measured by a calibrated spring. Different spindle shapes driven at variable rotational speeds allow a wide range of measurements of different disperse systems.

b. Falling or Rolling Ball Methods. A spherical ball falls or rolls in a temperature-controlled jacketed cylinder filled with the liquids whose viscosity is to be measured. The ball can roll in a straight or inclined cylinder through the liquid. True streamline flow of the sphere is essential for result reproducibility. A modification of the falling or rolling ball is achieved by using a perforated disc attached to a rod instead of the ball, and is referred to as a falling or rolling plunger.

c. Capillary Tube or Orifice Methods. These methods measure the relative viscosity either by measuring the pressure that drives the liquid flow through a capillary tube under gravity, or the rate of the liquid efflux through a small orifice. The two commercial instruments commonly used are the Ostwalt and the Saybolt apparatuses.

The Ostwalt apparatus consists of a reservoir for the liquid to be tested attached to a calibrated capillary pipette submerged in a temperature-controlled water bath. The liquid flows through the capillary pipette under gravity, and the time required for the meniscus to pass between the two calibrations is measured. For better accuracy, the results are compared against water as a reference standard liquid, and the capillary pipette is kept exactly vertical.

For disperse systems of high viscosity (104 centipoise or greater), the efflux method using the Saybolt apparatus is used. The Saybolt apparatus determines the relative viscosity by measuring the rate of efflux of the liquid under test through a small orifice. The viscosity result is reported as the number of seconds required for a specified volume of the liquid flow out.

The Saybolt apparatus consists of a jacketed calibrated vessel that can hold more than 60 mL with an overflow reservoir at the top and an efflux orifice at the bottom. The viscosity is determined by measuring the time required for 60 mL to flow into a calibrated vessel under gravity.

d. Surface Viscosity Methods. These methods measure both surface and interfacial viscosity by the use of either a two-dimensional capillary surface slit or by measuring the dampening rate of the oscillations of a disc in the surface of the test liquid.

The surface slit viscometer consists of a long narrow tube by which the difference of pressure at its two ends is kept constant by the use of surface slit barriers. The surface slit measurement is more difficult for interfacial viscosities. The oscillating disc method is preferred for measuring interfacial viscosities.

The oscillating disc apparatus consists of a disc fastened to a tension pendulum device in which the amplitude of oscillation is read out electronically by means of a gauge.

4. Measurement of Droplet or Particle Size and Distribution

Fine droplets or particles are described in terms of concentration, size, and size distribution. Some of the reasons for this type of characterization are primarily associated with preformulation, formulation, manufacturing convenience and cost, packaging and handling convenience, and product quality and stability.

The droplet, or particle size, of the dispersed phase in a disperse system is dependent on both the method of manufacture and formula used. The size of the droplet or particle can affect product appearance. Such particles cause difference in light scattering, absorption and reflection, rheological properties, and stability of the disperse system product. Droplet or particle size can be determined by microscopic methods, sedimentation methods, and optical methods, including light scattering, spectroturbidimetric, reflectance, and electrolyte displacement methods.

a. Microscopic Methods. Direct microscopic examination, while extremely laborious, is probably the most certain in results and should not be overlooked. Microscopic technique involves counting the droplets or particles of varying sizes and constructing a droplet/particle-size-distribution curve from the data (Fig. 11).

Alternatively, for emulsions, a photomicrographic record can be made, and the droplet sizes are determined from the print. Necessary dilutions and proper choice of enlargement facilitate measurement. The measurement of droplet sizes from a photomicrographic print is carried out with calipers or by the use of a transparent template. Statistically significant counts have been made with as few as 300 droplets, in which the error at any value of the cumulative distribution was less than 8% within a 95% confidence limit. Recently, automated image analyzers are often used in defining the size and number of droplets or particles measured microscopically.

Transmission and scanning electron microscopes are used for particle measurement below 1 μm diameter. The scanning electron microscope is used to determine size, shape, and surface characteristics of the particle, while the transmission microscope is

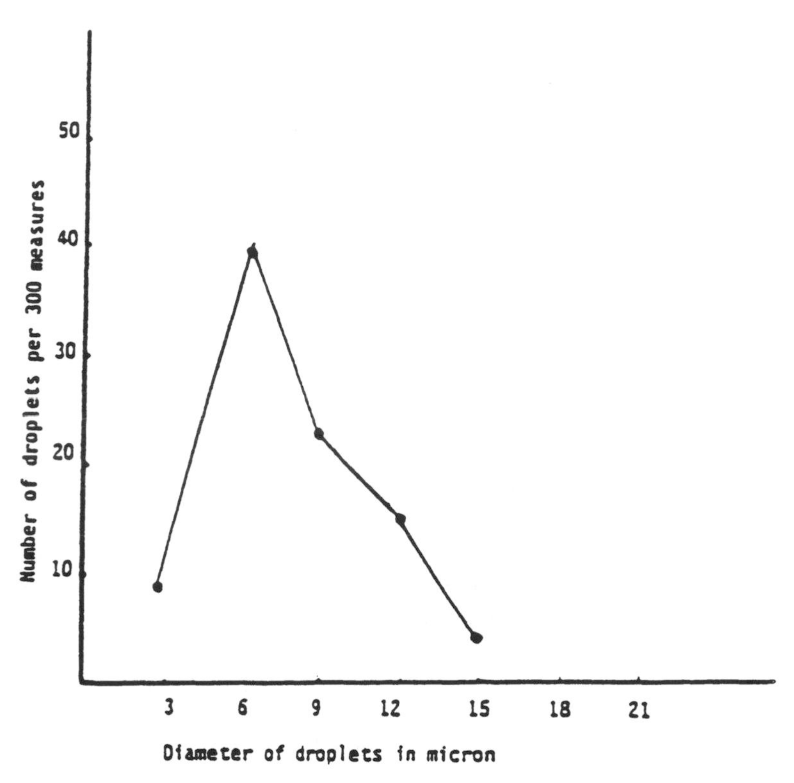

Fig. 11 Particle size distribution curve.

used when details of the size and projected shape of very small particles are required.

b. Sedimentation Methods. This method is used to measure emulsion creaming. The emulsion under test is placed in a large cylinder with a capillary side arm filled with the continuous phase. As creaming occurs, the change in density at the top of the emulsion in the large cylinder is reflected by a change in the level of disperse phase in the capillary side arm.

a. Electrolyte Displacement Methods. The electrolyte displacement method offers a fairly accurate technique for oil-droplet and particle size analysis of oil-in-water emulsions and suspensions, respectively. The electrolyte displacement method is usable only for oil-in-water emulsions where the continuous phase is or can be made electrically conducting.

To measure particle size electrically, the test sample is suspended in an electrolyte solution. The resultant suspension flows through a narrow orifice that has conductivity electrodes on both sides. Each time a droplet or particle passes through the orifice, the volume of electrolyte changes. As a consequence, the resistance between the two electrodes is changed. This change in resistance is converted to a voltage pulse whose magnitude is proportional to the droplet or particle size. If more than one droplet or particle passes the orifice at any time, the results can be corrected based on accepted statistical theory. A range of different-sized orifice tubes is available that allows this technique to measure droplets or particles with diameters less than 2 μm.

b. Optical Methods. Optical methods measure droplet or particle sizes by either measurement of reduction of light directly transmitted through the disperse system, as by spectroturbidimetric and nephelometric methods, or by light transmitted or scattered at a definite angle, usually 90° from the light path. The light-scattering technique has been widely used lately in order to study the droplet size of microemulsions. Optical methods assume a spherical droplet or particle shape and, as such, measure only the average size of all the particles. Consequently, this method is of little value in measuring the size distribution of the distribution of the diffused droplets or particles. The relationship between droplet or particle size and transmission, absorption, or scattering is usually determined empirically for the disperse system concerned, assuming that the scattering is proportional to the surface area of the droplets or particles.

5. Measurement of Electrical Conductivity

Measurements of electrical conductivity afford a convenient way of distinguishing between oil-in-water and water-in-oil emulsions. Oil-in-water emulsions, where water is the continuous phase, show high conductivity, whereas water-in-oil emulsions, where oil is the continuous phase, show little or no conductance. Measurements of an emulsion's electric conductivity are used to investigate degree of dispersion in dilution preformulation studies, and phase inversion and creaming in stability studies. The test is performed by measuring the current flow between two fixed platinum electrodes immersed in the emulsion. Oil-in-water emulsions pass a current of 10–13 mA, whereas water-in-oil emulsions pass a current of 0.1 mA or less.

a. Zeta Potential. Zeta potential is the work required to bring a unit charge from infinity to the edge of a fixed interface layer. Since the stability of an emulsion is so much a function of the electrical properties of the interface, zeta potential, in conjunction with other stability measurement, is important in emulsion stability studies and shelf-life prediction.

Zeta potential is obtained by direct measurement of the rate of migration of emulsion droplets using microelectrophoresis. Usually, the test is performed using the commercially available Zeta-Meter microelectrophoresis apparatus. This apparatus can determine electrophoretic mobility and zeta potential on the same emulsion sample.

b. Dielectric Constant. The dielectric constant is a measure of the work done by an external field in orienting molecules against the effect of their thermal agitation. The dielectric constant can be used as an index of extent of aggregation of disaggregation in disperse systems. An increase in dielectric constant indicates aggregation in the disperse system during stability studies.

For dilute disperse systems, the dielectric constant is calculated from Eq. (6)

$$\frac{\varepsilon - \varepsilon m}{\varepsilon + 2\varepsilon m} \phi \frac{\varepsilon p - \varepsilon m}{\varepsilon p + 2\varepsilon m} \tag{6}$$

where εm = dielectric constant of continuous phase; εp = dielectric constant of dispersed phase; ε = dielectric constant of dispersion systems; and ϕ = volume fraction of the dispersed phase.

For concentrated disperse systems, the dielectric constant is calculated from Eq. (7)

$$\frac{\varepsilon p - \varepsilon m}{\varepsilon m + \varepsilon p} = (1 - \phi) \quad \sqrt[3]{\frac{\varepsilon}{\varepsilon m}} \tag{7}$$

6. *Determination of Disperse System Type*

In medical and pharmaceutical practice, emulsion and suspension systems have been used since the earliest days for therapeutic purposes. Emulsion was defined previously as the dispersion of one immersible liquid in another continuous phase, frequently with the help of an emulsifying agent. Emulsions can be further classified into macro- and microemulsions. Emulsions are microemulsions when the dispersed liquid-liquid systems in which the particle size of the dispersed phase is less than 100 Å and as macroemulsions when the particle size of the dispersed phase is more than 100 Å. Suspensions are defined as the dispersion of solid in a continuous liquid phase. Several methods are used to identify the disperse system type: phase dilution, fluorescence, dye solubility, conductivity, and particle size methods.

a. Phase Dilution Method. This method is used to characterize the type of emulsion under test and depends on the fact that an emulsion is readily dilutable by the liquid that constitutes the continuous phase. To perform the test, several drops of the emulsion are placed on a glass slide under a low-powered magnification microscope. A drop of one emulsion component is added to each drop of the emulsion, stirred gently while being examined under the microscope. The continuous phase component will be the drop that blends readily with the emulsion drop.

b. Fluorescene Method. This method is used to identify whether the emulsion is oil-in-water or water-in-oil. Many oils fluoresce under ultraviolet light; thus, examining a drop of the emulsion under a fluorescent light microscope will characterize the emulsion type. The test is performed by examining a drop of the emulsion under a fluorescent light microscope. If the whole field fluoresces, the emulsion is water-in-oil; while in the oil-in-water type emulsion, only a few fluorescing dots are observed.

c. Dye Solubility Method. This method is used to identify the continuous phase of an emulsion. Particular water- or oil-soluble dyes are chosen (e.g., brilliant blue or red sudan III), which are soluble in one emulsion component and insoluble in the other.

d. Conductivity Method. Since aqueous systems are good conductors of electricity and since oil systems are poor conductors, a crude measure of electric conductance can characterize whether the emulsion continuous phase is either water or oil. The test is performed by dipping two electrodes attached to a simple electric circuit with a light indicator into a beaker that contains the test emulsion. A few crystals of sodium chloride are added and mixed gently into the emulsion. The light goes on if the continuous phase of the emulsion is aqueous in nature.

e. Droplet or Particle Size Method. In general, the use of microscopic, instrumental counting, and optical measurement techniques serves to characterize the disperse system under test. These methods were described in detail earlier.

7. *Particulate Matter in Injections*

Injectable solutions, including solutions composed of sterile solids intended for parenteral use, should be essentially free from particles that can be observed on visual injection. The USP describes two procedures for the determination of particulate matter to be used according to the volume of the injectable products to be tested. A filtration followed by microscopic examination test procedure is used in the case of injectable products that are labeled as containing more than 100 mL of a single-dose large-volume injection intended for administration by intravenous infusion. An electronic liquid-borne particle counter system, using a light-obscuration-based sensor, is used for small-volume injectables that are labeled as containing 100 mL or less, for single or multiple dose, either in solution or in solution constituted from sterile solids.

The large-volume injectable for single-dose infusion meets USP requirements if the units under test contain not more than 50 particles per mL that are equal to or larger and 10 μm and not more than 5 particles per mL that are equal to or larger than 25 μm in effective linear dimension.

In the case of small-volume injections, they meet USP requirements if the average number of particles contained is not more than 10,000 per container that are equal to or greater than 10 μm in effective spherical diameter and not more than 100 per container equal to or greater than 25 μm in effective spherical diameter.

8. Bacterial Endotoxins Test

Water for injection, transfusion, and infusion assemblies, and most of the antibiotic monographs in the USP allow the use of the bacterial endotoxin test to estimate the concentration of bacterial endotoxins that may be present in or on the article using limulus amebocyte lysate (LAL), which has been obtained from aqueous extracts of the circulating amebocytes of the horseshoe crab, *Limulus polyphemus*. USP endotoxin reference standard has a defined potency of 10,000 USP endotoxin units per vial. USP LAL reagent is prepared and characterized for use for gel-clot formation. However, as an alternative method, LAL reagent can be formulated to be used for turbidimetric or colorimetric procedures providing it complies with additional USP requirements. Since the form and amount per container of standard endotoxin and of LAL reagent may vary, constituting and/or dilution of contents should be as directed in the labeling. The pH of the test mixture of the specimen and the LAL reagent is adjusted within the range of 6.0–7.5 unless specified otherwise in the individual monograph. The pH may be adjusted using sterile, endotoxin-free sodium hydroxide or hydrochloric acid, or suitable buffers. Perform the test using the specified volumes of products, the endotoxin standard, negative controls, and a positive control. Not less than two replicate reaction tubes at each level of the dilution series for each product under test are conducted in parallel, whether the test is employed as a limit test or as a quantitative assay. A set of standard endotoxin dilution series should be included for each block of tubes, which may consist of a number of racks for incubation together under the same environmental conditions. Incubate each tube undisturbed for 60 ± 2 min at 37° ± 1°C and carefully remove for observation. A positive reaction is characterized by the formulation of a firm gel that remains when inverted through 180°. A negative result is characterized by the absence of such a gel or by the formulation of a viscous gel that does not maintain its integrity. The test should be considered invalid if the positive product control or endotoxin standard does not show the end-point concentration to be within a ± 1 twofold dilution from the label-claim sensitivity of the LAL reactant, or if any negative control shows a gel-clot end point. USP requirements are met if the concentration of endotoxin does not exceed that specified in the individual article monograph and the confidence limits of the assay do not exceed those specified.

9. Pyrogen Test

The pyrogen test is designed to measure the rise in temperature of rabbits following the intravenous injection of the sample test solution for products that can be tolerated by the test rabbit in a dose not to exceed 10 mL per kg when injected intravenously within a period of not more than 10 min. Special preparation directions and conditions of administration, such as in the case of antibiotics or biologics, should be followed.

Unless otherwise specified in the individual monograph, the ear vein of these rabbits is injected with 10 mL of the test solution at 37 ± 2°C per kg of body weight within 10 min after start of administration. Temperatures are recorded at 1, 2, and 3 hr subsequent to injection.

USP requirements are met if no rabbit shows an individual rise in temperature of 0.6°C or more above its respective control temperature and if the sum of the three rabbits' maximum temperature rises does not exceed 1.4°C. If these requirements are not met, an additional five rabbits should be tested. If not more than three of eight rabbits show individual rises in temperature of 0.6°C or more, and the sum of the eight individual rabbits maximum temperature rises does not exceed 3.7°C, the article under test meets USP requirements.

10. *Sterility Test*

Sterile pharmacopeial articles should comply with USP sterility test requirements. Alternative procedures may be used provided the results obtained are at least of equivalent reliability to the USP sterility test. However, in case of a dispute, the USP sterility test results are used.

USP sterility tests are either transfer to test media or membrane filtration techniques. Membrane filtration procedure is the method of choice for liquids and soluble powders processing bacteriostatic or fungistatic properties, as well as for oils, ointments, and creams that can be put into solution with nonbacteriostatic or nonfungistatic diluting fluids. Media, as well as diluting and rinsing fluids used for USP sterility tests, should be prepared following USP directions. Each autoclave load of each lot of medium should be tested for growth-promoting qualities by separating inoculating duplicate test containers of each media with 10–100 viable microorganisms of each strain listed in Table 25 and incubating at the temperature and for the length of time specified. The test media are satisfactory if clear evidence of growth appears in all inoculated media containers with seven days. Freshly prepared media are stored in the dark at 2–25°C if not used within two days. Media can be stored in sealed containers for one year, provided they are tested for growth promotion every three months and if the color indicator requirements are met.

The sterility test by direct transfer to media is performed by aseptically removing the specified volume of liquid from each test container to a vessel of culture medium according to Table 26. The vessel is mixed and incubated at the specified temperature for 14 days. Media should be visually inspected for growth on the 3rd, 4th, or 5th day; on the 7th or 8th day; and on the 14th day.

Sterility tests, using membrane filtration, are performed using volumes from liquid articles that are not less than the volumes listed in Table 26. A suitable membrane filter unit that uses a membrane with a nominal porosity of 0.45 μm, a diameter of approximately 47 mm, and a flow rate of 55–75 mL of water per minute at a pressure of 70 cm of mercury should be used. The entire unit is preferably assembled and sterilized with a membrane in place prior to use. Because of the diversity in the nature of articles to be tested and other factors affecting the conduct of the sterility test, it is important to follow the USP directions in performing the sterility test according to each article's requirements. Membranes are aseptically removed from the holder and cut in

Table 25 USP Sterility Tests Growth Promotion Microorganisms

Medium	Test Microorganisms	Incubation Temperature (°C)	Conditions
Fluid thioglycolate	*Bacillus subtilis* (ATCC No. 6633)[a]	30–35	—
	Candida albicans (ATCC No. 10231)	30 35	Aerobic
	Bacteroides vulgatus (ATCC No. 8482)[b]	30–35	—
Alternative thioglycolate	*Bacteroides vulgatus* (ATCC No. 8482)[b]	30–35	Anaerobic
Soy bean–casein digest	*Bacillus subtilis* (ATCC No. 6633)[a]	20–25	—
	Candida albicans (ATCC No. 10231)	20–25	Aerobic

[a]If a spore-forming organism is not desired, use *Micrococcus luteus* (ATCC No. 9341) at the incubation temperature indicated in the table.
[b]If a spore-forming organism is desired, use *Clostridium sporogenes* (ATCC No. 11437) at the incubation temperature indicated in the table.

half if only one membrane is used and immersed in an appropriate medium and incubated at the specified temperature for seven days. USP requirements are met if no growth is observed at the conclusion of the incubation period.

If microbial growth is found, but a review of the sterility testing facility, the environmental monitoring, the materials used, the testing procedure, or the negative controls indicates that an inadequate or faulty aseptic technique was used in the test itself, the first stage is declared invalid and may be repeated. If microbial growth is observed, but there is no evidence invalidating the first stage of the test, proceed to the second stage. The minimum number of specimens selected is double the number tested in the first stage. If no microbial growth is found, the article tested meets the requirements of the test for sterility. If growth is found, the result so obtained is conclusive that the article tested fails to meet the requirements of the test for sterility. If, however, it can be demonstrated that the second stage was invalid because of faulty or inadequate aseptic technique in the performance of the test, the second stage may be repeated.

E. Reconciliation

When the manufacturing process has been completed, the theoretical yields to be expected from the formulation at different stages of manufacture and the accountability calculations are checked for comparison with the practical and permissible yield limits. Such information is recorded on appropriate forms, as shown in Table 27, and any discrepancy must be reconciled if beyond process-allowable variation.

F. Auditing

Good manufacturing practices require that the manufacturing process be adequately documented throughout all stages of the operation. The history of each patch, from the

Table 26 Liquid Quantities for USP Sterility Test

Container content (mL)	Minimum volume taken from each container for each medium	Minimum volume of each medium		Number of containers per medium
		Use for direct transfer of volume taken from each container (mL)	Used for membrane or half membrane representing total volume from the appropriate number of containers (mL)	
Less than 100	1 mL, or entire contents if less than 1 mL	15	100	20 (40 if each does not contain sufficient volume for both media)
10 to less than 50	5 mL	40	100	20
50 to less than 100	10 mL	80	100	20
50 to less than 100, intended for intravenous administration	Entire contents	—	100	10
100–500	Entire contents	—	100	10
Over 500	500 mL	—	100	10

Table 27 Product Accountability

Product name: ______ Strength: ______ Size: ______ Lot # ______ Code # ______

I. Container accountability
  A. Containers issued to production
  B. Finished containers filled
  C. Excess containers returned to warehouse
  D. Samples
    1. Quality assurance
    2. Preservation
    3. Stability
  E. Rejects
  F. Containers accountability $= B + C + D + E = X$; deviation $= [(A - X)/A] \times 100 = \pm\ \%$.

II. Bulk accountability
  A. Weight or volume of bulk prepared
  B. Filled quantity = units sealed × average fill
  C. Samples
    1. Quality assurance
    2. Machine setup
    3. Other
  D. Bulk accountability $B + C = X$; deviation $= [(A - X)/A] \times 100 = \pm\ \%$.

Prepared by: ______________________ Date: ______________
Checked by: ______________________ Date: ______________
Quality assurance: ______________________ Date: ______________

starting materials, equipment used, and personnel involved in production and control until packaging is completed, should be recorded. Reserve samples are to be stored for at least two years beyond the labeled expiration date of each product. Areas for which records are kept are

- Individual components, raw materials, and packaging master formula cards
- Batch production
- Containers and labeling
- Packaging and labeling operations
- Laboratory control testing, in-process and finished
- Proper signing and dating by at least two individuals independently for each operation in the proper spaces
- Reconciliation of materials supplied with amount produced, taking into account allowable loss limits

Before quality assurance releases the product for distribution, it should evaluate the complete batch records of all in-process tests and controls and all tests of the final product to determine whether they conform to specifications.

A summary example of a quality assurance audit of a product to assure correct in-process operations is as follows:

- Approval of bulk for filling: lot number strength, size, signatures, working formula card and procedure, and corrections

- Approval of filled product for packaging: lot number, strength, size, signature, working formula card and procedure, biological control report, environmental control report, labeling, packaging forms, volume check, and visual inspection
- Approval of finished product in stock: lot number, strength, size, signatures, working formula card and procedure, analytical and microbiological results, biological control report, environmental control report, packaging and labeling inspection, reconciliation record, line clearance, and release for stock

G. Complaints

Good manufacturing practice regulations stipulate that the quality assurance team must record in a special file any complaints regarding the quality of a drug, including any change in its physical characteristics. Quality assurance must investigate thoroughly each complaint and route specimens or, in some instances, preservation and/or stability samples, to respective laboratories for appropriate testing. Appropriate testing should be performed, recorded, and filed with the original complaint. Reports of injuries or adverse reactions resulting from the use of a drug should be forwarded to the appropriate authorities. All complaint records should be retained for at least two years after distribution of the drug has been completed, or one year after the product's expiration date, whichever is shorter.

H. Stability

One of the primary objectives in pharmaceutical product development is to develop a product that meets the desired specification for the intended use and is stable during its expected shelf life. This goal requires that not only the active ingredient(s) be monitored, but that also other components such as antioxidants, preservatives, fragrance, and colors added to the disperse system be studied as well. It is also important to make sure that the disperse system chosen is stable to freeze-thaw conditions that may occur during shipping. To attain these goals, disperse systems are normally subjected to accelerated stress conditions that are used to predict shelf-life expiration dates.

The legal requirements for stability stem from the Federal Food, Drug, and Cosmetic Act and the Current Good Manufacturing Practices Regulations (CGMP) as published in the *Federal Register* on March 28, 1979, and Title 21, Code of Federal Regulations. The CGMPs state that there shall be assurances of stability of the finished product and suitable expiration date, based on appropriate stability studies and related to specific storage conditions placed on the label. Stability tests are to be performed in finished drug products' marketed containers using reliable and specific test methods. Effective September 28, 1979, all drug labels, prescription and nonprescription, are required to state the product's expiration date, the time after which the drug can no longer be considered within the legal potency requirement.

The United States Pharmacopeia/National Formulary (USP/NF) has required expiration dates on all monograph drugs since January, 1976. The USP/NF states that, for pharmacopeia products, the label shall bear an expiration date limiting the period during which the product may be expected to retain the full label potency of the active ingredient provided the product is stored as directed. A set of label storage temperatures is provided by the USP/NF to cover the various types of storage conditions; these directions include the following: store in a freezer not exceeding −10°C, store in a refrigerator held between 2 and 8°C, store in a cool place between 8 and 150°C, store at controlled room temperature between 15 and 30°C, store at room temperature (25°C ± 2), protect from excessive heat (any temperature above 40°C), protect from light, protect from freezing, protect from moisture, and keep tightly closed. The marketing container is to be used for all the stability studies and under the same conditions as storage and reconstitution time as specified on the label.

In addition to routine active(s) potency monitoring in disperse systems, minimum concentrations of each antioxidant are required to prevent rancidity, especially of unsaturated compounds; therefore, stability-indicating analytical procedures to monitor antioxidant and actives throughout the expected shelf life are required. On the other hand, chemical monitoring of preservative systems is not sufficient. Preservatives are added to minimize contamination with microorganisms during manufacture and consumer use through the expected shelf life. Therefore, it is necessary to monitor the preservative system chemically using stability-indicating analytical procedures and biologically using the USP microbial challenge test or other appropriate microbiological tests.

FDA stability guidelines recommend that at least three months of data generated from accelerated stability storage conditions be used in determining an expiry dating period. In addition, stability data from a minimum of three lots of the commercial drug product made according to production-scale procedures at the label storage condition should also be provided in support for a proposed expiration date. Tables 28 and 29 provide examples for stability protocols for oral suspension and topicals disperse system drug products.

I. Extent of Separation Measurement

In disperse systems, the extent of breakdown as a function of time is an important stability measurement. Ideally, in a stable emulsion, the droplets in the dispersed phase should maintain their size and remain suspended indefinitely. A stable disperse system should have an unchanging amount of dispersed phase in a given amount of continuous phase when subjected to an expected stress condition of temperature, agitation, and gravitational forces. However, the dispersed phase tends to agglomerate, coalesce, flocculate, or sediment even in stable emulsions, but to a far lesser degree than in unstable ones.

In agglomeration, the dispersed phase particles agglomerate to form a cluster of particles of coalesce to form larger droplets that lead to phase separation. In the case of flocculation, agglomeration can be reversed by shaking.

Creaming occurs when dispersed particles either settle or float with respect to the continuous phase and either the lower or upper portion, respectively, becomes more opaque or creamier.

Crystal growth in suspensions leads to instability. Phase inversion and creaming are two other indications of disperse system instability.

Several methods have been used to measure the extent of separation of disperse systems. Dielectric constant, particle size distribution, optical measurement, and counting methods are valuable tools to enable the product development scientist to decide whether a particular disperse system is more stable than another.

Volume of sediment and ease of redispersibility tests are used to measure the extent of separation of suspensions. The time during which redispersed suspension stays suspended is a measure of its stability, or even acceptability, as a product for marketing.

Table 28 Stability Protocol for Disperse System Oral Suspension Products

Time	Light	Storage conditions						
(months)	(600 Fc)	5°C	60%RH/25°C	37°C	50°C	−18°C	75%RH/40°C	Freeze-thaw
0	T_1[a]	T_2[b]	T_2	T_1	T_2	T_2	T_2	T_2
1	T_1	T_1	T_1	T_1	T_1	T_1	T_1	T_1
2	T_1	—	T_1	T_1	T_1	T_1	T_1	T_1
3	T_1	T_1	T_1	T_1	T_2	T_2	T_2	T_2
6	—	T_1	T_1	T_1	—	—	T_2	—
9	—	T_1	T_1	T_1	—	—	—	—
12	—	T_1	T_1	T_1	—	—	—	—
24	—	—	T_1	—	—	—	—	—
36	—	T_1	T_1	—	—	—	—	—
48	—	—	T_1	—	—	—	—	—
60	—	—	T_1	—	—	—	—	—

[a]T_1 = active(s)/degradation, preservative(s), visual/pH.
[b]T_2 = active(s)/degradation, preservative(s), visual/pH, specific gravity, and/or viscosity.

Table 29 Stability Protocol for Disperse System Topical Products

Time (months)	Light (600 Fc)	Storage conditions					
		5°C	60%RH/25°C	37°C	50°C	75%RH/40°C	Freeze-thaw
0	T[a]	T	T	T	T	T	T
1	T	T	T	T	T	T	T
2	T	T	T	T	T	T	T
3	T	T	T	T	T	T	T
6	—	T	T	T	—	T	—
9	—	T	T	T	—	—	—
12	—	T	T	T	—	—	—
24	—	—	T	—	—	—	—
36	—	—	T	—	—	—	—
48	—	—	T	—	—	—	—
60	—	—	T	—	—	—	—

[a]T = active(s)/degradation, preservative(s), visual (odor/color/appearance, etc.)/pH.

VI. REGULATIONS

Most governments promulgate regulations governing the manufacture, processing, quality, and distribution of finished dosage form pharmaceuticals. Examples of such regulations in the United States, EEC, and Japan are presented in Part 211, Title 21 of the Code of Federal Regulations, the *Rules Governing Medicinal Products in the European Community*, and Japan's *Provisions of the Pharmaceutical Affairs Law Concerning GMP*, respectively.

Disperse systems quality depends on many variables, such as the quality of the components or materials used, type of equipment, manufacturing, processing, handling, installation, testing, shipping, calibration, maintenance, training, experience of personnel, environmental conditions, and caliber of procedures used. The process, organizational structure, procedures, and resources that pharmaceutical manufacturers use to control these variables to produce product of consistent quality that meets defined specifications is called a quality system. While manufacturers may choose to evaluate their own quality systems, most governments have established a conformity assessment structure to regulate pharmaceutical products. In the United States the authority for duly appointed officers or employees of the FDA to enter and inspect establishments under the jurisdiction of the Federal Food, Drug and Cosmetic Act is in Section 704 of the Act (21 U.S.C. 374). Refusal to permit inspection upon presentation of official notice by appropriately identified FDA officers or employees pursuant to 21 U.S.C. 374, exposes any person responsible for such refusal to criminal penalties under 21 U.S.C. 331(F) and 333.

According to the FDA, assurance of product quality is derived from careful and systematic attention to a number of critical, important factors, including selection of quality components and methods, adaquate product and process design, and statistical control of the process through in-process and end-product testing. To clarify and identify those critical and important factors the FDA's office of Regulatory Affairs issued in the last few years numerous guidelines that cover most of the pharmaceutical industry's manufacturing and quality operations.

Those guidelines include *Guide to Inspection of Bulk Pharmaceutical Chemicals, Guide to Inspection of High Purity Water Systems, Guide to Inspections of Lyophilization of Parenterals, Guide to Inspections of Microbiological Pharmaceutical Quality Control Laboratories, Guide to Inspections of Validation of Cleaning Processes, Guide to Inspection of Dosage Form Drug Manufacturers—CGMPs, Guide to Inspection of Liquid Injectable Radiopharmaceuticals Used in Positron Emission Tomography (PET), Guide to Inspection of Oral Solid Dosage Forms Pre/Post Approval Issues for Development and Validation, Guide to Inspection of Sterile Drug Substance Manufacturers, Guide to Inspection of Topical Drug Products*, and *Guide to Inspections of Oral Solutions and Suspensions*.

Disperse systems products must be manufactured in accordance with the CGMP regulations, otherwise they are considered to be adulterated within the meaning the FD&C Act, Section 501(a)(2)(B). To assure compliance with these requirements, four of the above guidelines are explained in detail: *Guide to Inspection of Oral Solutions and Suspensions, Guide to Inspection of Dosage Form Drug Manufacturers—CGMPs, Guide to Inspection of Pharmaceutical Quality Control Laboratories*, and *Investigation Operations Manual*.

Guide to Inspection of Oral Solutions and Suspensions covers facility, equipment, raw materials, compounding, microbial quality, oral suspensions uniformity, product specifications, process validation, stability, and packaging. The FDA investigator is instructed under facility, to cover possible cross-contamination, microbial contamination, heating, ventilating and air conditioning systems including air recirculation, air filtration, and dust control. Equipment should be of sanitary design, identified, and detailed in drawings and SOPs. For the manufacture of suspensions, valves should be flushed and cleaning validation for equipment and transfer logs must be completed. The physical characteristics, particularly the particle size of the drug substance, are very important for suspensions. For oral suspensions, microbial and uniformity are of concern for the potential of contamination and segregation during manufacturing operation time, temperature and nitrogen purging, if needed, should be justified and validated. Disperse systems either for orals or injectables, especially those intended for pediatric use, should be protected from possible microbial contamination, especially *Gram-negative* or *Pseudomonas* sp. Assurance of uniformity, especially those preparations with low viscosity, is important to assure that segregation during manufacturing has not occurred. Final production specification should be adequate to assure that the product will meet the assigned expiration date. Physical characteristics, such as viscosity, particle size, and microbial limit, in addition to potency and degradation products should be part of the stability protocol. As with other pharmaceutical products, process validation including the predetermined specification at critical phases of the manufacturing operation is required, and for some product bio-equivalency studies are required. Stability studies should be performed in the marketed closure system with some units stored on their side or inverted to determine product incompatibility with the closure system. Cleaning of containers before packaging, testing for uniformity of fill, and calibration of measuring devices such as droppers must be performed.

Guide to Inspection of Dosage Form Manufacturers is intended to be a general guide to inspection of drug manufacturing to determine manufacturer compliance with drug CGMPRs. All manufacturers of prescription and over-the-counter drugs must comply with the drug CGMP requirements which are organization and personnel, buildings and facilities, equipment, components and product containers, production and process controls for tablets and capsule products, sterile products, ointments, liquids and lotions, packaging and labeling, holding and distribution, laboratory controls, control records, and returned drug products, as presented in 21 CFR, subparts B, C, D, E, F, G, H, I, J, and K, respectively.

Guide to Inspection of Pharmaceutical Quality Control Laboratories pertains to the quality control laboratory and product testing. It may include inspection for specific testing methodology or laboratory operation, or a complete assessment of a laboratory's conformance with CGMPs. Inspection guides are based on the team approach and detail FDA requirements in general and preapproval inspections. It covers failure or out-of-specification laboratory results including laboratory error, in-process–related or operator error and process-related or manufacturing process error, product failures, retesting, resampling, averaging results of analysis, blend sampling and testing, microbiological testing, sampling, laboratory records and documentation, laboratory standard solutions, analytical methods validation, laboratory equipment, raw material testing, in-process control and specifications, stability, computerized laboratory data acquisition systems, and laboratory management. The use of computerized laboratory data acquisition systems requirements is addressed in more detail in CGMP guidance documents, *Compliance Policy Guide Computerized Drug Processing: Input/Output Checking*, *Identification of "Persons" on Batch Production and Control Records*, *CGMP Applicability to Hardware and Software*, *Vendor Responsibility*, *Source Code for Process Control Application Programs*, and *Guide to Inspection of Computerized Systems in Drug Processing*.

Investigations Operations Manual (IOM) serves as the primary guide to FDA field investigators and inspectors on policies and procedures. The IOM has over 500 pages and contains 10 chapters: Administration, Organization, Federal and State Cooperation, Sampling, Establishment Inspection, Imports, Regulatory, Recall Procedures, Investigations, and Reference Materials.

SUGGESTED READING

Adams, P. B., Corning Research, Corning Glass Works, New York, 1973.

Adeyeye, C. M. and Price, J. C., *Drug Dev. Ind. Pharm.*, 17 (1991).

Barsh, H. G. and Sun, S.-T., *Anal. Chem.*, 63(12) (1991).

Becher, P. In Encyclopedia of Emulsion Technology, Vol. 1, Marcel Dekker, New York, 1983.

Becher, P. In: *Encyclopedia of Emulsion Technology*, Vol. 2, Marcel Dekker, New York, 1985.

Becher, P. In: *Encyclopedia of Emulsion Technology*, Vol. 3, Marcel Dekker, New York, 1987.

Benita, S., and Levy, M. Y., *J. Pharm. Sci.*, 82 (1993).

Caldwell, L. J., and Anderson, B. D., *Pharm. Res.*, 10 (1993).

Code of Federal Regulations, Part 211, Title 21, Bernam Press, Lonham, MD, 1995.

Deem, D., In: *Pharmaceutical Dosage Forms: Disperse Systems*, Vol. I (H. A. Lieberman, M. A. Rieger, and G. S. Banker, eds.), Marcel Dekker, New York, 1988.

Griffin, R. C., and Sacharow, S., *Drug and Cosmetic Packaging*, Noyes Data Corp., New Jersey, 1975.

Guide to Inspection of Dosage Forms, Drug Manufacturers - CGMPR's, The Division of Field Investigations, Office of Regional Operations, Office of Regulatory Affairs, Food & Drug Administration, Rockville, MD, 1993.

Guide to Inspection of Pharmaceutical Quality Control Laboratories, The Division of Field Investigations, Office of Regional Operations, Office of Regulatory Affairs, Food & Drug Administration, Rockville, MD, 1993.

Guide to Inspections of Oral Solutions and Suspensions, The Division of Field Investigations, Office of Regional Operations, Office of Regulatory Affairs, Food & Drug Administration, Rockville, MD, 1994.

Idson, B., *Drug Cosmetic Industry*, 143: 74 (1988).

Investigations Operations Manual, The Division of Field Investigations, Office of Regional Operations, Office of Regulatory Affairs, Food & Drug Administration, Rockville, MD, 1994.

Japan's Provision of the Pharmaceutical Affairs Law Concerning GMP, *International Drug GMP's*, Tokyo, Japan, 1990.

Lissant, K. J., *Emulsions and Emulsion Technology*, Marcel Dekker, New York, 1974.

Matsumoto, S., and Sherman, P., *J. Colloid Interface Sci.*, 33:294 (1970).

McClements, D. J., *Advances in Colloid and Interface Sci.*, 37: 33–72 (1991).

Radebaugh, G., In: *Pharmaceutical Dosage Forms: Disperse Systems*, Vol. 1, 2nd Ed., (H. A. Lieberman, M. A. Rieger, and G. S. Banker, eds.), Marcel Dekker, New York, 1996.

Rami Reddy, B., and Doyle , A. K., *Cosmet. & Toiletr.*, 99(Oct.): 67–72 (1984).

Rieger, M. M., *Surfactants in Cosmetics*, Vol. 16, Marcel Dekker, New York, 1985.

Rules Governing Medicinal Products in the European Community, GMP for Medicinal Products, Vol. IV, Commission of the European Communities, Brussels, Belgium, 1992.

Shah, D. O., *Macro- and Microemulsions Theory and Applications*, American Chemical Society, Washington, DC, 1985.

Sherman, P., *Industrial Rheology*, Academic Press, New York, 1970.

Sherman, P., *Soap Perf. Cosmet.*, 44:693 (1971).

Sugrne, S. I., *American Laboratory*, 64 (April) (1992).

Van Wazer, J. R., Lyons, J. W., Kim, K. Y., and Colwe, R. E., *Viscosity and Flow Measurement*, Interscience, New York, 1963.

Virtanen, S., Yliruusi, J., and Selkainaho, *E., Pharmazie*, 48(Jan.): 38–42 (1993).

12

Validation of Disperse Systems

Robert A. Nash

St. John's University, Jamaica, New York

I. INTRODUCTION

Validation has been made a requirement of current good manufacturing practices (CGMPs) for finished pharmaceuticals in the newly created Subpart L of article 21, Code of Federal Regulations (CFR) 211. Process validation has been defined, in 21CFR 210.3, as establishing, through documented evidence, a high degree of assurance that a specific process (i.e., dispersed systems) will consistently produce a product that meets its predetermined specifications and quality characteristics [1].

There is much confusion as to what process validation is and what constitutes process validation documentation. We use the term validation generically to cover the entire spectrum of CGMP concerns, most of which are, essentially, facility, equipment, component, methods, and process qualification. According to the official definition, the specific term process validation, when following qualification, should be reserved for the final stage(s) of the product and process development sequence and its subsequent manufacturing life cycle. The essential or key steps or stages of a successfully completed development program are presented in Table 1.

A. Validation of Pharmaceutical Systems

A number of texts and publications have been devoted to a discussion of the validation of both aseptic and solid pharmaceutical dosage forms [2–5]. In addition, the FDA has issued a number of guides and guidelines with respect to validation in general [6] as well as a number of specific topics, i.e., sterile drug products produced by aseptic processing, analytical data for methods validation, validation documentation inspection guide, cleaning validation, oral solid dosage forms: pre- and post-approval issues for development and validation, preapproval inspections/investigations, post-approval audit inspections, bulk pharmaceutical chemicals, and lyophilization of parenterals.

With specific reference to disperse systems, two guides have been issued separately by the FDA. They are the mid-Atlantic inspection guide for topicals [7] and the guide to inspection of oral solutions and suspensions [8].

Table 1 Key Stages of Successfully Completed Development Program

Developmental Stage	Batch size[a]
Product design	1X
Product characterization	1X
Product selection	1X
Process design	1X
Product optimization	10X
Process characterization	10X
Process optimization	10X
Process qualification	10X
Process qualification	100X
Process validation	100X
Process approval	100X

[a]Where 1X are laboratory-size formulation batches, 10X are laboratory-size pilot batches made for clinical and other studies in a GMP facility, and 100X are pilot production or full-scale batches in a GMP type of production facility. Process revalidation is 100X to 1000X.

In a well-designed validation program, most of the effort should be spent on facilities, equipment, component, methods, and process qualification. In such a program, the formalized, protocol-driven validation sequence provides the necessary process validation documentation required by the FDA to show product reproducibility and a manufacturing process in a state of control. Such a strategy is consistent with the Food and Drug Administration's (FDA's) preapproval inspection program directive [9].

B. Process Validation Options

The guidelines on general principles of process validation [6] mention three options. They are prospective process validation (also called premarket validation), retrospective process validation, and revalidation. There are actually four options, if concurrent process validation is also included.

1. Prospective Validation

A prospective validation is carried out prior to the distribution of either a new product or an existing product made under a completely revised manufacturing process where such revisions may affect product specifications or quality characteristics. The prospective approach features critical step analysis, in which the unit operations are challenged during the process qualification stage to determine, using either "worst case" analysis or a fractional factorial design [10], those critical process variables that may affect overall process performance. During formal, protocol-driven, prospective validation, critical process variables should be set within their operating ranges and should not exceed their upper and lower target limits during process operation. Output responses should be well within proposed finished-product specifications.

2. Retrospective Validation

Prospective validation is recognized in both CGMPs (21 CFR 211.110 b) and the process validation guidelines [6]. It involves using the accumulated in-process production and final product testing and control (numerical) data to establish that the product and its manufacturing process are in a state of control. Valid in-process results shall be consistent with drug product final specifications and shall be derived from previous acceptable process averages and process variability estimates where possible, and determined by the application of suitable statistical procedures, i.e., quality control charting, where appropriate [11].

The retrospective validation option is selected for established products whose manufacturing processes are considered to be stable and when, on the basis of economic considerations and resource limitations, prospective qualification and validation experimentation cannot be justified. Prior to undertaking either prospective or retrospective validation, the facilities, equipment, and subsystems used in connection with the manufacturing process must be qualified in conformance with CGMP requirements.

3. *Concurrent Validation*

Concurrent validation studies are carried out under a protocol during the course of normal production. The first three production-scale batches are usually monitored as comprehensively as possible. The evaluation of the results is used to establish the acceptance criteria and specifications of subsequent in-process control and final product testing. Some form of concurrent validation, using statistical process control techniques, such as quality control charting, may be used concurrently throughout the product manufacturing life cycle.

4. *Revalidation*

Revalidation is required to ensure that changes in process and/or in the process environment, whether introduced intentionally or unintentionally, do not adversely affect product specifications and quality characteristics [6]. There should be a quality assurance system (change control) in place that requires revalidation whenever there are significant changes in formulation, facilities, equipment, process, and packaging that may have an impact on product and manufacturing process performance [1]. Furthermore, when a change is made in the raw material supplier, the drug product manufacturer should be made aware of subtle, potentially adverse differences in raw material characteristics that may adversely affect product and manufacturing process performances, especially with respect to disperse systems.

Conditions requiring revalidation study and documentation are as follows:

- Change in a critical component (usually refers to active drug substance, key excipient, or primary packaging)
- Change or replacement in a critical piece of modular (capital) equipment
- Significant change in processing conditions that may affect subsequent unit operations and product quality
- Change in a facility and/or plant (usually location, site, or support system)
- Significant increase or decrease in batch size that affects the operation of modular equipment
- Sequential batches that fail to meet product and process specifications

In some situations process performance requalification studies may be required prior to undertaking specific revalidation assignments. With the exception of sterile products manufacture (including parenteral suspensions) periodic revalidation is not required at the present time. The performance and state of control of the product and its manufacturing process can be adequately covered during the annual product and process review.

C. Validation Master Plan, Protocol, and Committee

The creation of a master plan permits the development of an overview of the validation effort. It lays out in a logical sequence the activities and/or key elements to be performed vs. the approximate time schedule in a Gantt or PERT chart format. Once generated, the master plan establishes the critical path against which progress can be monitored [6].

The validation program starts with the design and development of raw materials and components, then is followed by the installation qualification (IQ) and operational qualification (OQ) of facilities, equipment, and systems, through performance and process qualification (PQ) stages and terminates in the protocol-driven, formal process validation program. Many of these activities move forward in series. However, by combining activities and elements in groups and functions in parallel where possible on independent tracks with respect to bulk pharmaceutical chemicals, analytical methods development, facilities, equipment, support systems, and the drug product design and manufacturing process development itself, valuable time can be saved before the individual elements or grouping of activities come together prior to the formal process validation program. Such a Gantt chart format presentation is shown in Fig. 1.

A validation protocol is a written plan describing the process to be validated, including production equipment and how validation will be conducted, including objective test parameters, product and/or process characteristics, predetermined specifications, and factors that will determine acceptable results. A successfully completed validation protocol forms the basis for the validation report following formal process validation studies. The following protocol is suggested in the guidelines *Validation of Manufacturing Processes* [4]:

- Purpose and prerequisites for validation
- Presentation of the whole process and subprocesses including flow diagram and critical-step analysis
- Validation protocol approvals
- Installation and operational qualifications including blueprints and drawings
- Qualification reports including methods, procedures, release criteria, calibration of test equipment, test data, summary of results, requalification options, and approvals

		Qualification stage		Validation stage	
Key elements	Design stage	Installation	Operational	Prospective	Concurrent
Facilities and equipment	Engineering phase →		Manufacturing start-up		
	↘ (Validation protocols) → ↗			(Batch records and validation documentation)	
Process and product including BPC and analytical methods	Developmental phase → (formula definition and stability testing)		Scale-up phase → (process optimization and pilot production)	QA and manufacturing (full production)	
		→ Timeline for new product introduction			

Fig. 1 Validation program Gantt chart.

- Product qualification test data from prevalidation batches
- Test data from formal validation batches
- Evaluation of test data, conclusions, and recommendations including the need for requalification and revalidation
- Certification and approval
- Summary report of findings with conclusions

The validation protocol and report may also include copies of the product stability report(s) or its summary plus validation documentation concerning cleaning and analytical methods.

The validation committee formal process validation assignments should be carried out by those individuals with both the necessary training and experience to perform such duties. The specifics of how a dedicated validation committee, group, or team is organized to conduct process validation assignments is beyond the scope of this chapter and should be left to the individual pharmaceutical company to establish [12].

Table 2 outlines an approach to core membership on the validation committee and the committee's specific missions and responsibilities with respect to new product and process development. Some companies may not have their own in-house engineering function. In such cases, traditional outside engineering services can be obtained. In other companies validation is a separate corporate function or concern, whereas in some companies other organizational functions, such as regulatory affairs, analytical development, computer services, and bulk pharmaceutical chemical operations, may be included in discussions and in carrying out certain validation assignments.

Process validation is more than three good manufactured batches and organized, documented commonsense. It is the key to successful liquid, semisolid dispersions, and solid dosage form design, development, and manufacture.

D. Process Validation—Order of Priority

Because of resource limitation, it is not always possible to validate an entire company's product line at once. With the obvious exception that a company's most profitable products should be given a high priority, it is advisable to draw up a list of product categories that are to be validated based on risk potential. Such an approach was developed separately by Nash [13] and Sharp [14].

Table 2 Specific Responsibilities of Each Organizational Structure Within the Scope of the Process Validation Concept

Engineering	Install, qualify and certify plant, facilities, equipment, and support systems.
Development	Design, optimize, and qualify manufacturing process within design limits, specification, and/or requirements. In other words, provide product development and process capability documentation.
Manufacturing	Operate and maintain plant facilities, equipment, support system, and the specific manufacturing process within its design limits, specifications, and/or requirements.
QC/QA	Help provide approvable validation protocols and assist with process validation by monitoring sampling, testing, challenging, and auditing, and by verifying the specific manufacturing process and its environment for compliance with design limits, specifications, and/or requirements.

Both approaches have been combined and summarized in Table 3. In each of the categories, high, moderate, and low risk disperse system dosage forms (suspensions, gels, aerosols, etc.) are present. Though disperse systems are more complex preparations than are simple solutions, other concerns, such as sterility, route of administration, and drug potency affect validation priorities to a greater extent.

E. Process Validation

Formal process validation trials are never designed to fail. Process validation failures, however, are often attributable to an incomplete understanding of the manufacturing process being evaluated. Often failures appear to be directly related to either a failure to establish process capability information (that is, what a process can or cannot do under a given set of processing conditions) or process qualification (or prevalidation) trials that are not properly defined for the later formal validation and production phases of the overall program.

1. Qualifications

Process qualification (or process capability design and testing) may be defined as studies that are carried out to establish the critical process parameters or operating variables that influence process output and the range of numerical data for each of these critical process parameters that result in acceptable process output.

Process qualification should proceed in the following stages or phases of the overall development program (Fig. 2):

Table 3 Validation Priorities

Risk level	Dosage forms
High risk (high priority)	Large-volume parenterals (LVPs) sterile solution greater than 50 mL
	Small-volume parenterals (SVPs) and sterile lyophilized powders
	SVPs sterile solutions and suspensions
	Sterile implants
	Sterile ophthalmic solutions and suspensions
	Sterile ophthalmic ointments and gels
	Sterile irrigating fluids
	Sterile otic solutions and suspensions
Moderate risk	Low-dose tablets and capsules
	Extended release tablets and capsules
	Transdermal delivery systems
	Intranasal delivery systems
	Oral inhalation aerosols
	Reconstituted powders
	Oral suspensions and emulsions
	High-dose tablets and capsules
Low risk (low priority)	Suppositories
	Troches and lozenges
	Oral solutions
	Mouthwashes, douches, and topical solutions
	Topical suspensions and lotions
	Topical ointments and creams
	Topical powders

- Preformulation studies with the active drug substance and key excipients (dispersing agents, stabilizers, and preservatives)
- Product design or formulation development of the prototype formula in laboratory equipment (1X)
- Preparation of the first scale-up batches in pilot-laboratory equipment (10X) for stability testing and possible clinical study.
- Further process scale-up to (100X) in production size equipment and evaluation (prevalidation batches).
- Formal, protocol-driven validation studies in production equipment (100X)

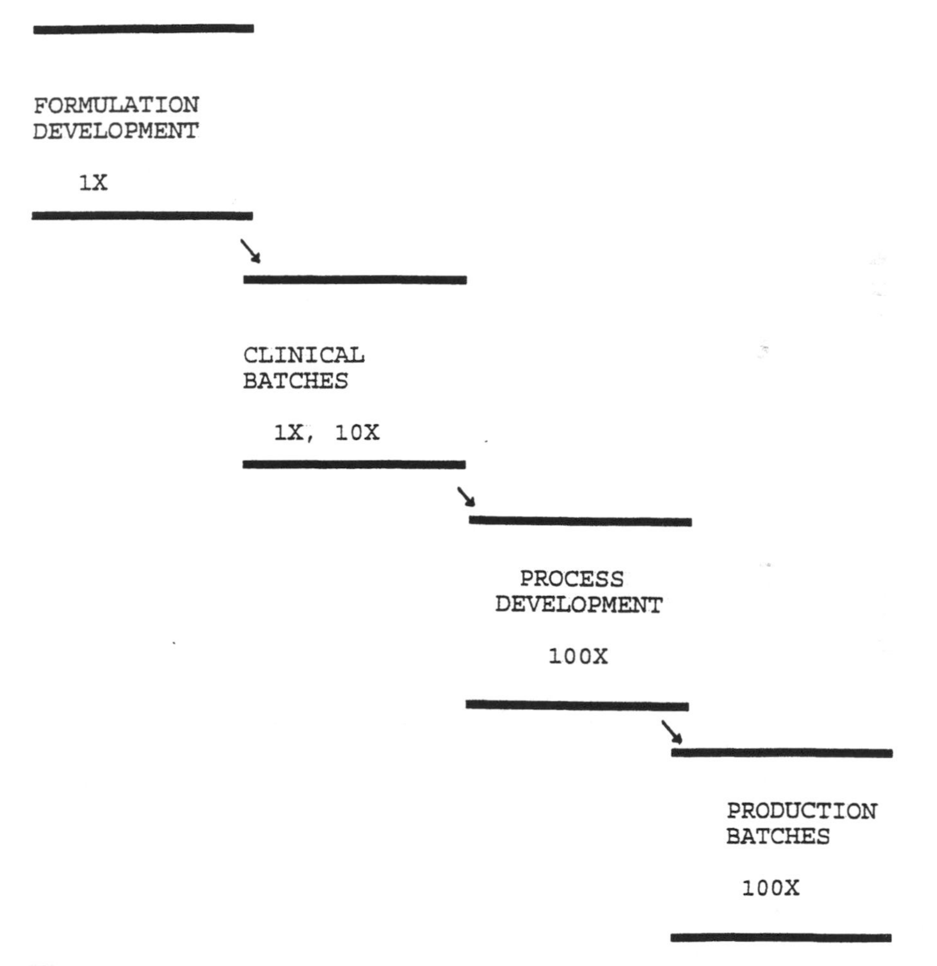

Fig. 2 Technology transfer stage.

Installation qualification procedures and documentation show that all important aspects of the installation of either the facility, support system, or piece of modular equipment, having been properly installed and calibrated, meet their design specifications and that the vendor or supplier's recommendations have been suitably considered.

Operational qualification follows IQ, and includes procedures and documentation that show that the facility, support system, or piece of modular equipment performed as intended throughout all anticipated operating ranges under a suitable load.

Performance qualification follows both IQ and OQ, and includes actual demonstrations during the course of the validation program that show that the facility, support system, or piece of modular equipment performed according to a predefined protocol and achieved process reproducibility and product acceptability.

Objectives for qualification of mixing equipment for dispersions may be handled by the following five-step procedure:

1. Establish normal operating conditions for the process i.e., rotational speed of agitator, temperature cycle(s), duration of mixing.
2. Determine how the normal operating conditions are controlled and what variations are likely to occur within a batch and among batches.
3. For continuous mixers, consider variations produced by feeders, i.e., flow variation, flow properties of input materials.
4. Establish the operating characteristics required to obtain appropriate and reproducible physical characteristics and product uniformity.
5. Reestablish normal operating conditions based on the results of steps 1–4.

2. Change Control

Change control procedures should be in place before, during, and after the completion of the formal validation program. A change control system maintains a sense of functionality as the process evolves and also provides the necessary documentation trail that ensures the process continues in a validated, operational state even when small noncritical adjustments and changes have been made to the process.

Such minor, noncritical changes in materials, methods, and machines should be reviewed by those responsible for carrying out elements of the validation function. They are development, engineering, production, and QA/QC personnel, who assure that all process integrity and process comparability have been maintained and documented before the specific requested change can be approved by the head of the quality control unit.

The change control system, based on an approved standard operating procedure(s), takes on added importance as the vehicle or instrument through which innovation and process improvements can be made more easily and more flexibly without formal review by the new drug application (NDA) and abbreviated new drug application (ANDA) reviewing function of the FDA. If more of the supplemental procedures with respect to the chemistry and manufacturing control sections of NDAs and ANDAs could be covered through annual review documentation procedures, process validation, with appropriate safeguards, could become more pro-innovative.

F. Cleaning Validation

According to section 211.67 of the CGMPs, equipment and utensils should be cleaned, maintained, and sanitized at appropriate intervals to prevent malfunction or contamination that would alter the safety, identity, strength, quality, or purity of the drug product. Written procedures should be established and followed for cleaning and maintenance of equipment, and these procedures should include, but not be limited to, the following:

- Assignment of responsibility for cleaning and maintaining equipment

- Maintenance and cleaning schedules and sanitizing schedules where appropriate
- Description in sufficient detail of methods, equipment, and material used in cleaning and maintenance operations, and the methods of disassembling and reassembling equipment as necessary to ensure proper cleaning and maintenance
- Removal or obliteration of previous batch identification
- Protecting clean equipment from contamination prior to use
- Inspection of equipment for cleanliness immediately before use.

Records should be kept of maintenance, cleaning, sanitizing, and inspection.

The objective of cleaning validation of equipment and utensils is to reduce the residues of one product below established limits so that the residue of the previous product does not affect the quality and safety of the subsequent product manufactured with the same equipment [15].

According to section 211.63 of the CGMPs, equipment used in manufacture, processing, packing, or holding of a drug product should be of appropriate design and adequate size, and suitably located to facilitate operations for its intended use and for its cleaning and maintenance. Some of the equipment design considerations include type of surface to be cleaned (stainless steel, glass, plastic), use of disposables or dedicated equipment and utensils (filters, etc.), use of stationary equipment (tanks, mixers, mills, etc.), use of special features (clean-in-place systems, steam-in-place systems) and identification of difficult-to-clean locations on the equipment (so-called hot spots and/or critical sites).

The specific cleaning procedure should define the amounts and the specific type of cleaning agents and/or solvents used. The cleaning procedure should give full details as to what is to be cleaned and how it is to be cleaned. The cleaning method should focus on worst-case conditions, such as potent drugs and the least-soluble, most difficult to clean formulations (i.e., creams and ointments). Cleaning procedures should identify time between processing and cleaning, the cleaning sequence, the equipment dismantling procedure, the need for visual inspection, and provisions for documentation.

The choice of a particular analytical method (HPLC, TLC, spectrophotometric, total organic carbon, pH, conductivity, etc.) and sampling technique chosen (direct surface by swabs and gauze or rinsing) depends on the residual limit to be established based on the sampling site, type of residue sought, and equipment configuration (critical sites vs. large surface area) considerations. The analytical and sampling methods should be challenged in terms of specificity, sensitivity, and recovery. The established residue limits must be "practical, achievable, verifiable and assure safety." The potency of the selected drug and the presence of degradation products, cleaning agents, solvents, and microorganisms should be taken into consideration.

In the case of ointments, creams, insoluble drugs, and propellants, the use of solvents in cleaning disperse systems is widely used. In the case of emulsified creams and lotions, the use of high salt (NaCl) concentration solutions in water to break emulsions has been shown to be an effective pretreatment cleaning step [16]. The use of cosolvents (alcohols, glycol ethers) to provide miscibility between cleaning solvents and water has been used to provide a purified-water USP final-rinse end point for tank, kettle, mill, mixer, homogenizer, etc., cleaning.

II. DISPERSE SYSTEMS

A dispersion is defined as a heterogeneous system in which the internal or dispersed phase is uniformly dispersed in the external phase or dispersing medium. It may be kept in a state of dispersion for prolonged periods with the aid of suitable dispersing agents.

Dispersions, as pharmaceutical dosage forms, are classified in Table 4 based on the physical states of matter (i.e., gas, liquid, solid) for both the internal and external phases of the disperse systems. In summary, the disperse system pharmaceutical dosage forms that require validation documentation include

- Oral inhalation and nasal aerosols
- Injectable, ophthalmic, oral, and topical suspensions
- Oral emulsions and topical emulsions, creams, and lotions
- Insoluble powders and granules for reconstitution with water to form suspensions
- Ophthalmic and topical ointments

Transdermal delivery patches, however, may be treated as a variant of extended release solid dosage forms (capsules and tablets).

There are a number of disperse systems that may be classified as colloidal dispersions, i.e., where the disperse phase has a particle size less than about 0.2 microns. In most cases, when properly prepared, visually these aqueous solutions appear to be clear or at worse turbid or hazy.

The active drug substances are solubilized in such systems with the aid of surfactants, complex proteins, lipoproteins, and/or pharmaceutically acceptable cosolvents. Such colloidally stable dispersions are referred to as liposomes, microemulsions, and micellular solubilized lipids in an aqueous dispersion. The difference in classification among the three is usally based on structural considerations. With respect to validation concerns, for the most part, colloidal systems may be treated as sterile and nonsterile simple aqueous solutions.

Table 5 shows the test parameters for pharmaceutical disperse systems.

A. Suspensions

A suspension is a disperse system in which solid, vehicle-insoluble particles (internal phase) are uniformly suspended by mechanical agitation and formulation design through-

Table 4 Classification of Pharmaceutical Dispersions

Internal phase	External phase	Dosage form
Gas	Gas	Mixture of gases (or propellants)
Gas	Liquid	Aerosol foam
Gas	Solid	Adsorbate
Liquid	Gas	Aerosol spray (fog)
Liquid	Liquid	Emulsion
Liquid	Solid	Absorbate (solvated powder)
Solid	Gas	Aerosol dust (smoke)
Solid	Liquid	Suspension
Solid	Solid	Mixture of powders or granules

Table 5 Test Parameters for Pharmaceutical Disperse Systems

Test parameter	Suspensions	Emulsions	Creams/ointments	Aerosols
Macroscopic appearance	Yes	Yes	Yes	No
Specific gravity	Yes	Yes	No	No
Viscosity	Yes	Yes	Yes	No
pH	Yes	Yes[a]	Yes[a]	No
Content uniformity	Yes	Yes	Yes	Yes
Sedimentation	Yes	No	No	No
Resuspendability	Yes	No	No	No
Particle size[b]	Yes	Yes	Yes	Yes
Release rate	Yes	Yes	Yes	Yes
Pressure	No	No	No	Yes

[a]Oil-in-water systems.
[b]Internal phase.

out the liquid vehicle (external phase). Most stable pharmaceutical suspensions are prepared by either flocculating (loose aggregates of individual particles) the vehicle-insoluble drug particles and suspending them in the liquid vehicle, where they may be easily resuspended with gentle agitation. Other suspensions may be prepared by suspending properly wetted, individual deflocculated particles in a structured vehicle that possesses infinite, very large viscosity at rest to overcome particle settling with time. Many pharmaceutical suspensions take advantage of both particle flocculation and structured vehicle techniques. The formulating aspects in validating suspensions have been discussed in elsewhere [17,18] and are covered in other sections of the present chapter.

There are three general classes of pharmaceutical suspensions. They are orally administered (sometimes referred to as mixtures), externally applied (topical lotions), and injectable (depot or parenteral) dosage forms.

Drug particle size is an important factor influencing product appearance, settling rates, drug solubility, in vivo absorption, resuspendability, and overall stability of pharmaceutical suspensions. The crystal growth with time of unprotected, slightly soluble drug particles and changes in their particle size distribution in suspension is a serious consideration.

Crystal growth and changes in particle size distribution can be largely controlled by using one or more of the following procedures and techniques:

- Selecting a narrow range of particle sizes (1–10 microns)
- Selecting the highest melting point crystalline form of the drug
- Using controlled precipitation techniques rather than high-energy milling to reduce particle size
- Using water-dispersible surfactant wetting agents to overcome free surface energy effects
- Using a protective colloid (gelatin, gum, cellulose derivative) to form a barrier around the particle
- Increasing the intrinsic viscosity of the suspending vehicle
- Avoiding temperature extremes during processing and storage

- Avoiding conditions of supersaturation of insoluble drug particles
- Slowing crystallization techniques instead of using shock cooling or flash evaporation
- Avoiding impurities, foreign substances, and exotic recrystallizing solvents
- Establishing reproducible, crystallizing conditions in order to reduce consistent material

During the preparation of physically stable pharmaceutical suspensions, a number of formulation components are used to keep the solid particles in a state of suspension (suspending agents), whereas other components are part of the liquid vehicle itself and have other functions in the dosage form. A list of excipients used in pharmaceutical suspensions is presented in Table 6.

B. Gels

Gels are semisolid disperse systems where polymers or long-chain molecules in the internal phase are capable of cross-linking and interacting with themselves to entrap the external phase within a weblike structure. The classic example of a gel is dissolved gelatin in water forming a highly viscous gel structure. Reversible gels are reproducible, semisolid, highly viscous structures at low temperatures, and colloidal solutions and/or liquids at higher temperatures. Most gels, whether aqueous or nonaqueous are prepared with the aid of heat and high shear agitation [19]. Suspensions, milks, and magmas of clays and inorganic hydroxides are sometimes referred to as gels because of their high viscosity.

A list of common pharmaceutical gelling agents for both aqueous and nonaqueous liquids is presented in Table 7. The most useful of the gelling agents listed in Table 7 are carbomers for aqueous solutions, colloidal silicon dioxide for both aqueous and nonaqueous liquids and stearate soaps for nonaqueous liquids.

Because of their importance as pharmaceutical suspending and gelling agents, six different carbomers are recognized in NF 18. They are carbomer 910, 934, 934P, 940, 941, and 1342. With the exception of carbomer 934 and 934P (which may be cross-linked with allyl ethers of sucrose) the other carbomers are essentially high molecular weight acrylic acid polymers cross-linked with allyl ethers of pentaerythritol.

Table 6 Excipients Used in Pharmaceutical Suspensions

Component	Agent	Examples
Suspending system (internal phase)	Wetting agents	Poloxamer, polysorbates, sodium lauryl sulfate
	Dispersants	Gelatin, lecithin
	Flocculants	Di- and trivalent anions and cations
	Thickeners	Cellulosics, clays, gums, polymers, soaps
Suspending vehicle (external phase)	pH control and buffers	Citrates, phosphates, and bases
	Osmotic agents	Mannitol, sodium chloride
	Humectants	Glycerin, propylene glycol
	Preservatives	Benzyl alcohol, parabens, sorbates
	Coloring agents	Flavors, fragrances, sweeteners
	Stabilizers	Ascorbic acid, bisulfite, edetate, butylated hydroxyanisole

Table 7 List of Pharmaceutical Gelling Agents

For Aqueous Liquids
Carbomers
Cellulosics
Colloidal silicon dioxide
Gelatin
Natural gums (algins, pectins, carrageenans, xanthans)
Nonionic surfactants
Starches
For Nonaqueous liquids, solvents, and oils
Clays and organoclays
Colloidal silicon dioxide
Low HLB surfactants
Modified cellulosics
Polyethylene

Carbomer 934P is highly purified 934 resin that is used in mucosal contact products. Carbomers NF are used in the range of 0.5–2.0% as topical gelling agents. They must be brought to neutrality (pH 6–8) with a suitable base (amine or hydroxide) when used to prepare gels with purified water USP, alcohol USP, and glycerin USP.

The procedure calls for a 0.5% concentration of carbomer NF to be dispersed in water by sprinkling it on the surface of purified water USP vigorous agitation to disperse the agent in the liquid. Agitation is stopped to permit deaeration before 0.5% of the neutralizer is added with very slow agitation. The gel forms instantaneously after the carbomer is neutralized.

Instant gelation causes several problems. They are contact microgel formation between neutralizer and carbomer, which may be difficult to redisperse and macro- and microbubble formation due to air entrapment in the highly viscous gel. Therefore, strict process control during agitation, pH neutralization, and ingredient addition is critical during scale-up.

Colloidal silicon dioxide NF is a material that has been used as a gelling agent in several nonaqueous liquids at concentrations between 2 and 10%. This relatively inert, low bulk density (0.1 g/cm^3), high specific surface area (250 m^2/g), inorganic material is capable of forming stable gels in a variety of organic liquids in the presence of high shear agitation and high temperatures, if required.

Stearate soaps are glycerinated gels for suppository manufacture and consist of dissolving 9 g of sodium stearate NF in 86 g of glycerin USP with agitation and heating between 115 and 120°C. Then 5 g of purified water USP are added with gentle agitation below 90°C. The mixture is cooled rapidly without agitation to form a rigid, nonaqueous gel. Aluminum and magnesium stearates have also been used to form gels in nonaqueous liquids [20].

C. Ointments

Ointments or ointment bases are semisolid disperse systems intended for topical application to the skin or certain mucous membranes. There are four general classes of ointment bases [21]. They are listed as follows:

- Water-soluble bases such as polyethylene glycol ointment USP
- Water-removable bases such as hydrophilic ointment USP and oil-in-water (O/W) emulsion bases
- Absorption bases such as hydrophilic petrolatum USP and water-in-oil (W/O) emulsion bases
- Hydrocarbon bases or oleaginous ointment bases such as white petrolatum USP and white ointment USP

The choice of the base selected for formulation and scale-up depends on factors such as the area of topical application, chemical and physical stability, and the aqueous solubility of the active drug substance itself. As a general rule, water-insoluble drugs are best dissolved or suspended in water-soluble and water-removable bases, whereas more water-soluble active drug substances are usually incorporated into absorption and/or hydrocarbon bases.

1. *Hydrocarbon Bases*

Modern hydrocarbon bases feature white petrolatum USP (prepared from Pennsylvania crude oils) and (white) mineral oils USP gelled with either microcrystalline (petrolatum) waxes, natural waxes, low molecular weight polyethylene, or colloidal silicon dioxide in order to achieve an appropriate viscosity range. In today's regulatory climate, formulating with the less pure petrolatum USP and yellow wax NF (beeswax) alternatives is undesirable.

Small amounts of oil-soluble (so-called low HLB) nonionic surfactants are added to the oleaginous blends to aid dispersion of active drug substance during product manufacture. The keys to effective process control of oleaginous ointment manufacture are as follows:

- Type and configuration of the agitation equipment selected for dispersion
- Speed of agitation and shear rates selected during the various phases and/or steps in the manufacturing sequence
- Temperature ranges selected for ingredient addition and agitation during the heat-up and cool-down phases of the manufacturing process.

2. *Water-Soluble Bases*

At the present time the only compendial formula for water-soluble bases is restricted to an admixture of 40% polyethylene glycol 3350 NF and 60% polyethylene glycol 400 NF. A small portion of the polyethylene glycol 3350 NF may be replaced by stearyl alcohol NF in the PEG base as a stiffening agent when small amounts of water are added to the base.

In official variations of water-soluble bases, nonionic surfactants, glycol cosolvents, and other polyethylene glycol derivatives may be added to increase the solubility of the active drug substance in the PEG base. The water content of most finished polyethylene glycol–based ointments range between 5 and 20%.

In preparing stable products in PEG water-soluble bases it is important to use highly purified grades of polyethylene glycols that are free of residual ethylene oxide and peroxides. In dispersing active drug substances in PEG water-soluble bases the order of addition, type and rate of agitation, and temperature ranges during processing, holding, and filling are important parameters for validation consideration and product manufacture.

D. Creams and Lotions

Creams are viscous, semisolid systems, either water-in-oil or oil-in-water emulsions, and for the most part are used topically. Lotions, which may be taken either orally (as emulsions or mixtures) or used topically, are either less viscous suspensions or O/W creams diluted with additional external phase (purified water USP) [21].

1. *Absorption Bases*

Hydrophilic petrolatum USP is a prototype absorption base that forms a W/O emulsion upon the incorporation of small amounts of purified water USP. The formula for hydrophilic petrolatum USP is as follows:

Cholesterol NF	3%	primary W/O emulsifier
Stearyl alcohol NF	3%	auxillary emulsifier
White wax NF	8%	stiffening agent
White petrolatum USP	86%	vehicle
	100%	by weight

Procedure: the stearyl alcohol and white wax are melted together at about 70°C. The cholesterol is then added and the resulting mixture is stirred until the cholesterol dissolves. The white petrolatum is then added and the resultant mixture is mixed until a uniform liquid blend is obtained. The mixture is then removed from the heat source and gently mixed until the liquid blend begins to congeal, between 45 and 55°C. The base is then filled into suitable containers before cooling to room temperature.

The base has the capability of absorbing an equal weight of purified water to form a W/O emulsion cream base. The purified water can be added at room temperature (25°C) or 45–55°C to form a stable W/O emulsion. A suitable preservative should be added to the purified water before forming the final emulsion blend.

Another simple W/O emulsion base consists of 30% preserved purified water USP, 6% sorbitan monostearate NF, and 64% white petrolatum USP.

2. *Water-Removable Bases*

Water removable O/W emulsions when diluted with external phase (purified water) are also referred to as vanishing creams or cold creams. Hydrophilic ointment USP is an example of a water-removable base with less than an ideal amount of external phase (purified water) present. The formula for hydrophilic ointment USP is given as follows:

Methyl paraben NF	0.025%	preservative
Propyl paraben NF	0.015%	preservative
Sodium lauryl sulfate NF	1.0%	primary emulsifier
Propylene glycol USP	12.0%	humectant and potentiator
Stearyl alcohol NF	25.0%	auxillary emulsifier
White petrolatum USP	25.0%	internal phase
Purified water USP	37.0%	external phase
	100.0%	by weight

Procedure: Stearyl alcohol and white petrolatum are heated to about 75°C in a suitable vessel. The parabens, sodium lauryl sulfate, and propylene glycol are dissolved in the purified water at 75°C in a separate vessel. The water phase is added to the oil phase (stearyl alcohol and white petrolatum). The resulting emulsion is mixed until it reaches the congealing point (40–50°C). The base is then filled into suitable containers before cooling to room temperature.

Typically O/W emulsion cream bases contain between 50 and 70% purified water as the external phase. Hydrophilic ointment USP is deliberately designed to permit the addition of 20–30% purified water based on the final weight of the finished cream. The makeup water usually contains a solution or suspension of the incorporated active drug substance.

Hydrophilic ointment USP is somewhere between anhydrous, self-emulsifying wax preparation and a finished, fully hydrated, low viscosity, O/W vanishing cream base.

More stable, more efficient O/W emulsion creams containing a combination of high- and low-HLB nonionic surfactants (Table 8) as substitutes for sodium lauryl sulfate and light mineral oil USP in place of all or a portion of the white petrolatum USP have been used extensively as water-removable cream bases for a large number of active drugs. A typical formula for an O/W emulsion cream base is given as follows:

Glyceryl monstearate NF	1.4%	emulsifier (HLB 2.7)
Poloxy 40 stearate NF	2.6%	emulsifier (HLB 16.9)
Cetyl alcohol NF	8.0%	auxillary emulsifier
Light mineral oil NF	20.0%	internal phase
Propylene glycol USP	8.0%	humectant & potentiator
Methyl paraben NF	0.15%	preservative
Propyl paraben NF	0.05%	preservative
Lactic acid USP		q.s. adjust pH to 5
Purified water USP	60.0%	external phase
	100.0%	by weight

(Note that the required HLB for mineral oil and cetyl alcohol to form an O/W emulsion is 11.8.)

Procedure: The high- and low-HLB emulsifiers and the cetyl alcohol are dissolved with mixing in the light mineral oil at 75°C in a suitable vessel. The parabens and the propylene glycol are dissolved with mixing in about half the final water (30% by weight) at 75°C in a separate vessel. The water phase is slowly added to the oil phase with

Table 8 HLB[a] Classification of Nonionic Surfactants Based on Their Water Dispersibility Characteristics

HLB value	Type	Water dispersibility
1–3	Antifoam agents	Nondispersible
3–6	W/O emulsifiers	Poor dispersion
6–9	Wetting agents	Milky dispersion
9–12	O/W emulsifiers	Stable milky dispersion
12–15	Detergency	Clear dispersion
15–20[b]	Solubilizers	Clear colloidal solution

HLB Concept (hydrophilic-lipophilic balance) of nonionic surfactants was introduced by W. C. Griffen in *J. Soc. Cosmet. Chem.*, 5:249–255 (1954).
[b]Limiting value for nonionic surfactants.

mixing at the same temperature (75°C) to form a viscous, primary W/O emulsion. Additional purified water is added with mixing during the cooling step to permit phase inversion to an O/W emulsion. The pH of the cream is adjusted to pH 5 with lactic acid and filled into suitable containers before final cooling to room temperature.

3. *Manufacture of Emulsified Cream and Lotions*

There are several methods for the manufacture of emulsion systems that are based on formulation, equipment, and requirements (pharmaceutical vs. cosmetic) of the product [22,23].

In one method, the external phase is prepared in the vessel in which the final emulsification is to take place and the internal phase is pumped in a thin stream with constant mixing into the external phase. This is a typical approach for a cosmetic product.

For another method, the two phases are prepared in two separate vessels and the emulsification is carried out in a third vessel or an in-line flow system. In a typical cosmetic approach, small quantities of each phase are added alternately to the emulsifying vessel until one phase (internal or external) is exhausted and the balance of the remaining phase is added at the end in a thin stream. In an in-line system the two phases are dispensed continuously, in their exact proportions, into the final vessel.

In the inversion method, the internal phase is prepared in the vessel in which the emulsification is to take place [24]. If an O/W emulsion is to be made, the oil phase is prepared in the emulsifying vessel. The external phase (purified water) is then added in a thin stream. Initially, a W/O emulsion is formed, (if the formulation contains two emulsifiers, low-HLB and high-HLB nonionic surfactants), which is thicker (viscous) and slightly off-white in appearance. When the water phase is added and its level approaches 50% of the total volume, the emulsion suddenly becomes less viscous and inverts to an O/W emulsion; the balance of the water (50–70% of the final volume) may be added rapidly. Phase inversion can yield a fine droplet emulsified internal phase (oils, etc.) without the need to use extensive agitation (colloid mills and homogenizers). This is a typical pharmaceutical approach, but pharmaceutical creams do not generally possess outstanding cosmetic "skin feel." The design features for pharmaceutical preparations are topical activity and product stability.

E. Aerosols

Aerosols consist of two-phase (gas and liquid) or three-phase (gas, liquid, and solid) disperse systems. Today, because of restrictions on chlorofluorohydrocarbon propellants, most topical aerosol systems have been converted to aqueous-alcohol pump-spray soltions.

The oral inhalation products consist of a suspension of the active ingredient(s), a cosolvent or dispersant in addition to a blend of propellents of the chlorofluorohydrocarbon type. Mixtures of propellants are frequently used to obtain desirable pressure, delivery, and spray characteristics. The use of chlorofluorohydrocarbon propellents for oral inhalation drug products has been granted a temporary extension through the end of the century [25]. They will eventually be replaced by fluorohydrocarbon propellants.

Aerosol containers are made of glass, plastic, metal, or combinations of these materials. Thermoplastics have ben used to coat glass and metals to improve their safety and corrosion resistance, and the physical stability of the disperse systems. Suitable metals for container construction include stainless steel, aluminum, and tin-plated steel.

Aerosols are prepared by either the "cold-fill" process or the "pressure-fill" method. In the cold-fill process, the concentrate cooled to a temperature below 0°C and the refrigerated propellant blend is measured into chilled open containers. The valve-actuator assembly is then crimped onto the container to form a pressure-tight seal. During the interval between propellant addition and crimping, sufficient volatilization of propellant occurs to displace the residual air from the container.

In the pressure-fill method, the concentrate is placed in the container, and either the propellant is forced under pressure through the valve orifice after the valve is sealed in place or the propellent is permitted to flow under the valve cap and then the valve assembly is sealed. In the pressure-fill method, provisions must be made to remove the air by vacuum.

Flow diagrams for the cold-fill process and the pressure-fill method are presented in Figs. 3 and 4.

The formula for a typical oral inhalation aerosol is given as follows:

Terbutaline sulfate	0.075 g	active (0.714%)
Sorbitan trioleate	0.105 g	dispersant
Trichlorofluoromethane	2.58 g	propellent
Dichlorotetrafluoroethane	2.58 g	propellent
Dichlorodifluoromethane	5.16 g	propellent
	10.5 g	total weight

Specific gravity equals 10.5g/7.5 mL or 1.4 g/cm^3.

There are 10.5g/300 actuations or 35 mg per actuation, which is equivalent to 0.714% times 35 mg or 0.25 mg per valve actuation. Since the delivery system is 80%

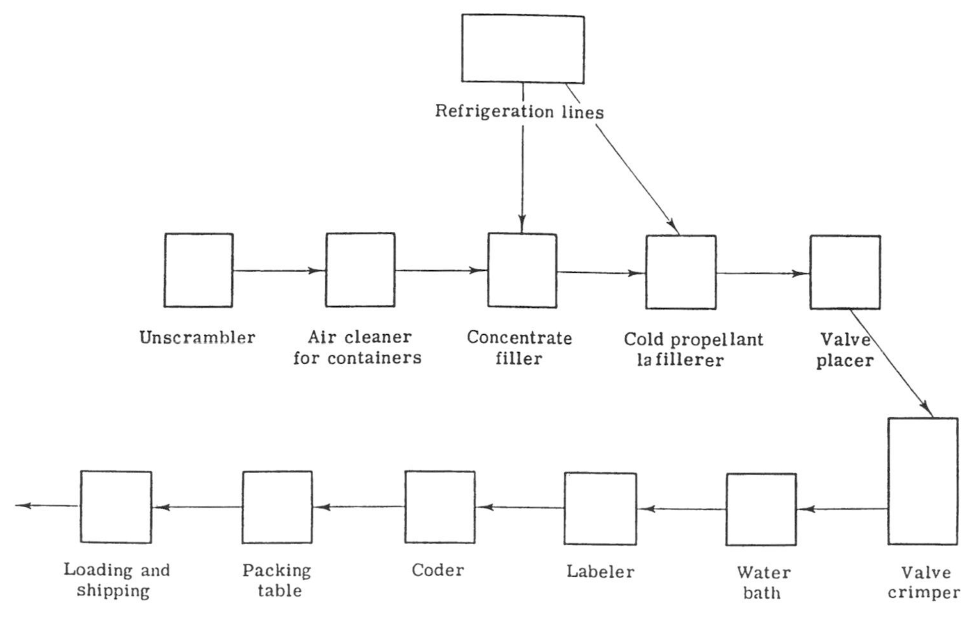

Fig. 3 Flow diagram of cold filling process for aerosols.

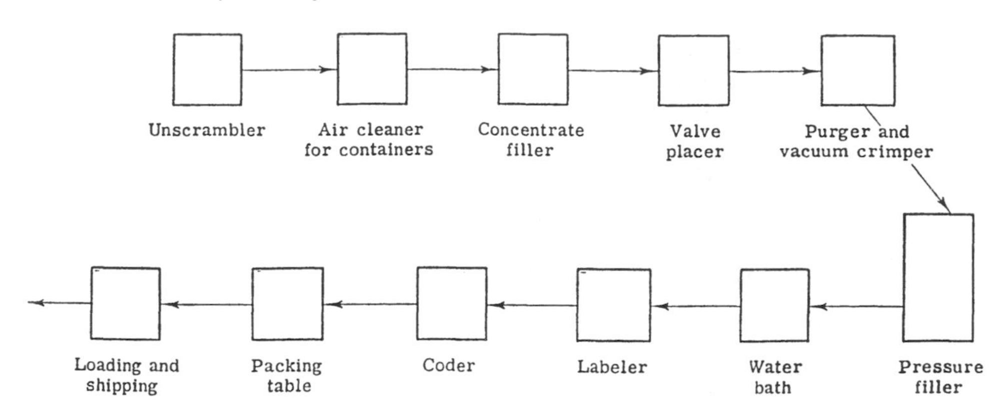

Fig. 4 Flow diagram of pressure filling process for aerosols.

efficient, the valve delivers 0.25 times 0.8 or 0.2 mg or terbutaline sulfate per actuation with a particle range between 1 and 6 microns.

Other dispersants that may be used in place of sorbitan trioleate include oleic acid, soya lecithin with and without small amounts of added alcohol (1–3%). Larger amounts of alcohol USP (30–38% by weight) are used to prepare solutions or liquid concentrates of the active drug substance that are immiscible with the liquid propellant blend. Since the actives and dispersant and propellant blends are essentially free of water, the use of a preservative is not necessary. Nevertheless, good manufacturing practice requires that the actives, dispersants, and propellents be free of pathogens. Other agents used in oral inhalation systems include 0.1% ascorbic acid as a stabilizer, and menthol and saccharin as flavor and taste agents.

1. Validation Requirements for Metered-Dose Inhalers

In no other pharmaceutical dosage form is the testing strategy more intense than for pressurized metered-dose inhalers (MDIs), also known as, oral inhalation aerosols (Fig. 5). The process parameters, quality attributes, and product specifications that should be present in a formal testing protocol for MDIs include the following items (taken from USP XXIII) [26]:

- Water content (921). Usually performed with 10 representative containers.
- Minimum or net fill weight (755). Usually performed with 10 representative containers.
- Total number of discharges per container (601). Usually performed with 10 representative containers.
- Leakage or weight loss test (601). Performed on 12 representative containers during long-term storage.
- Delivery rate test (601). Performed over a 5 sec period and repeated twice and reported in g/sec.
- Pressure test (601). Select not less than four containers and determine gauge pressure at 25°C.
- Metering performance test (601). Usually performed with 10 representative containers.
- Uniformity of unit spray content. Usually performed with not less than four filled containers.

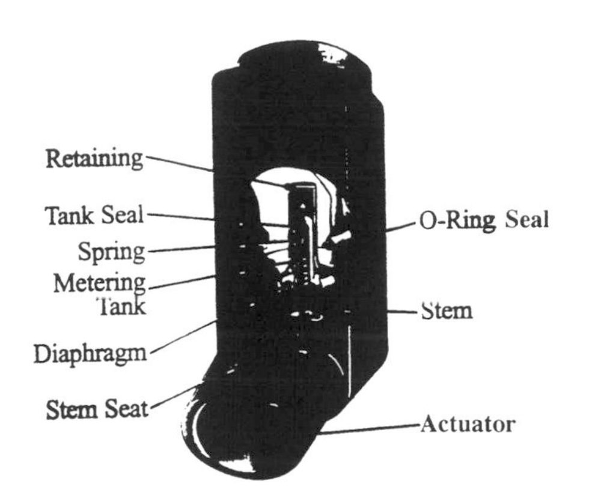

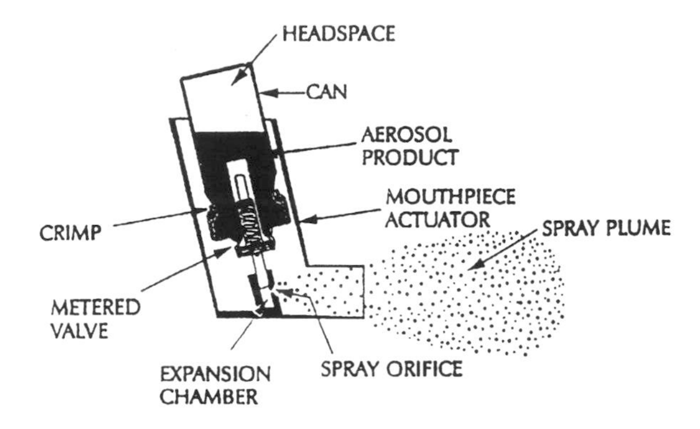

Fig. 5 Vital parts of the metered-dose inhaler aerosol system.

- Particle size (601). Check active ingredient by microscopic examination of sufficient fields. Particle size should be about 1–10 microns.
- Aerodynamic size distribution using a Cascade Impactor (601). Three different cascade impactors are described in the USP.
- Uniformity of dosage unit (905). Using 10–30 representative filled containers.
- Specific gravity (density). Check the final formula.
- Testing packaging components (cans, valves, liners, gaskets, etc.). Check for toxic extractables and heavy metals.

III. MANUFACTURING CONSIDERATIONS

A. Flow Diagram

A simplified flow diagram or chart showing all the unit operations (modules) or steps in the manufacturing process is constructed in a logical sequence. The major pieces of capital equipment to be used in each processing step and the stages or unit operations at which various ingredients are to be added to the process should be identified.

Several flow diagrams (Figs. 6 and 7) for the manufacture of disperse systems have been provided. The enclosed rectangles represent steps and or unit operations in the manufacturing process. The arrows represent transfer of material into and out of each step and/or unit operation in the process. The sequential arrangement of unit operations should be analogous to the major sequential steps in the operating instructions of the manufacturing process itself.

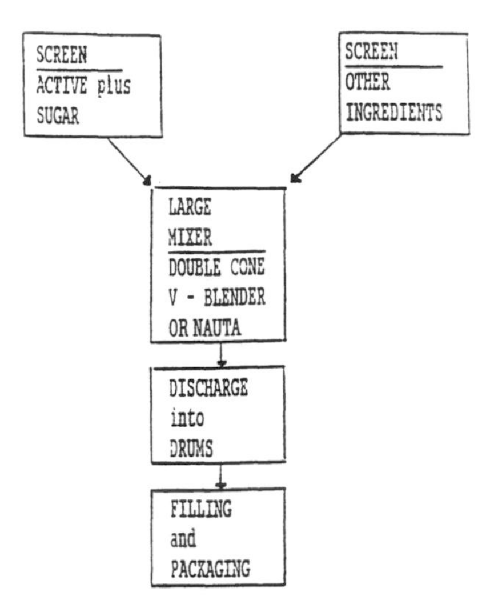

Fig. 6 Flowchart of powder for suspension manufacture.

Using the flow diagram or chart as a guide, a list of process or control variables are next drawn up for each unit operation or step in the manufacturing process.

B. Unit Operations for Disperse Systems

There are five unit operations that are fundamental to most disperse systems (suspensions, lotions, creams, ointments, and aerosols). They are as follows:

- Mixing of liquids
- Mixing of solids
- Mixing of semisolids
- Dispersing
- Milling and size reduction of solids and semisolids

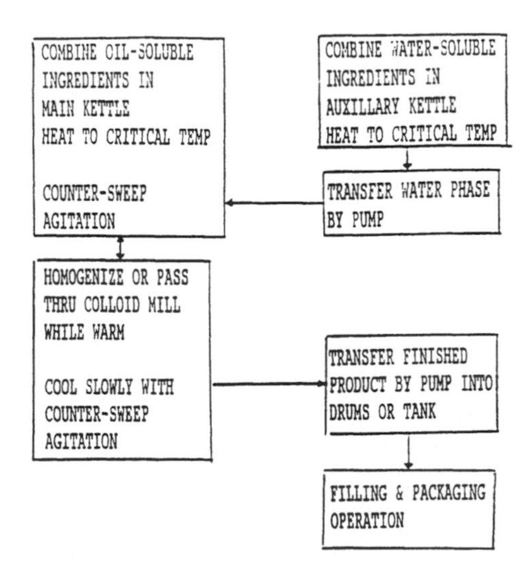

Fig. 7 Flowchart of cream, lotion, and ointment manufacture.

The essential aspects of each of these five unit operations is presented in the following outline format. Process variables that are designated by an asterisk are considered to have a critical effect on the control of the unit operation and the quality of the final product.

1. *Mixing of Liquids*

Equipment:	kettle and tank fitted with agitator
Process variables:	capacity of unit
	shape and position of agitation system
	order of addition
	rate of addition
	fill volume
	mixing speed of agitator*
	temperature of liquid*
	mixing time
Properties affected by variables:	appearance of liquid
	viscosity of liquid
Monitoring output:	potency
	appearance
	pH
	specific gravity
	viscosity

2. *Mixing and Blending of Solids*

Equipment:	blade mixers and tumblers
Process variables:	capacity of unit
	mixing speed of unit
	shape of unit and position of mixing elements within unit
	product load (fill volume)*
	order of addition of solids to unit
	mixing time*
Properties affected by variables:	particle size of solids
	blend uniformity
Monitoring output:	potency
	particle size analysis
	content uniformity of actives

3. *Mixing and Blending of Semisolids*

Equipment:	blade mixer and kneeder
Process variables:	type and capacity of unit
	shape of unit and position of mixing elements within unit
	product load (fill volume)*
	temperature*
	agitation speed
	mixing time*

Properties affected by variables:	homogeneity specific gravity viscosity
Monitoring output:	potency content uniformity viscosity density

4. *Dispersing*

Equipment:	homogenizers (pressure/piston), colloid mill (rotor/stator), or ultrasonic device
Process variables:	equipment option bore opening/clearance of stator/ power setting pressure/rotor speed/power consumption feed rate* temperature* dispersion time* order of mixing
Properties affected by variables:	particle size of solids viscosity of liquid
Monitoring output:	potency particle size distribution viscosity specific gravity

5. *Size Reduction of Solids and Semisolids*

Equipment:	end-runner mill, hammer mill, ball mill, colloid mill, micronizer (air jet mill)
Process variables:	mill type mill size mill speed/air pressure/clearance setting* product load (percent fill capacity) feed rate* inert atmosphere
Properties affected by variables:	particle size bulk density dissolution rate of solids
Monitoring output:	potency particle size analysis density/surface area dissolution rate/flow rate of solids

C. Filling and Packaging Operations

In the case of solid dosage forms (capsules and tablets) and simple, nonsterile solution products, filling and packaging operations are usually routine. With respect to these particular dosage forms, the critical aspects of filling and packaging are machinability (i.e., secure closures, and achieving proper count, volume, and weight), labeling, and maintaining the stability of the final product in the primary container-closure system during the expiry dating period of the product.

In the case of disperse systems (suspensions, lotions, creams, ointments, and aerosols) product settling and separation are often a concern during filling and packaging operations (getting product into bottles, tubes, jars, and cans). The following critical aspects of the filling and packaging operations must be evaluated and controlled during large-scale (100X) validation and manufacturing runs:

- Proper control of product temperature to aid product flow and maintain product consistency before and during filling and packaging operations
- Proper agitation in holding tanks and filling heads in order to main product uniformity and homogeneity during filling and packaging operations
- The use of air pressure and inert atmosphere to achieve product performance and maintain stability in the primary container-closure system

D. Particle Size Considerations

The mean particle diameter and the particle size distribution of disperse, insoluble drug substances represent the internal phase of most dispersions. With the exception of emulsions and creams, for which the drug particle substrates are essentially liquid microdroplets, the other disperse systems, i.e., suspensions, ointments, and aerosols, have internal phase microparticles that are usually solids.

Control of particle morphology (shape, polymorphic forms, crystalline substances, solvates, or amorphous materials) and particle size are important parameters to attain high-quality drug product manufacture and control procedures. In general, the smaller the particle size, the greater is its surface area, which favors broader topical coverage (skin and mucous membrane), increased solubility, and enhanced absorption. The particle size distribution for most disperse systems should be in the range of impalpability, i.e., 0.2–20 microns. Several excellent reviews of this important topic have been provided by Brittain et al. [27], Byrn et al. [28], and Rawle [29].

Maintaining control of particle morphology and particle size with respect to both raw materials and finished product specifications is important. Drug morphology and particle size are significant factors, because they influence product appearance, settling rates, drug solubility, topical absorption, redispersability, and overall product stability.

Coalescence of liquid microdroplets, crystal growth of solid particles, and unfavorable shifts in particle size distribution are strong indicators of either poorly formulated products or poor control of critical unit operations during the manufacturing process. These factors, listed in Table 9, are also important considerations during formal validation studies.

Table 9 Factors Influencing Particle Size Distributions of Actives During Process Manufacture

Order of addition of critical materials during milling and mixing operations
Addition rates of critical materials during milling and mixing operations
Agitation speed of mills and mixers during processing operations
Temperatures and ranges achieved during mixing operations
Residence time in mixers, mills, and tanks during processing operations

E. Rheology

Rheology (viscosity and consistency) is an extremely important quality characteristic of disperse systems. Rheology is simply defined as a study of fluid (gases and liquids) flow. There are important determinants of how well, for example, a liquid will suspend insoluble ingredients, namely its yield value and viscosity. Yield value is the minimum amount of applied stress (force) needed before fluid flow is initiated. This property defines how well a liquid system can suspend insolubles (the so-called structure vehicle concept). Unless the force of gravity operating on a suspended particle of a given mass exceeds the liquid's yield value, the insoluble particle in suspension will not settle [30,31]. Additional discussion of the topic is provided in Chapter 5 of Volume I of this series.

Viscosity is defined as a measurement of the applied stress per unit area required to maintain a certain flow rate or shear rate. In general, viscosity is the resistance to liquid flow, whereas fluidity is the reciprocal of viscosity or the coefficient of viscosity. The thicker the liquid, the higher the viscosity, the thinner the liquid, the lower the viscosity or the higher its fluidity. In a disperse system the viscosity of the finished product or external phase is generated by one or more of the following components:

- Suspended solids
- Blends of oils and waxes
- Dispersed clays, gums, cellulosics, and/or synthetic polymers
- High concentrations of dissolved solids (sugars) in water
- The presence of polyols and polyoxyethylene derivatives

Since viscosity is both temperature and structure dependent, in most cases viscosity often changes with time. A suitable rheological instrument (viscometer) can be used to measure indirectly the settling behavior and the structural characteristics of pharmaceutical dispersions (suspensions, lotions, etc.). Creams and ointments are best handled at elevated temperatures when they are in a liquid or semisolid state.

The viscometer should be properly calibrated to measure the apparent (non-Newtonian) viscosity of the disperse system at equilibrium at a given temperature to establish system reproducibility. Non-Newtonian viscosity, which is representative of most pharmaceutical dispersions, may be defined simply as a different viscosity at a different shear rate. Apparent viscosity, like pH, is an exponential term and therefore the log of the apparent viscosity is an appropriate way to report test results. A classification of approximate viscosity ranges for disperse systems is provided in Table 10.

F. Specific Gravity

Specific gravity or density of a disperse system is an important parameter. With the exception of propellant-containing aerosols, most dispersions have a density of about 1 g/cm^3 or less. A decrease in density of suspensions, creams, and lotions often indicates the presence of entrapped air within the structure of the disperse system. Density measurement should be made at a given temperature using well-mixed, uniform dispersions. Precision hydrometers have been conveniently used to measure the density of liquid dispersions.

G. pH Value

The pH of aqueous dispersions should be taken at a given temperature and only after equilibrium has been reached in order to minimize "pH drift" and possible electrode

Table 10 Approximate Range of Viscosities for Pharmaceutical Dispersions at Room Temperature

Consistency type	Approximate viscosity in cps at 25°C	Pharmaceutical examples
Mobile liquids	1–100	Water, solvents, colloidal solutions
Pourable liquids	100–1,000	Glycols, oils, syrups
Just pourable	1,000–10,000	Lotions, suspensions with structure
Soft, nonspreadable[a]	10,000–100,000	Thin gels
Soft, spreadable	100,000–300,000	W/O, O/W creams
Plastic flow, spreadable	300,000–1,000,000	Ointments, soft butters
Hard, spreadable	1,000,000–3,000,000	Suppositories, waxes, hard butters

[a]Nonspreadable is defined as movement after the mass has been put in a fixed position.

surface coating by the disperse phase. Electrolytes, such as potassium chloride, may be added to the aqueous external phase or dispersions to stabilize their pH readings. Neutral electrolytes, such as potassium chloride, however, may have adverse effects on the physical stability of aqueous dispersions.

H. Content Uniformity

Probably the single most important parameter governing product stability and process control of disperse systems is content uniformity and/or homogeneity. This is especially so, since unlike solutions, the internal phase is dispersed rather than solubilized throughout the external phase. Suspensions and lotions are influenced by gravity regarding the uniformity of the internal phase. In the case of solid dispersions (i.e., powders, creams, and ointments), the effects of gravity are less important. The latter systems, however, are more dependant on particle size, shear rate, and mixing efficiency in order to attain and maintain uniformity of the active drug component (usually the internal phase) in the disperse system.

Throughout the critical unit operations of the manufacturing process of disperse systems, mixing, blending, homogenizing, holding, and filling are all important for obtaining and maintaining content uniformity and/or homogeneity and meaningful control of the system.

The settling rates of disperse systems, such as suspensions, are governed by one or more of the following Stokes' Law factors:

- Particle size of the internal phase
- Particle density of the internal phase
- Density of the external phase
- Viscosity and structure of the external phase

Proper control of these four vital factors during validation studies and routine production batches assures success in attaining content uniformity and homogeneity of such disperse systems.

With respect to content uniformity testing requirements for disperse systems in USP XXIII, they are similar to conventional solid dosage forms (tablets and capsules) with the exception of pressurized metered-dose inhalers. In the latter case, the number of units outside the content uniformity limits and the limits themselves may be increased somewhat to accomodate the uniformity testing of aerosol spray products [26].

The usual sample size of creams, lotions, and ointments for potency content and content uniformity testing usually ranges between 0.5 and 1.5 g per sample for assay. With respect to content uniformity testing in finished tubes, after carefully opening the tube lengthwise with a sharp razor blade, the samples for testing are usually removed from the top, middle, and bottom segments of the tube. Variations on this procedure are used to test content uniformity and homogeneity of other disperse system products.

I. Preservative Effectiveness

Preservation against microbial growth is an important consideration with respect to the formulation, stability, and validation of disperse systems. This is especially true whenever one of the phases of the disperse system is aqueous. Concern for microbial growth in oleaginous preparations and organic propellant aerosols is of lesser importance. Changes in the physical stability of aqueous dispersions can be traced often to microbial contamination [32,33].

A well-preserved oral or topical suspension, lotion, or cream does not have to be sterile to prevent microbial growth. The use of small amounts of propylene glycol (5–15%) or disodium edetate (about 0.1%), or a decrease in the pH of the disperse system have often been used to increase the efficiency of the preservative system of the formulation without adversely affecting the physical stability of the aqueous disperse system. A list of useful preservatives for oral, topical, and injectable pharmaceutical dispersions may be found in the NF XVIII [21] and the *Handbook of Pharmaceutical Excipients* [34].

Incorporating a USP antimicrobial preservative effectiveness testing procedure or a microbial limit test into the formal validation of aqueous disperse systems may be useful [21]. Such testing procedures and the determination of bioburdens for validation and production batches can also be used to establish appropriate validated cleaning procedures for the facilities and equipment used in the manufacture of disperse systems.

J. Dissolution Testing

Dissolution testing of disperse systems is still evolving. In such products the drug substance, which is usually a solid and part of the internal phase, slowly dissolves into an aqueous test medium under controlled conditions (i.e., temperature, pH, and agitation rate). Dissolution testing is primarily used as a quality control procedure to determine product uniformity and secondarily as a means of assessing the in vivo absorption of the drug in terms of a possible in vitro/in vivo correlation. Dissolution test methods have been established in USP XXVIII for tablets, capsules, and transdermal delivery systems [21]. Presently, there is no official method for disperse systems.

With respect to the dissolution of suspensions and lotions, the best approach at the present time is to place a small, but known, amount of disperse system inside a secure Durapore (polyvinylidene fluoride) membrane (Millipore Products Div., Bedford, MA) pouch of a suitable porosity and submerge and suspend it "tea bag" fashion in a suitable dissolution medium using a USP Method I paddle apparatus. Optimum experimental conditions are required to achieve reproducible results. In other words, qualification of the dissolution test method is required.

With respect to creams and ointments, the Franz in vitro flow-through diffusion cell (Crown Glass Co., Inc., Somerville, NJ) has been modified by using a silicone rubber membrane barrier to simulate a percutaneous dissolution unit for testing purposes [35].

IV. PRACTICAL APPROACH FOR MANAGING THE VALIDATION OF DISPERSE SYSTEMS

With the exception of an aerosol, which is a unique pharmaceutical dosage form, the validation of disperse systems, such as suspensions, lotions, creams, and ointments can be handled in the same way, because their similarities, rather than their differences, are usually subjected to the validation exercise.

The common similarities among disperse systems are as follows:

- Particle size distribution of the drug itself
- Homogeneity or blend uniformity of the drug throughout the external phase of the specific disperse system
- Reproducibility and stability of the viscosity and/or density of the final product

The differences among disperse systems are as follows:

- Either aqueous or oleaginous external phases of the specific disperse system
- Varying apparent viscosity and/or density between and among the various suspension, lotion, cream, and ointment formulations

It should be remembered that validation concerns are primarily related to in-process control of the manufacturing operations, whereas the release, by the head of the quality control unit, of the final finished product in labeled packaging is in accordance with the requirements established in the final official product specifications. This does not preclude the same requirement, i.e., blend uniformity for the disperse system, meeting both an in-process control limit as well as a final product specification.

The primary focus of prospective validation is to identify the critical unit operations, their critical process variables, and reasonable control limits for these variables in order to establish in-process control of the manufacturing process. In this connection, fractional, factorial designed experiments are often effective in obtaining such information [10].

In the case of retrospective validation, the objective is to establish and maintain process control by a demonstration of reproducibility of the various manufactured batches primarily meeting their final product specifications. This can be shown effectively through the use of quality control charting. The same data set shown in Table 14 was used both to carry out the fractional factorial analysis in Table 13 and to construct the control chart shown in Fig. 8.

A. Fractional Factorial Analysis in Prospective Validation

Some factorial experimental designs are often used to determine the critical process variables during the prospective phase of product scale-up and process validation [11]. Depending on the number of variables to be evaluated, full factorial designs are often chosen to establish a polynomial equation in terms of the process variables studied for the purpose of predicting the outcome or response, i.e., blend uniformity in a disperse system.

The problem with the full factorial experimental design approach and its many variations is the number of experiments, trials, or runs that are required to establish the critical variables and to predict their influence on future process outcomes. In the case of four to six process variables, the trials often range between 12 and 64 experiments. Outside of the laboratory, such experimental designs are usually unacceptable during the important scale-up stage of the process.

The use of fractional factorial designs described by C. D. Hendrix and others is a more reasonable approach to prospective validation requirements during process scale-up [10]. In the case of the fractional factorial design, four to eight process variables for evaluation would require only six to 10 trials or runs, respectively. The objective of determining critical processing variables is maintained, but at the expense of producing a predictor equation. In the author's experience, predictor equations don't always work out during the scale-up of pharmaceutical processes.

An example of a fractional factorial design for a typical disperse system based on eight process variables in 10 trials is demonstrated to the reader in Tables 11–13. The number of processing variables taken for factorial analysis is usually reduced to no more than eight by the use of constraint analysis [12]. According to the concept of constraint analysis and the Pareto principle that supports its use, no more than a handful of variables are responsible for the overall quality of a process's output. Based on experience with similar products and the theories governing the preparation and control of disperse systems in general, the following randomly assigned processing variables (Table 12) at their suggested upper and lower process control limits were selected for prospective study. The fractional factorial design used to determine the critical process variables is presented in Table 11.

Fractional factorial analysis (Table 13) indicates that X_4, X_6, X_8, i.e., pH, apparent viscosity, and particle size, when active, are critical variables with respect to blend uniformity of the disperse system, since they deviate significantly from zero.

Table 11 Fractional Factorial Design (Eight Variables and Ten Trials, Random Assignment)

Trials	X_1	X_2	X_3	X_4	X_5	X_6	X_7	X_8	
1	–[a]	–	–	–	–	–	–	–	0/8
2	–	+[b]	–	–	–	–	–	–	1/8
3	–	–	–	–	+	–	–	+	2/8
4	–	+	–	–	–	+	+	–	3/8
5	+	–	+	–	+	–	+	–	4/8
6	+	+	+	+	–	–	–	–	4/8
7	+	–	–	+	+	+	–	+	5/8
8	–	+	+	+	–	+	+	+	6/8
9	+	–	+	+	+	+	+	+	7/8
10	+	+	+	+	+	+	+	+	8/8
(LCL vs. UCL)	5/5	5/5	5/5	5/5	5/5	5/5	5/5	5/5	

[a]Variable at the lower control limit (LCL).
[b]Variable at the upper control limit (UCL).

Table 12 Process Variables for Factorial Analysis

Processing variables (randomly assigned)	LCL (-)	UCL (+)
X_1—mill clearance setting (step 4)	1	4
X_2-moisture content suspending agent	5%	15%
X_3—processing temperature (step 3)	50°C	70°C
X_4—pH value (step 4)	5.0	7.0
X_5—processing time	2 hr	6 hr
X_6—apparent viscosity (step 4)	20,000 cps	200,000 cps
X_7—blender speed (step 5)	4,000 rpm	20,000 rpm
X_8—avg. particle size (active)	20 micron	40 micron

These three process variables (X_4, X_6, X_8) and their unit operations should be controlled prospectively during process scale-up and retrospectively during future product manufacturing operations.

The FDA's desire for "worst case" analysis of processing variables was carried out in both trials 1 and 10, respectively, at lower and upper control limits.

B. Quality Control Chart

The quality control chart is used to decide periodically whether a process is in statistical control. The use of such a chart facilitates the detection and possible elimination of assignable causes of process variation. The use of quality control charts is considered to be the best statistical tool available for establishing, monitoring, and verifying a validated product and/or process such as that used in the manufacture of disperse systems. The control chart, as devised by W. A. Shewhart (Bell Telephone Labs) and reported in *Economic Control of the Quality of Manufactured Products* (Van Nostrand, New York, 1931), is a graphic presentation on which the values of the quality characteristics (blend uniformity, pH, viscosity, etc.) under investigation are plotted sequentially.

Table 13 Determining Critical Process Variables Using Blend Uniformity Data

Trials	X_1	X_2	X_3	X_4	X_5	X_6	X_7	X_8	RSD value
1	-5.4	-5.4	-5.4	-5.4	-5.4	-5.4	-5.4	-5.4	5.4
2	-4.7	+4.7	-4.7	-4.7	-4.7	-4.7	-4.7	-4.7	4.7
3	-4.2	-4.2	-4.2	-4.2	+4.2	-4.2	-4.2	-4.2	4.2
4	-4.2	+4.2	-4.2	-4.2	-4.2	+4.2	+4.2	-4.2	4.2
5	+2.9	-2.9	+2.9	-2.9	+2.9	-2.9	+2.9	-2.9	2.9
6	+3.3	+3.3	+3.3	+3.3	-3.3	-3.3	-3.3	-3.3	3.3
7	+4.3	-4.3	-4.3	+4.3	+4.3	+4.3	-4.3	+4.3	4.3
8	-5.1	+5.1	+5.1	+5.1	-5.1	+5.1	+5.1	+5.1	5.1
9	+6.3	-6.3	+6.3	+6.3	+6.3	+6.3	+6.3	+6.3	6.3
10	+5.7	+5.7	+5.7	+5.7	+5.7	+5.7	+5.7	+5.7	5.7
(Avg. deviation)	-0.2	0.0	+0.1	+0.7	+0.1	+1.0	+0.4	+1.0	—

The control chart (Fig. 8) consists of a central line or grand average $\overline{\overline{X}}$ and a limit line above (upper control limit, UCL) and below (lower control limit, LCL) the central line. These two control limits represent $\pm$ 3 standard deviations about the grand average $\overline{\overline{X}}$.

The control chart can be used to determine the following processing effects:

- Non-random data (trending or drifting), a long series of data points that move gradually toward either the upper or lower control limit
- Inconsistent data (change in direction), sudden shifts in level from one side of the grand average to the other side
- Instability, an unstable data pattern with large fluctuations or wide swings and where data points are often found outside the upper and lower control limits.
- Random and consistent data (normal process noise), fluctuations or variability in the process with a random distribution of data points on both sides of the grand average and within the upper and lower control limits

C. Relative Standard Deviation

The relative standard deviation (RSD), formerly called the coefficient of variation (CV), is a popular and excellent way of estimating, at the same time, the central tendency $\overline{X}$ and the variability of the numerical data set. The RSD value represents the ratio of the standard deviation (SD) to the mean $\overline{X}$. Thus

$$\text{RSD (\%)} = \frac{\text{SD}}{\overline{\overline{X}}} \times 100 \tag{1}$$

The usual quality control charting technique consists of preparing two distinctly different charts, one for averages ($\overline{X}$) and the other for ranges R, i.e., the difference between the highest value and the lowest value in a given numerical data set.

Using the RSD value concept, one chart can be prepared covering both the influence of the central tendency $\overline{X}$ and the dispersion or variability R. In addition to the RSD control chart's robustness, which is a distinct advantage in statistical analysis, each data set (batch or manufacturing lot) is represented by one unique value, the RSD (%).

In order to prepare a control chart based on a single value, the mean of $\overline{X}$ value is obtained by averaging the RSD values of two consecutive batches and the R value represents the difference between RSD values of the two consecutive batches. In this way 10 batches produce nine $\overline{X}$ values and nine R values from the analysis of 10 individual RSD values for the 10 batches (Table 15). In order to construct the control limits for the RSD quality control chart, the average range R is multiplied by the scaling factor A_1, where the sample size of N equals two for consecutive batches. The scaling factor for calculating purposes equals 1.88. The RSD quality control chart analysis assumes that RSD values for batches are normally distributed, which is a reasonable assumption.

The construction of a control chart used to evaluate retrospectively blend uniformity data for 10 manufactured batches of a disperse system is presented in Tables 14 and 15 and Fig. 8. The average RSD values were obtained by averaging two consecutive blend uniformity values of two consecutive batches. The R range values were obtained by taking the difference of R values of two consecutive batches.

Table 14 Blend Uniformity Analysis Data Set[a]

Batch number	Avg. value	Std. dev. (±)	RSD value (%)
1	96.2	5.19	5.4
2	97.3	4.57	4.7
3	96.8	4.06	4.2
4	95.8	4.02	4.2
5	101.0	2.93	2.9
6	98.6	3.25	3.3
7	97.5	4.19	4.3
8	98.3	5.01	5.1
9	102.6	6.46	6.3[b]
10	97.1	5.53	5.7
(Avg.)	98.1	4.52	4.6

[a]The specification is based on USP XXIII. A RSD of not more than 6% for 10 samples and not more than 7.8% for 30 samples.
[b]RSD value was 6.6 after testing 30 samples.

With the exception of the last data point (6.0) in Fig. 8, all other values are within the upper and lower control limits. The analysis of the control chart, however, shows some interesting findings. The first six batches appear to trend toward the lower control limit (RSD value of 3.1), after which there is a change in direction with a trend toward the upper control limit (RSD value of 5.9) and beyond (moving out of control).

A cursory analysis of the control chart data appears to suggest the following basis for further study. Processing conditions and/or raw materials for the first six consecutive batches of product showed a favorable trend with respect to process optimization (i.e., lower RSD value). After which processing conditions and/or raw materials for the last four consecutive batches showed a dramatic change in direction and if continued unchecked they would trend out of process control. Such clear-cut data are not always available from control charts. Nevertheless, the use of RSD control charts is a valuable tool for providing retrospective validation documentation for disperse systems.

Table 15 Construction of Relative Standard Deviation Control Chart for Disperse System Blend Uniformity Testing

Avg. RSD values	R values	$N = 2$
5.1	0.7	
4.5	0.5	Upper control limit
4.2	0.0	UCL = 4.5 + 1.88 (0.72)
3.6	1.3	UCL = 5.9
3.1	0.4	Lower control limit
3.8	1.0	LCL = 4.5 − 1.88 (0.72)
4.7	0.8	LCL = 3.1
5.7	1.2	
6.0	0.6	
$\bar{\bar{X}} = 4.5$	$\bar{R} = 0.72$	

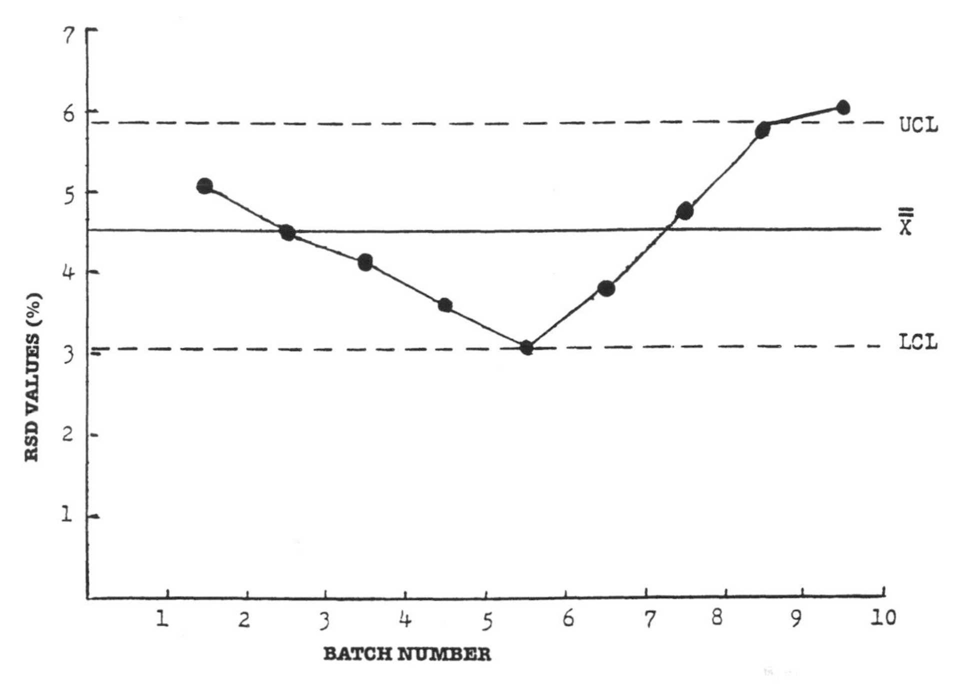

Fig. 8 Control chart for blend uniformity of a disperse system. The data points are plotted halfway between consecutive batches.

REFERENCES

1. CGMP: Proposed rule amendment of certain requirements for finished pharmaceuticals, 21CFR 210 & 211, *Federal Register* May 3, 1996.
2. *Validation of Aseptic Pharmaceutical Processes*, (F. J. Carleton and J. P. Agalloco, eds.,) Marcel Dekker, Inc. New York, 1986.
3. *Pharmaceutical Process Validation*, 2nd Ed., Revised and Expanded, (I. R. Berry and R. A. Nash, eds.), Marcel Dekker, Inc., New York, 1993.
4. *Validation of Manufacturing Processes,* Fourth European Seminar on Quality Control, Geneva, Switzerland, Sept. 1980.
5. *Validation in Practice*, (H. Sucker, ed.), Wissenschaftliche Verlagsegesellschaft GmBH, Stuttgart, Germany, 1983.
6. Guideline on General Principles of Process Validation, FDA, Rockville, MD, May 11, 1987.
7. Guide to Inspection of Topical Products, Mid Atlantic Region Inspection Guide, FDA, May 25, 1993.
8. Guide to Inspection of Oral Solutions and Suspensions, FDA, Rockville, MD, Aug. 1994.
9. FDA, *Pre-approval Inspection/Investigation Guidance Manual*, 7346, 832, Oct. 1, 1990.
10. C. D. Hendrix, What every technologist should know about experimental design, *CHEMTECH*, March. 167–174 (1979).
11. *Statistical Design and Analysis in Pharmaceutical Science, Validation, Process Control and Stability*, (S. C. Chow and J. P. Lin, eds.), Marcel Dekker, Inc., New York, 1995.
12. R. A. Nash, Process validation: A 17-year retrospective of solid-dosage forms, *Drug Dev. Ind. Pharm.*, 22:25–34 (1996).
13. R. A. Nash, The essentials of process validation, *Pharmaceutical Dosage Forms: Tablets*, Vol. 3, (H. A. Lieberman, L. Lachman, and J. B. Schwartz, eds.), Marcel Dekker, Inc., New York, 1990.

14. J. R. Sharp, Validation—How much is required? *PDA J. Pharm. Sci. Technol.*, 49:111-117 (1995).
15. Guide to inspection of validation of cleaning processes, *FDA Div. Field Investigations*, Rockville, MD, July, 1993.
16. P. Wray, Rutgers University, Piscataway, NJ, private communication.
17. R. A. Nash, *Pharmaceutical suspensions, Pharmaceutical Dosage Forms: Disperse System*, Vol. 2, 2nd Ed., (H. A. Lieberman, M. M. Rieger, and G. S. Banker, eds.), Marcel Dekker, Inc., New York, 1996.
18. N. K. Patel, L. Kennon, and R. S. Levinson, Pharmaceutical suspensions, *Theory and Practice of Industrial Pharmacy*, (L. Lachman, H. A. Lieberman, J. L. Kanig, eds.), Lea Febiger, Philadelphia, 1986.
19. D. L. Courtney, Physical Chemistry of Gels, *Am. Cosm. Perf.* 87:31-35 (1972).
20. J. S. Stephens, Flow properties of aluminum soap-hydrocarbon systems, *J. Pharm. Pharmac.*, 23:774-780 (1971).
21. USP XXIII NF XVIII United States Pharmacopeial Convention, Inc., Rockville, MD, 1994.
22. R. Y. Lockhead, Emulsions, *Cosm. and Toiletries*, 109 (May), (1994).
23. N. P. Redmond and R. W. Haltenberger, Design of a continuous manufacturing process for an oil-in-water cosmetic cream, *J. Soc. Cosmet. Chem.*, 23:637-655 (1972).
24. T. Forster and H. Tesmann, Phase inversion emulsification, *Cosm. and Toiletries*, 106 (Dec.) (1991).
25. M. A. Johnsen, Considerations in the Development of Aerosol Products, *Spray Technol.* 4 (Dec), (1994).
26. E. G. Lovering, Metered aerosol products: Dose uniformity over the entire contents, *Pharm. Forum*, 20:7495-7506 (1994).
27. H G. Brittain, et al., Physical characterization of pharmaceutical solids, *Pharm. Res.*, 8:963-973 (1991).
28. S. Byrn, et al., Pharmaceutical solids: A strategic approach to regulatory considerations, *Pharm. Res.*, 12:945-954 (1995).
29. A. Rawle, The importance of particle size analysis in the pharmaceutical industry, Malvern Instruments, Ltd., Worcester, UK.
30. K. Suzuki, Rheological study of vanishing creams, *Cosm. and Toiletries*, 91 (June), (1976).
31. H. T. Lashmar, J. P. Richardson, and A. Erbod, Correlation of physical parameters of an oil in water emulsion with manufacturing procedures and stability, *Int. J. Pharm.*, 125:315-325 (1995).
32. K. Klein, Improving emulsion stability, *Cosm. & Toiletries*, 99 (March), (1984).
33. M. M. Rieger, Stability testing of macroemulsions, *Cosm. & Toiletries*, 106(May), (1991).
34. *Handbook of Pharmaceutical Excipients*, 2nd Ed., (A. Wade and P. J. Weller, eds.), Am. Pharm. Assoc., Washington, DC, and The Pharmaceutical Press, London, UK (1994).
35. Workshop III report: Scaleup of liquid and semisolid disperse systems, *Pharm. Res.*, 11:1216-1220 (1994).

13

Drug Regulatory Affairs

John P. Tomaszewski and Dina R. Russello

Bayer Corporation, Morristown, New Jersey

I. INTRODUCTION

The purpose of this chapter is to provide a broad overview of regulatory affairs considerations within the pharmaceutical industry. Since the laws and regulations written to govern the industry are not product- or dosage-form specific, dispersion systems, as a dosage form category, are not covered specifically.

The information contained in this chapter applies equally to all pharmaceutical forms. Specific requirements for individual dosage forms emanate from the actual manufacture of the product and more aptly fall under the purview of quality control.

Regulatory affairs is distinct from quality control or quality assurance in that it deals with all legal and regulatory issues pertinent to pharmaceuticals, from the approval process, to labeling, to advertising, and is therefore involved with issues outside the scope of manufacturing, which tend to be more dosage-form-specific in nature.

All pharmaceutical dosage forms, including disperse systems, are equally impacted by the major areas of concern covered in this chapter: premarketing requirements, labeling, advertising, good manufacturing practices, and FDA inspections and, as such, this chapter should serve as good background for any regulatory issues concerning disperse systems.

II. HISTORY OF THE FDA

The United States Food and Drug Administration (FDA), as it exists today, did not come into being until passage of the 1906 Pure Food and Drug Act. However, its roots lie in the United States Department of Agriculture's (USDA's) Department of Chemistry, which, for the years preceding the 1906 Act, was mainly responsible for assuring the wholesomeness of America's food supply. The most prominent and influential figure in the early history of food and drug law was Dr. Harvey W. Wiley, chief of the USDA's Department of Chemistry who later became the first FDA commissioner. Dr. Wiley was instrumental in the creation of the FDA as well as in setting a standard of dedication and loyalty that can be seen in today's FDA. For example, while at the USDA, Dr. Wiley formed what came to be known as "Doctor Wiley's poison squad." This was a group of USDA employees who actually consumed great quantities of the substances, particularly preservatives, that were being added to the nation's food supply to observe what effect these substances had on humans. The experiments went on for five years and demonstrated that some of these preservatives were indeed harmful.

At the turn of the century, the pureness of the American food and drug supply was an important issue only for those involved with the respective industries, and only as far as it did not affect profit. The public at large was relatively unaware of the state of the American food supply until the "muckrakers" focused on industry problems. Perhaps the most influential of these muckrakers was Upton Sinclair, whose book *The Jungle* was greatly responsible for making the safety and purity of America's foods and drugs an important issue in the mind of the American public. Indeed, it was the alarm that the muckrakers caused among the public that helped spur Congress to pass the Pure Food and Drugs Act of 1906, despite the very strong lobbying of those interests opposed to any such legislation.

The 1906 Act was the very first legislation to give the federal government authority over the selling in interstate commerce of foods and drugs. It also gave the government the authority to create an agency responsible for regulating foods and drugs and, more important, the authority to promulgate regulations.

From its early beginnings in the Department of Agriculture, the FDA has grown into one of the most respected—as perceived by consumers—big government agencies, with almost 8000 employees. Through these decades, the FDA has maintained what it sees as its major mission—the protection of the American public from harmful and fraudulent foods and drugs. It is this precept on which the FDA bases all of its regulatory efforts, from the regulation of food and color additives to the approval of new drugs. Today's FDA not only oversees food and drugs, but it is also responsible for cosmetics, medical devices, and energy-emitting devices, and is developing new and more effective ways of assuring that all of these different products are safe and efficacious for use by the American public.

In recent years, particularly the last 35, the FDA has developed from a headquarters-only type operation to an effective field organization that can and does assure, at the manufacturing level, that the products it regulates are manufactured, packed, and held in a manner that assures that the products meet the qualities and standards they are supposed to have. The FDA has also begun to change its regulatory policy from a simple policing approach to a more proactive preventive strategy. This has been achieved through the promulgation of regulations that are open to industry comment and criticisms as well as the fostering of industry self-regulation.

III. EVOLUTION OF DRUG REGULATIONS

To understand the FDA's current philosophy on regulating drugs, one must go back through history to gain a perspective. In general, the FDA regulates all drugs under the same procedures regardless of dosage form, i.e., tablets, capsules, or liquids. The primary regulatory tool used by the FDA today is the new drug application (NDA), which allows the agency an opportunity to scrutinize the safety, efficacy, and quality of all new drugs and new indications for previously approved drugs. Once a drug has been approved, the FDA's main tool for assuring drug quality is inspection. Both of these tools are fairly recent developments in FDA history and date back to 1962. As it will be shown, prior to 1962, the FDA had no truly effective means to regulate drugs from the standpoints of either efficacy or quality. Today, it has at its disposal, a vast array of postmarketing surveillance regulations designed to ensure continued monitoring of the effect of drugs once marketing has begun.

A. The 1906 Act

The 1906 Pure Food and Drug Act represented the first comprehensive effort of the U.S. government to regulate drugs produced and marketed in the United States. In 1849, Congress passed a statute designed to prevent the importation of "adulterated and deleterious drugs and medicines," but drugs produced in the United States were not covered. The 1906 Act had no provisions regarding premarket approval of drugs, and manufacturers were free to market anything they wanted as long as the products were not adulterated or misbranded. The 1906 Act did require that standard drug products meet the requirements of the United States Pharmacopeia or the National Formulary, unless other standards were declared on the products' labels.

In terms of labeling, the 1906 Act stated that a product was misbranded if its labeling was "false or misleading in any particular." But in reference to therapeutic claims, this definition was qualified to the extent that the misstatement had to be shown to be fraudulent. In essence, this burdened the FDA with proving that the false therapeutic claims were purposely made by the seller. False or unsubstantiated claims made in good faith were immune from liability. In addition, the act had no standards regarding the safety or efficacy of drugs on the market. Manufacturers could use any substance they wanted to formulate drugs. As technology moved ahead and more and more new synthetic materials were developed and subsequently used in food and drug products, it became apparent to Congress that something had to be done to strengthen the FDA's authority to assure the safety of these new substances. Congress had begun to work on a new food and drug law in 1933, but it was not until the Elixir of Sulfanilamide tragedy that new legislation was finally passed in 1938.

Elixir of Sulfanilamide was manufactured and sold by the S. E. Massengill Company of Bristol, TN. It contained the active ingredient sulfanilamide with diethylene glycol as a solvent. Distribution of this drug resulted in at least 73 deaths from the toxic effects of diethylene glycol (antifreeze). The FDA could take no action on this deadly product on the basis of the 1906 Act; there were simply no provisions relating to safety of drugs. The FDA eventually took action against the product, forcing its removal from the market on the basis of a misbranding change, because it was falsely labeled as an elixir, although it contained no alcohol.

This episode made the inadequacies of the 1906 Act readily apparent and prompted Congress to pass the 1938 Federal Food, Drug, and Cosmetic Act (FD&C Act).

B. The 1938 Amendments

As early as 1933, Congress and those in the industry recognized that significant change was needed in the 1906 Act to adequately regulate the nation's drugs. This was due to the great progress in technology, which allowed a myriad of new synthetic chemical entities to be developed and, subsequently, used for their pharmacologic effects. Congress had been considering various legislative actions, but it was not until the Elixir of Sulfanilamide tragedy that they were forced into action. The 1938 Act was particularly important for two reasons: it established the definition of "new drug," and it required that all such drugs be shown to be safe prior to marketing. The act defined a "new drug" as one that was not generally recognized by experts as safe for use under the conditions prescribed, recommended, or suggested in the labeling. (Note that this definition did not include the term efficacy.) The concept of "general recognition" became one of great importance in the issue of whether or not a drug was "new." The 1938 Act required that for "new drugs," the manufacturer submit data to the FDA demonstrating that the product was indeed safe for use by humans. The burden of proof was placed on industry. Any new drug on the market had to be the subject of an "effective" NDA. Under the 1938 Act, and NDA was effective if it had not been "disapproved." In other words, explicit approval of the NDA was not required. NDAs at that point were required to include data regarding reports of investigation of the drug's safety, a list of components, methods of manufacture, and proposed labeling. The 1938 Act, however, did not affect how the FDA regulated the drugs that were brought to market prior to 1938. The FDA still had to rely on adulteration and misbranding charges as a way to control those pre-1938 "grandfathered" drugs. Significant change was limited to drugs coming to the market simply by the fact that the FDA was given the authority to closely scrutinize the safety of these new drugs.

C. The 1962 Amendments

As with the 1938 amendments, the next change in drug legislation was prompted by a catastrophe, this time involving thalidomide, a European sedative drug that was awaiting approval in the United States. While awaiting approval, it was discovered that the drug caused birth defects in the offspring of women taking it. In light of this situation, it again became apparent that greater control over drugs was needed not only in premarket approval but also in areas of postmarketing monitoring and manufacture. The 1962 amendments to the Food, Drug, and Cosmetic Act were important for two reasons: they required that all drugs on the market (even those approved before 1962) be effective for their labeled purposes, and they redefined adulteration by including the concept of good manufacturing practice (which is covered later). In order to properly review the drug products approved prior to 1962 for efficacy, the FDA established the drug efficacy study implementation (DESI) review. The purpose of this program was to review all drugs marketed between 1938 and 1962 for efficacy. Drugs for which new drug applications were filed after passage of the 1962 amendments were required to demonstrate efficacy in their NDAs.

D. The DESI Review

With the passage of the 1962 Drug Amendments the burden fell on the FDA to substantiate, on a retrospective basis, the efficacy of all drugs that were introduced into the market between 1938 and 1962. The first step, which was deemed the most important, was the review of the prescription drugs approved during that time period. The FDA enlisted the aid of the National Academy of Sciences/National Research Council (NAS/NRC) to conduct the first stage of the review. The NAS/NRC formed 30 expert panels to serve as the reviewing bodies for the categories of drugs assigned to them. Each panel was composed of six medical experts who were to review all data submitted as part of the review and make a recommendation as to the efficacy of the drug. The process began in earnest in 1966. Drugs were categorized by pharmacologic activity, i.e., anti-infectives, cardiovascular drugs, antidepressants, etc. The first order of business was to publish a notice in the *Federal Register* asking interested parties (namely manufacturers of drugs) to submit any data regarding the efficacy of the product. Aside from manufacturers' data a panel had access to FDA files on the drug and also reviewed the medical literature. After reviewing all pertinent data, the panel issued a report regarding its findings. Each drug ingredient had to be reviewed for each of the claims made for it. The panel classified the drug for each claim as follows:

- Probably effective—more data was needed to prove claim or the claim had to be modified to some extent
- Possibly effective—substantially more data was needed to support claim
- Ineffective—no data to support claim
- Ineffective as a fixed combination—combinations that were found not effective even though each individual ingredient was effective

By the middle of 1969 all of the NAS/NRC reports were completed and the FDA's evaluation of the panel's reports began. In total, the NAS/NRC found that approximately 60% of the products had at least one effective use. As the FDA published the NAS/NRC panel reports in the *Federal Register* (from 1968 to 1974), manufacturers and other interested parties were given the opportunity to comment on the findings and, if necessary, submit additional data in support of claims. Based on both the panels' reports and additional submitted data (if any), the FDA made a final determination as to each product's efficacy.

After reviewing more than 16,000 claims for some 3401 prescription drugs, the DESI review found 1099 drugs ineffective and 2302 effective for at least one use. The ineffective drugs were removed from the market, and unproved claims were removed from many of the 2302 effective drugs.

One significant result of the DESI review was that approximately 75% of the over-the-counter (OTC) drugs covered were found to be ineffective or lacking substantial evidence of efficacy. Many of the OTC drugs were on the market prior to 1938 and, therefore, enjoyed grandfather status under the 1938 amendments, meaning that a number of them had never been reviewed for safety. This in turn led to the establishment by the FDA of the OTC drug review. The OTC drug review was very similar to the DESI review, but much broader in scope. Since many of the OTC drugs had been on the market prior to 1938 and lacked adequate proof of safety, the OTC review evaluated safety as well as efficacy.

E. The OTC Review

At the start of the OTC review in 1972, there were approximately 250,000 different OTC drugs on the market. Again, the FDA's first step was to establish expert review panels whose charge was to make an initial review and evaluation of submitted data. Seventeen different panels were formed, and 250,000 drugs were categorized into approximately 55 therapeutic groups, and each panel was assigned to review certain groups. As with DESI, manufacturers submitted data to panels, and the panels reviewed and evaluated the data and published a report in the *Federal Register*, known as an Advanced Notice of Proposed Rulemaking (ANPR). With publication of the ANPR, manufacturers and other interested parties were provided the opportunity to respond and submit additional data. The FDA reviewed both the panels' reports and submitted data and published their findings as a tentative final monograph. Again time was allowed for comments and additional data before FDA promulgated the final monograph.

Ingredients reviewed under the OTC review were placed into one of three categories:

- Category I—those drugs found to be both safe and effective
- Category II—those drugs found to be either unsafe or ineffective.
- Category III those drugs for which data is insufficient to be found effective.

The final monographs are designed to include all requirements necessary to safely and effectively market OTC drugs, by specifying active ingredients and their dosages, indications, and warnings. The concept of the OTC review is interesting for two reasons: first, it was ingredient specific in that, except for a few instances, it did not specify particular dosage forms for any of the ingredients reviewed and, second, drugs reviewed by these expert advisory panels have gained a "general recognition" of safety and efficacy and, therefore, are not new drugs. Both of these circumstances are important to manufacturers because it allows any manufacturer to market a Category I OTC drug product, provided it is not a time-release dosage form, without prior FDA approval. This, of course, differs very greatly from the philosophy regarding prescription drugs, in which safety and efficacy are seen as extremely formulation dependent.

To date, the FDA has established over 90 separate rulemakings for OTC drugs. Of these, 45 rule-makings are in final monograph form, most notably the cold/cough ingredients, nighttime sleep aids, antacids, anorectal drug products, and topical first aid drug products. Significant pharmacologic categories not yet in final monograph form include internal analgesics, cough/cold combination products, laxatives, and oral health care drug products.

As these monographs are completed, the FDA will enter into a new phase of OTC drug regulation, which will include the reexamination of final monographs to assure that marketed products continue to be safe and effective. It may also turn out that due to new scientific information some ingredients may be found unsafe, and therefore subject to removal from the final monograph. In these cases, the FDA would have to reopen the monograph by publishing a proposed rule and allowing for public comment.

F. Adulteration and Misbranding

Since the 1906 Act, the FDA's primary means of regulating drugs was through use of the adulteration and misbranding charges, which were first defined on the federal level by the 1906 Act, and which have been strengthened and modified with each succeeding amendment to the basic act, i.e., the 1938 and 1962 amendments. In the 1906 Act the burden of proof was on the FDA; in addition, the agency had to show that, in cases of misbranding, the claims were fraudulent. The FDA's authority and the adulteration statute were once so weak that the agency needed a misbranding charge to remove the deadly Elixir of Sulfanilamide from the market.

The statutory requirements regarding drug adulteration and misbranding can be found in Sections 501 and 502 of the Food, Drug, and Cosmetic Act, respectively. The basic tenets of both sections, except the burden of proof requirement, have remained essentially unchanged since the 1906 Act. The adulteration clause reads

> A drug or device shall be deemed to be adulterated if it consists in whole or in part of any filthy, putrid or decomposed substance or if it has been prepared,

> packed or held under insanitary conditions, whereby it may have been contaminated with filth or where it may have been rendered injurious to health.

In the 1962 amendments, one very important change was made to the adulteration concept, which is that a drug could be deemed adulterated if "the methods used in, or the facilities or controls used for, its manufacture, processing, packing, or holding do not conform to or are not operated or administered in conformity with current good manufacturing practice." Thus were born the concepts of good manufacturing practices (GMPs), and the development of the FDA into an effective field-inspection force. This new clause gave the FDA the authority to promulgate regulations to define good manufacturing practices, which can be found in Title 21 of the Code of Federal Regulations (21 CFR 211). The GMPs are discussed in detail later in this chapter.

The misbranding clause states

> A drug or device shall be deemed to be misbranded if its labeling is false or misleading in any particular, if in package form it fails to bear a label containing (1) the name and place of business of the responsible firm and (2) an accurate statement of quantity of contents or if the required information is not prominently and conspicuously placed on the label.

As indicated, this definition has remained essentially unchanged from 1906, except that the burden of having to prove fraud has been removed from the FDA's shoulders in order to take legal action based on this clause.

IV. Rx VERSUS OTC DRUG REGULATION

This section deals with the regulatory differences between prescription (Rx) drugs and OTC drugs, not in terms of how a drug becomes one or the other, but in terms of how these drugs reach the market and how they are treated by the FDA upon marketing. The major topics covered include preclearance, labeling, and advertising requirements.

Over the past several years, many significant prescription drugs have been switched to over-the-counter status by manufacturers wishing to increase and maintain the profitability of the product. These recent switches have all been accomplished through the NDA route, in which manufacturers were required to demonstrate that the product was indeed safe and effective when actually used by the consumer under appropriate OTC labeling.

Recent examples of these significant Rx to OTC switches include the H_2 antagonists (Pepcid, Tagamet, Zantac) and the nonsteroidal anti-inflammatories (NSAIDS) (Naprosyn and Orudis).

A. Preclearance Requirements

Perhaps the biggest difference between the regulation of Rx and OTC drugs lies in the way drugs in these areas are reviewed prior to marketing, and what is required of these drugs to prove that they are safe and effective.

Since the enactment of the 1962 Drug Amendments, all Rx drugs that come on the market are considered "new drugs" and, therefore, are subject to stringent preclearance requirements, including the submission of a NDA. In the area of OTC drugs, there are virtually no "new drugs." Instead, the underlying purpose of the OTC review was to confer "not new" status on these drugs by establishing them as "generally recognized as safe and effective."

As background it is helpful to review the definitions of "drug" and "new drug" as delineated in the Food, Drug, and Cosmetic Act. In the Definitions section of the FD&C Act, drug is defined under Sec. 201 (g) as follows:

> The term 'drug' means: (A) articles recognized in the official United States Pharmacopeia, official Homeopathic Pharmacopeia of the United States or official National Formulary or any supplement to any of them; and (B) articles intended for use in the diagnosis, cure, mitigation, treatment, or prevention of disease in man or other animals; and (C) articles intended for use of the body of man or other animals; and (D) articles intended for use as a component of any articles specified in clause (A), (B) or (C); but does not include devices or their components, parts, or accessories.

Although this definition appears relatively straightforward, much debate and ultimately many court cases have arised based on the "intended" language in subsection (B). The courts have pretty much held that the intention of a product can be gleaned from label and labeling statements made on the product or from statements made in promotional material and advertising.

A "new drug" as defined in Section 201 (p) is as follows:

> (1) Any drug (except a new animal drug or an animal feed bearing or containing a new animal drug) the composition of which is such that such drug is not generally recognized among experts qualified by scientific training and experience to evaluate the safety and effectiveness of drugs as safe and effective for use under the conditions prescribed, recommended, or suggested in the labeling thereof, except that such a drug not so recognized shall not be deemed to be a "new drug" if at any time prior to the enactment of this Act it was subject to the Food and Drugs Act of June 30, 1906, as amended and, if at such time, its labeling contained the same representations concerning the conditions of its use; or
> (2) Any drug (except a new animal drug or an animal feed bearing or containing a new animal drug) the composition of which is such that such drug, as a result of investigations to determine its safety and effectiveness for use under such conditions has become so recognized but which has not, otherwise than in such investigations, been used to a material extent or for a material time under such conditions.

In essence this means that any drug not generally recognized as safe and effective and which has not been used for a material extent or a material time is a "new drug."

To further define "new drug" the FDA has promulgated definitions in the Code of Federal Regulations title 21 Part 310.3 (h) that points out circumstances under which the newness of a drug may arise. These include the use of new ingredients, new chemical entities (NCEs), new combinations of drugs, use of a drug for a new indication, and the use of a new dosage form, i.e., going from a solid oral dosage form to a liquid oral dosage form. If a drug fits into any of these categories of "newness," it is considered a "new drug" and is, therefore, subject to the new drug requirements of the FD&C Act, which are found in Section 505.

Bringing a new drug to market is a multistep process. The first order of business is to ascertain whether or not in a human population the proposed drug is safe and effective. To determine this, one must test the drug in humans. But this poses a dilemma. According to the FD&C Act, in order to legally ship a new drug in interstate commerce (drugs being tested in all probability will be shipped in interstate commerce), it must be covered by an approved NDA since the drug is only being tested, it does not have an NDA and therefore it cannot be shipped. However, the FD&C Act in Section 505, allows for this circumstance by including a section on "Investigational New Drugs," which can be found at Section 505 (i): "The Secretary shall promulgate regulations for exempting from the operation of the foregoing subsections of this section drugs intended solely for investigational use."

The specific regulations promulgated for investigational new drugs are found in the Code of Federal Regulations (CFR) Title 21 part 312 and are discussed in detail below.

1. Investigational New Drugs

Investigational new drugs (INDs) are addressed in Section 212.1 of the CFR, which requires that drugs shipped for purposes of investigation meet two conditions: (1) that the label of such a drug bears the statement "Caution: New drug–limited by Federal (or United States) law to investigational use"; and (2) that the person responsible for this new drug submit to the FDA a completed and signed "Notice of Claimed Investigational Exemption for a New Drug." This document must be submitted to the FDA 30 days before any study is planned to begin and must be submitted in triplicate. This information required for the claimed exemption can be found on what is known as an FD Form 1571, which is available from the FDA upon request. The requirements stated on this form can also be found in 21 CFR at part 312.1.

The statutory authority for promulgation of regulations governing investigational drugs lies in section 505 (i) of the Food, Drug, and Cosmetic Act. The requirements of this section are threefold. First, before any clinical testing of a new drug begins, the sponsor (person or company undertaking the study) must report to the FDA any data on preclinical studies that justify using the drug in humans for a particular purpose.

Second, the sponsor must obtain and submit to the FDA signed agreements from the clinical investigators that persons receiving the drug will be under the direct supervision of the investigator. Third, records of data must be established and maintained by the sponsor and made available to the FDA so that an appropriate evaluation of the safety and efficacy of the drug can be made. From this statement in the act, the FDA has promulgated specific regulations detailing exactly how one is expected to comply with this statute. These regulations appear in section 312.1 of 21 CFR and, as noted, go into much greater depth regarding mechanisms of compliance.

The basic purpose of regulating INDs is to assure the safety of the subjects being tested by informing the FDA of the intent to test a new drug on humans and by allowing the FDA an opportunity to review and evaluate the safety of the compounds as well as to assess the scientific and medical rationale behind the specific use of the drug.

In order to conduct an appropriate review of the proposed IND testing, certain basic information is required:

- Name and description of the drug, as well as a list of components and quantitative composition for the drug
- Information as to the methods, facilities, and controls used to make the drug
- All information regarding any preclinical studies conducted on the drug that indicates that it would be "reasonably safe" to administer to humans
- A list of the clinical investigators, including a summary of their training and experience
- The protocol to be followed in the clinical study
- A statement that the proposed clinical study has been reviewed by an institutional review board to assure protection of the subjects

By the nature of the information required by an IND application, it is readily apparent that the main purpose of the document is to protect the safety of any individual who participates in the clinical evaluation of a new drug. However, the IND submission also provides the FDA the opportunity to review and evaluate the technical parameters relating to the drug, particularly manufacturing and analytical procedures.

Thirty days after an IND application is submitted, a firm can legally begin to ship the new drug in interstate commerce for the purpose of experimentation. All drugs tested as an IND must bear on their label the statement "FOR INVESTIGATIONAL USE ONLY."

The clinical evaluation of investigational drugs is divided into three distinct phases referred to as phase 1, phase 2, and phase 3. Phase 1 clinical trials are conducted in limited populations, usually healthy, adult males, and are designed primarily to assess the toxicology and pharmacokinetic effects of various doses of the drug. Phase 2 studies are conducted in somewhat larger populations of individuals, who have the disease or condition the drug is intended to effect. The drug is evaluated at various dose levels for an extended period of time. Phase 2 studies can accumulate anywhere from 100 to 500 patients. Phase 3 studies are the pivotal efficacy studies conducted in large populations designed to adequately assess the safety and efficacy of the drug under what would be conditions of actual use.

Once all necessary clinical data are obtained through testing conducted while a drug is an IND, a NDA can be submitted to the FDA. On March 19, 1987, the FDA published major new regulations pertaining to investigation new drugs (*Federal Register* Vol. 52, pp. 8797–8857). The main purpose of the new regulation is to allow large-scale treatment of persons with diseases that are particularly severe or life-threatening. Prior to these new regulations, promising new drugs for certain diseases required many years of testing before they became available for use by the population that could have most benefited from them. Because they were "investigational," only a small number of test subjects could use them, but in the mid-1980s, when the AIDS problem came to the forefront, demand for many promising new drugs without fully established safety profiles in humans increased. Individuals suffering and dying from AIDS were willing to be treated with unproven drugs. Recognizing the needs of these patients and other patients with severe life-threatening diseases, the FDA sought to make these investigational drugs more readily available. There are certain criteria by which these drugs are evaluated to determine if they are eligible to become a treatment IND, with emphasis on the severity/mortality profile of the particular disease.

2. New Drug Applications

As discussed above, any new drug that was marketed after 1962 is required by the Food, Drug, and Cosmetic Act to have an effective new drug application. The statutory mandate for this can be found in Section 505 of the Act. The Act requires that the NDA document the following:

- Full reports of investigations that show whether or not the drug is safe and effective (those conducted under the IND)
- Full list of articles used as components of such drug
- Full statement of the composition of the drug
- Full description of methods and facilities used to manufacture the drug
- Samples of the drug
- Labels and labeling for the drug
- Certification as to patent status

The last item was incorporated into the NDA process with the passage of the Price Competition/Patent Term Restoration Act of 1984. This amendment to the Food, Drug, and Cosmetic Act allows extension for the patent life of a drug if it can be shown that some amount of marketing time was lost during the FDA review process. Under NDA procedures, approval for marketing drugs can require up to six years. Since patents are valid for only 17 years, the review process left only about 10 years (on average) for marketing exclusivity, which in the long run compromised the innovating drug firms from a financial standpoint and to some extent thwarted research and development. Under the provisions of the Patent Term Restoration Act, exclusive marketing time can be restored to the innovator based on the amount of time the drug was in the approval process. The maximum extension that can be granted is five additional years patent life, allowing for up to 14 years marketing exclusivity. There is another important aspect to the Patent Term Restoration Act that is discussed in the section on abbreviated new drug applications.

The specific regulations covering the details for submission of a NDA can be found in 21 CFR 314. This section of the CFR covers all aspects of the submission from the format (i.e., paper size, type size, pagination system, etc.), to the FDA's action regarding the submission. The regulation also outlines steps a company can take if it disagrees with the findings of the FDA regarding the disposition of the NDA.

The data-reporting requirements for a NDA are very extensive and can take many years to accumulate. The basic requirements of a NDA are as follows:

Application form [FD form 356 (h)].
Index and summary of entire application.
Technical section.
Chemistry—All information on the drug substance and drug product chemical characteristics, stability, name and address of manufacturer, method of synthesis, analytical methods to assure identity, strength, quality, and purity of the drug substance.
Nonclinical pharmacology and toxicology—Reports of all studies conducted to assess pharmacologic activities and toxicological effects of the drug.
Human pharmacology and bioavailability—A description and report of the results of all bioavailability studies and pharmacokinetic studies.

Microbiology section—Only required for anti-infective drugs

Clinical data—A description and analysis of each controlled clinical study pertinent to the proposed use of the drug including a summary of data demonstrating efficacy and support for the dosage/administration required of the drug. Summary of the safety information including adverse effects and drug interactions.

Statistics—A description of the statistical evaluation of the clinical data supporting safety and efficacy.

Samples—Proposed product labeling must also be included.

Case report forms and tabulations—Required for all clinical studies submitted in support of the drugs claimed safety and efficacy.

3. *Abbreviated New Drug Applications*

For generic versions of new drugs, so-called me-too drugs, the statute allows for the submission of an abbreviated new drug application (ANDA). Prior to enactment of the Price Competition/Patent Term Restoration Act of 1984, ANDAs were allowed only for those drugs approved before 1962. The new drugs approved after 1962 were, by law, not amenable to subsequent ANDA approvals, in essence granting a lifetime of exclusivity to innovators of post-1962 new drugs. This obviously was not acceptable to generic drug manufacturers nor to Congress, which was concerned about the rising costs of drugs. Congress reasoned that if more generics were made available, overall health-care cost would decrease. To achieve this end, Congress pass the price competition portion of the 1984 amendment to open up all post-1962 new drugs to generic marketing, provided that the generic manufacturer could certify that the new drug was not longer on patent.

Drugs that are available for generic copies are so-called listed drugs. The FDA is required under the Price Competition/Patent Term Restoration Act to publish an official list of drugs that are amenable to generic marketing. These drugs are published by the agency in what is known as the "orange book," which lists all drugs approved by the FDA, their dosage form, therapeutic equivalence rating, (which dictates what type equivalence data is required), NDA number, approval date, patent expiration, and any exclusivity period. The orange book is used not only by generic manufacturers, to determine what drugs can be copied, but also by individual state boards of pharmacy to determine what drugs are appropriate for generic substitution. The FDA updates the orange book on a quarterly basis, and issues a new version each year.

The requirements for an ANDA application are not as extensive as those for a NDA because clinical data on efficacy and safety are not required. What is required, however, is that the generic drug be shown to be bio-equivalent to the innovator drug through either human or in vitro testing. The requirements can be as simple as dissolution testing or as complex as human pharmacokinetic studies. Regarding the bulk of the document, information on chemistry, manufacturing, and a certification that the drug will be manufactured according to good manufacturing practices must be included.

The FDA considers a generic drug to be bio-equivalent to the innovator drug if the "rate and extent" of absorption of the drugs are not significantly different. The FDA definition of "significantly different" has changed over time and in some cases has changed on a drug-by-drug basis. The agency's current thinking is that no significant difference means no more than a 20% difference between drugs within a 90% confidence interval.

4. *Submission Guidelines*

To help firms comply with the many provisions of the application requirements, including the complicated format requirements, the FDA makes available guidelines for each specific section of an NDA. These guidelines can be obtained from the FDA's Docket Management Branch under Docket No. 85D-0248. These guidelines are very detailed and provide an excellent guide to what documents the FDA will review and how it will review them.

As far as approval timing is concerned, the Act mandates that the FDA rule on either the acceptance or rejection of a submission within 180 days of receipt. However, in the case of NDAs this is seldom the case. What usually happens is that within the initial 180-day time frame, the FDA will inform the firm that the NDA submission is lacking in some respect or that additional information is needed and will issue a "nonapprovable" letter. This letter assures FDA compliance with the statutory 180-day response requirement and allows the FDA necessary time for in-depth review, as the submitting firm is generating additional data. Once the FDA is satisfied that all data needed to review the NDA are on hand and the submission is complete, it will issue an "approvable letter." The date of this letter then becomes the official submission date and the 180-day clock starts ticking. The time span between the "unapprovable letter" and the "approvable letter" can sometimes be several years.

Over the past several years significant changes have been made to the FDA's authority and responsibility concerning drug approvals, which have served to improve the FDA's activities in terms of review times and the number of applications actually reviewed and approved. These changes include user fees, the refuse to file policy, preapproval inspections, and advisory committee review, and are discussed in some detail below.

5. *Prescription Drug User Fee*

The Prescription Drug User Fee Act of 1992 (PDUFA) was signed into law on October 29, 1992. This law requires pharmaceutical companies to pay for certain new drug and biologic applications, certain marketed drug products and certain facilities where these products are made. In return, the FDA is required to meet user fee performance goals, which include the elimination of its 1992 backlog of overdue applications, the shortening of review times, and the initiation of programs to improve and ensure the quality and integrity of the review process. This law was made effective on September 1, 1992, and sunsets on October 29, 1997, at which time Congress must renew the act or it will expire. The amount of revenue to be collected over these five years is based on the amount the FDA estimates it will need to achieve its stated performance goals. For 1993, the total revenue and corresponding fee rates were established in the PDUFA. For subsequent years, however, the total revenue and rates have been estimated. The FDA is authorized to increase the total revenue on a yearly basis to cover the greater of (1) the total percentage increase that occurred during the preceding fiscal year (FY) in the consumer price index; or (2) the total percentage pay increase for that FY for federal employees. The actual revenue and corresponding fee rates must be calculated and published in the *Federal Register* in December of each year. These new rates are retroactive, each year, to October 1 of the previous year. In other words, the revenue collected for 1993 applied to applications, establishments, and products that were subject to fees during October 1, 1992, through September 30, 1993. The actual user fee revenue for 1993–1997 are shown in Table 1. The total fee revenue for each year is collected in equal amounts from each of the three categories of user fees. Thus, in 1993, the total fee revenue was $36 million and one third, or $12 million, was collected each from application, product, and establishment fees.

6. Application Fees

Under the PDUFA, the FDA is authorized to collect fees for certain types of human drug applications that are submitted on or after September 1, 1992. These applications, listed below, cover essentially all NDAs and supplements as well as all product license applications (PLAs) and PLA amendments. Application fees are due for all such applications even if the application is requesting approval to market a product over the counter.

Applications subject to user fees are

- Applications or supplements submitted under §505(b)(1) of the FD&C Act
- Applications or supplements submitted under §505(b)(2) of the FD&C Act after September 30, 1992, which request approval of a molecular entity, which is an active ingredient or new indication for use not previously approved under a 505(b) application
- Applications or amendments for initial certification or initial approval of an antibiotic drug under §507
- Applications or amendments for licensure of a new biologic under §351 of the Public Health Service Act

Applications for the products listed below are not subject to user fees

- Generic drugs approved under section 505(j)
- Whole blood or a blood component for transfusion
- Crude allergenic extracts
- In vitro diagnostic biologics licensed under section 351 of the Public Health Service Act
- Large-volume parenterals with an NDA or supplement, or PLA or amendment, approved before September 1, 1992

Different fees are stipulated for the above-noted applications based on whether or not they contain clinical (other than bioavailability or bioequivalence) data. There are three categories of applications for purposes of user fees. They are (1) full applications with clinical data; (2) full applications without clinical data; and (3) supplements or amendments with clinical data. In 1993, the fee for full applications with clinical data was

Table 1 Total Revenue by User Fee Category for Fiscal Years 1993–1997 (in dollars)

	1993	1994	1995	1996	1997
Application	12×10^6	18,761,400	25,805,000	26,660,400	28,932,400
Product	12×10^6	18,761,400	25,805,000	26,660,400	28,932,400
Establishment	12×10^6	18,761,400	25,805,000	26,660,400	28,932,400
Total	36×10^6	56,284,200	77,415,000	79,981,200	86,797,200

$100,000. Per the PDUFA, the fee for full applications without clinical data must be one half the fee for full applications with clinical data. Thus, in 1993, the fee for full applications without clinical data was $50,000. For supplements or amendments with clinical data, the fee stipulated is also one half the fee for full applications with clinical data, and thus was $50,000 in 1993. There is no fee for supplements and amendments without clinical data. Table 2 shows the actual application fee rates for 1993–1997.

7. *Product Fees*

Pharmaceutical companies are subject to product fees for each final dosage form of each prescription drug product, which they have listed per §510 of the Food, Drug, and Cosmetic Act, when the company has one or more human drug applications (or supplements, or PLAs or amendments) pending with FDA on or after September 1, 1992. This fee is due, as long as the product remains listed, even if it is not currently marketed. A prescription drug product is a drug of specific strength, in a final dosage form, that is marketed through an approved human drug application. A final dosage form is defined as a "finished dosage form, which is approved for administration to a patient without further manufacturing," where each different form of a drug product, even if the same strength, is considered a final dosage form for which a fee is payable. This fee is due once per year for each drug product when the product is first listed and then in December for upcoming years. Certain products, however, are exempt from annual product fees. Drug products not subject to user fees are

- Generic drugs approved under section 505(j)
- Drug products approved under section 505(b)(2) of the FD&C Act
- Whole blood or a blood component for transfusion
- Bovine blood products for topical application licensed before September 1, 1992
- Crude allergenic extracts
- In vitro diagnostic biologics licensed under section 351 of the Public Health Service Act
- Large-volume parenterals with an NDA or supplement, or PLA or amendment, approved before September 1, 1992
- Products that may be dispensed without a prescription (OTC drug products)
- Products that are never dispensed as finished pharmaceuticals (i.e., require further manufacturing)

Table 2 Application Fee Rate Schedule for 1993–1997

	1993	1994	1995	1996	1997
Full applications with clinical safety or efficacy data	$100,000	$162,000	$208,000	$204,000	$205,000
Full applications without clinical data	$50,000	$81,000	$104,000	$102,000	$102,500
Supplements (or amendments) with clinical data	$50,000	$81,000	$104,000	$102,000	$102,500

Product fees are not required for products marketed over the counter even if they have an approved new drug application and an application fee was paid to review that application. The actual product fee rates for 1993–1997 are as follows: 1993, $6000; 1994, $9400; 1995, $12,200; 1996, $12,600; and 1997, $13,200.

8. *Establishment Fees*

Under the PDUFA, companies are required to pay annual establishment fees for their facilities that manufacture prescription drugs if they have at least one human drug application (or supplement, or PLA or amendment) pending with FDA on or after September 1, 1992. For purposes of assessing user fees, an establishment is defined as "either a foreign or domestic place of business where at least one prescription drug product is manufactured in final dosage form, and/or which is under the management of a person listed as the applicant in a human drug application for a prescription drug product." One establishment, however, may consist of several buildings as long as they are less than five miles apart. Thus, establishments where no prescription drug products are manufactured but for which a NDA (or supplement, or PLA or amendment) for a prescription drug product has been submitted are also subject to an establishment fee. Facilities where only packaging of a drug product is performed are exempt from establishment fees. Establishment fees are billed by December 31 of each year and are due within 30 days, or by January 31 of the next year. Manufacturing of the products listed below, even though they may be prescription drug products or they may have approved human drug applications, are exempt from the establishment fee:

- Over-the-counter drug products
- Generic drug products
- Drug products approved under §505(j), a generic drug, or 505(b)2, a paper NDA drug
- Whole blood or blood components for transfusion
- Bovine drug products for topical application licensed before September 1, 1992
- Crude allergenic extract products
- In vitro diagnostic biologic products licensed under section 351 of the Public Health Service Act
- Large-volume parental products manufactured under applications or supplements approved before September 1, 1992

The actual establishment fee rates for 1993–1997 and the estimated establishment fee rate schedule for 1997 are as follows: 1993, $60,000; 1994, $93,800; 1995, $129,000; 1996, $135,300; and 1997, $115,700.

9. *Waivers and Fee Reductions*

The FDA can grant waivers or reductions in any user fee if

- Such a waiver or reduction is necessary to protect the public health
- The assessment of the fee would present a significant barrier to innovation (because of limited resources available to such person (the applicant) or other circumstances)
- The fees to be paid will exceed the FDA's anticipated present and future costs incurred in reviewing the application

- Assessment of the fee for an original or supplement application would be inequitable because an application (e.g., an ANDA) for a product containing the same active ingredient filed by another person cannot be assessed a fee. The fee would be inequitable because the product is similar to certain generic drugs that are not part of the user fee program.

Anyone can request a waiver or reduction in a user fee, but the request must be made in writing 90 days before the fee is due.

The FDA has stated that waivers and/or reductions would most likely be granted to companies with less than $10 million in annual gross revenues with no corporate parent or funding source with annual gross revenues of $100 million or less.

10. Logistics

In December of each year, the FDA establishes total fee revenues; sets application, establishment, and product fee rates for the following year; and publishes this information in the *Federal Register*. Also in December, the FDA prepares and mails invoices to companies for product and establishment fees assessed for the upcoming year. Products and establishments, which become subject to fees during the year, are billed at that time. These fees are due within 30 days. For application fees, applicants are required to pay 50% of the appropriate fee upon submission of the application. This fee, however, is not to be included with the application but instead should be sent to the address shown below. Effective January 1, 1994, Form FDA 3397 must accompany each application submitted. This form, entitled User Fee Cover Sheet, helps the FDA determine what, if any, kind of application fee is required. Failure to complete and include the form results in processing delays, and failure to submit the appropriate fee results in a refusal-to-file (RTF) notification. Should a RTF be issued for any reason, the FDA will refund one-half the amount paid at the time of submission. Thus in 1993, the submission of a full application with clinical data would have required a $50,000 fee at the time of submission. If the application wound up as a RTF, the FDA would refund only half of that, or $25,000. The FDA feels that refunding only 25% of the full application fee is necessary to encourage higher quality submissions and to pay for the work done to determine if the application could be filed. If an application were withdrawn for any reason, the remaining 50% of the application fee is still due. The FDA can, however, waive the rest of the fee if "no substantial work was yet performed on the submission." When an application is accepted for filing, the remaining 50% of the application fee is due when the applicant receives an action letter (approvable, approved, not approved) from the FDA. Per 21 CFR 314.101 and 601.2 and 601.3, the FDA should take no longer than 60 days to determine the status of filing an application. Per the FDA's performance goals stated under the PDUFA, the FDA should take no longer than 6–12 months (depending on whether the application is deemed standard or priority) to issue action letters.

All checks for application, product, and establishment fees must be made payable to and mailed to

Food and Drug Administration
P.O. Box 777-W7745
Philadelphia, PA 19175-7745

11. *Performance Goals*

In July of 1993, the average time to gain approval for marketing a new drug product was 22 months. Pursuant to the PDUFA, the FDA is required to complete review of new applications in 6 or 12 months, depending on the type of application. A standard application, which involves a drug similar to one already marketed, must be reviewed within 12 months. A priority application, which involves a drug offering significant advances over existing treatments, must be reviewed within 6 months. In order to meet these goals, the FDA plans to hire about 700 new drug reviewers and support staff with the approximately $325 million they expect to collect under the user fee act (during 1993–1997). Consequently, the FDA must prepare and submit to the Committee on Energy and Commerce of the House of Representatives and the Committee on Labor and Human Resources of the Senate an annual report stating the agency's progress in achieving the goals provided in the act and their plans for meeting these goals in the future. In summary, the Prescription Drug User Fee Act allows industry to supplement existing FDA appropriations to be used solely for improving the drug review process so that new drug products can be brought to market faster. It's the FDA's challenge to make this a reality.

12. *Refusal to File*

In conjunction with the Prescription Drug User Fee Act, the FDA initiated and refined several other policies having a direct impact on their drug review and approval activities. The most significant and directly relatable to the PDUFA approach is the agency's policy on refusing to file applications found deficient. As the PDUFA imposed strict and aggressive time lines on the FDA's approval activity, the agency needed some way to clear the backlog of approvals and also some mechanism by which they could control the number of new applications being submitted.

In July of 1993, the agency formally issued a guidance document relating to the refuse-to-file policy, spelling out exactly what would constitute a RTF action and how the actions would be carried out.

The first step in the drug review process that the agency now undertakes is to have each reviewing group in the division review the application in context of specific checklists to determine if the appropriate information required for the FDA to conduct its review is in the document. The FDA conducts this checklist review within 45 days of submission and notifies the applicant at this point as to whether or not the application has been filed. If it can be filed, the agency then proceeds with its review and the normal review clock starts.

If the agency refuses to file an application, it issues a detailed letter to the applicant stating the exact reason for the refusal. The RTF letter also provides for an appeal process whereby the applicant can appeal the RTF to higher FDA management. In most cases, soon after the RTF letter is issued, the applicant will have a meeting with the FDA reviewing division where the NDA is discussed in detail and a plan is agreed to that will address the RTF issues.

A RTF action has a significant impact on the applicant, not only from the review-time point of view, but from the immediate financial point of view as well. Under the PDUFA, the applicant has already paid half of the review fee, and can forfeit half of the initial payment because of the RTF action. In most cases, one-fourth the total fee is lost on an RTF action.

Although each reviewing division has for the most part developed specific checklists on which they base their RTF findings, there are three basic circumstances that will invoke an RTF action. These are

- Omission of a section of the NDA required under federal regulations, or the presentation of the information in such a haphazard manner that renders it incomplete
- Clear failure to provide evidence of effectiveness
- Omission of critical data

13. Preapproval Inspections

Preapproval inspections (PAIs) are another FDA initiative that have become more evident and more meaningful since the advent of the PDUFA. Preapproval inspections have now become an integral part of the NDA approval process. PAIs focus not only on the finished dosage form manufacturing preparations but the bulk raw material active ingredient operations as well.

The reviewing division requests a preapproval inspection not long after the application is accepted for filing. The division requests the FDA district office, in which the manufacturing plant is located, to conduct a GMP inspection of the proposed manufacturing site, focusing on the specific product covered by the application. The district office uses the "field copy" of the Chemistry, Manufacturing and Controls (CMC) Section of the NDA as the basis on which it conducts its inspection. The field copy of the CMC Section is provided to the appropriate FDA district office by the applicant. This requirement came into effect with the passage of the Generic Drug Enforcement Act, which stemmed from the generic drug scandals of the mid-1980s. Under FDA policy, PAIs are usually requested by the division within 45 days of the NDA acceptance and take place usually six to seven months after the application is filed.

The main purpose of a PAI is to ensure that the applicant is capable of producing the drug product under the conditions it established in the CMC section of the NDA. Specifically the inspection focuses on

- The verification of the accuracy and completeness of the information in the CMC
- The evaluation of the manufacturing process of the preapproval batches (batches used for clinical studies)
- Evaluation of the manufacturers overall compliance to the GMPs
- Collection of appropriate samples if required

The FDA usually makes a practice of conducting these PAIs after some cursory review of the safety and efficacy portions of the NDA have been completed to assure that valuable district office inspection time is reasonably spent.

Although PAIs were initially used in circumstances where a new chemical or molecular entity was the subject of the application, the FDA has expanded their use to help them control the ebb and flow of approvals and to help them meet their review time targets under the user fee requirements. Often times the agency uses adverse PAI findings to have an application removed from its review queue, so it thereby does not have an adverse impact on its review time obligation.

14. Advisory Committees

Perhaps one of the most influential factors during recent years relating to the FDA's drug review activities has been the increase in use of the FDA advisory committees. These committees have been in existence for a good number of years, but it has only been since the passage of the PDUFA that the agency has come to rely on the committees to help make prompt decisions regarding drug approvability. Today, virtually no new drug gets approved without advisory committee review.

The FDA's Center for Drug Evaluation and Research relies on approximately 16 standing advisory committees, each focused on a specific pharmacologic class of drugs. These committees cover the following drug classes:

- Gastrointestinal drugs
- Antiviral drugs
- Cardiovascular and renal drugs
- Peripheral and CNS drugs
- Psychopharmacologic drugs
- Oncologic drugs
- Pulmonary-allergy drugs
- Medical imaging drugs
- Anti-Infective drugs
- Dermatologic drugs
- Endocrinologic and metabolic drugs
- Fertility and maternal health drugs
- Arthritis drugs
- Anesthetic and life supports drugs
- Drug abuse

These committees are composed of experts in the particular fields of drugs covered, primarily physicians. Other qualified experts are also members of the committees, and these include epidemiologists, biostatisticians, pharmacologists, and toxicologists. Many committees also have a voting member nominated by consumers to ensure that the interests of the American consumer can be considered as part of the process.

The role of an advisory committee is to provide the FDA with scientific input on issues related to drug approval. This not only includes the review of safety and efficacy data, but also the overall benefit/risk equation, specific labeling information and recommendations, and the appropriateness of Rx to OTC switches. In some instances, an advisory committee is also assigned the primary review responsibility of selected portions of the NDA. The committees are provided with all necessary information in advance of the actual meeting, and at the meeting respond to specific questions the FDA has generated. Decisions of the committee are usually made by vote. The FDA is not legally bound by the vote of the committee, but in the great majority of cases the majority vote of the committee is followed.

Advisory committee meetings are usually held two to three times a year and the announcement of each meeting is made in the *Federal Register* at least 15 days in advance of the meeting date. The *Federal Register* announcement also specifies the agenda for the meeting, i.e., what drug or NDA is being considered and what type of meeting it will be. There are several different ways that a advisory committee can conduct the meeting; these are

- In an open public hearing that includes public participation
- In an open committee discussion with no public participation, but open to the public
- In a closed presentation of data where only the sponsor is present
- In closed committee deliberations involving only committee members and FDA representation

Most announced meetings include all segments listed above and cover more than one topic or drug entity. As the FDA drug review burden increases and more drugs are considered for OTC status, the use and reliance on these advisory committees will no doubt increase as well.

15. OTC Drug Review

For the majority of drugs sold over the counter, the preclearance requirements are very different. There are occasions when a totally new chemical entity is developed that would be suitable for OTC marketing except that it requires a full NDA prior to marketing. In the case of NCEs, there is no statutory or regulatory requirement that they be initially restricted to prescription marketing, but they usually are. This is to ensure that the use of these NCEs can be closely monitored for safety in a real population. NDA regulations require very close monitoring of newly approved drugs to ensure that there are no unexpected side effects or toxicity problems, which sometimes only surface in a general population as opposed to a more narrow clinical study population. For the most part, drugs that are sold OTC have been marketed thus for a long period of time, satisfying the "material time" aspect of the statutory definition, and have been reviewed under the OTC drug review. As explained earlier, once placed in Category I (safe and effective), an ingredient is conferred general recognition as safe and effective and any manufacturer can market a product with that ingredient, without prior FDA approval, provided the dosage and labeling recommendations in the particular OTC monograph are followed. The manufacturer does not have to seek FDA review and evaluation of the product.

If a new indication or a new dosage form or dosage regimen is developed for an OTC drug, then some type of FDA approval will be required. With regard to new or revised OTC drugs, the options available to gain market clearance include filing of an NDA or petitioning to amend the monograph. These options require, to varying degrees, demonstration through clinical study that the new indication or dosage is safe and effective. With the petition mechanism, reliance can be placed on data in the literature. The filing of an NDA requires that a firm conduct some kind of essential clinical study to support its position.

B. Labeling

Although the requirements for labeling all drug products are found in Section 502 of the Food, Drug, and Cosmetic Act under "Misbranded Drugs and Devices," there are major differences in the way Rx and OTC drugs are labeled. Besides the FD&C Act, drug labeling is also governed by the Fair Packaging and Labeling Act (FPLA) and the Poison Prevention Packaging Act (PPPA). The requirements of the FPLA in the case of drugs are identical to those of the FD&C Act. The PPPA requires that all prescription drugs in packages intended for home use and certain OTC drugs, such as acetaminophen and aspirin, be packaged in child-resistant packages. For products available in multiple package sizes, one package size can be made exempt by a manufacturer if it is prominently labeled, "NOT FOR HOUSEHOLDS WITH CHILDREN."

The basic labeling requirements spelled out in Section 502 for all drugs must include

- Name and place of business of manufacturer, packer, or distributor
- Accurate statement of quantity of contents
- The established name of the drug and whether or not active; the established name and quantity of any alcohol, bromide, ether, chloroform, etc.
- Adequate directions for use
- Adequate warnings

The Act also requires that habit-forming drugs bear a warning, "Warning—May Be Habit Forming" and that alcohol, whether an active ingredient or not, be quantitatively declared on the label. Section 503 of the Act specifically requires that prior to dispensing, all prescriptions drugs must be labeled "Caution: Federal Law prohibits dispensing without prescription."

The Act itself does not require the labeling of either a lot number or an expiration date; these requirements are found in regulations promulgated by the FDA under authority given to it in Section 502. The specific labeling regulations can be found in 21 CFR; general labeling regulations at 21 CFR Part 201 Subpart A, specific requirements for Rx drugs at Subpart B, and OTC drugs at Subpart C. The sections dealing with expiration date and lot numbers are 201.17 and 201.18, respectively. Also appearing in 21 CFR 369 is a section relative to warning statements for specific OTC drugs. This section was promulgated as an aid to manufacturers in complying with Section 502(f) of the Act, which requires adequate warnings. This section provides warning statements the FDA considers appropriate in meeting the legal requirements of 502(f).

The regulations require that all mandatory labeling be prominent and conspicuous; the FPLA (16 CFR) specifically spells out type size requirements for the net quantity of contents declaration.

The major difference between Rx and OTC labels is in terms of what consumers see when they purchase and use the drugs. In the case of Rx drugs, for which the responsibility for safely and effectively using the drug is at the direction of a physician, the consumer will only see the pharmacy's label, which merely includes the name of the drug, the dosage directions, and physician's name. All other labeling, including the insert, is available to the pharmacist and the doctor. The insert in particular contains all the information required by a physician to properly prescribe the drug, and indeed 21 CFR 201.57 spells out in great detail the exact content and format for inserts. OTC drugs are a different matter. The actual label of the product must contain all information needed for the consumer himself to safely and effectively use the product. This includes a statement on indications, directions, and warnings. Thus, a prime consideration as to what is and what is not appropriate for over-the-counter use is whether or not appropriate labeling can be written for the drug.

Labeling appearing on Rx products is approved by the FDA during the NDA review process. The manufacturer is required to submit data in support of all labeling statements, including indications, warnings, adverse reactions, and side effects. Labeling for OTC drugs has been and continues to be reviewed and approved by the FDA on a drug category basis as part of the OTC drug review. The labeling sections in the OTC monographs dictate exact language to be used for statement of identity, indications, dosage, and warnings.

The fundamental differences between the labeling requirements of Rx and OTC drugs are due in part to the fact that OTC drug labels need to provide all information necessary for the consumer to safely and effectively self-administer these products. However, because of the nature of the drugs themselves, the labels of these products are not as extensive as one might think. OTC drugs, by nature, must be extremely safe to use and they should be used only for conditions that are self-diagnosable and self-limiting.

With these considerations as limiting factors, OTC drug labels need only bear information directly relating to three issues. A simple statement of indications, a proper dosage regimen, and adequate warnings. This information can be accommodated on the actual container label itself. The FDA is currently focusing much attention on OTC drug labels, and in the case of recent OTC switches has required that products bear a "drug facts" box statement on the back panel. This statement is similar in concept to the nutritional facts box on food products, where important information is presented in standard form for easier consumer reading. The FDA is examining whether or not adopting a drug facts box for all OTC drug labeling would be appropriate.

The OTC drug industry through its major trade association, the Nonprescription Drug Manufacturers Association (NDMA) is recommending labeling revision that will require: (1) a specific order of information (i.e., active ingredient, uses, directions for use, warnings, and inactive ingredients); (2) specific subheadings under warnings (i.e., "Consult a doctor before use if . . .", "Consult a doctor after use if . . ."); and (3) simplification of label language. Although it remains to be seen which type of labeling the FDA will require, it is certain that OTC labeling will change from its present form.

Prescription drugs, on the other hand, must bear labeling inclusive of all information necessary for the physician to properly prescribe and supervise the use of the product. All the information required for a physician to safely and effectively administer Rx drugs cannot fit on the immediate container label. The specific requirements for Rx drug labeling regarding content and format are at 21 CFR 201.57. The additional information required for Rx drugs is as follows:

- Structural formula of the drug
- Clinical pharmacology
- Indications and usage
- Dosage and administration
- Contraindications
- Warnings
- Precautions
- Drug interactions
- Use in pregnancy
- Pediatric use
- Drug abuse and dependence
- Overdosage
- Dosage and administration
- How supplied

The information that constitutes the text under each of these headings is, for the most part, generated from the toxicological and clinical efficacy studies conducted for INDs as part of the approval process.

Over-the-counter drugs, as indicated previously, must bear all information necessary for the consumer to safely and effectively use the product. Due to the great variety of OTC drugs on the market, as well as the great differences in the wording used on the labels of these drug products, the FDA wanted to bring some sort of consistency to the OTC label so that consumers could make appropriate choices in deciding what OTC medication was best for their particular need. This, of course, was another impetus for the OTC drug review. As mentioned earlier, the OTC monographs, which are the final result of the review process, specify the critical labeling to be used on OTC drugs. The critical labeling, as the FDA sees it, includes a statement of identity, indications, directions for use, and warnings, which are all related to how safely and effectively the consumer can use the product. For the most part, the labeling used on OTC drugs has to be identical to the labeling specified in the monograph in terms of dosage amounts, dosing intervals, and maximum daily dose. Regarding indications, the FDA has revised its "exclusivity" policy to allow for flexibility in this area. Manufacturers are free to use indication language different from that specified in the monograph, provided that the alternative language is synonymous with the monograph language. The FDA has stated that the monograph labeling is the benchmark in determining whether or not alternative language is synonymous with the monograph language. An interesting aspect of what has been called the "flexibility policy" is that, if the monograph specified indication statement is used, the statement on the label may be headed "FDA APPROVED USES." This represents the first time that the FDA has allowed labeling to state either implicitly or explicitly that a product is sanctioned by them.

Another difference between OTC and Rx drug labels is that the majority of OTC labels bear inactive ingredient labeling. Although not required by regulation, inactive ingredient labeling was adopted by OTC drug manufacturers at the behest of the NDMA, the trade organization representing OTC drug manufacturers, so that consumers can make more informed decisions as to the medicines they use.

C. Advertising

Another major area in which OTC and Rx drug regulation differs is advertising. As a general rule, all advertising is regulated by the Federal Trade Commission (FTC). However, this is not the case for Rx drugs. Section 502(A) of the Food, Drug, & Cosmetic Act specifically exempts prescription drugs from Sections 12 through 17 (advertising regulations) of the Federal Trade Commission Act. Section 502(A) also states that no regulation issued under this section shall require prior approval of the content of any advertising. In reality, Rx drug advertising is actually reviewed by the agency prior to release on the basis of requirements of Section 505(B), the new drug approval requirements, under the heading of labeling. This promotional material is reviewed by the FDA's Division of Drug Marketing, Advertising and Communications. Technically, labeling is any printed or graphic material that accompanies the product and, therefore, covers all kinds of promotional material. The FD&C Act specifically requires that all Rx drug advertising include the established name of the drug, the quantitative active ingredient formula, and a brief summary relating to side effects, contraindications, and effectiveness. Since this is information that would be difficult to include in a 30-second, or even a one-minute television commercial, this requirement, in effect, prohibited television advertising of prescription drugs. Even if a commercial was developed that could include the information required by Section 502(A), the FDA felt that the information would not be understandable to the consumer and therefore not be appropriate. Over recent years this thinking has changed dramatically and it is now quite common to see Rx drug advertising on television as well in the lay print media.

Unlike prescription drugs, OTC drugs are not exempt from the Federal Trade Commission Act, and all the rules and regulations regarding truth in advertising apply. The Federal Trade Commission does not have the authority to prereview or approve advertising and, therefore, acts on advertising after it has been disseminated. For the most part, the FTC will act on advertising only upon a complaint or if it feels the advertising is fraudulent and of major impact on the public. The regulation of OTC drug advertising is mostly a self-regulatory system; the National Advertising Division (NAD) of the Council of Better Business has been accepted by advertisers as an impartial judge. Although all aspects of the NAD review process are voluntary, advertisers do respond to NAD requests and, for the most part, adhere to its findings. Although the FTC has authority over OTC drug advertising, the FDA does have some jurisdiction under the general provision regulations for OTC drugs 21 CFR 330(d). This section states that the advertising for an OTC drug product should prescribe, recommend, or suggest its use only under the conditions stated in the labeling. In effect, a drug can be considered misbranded if it advertises a use that is not on its label.

V. DRUG PRODUCTION

After gaining approval to market a drug product, the next most critical aspect of regulation for the pharmaceutical company is the manufacture of that drug product. This section deals mainly with the second most significant drug regulatory tool FDA has in its arsenal--the current good manufacturing practices regulation, better known as the CGMPs or, simply, GMPs.

A. Good Manufacturing Practices

The GMPs, as codified regulations, did not come into existence until the passage of the 1962 amendments to the Food, Drug, and Cosmetic Act. The statutory basis for the GMPs can be found in the adulteration section of the FD&C Act, Section 501, which was amended in 1962 to define an adulterated drug as one in which the "methods used in, or the facilities or controls used for, its manufacture, processing, packing or holding do not conform to or are not operated or administered in conformity with current good manufacturing practice." The Act does not define, however, what current good manufacturing practice is and it was left up to the FDA as well as industry to establish GMPs that would be both effective in achieving its purpose and workable from the standpoint of industry compliance. This task of establishing the GMPs was left up to the FDA through its regulation promulgation procedures. Interaction and cooperation with industry ensured that the GMPs were indeed workable standards that could be introduced as advances in manufacturing control technology evolved. The GMPs were intended to reflect the current state of the art in pharmaceutical manufacturing, practices that were actually in use by industry so as to not impose standards that industry could not adhere to. One important feature of the GMPs was that it mandated documentation of certain critical manufacturing steps and control operations so that the compliance could be reflected in these documents; in the FDA's thinking, "if you didn't document it, you didn't do it." To maintain current status of the GMPs the original 1962 version was updated in 1978 to include new developments in manufacturing and control. One of the more critical aspects of the 1978 GMPs was the concept of "validation" from both the manufacturing process standpoint, as well as the testing methodology standpoint. The GMPs cover every aspect of the pharmaceutical plant operations, from building construction and personnel, up to and including warehousing and distribution of finished product. The GMPs can be found at 21 CFR 211. What follows is a brief outline of GMP requirements from various Subparts of 21 CFR Part 211.

Subpart A—General Provisions

This section includes a statement of the scope of the regulations, as well as definitions of terms. An important point made in the statement of scope is that these GMPs are minimum requirements. In other words, manufacturers are encouraged to go beyond the GMPs if they desire to add even more confidence in the quality of the products they produce.

This section also stipulates that the definitions found in Part 210.3 apply to Part 211. Section 210.3 defines such terms as batch, active ingredient, inactive ingredient, lot, theoretical yield, actual yield, and representative sample.

Subpart B—Organization and Personnel

This subpart spells out requirements for the establishment of a quality control unit, the qualifications and responsibilities of plant personnel, and of consultants who may advise on manufacturing or quality issues. General requirements for personnel and consultants are that they shall have "education, training, and experience, or any combination thereof to perform their jobs." This section deals specifically with the responsibilities of a quality control unit and states that such unit shall have the responsibility and authority to approve or reject any and all components and procedures of drug production, including the finished product. This section requires that there be adequate laboratory facilities available to the quality control unit to test components and finished drug products and that the quality unit's responsibilities and procedures shall be in writing; such written procedures shall be followed.

Subpart C—Buildings and Facilities

This section covers all aspects of the physical manufacturing site including building design and construction, lighting, ventilation, air filtration, heating and cooling, plumbing, sewerage, washing and toilet facilities, sanitation, and maintenance. Regarding design and construction of the plant, this section requires that the building shall have adequate space for operations to help prevent mix-up and that there shall be specifically defined and/or separate areas for the firm's operations to prevent cross-contamination or mix-up. The specifically defined or separate areas shall be as follows: receipt and quarantine of incoming components, holding of rejected components, storage of in-process components of manufacturing and processing operations, packaging and labeling operations, quarantine storage before release of drug products, storage of drug products after release, control and laboratory operations, and aseptic processing. This section also requires that operations relating to the manufacture of penicillin products shall be performed in totally separate facilities. In order to assure compliance with these provisions, some manufacturers have consulted with the FDA prior to building new facilities.

Subpart D—Equipment

The major emphasis of this section is on the adequacy, cleaning, and maintenance of production equipment and the requirement that written procedures be established and followed. The section requires documentation of all maintenance, cleaning, sanitizing, and inspection activities for equipment.

Subpart E—Control of Components and Drug Product Containers and Closures

As indicated, this subsection details all aspects of receipt, storage, testing, and release or rejection of all components used in drug product manufacturing. The sections of this subpart specifically require that each shipment of components be identified with a distinctive code, that it be quarantined until tested and released, and that the quality control unit sample and test or examine the lot where appropriate. A specific section spells out in detail the sampling criteria and methods that are to be used by the quality control unit in relation to incoming components. This section also requires that at least one identity test be performed on each incoming shipment and that specific identity tests, if they exist, be used. The regulations do permit manufacturers to rely on suppliers' protocols for components as documentation of purity, strength, and quality of the material, provided the manufacturer establishes the reliability of the suppliers' analyses through periodic validation of the suppliers' test results. Of course, appropriate written specifications must be established and met for each component used in drug production.

Subpart F—Product and Process Controls

Requirements for all phases of in-process procedures and documentation are included under this subpart. Section 211.100, e.g., requires that written procedures for production and process controls be established, drafted, reviewed, and approved by appropriate organizational units and reviewed and approved by the quality control unit. These procedures must be followed and documented at the time of performance. This subsection also requires that actual yields be determined and compared with theoretical yields at appropriate manufacturing phases and that equipment be identified as to its contents and identified on the batch production records as being used in production of specific product lots. Another section provides requirements for the sampling and testing of in-process controls and tests. This section also requires that control procedures be established to "monitor the output and to validate the performance of those manufacturing processes." It is with this authority that the FDA initiated and encouraged the emphasis on process validation that is now such a major part of all pharmaceutical operations.

When this requirement was first included in the 1978 revision to the GMPs, industry as well as the FDA did not really know what would constitute compliance with the validation requirement. With the help of industry, the FDA developed guidelines spelling out the steps and procedures determined to be appropriate for validation. Validation has become an integral part of any pharmaceutical process and is taken into account at the very inception of any new drug product or process.

The FDA has published guidelines on validation that covers all types of processes from tablet manufacture to ointment manufacture to manufacture of sterile products. The guidelines are available from FDA upon request.

Subpart G—Packaging and Labeling Control

This section basically covers the requirements for labeling and packaging operations, including the need to sample and test or examine all incoming labeling and packaging material, as well as all finished drug products. This is the section that requires the identification of each production batch with a lot number, one that would permit determination of the manufacture and control history of the product, as well as the declaration of an expiration date. According to Section 211.137, the expiration date shall be related to any storage condition stated on the label.

As of November 5, 1982, Subpart G also incorporates regulations pertaining to tamper-resistant packaging. These regulations were promulgated in wake of the tampering incidents of the early 1980s. It also is unique in that it is one of the few sections of the GMPs that deal directly with OTC drugs.

Subpart H—Holding and Distribution

This subpart requires that written procedures shall be established and followed for warehousing and distribution of finished drug products. Distribution procedures require that there be a method where by the oldest stock is distributed first and a system by which the distribution of each lot of any product can be readily determined to facilitate a recall, if necessary.

Subpart I—Laboratory Controls

This subpart covers all aspects of testing for all components and finished drug products, including stability testing, in-process testing, and testing for release of finished drug products. Laboratory controls shall include determination of conformance to appropriate written specifications for acceptance of each lot of raw materials, in-process materials, and finished drug products. Also required is the establishment of a written program to periodically calibrate appropriate instruments, apparatus, gauges, and recording devices. This section also requires that the accuracy, sensitivity, reliability, and reproducibility of test methods be validated.

Regarding stability testing, a written testing program shall be developed for each drug product that specifies sample size, test intervals, and storage conditions for samples, reliable and meaningful test methods, testing of the drug product in the same container-closure system as that in which the product is marketed, and the testing of an adequate number of batches to determine an appropriate expiration date. The regulation specifically allows for the use of accelerated stability studies (storage of samples at high temperature and high relative humidity) for shorter periods of time to set tentative expiration dates, provided, of course, that full real-time room temperature studies are being conducted to verify the tentative date.

This section also requires that samples of all finished drug product lots and lots of raw material active ingredients be reserved for future testing, if necessary. The majority of samples are to be retained until one year after the labeled expiration date.

Subpart J—Records and Reports

This subsection covers general and specific requirements for all records and reports required to be maintained under the GMPs. General requirements include maintenance of all records for one year after the expiration date of the finished product and the provision that all such records be available for authorized inspection and copying. It should be noted that the GMPs do not make a distinction between Rx and OTC drugs in this respect and that failure to make OTC drug records available for inspection is a violation of GMPs. This is in conflict with the Food, Drug, and Cosmetic Act, which only specifies that records for prescription drugs must be available for inspection, and has often created controversy when FDA inspects OTC drug manufacturers.

General requirements also include provisions for an annual review of all records, including a review of every batch record, to ensure that all procedures and processes are operating adequately to maintain acceptability of finished drug products.

This section covers, in great detail, specific records and reports for the following:

- Equipment cleaning and use
- Component, drug product container, closure, and labeling records
- Master production and control records
- Batch production and control records
- Production record review
- Distribution records
- Complaint files

All these previously listed records must be dated and signed by the performing individual and independently checked, dated, and signed by a second individual. They must also be independently checked and approved by the quality assurance unit before the finished product is released for distribution.

Subpart K—Returned and Salvaged Drug Products

These sections allow manufacturers to salvage returned drug products provided appropriate examination, testing, and reconditioning procedures are undertaken and documented to ensure that the salvaged products meet all appropriate standards and specifications.

B. Inspections

The third major drug regulatory tool at the FDA's disposal is plant inspection. Under Section 704 of the Food, Drug, and Cosmetic Act, the FDA was given authority by Congress to enter and inspect any factory, warehouse, establishment, or vehicle in which foods, drugs, devices, or cosmetics are manufactured, processed, packed, or held. In terms of drugs, the FDA's emphasis of the investigational activities is directed toward compliance with good manufacturing practices, and the FDA investigator is authorized to examine each phase of a plant's operation to ensure that it is operating in conformance with GMPs in terms of procedures as well as documentation.

Drug manufacturers can be certain of a FDA inspection at least once very two years. FDA is required by the Food, Drug, and Cosmetic Act, Section 501(h), to inspect drug manufacturing facilities (including repackers and relabelers) once every two years. If the FDA does not inspect a facility at least once every two years, it is in violation of the act. For this reason, the FDA is very careful about its "statutory obligation" inspection program and succeeds in meeting its obligation under the law. Throughout the FDA's inspectional history since the 1962 Drug Amendments, there have been very few instances in which the FDA has failed to inspect a drug firm in the mandated two-year period. Drug firms would be most happy if the agency inspected only once every two years. In reality, drug manufacturing firms can expect a FDA visit at least once year, for any number of reasons, i.e., complaint investigation, surveillance sampling, or recall investigation.

Although Section 704 gives the FDA broad inspectional authority over regulated products, it does provide several caveats that must be considered by the FDA in the course of its activities. The investigator must first present a written notice to the agent in charge of the facility before he can legally enter it. Refusal to permit inspection by a FDA investigator, after receipt of a written notice of inspection (FDA Form 482), is a violation of federal law.

Section 704 does not require that the FDA give prior notice of an inspection, and GMP inspections are always unannounced. The Act does require, however, that the inspection is conducted at reasonable times and in a reasonable manner. This does not mean, though, that the manufacturer can delay an inspection by stating that it is a bad time for an inspection. Courts have stated that a reasonable time is any time that a facility is in operation, whether it is in the middle of the day or the third shift. If a facility is operating at three o'clock in the morning, it is considered a reasonable time for the FDA to inspect.

Regarding prescription drugs in particular, Section 704 provides the FDA very broad authority. The section allows the investigator access to "all things therein (including records, files, papers, processes, controls, and facilities)," and also gives the FDA access to research data pertaining to new drugs and new devices subject to provisions of Section 505 of the Act. This authority does not extend to nonprescription drugs. The FDA does not have authority to review financial or sales data except for shipment data.

The following describes the steps that normally take place during a FDA inspection conducted as part of the FDA's statutory obligation program to assess compliance with the GMPs.

First, the firm's name comes up on the FDA computer as requiring the periodic inspection, and the firm is assigned to an investigator. The investigator prepares for the inspection by reviewing the official FDA establishment file. This file includes all previous establishment inspection reports (EIRs), which are documents written by the investigators detailing the inspection, including the problems encountered. The investigator looks for problem areas or for areas that have not been inspected for a while. As part of the establishment file, a listing of all types of products manufactured by the firm is included. The investigator oftentimes researches a particular drug or dosage form to cover in order to be better prepared to examine the drug products he will inspect. The file also gives the investigator a good idea as to the structure of the firm in terms of individual responsibility and operations. Once the investigator has done the preinspectional research, she is ready for the inspection.

When the investigator enters a firm, her first request is to see the person in charge of the entire facility in order to issue a written notice of inspection, the FD482. The investigator is seeking the most responsible official of the firm, who is often the plant manager. On occasion, in smaller firms, the FD482 is given to the president of the firm. Once the FD482 is issued, it is a violation of the FD&C Act to refuse inspection.

After the notice is issued, the first thing the investigator does is ask for a brief history of the firm, including chain of command, so that she can determine if anything has changed since the last inspection. The investigator then inquires about the responsibility of each of the corporate officers and individuals in charge of plant operations. This is a routine procedure performed by the investigator and helps to determine who is responsible for any violations encountered during the inspection, from both the commission and the correctional points of view.

Oftentimes the investigator asks to see a current list of products manufactured in order to determine whether or not any new products are being produced.

After this initial phase, the actual plant inspection begins. The investigator most likely walks through the facility, starting at the receiving dock and working his way through the plant, as would a product. She familiarizes herself with the general procedures and record-keeping aspects of each department and reviews, in general, the adequacy of the records being maintained.

During physical inspection, which usually follows the flow of product through the facility, the investigator examines receipts, quarantine, sampling, and storage of incoming materials. The weighing rooms and production facilities, finished product storage, sampling, packaging areas, and the label storage and handling procedures are also examined. At each department, the investigator may delve into specific areas to determine the accuracy and adequacy of operations and documentation. For instance, the investigator may randomly select a particular product label and examine its history from receipt, sampling, inspection, storage, issuance, and reconciliation.

The initial walk-through of the plant also includes inspection of the quality control laboratories, where the investigator can become familiar with all laboratory procedures from handling of samples, to reagent preparation, to a review of individual laboratory technician notebooks. Retention and stability sample areas are also reviewed to ensure acceptability.

As the investigator observes conditions of practices that are considered inadequate or inappropriate, in terms of the GMPs, he is required to point them out immediately, so that the firm is made aware of any deficiencies and can make corrections, if necessary. The investigator also presents all observations at the close of the inspection. Even if a condition was noted and corrected during the inspection, it still appears on the investigator's official list of observations (FD483) issued at the end of the inspection.

After the walk-through, the investigator examines the record-keeping and documentation procedures of the firm. Specific products and lots are scrutinized from beginning to end. The first step in the procedure is review of a specific batch record, which leads to a review of all records concerned with that particular batch. The investigator works back through the records of incoming raw materials, covering all aspects from receipt to final product distribution records. Review of these records undoubtedly leads to more questions and may take the investigation in many different directions.

The last phase of the inspection is the exit interview, at which point the list of observations (FD483) is presented to management. This document lists, usually in order of significance, instances or circumstances that, in the opinion of the investigator, violate GMPs. It is best at this point not to respond directly to the investigator, except to gain clarification of specific points, but to indicate that the firm will respond in writing to each of the points listed. The purpose of the exit interview is to ensure that top management is made aware of the investigator's findings so that appropriate action can be taken; for this reason, the investigator will insist on presenting the FD483 to the highest ranking individual possible. If this is not possible, the investigator can mail copies of the FD483 to the president or chief executive officer of the firm.

After the inspection, the investigator prepares a narrative report of the inspection, detailing all information gathered during the inspection and documenting each of the points listed on the FD483. This establishment inspection report will specifically address each of the following:

- History of business
- Persons interviewed and administrative procedures
- Individual responsibility
- Firm's training program
- Raw materials
- Operations
- Objectionable conditions or practices
- Manufacturing codes
- Consumer complaints
- Recall procedures
- Promotion and distribution
- Discussion with management

The EIR is endorsed by the investigator's supervisor as either "no action indicated" or "further action indicated." This decision is sent to the district's compliance branch, which evaluates the report and recommends appropriate regulatory or legal action. Depending on the nature of the problems uncovered by the investigator, legal action can consist of any of the following: issuance of a "warning letter," seizure of the product, or criminal prosecution. In recent years, the FDA's emphasis has been on industry voluntary compliance. The action a firm takes immediately after receiving an FD483 can have an influence on the agency's actions, and in some cases can prevent the issuance of a "warning letter." It is important for a firm to respond immediately in writing to any point on a FD483, providing the FDA with a plan for corrective action with specific documentation, if available.

VI. SUMMARY

As evidenced in this chapter, one can see that the FDA is an ever-changing organization, and that the rules and regulations pertaining to the manufacture and marketing of pharmaceutical products are also ever-changing. There is no doubt that in the future, as technology and product categories develop, regulatory requirements will also evolve. Specific regulations for specific product types may be established, such as specific GMPs for disperse systems. As the GMPs and federal law now stands, they are not only broad enough to adequately cover all the various types of drug products currently on the market, but they are also flexible enough to permit changes that will continue to ensure safe and effective drug products for consumption by the American public.

SUGGESTED READING

Adair, F. W., Process validation—Biological consideration, *Proc., Proprietary Assoc. Manufacturing Controls Seminar*, 1979, pp. 4–16.

Avallone, H. L., Process validation for drug manufacturing, *FDA By-Lines*, 10:171–178 (1980).

Bachrach, E. E., The FDA's flexibility policy for labeling over-the-counter drugs: A review and analysis, *Food Drug Cosmet. Law J.*, 42:184–191 (1987).

Buyers, T. E., Overview of the regulations, *Food Drug Cosmet. Law J.*, 34: 465–469 (1979).

Celeste, A. C., The inevitable FDA inspection, *Food Drug Cosmet. Law J.*, 34:32–39 (1979).

Establishment of prescription drug user fee revenues and rates for fiscal year 1993, *Federal Register*, 58(237), Dec. 13, 1993.

Establishment of prescription drug user fee revenues and rates for fiscal year 1993, *Federal Register* 59(236), Dec. 9, 1994, pp. 63808-63810.

FDA Backgrounder, User Fees, BG 93-2, July 21, 1993.

Filter validation symposium, *J. Parenter, Drug Assoc.*, 33:246–282 (1979).

Gardener, R. Y., Current concepts for the microbiological control of nonsterile drug products: A report from the PMA quality control and biologics sections, *Pharm. Technol.*, 7:54–55, 58 (1983).

Gilston, S. M., FDA's Robert Temple: Examining the drug review process, Part 1, *Med. Adv. News*, pp. 10, 18–20, 23–24 (1984).

Gilston, S. M., FDA: The year in review, *Med. Adv. News,* 6:15–16, 18 (1987).

Halperin, J. A., FDA: Pharmaceutical regulations in the 1980's, Predictions, *Drug. Intel. Clin. Pharm.*, 16:320–324 (1982).

Herzog, J., Recurrent criticisms: A history of investigations of the FDA, *Med. Mktg. Media,* 13:41–45 (1978).

Loftus, B. T., Validation and stability, *J. Parenter. Drug Assoc.*, 32:269–272 (1978).

Mead, W. J., Process validation, *Cosmet. Toilet.*, 93:19–21 (1978).

Prescription Drug User Fee Act, Public Law 102-571.

Stenecher, R., Process validation and reliability. *Proc., Proprietary Assoc. Manufacturing Controls Seminar*, 1979, pp. 32–40.

User Fee Correspondence 1, PHS, DHHS, FDA Feb. 1, 1993.

User Fee Correspondence 2, PHS, DHHS, FDA, July 16, 1993.

User Fee Correspondence 3, PHS, DHHS, FDA, Aug. 5, 1993.

User Fee Correspondence 4, PHS, DHHS, FDA, Jan. 4, 1994.

Index